Biomanufacturing

Scrivener Publishing
100 Cummings Center, Suite 541J
Beverly, MA 01915-6106

Advances in Production Engineering

Editor: Dr. Sandip Kunar (sandip.sandip.kunar@gmail.com)

"Advances in Production Engineering" addresses recent developments and research issues related to the production engineering which necessitates the development of new materials and manufacturing processes in order to produce high-quality products at lower prices and in less time. This series aims to provide a scientific platform for researchers, practitioners, professionals, and academicians to discuss the most recent technological developments in metals, polymers, ceramics, composites, biomaterials, nanomaterials, special materials, metals, micro-forming, powder metallurgy, ceramics processing, non-traditional machining, high speed machining, micro and nanomachining, and laser processing. Tribological analysis, friction behavior, modelling, and optimization techniques in materials, machining, and manufacturing are also covered in the series.

Publishers at Scrivener
Martin Scrivener (martin@scrivenerpublishing.com)
Phillip Carmical (pcarmical@scrivenerpublishing.com)

Biomanufacturing

Transforming Production of Biologics and Biopharmaceuticals

Edited by

Sandip Kunar

Department of Mechanical Engineering, Aditya University, Surampalem, Andhra Pradesh, India

Jagadeesha T.

Department of Mechanical Engineering, National Institute of Technology Calicut, Kozhikode, Kerala, India

and

D. Sathis Kumar

School of Pharmacy, Aditya University, Surampalem, Andhra Pradesh, India

Scrivener Publishing

WILEY

This edition first published 2026 by John Wiley & Sons, Inc., 111 River Street, Hoboken, NJ 07030, USA and Scrivener Publishing LLC, 100 Cummings Center, Suite 541J, Beverly, MA 01915, USA

For more information about Scrivener publications please visit www.scrivenerpublishing.com.

Wiley Global Headquarters
111 River Street, Hoboken, NJ 07030, USA

For details of our global editorial offices, customer services, and more information about Wiley products visit us at www.wiley.com.

The manufacturer's authorized representative according to the EU General Product Safety Regulation is Wiley-VCH GmbH, Boschstr. 12, 69469 Weinheim, Germany, e-mail: Product_Safety@wiley.com.

Library of Congress Cataloging-in-Publication Data

ISBN 978-1-394-44051-1

Cover image: Generated with AI using Adobe Firefly
Cover design by Russell Richardson

Set in size of 11pt and Minion Pro by Manila Typesetting Company, Makati, Philippines

Printed in the USA

10 9 8 7 6 5 4 3 2 1

Dedicated to the Loving Memory of My Parents

Mr. Haripada Kunar
and
Ms. Alaka Bera Kunar

With love and gratitude, also dedicated to

My Wife
Ms. Anusree Mal Kunar

and

My Daughter
Ms. Trinetra Kunar

Contents

Preface

Biomanufacturing stands at the forefront of a new era in the life sciences, reshaping the conception, development, and delivery of advanced biologics and biopharmaceuticals to address global health challenges. Biomanufacturing uses living systems, such as genetically modified cells and microorganisms, to create complex molecules like monoclonal antibodies, therapeutic proteins, vaccines, and cell and gene therapies, in contrast to conventional pharmaceutical production, which depends on chemical synthesis. Significant technological developments, such as single-use systems, continuous processing, real-time analytics, artificial intelligence, and machine learning, have coincided with the move toward biologics. These developments are speeding up production schedules, enhancing product quality, and reducing manufacturing costs. Simultaneously, growing biotechnology investment and changing regulatory frameworks have accelerated innovation and scalability. But these developments also bring about new challenges, such as maintaining strict adherence to Good Manufacturing Practices (GMP) and ensuring the stability and consistency of biological products.

This book provides a thorough examination of the dynamic and multifaceted subject of biomanufacturing, emphasizing how engineering, science, and policy converge to enable the production of life-saving treatments.

The book comprises twenty-eight chapters.

Chapter 1 discusses the transformations in biomanufacturing, fundamentals of bioprocessing and biomanufacturing, innovations in bioprocessing, the need for a bio-based economy, biological systems, progress in biomanufacturing, case studies of bioprocesses, overview of biomaterials, manufacturing process for porous implants, investment and strategic alliances to boost commercial biomanufacturing, challenges and limitations in bioprocessing and biomanufacturing, and future prospects of biomanufacturing. Chapter 2 presents an in-depth examination of the key role of mathematical modelling and simulation in the design and optimisation of contemporary biomanufacturing systems.

Chapter 3 explains a gateway into various materials that can be used for unlocking the full potential of bioprinting. Chapter 4 examines key bioprinting techniques, including inkjet, extrusion-based, and laser-assisted methods, evaluating their printability, resolution, and biocompatibility. Chapter 5 highlights the methods for aligning lean strategies with good manufacturing practice standards to achieve process efficiency without compromising safety or effectiveness in bioproduction.

Chapter 6 addresses industry barriers, accelerates innovation, and improves global access to life-saving drugs. Chapter 7 illuminates future changes such as producing biomedical goods out-of-town, providing treatment as needed, and strengthening healthcare systems against climate change. Involving insights from several fields, the chapter functions as an all-inclusive resource for people interested in making biopharmaceutical production more sustainable and progressive.

Chapter 8 emphasizes additive manufacturing (AM) as a fundamental technology for future biomedical innovations, providing avenues for personalized and regenerative healthcare that enhance patient outcomes and surgical success rates. Chapter 9 surveys the scientific principles and engineering practice that underpin biomedical functionally graded materials, summarizes the state of the art across hard- and soft-tissue applications, and outlines challenges that must be overcome before widespread clinical adoption.

Chapter 10 considers how genomics-driven bio-manufacturing can replace traditional petrochemical processes through advanced microbial engineering, renewable feedstocks, and technological innovation, while addressing ongoing scalability and regulatory challenges. Chapter 11 investigates the optimized utilization of agro-industrial wastes, which are abundant in biological material and are often treated as refuse, converting them into feasible substrates through microbial bioprocesses.

Chapter 12 analyzes a broad base of information about automation, AI-based processes, and lean technology involved in biomanufacturing. Chapter 13 discusses contemporary difficulties, such as material constraints, regulatory barriers, and problems with biological integration. It concludes by outlining the future directions of the process and highlighting its revolutionary potential in regenerative healthcare and personalized medicine.

Chapter 14 assesses the improvements in lean biomanufacturing processes with AI applications in both the upstream and downstream, including inventory of raw materials, bottle release, and so forth. Chapter 15 surveys the design, classification, and functionality of emerging bio-inks for applications in tissue engineering and scalable biomanufacturing.

It provides an overview of natural, synthetic, and hybrid hydrogels, highlighting their rheological properties, printability, and cell compatibility.

Chapter 16 examines an inclusive overview of the current landscape and future prospects of AI-ML-driven bioprocess optimization, offering valuable insights for practitioners, researchers, and decision-makers in the field of biomanufacturing. Chapter 17 details the changing landscape of RNA-based biologics, the centrality of biomanufacturing therein in the swift development, scale-up, and delivery to the world. This chapter also presents technological inventions like microfluidics, single-use bioreactors, and real-time analytical tools, which are revolutionizing RNA biomanufacturing.

Chapter 18 provides a broad view of rapid response biomanufacturing infrastructure that allows making and releasing biologics, such as vaccines, monoclonal antibodies, and anti-viral proteins, within a short period. Chapter 19 reviews the philosophy behind the techniques and technologies, and the practical realization of real-time analytical applications and control systems of downstream bioprocessing. It discusses Process Analytical Technology (PAT), spectroscopy-based sensors (e.g., NIR, Raman, UV-Vis), and soft-sensor models to ensure that processing conditions are controlled to monitor critical quality attributes (CQA) in purification, filtration, and formulation.

Chapter 20 traces how designed experiments (DoE) have been used in the process development of biopharmaceuticals, with their role in defining critical process parameters (CPPs) and the interactions among them to improve product quality, productivity, and process robustness. Chapter 21 outlines the essence, enabling technologies, and practical applications of zero-waste systems in the biopharmaceutical and biologics manufacturing industry.

Chapter 22 focuses on the progress in terms of the use of fewer auxiliary chemicals, and the development of metabolic engineering to maximize the ends whilst minimizing wastes and dangerous by-products. Chapter 23 lays out a comprehensive outline of the merits, advancements, and struggles of the establishment of the regulatory framework with quality by design (QbD) in biologics to ensure the robustness and scalability of bioproduction.

Chapter 24 considers the technical underpinnings of AI-based maintenance, such as data ingestion systems, anomaly detection, conditioning monitoring, and predictive modelling for biomanufacturing equipment. Chapter 25 lays out common working hazards, such as environmental safety concerns and some engineering control measures put in place to mitigate them. The combination of process safety, levels of biosafety, engineering controls, and digital technologies of risk prediction used in the chapter is focused on the ability to meet the international regulatory standards.

Chapter 26 provides insight into the functions of solvent recovery, single-use systems, and waste valorization in reducing operational expenses and carbon emissions. Regulatory factors and empirical case studies from industrial biomanufacturing are analyzed to demonstrate practical use. This chapter proposes a framework for assessing and implementing green Downstream processing (DSP) technologies, thereby advancing sustainable and circular biomanufacturing in accordance with global environmental and policy goals. Chapter 27 discusses how transdermal drug delivery faces challenges with biologics due to their large molecular size and the skin's barrier. Next-gen nanocarriers like polymeric nanoparticles, liposomes, and metallic nanoparticles are being developed to overcome this, enabling targeted, efficient delivery. These systems improve bioavailability and reduce toxicity but face issues like low drug loading and manufacturing complexity. They show promise in treating diseases like cancer and autoimmune disorders.

Chapter 28 highlights emerging trends, such as the utility of artificial intelligence and automation facilitated bioprocess refinement, the trending movements toward circular bioeconomy models and the biomanufacturing reliability in personalized healthcare and sustainable Chemistry. This chapter also describes the revolutionary impact of biocatalysis and metabolic engineering in delivering eco-friendly and cost-effective solutions for industrial biotechnology.

This book provides a thorough resource for scientists, engineers, regulatory specialists, and students alike by looking at the historical background as well as the innovative methods influencing the creation of biologics today and in the future. Biomanufacturing is more than just a technological advancement; it is a vital component of contemporary medicine that has the potential to revolutionize patient care globally as the needs of global health become more pressing and personalized treatment becomes a reality.

The Editors
June 2026

Acknowledgements

This book is the result of collaborative effort, dedicated research, and the generous support of many individuals and organizations.

We would first like to express our sincere gratitude to all authors, whose work continues to push the boundaries of biomanufacturing. Their pioneering innovations in cell biologics, upstream and downstream processing, and regulatory science laid the foundation for much of the content in this book.

Special thanks go to our colleagues and editors in the field of bioprocessing and pharmaceutical sciences, whose insights, critical feedback, and encouragement helped shape the structure and scope of this work. We are also indebted to the academic institutions and industry partners that provided access to valuable resources, case studies, and real-world examples.

To our editors, Martin Scrivener, and the team at Scrivener Publishing, thank you for your professionalism, editorial guidance, and commitment to quality throughout the publishing process.

Finally, we are deeply grateful to our family and friends for their unwavering support, patience, and encouragement during the long hours of writing and research.

The Editors

Part 1

FOUNDATIONS OF BIOMANUFACTURING

1

Introduction to Biomanufacturing

Sandip Kunar[1]*, Jagadeesha T.[2], Anusha Mylavarapu[1], Ajithkumar Sitharaj[3], Gurudas Mandal[4] and Satishkumar P.[5]

[1]Department of Mechanical Engineering, Aditya University, Surampalem, Andhra Pradesh, India
[2]Department of Mechanical Engineering, National Institute of Technology Calicut, Kozhikode, Kerala, India
[3]Department of Mechanical Engineering, KPR Institute of Engineering and Technology, Coimbatore, Tamil Nadu, India
[4]Department of Metallurgical Engineering, Kazi Nazrul University, Asansol, West Bengal, India
[5]Department of Mechanical Engineering, Rathinam Technical Campus, Coimbatore, Tamil Nadu, India

Abstract
The process of creating economically significant biomaterials and biomolecules using biological systems, such as cells, enzymes, or organisms, is known as biomanufacturing. Pharmaceuticals, food and beverage, chemicals, and biofuels are among the businesses that employ these materials. The process for creating solid and porous biomedical implants was thoroughly studied to ensure their effective use. Technologies such as electrospinning, additive manufacturing, and powder metallurgy have been identified as promising approaches for creating porous, mechanically adjustable ceramic and metallic implants for use in medicine. The chapter discusses the transformations in biomanufacturing, fundamentals of bioprocessing, and biomanufacturing, innovations in bioprocessing, need of bio-based economy, biological systems, progress in biomanufacturing, case studies of bioprocesses, overview of biomaterials, manufacturing process for porous implants, investment and strategic alliances to boost commercial biomanufacturing, challenges and limitations in bioprocessing and biomanufacturing, and future prospects of biomanufacturing.

Keywords: Biomanufacturing, biomaterials, implants, biological systems, bioprocessing, biofactory

**Corresponding author*: sandip.sandip.kunar@gmail.com

Sandip Kunar, Jagadeesha T. and D. Sathis Kumar (eds.) Biomanufacturing: Transforming Production of Biologics and Biopharmaceuticals, (3–46) © 2026 Scrivener Publishing LLC

1.1 Introduction

Since its origin, biomanufacturing, the technique of harnessing live organisms to create useful products has advanced significantly. Biomanufacturing is the process of producing commercial goods for food, energy, material, pharmaceutical, and agricultural industries using biological systems such as enzymes, bacteria, cells, tissues, plants, and animals [1]. Technological developments, scientific breakthroughs, and a growing understanding of biological systems have made it possible to optimize biomanufacturing procedures and produce new applications throughout time. In the earlier days, the food and beverages were prepared using microbes like yeast and bacteria. For example, the yeast is utilized in the fermenting process to ferment sugars and generate alcohol. Similar to this, bacteria are applied in the making of cheese and yogurt; it is used to transform milk into their respective products.

By the development of genetically modified organisms and capable of producing particular proteins or other chemicals, the application of recombinant DNA technology in the 1970s and 1980s transformed biomanufacturing. The first recombinant protein, human insulin was invented in the early 1980s to treat diabetes on an industrial scale. The biomanufacturing technology continued to advance throughout the 1990s and 2000s. Advances in directed evolution and the rapid screening of hundreds of potential drug candidates were made possible by high-throughput screening techniques. The protein engineering made it possible to boost biomanufacturing methods and produce new enzymes and other proteins with particular roles. Biomanufacturing is now a significant part of the biotechnology and pharmaceutical sectors. Biomanufacturing techniques are exploited to generate many of the vaccines and medications required to prevent and treat diseases, and new uses for biomanufacturing are frequently being explored. The application of synthetic biology, which involves creating new biological systems with particular roles, is another rapidly developing the field of biomanufacturing. The biomanufactured new organisms can perform the specified tasks, including removing ecological pollutants; this technology has capability to completely transform biomanufacturing.

The most significant biomanufactured products and their applications are discussed:

- Insulin: Blood sugar levels are regulated by biomanufactured insulin to those with diabetes.

- Vaccinations: The production of vaccinations, such as the COVID-19, HPV, and flu vaccines, is done using biomanufacturing.
- Enzymes: Biomanufacturing creates enzymes that are utilized in several industrial processes, including food processing, laundry detergents, and biofuels.
- Antibodies: Monoclonal antibodies are produced *via* biomanufacturing and are utilized to cure many diseases, including autoimmune disorders and cancer.
- Biofuels: Biofuels such as ethanol and biodiesel are produced through the process of biomanufacturing.
- Hormones: Hormones including growth hormone, erythropoietin, and follicle-stimulating hormone are generated through biomanufacturing.
- Industrial chemicals: Biomanufacturing is the process used to create industrial chemicals, including solvents, bio-based chemicals, and biodegradable plastics.
- Amino Acids: Amino acids used in food additives and animal feed are produced through biomanufacturing.
- Nutraceuticals: Vitamins, probiotics, and prebiotics are examples of nutraceuticals that are made using biomanufacturing.
- Personal care products: Products including lotions, creams, and shampoos are made by biomanufacturing.
- Food: Dairy items, fermented edibles, and various foods are produced using microorganisms.
- Bio-based fibers: Environmentally sustainable alternatives are used to synthetic fibers in clothing and other industries.
- Biopesticides: Products for pest management made from natural sources are called biopesticides.
- Biofertilizers: Additives to soil that support the health and growth of plants.

Biomanufacturing encompasses metal biorefineries, regenerative medicine, and a range of chemical products [2, 3, 25]. These products may be reformed through different methods, including recombinant DNA technologies, synthetic biology and protein and metabolic engineering, or they may be applied in their unaltered form [4].

Latest studies explore the rapid demand of bone and joint replacements. According to estimates, most people over 45 years with bone weakness, inflammation of the joints, and cartilage deterioration, which leads to bone degradation under conditions of constant cyclic loading [5, 6]. This is the primary reason for bone joint loss due to the friction and deterioration of the physical and mechanical characteristics of normal human bone. The best solution to these issues is total joint replacement (TJR), in which artificial implants are used to replace load-bearing joints [7, 8]. Artificial implants are made of biomaterials and consist of all or a portion of a live structure that substitutes a natural function to enhance the quality of the human body. Three elements determine how stable implants are: (i) the mechanical, chemical, and morphological characteristics of biomaterials; (ii) their biocompatibility; and (iii) their nontoxicity [9, 10]. The reactions of implanted biomaterials indicate that titanium alloy, magnesium-based alloys, and stainless steel are the main prevalent metallic biomaterials. It is important to note that the stiffness values of SS, Co, and Ti alloys are significantly higher than those of cortical bone. They should have respectable chemical and mechanical qualities. Foreign body reaction, acute and chronic inflammation, and blood–material interaction are examples of host reactions that typically happen in the biomaterial implantation. A wide range of implant biomaterials close to the implant site can cause different kinds of reactions.

The changes in biomanufacturing, bioprocessing fundamentals, biomanufacturing innovations, the need for a bio-based economy, biological systems, biomanufacturing progress, case studies of bioprocesses, biomaterials overview, the manufacturing process for porous implants, investment and strategic alliances to increase commercial biomanufacturing, bioprocessing and biomanufacturing challenges and limitations, and biomanufacturing's future prospects are all covered in this chapter.

1.2 Transformations in Biomanufacturing

The technological advancements that paved the way for biomanufacturing usually occurred concurrently and led to new revolutions. Revolutions in biomanufacturing are explained as follows [11]:

(i) Premodern Biomanufacturing

Biomanufacturing was certainly used to transform one sort of food into another thousands of years ago, before people were aware of fermentation. According to documents dating back over 9,000 years, the ancient Chinese may have made wine with rice, honey, and fruits as raw ingredients. Comparable accounts have been found in other texts that detail the Sumerian and Babylonian brewing of beer around 6000 BC [12], as well as the baking of bread in Egypt. Prior to

Louis Pasteur popularizing fermentation, it was employed in several processes, including pickling food and making cheese, curd, beverages, and culinary sauces. These fermentation methods could be categorized as premodern biomanufacturing methods since they employed naturally existing, readily cultivable mixed microorganisms that could create desired products from simple feedstocks and included solid state fermentation.

(ii) Biomanufacturing 1.0

The production of primary metabolites is the main focus of the bio-manufacturing 1.0 revolution. Biomanufacturing 1.0 may have begun in the 1910s when Chaim Weizmann identified Clostridium acetobutylicum and established the fermentation processes for ethanol, butanol, and acetone. Pure microbial cultures were used in an anaerobic liquid fermentation system for the first time in biomanufacturing 1.0. The demands of the market acted a significant role in the progress of fermentation technology. The major goal of biomanufacturing 1.0 was to produce primary metabolites. The most fundamental metabolic byproducts are needed for normal growth and reproduction. Acetone, n-butanol, glycerol, nucleosides, carbohydrates, organic acids, amino acids, and so forth are examples of these. Similarly, to meet market demand and reduce production costs, the leading glutamate producers shifted from chemical synthesis to microbial fermentation for the commercial production of monosodium glutamate [13]. The amino acid that is most widely synthesized, is monosodium glutamate.

(iii) Biomanufacturing 2.0

The primary focus of this biomanufacturing revolution was the generation of secondary metabolites, which began during World War II with the fermentation of penicillin, tetracycline, and streptomycin. Unlike primary metabolites, which are produced by a limited number of species within a taxonomic group, secondary metabolites are not required for an organism's growth, development, or reproduction. However, being a crucial component of natural defense, they provide a selection advantage to the producing organism, enhancing its chances of survival. These secondary metabolites are commercially useful to humans and are used as components in medications, flavors, fragrances, and colorings. Biomanufacturing 2.0 built upon the previous revolution in terms of the employment of specialized mutant microbial strains and aerobic submerged fermentation techniques.

(iv) Biomanufacturing 3.0

The transition in the 1980s from the synthesis of primary or secondary metabolites to large proteins crucial for catalysis and medicine led to the creation of Biomanufacturing 3.0. The two main technologies that this revolution concentrated on were improved cell cultures and recombinant DNA technology. After Stanley Cohen and Herbert Boyer created the first *in vitro* recombinant DNA plasmid in 1973, other for-profit initiatives using the same technology began to appear. These businesses concentrated on high-value pharmaceutical proteins since they are difficult to extract from their natural sources and costly to synthesize. Biomanufacturing 3.0 made rapid progress with the sequential production of recombinant proteins such as growth hormone and human insulin. During this time, customized mammalian cell cultures were used to produce human erythropoietin, a protein hormone that increases the rate of RBC synthesis, in a low-cost with higher-quality manner. Additionally, the research of antibody-drug conjugates for the treatment of cancer gained momentum [14]. Furthermore, a crucial element of biomanufacturing 3.0 was the creation of recombinant enzymes for commercial and scientific applications. New England Biolabs was the first company to be established in 1974 for the commercial production of enzymes needed for recombinant DNA technology, such as restriction enzymes and DNA polymerases. The main challenges of this production encountered the limited use and poor stability of enzymes were overcome during the 1980s to 1990s, which included immobilization of enzymes [15, 16]. Later, in the 1980s, a few more companies were founded, including Novozymes and Genencor, which used technology for structure-based protein engineering and focused on the mass production of enzymes used in food, chemical, detergent, and textile industries. The third and current wave of enzymes focused on creating the customized enzymes with major chemical changes starting in the middle and late 1990s [17, 18].

(v) Biomanufacturing 4.0

The production of biology-driven organisms, *in-vitro* bio-systems for artificial starch synthesis, metabolic engineering for iso-butanol production, and human tissues are the focus of a different category known as biomanufacturing 4.0. The world's most urgent issues, including those pertaining to food safety, energy generation, water supply, and sustainability, may be resolved by the bio-manufacturing 4.0 revolution. Additionally, in the biomanufacturing industry, implementing digital technology and artificial intelligence has become essential and a competitive advantage. Consequently, many biological product manufacturing firms today depend on more robots than humans. The complete implementation and execution of robotics is yet a ways off [19]. In recent years, the biomanufacturing industry has grown more and more dependent on the advancement of Internet of Things (IoT) technologies. The concepts of a data-driven, networked society have replaced those of industry. Industry 4.0 encompasses the biomanufacturing industry, which is characterized by the application of cyber-physical-based systems. The massive revolution brought about by Industry 4.0—which is fueled by big data, cloud computing, and digital transformation—has altered our daily lives. The current biomanufacturing revolution, referred to as "Biomanufacturing 4.0," is focused on producing a broad range of novel, sustainable, effective, and safe products associated with pharmaceuticals, regenerative medicine, renewable energy sources, artificial food, and the integration of various biological tools from synthetic biology to metabolic engineering, among other things [20].

(vi) Biomanufacturing 5.0

Biomanufacturing 5.0 signifies the latest development in production powered by biotechnology, incorporating cutting-edge technologies such as artificial intelligence, machine learning, robotics, and the Internet of Things (IoT) to improve accuracy, efficiency, and sustainability. In contrast to previous generations, biomanufacturing 5.0 focuses on real-time data analysis, digital twins, and intelligent automation to enhance each phase of the production process, from cell culture to product purification. This method not only speeds up innovation but also promotes personalized medicine and eco-friendly manufacturing practices. Among other fascinating advancements in chemical synthesis and assembly that are employed to biological processes, components, and systems, synthetic biology may now modify already-existing biological cells, produce synthetic cells, transplant synthetic genomes, and produce chemical models. Self-regulating biological self-organization and self-production processes are successfully added by synthetic biology, and these systems are of interest to artificial intelligence [21].

1.3 Industrial Biotechnology and Biomanufacturing from a Historical Perspective

Industrial biotechnology has its origins in ancient civilizations, when bread, beer, and wine were produced using early biotechnological techniques like fermentation. The foundation for contemporary biotechnological applications was established by these procedures, which were conducted without a scientific understanding of the microorganisms involved. A scientific foundation for comprehending and working with microbes was established by the 19th-century development of microbiology, which was led by trailblazers such as Louis Pasteur and Robert Koch [22]. Significant developments during the 20th century accelerated the growth of industrial biotechnology. The pharmaceutical industry underwent a sea change in 1928 when Alexander Fleming discovered penicillin, and it was mass produced during World War II. Large-scale fermentation methods were developed at this time, making it possible to generate vitamins, antibiotics, and organic acids [23]. The commercialization of genetic engineering initiated in 1976 with the founding of the first biotechnology companies, such as Genentech. Researchers invented the ability to alter the genetic material of organisms, leading to developments in recombinant DNA technology during the 1970s, which essentially changed the discipline. This method empowered the development of inherently modified microorganisms by generating human insulin, various therapeutic proteins and growth hormones. A significant milestone demonstrates the potentiality of biotechnology to generate complex biological molecules, which was the authorization of recombinant human insulin in 1982 [24].

The domain of industrial biotechnology started the quick growth during the late 20th and early 21st centuries. Developments in proteomics, genomics and systems biology accelerated more accurate genetic modifications, providing a deeper perception into cellular functions. The pace of innovation was further boosted by the initiation of high-throughput screening techniques and automation. During this period, synthetic biology developed, which implies the design and development of unique biological components and systems by integrating the concepts from biology and engineering [1]. One of the most substantial transformations is the growth of biomanufacturing, remarkably within the pharmaceutical industry. The biomanufacturers produce the biopharmaceuticals such as cellular therapies, monoclonal antibodies and vaccines utilizing living cells. These medicines are essential for recovering various diseases, including autoimmune disorders, cancer, and infectious diseases due to their intricate specificity and nature. As the field continues to grow, several key trends are forming its future. The addition of digital technologies such as machine learning (ML), Internet of Things (IoT), and artificial intelligence (AI) is changing biomanufacturing and bioprocessing. These technologies augment the product quality and productivity by facilitating optimization and real-time monitoring of manufacturing processes. Furthermore, advancements in synthetic biology and metabolic engineering are growing the range of products that can be progressed through biotechnological methods [25]. The chronological context of industrial biotechnology focuses the remarkable progress made over the past century. The advancements in engineering science and technology, the field has experienced the continuous evolution, from the early days of fermentation to the sophisticated biomanufacturing processes for present days. This progress has not only converted industries but has also contributed to addressing global challenges such as environmental sustainability, healthcare, and energy [26].

1.4 Fundamentals of Bioprocessing and Biomanufacturing

The main aspects of industrial biotechnology are biomanufacturing and bioprocessing, which develop the biological systems to produce goods with a market assessment. The production of biological materials, such as biochemicals, biofuels, and medicines can be optimized by a variety of technologies that fall under these categories. To fully utilize these methods, it is mandatory to realize their foundations [27]. Upstream processing is the first step among several crucial steps in bioprocessing. The preparation and growing of the microorganisms that will provide the intended product are involved in this step. The efficiency of the production process is decided by the choice of appropriate cell lines or microorganisms. To enhance these biological systems' productivity, genetic engineering techniques are commonly used [28]. The next step is fermentation in the bioprocessing process. The intended product is formed during fermentation by generating cells or microorganisms in bioreactors under carefully monitored conditions. The ideal growth conditions, such as pH, nutrition supply, temperature, oxygen levels, bioreactor design and operation are important. The scalability and efficiency of bioprocessing have been significantly augmented by the developments in bioreactor technology, such as the creation of continuous fermentation systems and single-use bioreactors [29].

The recovery and purification of the product from the fermentation broth create the downstream processing, which comes after fermentation. A number of procedures are involved at this stage, including purification, product isolation, and cell separation. To get the product's concentration and purity, the techniques like chromatography, filtering, centrifugation, and crystallization are commonly utilized. The product's nature and specifications for its intended use decide, which downstream processing techniques are applied [30]. By integrating bioprocessing into a unified production system, biomanufacturing promises the smooth transfer of information and materials from raw materials to final products. Upstream and downstream methods must be directed as part of this integration, and quality control techniques must be put in place to ensure the consistency and safety of the final product. Real-time monitoring and management of biomanufacturing methods have become simpler with the use of sophisticated control systems and process analytical technologies (PAT), increasing efficiency and decreasing adaptability [31]. The adoption of continuous manufacturing techniques is one of the major improvements in biomanufacturing. Continuous manufacturing functions in a steady-state mode, where raw materials are frequently fed into the system and products are constantly collected, in contrast to typical batch processing. The lower prices, shorter production times, and higher-quality products are just a few advantages of this strategy. The potentiality of continuous manufacturing to transform biomanufacturing has been exposed by its effective application in the production of biopharmaceuticals, including monoclonal antibodies and vaccines [32]. The creation of single-use technology, which makes the use of disposable bioreactor components, is another significant breakthrough. These technologies provide several advantages, such as decreased capital investment, a decreased danger of cross-contamination and greater flexibility. Particularly, single-use bioreactors have become more and more popular in the biopharmaceutical sector, where they are applied to produce therapeutic proteins and vaccines in tiny to medium-sized batches [33].

1.5 Innovations in Bioprocessing

The contemporary advancements in technology and an enhanced understanding of biological processes have resulted in significant innovations within the field of bioprocessing. These improvements have progressed the scalability, efficiency and sustainability of bioprocesses, thereby improving their applicability in industrial environments. Key areas of innovation consist of novel bioreactor designs, synthetic biology, the implementation of process analytical technologies (PAT) and advanced fermentation techniques.

(i) Advanced Methods of Fermentation

The core principle of bioprocessing is fermentation, which involves cultivating microorganisms or cells to manufacture the specific products. Recent developments in fermentation methods aim to enhance the productivity and yield through the optimization of various conditions. For example, the high-cell-density fermentation pointedly increases the product concentration by allowing the growth of numerous cells within a limited volume. This method is particularly beneficial in the production of biopharmaceuticals, where high yields are vital for economic feasibility.

Another important innovation is the growth of fed-batch and continuous fermentation methods. Fed-batch fermentation allows for extended production periods and increased cell densities by methodically introducing nutrients into the bioreactor. Conversely, the continuous fermentation keeps cells in a productive state for extended durations by perpetually supplying fresh medium while simultaneously eliminating spent medium and cells. In comparison to traditional batch fermentation, the continuous methods offer numerous advantages, including improved product uniformity, reduced downtime, and enhanced output.

(ii) Novel Bioreactor Designs

The productivity of bioprocessing is pointedly affected by the design and operation of bioreactors. Recent developments in bioreactor technology have focussed on oxygen transfer, improving mixing, and temperature regulation—all vital factors for product development and cellular growth. Due to their lower initial investment, adaptability, and reduced risk of cross-contamination, single-use bioreactors made from disposable plastic materials have become progressively popular. These bioreactors are mostly beneficial for processes that require stringent sterility and for production on a small to medium scale. Another important innovation is the use of perfusion bioreactors, where cells continue within the bioreactor while the medium is incessantly refreshed. This setup is ideal to produce biologics such as monoclonal antibodies and viral vectors, as it supports extended production durations and high cell densities. Additionally, the incorporation of sensors and advanced control systems in bioreactors enables adjustment of process parameters and real-time monitoring, thereby improving the productivity and product quality.

(iii) Metabolic Engineering and Synthetic Biology

Metabolic engineering and synthetic biology have transformed the bioprocessing by empowering the formation of unique cell lines and microbial strains tailored for specific industrial applications. Through genetic alterations, it is possible to enhance the metabolic pathways to progress the production of desired molecules, reduce byproduct formation, and improve cell growth and stability. For example, the modified yeast and bacteria create biofuels, bioplastics, and pharmaceuticals more skilfully than their wild-type counterparts. CRISPR-Cas9 technology acts as a controlling technology for precise genome editing, allowing for the targeted alteration of genes that perform a significant role in metabolic pathways. This technique has empowered the identification of novel biosynthetic pathways for valuable compounds and has improved the development of microbial strains with enhanced production capabilities. The integration of metabolic engineering and synthetic biology extends the range of organically formed products, ensuring that bioprocessing remains at the forefront of innovation.

(iv) Sustainable Bioprocessing

Sustainability performs a substantial role in modern bioprocessing with initiatives, focused on reducing ecological impact and augmenting the economic viability of biotechnological production. One of the developments in this area is the utilization of renewable feedstocks as substrates for microbial fermentation, including industrial byproducts and agricultural residues. This technique supports a circular economy by reducing the dependence on fossil fuels and enhancing the value of waste materials. Additionally, the progress of more effective and less energy-consuming separation and purification techniques has further boosted the sustainability of bioprocessing. For example, the membrane technologies provide a low-energy alternative to conventional filtration and chromatography methods, thereby diminishing the overall carbon footprint of the production process. The implementation of green chemistry principles and the design of bioprocesses that aim to reduce the waste generation and energy use are also vital for meeting the sustainability objectives.

1.6 Need to Shift to a Bio-Based Economy

The decline of fossil fuel resources is increasing the demand for biomass-based fuels and commodity chemicals. Therefore, new manufacturing methods that allow biomass to be converted into bulk and fine compounds in an environmentally responsible manner must be developed. The chemical industry is particularly affected by this paradigm shift because a variety of components are commonly created from materials with a petroleum base. In this case, organic solvents are great examples. The rise in demand is starting to affect the price of crude oil, which might have catastrophic consequences for world economies.

As our understanding of the many basic biological molecules grows, biotechnology advances. The petroleum-based economy is shifting gradually to the biofuels. It must improve the policies, financing, and legislation that nations are implementing. Biomanufacturing will then promote the engineering of living things to produce bioproducts. The bio factories will produce hydrogen, ethanol, butanol, bioenergy, and other biofuels in addition to raw materials for the building, automotive, and industrial sectors [34].

1.7 Biological Systems as a Biofactory

1.7.1 Bacterial Biofactory

A microbial cell factory, quick growth, and access to tools for gene and genome engineering, the well-established high cell density culture techniques, and an abundance of metabolic engineering systems are just a few advantages of E. coli, the most researched organism [35]. Its limited metabolic engineering systems are just one of its many drawbacks. Pseudomonas putida KT2440, a TOL plasmid-cured strain of Pseudomonas putida mt-2 is a frequently used strain. A novel bacteria called Pseudomonas putida is being assessed for use as an industrial host [36, 37]. Because it has specialized mechanisms that allow it to break down a wide range of dangerous substances in highly contaminated environments, it is very common [38]. The capacity of Pseudomonas species to create medium-chain-length polyhydroxyalkanoates makes them one of the most efficient biopolymer producers (PHAs) [39]. Numerous studies have been conducted on bacillus subtilis, an aerobic bacterium, because of its capacity to synthesize a variety of heterologous proteins. B. subtilis has demonstrated the productive expression of several enzymes, including lichenase, amylases, xylanases, and galactosidase. Through fermentation processes, bacillus subtilis can produce bacterial cellulases, proteases, and other enzymes [40].

1.7.2 Yeast Biofactory

Yeast has been used to produce wine, beer, and bread since the dawn of human civilization. It has, however, recently become one of the most well-known organisms that work as a sort of cell factory to produce various chemicals, medications, food ingredients, etc. [41]. Yeast cells are quite interesting since they are often utilized as model organisms to improve our understanding of human cells. Pichia pastoris, saccharomyces cerevisiae schizosaccharomyces, kluyveromyces lactis, and hansenula polymorpha are among the various yeast types employed in biomanufacturing [42]. Yeast is used to produce a wide range of goods, such as biofuels [43], and virus-like particles (VLPs), which are used to make vaccines [44], biopharmaceuticals, etc. The eukaryotic microbe yeast can produce post-translationally altered products, such as N-glycosylated molecules [45], which sets it apart from other eukaryotic organisms. In addition to their ability to be successfully grown at high concentrations, yeasts also provide scalable fermentation, rapid growth, a lower risk of contamination, easy genetic modification, and a more economical process.

1.7.3 Insect Cell Biofactory

In vitro insect cell culture has been extensively utilized since approximately 1950. Due to the ability to modify baculoviruses through biotechnology and use them as vectors, the infection of lepidopteran cell lines within a lytic system has facilitated the expression of heterologous proteins. The insect cell baculovirus expression vector system, or IC-BEVS, serves as a crucial facility to produce recombinant proteins and viral particles, which are essential for the development of valuable products. This underscores the importance of cell culture and gene transfer vectors [46, 47]. With superior glycosylation characteristics over other model systems, BEVS was created as a substitute for mammalian expression methods, and insect cells have produced a vast array of proteins [48]. During the initial phases of drug research, this is a common mode of expression. The yield would be lower than with a mammalian system due to the variable glycosylation pattern [49], even though desired commodities might be produced on a big scale using its vector technology.

1.7.4 Mammalian Cell Culture

Instead of using recombinant DNA (rDNA) in bacterial culture, certain biotechnology products, such viral vaccines, must be made using mammalian cells. The reason for this is because many complex proteins include glycan molecules through post-translational modifications. Biopharmaceuticals including enzymes and hormones, as well as immunobiological products like interleukins, lymphokines, and monoclonal antibodies, can be produced using rDNA technology in mammalian cell culture [50]. First-generation vaccines and replacement hormones were the first biopharmaceuticals to hit the market. Second-generation vaccines, recombinant proteins, and monoclonal antibodies came next. The COVID-19 pandemic has prompted major biotechnology and pharmaceutical companies to create a number of licensed or in-development COVID-19 vaccines or antiviral medications. Since cell culture is such an important part of drug development and manufacture, there is a greater need for cell culture products [51].

1.7.5 Biosensors

A biosensor is a measurement tool made up of an electrical system, transducer, base material, and bioreceptor [52]. To identify biological, organic, and inorganic chemical substances, it uses certain metabolic processes to generate electrical, thermal, or optical signals. The following are the many kinds of biosensors made from different organisms [53].

(i) Acetylcholinesterase Inhibition-Based Biosensors
Since the 1960s, organophosphorus pesticides (OPs) have been shown to be the most effective in maintaining agricultural productivity due to their high pest toxicity, ease of manufacture, and low production costs. Overuse of OP, however, could result in bioaccumulation in the environment, which would be a serious environmental issue. Furthermore, OPs phosphorylate acetylcholinesterase, irreversibly inhibiting it and potentially leading to neurological problems [54]. Researchers have developed a number of acetylcholinesterase inhibition-based biosensors to identify OP residue.

(ii) Microalgae Biosensors
Anthropogenic contamination is unavoidable when the industrial and agricultural sectors incorrectly discharge chemical waste into the soils, water, and air. A sustainable approach employing microalgae biosensors could be an advantageous option for the environment because of their availability, affordability, and disposability [55]. Additionally, a novel fluorometric biosensor chip based on the pseudokirchneriella subcapitata strain was created [56], that can rapidly identify a range of herbicides. Diuron, simetryn, simazine, and atrazine are the four herbicides that this biosensor chip is most sensitive to. Using a non-genetically modified strain of chlorella vulgaris, a bioluminescent microalgae biosensor was created [57], for the rapid detection of both light metals, like sodium, aluminum, and lithium, and heavy metals, like copper, lead, and cadmium.

(iii) Yeast-Based Biosensors
Indeed, a few cellular characteristics and molecular processes, such as numerous signalling pathways that are critical for identifying and reacting to environmental stimuli, are shared by yeast and mammalian cells. Interest in researching biosensing technologies has increased because of these advantages of yeast. For hormone detection, a fluorometric biosensor would be employed to accurately identify androgens, progestogens, and estrogens [58].

1.8 Progress in Biomanufacturing

It is believed that antibody-presenting viral nanoparticles (VNPs) hold great promise as a therapeutic and diagnostic tool. To exhibit nanobodies [59], two representative members of the genus Potyvirus—Tobacco etch virus and Zucchini yellow mosaic virus—can be used to produce genetically encoded viral nanoparticles that act as nanoscaffolds.

CdS quantum dots could be biomanufactured by employing directed evolution to develop a strain of stenotrophomonas maltophilia (SMCD1) that can make CdS nanocrystals extracellularly to aid in purification. Utilizing the biomanufacturing technique demonstrated here, the CdS semiconductor nanocrystals exhibit photoluminescent properties [60], and a size-dependent band gap. Although it will take advancements in our ability to predictably engineer plant biology, it is feasible to construct plants in space for the ideal biomanufacturing platform that can produce medications and biomaterials on demand. Small amounts of seeds can be kept viable at room temperature for many years, unlike microbes, which need to be preserved at below-freezing temperatures. Because they can repair their own carbon through photosynthesis mechanisms, plants just require CO_2 to grow [61]. Mannitol, a readily available component of seaweed and a promising feedstock for biomanufacturing, has a higher energy content than glucose. Mannitol can be converted by engineered microbes into several different bioproducts [62]. The ability of computers or machines under computer control to accomplish tasks that are comparable to those carried out by humans is known as artificial intelligence (AI). To attain the necessary titres, rates, and yield, a fully working AI would identify the best compounds to make, offer possible production routes, choose the ideal pathway configuration, and evaluate experimental data to correct the attempt. Machine learning in bioinformatics [63], refers to the use of machine learning methods in bioinformatics fields like as text mining, systems biology, microarrays, proteomics, evolution, and genomics. CRISPR/Cas has become a novel tool for metabolic pathway engineering in certain microorganisms [64], by altering the metabolic flux in the direction of the production of important biomolecules.

The exciting and novel field of biofabrication is being studied by tissue engineering and regenerative medicine. Biofabrication employs automated processes, such as additive manufacturing technology, to produce 3D biomaterial-based cell culture systems. Biofabrication is characterized by biological products, bio-inspired manufacturing processes, and cell-based building blocks. Because of this, most bio-products made by biofabrication are stem cell and biomaterial hybrids, including scaffolds, microcarriers, and microgels [65].

A branch of industrial biotechnology known as "biomanufacturing" is related to biological systems to generate high-value goods like biochemicals, biofuels and medicines. Improvements in technology and the fundamental understanding of biological mechanisms have driven important progress in this field. Key areas of development include single-use technologies, synthetic biology, the integration of automation with digital tools, and continuous biomanufacturing.

(i) Continuous Biomanufacturing

Traditional batch processing has transformed to continuous biomanufacturing. In continuous biomanufacturing, the inputs and outputs are in a constant flow as the production process remains uninterrupted. Enhanced productivity, consistent product quality, and reduced manufacturing times are merely a few advantages of this approach. The steady-state characteristics of continuous processes consent for the greater control over production conditions, which in turn increases lower costs. The production of biopharmaceuticals is an important application of continuous biomanufacturing. The reliable and highly effective production of therapeutic proteins have been enabled by continuous upstream processing methods, such as perfusion culture systems. To ensure an efficient production flow, downstream processes like filtration and continuous chromatography have also been advanced to complement continuous upstream systems. It presents the extensive benefits over traditional batch methods and has demonstrated the potentiality in the production of the biologics and monoclonal antibodies.

(ii) Integration of Digital Tools and Automation

The integration of digital tools and automation has pointedly improved the reliability and efficiency of biomanufacturing processes. Sophisticated control systems, data analytics, and sensors facilitate the real-time production monitoring and optimization. These technologies enable the precise control of input variables such as nutrient levels, temperature, and pH, ensuring optimal conditions for product synthesis and cell growth. Biomanufacturing is leveraging artificial intelligence (AI) and machine learning (ML) to envisage and improve the process performance. These tools can analyze the large datasets during the generation production time to identify trends and patterns that allow for proactive adjustments. AI-driven predictive maintenance decreases downtime and aids prevent equipment failures, further improving operational efficiency.

The perception of the "digital twin" is also being exploited in biomanufacturing. A digital twin, which is a virtual representation of a physical manufacturing system, is formed using advanced modeling techniques and real-time data. This digital model can be applied to test new perceptions, identify issues, and optimize and simulate production processes without disrupting ongoing operations. Digital twins hold the potential to improve scalability, industrial efficiency, and process development.

(iii) Sustainable Biomanufacturing

A substantial aspect of modern biomanufacturing is sustainability. The primary objectives are to improve the economic feasibility of biotechnological products and to diminish the environmental impact of production methods. Innovations in this area include waste reduction strategies, the use of renewable feedstocks, and energy-efficient production techniques. One approach to sustainable biomanufacturing includes waste and by-products as feedstocks for microbial fermentation. Substrates for microbial growth can comprise industrial byproducts, food waste, and agricultural residues, which helps to decrease the reliance on fossil fuels and substitute a circular economy. This method not only decreases the overall carbon footprint of biomanufacturing but also enhances the value to waste materials. Energy efficiency is another significant element of sustainable biomanufacturing. By advanced separation and purification methods, optimal process conditions and developing more efficient bioreactor designs, energy consumption can be significantly minimized. For instance, the membrane technologies improve the sustainability of biomanufacturing procedures by offering a low-energy substitute to chromatography techniques and traditional filtration.

1.9 Case Studies of Innovative Bioprocesses

Case studies focusing on advanced bioprocesses provide the remarkable understandings into the benefits and applications of industrial biotechnology. These instances demonstrate the impact of biotechnological improvements on industrial operations by showcasing successful applications across various fields.

(i) Insulin Production through Recombinant DNA Technology

The application of recombinant DNA technology for the production of human insulin stands out as one of the most important case studies. Before the initiation of biotechnology, the insulin was extricated from the pancreases of cows and pigs for diabetes treatment. In 1982, Genentech and Eli Lilly effectively generated the first recombinant human insulin by applying genetically modified E. coli bacteria, resulting in insulin that mirrors the production of the human pancreas. This breakthrough reduced the likelihood of allergic reactions and improved the availability of insulin, thereby altering diabetes treatment. The use of recombinant DNA technology not only improved the efficiency of insulin production but also enhanced its quality and safety. This case study represents how biotechnology can transform the pharmaceutical production and enhance the patient outcomes.

(ii) Bioethanol Production from Lignocellulosic Biomass

The production of bioethanol from lignocellulosic biomass, which assists as a sustainable substitute to fossil fuels, signifies an important case study. Traditional bioethanol production depends on the food crops such as sugarcane and corn, leading to concerns regarding food security and land use. Lignocellulosic biomass, agricultural waste, non-food plants, and sourced from wood offer a more sustainable feedstock for bioethanol production. Researchers have developed advanced bioprocessing methods to proficiently convert lignocellulosic biomass into fermentable sugars and subsequently into bioethanol. One of these methods includes the pre-treating the biomass to break down its complex structure, followed by the release of sugars through enzymatic hydrolysis. These sugars are then fermented into ethanol utilizing genetically yeast and modified bacteria. The successful commercialization of this process by companies such as DuPont and POET-DSM has presented the viability of producing bioethanol from lignocellulosic sources.

(iii) Production of Plant-Based Proteins

Microbial fermentation has been applied to generate the plant-derived proteins in reaction to the increasing demand for sustainable food options. Industrial biotechnology has been applied by companies such as Impossible Foods and Beyond Meat to develop meat substitutes that imitate the texture and taste of actual meat. Heme proteins generated through yeast fermentation are combined into plant-based components such as pea and soy protein to improve these products. Genetically modified yeast strains are used in the fermentation process to create heme, the chemical that gives meat its distinctive flavor and color. This invention has addressed ethical and environmental issues related to cattle rearing in addition to offering a sustainable substitute for animal flesh. The popularity of plant-based meat substitutes shows how biotechnology can revolutionize the food industry.

1.10 Overview of Biomaterials

The value and usefulness of metallic materials and their composites have been clear since the early 19th century. Significant progress has been made in the field of synthetic materials to understand how biomaterials interact with the human body and how to use them effectively [66, 67].

Since then, biomaterials—which are artificial—have been widely employed to replace or restore the many biological functions of human tissue. Because they constantly come into contact with bodily fluids, they aid in enhancing everyday human activities. The designed material should be examined for the same or distinct qualities required for its precise use through engineering design. When designing biomaterials, one crucial factor, that is, biocompatibility should be considered together with the material's chemical, physical, and mechanical properties, including whether it is bioactive, bioinert, or biodegradable [68].

In recent years, biotherapy employing these biomaterials has gained more traction in the prevention of several deadly illnesses, including cancer, heart infections, and AIDS. Combining basic science with practical applications, biotherapy has the potential to be a powerful treatment for fatal and incurable illnesses [69–71]. Biotherapy is mostly linked to the prevention of fatal illnesses using biotechnology or bio-based medications, including the identification of therapeutic targets through pathological analysis and the development of biotechnology related to immunotherapy, liver, etc. While immunotherapy is the use of vaccinations and antibodies to prevent disease, gene therapy is the intentional modification of gene expression in a specific tissue for the prevention of pathological diseases. The main conditions that can be treated with stem cell treatment are malignancies, hematological illnesses, cardiovascular disorders, etc. [72–75]. Furthermore, bone marrow transplantation is becoming a widely accessible stem-cell treatment. Tissue engineering is a rapidly expanding area of biomedical technology that makes it easier to regenerate and repair damaged tissues.

Biomaterials can be divided into two main classes based on where they come from: natural and synthetic (man-made). Natural biomaterials, such as gelatin, hyaluronic acid, heparin, and alginate, have been used for many years. Furthermore, because they are needed in big quantities, synthetic biodegradable polymers are currently being considered because of their wide-scale manufacture and simplicity of manipulation [76].

1.10.1 Biomaterials Classification

Categorization of biomaterials: Although many natural materials and their alloys are used in industry, only a small number of them demonstrate biocompatibility; as a result, they have outstanding bio-functionality and may be used as implantation materials in the future [77]. Table 1.1 explains some important traits and applications of different biomaterials.

1.10.2 Features of Biomaterials

The kind of medical application determines the requirements for biomaterial design and selection criteria. To have a strong use in the biomedical area for a longer period of time without immune rejection, the biomaterial needs to possess certain special qualities. Here are some descriptions of these traits [78, 79].

1. Exceptional biocompatibility
2. Adequate mechanical characteristics
3. Superior physical and chemical attributes
4. Sufficient wear resistance
5. Adequate rust resistance
6. Osseointegration (For bone implants)

Table 1.1 Important applications of biomaterials.

Sl. no.	Materials	Applications
1	Metals (such as stainless steel, Ag, Au, Ti and its alloy, etc.)	Weight-bearing implants include wires, joint replacements, screws, dental root implants, bone plates and pins, etc.
2	Ceramics, such as hydroxyapatite, zirconia, and alumina	Covering for dental and orthopedic implants, medical sensors, load-bearing implants, etc.
3	Polymers (such as polyester, silicon, and nylon)	Heart valves, blood vessels, artificial hearts, hip joints, contact lens sockets, etc.
4	Composites/Bio composites	Dental implants, bone cement, and joint replacements.

1.10.3 Applications of Biomaterials

Biomaterials are used in a wide range of industries, including pharmaceutical firms, food processing facilities, fashion design, and other home products. Biomaterials are widely used in the medical sciences for the creation of tissue scaffolds, implants, prostheses, and dental fixtures.

(i) Cardiovascular biomaterials

Certain heart conditions have a significant negative impact on people's quality of life and cannot be adequately treated with medication alone. As a result, the patient urgently needs supportive therapy that aids in improving their health. The usage of "biomaterials" is one such new treatment. These biomaterials can aid in the effective construction and implantation of vascular grafts and heart valves in cardiac diseases. Cardiovascular biomaterials (CB) are materials that are used as implants in conditions affecting the heart and that efficiently work with biological machinery. These days, there is a high need for cardiovascular biomaterials because of their success. Metals, polymers, and biological materials are the three groups into which these cardiovascular biomaterials can be divided. The type of impairment or injury determines the properties of the cardiovascular biomaterial utilized. For example, polymers have been the therapy of choice for cardiac damage; however, blood compatibility may provide issues. To solve this issue, certain novel technologies, such as surface changes, are now being investigated and used.

(ii) Blood vessel prostheses/vascular grafts

Blood vessel prostheses are the most widely used device for treating damaged or compromised blood vessels. Furthermore, vascular bypass, sometimes referred to as vascular graft technology, is a surgical technique that reopens blood channels to allow blood to freely flow from one location to another, reducing the effects of obstruction. It's a common procedure that involves avoiding the blocked blood channel or vessels to maintain normal blood flow between two healthy locations. The best material for this vascular bypass operation is either the patient's own vein (autograft) or a donor vein (allograft). In addition to them, polytetrafluoroethylene, or PTFE (teflon), and polyethylene terephthalate, or dacron, are also widely employed.

(iii) Heart valves

Nowadays, patients with valvular heart disease are treated by having an artificial heart valve implanted. This condition arises when the functioning of any one of the four heart valves is compromised. These prosthetic heart valves are divided into two categories based on how they work: biological and mechanical. PTFE, Dacron, and other synthetic materials or components of biological origin are used to make biological valves. Conversely, metals, polymers, and ceramics—such as silicone, titanium, stainless steel, and pyrolytic carbon—are used to make mechanical heart valves. Transcatheter aortic valve implantation has recently been accomplished using biological material (pig heart valve) and metals, ceramics, and polymers (titanium, stainless steel, and Dacron) [80].

(iv) Stents

The clogged blood channels supplying the heart with oxygenated blood are opened or bypassed by a stent. These stents, which can be either plastic or metallic, are placed into the blood vessel lumen. Stents come in various varieties, including urethral, coronary, prostatic, vascular, esophageal, and biliary stents, depending on the needs [81]. These stents can be divided into three groups based on their distinct physical and functional characteristics: drug-eluting stents (DES), bioabsorbable stents, and bare-metal stents (BMS).

(v) Artificial ligaments and tendons

The musculoskeletal system is supported and given stiffness by specific connective tissues, such as ligaments and tendons. Artificial tendon and ligament replacement is urgently needed when these tendons and ligaments are destroyed as a result of an accident or sports injury. However, it should be remembered throughout the implantation procedure that these synthetic tendons and ligaments need to share structural, functional, and biochemical traits with their natural counterparts. In order to withstand the body weight immediately following implantation, the biomaterial employed to create these artificial tendons and ligaments needs to possess sufficient mechanical strength [82].

(vi) Bone cement

A biomaterial called bone cement is utilized to bridge the space between the prosthetic and the bone. This bone cement is essentially made up of two parts: a liquid (MMA monomer) that serves as an inhibitor and a powder (pre-polymerized PMMA/PMMA copolymer beads) that functions as an initiator. The process of free radical polymerization is started when the initiator is added to the accelerator, changing the cement's viscosity and turning it into a hard substance [83]. The biomaterial utilized to fill in the gaps must mostly consist of bioresorbable material that promotes the growth of new bone in the injured location. CaP-based bioceramics are the most widely utilized biomaterial for bone regeneration because of their superior biocompatibility, osseointegration, and osteoconduction properties [84]. Hip implants are the most typical instance of CaP.

1.11 Manufacturing Process for Porous Implants

A popular fabrication approach for creating porous implants is the powder metallurgy method, which provides fine control over mechanical characteristics and porosity. Space-holder substances like ammonium bicarbonate or sodium chloride are integrated with metal particles like its alloys or titanium in this method. The combination experiences a sintering process, where high temperatures fuse the metal particles without melting them, after they have been crushed into the desired shape. The space-holder materials decompose during the sintering process, resulting in a porous structure in their place. This controlled porosity is vital for promoting osseointegration, as it allows for vascularization and bone ingrowth. Moreover, other methods that provide unique advantages regarding interconnectivity, customization, and pore size distribution, such as electrospinning, additive manufacturing and freeze casting are presently under investigation. Among these approaches, additive manufacturing is progressively favored due to its ability to create intricate geometries and patient-specific designs with controlled internal architectures, thereby improving the biological and mechanical performance of the implant.

1.11.1 Powder Metallurgy

A popular production method for generating porous implants is powder metallurgy, which provides the exact control over the material's microstructure and porosity. This procedure integrates metal powders, compacts them into the essential shape, and then sinters the compact at temperatures much below the underlying material's melting point. Particle bonding is controlled by the sintering process while maintaining the porosity structure, which is energetic for biomedical applications. Because it improves tissue integration, assists the bone growth, and increases the mechanical compatibility between the natural bone and implant, porosity in implants is vital. By adapting the powder properties, sintering parameters and compaction pressure, the pore interconnectivity and degree of porosity can be customized. Powder metallurgy regularly uses materials like its alloys and titanium because of its exceptional corrosion resistance, biocompatibility and mechanical strength. All things considered, powder metallurgy is an adaptable and economical technique for generating intricately shaped, porous implants that satisfy the functional needs of orthopedic and dental applications [85].

1.11.2 Additive Manufacturing

In the field of biomedical engineering, additive manufacturing (AM) has developed as a groundbreaking production technique, particularly for the development of porous implants. Unlike traditional subtractive manufacturing methods, additive manufacturing (AM) empowers the construction of highly tailored structures layer by layer from digital blueprints. This ability, which emulates the mechanical properties of natural bone, is particularly valuable for manufacturing implants with controlled porosity, an energetic factor for promoting osteointegration, enhancing nutrient flow, and dropping stress shielding. Various methods such as Selective Laser Melting (SLM), Fused Deposition Modeling (FDM), and Electron Beam Melting (EBM) are commonly applied to generate polymeric and metallic implants with changing pore shapes, sizes, and distributions. Due to their mechanical strength and biocompatibility, materials like cobalt-chromium alloys, biodegradable polymers and titanium alloys are often utilized. AM's design versatility allows for the making of implants featuring gradient porosity and interconnected pore networks, both of which are vital for long-term stability and vascularization. Despite challenges such as complying with regulatory standards, optimizing post-processing procedures, and ensuring consistent quality, additive manufacturing remains to boost improvements in implant design. Its capacity to generate functionally graded, the patient-specific porous implants signify an important advancement in the fields of personalized and regenerative medicine [86].

1.11.3 Electrospinning

For producing porous implants, electrospinning is a popular production method, especially in the fields of regenerative medicine and tissue engineering. By exposing a polymer solution or melt to a high-voltage electric field, fine, and continuous fibers are formed. These fibers are then assembled on a grounded surface to generate a nonwoven fibrous mat. The resultant structure offers a highly porous architecture with linked pores and a high surface area, closely resembling the extracellular matrix (ECM) of real tissues. Electrospun scaffolds are perfect for promoting cell adhesion, proliferation, and differentiation because of their properties. Implants that are both biocompatible and biodegradable can be made using a variety of natural and synthetic polymers, including collagen, gelatin, polycaprolactone (PCL), and polylactic acid (PLA). Furthermore, by modifying processing parameters such as voltage, solution concentration, flow rate, and collector type, the porosity and fiber diameter may be precisely regulated. Although changes like multi-layering or the addition of nanoparticles can improve their mechanical qualities for possible use in bone regeneration, electrospinning is especially useful for producing porous implants used in soft tissue applications, such as skin, nerve, and vascular grafts. cost-effectiveness, the capacity to create ECM-like structures that support tissue regeneration, and scaling up and attaining consistent thickness in larger constructs, despite certain limits [87].

1.12 Investment and Strategic Alliances to Boost Commercial Biomanufacturing

1.12.1 Partnerships for Product Development

The development and uptake of biomanufacturing techniques can be greatly accelerated by strategic alliances. Platform businesses like Amyris, Zymergen, and Ginkgo Bioworks present fresh chances for cross-industry application co-development. Significant scale and cost advantages for new development work are provided by collaborating with and engineering on platforms created by these kinds of businesses. Furthermore, industry experts can assist in guiding development towards the most crucial key performance indicators (KPIs) for product success for application-focused biomanufacturing startups founded on substantial experience in a specific technology area, such as Modern Meadow, Impossible Foods, and Bolt Threads.

(i) Cell development toolkits

Platform businesses make use of economies of scale inside a certain technology ecosystem that can support a wide range of applications. As a stand-in, the majority of computer software applications are created using a standard toolkit on a distinct platform, and the software is separated from the application. Similarly, platform firms are making the bioeconomy more accessible as basic biological unit activities become more standardized. For instance, Ginkgo Bioworks is creating a platform for cell programming to facilitate the engineering of biology. To facilitate the development of new products in close collaboration with product-oriented companies that lack in-house capabilities for large-scale biological engineering, their platform compiles an exhaustive list of the infrastructure, personnel, and tools required by businesses to design and develop microbes for new applications.

To reduce the barrier to creating new biomanufacturing apps, Ginkgo has launched Cell Development Kits (CDKs), which are the biological counterpart of software development kits used to create applications for iOS and Android. Pre-engineered cells are optimized for absorbing DNA libraries, instruments for creating and printing new DNA sets, specialized machinery for creating and cultivating the engineered organisms, and the automation, know-how, and infrastructure required to assess and validate bioprocess performance are all included in cell development kits. These platforms provide substantial cost savings for biomanufacturing when compared to the substantial time, money, and human commitment needed for new biomanufacturing companies to start producing material to evaluate novel applications. Technology pioneers find it particularly alluring to test and develop early theories without having to make significant financial investments. Larger businesses also used these kinds of technologies to investigate novel biomanufacturing approaches without having to worry about setting up labs and staff. CDKs are anticipated to shorten development durations from years to a few months in addition to saving money.

(ii) Sustainable product performance

Currently, the rate of application testing and product development required to bring a product to market does not keep up with the pace of upstream biomanufacturing advances, which refers to the biological manufacture or fermentation of a molecule. Although they are not always the most qualified to guide the subsequent procedures to purify, isolate, and manufacture a completed product from that molecule, synbiotechnology pioneers frequently contribute extensive technical experience to the manufacturing of a class of molecules. Technology innovators have a great chance to get the input they need on key upstream KPIs for successful deployment through strategic alliances with established product firms, while traditional manufacturers can quickly embrace a biomanufacturing approach. Genomatica and Lululemon announced their collaboration in 2020 to create a sustainable bio-nylon.

The $10 billion global nylon industry is based on the petrochemical derivative caprolactam [88]. Genomatica and Lululemon are collaborating to create caprolactam sustainably by fermenting plant sugars in a microbe. This presents a chance to get rid of the 60 million tons of nitrous oxide, a dangerous greenhouse gas that is normally produced by the production of petrochemicals. Genomatica has a great deal of experience developing biomaterials, and it is likely to hasten the use of sustainable biomaterials going forward while gaining from the product knowledge of seasoned producers like Lululemon. The production of bio-nylon uses a lot less energy and water than petrochemical alternatives, which gives it a competitive sustainability edge. It also dramatically lowers greenhouse gas emissions. The intrinsic sustainability of the bionylon production process turns into a "performance" indicator that has retail value for customers of Lululemon's premium goods even in the absence of improved material performance. Leveraging the sustainability advantage targeted at high-margin and luxury consumers may be a practical way to offset the upfront costs associated with the development and uptake of biomanufactured products.

(iii) Collaborations in manufacturing technologies

More companies are finding it economically feasible to use sustainable and cutting-edge materials as the cost of biomanufacturing applications keeps falling. Pharmaceuticals are often manufactured on a smaller scale (less than 1000-liter tanks), with greater manufacturing costs that are offset by the large profit margins. The production expenses of commodity chemicals, like bioethanol, are spread across larger batch sizes, allowing them to attain positive economics at large volumes (>200-kiloliter tanks). Until a market is sufficiently established to sustain scaled-up production, many potential biomanufacturing applications find it difficult to reconcile the high expense of producing at comparatively smaller levels. As the efficiency of these bio-transformations increases due to advancements in biological sciences, biomanufacturing costs are anticipated to continue declining. Partnerships with manufacturing technology partners can also result in significant synergies, even though they can also hasten the market acceptance of specific applications.

Partnerships that combine advances in synthetic biology with current manufacturing techniques, for example, are crucial to accelerating product development and testing required to commercialize these difficult applications, such as biomaterials. Modern Meadow is a synthetic biology startup platform that uses protein engineering and design to create sustainable performance materials through biofabrication. Today, Modern Meadow has concentrated its technological efforts on creating Bio-Alloy, a patented application platform that combines biopolymers and tailored proteins. In comparison to conventional leather, the Bio-Alloy technology offers several mechanical and thermal benefits while lowering greenhouse gas emissions by 60–85%, making it possible to make a strong entry into the materials market. To further speed up the implementation and uptake of the latest generation of performance biomaterials, Modern Meadow established a joint venture collaboration, BioFabbrica, with renowned Italian textile and materials supplier Limonta this year. BioFabbrica is prepared to provide sustainable biofabricated material solutions that satisfy the distinct performance and design requirements of brands by fusing Limonta's vast experience in state-of-the-art materials manufacturing techniques with Modern Meadow's patented Bio-Alloy Application Platform.

1.12.2 Expanding the Implementation of Novel Solutions

A large portion of the biomanufacturing capacity available today is focused on pharmaceuticals, which creates a difficult pricing structure for novel low-cost biomanufacturing applications. The infrastructure created for pharmaceutical applications is more advanced than downstream capabilities to chemically and physically separate targets for applications in alternative biomanufacturing domains. To guarantee equivalency and/or compatibility with the intended use, early prototyping is generally crucial in the commercialization of biomanufactured goods. If necessary, this would enable more upstream engineering cycles. For technology pioneers, the absence of this early prototyping turns into a significant hazard in bioengineering initiatives. For instance, gating to biomanufacturing applications is the availability of commercial demonstration scale for non-pharmaceutical uses. However, for early stage biomanufacturing goods, producing an initial prototype batch or maiden commercial production run presents a considerable logistical and financial hurdle. While historical leaders in fermentative manufacturing, such as DSM and Cargill, have achieved notable success in these kinds of applications, technology pioneers introducing novel manufacturing concepts to the market face a capital-intensive process in bearing the costs of development and manufacturing scale. Academic labs can now demonstrate small-scale processes, although their equipment and capabilities are frequently restricted. Large-scale solutions can also be provided by contract manufacturing companies, although working with these companies can be costly and necessitates pre-existing business need. The significance of having an in-house manufacturing footprint is also being diminished by advancements in the automation and sophistication of upstream process development technology (such as Sartorius' ambr250, Culture Biosciences, and Univercell), which allow for faster technology transfer across production.

(i) The biomanufacturing co-operative

Pre-competitive bioprocess demonstration facilities that are built with the flexibility of applications now possible are necessary, as evidenced by the dispersed status of pilot and prototyping capacities for biomanufacturing applications. Technology pioneers may be able to speed up application testing and prototyping through public-private initiatives like "biomanufacturing cooperatives," in which businesses jointly fund and co-design, hence raising the success rate of biomanufacturing applications overall. Applications in biomanufacturing have many upstream equipment requirements, which might facilitate market segment and biosecurity application variety. To cover the required range of applications and overcome a big obstacle to biomanufacturing prototyping, downstream capabilities are far more varied and demand careful design and substantial expenditure. The features of such a model are examined.

(ii) Co-operative design needs

Today, the majority of biomanufacturing prospects come from growing microorganisms in a stirred tank to carry out a biotransformation on inexpensive feedstock. Microbial fermentation is now the first step in the production of proteins for meatless burgers, compounds like caprolactam for bio-nylon, and the plasmid DNA starting material for COVID-19 vaccines.

Tank size, particular process control requirements, and regulatory compliance (such as oxygen feeds, solvent rating, and good manufacturing practice compliance) can vary depending on the application; nonetheless, the main capital components and capabilities are comparable. Although the biomanufacturing ecosystem shares the engineering biology toolbox, each product application has different processing requirements in order to meet the critical performance indicators for product success. As biological engineering capabilities mature, the pipeline of upstream bioprocess development to biologically create desired items is growing and speeding up. The downstream unit procedures necessary to separate the product and get it ready for product application cannot be generally described in the same way. A wide range of chemical, physical, and thermal unit activities are required to extract and process target products from biological systems for several uses outside of the biological component of biomanufacturing [89]. To guarantee minimal acceptable performance against fundamental product KPIs and prevent product integration issues following significant time and financial commitment, downstream development must be started as early as feasible in the biomanufacturing development cycle. The capacity to evaluate the newly obtained raw input in the product context—through nutritional research of novel ingredients, stress testing of biomaterials, and sensory testing of flavor and fragrances—is essential for the successful integration of biomanufacturing applications. Only once, all upstream and downstream unit procedures have been designed to purify and isolate the finished product are these application testing techniques feasible.

A lot of startups squander time and money building up a specialized manufacturing scale before confirming the targeted application's financial feasibility. To boost the rate of commercial biomanufacturing achievements, scaling partnerships present a chance to broaden the biomanufacturing ecosystem and combine complementary knowledge and technology. Collaborative initiatives, would pool the resources and knowledge required to support the deployment of biomanufacturing products. A significant obstacle to tech pioneers' commercialization of novel biomanufacturing applications will be removed with investment in the development of state-of-the-art, collaborative spaces and manufacturing resources, which will also hasten bioeconomic growth.

1.13 Challenges and Limitations in Bioprocessing and Biomanufacturing

To fully realize the potential of industrial biotechnology, several obstacles and constraints still need to be overcome, even considering the many developments in bioprocessing and biomanufacturing. These difficulties can be roughly divided into four categories: environmental, regulatory, economic, and technical.

(i) Challenges in Process Integration and Automation

There are advantages and disadvantages to incorporating new technologies like machine learning (ML) and artificial intelligence (AI) into bioprocessing and biomanufacturing processes. Although these technologies can improve predictive maintenance and process optimization, their deployment necessitates a large investment and level of knowledge. Extensive data gathering, model training, and validation are all necessary to ensure that AI and ML models are accurate, dependable, and applicable to actual bioprocesses. System integration and interoperability issues are also brought about by the automation of bioprocessing and biomanufacturing. Complex software and hardware solutions are needed to guarantee that various production system components, including bioreactors, sensors, and control systems, operate in unison. Furthermore, it is essential to keep automated systems flexible and adaptive in order to support various bioprocesses and production scales.

(ii) Environmental Challenges

Industrial biotechnology presents environmental challenges even as it provides more sustainable alternatives to conventional chemical processes. To reduce the impact on the environment, biowaste disposal, including fermentation wastes and discarded microbial cultures, must be controlled. High concentrations of organic matter and nutrients found in biowaste have the potential to pollute and eutrophicate water basins if improperly handled. Another important issue is feedstock sustainability. The sourcing and supply chain of biomass must be controlled to prevent adverse environmental effects, even though using renewable biomass as fuel is more sustainable than using fossil-based resources. For instance, if biomass feedstocks are not managed sustainably, their large-scale cultivation may result in deforestation, biodiversity loss, and changes in land use.

(iii) Economic Challenges

There are important financial barriers due to the high costs related to the raw materials, process development and bioreactor equipment. Bioreactors and other biomanufacturing apparatus often require the extensive initial investments. Moreover, the overall economics of bioprocessing can be profoundly influenced by the costs of raw materials such as growth media and feedstocks. Considerable expenses are also associated to optimization and process development, specifically in the creation of cell lines and microbial strains. The iterative nature of strain generation, which includes multiple rounds of genetic testing and modifications, further increases the financial resources and time required. Furthermore, the regulatory demands for biopharmaceuticals and other biotechnological products require the rigorous testing and validation to confirm the safety and effectiveness, leading to additional costs.

(iv) Technical Challenges

Enhancing the microbial strains and cell lines implies a significant technological challenge in bioprocessing. The optimization of metabolic pathways and comprehensive genetic engineering are crucial to attain higher efficiency, yet these systems can be intricate and time-consuming. Additionally, maintaining the steadiness of genetically modified strains in industrial environments is both challenging and critical. Strain instability can lead to productivity losses and variations in product quality. Another key technical challenge is the scalability of bioprocesses. While laboratory-scale processes can be precisely controlled and optimized, scaling them up to industrial levels often explores the variability and inefficiencies. At larger scales, the issues such as oxygen transfer, mixing, and heat removal become more noticeable, necessitating the advanced bioreactor designs and control systems.

1.14 Conclusion

Industrial biotechnology has been changed bioprocessing and biomanufacturing, leading to more active and sustainable production techniques. This study indicates advancements, key applications, and future prospects in the field, showcasing biotechnology's widespread impact across various industries. The historic development of industrial biotechnology reveals the evolution from traditional fermentation systems to advanced biomanufacturing approaches. The development of microbial strains has been pointedly improved through the integration of synthetic biology, genetic engineering, and genome editing technologies such as CRISPR-Cas9. This progress has empowered the efficient and precise production of bio-inspired materials. Innovations in bioprocessing, including new fermentation methods, novel bioreactor designs, and process analytical technologies (PAT) have enhanced the efficiency of biotechnological production. Continuous biomanufacturing and single-use technologies have made biomanufacturing more effective and accessible, while also improving flexibility, lowering contamination risks, and reducing costs. Industrial biotechnology is used in various sectors, including pharmaceuticals, food and beverage, biofuels, agriculture, and environmental management. Bioprocessing has facilitated the creation of sustainable biofuels, environmentally friendly bioplastics, life-saving biopharmaceuticals, addressing urgent global challenges such as healthcare, energy sustainability, and plastic pollution. Notable instances that demonstrate the practical and transformative potential of biotechnological innovations include the bioethanol from lignocellulosic biomass, production of insulin, and bioplastics derived from microbial fermentation. Despite these developments, numerous challenges and limitations remain. Technical hurdles, such as the improvement of microbial strains and the expansion of bioprocesses, necessitate ongoing research and development.

The sustainable bioprocessing and circular economy will continue to be significant areas of concentration, propelling initiatives to reduce their negative effects on the environment and increase resource efficiency. The sustainability of biotechnological production will be improved by the creation of green chemistry concepts, biorefinery platforms, and renewable feedstocks. New markets and prospects for industrial biotechnology will be created by emerging applications in specialty chemicals, sophisticated materials and food technology. The advancement of industrial biotechnology will require interdisciplinary and collaborative approaches. Innovative bioprocess development and commercialization will be facilitated by collaborations among academic institutions, industry, and regulatory agencies. The next generation of biotechnologists will be trained through interdisciplinary research and education programs that promote creativity and teamwork. In conclusion, by offering effective and sustainable manufacturing methods, industrial biotechnology has the potential to completely transform a number of industries. Bioprocessing and biomanufacturing have the potential to significantly improve healthcare, energy, food, and environmental management by tackling present issues and utilizing new trends. Industrial biotechnology's ongoing progress will be essential to tackling global issues and advancing sustainable development, which will lead to a more prosperous and sustainable future.

It has been noted in recent years that these biodegradable polymers have been widely used in biotherapy because of their biocompatibility and biodegradability. However, the higher standard for these biomaterials has been established following the development of recently developing fields like biotechnology and medical technology. Perfect biodegradable polymers with special properties are quite valuable. Synthetic biodegradable polymers have been discovered to have lower immunogenicity and to be highly amenable to chemical alteration when compared to natural biodegradable polymers. As a result, they may present new opportunities in the future.

In summary, the development of biomanufacturing has been a protracted and intricate process, with numerous significant discoveries and breakthroughs for the creation of goods in the biotechnology, pharmaceutical, food processing, and industrial areas, as well as new biomanufacturing applications, are constantly being explored. Many organisms have the potential to be biofactories, whether they be whole organisms, their cells, or extracts of bacteria, yeast, algae, plants, animals, and insects. The possibilities for biomanufacturing are almost endless with further study and advancement. Numerous sectors could undergo a transformation because of biomanufacturing. Biomanufacturing has made it possible to generate more economical, sustainable, and efficient goods that could benefit both the human health and environment. Biomanufacturing has the potentiality to play a bigger role in our economy and daily lives with extra study and development.

1.15 Future Prospects in Biomanufacturing

As a result of advancement in technological advancements, scientific research, and an increasing demand for sustainable solutions, industrial biotechnology is anticipated to experience considerable growth and change in the coming years. Various innovative concepts are persuading the areas of biomanufacturing and bioprocessing, offering potential enhancements in scalability, sustainability, and efficiency.

(i) Synthetic Biology and Genome Engineering
One of the most important areas for the future of industrial biotechnology is synthetic biology. This discipline combines engineering principles with biological concepts to generate and develop innovative biological tools, components, and systems. Recent advancements in synthetic biology have empowered the production of tailored microbial strains and cell lines with optimized metabolic pathways for the production of various products, including biochemicals, biofuels, pharmaceuticals, and food ingredients. The field of genetic engineering has been reformed by CRISPR-Cas9 and other genome-editing technologies, which facilitate precise modifications to cellular and microbial genomes. These methods improve the productivity and efficiency of bioprocesses by optimizing metabolic pathways and eliminating undesirable byproducts.

(ii) Incorporation of Artificial Intelligence and Machine Learning
The bioprocessing and biomanufacturing sectors are on the border of an important transformation because of the integration of artificial intelligence (AI) and machine learning (ML). The vast amounts of data created during production can be analyzed by AI and ML algorithms to identify trends, improve workflows, and envisage outcomes. By facilitating control of bioprocesses and real-time monitoring, these systems improve the productivity, lessen variability, and augment product quality. Moreover, AI-driven predictive maintenance can anticipate the equipment failures and schedule repairs proactively, minimizing downtime and ensuring uninterrupted production. Furthermore, AI-enabled process optimization can reduce the resource and energy consumption, thereby enlightening the sustainability of bioprocessing. It is projected that the application of AI and ML in industrial biotechnology will improve and exert a more important impact as these technologies change.

(iii) Industry 4.0 Technology Adoption

Bioprocessing and biomanufacturing are developing owing to the influence of Industry 4.0. Key components of Industry 4.0 being combined into biotechnological production include automation, improved data analytics, and the Internet of Things (IoT). Real-time information regarding process conditions can be attained from IoT-enabled equipment and sensors, facilitating precise control and optimization. Digital twins, which are virtual representations of actual production systems, are applied to improve bioprocesses. These digital models can assess various scenarios, identify potential issues, and refine process design without disrupting real operations. With the adoption of Industry 4.0 technologies, biomanufacturing is projected to become increasingly efficient, flexible, and scalable, thereby improving its ability to reply to changing production necessities and market demands.

(iv) Bioprocessing Sustainability and the Circular Economy

An important concern for the future of industrial biotechnology is sustainability. Efforts are currently being made to develop environmentally friendly and efficient bioprocesses. This includes the generation of biochemicals, bioplastics and biofuels from renewable feedstocks, such as agricultural byproducts and waste materials. Moreover, the application of green chemistry is a growing trend, which aims to reduce the use of harmful substances and decrease waste production. Bioprocessing is embracing the perception of a circular economy, which promotes the recycling and repurposing of waste resources within production processes. One efficient method to diminish waste and confirm a sustainable supply of raw materials is to apply the waste biomass as a feedstock for microbial fermentation.

(v) Teamwork and Multidisciplinary Methods

Interdisciplinary and collaborative methods that integrate knowledge from various fields, including chemistry, biology, data science, and engineering, will configure the future of industrial biotechnology. The development of new technologies, the progress of research, the safe and effective application of bioprocesses depend on the alliance among regulatory bodies, academia, and industry. Public-private partnerships can accelerate the development and commercialization of biotechnological novelties by enabling the sharing of resources, infrastructure, and expertise. Preparing the next generation of biotechnologists and nurturing a culture of innovation also deliberately relies on educational initiatives and interdisciplinary research.

AI disclosure statement: The author acknowledges the use of AI-based tools for minor language editing or grammatical correction. However, all intellectual contributions, analyses, and conclusions presented in this chapter are entirely the author's own.

References

1. Zhang, Y.H.P., Sun, J., Ma, Y., Biomanufacturing: history and perspective. *J. Ind. Microbiol. Biotechnol.*, 44, 4–5, 773–784, 2017.
2. Roh, K.H., Nerem, R.M., Roy, K., Biomanufacturing of therapeutic cells: state of the art, current challenges, and future perspectives. *Annu. Rev. Chem. Biomol. Eng.*, 7, 1, 455–478, 2016.
3. Murray, A.J., Zhu, J., Wood, J., Macaskie, L.E., A novel biorefinery: Biorecovery of precious metals from spent automotive catalyst leachates into new catalysts effective in metal reduction and in the hydrogenation of 2-pentyne. *Miner. Eng.*, 113, 102–108, 2017.
4. Vasic Racki, D., History of industrial biotransformations–dreams and realities, in: *Industrial biotransformations*, pp. 1–36, Wiley Online Library, Weinheim, Germany: Wiley-VCH Verlag GmbH & Co. KGaA, 2006.
5. Natarajan, S., Biomimetic, bioresponsive, and bioactive materials edited by Matteo Santin and Gary J. Phillips. *Mater. Manuf. Process.*, 31, 7, 976–977, 2016.
6. Geetha, M., Singh, A.K., Asokamani, R., Gogia, A.K., Ti-based biomaterials, the ultimate choice for orthopaedic implants—a review. *Prog. Mater. Sci.*, 54, 3, 397–425, 2009.
7. Gao, C., Peng, S., Feng, P., Shuai, C., Bone biomaterials and interactions with stem cells. *Bone Res.*, 5, 1–33, 2017.
8. Prakash, C., Kansal, H.K., Pabla, B.S., Puri, S., Aggarwal, A., Electric discharge machining—a potential choice for surface modification of metallic implants for orthopedic applications: a review. *Proc. Inst. Mech. Eng., Part B: J. Eng. Manuf.*, 230, 2, 331–353, 2016.
9. Liu, X., Chu, P.K., Ding, C., Surface modification of titanium, titanium alloys, and related materials for biomedical applications. *Mater. Sci. Eng. R Rep.*, 47, 3–4, 49–121, 2004.
10. Bartolo, P., Kruth, J.P., Silva, J., Levy, G., Malshe, A., Rajurkar, K., Mitsuishi, M., Ciurana, J., Leu, M., Biomedical production of implants by additive electro-chemical and physical processes. *CIRP Ann. Manuf. Technol.*, 61, 2, 635–655, 2012.
11. Bhatt, I., Mishra, S.K., Paul, P.K., Challenges and Opportunities in Biomanufacturing, in: *Biomanufacturing for Sustainable Production of Biomolecules*, pp. 349–359, Springer Nature Singapore Pte Ltd., Singapore, 2023.

12. McGovern, P.E., Zhang, J., Tang, J., Zhang, Z., Hall, G.R., Moreau, R.A., Nunez, A., Butrym, E.D., Richards, M.P., Wang, C.S., Cheng, G., Fermented beverages of pre- and proto-historic China. *Proc. Natl. Acad. Sci.*, 101, 51, 17593–17598, 2004.
13. Sano, C., History of glutamate production. *Am. J. Clin. Nutr.*, 90, 3, 728S–732S, 2009.
14. Goto, M., Akai, K., Murakami, A., Hashimoto, C., Tsuda, E., Ueda, M., Kawanishi, G., Takahashi, N., Ishimoto, A., Chiba, H., Sasaki, R., Production of recombinant human erythropoietin in mammalian cells: host-cell dependency of the biological activity of the cloned glycoprotein. *Biotechnology*, 6, 1, 67–71, 1988.
15. Bornscheuer, U.T., Huisman, G.W., Kazlauskas, R.J., Lutz, S., Moore, J.C., Robins, K., Engineering the third wave of biocatalysis. *Nature*, 485, 7397, 185–194, 2012.
16. Liese, A. and Hilterhaus, L., Evaluation of immobilized enzymes for industrial applications. *Chem. Soc. Rev.*, 42, 15, 6236–6249, 2013.
17. Siegel, J.B., Smith, A.L., Poust, S., Wargacki, A.J., Bar-Even, A., Louw, C., Shen, B.W., Eiben, C.B., Tran, H.M., Noor, E., Gallaher, J.L., Computational protein design enables a novel one-carbon assimilation pathway. *Proc. Natl. Acad. Sci.*, 112, 12, 3704–3709, 2015.
18. Ye, X., Zhang, C., Zhang, Y.H.P., Engineering a large protein by combined rational and random approaches: stabilizing the Clostridium thermocellum cellobiose phosphorylase. *Mol. Biol. Syst.*, 8, 6, 1815–1823, 2012.
19. Nahavandi, and Saeid, Industry 5.0—A human-centric solution. *Sustainability*, 11, 16, 4371, 2019.
20. Majeed, Z., Mubashir, M., Chiong, M.C., Transition of Bio-manufacturing Industry from 4.0 to 5.0: Advances and Challenges, in: *The Prospect of Industry 5.0 in Biomanufacturing*, pp. 115–133, CRC Press (Taylor & Francis Group), Boca Raton, FL, USA, 2021.
21. Damiano, L. and Stano, P., Synthetic biology and artificial intelligence. Grounding a cross-disciplinary approach to the synthetic exploration of (embodied) cognition. *Complex Syst.*, 27, 199–228, 2018.
22. Smart, N.J., *Lean biomanufacturing: Creating value through innovative bioprocessing approaches*, pp. 1–384, Woodhead Publishing Limited, Cambridge, UK, 2013.
23. Gottschalk, U., Brorson, K., Shukla, A.A., The need for innovation in biomanufacturing. *Nat. Biotechnol.*, 30, 6, 489–492, 2012.
24. Pandey, K., Pandey, M., Kumar, V., Aggarwal, U., Singhal, B., Bioprocessing 4.0 in biomanufacturing: paving the way for sustainable bioeconomy. *Syst. Microbiol. Biomanuf.*, 4, 2, 407–424, 2024.
25. Clomburg, J.M., Crumbley, A.M., Gonzalez, R., Industrial biomanufacturing: the future of chemical production. *Science*, 355, 6320, aag0804, 2017.
26. Rollin, J.A., Tam, T.K., Zhang, Y.H.P., New biotechnology paradigm: cell-free biosystems for biomanufacturing. *Green Chem.*, 15, 7, 1708–1719, 2013.

27. Nielsen, J., Tillegreen, C.B., Petranovic, D., Innovation trends in industrial biotechnology. *Trends Biotechnol.*, 40, 10, 1160–1172, 2022.
28. Cordova, L. and Lee, K., Cells as Biofactories: Parallels Between Biopharmaceutical Manufacturing and Industrial Biotechnology. *Ind. Biotechnol.*, 19, 4, 208–217, 2023.
29. Kelley, B., Industrialization of mAb production technology: the bioprocessing industry at a crossroads. *Mabs*, 1, 5, 443–452, 2009.
30. Yu, L.P., Wu, F.Q., Chen, G.Q., Next-generation industrial biotechnology-transforming the current industrial biotechnology into competitive processes. *Biotechnol. J.*, 14, 9, 1800437, 2019.
31. Straathof, A.J., Wahl, S.A., Benjamin, K.R., Takors, R., Wierckx, N., Noorman, H.J., Grand research challenges for sustainable industrial biotechnology. *Trends Biotechnol.*, 37, 10, 1042–1050, 2019.
32. You, C. and Zhang, Y.H.P., Cell-free biosystems for biomanufacturing, in: *Future trends in biotechnology*, pp. 89–119, Springer Science+Business Media, Dordrecht, Netherlands, 2013.
33. São Pedro, M.N., Silva, T.C., Patil, R., Ottens, M., White paper on high-throughput process development for integrated continuous biomanufacturing. *Biotechnol. Bioeng.*, 118, 9, 3275–3286, 2021.
34. Brumbley, S.M., Purnell, M.P., Petrasovits, L.A., Nielsen, L.K., Twine, P.H., Developing the sugarcane biofactory for high-value biomaterials. *Int. Sugar J.*, 1297, 5, 2007.
35. Ibrahim, T.N.B.T., Abas, A.B., Razak, N.F.A., Escherichia coli Cell Factory for Synthesis of Biomolecules, in: *Biomanufacturing for Sustainable Production of Biomolecules*, pp. 141–163, Springer Nature Singapore Pte Ltd., Singapore, 2023.
36. Wigneswaran, V., Pseudomonas putida as a microbial cell factory, Department of Systems Biology, Technical University of Denmark, pp. 1–123, 2016.
37. Weimer, A., Kohlstedt, M., Volke, D.C., Nickel, P.I., Wittman, C., Industrial biotechnology of Pseudomonas putida: advances and prospects. *Appl. Microbiol. Biotechnol.*, 104, 7745–7766, 2020.
38. Faizal, I., Dozen, K., Hong, C.S., Kuroda, A., Takiguchi, N., Ohtake, H., Takeda, K., Tsunekawa, H., Kato, J., Isolation and characterization of solvent-tolerant Pseudomonas putida strain T-57, and its application to biotransformation of toluene to cresol in a two-phase (organic-aqueous) system. *J. Ind. Microbiol. Biotechnol.*, 32, 542–547, 2005.
39. Kamaludin, N.H. and Feisal, N.A.S., Biomanufacturing for Sustainable Production of Biomolecules: Pseudomonas putida Cell Factory, in: *Biomanufacturing for Sustainable Production of Biomolecules*, pp. 175–188, Springer Nature Singapore Pte Ltd., Singapore, 2023.
40. Mishra, S.K., Bhatt, I., Paul, P.K., Bacillus subtilis Cell Factory, in: *Biomanufacturing for Sustainable Production of Biomolecules*, pp. 165–173, Springer Nature Singapore Pte Ltd., Singapore, 2023.

41. Nielsen, J., Yeast systems biology: model organism and cell factory. *Biotechnol. J.*, 14, 9, e1800421, Germany, 2019.
42. Mittal, M., Varshney, A., Singh, N., Saini, A., Mani, I., Yeast Cell Factory for Production of Biomolecules, in: *Biomanufacturing for Sustainable Production of Biomolecules*, pp. 211–251, Springer Nature Singapore Pte Ltd., Singapore, 2023.
43. Shi, S., Si, T., Liu, Z., Zhang, H., Ang, E.L., Zhao, H., Metabolic engineering of a synergistic pathway for n-butanol production in Saccharomyces cerevisiae. *Sci. Rep.*, 6, 1, 25675, 2016.
44. Kim, H.J. and Kim, H.J., Yeast as an expression system for producing virus-like particles: what factors do we need to consider? *Lett. Appl. Microbiol.*, 64, 2, 111–123, 2017.
45. Han, M. and Yu, X., Enhanced expression of heterologous proteins in yeast cells via the modification of N-glycosylation sites. *Bioengineered*, 6, 2, 115–118, 2015.
46. Ikonomou, L., Schneider, Y.-J., Agathos, S.N., Insect cell culture for industrial production of recombinant proteins. *Appl. Microbiol. Biotechnol.*, 62, 1–20, 2003.
47. Palomares, L.A., Estrada, S., Ramirez, O.T., Principles and applications of the insect cell baculovirus expression vector system, in: *Cell culture technology for pharmaceutical and cell-based therapies*, S.S. Ozturk and W.-S. Hu (Eds.), pp. 627–692, Taylor and Francis, New York, 2006.
48. Kost, T.A., Condreay, J.P., Jarvis, D.L., Baculovirus as versatile vectors for protein expression in insect and mammalian cells. *Nat. Biotechnol.*, 23, 5, 567–575, 2005.
49. Kaur, R., Kumar, A., Mandal, D., Insect Cell Factory for Production of Biomolecules, in: *Biomanufacturing for Sustainable Production of Biomolecules*, pp. 283–296, Springer Nature Singapore Pte Ltd., Singapore, 2023.
50. Verma, A., Verma, M., Singh, A., Animal tissue culture principles and applications, in: *Animal Biotechnology*, pp. 269–293, Amsterdam, Netherlands: Academic Press (Elsevier), 2020.
51. Arora, P., Singh, V., Kumar, A., Mammalian Cell Culture: An Edge to Biopharmaceutical Industry, in: *Biomanufacturing for Sustainable Production of Biomolecules*, pp. 297–313, Springer Nature Singapore Pte Ltd., Singapore, 2023.
52. Tavakoli J and Tang, Y., Hydrogel based Sensors for biomedical applications: an updated review. *Polymers (Basel)*, 9, 8, 364, 2017.
53. Lee, Y.Z., Low, S.C., Ngan, C.L., Biosensor for Detecting Biomolecules, in: *Biomanufacturing for Sustainable Production of Biomolecules*, pp. 87–122, Springer Nature Singapore Pte Ltd., Singapore, 2023.
54. McHardy, S.F., Wang, H.Y.L., McCowen, S.V., Valdez, M.C., Recent advances in acetylcholinesterase Inhibitors and Reactivators: an update on the patent literature (2012-2015). *Expert Opin. Ther. Pat.*, 27, 4, 455–476, 2017.

55. Purohit, J., Chattopadhyay, A., Singh, N.K., Green synthesis of microbial nanoparticle: approaches to application, in: *Microbial nanobionics*, R. Prasad (Ed.), pp. 35–60, Springer Nature, Basel, 2019.
56. Kashem, M.A., Kimoto, K., Iribe, Y., Suzuki, M., Development of microalgae biosensor chip by incorporating microarray oxygen sensor for pesticides sensing. *Biosensors*, 9, 4, 133, 2019.
57. Wong, L.S., Judge, S.K., Voon, B.W.N., Tee, L.J., Tan, K.Y., Murti, M., Chai, M.K., Bioluminescent microalgae-based biosensor for metal detection in water. *IEEE Sens. J.*, 18, 5, 2091–2096, 2017.
58. Chamas, A., Pham, H.T.M., Jahne, M., Hettwer, K., Uhlig, S., Simon, K., Kunze, G., Simultaneous detection of three sex steroid hormone classes using a novel yeast based biosensor. *Biotechnol. Bioeng.*, 114, 7, 1539–1549, 2017.
59. Xin, B., Zhong, C., Wang, Y., Integrating the marine carbon resource mannitol into biomanufacturing. *Trends Biotechnol.*, 2282, 5, 2023.
60. Yang, Z., Lu, L., Berard, V.F., He, Q., Kiely, C.J., Berger, B.W., McIntosh, S., Biomanufacturing of CdS quantum dots. *Green Chem.*, 17, 7, 3775–3782, 2015.
61. Mortimer, J.C. and Gilliham, M., SpaceHort: redesigning plants to support space exploration and on-earth sustainability. *Curr. Opin. Biotechnol.*, 73, 246–252, 2022.
62. Marti, M., Merwaiss, F., Butkovic, A., Daros, J.A., Production of potyvirus-derived nanoparticles decorated with a nanobody in biofactory plants. *Front. Bioeng. Biotechnol.*, 502, 2022.
63. Wani, A.W., Khalangre, A.B., Rajan, R., Mirza, A.A., Yaseen, I., *et al.*, Artificial Intelligence-Assisted Production of Biomolecules, in: *Biomanufacturing for Sustainable Production of Biomolecules*, pp. 123–140, Springer Nature Singapore Pte Ltd., Singapore, 2023.
64. Ingle, K.P., Singh, A., Sahni, T., Aakanksha, K., Kumar, S., Geethika, P., Wani, A.W., Genome Editing in the Synthetic Biology for Sustainable Production of Biomolecules, in: *Biomanufacturing for Sustainable Production of Biomolecules*, pp. 315–329, Springer Nature Singapore Pte Ltd., Singapore, 2023.
65. Xu, Y., Chen, C., Hellwarth, P.B., Bao, X., Biomaterials for stem cell engineering and biomanufacturing. *Bioact. Mater.*, 4, 366–379, 2019.
66. Chen, Q. and Thouas, G.A., Metallic implant biomaterials. *Mater. Sci. Eng. R*, 87, 1–57, 2015.
67. Patel, N.R. and Gohil, P.P., A Review on Biomaterials: Scope, Applications & Human Anatomy Significance. *Int. J. Emerging Technol. Adv. Eng.*, 2, 2250–2459, 2012.
68. Navarro, M., Michiardi, A., Castaño, O., Planell, J.A., Biomaterials in orthopaedics. *J. R. Soc. Interface*, 27, 1137–1158, 2008.

69. Jieheng, W., Angang, Y., Weihong, W., Current status of bench-to-bedside translation of cancer therapeutic antibodies. *Chin. J. Cancer Biother.*, 3, 023, 2014.
70. Abdelbary, H., Brown, C.W., Werier, J., *et al.*, Using targeted virotherapy to treat a resistant ewing sarcoma model: from the bedside to the bench and back. *Sci. World J.*, 171439, 1–11, 2014.
71. Uckun, F.M., Monoclonal antibody (mAb)-based biotherapy options for b-lineage non-Hodgkin's lymphoma (NHL). *Trans. Med.*, 3, e118, 2013.
72. Segers, V.F. and Lee, R.T., Stem-cell therapy for cardiac disease. *Nature*, 451, 937–42, 2008.
73. Huang, X., Liu, D., Liu, K., *et al.*, Haploidentical hematopoietic stem cell transplantation without in vitro T- cell depletion for the treatment of hematological malignancies. *Bone Marrow Transplant.*, 8, 291–7, 2006.
74. Kim, S.U. and De Vellis, J., Stem cell-based cell therapy in neurological diseases: a review. *J. Neurosci. Res.*, 87, 2183–200, 2009.
75. Childs, R., Chernoff, A., Contentin, N., *et al.*, Regression of metastatic renal-cell carcinoma after non-myeloablative allogeneic peripheral-blood stem cell transplantation. *N. Engl. J. Med.*, 343, 750–8, 2000.
76. Tian, H., Tang, Z., Zhuang, X., *et al.*, Biodegradable synthetic polymers: preparation, functionalization and biomedical application. *Prog. Polym. Sci.*, 37, 237–80, 2012.
77. Jenkoa, M., Sekb, M.G., Godeca, M., Hodnikc, M., Batica, B.S., Donika, C., Grantd, J.T., Dolinar, D., Surface chemistry and microstructure of metallic biomaterials for hip and knee endoprostheses. *Appl. Surf. Sci.*, 427, 584–593, 2018.
78. Kadam, A.G., Pawar, S.A., Abhang, S.A., A Review on Finite Element Analysis of Different Biomaterials used in Orthopedic Implantation. *Int. Res. J. Eng. Technol.*, 4, 2192–2195, 2017.
79. Vesque, J.L., Hermawan, H., Dube, D., Mantovani, D., Design of a pseudo-physiological test bench specific to the development of biodegradable metallic biomaterials. *Acta Biomater.*, 4, 284–295, 2008.
80. Kostrzewa, B. and Rybak, Z., History, present and future of biomaterials used for artificial heart valves. *Polim. Med.*, 43, 3, 183–189, 2013.
81. Jaganathan, S.K., Supriyanto, E., Murugesan, S., Balaji, A., Asokan, M.K., Biomaterials in cardiovascular research: applications and clinical implications. *Biomed Res. Int.*, 459–465, 2014.
82. Kuo, C.K., Marturano, J.E., Tuan, R.S., Novel strategies in tendon and ligament tissue engineering:advanced biomaterials and regeneration motifs. *Sports Med. Arthrosc. Rehabil. Ther. Technol.*, 2, 20, 2010.
83. Havelin, L.I., Espehaug, B., Vollset, S.E., Engesaeter, L.B., The effect of the type of cement on early revision of Charnley total hip prostheses: A review of eight thousand five hundred and seventy-nine primary arthroplasties from the Norwegian Arthroplasty Register. *J. Bone Joint Surg. Am.*, 77, 10, 1543–1550, 1995.

84. Eliaz, N. and Metoki, N., Calcium phosphate bioceramics: a review of their history, structure, properties, coating technologies and biomedical applications. *Materials (Basel, Switzerland)*, 10, 4, 2017.
85. Dutta, S., Devi, K.B., Roy, M., Processing and degradation behavior of porous magnesium scaffold for biomedical applications. *Adv. Powder Technol.*, 28, 12, 3204–3212, 2017.
86. Bose, S., Ke, D., Sahasrabudhe, H., Bandyopadhyay, A., Additive manufacturing of biomaterials. *Prog. Mater. Sci.*, 93, 45–111, 2018.
87. Muerza-Cascante, M.L., *et al.*, Melt electrospinning and its technologization in tissue engineering. *Tissue Eng. Part B Rev.*, 21, 2, 187–202, 2014.
88. Cumbers, J., Bio-nylon is the new green: How one company is fermenting a $10 billion market, Forbes, https://www.forbes.com/sites/johncumbers/2020/02/11/bio-nylon-is-the-new-green-how-one-company-is-fermentingyour-futurematerials/?sh=72cd6e284030.
89. Community data, Forum consultation, Policy Department, Ginkgo Bioworks, 2021.

2

Mathematical Modeling and Simulation Techniques for Biomanufacturing

Jagadeesha T.[1]* **and Sandip Kunar**[2]

[1]*Department of Mechanical Engineering, National Institute of Technology Calicut, Kozhikode, Kerala, India*

[2]*Department of Mechanical Engineering, Aditya University, Surampalem, Andhra Pradesh, India*

Abstract

This book chapter presents an in-depth examination of the key role of mathematical modeling and simulation in the design and optimization of contemporary biomanufacturing systems. The focus is placed on the integration of mechanistic models, kinetic equations, thermodynamic balances, stoichiometric approaches, and transport phenomena to explain and predict biological behavior across a variety of scales. Sophisticated simulation techniques such as computational fluid dynamics (CFD), finite element modeling (FEM), and multiscale modeling are described to analyze complicated interactions such as nutrient transport, fluid flow, and scaffold mechanics. The application of dynamic simulation methodology using ordinary differential equations and digital twins and machine learning-based predictive analytics for real-time monitoring of processes is also discussed in the chapter. Mathematical science complemented with engineering significance positions modeling and simulation as the foundation for scalable, robust, and smart biomanufacturing. This chapter identifies spaces of considerable potential in the future through cyber-physical integration, biologically realistic virtual models, and simulation together with high-throughput experimental data.

***Keywords*:** Mathematical modeling, biomanufacturing, simulation, digital twins, multiscale modeling, tissue engineering

**Corresponding author*: jagdishsg@nitc.ac.in

Sandip Kunar, Jagadeesha T. and D. Sathis Kumar (eds.) Biomanufacturing: Transforming Production of Biologics and Biopharmaceuticals, (47–66) © 2026 Scrivener Publishing LLC

2.1 Introduction

Biomanufacturing is truly a multidisciplinary and dynamically evolving and emerging active research area; it combines biology, chemistry, and engineering in designing and manufacturing biologically significant products critical to medicine, agriculture, food, and industry. Basically, biomanufacturing utilizes living entities cells or biologically active molecules in most instances to transform such products as therapeutic proteins, vaccines, antibodies, enzymes, and bio-based materials [1]. This technology differs from traditional manufacturing by virtue of inherent complexity, biological heterogeneity, and the close reliance on living organisms, adding an element of non-linearity and uncertainty not commonly experienced in most traditional mechanical systems. In the last twenty years, biomanufacturing has evolved from experimental-scale operations in research labs to commercial-scale operations. It is not easy to scale up the process to production line because of so many practical and legal challenges. In contrast to traditional manufacturing processes regulated by deterministic and linearly interpreted physical laws, biomanufacturing is based on biochemical kinetics, cellular metabolism, and environmental interactions, whose behavior is dynamic and nonlinear [2]. Successful translation of laboratory-scale processes to industrial production not only demands empirical knowledge but accurate predictive ability. That is where mathematical modeling and simulation come in.

Mathematical modeling in biomanufacturing plays several critical steps and procedures. It enables system behavior prediction, optimization of process parameters, critical control points identification, and simulation of process response under different operating conditions. Models may be simple empirical correlations or highly complex mechanistic models of mass transport, reaction kinetics, gene expression, and metabolic flux distributions [3]. By the application of experimental data, along with engineering theory, models provide alternatives to mimic biological processes in silico, so that there is less dependence on lengthy and expensive experimental iterations. Simulation software enables engineers and researchers to analyze alternatives to design choice, process variations, and system disturbances without physically carrying out experiments. For example, bioreactor models that mimic nutrient uptake, oxygen transfer, pH fluctuation, and cell growth kinetics are used to guide process design and control [4]. In addition, there is a requirement for computational models for the assessment of cell culture dynamics, tissue scaffold interactions, and microenvironmental gradients for tissue engineering purposes. Thus, modeling and simulation play a great role in risk reduction, process standardization, and regulatory compliance for biomanufacturing [5].

Bio manufacturing has gained its momentum due to advances in biology, computational fluid dynamics (CFD), finite element methods (FEM), and machine learning. Advanced characterization techniques and instrumentation push biomanufacturing modeling to unprecedented levels. Hybrid strategies such as these, enable the examination of multiscale effects—molecular and cellular scale through the macroscale process scale—to reconcile the complexity of the biological system with engineering control [6]. For instance, in perfusion bioreactors used for tissue construction, cell growth models are coupled with fluid mechanics models for enhancing the delivery of nutrients and mechanical forces. Similarly, in additive biomanufacturing or 3D printing of biological tissues, it is important to model cell viability, material response, and extrusion process under changing thermal and shear stresses to enable structural accuracy and biological functionality. The destiny of biomanufacturing is inseparably joined with predictive modeling in the future. As regulatory bodies like the FDA demand quality-by-design (QbD) and process analytical technology (PAT) philosophies, usage of validated mechanistic-in-nature models has come to the forefront [7]. Additionally, in an era of personalized medicine, with patient-specific treatments and products on the horizon, modeling and simulation can readily cut down development time and costs.

The purpose of this chapter is to present a detailed analysis of mathematical modeling and simulation's use within biomanufacturing. It evolves from basic concepts, history, to current process methods and technologies. Highlighting the utilization of mathematical approaches for modeling cell systems, bioreactor performance, and tissue engineering constructs will be the focus. Case studies and examples will be presented for successful applications. Lastly, future challenges and directions are mentioned, pointing out areas where modeling is continuing to evolve to match expectations of this evolving, multidisciplinary field.

2.2 Biomanufacturing Processes with Examples

Biomanufacturing involves a wide array of operations that transform biological materials to products of choice for commercial sale. Most operations progress through three main phases: upstream processing, downstream processing, and formulation/final product development. Each phase is distinguished by particular objectives, technologies, and biological interactions, and hence by distinctive design, control, and optimization strategies

2.2.1 Upstream Processing

Upstream processing starts with the choice and cultivation of the right biological system—bacterial, yeast, fungal, insect, or mammalian cells. These are optimized to produce the protein, metabolite, or structural tissue component sought. Inoculum growth in shake flasks or spinner flasks is scaled up to seed and production bioreactors. Classic example of an oldie but goldie is Chinese Hamster Ovary (CHO) cell manufacture of monoclonal antibodies, whose productivity and cell viability are maintained by media composition control, dissolved oxygen, and pH [8]. A typical example is microbial fermentation to produce bioethanol or lactic acid by saccharomyces cerevisiae or lactobacillus species, whose substrate feeding profiles are optimized for yield and reduced for by-product formation.

2.2.2 Bio Reactor Operations

Bioreactor operation is the foundation of upstream processing. These may be continuous, fed batch, or batch bioreactors based on the nature of the product and the microorganism. Fed-batch reactors are common in biopharmaceutical production because they can meet both the cell density and quality product requirements [9]. Continuous perfusion systems are, however, gaining popularity to produce labile proteins and for applications in regenerative medicine where long-term stability in culture is desired. For instance, perfusion bioreactors have a pivotal position in the osteoblast and chondrocyte culture in tissue engineering when there is a need of continuous removal of waste and supply of nutrients to mature tissues [10].

2.2.3 Downstream Processing

Downstream processing entails recovery and purification of the product from the multi-component biological system. It generally includes cell separation (e.g., centrifugation or filtration), protein capture (e.g., affinity chromatography), and polishing (e.g., ion exchange or size-exclusion chromatography). Ultrafiltration and diafiltration are usually applied for the production of enzymes to desalt and concentrate the product without inactivating it [11]. An excellent example would be the purification of insulin produced in recombinant E. coli, where inclusion bodies are solubilized, refolded, and purified using sequential chromatography steps.

2.2.4 Cell and Tissue Based Biomanufacturing

In biomanufacturing based on cells and tissues, it is quite a different process. Instead of purifying a molecule, the final product could be the entire cellular construct. In 3D printing of cells, for instance, the procedure starts with bioink preparation—cells in hydrogels such as alginate or gelatine methacrylate and continues with layer-by-layer deposition to create functional tissues like vascular grafts or skin patches [12]. Post-treatment includes crosslinking, bioreactor incubation, and mechanical or biochemical stimulation for maturation. Scaffold-based methods use biomaterial matrices seeded with stem cells that differentiate *in situ* under directed biochemical gradients and mechanical loads mimicking *in vivo* environments.

2.2.5 Integration of Process and Control

Process integration and control are needed in every step of the processes. Parameters such as temperature, pH, oxygen transfer rate, and agitation need to be regulated in very narrow ranges to provide consistent products [13]. These are regulated with high-level process control systems employing real-time feedback and monitoring in up-to-date plants. Integration with simulation packages allows predictive tuning, e.g., regulation of feeding regimen based on modeled growth of cells or metabolic changes.

2.2.6 Modular Biomanufacturing

Another emerging area is modular biomanufacturing that enables decentralized and flexible manufacturing platforms. This is being experimented with for on-site production of personalized therapies like CAR-T cells, where modular cleanroom modules and closed-system bioreactors enable fast, GMP-conform manufacture without the need for central infrastructure [14]. These operations are dependent on model-based digital twins for predicting yield, sterility assurance, and patient-specific variation.

2.3 Mathematical Modelling in Biomanufacturing

Mathematical modeling helps in prediction, and optimization of biomanufacturing processes. While traditional manufacturing systems are linear, single scale, and time-invariant because they utilize non-living systems and static responses, biomanufacturing is nonlinear, multiscale, and time-varying because it employs living systems and dynamic biological responses. Mathematical models here attempt to characterize the intricate interconnection among biological reactions, transport effects, and bioreactor dynamics in a quantitative structured description [15]. These models vary from empirical correlations to mechanistic hypotheses, and most of them combine several fields of biology and engineering.

2.3.1 General Form of Bioprocess Models

The common shape of a model for a biomanufacturing process can be given by a system of ordinary differential equations (ODEs) or partial differential equations (PDEs) that define mass balances, energy balances, and reaction kinetics [16]. A typical mass balance for a substrate in a perfectly mixed batch reactor can be given as:

$$\frac{dS}{dt} = -\frac{1}{Y_{X/S}} \cdot \mu(S) \cdot X \tag{2.1}$$

Where, S is the substrate concentration (g/L), X is the biomass concentration (g/L), μ(S) is the specific growth rate (1/h) and $Y_{X/S}$ is the yield coefficient of biomass on substrate (g biomass/g substrate).

For biomass, equation can be written as

$$\frac{dX}{dt} = \mu(S) \cdot X \tag{2.2}$$

From this, growth rate equation becomes

$$\mu(S) = \mu_{\max} \cdot \frac{S}{K_S + S} \tag{2.3}$$

Where, $\mu_{\max}$ is the maximum specific growth rate K_S is the half-saturation constant (g/L) and bio manufacturing process involves product formation, the model suggested is Luedeking-Piret equation, which can give insight into product accumulation.

$$\frac{dP}{dt} = \alpha \cdot \mu(S) \cdot X + \beta \cdot X \tag{2.4}$$

Where, P is the product concentration (g/L). α and β are growth-associated and non-growth-associated product formation coefficients.

2.3.2 Kinetic Models Used in Biomanufacturing

For more sophisticated biological systems, such as mammalian cell cultures where recombinant proteins are being produced, models include several substrates (glucose, glutamine), inhibitors (ammonia, lactate), and metabolic switching [17].

$$\mu(S) = \mu_{max} \cdot \frac{S}{(K_S + S)} \frac{K_I}{K_I + I} \tag{2.5}$$

Where I is inhibitory by-product and K_I is the inhibition constant and structural kinetic model is divided into active and inactive fractions which gives dynamics cell viability.

$$\frac{dX_a}{dt} = X_a(\mu - k_d) \tag{2.6}$$

$$\frac{dX_d}{dt} = k_d X_a \tag{2.7}$$

k_d is known as death rate constant of order one.

2.3.3 Transport Phenomena and PDE Models

Spatial gradients of nutrients, oxygen, and waste products such as those found within tissues or biofilms must be described through PDEs in non-ideal systems: In a general equation regarding mass transport within a porous scaffold

$$\frac{\partial C}{\partial t} = D\nabla^2 C - R(C, X) + \mathrm{u} \cdot \nabla C \tag{2.8}$$

Where, C is the concentration of species like oxygen, D is the diffusivity, $R(C, X)$ reaction term, u is the velocity of the fluid, ∇^2 is Laplacian operator and product $\mathrm{u} \cdot \nabla C$ represents convective transfer term.

$$\frac{\partial C_O}{\partial t} = D_O \nabla^2 C_O - q_O X + u \cdot \nabla C_O \tag{2.9}$$

C_O is oxygen concentration D_O is oxygen diffusivity, q_O specific oxygen uptake rate and X is local cell density.

2.3.4 Thermodynamic and Stoichiometric Models

Biomanufacturing processes frequently depend on cellular metabolism to transform substrates into important products like biofuels, enzymes, drugs, or structural biomaterials. To effectively design and optimize these processes, it is important to formulate models that represent the conservation principles of cellular functions [18]. Stoichiometric and thermodynamic models are critical analytical tools for this field, especially for genome-scale modeling, pathway analysis, and metabolic engineering.

i) Stoichiometric Modelling Framework
The basic mass balance for a metabolite within a biochemical network under pseudo-steady state (where intracellular metabolite concentration is nearly constant) is expressed as:

$$S \cdot v = 0 \tag{2.10}$$

S is the stoichiometric matrix with order $\mathrm{m} \times \mathrm{n}$ where m represents metabolites and n represents reaction and v is the rate of reaction fluxes.

ii) Flux Balance Analysis (FBA) FBA is a strong linear programming method applied for the prediction of metabolite flux in a metabolic network. The purpose is to maximize or minimize a specific biological process, typically biomass production, product yield, or ATP generation. FBA model can be expressed as

$$\text{Maximize} \quad Z = c^T v \tag{2.11}$$

$$\text{subject to} \quad S \cdot v = 0 \tag{2.12}$$

$$v_{min} \leq v \leq v_{max} \tag{2.13}$$

iii) Thermodynamic Constraints

These stoichiometric models are also augmented by incorporating thermodynamic viability into them, i.e., the fluxes of metabolites predicted should conform to the principles of thermodynamics.

If the Gibbs free energy is negative, then the reaction is said to thermodynamically feasible.

$$\Delta G < 0 \tag{2.14}$$

$$\Delta G = \Delta G^{\circ} + RT\ln\left(\frac{\prod[\text{products}]^{\nu_p}}{\prod[\text{reactants}]^{\nu_r}}\right) \tag{2.15}$$

Where ΔG° is standard Gibbs free energy change, R is the universal gas constant, T is absolute temperature, ν_p and ν_r are stoichiometric coefficients of products and reactants respectively.

This connection to thermodynamic properties involves directions of reaction and metabolite concentration. Thermodynamic constraints may be incorporated into linear or mixed-integer models when metabolite concentration ranges are known or estimated to rule out infeasible patterns of flux [21]. In thermodynamically constrained flux balance analysis, inequalities are introduced so that reactions cannot move in the opposite direction of their thermodynamic potential.

$$\Delta G_j \leq 0 \quad \text{for all } j \in \text{active reactions} \tag{2.16}$$

Extra binary variables are then added to provide for directionality of reaction, resulting in MILP structures.

2.4 Simulations in Biomanufacturing

Simulation of biomanufacturing is an authoritative method to anticipate, examine, and maximize intricate biological and biochemical processes engaged in biopharmaceutical, biofuel, enzyme, and regenerative material manufacture. In contrast with traditional manufacturing, biomanufacturing is controlled by nonlinear time-dependent and multiscale dynamics, with living cells or enzymes as active entities often utilized [19]. Virtual experimentation through simulation models reduces expense, facilitates

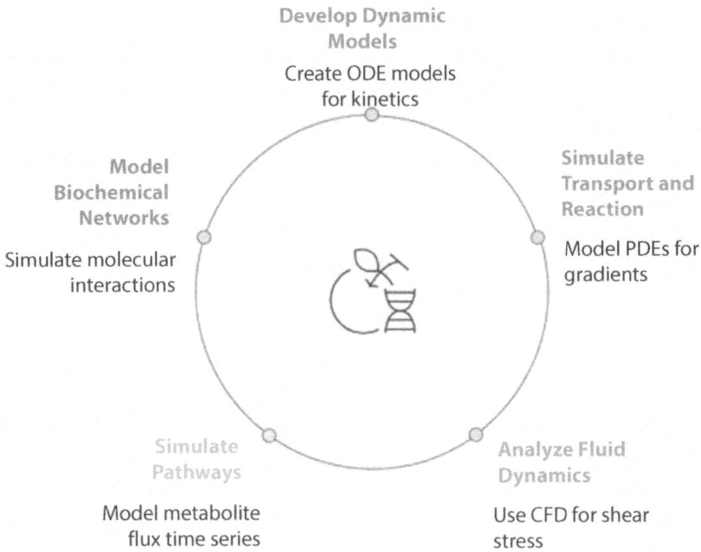

Figure 2.1 Application area and examples in simulation of biomanufacturing.

scale-up design, and allows for online observation of the process. The main objectives of simulations in biomanufacturing are

a) Process Design & Optimization: Designing bioreactor configurations, medium, and inoculum strategy.
b) Scale-Up Studies: Simulation of the behavior of mixing, oxygen transfer, and heat generation on scaled-up from bench to industrial scale.
c) Real-Time Monitoring: Application of digital twins and soft sensors for adaptive control.
d) Cost and Waste Minimization: Reducing batch failures, cycle time, and resource utilization.

Application area and typical examples are shown in Figure 2.1.

2.4.1 Bio Reactor Simulations Using CFD (Computational Fluid Dynamics)

Bioreactors are the heart of biomanufacturing systems, providing the physical and chemical environment for biological reactions. Their performance is governed by complex interactions among fluid flow, mass transfer, heat transport, and biochemical reactions [20]. Computational fluid dynamics (CFD) offers a robust simulation approach to analyze these phenomena in silico, allowing for design optimization and process scale-up without the cost and time associated with physical experiments.

CFD uses numerical solution of Navier-Stokes equations supplemented by transport equations for species, energy, and occasionally population balances.

Conservation of the mass is given by

$$\frac{\partial \rho}{\partial t} + \nabla \cdot (\rho \vec{v}) = 0 \tag{2.17}$$

ρ is the density of fluid and v is the fluid velocity.

Conservation of the momentum is given by

$$\frac{\partial (\rho \vec{v})}{\partial t} + \nabla \cdot (\rho \vec{v} \vec{v}) = -\nabla p + \nabla \cdot \tau + \rho \vec{g} \tag{2.18}$$

p is the pressure, τ is the shear stress and g is acceleration due to gravity.

Transport equation is given by

$$\frac{\partial C_i}{\partial t} + \nabla \cdot (\vec{v} C_i) = \nabla \cdot (D_i \nabla C_i) + R_i \tag{2.19}$$

Where C_i is concentration of species i (mol/m^3), D_i is diffusion coefficient (m^2/s) and is R_i is source or sink term (reaction rate).

Energy equation can be written as

$$\rho C_p \left(\frac{\partial T}{\partial t} + \vec{v} \cdot \nabla T \right) = \nabla \cdot (k \nabla T) + Q \tag{2.20}$$

where C_p is specific heat capacity (J/kgK), T is temperature (K), k is thermal conductivity (W/mK) and Q is volumetric heat generation (W/m^3).

There are many models that are used in commercial software such as Turbulence: calculated using Reynolds-Averaged Navier-Stokes (RANS) models like k-ε or k-ω models, or more advanced techniques like Large Eddy Simulation (LES) for higher accuracy. Multiphase Flow: For gas-liquid mixtures (e.g., air sparged bioreactors), Euler-Euler or Volume of Fluid (VOF) models are used. Non-Newtonian Fluids: Most biological fluids (e.g., cell cultures, fermentation broths) are shear-thinning and need rheological models like Power Law or Carreau-Yasuda models.

2.4.2 Multi-Scale Modeling and Simulation

Multiscale modeling and simulation provide an integrated and predictive platform for the capture of interactions between phenomena at various organization levels. These models are essential in optimizing product yield, guaranteeing quality, and establishing robust systems in such areas as tissue engineering, biologics production, and cell-based treatments.

On the molecular level, enzyme-substrate interactions, gene expression control, and signal transduction are simulated by ordinary differential equations (ODEs) for time-dependent biomolecule concentration changes [21]. One of the most standard methods is Michaelis–Menten kinetics:

$$\frac{d[S]}{dt} = -\frac{V_{max}[S]}{K_m + [S]} \tag{2.21}$$

$$\frac{d[P]}{dt} = \frac{V_{max}[S]}{K_m + [S]} \tag{2.22}$$

S is substrate concentration, P is product concentration, V_{max} is maximum reaction velocity, K_m is constant which is popularly known as Michaelis constant

Gene regulatory networks are also modelled additionally using

$$\frac{d[\text{mRNA}]}{dt} = \alpha - \delta[\text{mRNA}] \tag{2.23}$$

$$\frac{d[\text{Protein}]}{dt} = \beta[\text{mRNA}] - \gamma[\text{Protein}] \tag{2.24}$$

Where, α and β are known as transcription and translation rates, δ and γ are known as degradation constants.

Cell population dynamics are commonly described using agent-based models (ABMs), cellular automata, or compartmental systems of ODEs. An example of logistic growth model for biomass growth is expressed as:

$$\frac{dX}{dt} = \mu_{\text{max}} X\left(1 - \frac{X}{X_{\text{max}}}\right) \tag{2.25}$$

where X is cell density, μ_{max} is the maximum specific growth rate, and X_{max} is the carrying capacity. Cell mechanics and migration in scaffolds or bioreactors can also be modeled using force-based models:

$$\frac{d^2\vec{x}}{dt^2} = \overrightarrow{F_{\text{adh}}} + \overrightarrow{F_{\text{drag}}} + \overrightarrow{F_{\text{chemotaxis}}} + \overrightarrow{F_{\text{repulsion}}} \tag{2.26}$$

where F terms denote adhesion, fluid drag, chemical attraction, and repulsive forces between cells.

At the tissue level, deformation, stress distribution, and cell growth are described by continuum mechanics. Mass transport is modeled through reaction-diffusion equations

$$\frac{\partial C}{\partial t} = D\nabla^2 C - R(C, X) \tag{2.27}$$

$$\nabla \cdot \sigma + \vec{f} = 0 \tag{2.28}$$

$$\frac{\partial C_O}{\partial t} + \vec{v} \cdot \nabla C_O = D_O \nabla^2 C_O - q_O X \tag{2.29}$$

where C is concentration of a nutrient or signal, D is diffusion coefficient, and R is reaction or consumption rate dependent on both concentration and cell density X. Cauchy momentum equation is to express mechanical equilibrium.

2.4.3 Finite Element Modeling in Scaffold and Tissue Engineering

In tissue engineering and regenerative medicine, scaffolds are of prime importance in facilitating cellular attachment, growth, and differentiation. Besides offering structure, they need to ensure the maximum transport of nutrients, oxygen, and waste and the capacity for biological and mechanical stresses within the body [22]. Finite Element Modeling has proven to be a highly useful means of simulating and optimizing such intricate structures.

i) Simulation of Mechanical Behavior of Scaffolds

The mechanical behavior of scaffolds—specifically, their stress-strain response, load transfer, and deformation under physiologic loading—can be studied *via* FEM. The scaffold is divided into small elements, each of which follows equations of photoelasticity or elasticity.

$$\nabla \cdot \sigma = \vec{0} \tag{2.30}$$

σ is the Cauchy stress tensor.

$$\sigma = C : \varepsilon \tag{2.31}$$

C is the fourth-order elasticity tensor, ε is the strain tensor, u is the displacement vector

$$\varepsilon = \frac{1}{2}\left(\nabla\vec{u} + \overrightarrow{\nabla u^T}\right) \tag{2.32}$$

Force, displacement and rates are related in FEM using the following equation.

$$K\vec{u} = \vec{f} \tag{2.33}$$

ii) Nutrient and Oxygen Diffusion Modeling

In tissue scaffolds, adequate transport of nutrients and oxygen is required for cell survival, especially in large or complex structures. FEM may be applied to model the mass transport equations, typically by Fick's law of diffusion.

$$\frac{\partial C}{\partial t} = D\nabla^2 C - R(C, X) \tag{2.34}$$

C is the concentration of the nutrient (e.g., oxygen or glucose), D is the effective diffusion coefficient and $R(C,X)$ is the rate of consumption, which may depend on cell density X.

iii) Poroelastic and Fluid-Structure Interaction (FSI) Models
For more realistic situations in which fluid flow within the scaffold influences mechanical behavior (e.g., when applying perfusion bioreactor), poroelasticity models are used. The BIOT theory of poroelasticity integrates mechanical deformation and interstitial flow of fluids

$$\nabla \cdot \sigma + \alpha \nabla p = 0 \tag{2.35}$$

$$\frac{\partial \epsilon_v}{\partial t} + \nabla \cdot \vec{q} = Q \tag{2.36}$$

p is pore pressure, ϵ_v is volumetric strain, $\vec{q}$ is Darcy's fluid flux, Q is the source/sink term, α is the Biot coefficient.

2.4.4 Dynamic Simulation of Bioprocess Kinetics

Bioprocessing operations like microbial fermentation, mammalian cell culture, and enzymatic catalysis are frequently conducted under dynamic, non-steady-state conditions. For design, monitoring, and optimization of these systems, dynamic simulations based on kinetic models in the form of ODEs are of primary significance [23]. These simulations enable the provision of facilities for predicting system dynamics within a time interval as well as implementing real-time control measures so that robustness, productivity, and quality are maintained.

i) Modeling Time-Dependent Bioprocesses Using ODEs
The levels of biomass, substrate, product, and by-products vary with time in dynamic systems. A generic unstructured model applies first-order ODEs to define the rates of such variations [24]. The biomass growth is governed by

$$\frac{dX}{dt} = \mu(S)X \tag{2.37}$$

Substrate equation can be expressed as

$$\frac{dS}{dt} = -\frac{1}{Y_{X/S}}\mu(S)X \tag{2.38}$$

Growth rate expression is given as

$$\mu(S) = \mu_{max} \frac{S}{K_S + S} \tag{2.39}$$

Where, μ_{max} is the maximum specific growth rate, K_S is the half-saturation constant (Monod constant). Numerical methods are adopted to solve this property.

ii) Integration of Inhibition, Maintenance, and Product Decay

Later models can have substrate or product inhibition and maintenance energy requirements: With Substrate Inhibition (Haldane kinetics):

$$\mu(S) = \mu_{max} \frac{S}{K_S + S + \frac{S^2}{K_I}} \tag{2.40}$$

where K_I is constant.

These extended models may be employed to simulate behavior in over fed, poisoned, or high-productivity systems.

$$\frac{dP}{dt} = Y_{P/X}\mu(S)X - k_d P \tag{2.41}$$

where K_d is known as decay constant.

2.4.5 Digital Twins and Data-Driven Simulation Approaches

Digital Twins in biomanufacturing is an innovation method that safely unites physical bioprocesses with their digital twins in real-time simulation, prediction, and control. In contrast to disconnected classic mechanistic models, digital twins combine first-principles-based equations with data-driven machine learning models and sensor feedback into a closed predictive system [25]. The process is shown in Figure 2.2.

Mathematically, Digital Twins can be represented as

$$\frac{dx}{dt} = f(x(t), u(t), \theta) + \varepsilon(t) \tag{2.42}$$

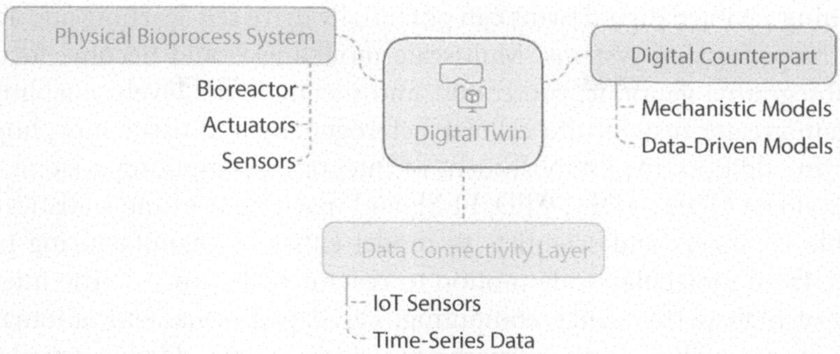

Figure 2.2 Main three components of digital twins in biomanufacturing.

Where, x(t) represents mechanistic and data driven functions, θ is model parameters ε(t) represents error in models. Uncertainty in model is inherent and is captured by data driven portion of the equation. Neural networks, Support vector regression and Gaussian Processes are commonly used. Model predicative controls are also used.

2.5 Conclusions and Future Opportunities in Biomanufacturing

The union of mathematical modeling and simulation with biomanufacturing has revolutionized the field of biotechnology by enabling predictive capability, process optimization, and design guidance on sizes of biological systems from microscopic to macroscopic. From kinetic modeling at the grassroots level to complex computational practices like CFD, FEM, and digital twins, the field has become an integrated engineering science unifying biology with mechanics, thermodynamics, transport phenomena, and data analysis. This chapter has extensively discussed the development and progression of mathematical modeling paradigms, emphasizing their application in microbial and mammalian systems, scaffold-based tissue engineering, and bioreactor design. The progression from empirical toward mechanistic, multiscale, and data-driven paradigms has made it possible to better understand intricate interactions that regulate cell growth, product synthesis, nutrient transport, and bioprocess control.

As the biomanufacturing industry trends towards greater precision, sustainability, and scalability, the horizon calls for its integration of simulation platforms with real-time process data. Artificial intelligence and machine learning-enabled digital twins can potentially drive self-learning and adaptive bioprocessing systems. Multiscale modeling would become focused on integrating genomic, proteomic, and metabolomic levels, enabling us to gain greater insight into cellular heterogeneity and tissue morphogenesis. In addition, the establishment of integrated computing systems that consolidate ODEs, PDEs, CFD, FEM, and machine learning statistics will enable engineers and scientists to model entire biomanufacturing pipelines, from molecular-scale motion to reactor-scale process. The integration of high-performance computing, cloud platforms, and automation with sensors will make them more pervasive and embedded in actual production. Ultimately, the future of biomanufacturing modeling is as much about enabling adaptability, scalability, and increased biological relevance as it is about enhanced accuracy. The pathway from abstractions to living digital factories is no longer a pipe dream but increasingly real, and those with a solid foundation in mathematical rigor as well as comprehension of biology will spearhead the next generation of bio-industrial innovation.

AI disclosure statement: The author acknowledges the use of AI-based tools for minor language editing or grammatical correction. However, all intellectual contributions, analyses, and conclusions presented in this chapter are entirely the author's own.

References

1. Orth, J.D., Thiele, I., Palsson, B.O., What is flux balance analysis? *Nat. Biotechnol.*, 28, 245, 2010.
2. Henry, C.S., Broadbelt, L.J., Hatzimanikatis, V., Thermodynamics-based metabolic flux analysis. *Biophys. J.*, 92, 1792, 2007.
3. Beard, D.A., Babson, E., Curtis, E., Qian, H., Thermodynamic constraints for biochemical networks. *J. Theor. Biol.*, 228, 327, 2004.
4. Mahadevan, R. and Schilling, C.H., The effects of alternate optimal solutions in constraint-based genome-scale metabolic models. *Metab. Eng.*, 5, 264, 2003.
5. Sandhya, S. and Viswanathan, K.S., Modelling and simulation approaches in biomanufacturing. *Biochem. Eng. J.*, 145, 109680, 2019.
6. Singh, V. and Kumar, A., Computational fluid dynamics in bioreactor design and scale-up. *Chem. Eng. Res. Des.*, 150, 486, 2019.
7. Borys, M. and Bazan, J., Multiscale modeling for bioprocess optimization: A review. *Biotechnol. Progr.*, 35, e2917, 2019.
8. Vo, T.D. and Lee, K.M., Michaelis–Menten kinetics: A review of models and applications. *J. Biotechnol.*, 299, 21, 2019.

9. Mandal, S. and Das, S., Finite element modeling of scaffolds for tissue engineering applications. *J. Biomed. Mater. Res. A*, 108, 2201, 2020.
10. Smith, R.L. and Dillard, J.G., Simulation of scaffold mechanical behavior under physiological loading. *J. Mech. Behav. Biomed. Mater.*, 103, 103570, 2020.
11. Chaudhury, P. and Mahadevan, R., Dynamic simulation of microbial fermentation processes using ODE models. *Biochem. Eng. J.*, 98, 103745, 2015.
12. Wang, Y. and Huang, M., Modeling and simulation of biomass growth kinetics in bioprocesses. *J. Biotechnol.*, 150, 99, 2010.
13. Tao, Y. and Zhang, Y., Digital twins for biomanufacturing: Concepts and applications. *Comput. Chem. Eng.*, 130, 106617, 2019.
14. Lee, S. and Park, J., Multiscale computational modeling in tissue engineering and regenerative medicine. *Biotechnol. Adv.*, 37, 107443, 2019.
15. Patel, V. and Shah, S., Finite Element Modeling of scaffold mechanics for tissue regeneration. *J. Biomed. Mater. Res. A*, 108, 2201, 2020.
16. Kumar, A. and Singh, R., Simulation-based optimization of bioprocesses for enhanced productivity. *Biochem. Eng. J.*, 145, 109680, 2019.
17. Zhao, Y. and Wang, J., CFD modeling for scale-up of bioreactors. *Chem. Eng. Sci.*, 203, 16, 2019.
18. Li, H. and Chen, Y., Thermodynamic analysis of biochemical reaction networks. *J. Theor. Biol.*, 254, 48, 2008.
19. Jain, R. and Singh, S., Incorporating thermodynamic constraints in flux balance analysis: MILP approaches. *Metab. Eng.*, 12, 220, 2010.
20. Brown, J.M. and Lee, H.K., Multiscale modeling of metabolic networks. *Curr. Opin. Biotechnol.*, 23, 321, 2012.
21. Guo, J. and Chen, W., Soft sensor development using digital twins in biomanufacturing. *Biotechnol. Bioeng.*, 117, 229, 2020.
22. Davis, M. and Johnson, P., Application of ODE models to bioprocess control. *Biochem. Eng. J.*, 110, 89, 2016.
23. Thompson, L. and Garcia, M., Real-time adaptive control in bioprocessing using machine learning. *Comput. Chem. Eng.*, 140, 106997, 2020.
24. Lee, J.M. and Kim, Y.H., Mathematical modeling and dynamic simulation of microbial growth kinetics in bioprocesses. *J. Biotechnol.*, 150, 99, 2010.
25. Gupta, R. and Mehta, S., Dynamic simulation of bioprocesses using ODE-based kinetic models. *Biochem. Eng. J.*, 98, 103745, 2015.

3

Potential Material for Bio-Printing Applications

Amey Dukle[1], Sirisha Puttabanthi[2] and Mamilla Ravi Sankar[1]*

[1]*Dept. of Mechanical Engineering, Indian Institute of Technology Tirupati, Andhra Pradesh, India*
[2]*Department of Anatomy, Sri Balaji Medical College Hospital & Research Institute, Renigunta, AP, India*

Abstract
Bioprinting is an advanced additive manufacturing technique that can be used for fabrication of artificial organs, tissues, scaffolds, and platforms for new drug discovery and testing. However, despite the hardware capabilities, the exiting materials often lack the capabilities that are essential for unlocking the full capabilities of bioprinting. To address these issues, various potential materials are being developed for use in bioprinting applications. Natural polymers are the most used materials due to inherent bioactivity and biocompatibility. Synthetic polymers are also widely used for their easy availability and tuneable mechanical properties. Composite bioinks are used to develop materials that closely mimic the native ECM architecture and obtain favorable mechanical properties. This chapter lists the various factors that need to be considered for bioink selection and various materials that have potential for use in bioprinting applications. Various properties of the different bioink materials, their advantages and limitations have been discussed. Various emerging materials that are gaining prominence in the field have also been critically analyzed. The chapter aims to provide a gateway into various materials that can be potentially used for unlocking the full potential of bioprinting.

***Keywords*:** Biomaterials, bioactive, bioprinting, tissue engineering

**Corresponding author*: evmrs@iittp.ac.in

Sandip Kunar, Jagadeesha T. and D. Sathis Kumar (eds.) Biomanufacturing: Transforming Production of Biologics and Biopharmaceuticals, (67–90)

3.1 Introduction

Bioprinting is an additive manufacturing technique that is used to fabricate three-dimensional biological constructs through deposition of biomaterials that are loaded living cells, and growth factors in a layer by layer [1, 2]. While traditional 3D printing technologies tend to use plastics or metals as their feedstock material, bioprinting focuses on recreating complex tissue architectures that closely mimics the native extracellular matrix (ECM) [3, 4]. Through integration of the principles from tissue engineering and regenerative medicine, the technology allows for the creation of customized biological scaffolds that support cell growth and tissue regeneration [5, 6]. The precision and versatility of bioprinting have opened new frontiers in personalized medicine, drug testing, and organ transplantation, thereby demonstrating its potential to revolutionize the healthcare sector [7, 8].

Traditional organ transplants suffer from donor shortages, immunological rejection, and long waiting times [9, 10]. Bioprinting offers a promising alternative by fabricating patient-specific tissues using autologous cells, thereby reducing immune complications and enhancing treatment outcome [11, 12]. Bioprinted tissue models also hold potential to advance drug discovery and toxicity testing by providing predictive platforms that replicate *in vivo* conditions in more accurate manner compared to two-dimensional cultures or animal models [11, 12]. Through these innovations, bioprinting presents itself as a more effective, safe, and personalized therapeutic strategy that can potentially reduce healthcare costs and improve patient outcomes [13].

There are various bioprinting strategies that are in use. The most suitable technology is selected based on the target tissues and the materials to be used. Inkjet bioprinting is one of the first developed bioprinting techniques. The technique is used for the fabrication of 3D cell culture samples [14]. Its operation mechanism is like traditional inkjet printers but uses bioinks containing cells and biomaterials, allowing precise droplet deposition with high speed and resolution [15]. The technique offers low cost and scalability making it suitable for industry applications. However, the need for low viscosity bioinks and poor cell viability due to shear stresses encountered during printing. make it unsuitable for organ printing applications [12, 16]. Extrusion-based bioprinting is the commonly used technique due to its capability to print materials with wide range of viscosities and 3D print large geometries. The printing setup is comprised of a barrel loaded with bioink. a pneumatic or mechanical plunger and a nozzle through which the bioink is extruded [17]. This approach is suitable for wide range of tissue constructs with low cost and high efficiency [18, 19]. Photopolymerization based bioprinting is another widely used technology. It produces the objects through photopolymerization of photosensitive bioinks in layer-by-layer manner [20]. This technology is advantageous for fabricating scaffolds with intricate architectures due to its high resolution. Each of these bioprinting modalities addresses specific material and cellular requirements, enabling tailored fabrication approaches suited to the mechanical, biological, and architectural demands of diverse tissues [20, 21]. Figure 3.1 provides overview of workflow in bio-printing processes and various bio-printing processes.

The selection of materials for bioprinting is the most important factor that determines the printing performance, mechanical, and biological properties of the printed scaffold. One of the vital properties to be considered is biocompatibility [22]. It ensures that materials support cell survival and function without causing adverse immune reactions. Bioactivity is another vital property in biomaterials [23]. Bioactivity is the capability of the biomaterial to support and promote cell adhesion, proliferation, and differentiation [24, 25]. Rheological characteristics such as viscosity and shear-thinning behavior, also play a vital role in bioprinting. Materials must exhibit appropriate rheological properties to allow smooth extrusion or deposition while maintaining shape fidelity post-printing [24, 26]. Often overlooked in the past, environmental sustainability has become a key consideration in material development. Efforts are being made to

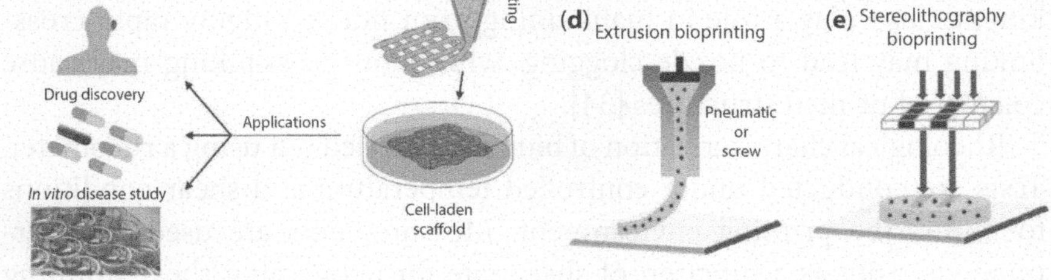

Figure 3.1 Overview of (a) workflow of the bioprinting processes, (b) inkjet bioprinting, (c) laser assisted bioprinting, (d) extrusion bioprinting, (e) stereolithography bioprinting. (Reproduced with permission from [8]).

encourage the development of bioinks derived from renewable, biodegradable sources that minimize ecological impact throughout their life cycle [14, 27].

This chapter presents a comprehensive exploration of the materials integral to bioprinting technology. It details the important factors that need to be considered for successful scaffold fabrication and biological integration. Various materials such as natural polymers, synthetic polymers, and composite materials have been discussed in detail. The chapter also introduces emerging material innovations, including stimuli-responsive polymers and biowaste-derived bioinks and their potential to overcome existing limitations and expand the horizons of bioprinting applications.

3.2 Factors Affecting Bioink Functionality

Bioink is a complex material that must possess suitable physicochemical, biological, and operational properties for ensuring its usability for tissue engineering application. The functionality of a bioink is determined by its rheological properties, shelf life, transportability, and cost. The properties of an ideal bioink for bioprinting are given below.

3.2.1 Rheological Properties

Rheology behavior corresponds to the flow and deformation properties during printing. It directly affects the printability, structural fidelity, and cell viability [28, 29]. For extrusion-based bioprinting, bioinks should exhibit shear-thinning behavior [30, 31]. Shear thinning behavior is characterized with reduction in viscosity with increasing shear rate [32]. This behavior facilitates smooth extrusion through the nozzle while maintaining integrity of the bioink network [33]. Post-extrusion, the bioink must again display high viscosity to maintain the printed shape. Crosslinking kinetics also play a role in maintaining print fidelity; highly rapid crosslinking may lead to nozzle clogging, while slow crosslinking may cause collapse of printed structures [34].

Rheological characterization of bioinks is carried out using a rheometer. Tests are conducted under controlled temperature and shear conditions to mimic the printing environment. Measurements are used to determine viscosity as a function of shear rate for confirming shear-thinning behavior [34, 35]. Oscillatory shear tests are conducted to gain insight into the viscoelastic properties of the bioink by measuring the storage modulus (G') and loss modulus (G''). The storage modulus (G') represents the elastic or solid-like component of the bioink. It indicates the ability to store energy during deformation. A higher G′ suggests a more solid-like behavior, contributing to shape retention after deposition. The loss modulus (G″) represents to the viscous or liquid-like component. For bioprinting applications, a balance between G′ and G″ is required. The ratio of these moduli is expressed as the loss modulus ($\tan \delta = G''/G'$) that provides an index of the relative dominance of viscous versus elastic behavior [29]. Amplitude sweep tests are used to identify the linear viscoelastic region (LVER). Frequency sweep tests assess the stability of viscoelastic behavior over different deformation rates, providing information relevant to both extrusion and post-deposition performance. These rheological data allow for the design and optimization of bioinks with controlled flow properties leading to reproducible printing outcomes [19, 30].

3.2.2 Shelf Life and Stability

Various factors such as polymer degradation, microbial contamination, and instability of incorporated bioactive agents have significant impact on the shelf life of the developed bioink. Natural polymers such as gelatin or alginate degrade due to hydrolysis or enzymatic activity thereby impacting their printability and mechanical performance. Biomolecules and cell laden bioinks have a shorter shelf life and also require special handling and storage precautions. Strategies to extend shelf life include lyophilization, storage at low temperatures, and incorporation of stabilizing agents. Maintaining consistent properties over the shelf life is essential to ensure predictable printing performance [36].

3.2.3 Transportability

Transportability is another crucial factor for centralized bioink manufacturing and supply to remote printing facilities. Factors such as temperature sensitivity, vibration during transit, and exposure to light or oxygen can possibly compromise bioink quality. Cold-chain logistics are often required for cell-laden formulations. Another approach used to improve the shelf life is through encapsulation technologies and use of protective additives. Transportability is an important factor that will determine the commercial and large-scale adoption of any bioink. The bioink must be designed such that it can seamlessly be transported using the existing global pharmaceutical supply chains.

3.2.4 Cost

The cost of bioink is another factor that impacts its accessibility for research and clinical applications. Highly purified biomaterials, growth factors, and nanoparticles can significantly increase the overall cost of the bioink. The number of processing steps such as sterilization, purification, and quality control can also increase the cost. Scalable manufacturing processes are necessary for transitioning from laboratory to clinical or commercial production.

3.3 Natural Polymers for Bioink Preparation

Natural polymers possess various characteristics that make them highly advantageous for bioprinting applications [37]. These materials have inherent biocompatibility and bioactivity that supports cell adhesion, proliferation, and tissue regeneration [38]. One of the main advantages is the high level of mimicry, naturally found in the extracellular matrix (ECM), thereby providing cells with an environment conducive to maintaining physiological functions [39, 40]. Their ability to support cell differentiation and migration is directly attributed to their molecular composition and structure that promotes cell-material interaction [41]. Natural polymers are also biodegradable in nature that eliminates the need for revision surgery and allows for dynamic scaffold remodeling and seamless integration with host tissues after implantation [41, 42].

The degradation property also allows for temporal control over scaffold stability and allows for programming of scaffold resorption that matches the tissue regeneration rate . Common examples of natural polymers used in bioprinting include collagen, gelatin, alginate, and hyaluronic acid. Collagen is the most abundant protein in the extracellular matrix (ECM) and plays a crucial role in bioprinting due to its excellent structural and biological characteristics. It possesses a triple-helix based molecular structure that provides mechanical strength and stability making it especially suitable for engineering load-bearing tissues such as cartilage and skin [43, 44]. Collagen-based scaffolds can mimic the native cellular environments thereby enhancing cell adhesion, proliferation, and differentiation [45, 46]. This ability to closely mimic the natural ECM efficiently supports tissue regeneration by directing native cell behaviors and facilitating matrix remodeling [43, 47]. Figure 3.2 schematically illustrates the workflow for 3D bioprinting of collagen.

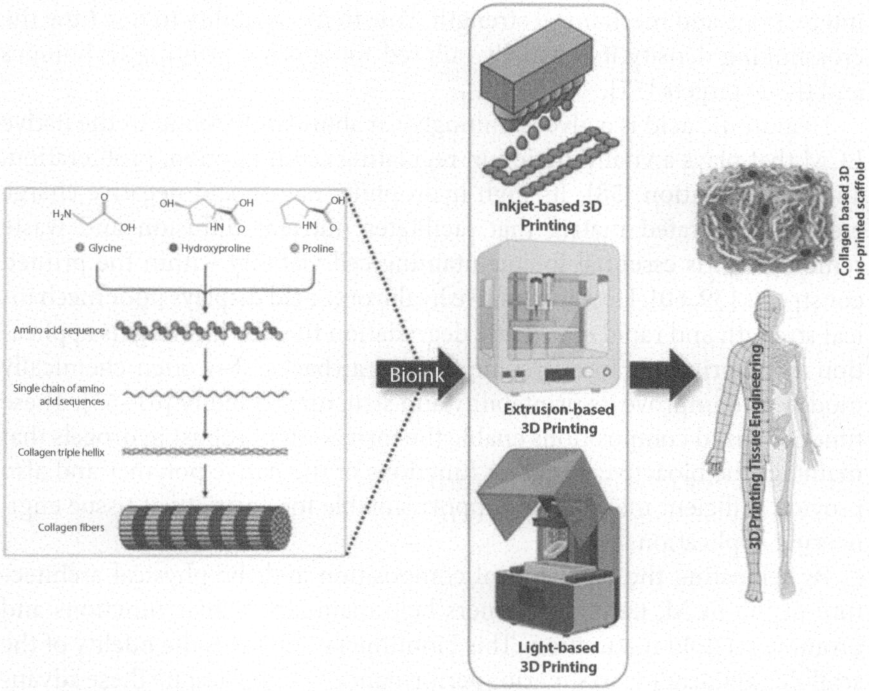

Figure 3.2 Schematic illustration of the workflow for 3D bioprinting of collagen. (Reproduced with permission from [43]).

Gelatin is extracted by the partial hydrolysis of collagen that promote cell adhesion and proliferation. It is widely used in bioprinting applications due to its easy availability and gelation properties [48]. Gelatin is chemically modified into gelatin methacrylate (GelMA) to introduce photocross-linkable groups [49, 50]. These groups enhance mechanical properties and allow for precise control over scaffold architecture after printing [51]. These modifications expand gelatin's application to various tissue types with high bioactivity that closely resembles ECM components [48, 52]. Alginate is another popular biopolymer that is sourced from seaweed. It is widely used bioprinting for its rapid gelation through ionic cross-linking using divalent cations such as calcium [53, 54]. This rapid crosslinking has been used in extrusion-based printing processes. It also maintains excellent shape fidelity during printing making it suitable for printing of various scaffolds [55, 56]. Despite this advantage, alginate lacks inherent cell-adhesion sites, which limits its bioactivity and hinders cell attachment function. To address these limitations, alginate is commonly blended with polymers like gelatin or functionalized with peptides to enhance cellular interactions and mechanical strength. Due to its capability to fine tune the crosslinking density it is usually tailored for specific printing techniques and tissue targets [57].

Hyaluronic acid is a glycosaminoglycan abundantly found in the native ECM that plays a vital function in regulating cell migration, proliferation, and differentiation [58]. Its high hydrophilic nature and negative charge creates a hydrated matrix that facilitates nutrient diffusion and waste removal that is essential for maintaining cell viability within the printed constructs [59, 60]. However, native hyaluronic acid displays poor mechanical strength and rapid enzymatic degradation thereby limiting its application in bioprinting. To overcome these drawbacks, it is often chemically modified to improve its printability and structural stability [61–63]. These functionalized compositions enable the formation of robust hydrogels that maintain the bioactive signaling functions of the native polymer and also provide sufficient mechanical support suitable for a variety of tissue engineering applications [64].

By recreating the biochemical composition and the physical architecture of the ECM, natural polymers help maintain cellular functions and promote scaffold maturation. This biomimicry enhances the fidelity of the scaffolds and leading to superior performance *in vivo*. Despite these advantages, natural polymers face notable challenges that need to be addressed. One significant issue is batch-to-batch variability inherent to biologically sourced materials that may lead to inconsistent physicochemical properties and unpredictable printing and biological outcomes. Careful optimization strategies need to be adopted to preserve biocompatibility and bioactivity of the materials and make it suitable for tissue engineering applications [41, 65].

3.4 Synthetic Polymer for Bioink Preparation

Synthetic polymers offer several significant advantages compared to natural polymers in the field of bioprinting. One of the primary factors is their controlled and reproducible physicochemical properties and the absence of batch-to-batch variability that may arise in natural polymers due to biological origin [41]. Synthetic polymers are manufactured through controlled chemical processes that ensure consistent composition and quality. The ability to fine-tune the properties such as molecular weight, chain architecture, and functional group density further enables the design of bioinks with specific rheological and degradation characteristics suited to specific tissue engineering requirements. Synthetic polymers also provide the capability to tailor mechanical strength and degradation rates with a high degree of precision. This tunability is crucial for ensuring scaffold integrity during implantation and supporting tissue regeneration without premature collapse or prolonged persistence that could hinder remodeling [66, 67].

One widely used synthetic polymer is polyethylene glycol (PEG) due to hydrophilic and biocompatible behavior. PEG's chemical structure allows functionalization with various cross-linkable groups which makes it an ideal candidate for production of hydrogels with tuneable mechanical properties and degradation profile [68, 69]. The material is non-immunogenic in nature making it suitable for implantation without causing an immunogenic host responses [69, 70]. Polycaprolactone (PCL) is another synthetic polymer that is used in bioprinting of hard tissue constructs [71]. PCL is a biodegradable polymer with a slower degradation rate and sufficient mechanical strength to provide necessary support for load-bearing applications. Its high mechanical strength ensures scaffold integrity is maintained during the remodeling process. Due to the low melting temperature of 60°C and solubility in common solvents, PCL based inks are used in various 3D printing setups [71, 72].

Polylactic acid (PLA) and polyglycolic acid (PGA) are widely used synthetic polymers due to their tunable degradation behavior. PLA has also received US FDA approval for the fabrication of 3D printed devices and scaffolds. These polymers degrade through hydrolysis into non-toxic metabolites which eliminates the risk associated with *in-vivo* scaffold degradation [73, 74]. Their FDA approval for biomedical applications is symbolic of its biocompatibility and safety for clinical use. These capabilities make PLA and PGA a promising material for wide range of bioprinting applications such as soft tissues as well as hard tissues [75, 76].

Despite these advantages, most synthetic polymers lack intrinsic bioactivity limiting their capacity to promote cellular adhesion, proliferation, and differentiation Unlike natural polymers that inherently mimic extracellular matrix components synthetic materials are biologically inert and do not play an active role in bone regeneration process [77, 78]. This shortcoming has led to a need for bioactivity enhancement of synthetic polymer-based scaffolds through strategies such as surface functionalization or blending with natural polymers to incorporate bioactive components that facilitate cell-scaffold interactions [73].

3.5 Composite Materials for Bioink Preparation

Primarily bioinks should fabricate scaffolds with a conducive microenvironment for tissue regeneration. To achieve the properties, bioink should display favorable printability, mechanical integrity, and biological functionality [77, 79]. Single component bioinks typically fail to meet the functional requirements which necessitates development of multi-component composite bioinks by combining polymers, ceramics, nanoparticles, or bioactive agents for bioprinting applications. These composite materials are designed such that they achieve the desired mechanical properties, degradation behavior, bioactivity, and print fidelity [80, 81]. Composite bioinks are also used to improve the overall treatment outcome. Drug-loaded composites allow localized delivery of therapeutic agents from printed constructs. Incorporation of growth factors within composite bioinks supports tissue regeneration and cell differentiation [82, 83].

One of the most used composite materials is Polymer–ceramic composites used for bone tissue engineering applications (Figure 3.3) [84]. The human bone is a composite of collagen (a polymer) and hydroxyapatite (a bioceramic); to achieve the properties mimicking the human bone the scaffold should closely match the human bone composition [83, 85]. Various hydroxyapatite (HA)–polymer composites that closely mimic the human bone are widely explored for bone tissue engineering. Hydroxyapatite

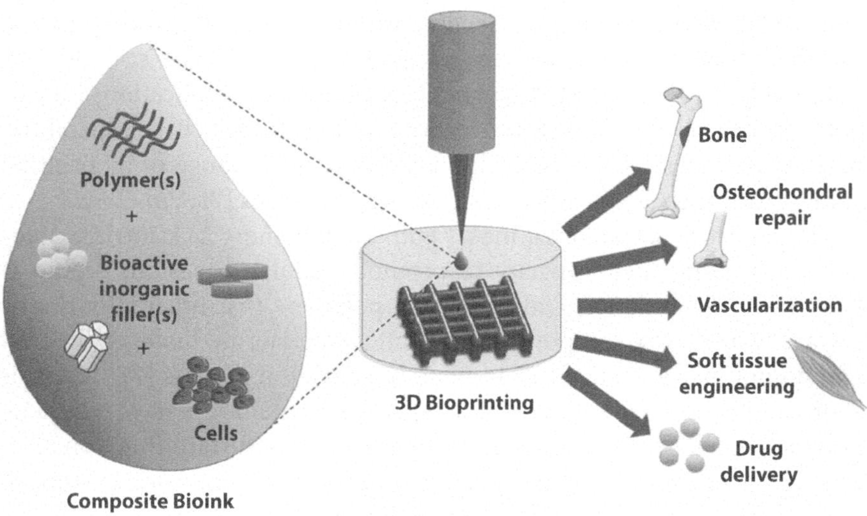

Figure 3.3 Schematic illustration of the properties and uses of composite bioinks. (Reproduced with permission from [80]).

provides the scaffold with osteoconductivity and mechanical strength, whereas polymers provide the bioink with biocompatibility and printability. Common polymers used in such composites include gelatin methacrylate (GelMA), alginate, and polyethylene glycol derivatives [83, 84]. These composites can mimic the extracellular matrix and bone mineral content thereby leading to faster tissue regeneration.

Another widely used type of bioinks is nanoparticle-reinforced polymeric bioinks. Various nanoparticles, such as silica, gold, or carbon-based nanomaterials are added to polymeric matrix to enhance the mechanical strength, print resolution, and biological response of the biomaterial. Nanosilica–alginate composites have shown better shear-thinning behavior as well as better extrusion control. Various composite bioink strategies are emerging in recent times. Few notable strategies are detailed in the next section.

3.6 Emerging Materials in Bioprinting

Emerging materials in bioprinting include stimuli-responsive biomaterials and biowaste-derived biomaterials for biomedical applications. Stimuli-responsive biomaterials show change in physical or chemical properties in response to external stimuli such as temperature, pH, or magnetic fields that allow for controled drug delivery, cell release, and dynamic tissue engineering environments. Biowaste-derived biomaterials that are obtained from agricultural or marine waste provide a sustainable and cost-effective source of biopolymers and bioactive compounds. These materials reduce environmental impact making them a promising candidate for scaffold fabrication and regenerative medicine application.

3.6.1 Stimuli-Responsive Smart Biomaterials

Stimuli-responsive polymers are an innovative class of biomaterials that can alter their physical or chemical properties in response to external environmental stimuli such as temperature, pH, light, or magnetic fields [86–88]. These materials allow for the fabrication of smart scaffolds that can adapt to evolving biological conditions and enhance the treatment outcomes.

Ultrasound stimulation represents a promising approach for creating stimuli–responsive bioinks. Paci *et al.*, in their study developed a printable alginate/Pluronic-based bioink including piezoelectric barium titanate nanoparticles (diameter 60 nm) for the 3D bioprinting of muscle cell-laden hydrogels. The study aimed to investigate the effects of the combination of piezoelectric nanoparticles with ultrasound stimulation on early myogenic differentiation of the printed structures. The combination of piezoelectric nanoparticles and ultrasound stimulation upregulated the expression of MYOD1, MYOG, and MYH2 genes thereby contributing to faster regeneration. These results suggest that the piezoelectric nanoparticles boosts myogenesis in its early phases [89]. Light responsive biomaterials are another class of biomaterials that have shown significant potential. The study by Ferreras *et al.*, developed a wound healing patch based on 4D printing Three distinct GBM such as graphene oxide (GO), reduced graphene oxide (rGO), and graphene platelets (GP), were used for the fabrication of the wound healing patch. The investigations revealed that rGO exhibited the highest photothermal efficiency and antibacterial properties when irradiated, even at a concentration as low as 0.05 mg/mL, without compromising human fibroblast viability. These findings suggest that light responsive biomaterials have promising therapeutic capabilities in drug delivery and anti-bacterial wound dressings [90].

Integration of stimuli-responsive scaffolds allows for real-time monitoring and drug delivery in tissue engineering. Stimuli scaffolds can detect physiological parameters such as pH, oxygen levels, or metabolic markers, by allowing for therapeutic release based on live data [86, 91]. This closed-loop system enhances the scaffold's responsiveness to the biological microenvironment and promotes optimized tissue regeneration outcomes [87]. Despite these promising advancements, stimuli-responsive and smart materials currently face challenges related to biocompatibility and scalability. Some smart polymers are prone to immune rejection and also immune responses or cytotoxicity due to their synthetic components or degradation byproducts [91, 92]. The complexity of fabricating stimuli-sensitive bioinks remains a significant hurdle. Scale-up production with consistent quality and reproducibility is a challenge that limits its clinical translation.

3.6.2 Biowaste Derived Sustainable Bioinks

Biowaste-derived materials represent a promising and sustainable source of bioinks for bioprinting [93, 94]. These materials are obtained from residues that are produced due to agricultural, food processing, and industrial activity. Due to its biological origin, biowaste acts as a rich source for biological components such as growth factors, and bioactive molecules. The inherent presence of these critical biochemical elements allows biowaste-derived bioinks to promote cell-material interactions with the printed constructs leading to faster tissue regeneration [95, 96]. Biowaste-derived bioinks also display significant improvement in cell differentiation and extracellular matrix remodeling processes leading to faster regeneration. The presence of tissue-specific biochemical signals within these materials supports lineage-specific cellular behaviors, accelerating processes such as osteogenesis and angiogenesis based on the biowaste used [95, 97].

Despite these biological advantages, significant challenges can be faced with respect to the standardization, sterilization, and printability of biowaste-derived bioinks. The biological matter has inherent variability in its composition based on the age, diet, and geographical location of the biowaste source [98, 99]. This leads to difficulties in achieving consistent composition with reproducible biological and printing outcomes. To overcome these challenges new strategies, need to be developed for efficient extraction and sterilization of biowaste for bioink preparation [100, 101].

Biowaste-derived bioinks have demonstrated success in the fabrication of cardiac, liver, and cartilage constructs. For cardiac tissue engineering, decellularized cardiac ECM obtained from biowaste has improved cardiomyocyte functionality and vascularization within printed scaffolds, enhancing their physiological performance [102, 103]. Liver-derived biowaste materials are rich in hepatic matrix. These components support hepatocyte viability and specific liver functions, contributing to the advancement of functional hepatic tissue models suitable for disease research and potential transplantation [104]. Biowaste bioinks prepared using animal cartilage waste have shown to provide suitable mechanical and biochemical environments that promote chondrocyte encapsulation and matrix deposition thereby promoting faster regeneration of load-bearing articular cartilage [105, 106]. Various biowaste materials have also shown potential in bone scaffold fabrication. Figure 3.4 shows the various biowastes that can be used in preparation of bioink for bone tissue engineering.

Biowaste derived bioinks present a paradigm shift in regenerative medicine by aligning technological innovation with sustainable development goals and reduce environmental impact. This approach also holds the potential to make treatments more affordable and accessible by currently faced challenges such as addressing donor shortages and minimizing immune complications. The convergence of sustainability and bioprinting thus promises a future where high-quality, patient-specific therapies that contribute to global health equity and environmental stewardship simultaneously.

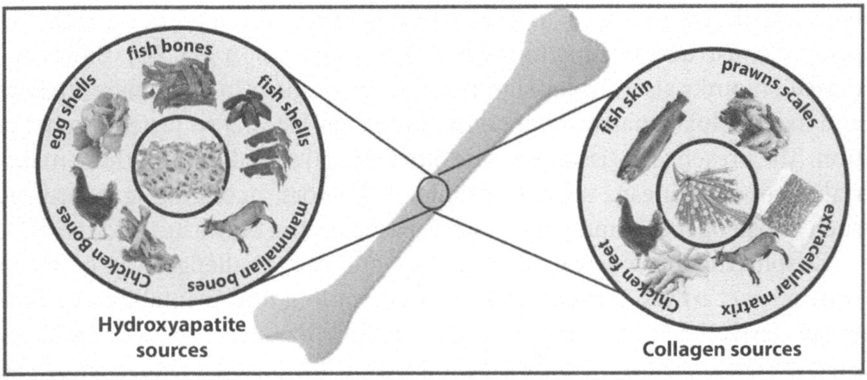

Figure 3.4 Biowaste sources for bioink preparation for bone tissue engineering. (Reproduced under CC-BY-4.0 from [97]).

3.7 Conclusions

Bioprinting technologies hold the potential to transform tissue engineering by tackling the challenges of organ shortages and prolonged transplant waiting times, offering a more efficient pathway for organ fabrication. Current technical equipment is already capable of producing highly precise and complex artificial organ structures. However, the available bioinks have yet to fully meet the requirements necessary to realize this goal. Advancements in bioink development must focus on replicating the intricate biochemical and biomechanical cues of native tissues while maintaining printability and structural stability. This requires integrating material science innovations with cell biology to design bioinks that not only support high cell viability but also promote tissue-specific differentiation and function. Efforts should be directed towards creating composite bioinks with tunable rheology, bioactivity, and degradation profiles tailored to specific tissue applications.

The successful translation of bioprinting from laboratory research to clinical organ fabrication will depend on overcoming limitations in bioink performance, standardizing testing protocols, and ensuring scalability under regulatory frameworks. By aligning technological capabilities with biologically advanced bioinks, bioprinting can move closer to producing functional, patient-specific organs, ultimately reducing dependence on donor organs and improving healthcare outcomes.

AI disclosure statement: The author acknowledges the use of AI-based tools for minor language editing or grammatical correction. However, all intellectual contributions, analyses, and conclusions presented in this chapter are entirely the author's own.

References

1. Gungor-Ozkerim, P.S., Inci, I., Zhang, Y.S., Khademhosseini, A., Dokmeci, M.R., Bioinks for 3D bioprinting: an overview. *Biomater. Sci.*, 6, 2018, https://doi.org/10.1039/C7BM00765E.
2. Ramesh, S., Harrysson, O.L.A., Rao, P.K., Tamayol, A., Cormier, D.R., Zhang, Y., Rivero, I.V., Extrusion bioprinting: Recent progress, challenges, and future opportunities. *Bioprinting*, 21, e00116, 2021, https://doi.org/10.1016/j.bprint.2020.e00116.
3. Vijayavenkataraman, S., 3D Bioprinting: Challenges in Commercialization and Clinical Translation. *J. 3D Print. Med.*, 7, 2023, https://doi.org/10.2217/3dp-2022-0026.
4. Gu, Z., Fu, J., Lin, H., He, Y., Development of 3D bioprinting: From printing methods to biomedical applications. *Asian J. Pharm. Sci.*, 15, 529–557, 2020, https://doi.org/10.1016/j.ajps.2019.11.003.
5. Huang, Y., Zhang, X., Gao, G., Yonezawa, T., Cui, X., 3D bioprinting and the current applications in tissue engineering. *Biotechnol. J.*, 12, 2017, https://doi.org/10.1002/biot.201600734.
6. Zhang, P., Hu, L., Tian, Y., Zhang, Z., Zhang, P., Yang, Y., Li, S., Ma, J., Inhibition of Ca2+-dependent protein kinase C rescues high calcium–induced pro-arrhythmogenic cardiac alternans in rabbit hearts. *Pflugers Arch.*, 473, 1315–1327, 2021, https://doi.org/10.1007/s00424-021-02574-7.
7. Derakhshanfar, S., Mbeleck, R., Xu, K., Zhang, X., Zhong, W., Xing, M., 3D bioprinting for biomedical devices and tissue engineering: A review of recent trends and advances. *Bioact. Mater.*, 3, 144–156, 2018, https://doi.org/10.1016/j.bioactmat.2017.11.008.
8. Mandrycky, C., Wang, Z., Kim, K., Kim, D.-H., 3D bioprinting for engineering complex tissues. *Biotechnol. Adv.*, 34, 422–434, 2016, https://doi.org/10.1016/j.biotechadv.2015.12.011.
9. Hunsberger, J., Neubert, J., Wertheim, J.A., Allickson, J., Atala, A., Bioengineering Priorities on a Path to Ending Organ Shortage. *Curr. Stem Cell Rep.*, 2, 118–127, 2016, https://doi.org/10.1007/s40778-016-0038-4.
10. Ozbolat, I.T. and Yu, Y., Bioprinting Toward Organ Fabrication: Challenges and Future Trends. *IEEE Trans. Biomed. Eng.*, 60, 691–699, 2013, https://doi.org/10.1109/TBME.2013.2243912.
11. Satpathy, A., Datta, P., Wu, Y., Ayan, B., Bayram, E., Ozbolat, I.T., Developments with 3D bioprinting for novel drug discovery. *Expert Opin.*

Drug Discov., 13, 1115–1129, 2018, https://doi.org/10.1080/17460441.2018.1542427.

12. Peng, W., Datta, P., Ayan, B., Ozbolat, V., Sosnoski, D., Ozbolat, I.T., 3D bioprinting for drug discovery and development in pharmaceutics. *Acta Biomater.*, 57, 26–46, 2017, https://doi.org/10.1016/j.actbio.2017.05.025.
13. Yaneva, A., Shopova, D., Bakova, D., Mihaylova, A., Kasnakova, P., Hristozova, M., Semerdjieva, M., The Progress in Bioprinting and Its Potential Impact on Health-Related Quality of Life. *Bioengineering*, 10, 910, 2023, https://doi.org/10.3390/bioengineering10080910.
14. Donderwinkel, I., van Hest, J.C.M., Cameron, N.R., Bio-inks for 3D bioprinting: recent advances and future prospects. *Polym. Chem.*, 8, 4451–4471, 2017, https://doi.org/10.1039/C7PY00826K.
15. Li, X., Liu, B., Pei, B., Chen, J., Zhou, D., Peng, J., Zhang, X., Jia, W., Xu, T., Inkjet Bioprinting of Biomaterials. *Chem. Rev.*, 120, 10793–10833, 2020, https://doi.org/10.1021/acs.chemrev.0c00008.
16. Angelopoulos, I., Allenby, M.C., Lim, M., Zamorano, M., Engineering inkjet bioprinting processes toward translational therapies. *Biotechnol. Bioeng.*, 117, 272–284, 2020, https://doi.org/10.1002/bit.27176.
17. Tian, S., Zhao, H., Lewinski, N., Key parameters and applications of extrusion-based bioprinting. *Bioprinting*, 23, e00156, 2021, https://doi.org/10.1016/j.bprint.2021.e00156.
18. Ouyang, L., Pushing the rheological and mechanical boundaries of extrusion-based 3D bioprinting. *Trends Biotechnol.*, 40, 891–902, 2022, https://doi.org/10.1016/j.tibtech.2022.01.001.
19. Amorim, P.A., d'Ávila, M.A., Anand, R., Moldenaers, P., Van Puyvelde, P., Bloemen, V., Insights on shear rheology of inks for extrusion-based 3D bioprinting. *Bioprinting*, 22, e00129, 2021, https://doi.org/10.1016/j.bprint.2021.e00129.
20. Torras, N., Zabalo, J., Abril, E., Carré, A., García-Díaz, M., Martínez, E., A simple DLP-bioprinting strategy produces cell-laden crypt-villous structures for an advanced 3D gut model, https://doi.org/10.1101/2022.02.09.479715.
21. Goodarzi Hosseinabadi, H., Nieto, D., Yousefinejad, A., Fattel, H., Ionov, L., Miri, A.K., Ink material selection and optical design considerations in DLP 3D printing. *Appl. Mater. Today*, 30, 101721, 2023, https://doi.org/10.1016/j.apmt.2022.101721.
22. Ji, S. and Guvendiren, M., Recent Advances in Bioink Design for 3D Bioprinting of Tissues and Organs. *Front. Bioeng. Biotechnol.*, 5, 2017, https://doi.org/10.3389/fbioe.2017.00023.
23. Camacho, P., Busari, H., Seims, K.B., Tolbert, J.W., Chow, L.W., Materials as Bioinks and Bioink Design, in: *3D Bioprinting in Medicine*, pp. 67–100, Springer International Publishing, Cham, 2019, https://doi.org/10.1007/978-3-030-23906-0_2.

24. Zhang, T., Zhao, W., Xiahou, Z., Wang, X., Zhang, K., Yin, J., Bioink design for extrusion-based bioprinting. *Appl. Mater. Today*, 25, 101227, 2021, https://doi.org/10.1016/j.apmt.2021.101227.
25. Zhang, W., Recent Progress in Bioprinting: From Bioink Design to Applications. *Bioengineering*, 9, 785, 2022, https://doi.org/10.3390/bioengineering9120785.
26. Lee, J., Oh, S.J., An, S.H., Kim, W.-D., Kim, S.-H., Machine learning-based design strategy for 3D printable bioink: elastic modulus and yield stress determine printability. *Biofabrication*, 12, 035018, 2020, https://doi.org/10.1088/1758-5090/ab8707.
27. Qiao, T., He, C., Xia, P., Liu, G., Sun, Y., Sun, M., Wang, Y., Cheng, Y., Yu, M., He, Y., Bioink design for organ-scale projection-based 3D bioprinting. *Nat. Protoc.*, 2025, https://doi.org/10.1038/s41596-025-01221-0.
28. Zhang, Z., Jin, Y., Yin, J., Xu, C., Xiong, R., Christensen, K., Ringeisen, B.R., Chrisey, D.B., Huang, Y., Evaluation of bioink printability for bioprinting applications. *Appl. Phys. Rev.*, 5, 2018, https://doi.org/10.1063/1.5053979.
29. Yin, J., Zhao, D., Liu, J., Trends on physical understanding of bioink printability. *Bio-Des. Manuf.*, 2, 50–54, 2019, https://doi.org/10.1007/s42242-019-00033-y.
30. Cooke, M.E. and Rosenzweig, D.H., The rheology of direct and suspended extrusion bioprinting. *APL Bioeng.*, 5, 2021, https://doi.org/10.1063/5.0031475.
31. Bercea, M., Rheology as a Tool for Fine-Tuning the Properties of Printable Bioinspired Gels. *Molecules*, 28, 2766, 2023, https://doi.org/10.3390/molecules28062766.
32. Townsend, J.M., Beck, E.C., Gehrke, S.H., Berkland, C.J., Detamore, M.S., Flow behavior prior to crosslinking: The need for precursor rheology for placement of hydrogels in medical applications and for 3D bioprinting. *Prog. Polym. Sci.*, 91, 126–140, 2019, https://doi.org/10.1016/j.progpolymsci.2019.01.003.
33. Chopin-Doroteo, M., Mandujano-Tinoco, E.A., Krötzsch, E., Tailoring of the rheological properties of bioinks to improve bioprinting and bioassembly for tissue replacement. *Biochim. Biophys. Acta (BBA) - Gen. Subj.*, 1865, 129782, 2021, https://doi.org/10.1016/j.bbagen.2020.129782.
34. Paxton, N., Smolan, W., Böck, T., Melchels, F., Groll, J., Jungst, T., Proposal to assess printability of bioinks for extrusion-based bioprinting and evaluation of rheological properties governing bioprintability. *Biofabrication*, 9, 044107, 2017, https://doi.org/10.1088/1758-5090/aa8dd8.
35. Gillispie, G.J., Copus, J., Uzun-Per, M., Yoo, J.J., Atala, A., Niazi, M.K.K., Lee, S.J., The correlation between rheological properties and extrusion-based printability in bioink artifact quantification. *Mater. Des.*, 233, 112237, 2023, https://doi.org/10.1016/j.matdes.2023.112237.
36. Murphy, J.F., Lavelle, M., Asciak, L., Burdis, R., Levis, H.J., Ligorio, C., McGuire, J., Polleres, M., Smith, P.O., Tullie, L., Uribe-Gomez, J., Chen, B., Dawson, J.I., Gautrot, J.E., Hooper, N.M., Kelly, D.J., Li, V.S.W., Mata, A.,

Pandit, A., Phillips, J.B., Shu, W., Stevens, M.M., Williams, R.L., Armstrong, J.P.K., Huang, Y.Y.S., Biofabrication and biomanufacturing in Ireland and the UK. *Bio-Des. Manuf.*, 7, 825–856, 2024, https://doi.org/10.1007/s42242-024-00316-z.

37. Liu, F., Chen, Q., Liu, C., Ao, Q., Tian, X., Fan, J., Tong, H., Wang, X., Natural Polymers for Organ 3D Bioprinting. *Polymers (Basel)*, 10, 1278, 2018, https://doi.org/10.3390/polym10111278.
38. Fatimi, A., Okoro, O.V., Podstawczyk, D., Siminska-Stanny, J., Shavandi, A., Natural Hydrogel-Based Bio-Inks for 3D Bioprinting in Tissue Engineering: A Review. *Gels*, 8, 179, 2022, https://doi.org/10.3390/gels8030179.
39. Cui, R., Li, S., Li, T., Gou, X., Jing, T., Zhang, G., Wei, G., Jin, Z., Xiong, X., Qu, S., Natural polymer derived hydrogel bioink with enhanced thixotropy improves printability and cellular preservation in 3D bioprinting. *J. Mater. Chem. B*, 11, 3907–3918, 2023, https://doi.org/10.1039/D2TB02786K.
40. Benwood, C., Chrenek, J., Kirsch, R.L., Masri, N.Z., Richards, H., Teetzen, K., Willerth, S.M., Natural Biomaterials and Their Use as Bioinks for Printing Tissues. *Bioengineering*, 8, 27, 2021, https://doi.org/10.3390/bioengineering8020027.
41. Khoeini, R., Nosrati, H., Akbarzadeh, A., Eftekhari, A., Kavetskyy, T., Khalilov, R., Ahmadian, E., Nasibova, A., Datta, P., Roshangar, L., Deluca, D.C., Davaran, S., Cucchiarini, M., Ozbolat, I.T., Natural and Synthetic Bioinks for 3D Bioprinting. *Adv. NanoBiomed Res.*, 1, 2021, https://doi.org/10.1002/anbr.202000097.
42. Puertas-Bartolomé, M., Mora-Boza, A., García-Fernández, L., Emerging Biofabrication Techniques: A Review on Natural Polymers for Biomedical Applications. *Polymers (Basel)*, 13, 1209, 2021, https://doi.org/10.3390/polym13081209.
43. Debnath, S., Agrawal, A., Jain, N., Chatterjee, K., Player, D.J., Collagen as a bio-ink for 3D printing: a critical review. *J. Mater. Chem. B*, 13, 1890–1919, 2025, https://doi.org/10.1039/D4TB01060D.
44. Stepanovska, J., Supova, M., Hanzalek, K., Broz, A., Matejka, R., Collagen Bioinks for Bioprinting: A Systematic Review of Hydrogel Properties, Bioprinting Parameters, Protocols, and Bioprinted Structure Characteristics. *Biomedicines*, 9, 1137, 2021, https://doi.org/10.3390/biomedicines9091137.
45. Griffanti, G., Rezabeigi, E., Li, J., Murshed, M., Nazhat, S.N., Rapid Biofabrication of Printable Dense Collagen Bioinks of Tunable Properties. *Adv. Funct. Mater.*, 30, 2020, https://doi.org/10.1002/adfm.201903874.
46. Marques, C.F., Diogo, G.S., Pina, S., Oliveira, J.M., Silva, T.H., Reis, R.L., Collagen-based bioinks for hard tissue engineering applications: a comprehensive review. *J. Mater. Sci. - Mater. Med.*, 30, 32, 2019, https://doi.org/10.1007/s10856-019-6234-x.
47. Osidak, E.O., Kozhukhov, V.I., Osidak, M.S., Domogatsky, S.P., Collagen as Bioink for Bioprinting: A Comprehensive Review. *Int. J. Bioprint.*, 6, 270, 2020, https://doi.org/10.18063/ijb.v6i3.270.

48. Asim, S., Tabish, T.A., Liaqat, U., Ozbolat, I.T., Rizwan, M., Advances in Gelatin Bioinks to Optimize Bioprinted Cell Functions. *Adv. Healthcare Mater.*, 12, 2023, https://doi.org/10.1002/adhm.202203148.
49. Yin, J., Yan, M., Wang, Y., Fu, J., Suo, H., 3D Bioprinting of Low-Concentration Cell-Laden Gelatin Methacrylate (GelMA) Bioinks with a Two-Step Cross-linking Strategy. *ACS Appl. Mater. Interfaces*, 10, 6849–6857, 2018, https://doi.org/10.1021/acsami.7b16059.
50. Rajabi, N., Rezaei, A., Kharaziha, M., Bakhsheshi-Rad, H.R., Luo, H., RamaKrishna, S., Berto, F., Recent Advances on Bioprinted Gelatin Methacrylate-Based Hydrogels for Tissue Repair. *Tissue Eng. Part A*, 27, 679–702, 2021, https://doi.org/10.1089/ten.tea.2020.0350.
51. Liu, W., Heinrich, M.A., Zhou, Y., Akpek, A., Hu, N., Liu, X., Guan, X., Zhong, Z., Jin, X., Khademhosseini, A., Zhang, Y.S., Extrusion Bioprinting of Shear-Thinning Gelatin Methacryloyl Bioinks. *Adv. Healthcare Mater.*, 6, 2017, https://doi.org/10.1002/adhm.201601451.
52. Lee, H., Kim, J., Choi, Y., Cho, D.-W., Application of Gelatin Bioinks and Cell-Printing Technology to Enhance Cell Delivery Capability for 3D Liver Fibrosis-on-a-Chip Development. *ACS Biomater. Sci. Eng.*, 6, 2469–2477, 2020, https://doi.org/10.1021/acsbiomaterials.9b01735.
53. Shams, E., Barzad, M.S., Mohamadnia, S., Tavakoli, O., Mehrdadfar, A., A review on alginate-based bioinks, combination with other natural biomaterials and characteristics. *J. Biomater. Appl.*, 37, 355–372, 2022, https://doi.org/10.1177/08853282221085690.
54. Jia, J., Richards, D.J., Pollard, S., Tan, Y., Rodriguez, J., Visconti, R.P., Trusk, T.C., Yost, M.J., Yao, H., Markwald, R.R., Mei, Y., Engineering alginate as bioink for bioprinting. *Acta Biomater.*, 10, 4323–4331, 2014, https://doi.org/10.1016/j.actbio.2014.06.034.
55. Piras, C.C. and Smith, D.K., Multicomponent polysaccharide alginate-based bioinks. *J. Mater. Chem. B*, 8, 8171–8188, 2020, https://doi.org/10.1039/D0TB01005G.
56. Axpe, E. and Oyen, M., Applications of Alginate-Based Bioinks in 3D Bioprinting. *Int. J. Mol. Sci.*, 17, 1976, 2016, https://doi.org/10.3390/ijms17121976.
57. Pahlevanzadeh, F., Mokhtari, H., Bakhsheshi-Rad, H.R., Emadi, R., Kharaziha, M., Valiani, A., Poursamar, S.A., Ismail, A.F., RamaKrishna, S., Berto, F., Recent Trends in Three-Dimensional Bioinks Based on Alginate for Biomedical Applications. *Materials*, 13, 3980, 2020, https://doi.org/10.3390/ma13183980.
58. Noh, I., Kim, N., Tran, H.N., Lee, J., Lee, C., 3D printable hyaluronic acid-based hydrogel for its potential application as a bioink in tissue engineering. *Biomater. Res.*, 23, 2019, https://doi.org/10.1186/s40824-018-0152-8.
59. Petta, D., D'Amora, U., Ambrosio, L., Grijpma, D.W., Eglin, D., D'Este, M., Hyaluronic acid as a bioink for extrusion-based 3D printing. *Biofabrication*, 12, 032001, 2020, https://doi.org/10.1088/1758-5090/ab8752.

60. Chen, H., Xue, H., Zeng, H., Dai, M., Tang, C., Liu, L., 3D printed scaffolds based on hyaluronic acid bioinks for tissue engineering: a review. *Biomater. Res.*, 27, 2023, https://doi.org/10.1186/s40824-023-00460-0.
61. Antich, C., de Vicente, J., Jiménez, G., Chocarro, C., Carrillo, E., Montañez, E., Gálvez-Martín, P., Marchal, J.A., Bio-inspired hydrogel composed of hyaluronic acid and alginate as a potential bioink for 3D bioprinting of articular cartilage engineering constructs. *Acta Biomater.*, 106, 114–123, 2020, https://doi.org/10.1016/j.actbio.2020.01.046.
62. Hauptstein, J., Böck, T., Bartolf-Kopp, M., Forster, L., Stahlhut, P., Nadernezhad, A., Blahetek, G., Zernecke-Madsen, A., Detsch, R., Jüngst, T., Groll, J., Teßmar, J., Blunk, T., Hyaluronic Acid-Based Bioink Composition Enabling 3D Bioprinting and Improving Quality of Deposited Cartilaginous Extracellular Matrix. *Adv. Healthcare Mater.*, 9, 2020, https://doi.org/10.1002/adhm.202000737.
63. Sekar, M.P., Suresh, S., Zennifer, A., Sethuraman, S., Sundaramurthi, D., Hyaluronic Acid as Bioink and Hydrogel Scaffolds for Tissue Engineering Applications. *ACS Biomater. Sci. Eng.*, 9, 3134–3159, 2023, https://doi.org/10.1021/acsbiomaterials.3c00299.
64. Lee, J., Lee, S.-H., Kim, B.S., Cho, Y.-S., Park, Y., Development and Evaluation of Hyaluronic Acid-Based Hybrid Bio-Ink for Tissue Regeneration. *Tissue Eng. Regener. Med.*, 15, 761–769, 2018, https://doi.org/10.1007/s13770-018-0144-8.
65. Rijal, G., Bioinks of Natural Biomaterials for Printing Tissues. *Bioengineering*, 10, 705, 2023, https://doi.org/10.3390/bioengineering10060705.
66. Kumar, S., Synthetic polymer-derived single-network inks/bioinks for extrusion-based 3D printing towards bioapplications. *Mater. Adv.*, 2, 6928–6941, 2021, https://doi.org/10.1039/D1MA00525A.
67. Abelardo, E., Synthetic material bioinks, in: *3D Bioprinting for Reconstructive Surgery*, pp. 137–144, Woodhead Publishing (Elsevier), Oxford, UK, 2018, https://doi.org/10.1016/B978-0-08-101103-4.00009-0.
68. Chang, S.-Y., Ching, T., Hashimoto, M., Bioprinting using PEGDMA-based hydrogel on DLP printer. *Mater. Today Proc.*, 70, 179–183, 2022, https://doi.org/10.1016/j.matpr.2022.09.017.
69. Piluso, S., Skvortsov, G.A., Altunbek, M., Afghah, F., Khani, N., Koç, B., Patterson, J., 3D bioprinting of molecularly engineered PEG-based hydrogels utilizing gelatin fragments. *Biofabrication*, 13, 045008, 2021, https://doi.org/10.1088/1758-5090/ac0ff0.
70. Xin, S., Chimene, D., Garza, J.E., Gaharwar, A.K., Alge, D.L., Clickable PEG hydrogel microspheres as building blocks for 3D bioprinting. *Biomater. Sci.*, 7, 1179–1187, 2019, https://doi.org/10.1039/C8BM01286E.
71. Borkar, T., Goenka, V., Jaiswal, A.K., Application of poly-ε-caprolactone in extrusion-based bioprinting. *Bioprinting*, 21, e00111, 2021, https://doi.org/10.1016/j.bprint.2020.e00111.

72. Zhang, K., Fu, Q., Yoo, J., Chen, X., Chandra, P., Mo, X., Song, L., Atala, A., Zhao, W., 3D bioprinting of urethra with PCL/PLCL blend and dual autologous cells in fibrin hydrogel: An *in vitro* evaluation of biomimetic mechanical property and cell growth environment. *Acta Biomater.*, 50, 154–164, 2017, https://doi.org/10.1016/j.actbio.2016.12.008.
73. Dukle, A. and Sankar, M.R., 3D-printed polylactic acid scaffolds for bone tissue engineering: Bioactivity enhancing strategies based on composite filaments and coatings. *Mater. Today Commun.*, 40, 109776, 2024, https://doi.org/10.1016/j.mtcomm.2024.109776.
74. Zhou, H., Lawrence, J.G., Bhaduri, S.B., Fabrication aspects of PLA-CaP/PLGA-CaP composites for orthopedic applications: A review. *Acta Biomater.*, 8, 1999–2016, 2012, https://doi.org/10.1016/j.actbio.2012.01.031.
75. Kalirajan, C., Dukle, A., Nathanael, A.J., Oh, T.-H., Manivasagam, G., A Critical Review on Polymeric Biomaterials for Biomedical Applications. *Polymers (Basel)*, 13, 3015, 2021, https://doi.org/10.3390/polym13173015.
76. Kitchens, W.H., Legal Issues Involved in Tissue Engineering: FDA Regulation of Tissue Engineering, in: *Synthetic Biodegradable Polymer Scaffolds*, pp. 33–49, Birkhäuser Boston, Boston, MA, 1997, https://doi.org/10.1007/978-1-4612-4154-6_3.
77. D.A.A.D.S.B.S.S.M.A.M.P.S., Kumari, S., Katiyar, S., Design strategies for composite matrix and multifunctional polymeric scaffolds with enhanced bioactivity for bone tissue engineering. *Front. Chem.*, 2022, https://doi.org/10.3389/fchem.2022.1051678.
78. Saniei, H. and Mousavi, S., Surface modification of PLA 3D-printed implants by electrospinning with enhanced bioactivity and cell affinity. *Polymer (Guildf)*, 196, 122467, 2020, https://doi.org/10.1016/j.polymer.2020.122467.
79. V.G.M.R.M.S.A.T.L.A.M.F., Veronesi, F., Giavaresi, G., Bioactivity and bone healing properties of biomimetic porous composite scaffold: *in vitro* and *in vivo* studies. *J. Biomed. Mater. Res. A*, 2015, https://doi.org/10.1002/jbm.a.35433.
80. Heid, S. and Boccaccini, A.R., Advancing bioinks for 3D bioprinting using reactive fillers: A review. *Acta Biomater.*, 113, 1–22, 2020, https://doi.org/10.1016/j.actbio.2020.06.040.
81. Maan, Z., Masri, N.Z., Willerth, S.M., Smart Bioinks for the Printing of Human Tissue Models. *Biomolecules*, 12, 141, 2022, https://doi.org/10.3390/biom12010141.
82. Rasouli, R., Sweeney, C., Frampton, J.P., Heterogeneous and Composite Bioinks for 3D-Bioprinting of Complex Tissue. *Biomed. Mater. Devices*, 3, 108–126, 2025, https://doi.org/10.1007/s44174-024-00171-7.
83. Ravanbakhsh, H., Bao, G., Luo, Z., Mongeau, L.G., Zhang, Y.S., Composite Inks for Extrusion Printing of Biological and Biomedical Constructs. *ACS Biomater. Sci. Eng.*, 7, 4009–4026, 2021, https://doi.org/10.1021/acsbiomaterials.0c01158.

84. Han, Y., Wei, Q., Chang, P., Hu, K., Okoro, O.V., Shavandi, A., Nie, L., Three-Dimensional Printing of Hydroxyapatite Composites for Biomedical Application. *Crystals (Basel)*, 11, 353, 2021, https://doi.org/10.3390/cryst11040353.
85. Kim, W. and Kim, G., Collagen/bioceramic-based composite bioink to fabricate a porous 3D hASCs-laden structure for bone tissue regeneration. *Biofabrication*, 12, 015007, 2019, https://doi.org/10.1088/1758-5090/ab436d.
86. Dutta, S. and Cohn, D., Temperature and pH responsive 3D printed scaffolds. *J. Mater. Chem. B*, 5, 9514–9521, 2017, https://doi.org/10.1039/C7TB02368E.
87. Bandiera, A., Catanzano, O., Bertoncin, P., Bergonzi, C., Bettini, R., Elviri, L., 3D-printed scaffold composites for the stimuli-induced local delivery of bioactive adjuncts. *Biotechnol. Appl. Biochem.*, 69, 1793–1804, 2022, https://doi.org/10.1002/bab.2245.
88. Tariq, A., Arif, Z.U., Khalid, M.Y., Hossain, M., Rasool, P.I., Umer, R., Ramakrishna, S., Recent Advances in the Additive Manufacturing of Stimuli-Responsive Soft Polymers. *Adv. Eng. Mater.*, 25, 2023, https://doi.org/10.1002/adem.202301074.
89. Paci, C., Iberite, F., Arrico, L., Vannozzi, L., Parlanti, P., Gemmi, M., Ricotti, L., Piezoelectric nanocomposite bioink and ultrasound stimulation modulate early skeletal myogenesis. *Biomater. Sci.*, 10, 5265–5283, 2022, https://doi.org/10.1039/D1BM01853A.
90. Ferreras, A., Matesanz, A., Mendizabal, J., Artola, K., Nishina, Y., Acedo, P., Jorcano, J.L., Ruiz, A., Reina, G., Martín, C., Light-Responsive and Antibacterial Graphenic Materials as a Holistic Approach to Tissue Engineering. *ACS Nanosci. Au*, 4, 263–272, 2024, https://doi.org/10.1021/acsnanoscienceau.4c00006.
91. Yuan, X., Zhu, W., Yang, Z., He, N., Chen, F., Han, X., Zhou, K., Recent Advances in 3D Printing of Smart Scaffolds for Bone Tissue Engineering and Regeneration. *Adv. Mater.*, 36, 2024, https://doi.org/10.1002/adma.202403641.
92. Gelmi, A. and Schutt, C.E., Stimuli-Responsive Biomaterials: Scaffolds for Stem Cell Control. *Adv. Healthcare Mater.*, 10, 2021, https://doi.org/10.1002/adhm.202001125.
93. Kannan, M.B. and Ronan, K., Conversion of biowastes to biomaterial: An innovative waste management approach. *Waste Manage. (Oxford)*, 67, 67–72, 2017, https://doi.org/10.1016/j.wasman.2017.05.045.
94. Albihn, A., Recycling Biowaste – Human and Animal Health Problems. *Acta Vet. Scand.*, 43, S69, 2002, https://doi.org/10.1186/1751-0147-43-S1-S69.
95. Xu, C., Nasrollahzadeh, M., Selva, M., Issaabadi, Z., Luque, R., Waste-to-wealth: biowaste valorization into valuable bio(nano)materials. *Chem. Soc. Rev.*, 48, 4791–4822, 2019, https://doi.org/10.1039/C8CS00543E.
96. Mishra, B., Mohanta, Y.K., Reddy, C.N., Reddy, S.D.M., Mandal, S.K., Yadavalli, R., Sarma, H., Valorization of agro-industrial biowaste to

biomaterials: An innovative circular bioeconomy approach. *Circ. Econ.*, 2, 100050, 2023, https://doi.org/10.1016/j.cec.2023.100050.

97. Dukle, A. and Ravi Sankar, M., Sustainable 3D printing of bone scaffolds using animal biowaste feedstocks. *Next Sustainability*, 5, 100099, 2025, https://doi.org/10.1016/j.nxsust.2025.100099.
98. Shibru, M.G., Ali, Z.M., Almansoori, A.S., Paunovic, J., Pantic, I.V., Corridon, P.R., Slaughterhouse Waste: A Unique and Sustainable Source for dECM-Based Bioinks. *Regener. Med.*, 19, 113–118, 2024, https://doi.org/10.2217/rme-2023-0194.
99. Madadi, M., Amiri, H., Pan, J., Song, G., Liu, D., Gupta, V.K., Aghbashlo, M., Tabatabaei, M., Food loss and waste valorization offers a sustainable source of biopolymers in bioinks for 3D printing. *Nat. Food*, 6, 323–330, 2025, https://doi.org/10.1038/s43016-025-01146-7.
100. Barua, E., Deb, P., Das Lala, S., Deoghare, A.B., Extraction of Hydroxyapatite from Bovine Bone for Sustainable Development, pp. 147–158, 2019, https://doi.org/10.1007/978-981-13-9977-0_10.
101. Dukle, A., Dua, R., Deepati, A.K., Sankar, M.R., Repurposing Biowaste for Biomedical Applications: Eggshell containing Injectable Bone Substitution Cements for Bone Regeneration. *Hybrid Adv.*, 100425, 2025, https://doi.org/10.1016/j.hybadv.2025.100425.
102. Basara, G., Ozcebe, S.G., Ellis, B.W., Zorlutuna, P., Tunable Human Myocardium Derived Decellularized Extracellular Matrix for 3D Bioprinting and Cardiac Tissue Engineering. *Gels*, 7, 70, 2021, https://doi.org/10.3390/gels7020070.
103. Das, S. and Jang, J., 3D Bioprinting and Decellularized ECM-Based Biomaterials for *in Vitro* CV Tissue Engineering. *J. 3D Print. Med.*, 2, 69–87, 2018, https://doi.org/10.2217/3dp-2018-0002.
104. Lee, H., Han, W., Kim, H., Ha, D.-H., Jang, J., Kim, B.S., Cho, D.-W., Development of Liver Decellularized Extracellular Matrix Bioink for Three-Dimensional Cell Printing-Based Liver Tissue Engineering. *Biomacromolecules*, 18, 1229–1237, 2017, https://doi.org/10.1021/acs.biomac.6b01908.
105. Stone, R.N., Reeck, J.C., Oxford, J.T., Advances in Cartilage Tissue Engineering Using Bioinks with Decellularized Cartilage and Three-Dimensional Printing. *Int. J. Mol. Sci.*, 24, 5526, 2023, https://doi.org/10.3390/ijms24065526.
106. Visscher, D.O., Lee, H., van Zuijlen, P.P.M., Helder, M.N., Atala, A., Yoo, J.J., Lee, S.J., A photo-crosslinkable cartilage-derived extracellular matrix bioink for auricular cartilage tissue engineering. *Acta Biomater.*, 121, 193–203, 2021, https://doi.org/10.1016/j.actbio.2020.11.029.

4

3D Bioprinting in Precision Biomanufacturing: From Design to Functional Tissue Constructs

Ajithkumar Sitharaj[1]*, Idhayaraja S.[2], Arulmurugan B.[1] and Sandip Kunar[3]

[1]Department of Mechanical Engineering, KPR Institute of Engineering and Technology, Coimbatore, Tamil Nadu, India
[2]Department of Manufacturing Engineering, PSG College of Technology, Coimbatore, Tamil Nadu, India
[3]Department of Mechanical Engineering, Aditya University, Surampalem, Andhra Pradesh, India

Abstract

3D bioprinting provides unparalleled accuracy in arranging cells, biomaterials, and bioactive signals to construct functional tissues. This chapter examines key bioprinting techniques, including inkjet, extrusion-based, and laser-assisted methods, evaluating their printability, resolution, and biocompatibility. The classification and formulation of bioink are examined, focusing on rheology, cross-linking, and compatibility factors that affect structural integrity and maturation. Computational tools, such as CAD/CAM and finite element analysis, are emphasized for enhancing scaffold design and mechanical efficacy. Applications encompass skin, bone, and cartilage regeneration, as well as vascularized constructs, organoids, and *in vitro* models for disease and drug screening. The incorporation of bioprinting into workflows compliant with Good Manufacturing Practice (GMP), along with automation, real-time monitoring, and AI-driven optimization, is examined for translational readiness. Emerging trends include 4D bioprinting, CRISPR-enabled constructs, and decentralized manufacturing. These advancements establish 3D bioprinting as a fundamental element of future regenerative medicine, biopharmaceutical manufacturing, and individualized therapeutic approaches.

**Corresponding author*: ersajithkumarbe@gmail.com

Sandip Kunar, Jagadeesha T. and D. Sathis Kumar (eds.) Biomanufacturing: Transforming Production of Biologics and Biopharmaceuticals, (91–112)

Keywords: 3D bioprinting, bioinks, tissue engineering, precision biomanufacturing, regenerative medicine

4.1 Introduction

Precision biomanufacturing is a shift in how biologics and biopharmaceuticals are manufactured and developed. Traditionally, biologics and biopharmaceuticals have been manufactured using general manufacturing practices to manage batch processes and generalized formulations. Precision biomanufacturing employs new techniques like computational modeling, automation, high-throughput screening, and real-time monitoring to allow for complex biological products to be manufactured with great specificity, consistency, and scalability [1]. Rather than batch processing, the core focus of precision biomanufacturing is tailoring the processes to the individual patient or therapeutic indication to provide effective treatment that is less variable. This shift is being propelled by the growth of patient-centric approaches to medicine, next-generation therapeutics like gene and cell therapy, and the demand for rapid production that can respond to emerging health challenges [2]. Digital technologies, artificial intelligence, and biodesign principles are also contributing to the advancement in precision manufacturing systems, which will be a critical approach in the future of biologics [3].

3D bioprinting has become a groundbreaking technology within precision biomanufacturing technology, that can produce biologically functional constructs using living cells, biomaterials and bioactive molecules in layer-by-layer increments. 3D bioprinting is based on the conventional additive manufacturing technologies that have developed over the last two decades to effectively navigate the technical challenges inherent within biological systems. The modalities have transitioned beyond basic inkjet, extrusion, and laser-assisted printing, and spanned unique demands such as spatial control over cells and tissue architecture and bioink composition. Consequentially, 3D bioprinting has matured from basic scaffold production to individual tissue constructs such as skin and cartilage, and complex tissue analogs such as vascularized constructs and the liver [4, 5]. These bioprinted constructs are now increasingly presented for applications in regenerative medicine, as well as *in vitro* models for studying disease, drug screening, and toxicology. The combination of the recent advances in bioink composition development, biocompatible materials development and engineered cellular populations and tissues are providing a much broader service to interdisciplinary fields of academic, clinical, and industrial applications and impact.

3D bioprinting is relevant to biologics and biopharmaceuticals because of its ability to create tightly controlled and physiologically relevant microenvironments for production and application of therapeutic products. In the upstream biomanufacturing context, bioprinting enables the creation of tissue-like scaffoldings and microenvironments which ultimately enhance cell proliferation, differentiation, and protein expression that result in biologic products of higher yield and quality. In the downstream and clinical application areas, it enables the creation of customized delivery systems and tissue implants that have direct interfaces with the biological system for improved bioavailability and efficacy of biopharmaceutical agents. Additionally, 3D bioprinting allows for development of personalized therapeutic platforms using patient-derived cells and customized architecture, overlapping in potentially significant ways with the goals of precision medicine. Concurrently, as biologics, such as monoclonal antibodies, therapeutic proteins, and cell therapies increase in prominence as therapy options, adoption of bioprinting technologies presents a new avenue for improving development, regulation, and effectiveness in therapeutics [6, 7]. This chapter will provide a detailed and timely overview of the relevance of 3D bioprinting to precision biomanufacturing, while focusing on its role in the production and delivery of biologics and biopharmaceuticals.

4.2 Core Technologies and Mechanisms of 3D Bioprinting

3D bioprinting, like all additive manufacturing, builds living tissues, in part, by building bioinks, which are living cells and biomaterials and bioactive substances, one layer at a time. The goal is to create biological structures that imitate their cellular microenvironments, mechanical properties, and structure from actual tissues. As a result, several methods of building such tissues have been developed, with each method being defined by the properties of bioink, required resolution, cell type, and intended application. Table 4.1 provides a summary of the main 3D bioprinting methods, with details of their principles/operational mechanisms, merits, limitations, and examples of standard uses.

Inkjet bioprinting is a non-contact, drop-on-demand (DoD) technology that uses thermal, piezoelectric, or electromagnetic actuation to eject discrete droplets of bioink onto a substrate. The different benefits of inkjet

Table 4.1 Comparison of major 3D bioprinting modalities [8, 9, 11].

Modality	Working principle	Resolution	Suitable bioink type	Advantages	Limitations
Inkjet Bioprinting	Drop-on-demand *via* thermal/piezoelectric force	~50–100 μm	Low-viscosity hydrogels, cells	High speed, low cost, good cell viability	Limited to low-viscosity inks
Extrusion-Based Printing	Continuous extrusion *via* pressure/mechanical	~100–200 μm	Medium/high-viscosity hydrogels	Broad material range, high cell density	Lower resolution, nozzle clogging
Laser-Assisted Printing	Laser-induced forward transfer (LIFT)	~10–50 μm	High-precision bioinks, ECM-based	High resolution, nozzle-free, precise control	Expensive, complex setup, lower throughput

bioprinting are high printing speed, low materials cost in relation to other bioprinting methods, and its ability to drop small volumes of bioink accurately to create cell and biomolecule patterns in two-dimensional and very thin three-dimensional constructs. In addition, inkjet bioprinting is limited to primarily low viscosity bioinks and low cell densities, and heat or shear stress resulting from similar constructions of deposited bioink could alter cell viability [8]. Inkjet bioprinting is typically employed to deposit discrete droplets in drug screening platforms, tissue patterning, and microarray construction.

Extrusion bioprinting is the most widely utilized modality in tissue engineering and biomanufacturing due to material versatility and the ability to print viscous hydrogels and cell-laden matrices. This technology is reliant on pneumatic, mechanical or screw-based systems to extrude bioink continuously through a nozzle and forms continuous filaments to form 3D structures in a layer-by-layer fashion [9]. This modality allows for both higher cell concentrations and a more diverse range of biomaterials including thermoresponsive bioinks, and composite bioinks. In general, extrusion bioprinting has lower resolution than inket or laser-based modalities; yet is well-suited for printing bulk tissues, scaffold-based constructs, and bioprinting tissues that require greater mechanical intransigence, such as bone and cartilage.

Laser-assisted bioprinting (LaBP) is a nozzle-free, high-resolution method utilizing laser pulses and the laser-induced forward transfer (LIFT) mechanism to micropattern bioink droplets, mitigating shear stress and clogging. This method is ideal for sensitive cells such as stem cells and vascular tissues and offers higher resolution and faster results than inkjet or direct-write bioprinting methods. However, LaBP is less widely used in bioprinting applications because the method is complex, costly, and has lower throughput than other methods. Sorkio *et al.* [10], used LaBP to fabricate three-dimensional cornea tissue from human embryonic stem cell derived limbal epithelial stem cells (hESC LESCs) and adipose derived stem cells (hASCs) in laminin and collagen I bioinks. The constructs exhibited high viability, marker expression and native like architecture. These templates exhibited a multilayered cornea structure, moderate

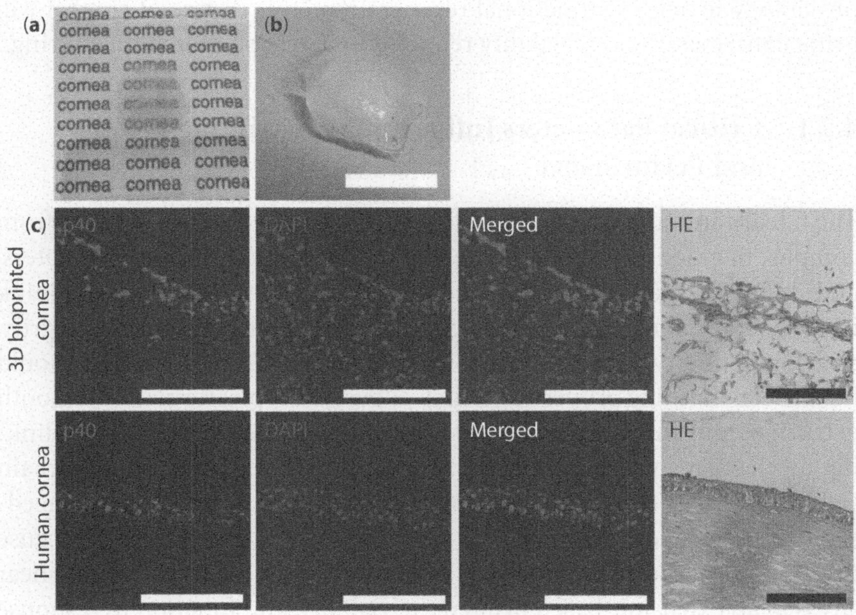

Figure 4.1 Fabrication of a 3D corneal construct using laser assisted bioprinting (LaBP) with human embryonic stem cell derived limbal epithelial stem cells (hESC LESCs) and human adipose derived stem cells (hASCs). (a) Bioprinted 3D cornea on a PET substrate showing moderate transparency. (b) Printing on a non-transparent Matriderm® substrate prevented shrinkage during culture. (c) Comparison of the bioprinted construct with native human cornea. Immunofluorescence staining reveals multilayered hESC LESCs expressing the progenitor marker p40 (red) on the 3D scaffold two days post printing. Hematoxylin and eosin staining demonstrates structural integrity of the bioprinted tissue. Scale bars: 10 mm (b), 200 μm (c) [10]. Image available under an open access license.

transparency, close resemblance to native tissue (indicated in Figure 4.1), and stromal layers were able to integrate with porcine cornea when placed *ex vivo*, suggesting cells with regenerative capability were able to migrate into the tissue.

Conventional methods for tissue engineering often involve manually plating cells on scaffolds *via* seeding, and then subsequently culturing in bioreactors to encourage tissue development. While important for figure applications, these methods do not enable precise spatial control over cell placement, are laborious, and prevent users from make consistent tissue architectures. 3D bioprinting provides a programmable, automated, and highly reproducible method to create tissues with complex geometry and hierarchical organization with exact placement of cells. Bioprinting enabled additional advantages including the incorporation of multiple cell types, gradients for growth factors, and vascularized channels, all in a single construct, features that are critical for recreating its native tissue like physiology to achieve functional regeneration. This degree of control will bring consistency and scalability required in precision biomanufacturing.

4.2.1 Critical Parameters Influencing Bioprinting Fidelity and Performance

The fidelity and functional efficacy of 3D bioprinted constructs result from complex interactions among properties of the bioink, printing parameters, and cellular responses. Perhaps the most important property of a bioink is its rheological response, which governs printability, structural integrity, and compatibility with a specific bioprinting area. An ideal bioink should have shear-thinning properties, which allow for droplet formation, smooth extrusion, and shape retention post-printing. Furthermore, the crosslinking mechanisms will dictate the ability of the printed structure to remain stable in 3D, with ionic, thermal, and photopolymerization strategies needing to be matched to the printing speed and resolution [12, 13]. Because printing may lead to mechanical stresses on cells, such as during shear stress in an extrusion-based printer, the viability, and function in response to the stresses is also affected.

Furthermore, nozzle diameter, dispensing pressure, substrate temperature, and print speed all affect spatial resolution, layer registration, and construct integrity. Understanding how to optimize these parameters for each cell type, and application, is ultimately important for both geometric accuracy and biological veracity. For precision biomanufacturing, repeatability, sterility, scalability, and in line quality monitoring become important to comply with clinical and regulatory requirements [14]. New approaches, such as adaptive control, machine learning based control and sensor integrated systems will allow for real time adaptation and enhance the consistency, rate, and functional properties of bioprinted products.

4.3 Bioinks: The Core of Functional Bioprinting

Bioinks constitute the primary components of 3D bioprinting, serving as both the structural matrix and biological environment required for cell survival, proliferation, and new tissue formation. The success of any bioprinted construct depends substantially on the bioink's characterization and performance [15]. A bioink should have exceptional biocompatibility, appropriate mechanical strength, controllable degradation rates, and appropriate rheological properties to afford printability while providing mechanical integrity. The bioink must support cellular activities like adherence, differentiation, and signaling. As 3D bioprinting approaches more complex and clinically relevant applications, the design and engineering of bioinks has also become more advanced, allowing the production of functional, tissue-mimetic constructs with a high level of precision and fidelity.

4.3.1 Classification of Bioinks

Based on the source of bioinks and their composition, bioinks can be considered natural or synthetic. Table 4.2 summarizes the types and properties. Natural bioinks (alginate, gelatine, collagen, hyaluronic acid, chitosan, and fibrin) are biologically sourced material with biocompatibility and bioactivity and can replicate components of the extracellular matrix (ECM) to promote cell adhesion and signaling [16]. Natural bioinks have some drawbacks due to variability between batches, low mechanical strength, and limited control on the rate of degradation. Synthetic bioinks (for example, polyethylene glycol (PEG), Pluronic F127, and polycaprolactone (PCL)) lack native biological cues, but allow for accurate control of stiffness, degradation and crosslinking. The bioactivity can be improved with functionalization by peptides or proteins. The decision regarding which bioink to use will depend on the *in-situ* target tissue, mechanical requirements, and cell compatibility [17].

In addition to simply grouping bioinks by classification, bioinks can also be grouped according to structural formulation. Bioinks derived from extracellular matrix (ECM) by decellularizing tissues are pure bioinks and have the advantage of retaining native proteins and signaling molecules and

Table 4.2 Classification and properties of common bioinks [16–18].

Bioink type	Materials	Key properties	Applications
Natural Bioinks	Collagen, Gelatin, Alginate, Fibrin	High biocompatibility, cell-adhesion sites	Skin, cartilage, vascular tissue
Synthetic Bioinks	PEG, Pluronic F127, PCL	Tunable mechanics, reproducibility	Bone scaffolds, hybrid constructs
ECM-Derived Bioinks	Decellularized organ/tissue ECM	Tissue-specific signals, native microenvironment	Organ-specific models, disease modeling
Composite Bioinks	GelMA + nanoclay, Alginate + HA	Balanced bioactivity and mechanical strength	Osteochondral, skin–bone interfaces
Smart/ Responsive Bioinks	pH/ temperature/ light-sensitive hydrogels	Stimuli-responsive behavior	4D bioprinting, smart implants

therefore ideal to work with for tissue specific and patient tailored applications [18]. Hydrogel bioinks, including gelatin methacrylate (GelMA), alginate, and agarose, also exhibit a host of advantages such as high-water content, tunable porosity, and biocompatibility. Composite bioinks introduce one or more synthetic components, which could be a mixture of both natural and synthetic materials, or might include nanoparticles or nanofibers, which can enhance the mechanical properties of bioinks, increase printability, and also improve biofunctionality. Sanchez Rubio *et al.* [19], highlighted the variability of composite hydrogels when using Alginate Gelatin (AlgGel) and Alginate Gelatin Methacrylate (AlgGelMA), and their capability for *in vitro* tissue modeling. AlgGel formed soft hydrogels (1.5–4.5 kPa) that were observed to soften over time, while AlgGelMA created material that had stiffer constructs (6–40 kPa) which were cured *via* UV crosslinking. THP 1 cells in AlgGel formed larger clusters of cells, while hMSCs in AlgGelMA showed a greater extent of invasion and matrix remodeling. The morphological and mechanical differences are demonstrated in Figure 4.2.

The use of living cells within bioinks, known as cell-laden bioinks, is what separates bioprinting from scaffold printing. These bioinks need to

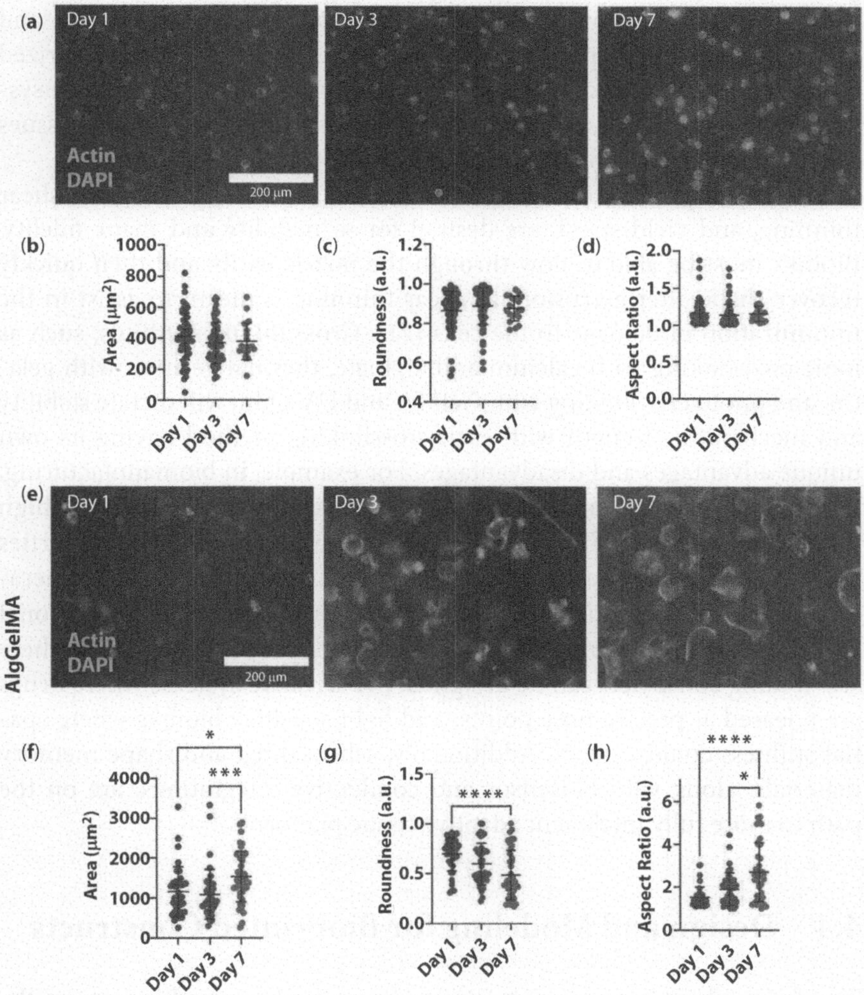

Figure 4.2 Human mesenchymal stem cell (hMSC) response in bioprinted AlgGel and AlgGelMA bioinks over 7 days. (a, e) Representative brightfield images of hMSCs in AlgGel (2%/8%) (a) and AlgGelMA (1%/9%) (e) on days 1, 3, and 7. (b–d) Quantification of cell area (b), roundness (c), and aspect ratio (d) in AlgGel constructs. (f–h) Corresponding metrics for AlgGelMA. Data are presented as individual points with mean ± SD. Statistical significance was assessed using the Kruskal–Wallis test [19]. This figure is available under an open access license.

sustain the viability of the cells, diffusion of nutrients, and phenotypic stability through the printing and maturation phases. Biocompatibility is important here, particularly in how the cells react to being exposed to shear stress, thermal gradient, or cytotoxic crosslinking agents [20]. Important factors influencing cell-laden bioinks include the method of gelation, the properties of the ink, pH, osmolarity, and reactive species. Cell-laden bioinks should also facilitate cell adhesion, proliferation, and matrix remodeling, in particular for soft tissue, cartilage, and vascularized tissue applications. Recent advances in encapsulation and co-culture systems have made it possible to print multicellular and vascularized tissues to provide more physiological relevance in bioprinting.

The rheological characteristics of bioinks, including viscosity, shear thinning, and yield stress, are desired for printability and shape fidelity. Bioinks must be able to flow through the nozzle easily and then quickly recover shape after extrusion, as shear thinning is meant to assist in the minimization of damage to the cells [21]. Crosslinking methods, such as ionic crosslinking with calcium and alginate, thermal gelation with gelatin, and photo crosslinking with GelMA and UV light, will dictate stability and mechanical strength with each crosslinking method having its own unique advantages and disadvantages. For example, in biomanufacturing, the ability to rapidly and controllably crosslink supports precision in high throughput printing. Stimuli responsive bioinks change their properties based on their biological context, which could include changes in temperature, pH, light, magnetic fields, or enzymes that support four dimensional bioprinting [22]. Examples include, thermoresponsive hydrogels where self-folding constructs can be designed, pH sensitive systems where drugs are released at precise time points, and light sensitive bioinks where spatial stiffness could control. Additionally, self-healing, and shape memory materials, along with biosensor and conductive integrations, are on the path to more fully intelligent adaptive tissue platforms.

4.4 Design and Modeling for Bioprinted Constructs

The design phase in ternary bioprinting is a critical step and focuses on the geometry and functions of the final constructs. Manufacturing of constructs is radically different from bioprinting because bioprinted constructs must consider living cells, soft biomaterials, and *in situ* remodeling. To be successful, the design must incorporate computational design, geometric optimization, and predictive modeling that combine engineering principles with biological complexity. Programs such as computer aided design (CAD), computer aided manufacturing (CAM), simulation software, and multi-material design frameworks are especially useful for making biomimetic structures with spatial and functional fidelity [23]. CAD allows for the technical modeling of complicated structures (i.e., the anatomic detail of patient-specific anatomy reconstructed using CT or MRI images or models). Models can be segmented, manipulated, and enhanced for end use in applications, such as scaffolds for bone, cartilage, or some organ. CAM software uses digital models to generate print layers and to inform the printer about how to create those layers. Integrated CAD/CAM workflows can account for application with multiple materials while defining spatial patterns for cells, vascular channels, and complex arrangements; if adhering to a parametric modeling framework, porosity and density can be defined and quickly changed to fit overall design objectives [24].

Scaffold architecture markedly affects biological and mechanical performance. Key scaffold features such as pore size, shape, distribution, interconnectivity, and wall thickness are key determinants of cell migration, nutrient diffusion, vascularization, and strength. High-permeability, interconnected structures will accommodate soft tissue regeneration and subsequent vascularization, while dense, stiff structures will bear load, such as bone tissue scaffolds [25, 26]. Optimal pore size will range from tissue to tissue, but generally, bone has pores that range from 100 to 300 micrometers in size, while cartilage or skin have pores that range from 20 to 100 micrometers. Three unique design approaches, triply periodic minimal surfaces (TPMS), Voronoi tessellations, and random pores, can create unique three-dimensional (3D) pore networks similar in structure to the natural extracellular matrix (ECM). In bioprinting, scaffold shape is controlled by the device design in combination with the spatial freedom afforded by deposition path programming and layer by layer deposition patterning and spatial registration. All 3D printed biological scaffolds can reflect the designers' control of mechanical property versus architectural characteristics by providing defined structural heterogeneity in the one construct. Taking a holistic approach to scaffold design should always balance mechanical integrity with biological function to ensure scaffolds provide immediate protection in load-bear applications, while adapting through the remodeling process to meet the long-term biological.

4.4.1 Finite Element Modeling and Mechanical Simulation

Finite element modeling (FEM) and other numerical simulation programs can be easily ignored in bioprinting with good experimental practices. Using FEM allows virtual testing of scaffold designs by simulating stress strain response with deformation and fluid flow or diffusion, allowing materials, architecture and printing specifications that provide tissue specific performance [27]. Bone scaffolds typically used for simulation would focus on the compressive strength and fatigue resistance of the scaffolds while soft tissues focus on the elasticity, viscoelasticity of soft biological tissues. When linked with experimental testing to set up a calibration/validation of the testing, design processes and material selection can be done in FEM. Advanced technologies in the biomanufacturing space have the ability to use simulation in pre-print design optimization and adjust designs and printing conditions with real time feedback [28].

Noble *et al.* [29], assessed a collagen based bioink for tissue engineered heart valves (TEHV) using implantation in rats and biaxial testing with FEM. Twelve-week explants approached native strength but lacked anisotropy, with FE analysis revealing differences in diastolic stress strain patterns. As shown in Figure 4.3, native valves exhibited higher stress in the leaflet belly and lower strain at the commissures during diastole, indicating the

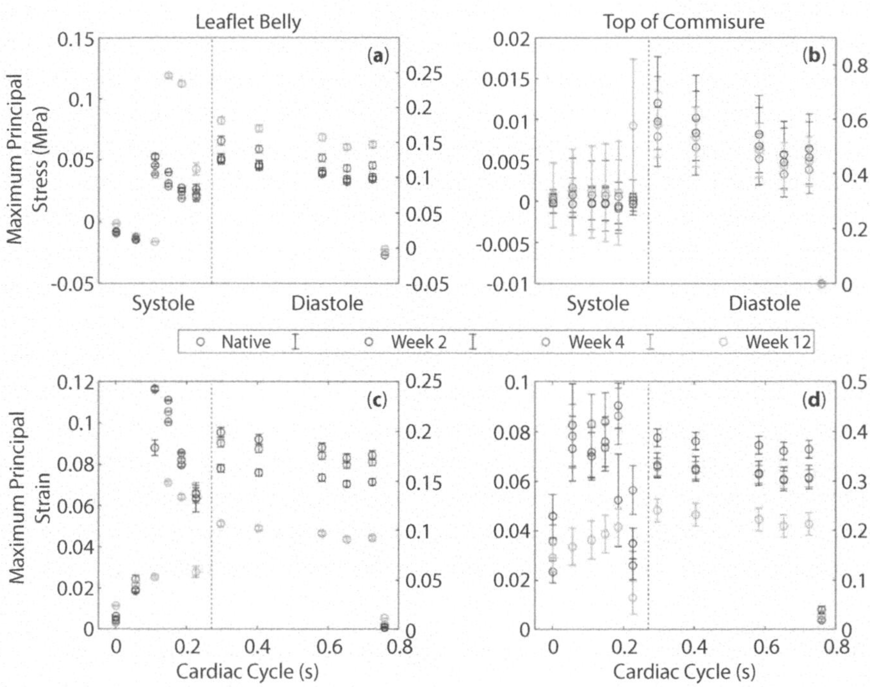

Figure 4.3 Maximum principal stress and strain at the commissure tips and leaflet belly for implant and explant valves across the cardiac cycle [29]. It is available under an open-access license.

absence of native like anisotropy in the explants. These findings highlight FEM's role in optimizing scaffold design and conditioning for improved *in vivo* performance.

The fabrication of functional tissue constructs often requires integration across several length scales and materials, to emulate the inherently complex architecture of native tissues. Multiple scale design includes features from nanoscales to macroscale; for example, nanoscale textures on surfaces influence the adhesion of cells, and macroscale features determine the load bearing capacity of the construct. Multiple material bioprinting enables the deposition of different bioinks, with distinct mechanical, biological, and chemical properties, in a single construct [30]. This is particularly important for the engineering of heterogeneous tissues such as osteochondral interfaces, where soft cartilage and hard bone need to functionally interact with one another. The design of multi material constructs requires sophisticated mapping with design software to identify areas of material in a construct, establish print paths and sequence of extrusions. Gradient based designs are increasingly being utilized to create constructs with gradual or smooth changes in properties, that mimic the transition found in natural tissues. This is a large step toward potential bioprinting capabilities to produce highly biomimetic and functionally graded tissue analogues that integrate well into their host environments and can perform biological functions in real-time.

4.5 Bioprinting Functional Tissues and Constructs

The final aim of three-dimensional bioprinting is to make biologically functional tissue constructs for restoring damaged tissue, modeling disease *in vitro*, or supporting drug discovery [31]. It is essential to be able to deposit cells and biomaterials in precise locations and also be able to duplicate tissue architecture, mechanical properties, and biochemical signals that are found in native tissues. Improvements in bioinks and printing technologies and designing multi-material constructs are making it possible to build biologically and physiologically relevant complex tissue constructs. Functional constructs will also have cells that survive, mature the tissues, and retain an ability to integrate with a host system as an *in vivo* functional construct. Table 4.3 provides a list of bioprinting applications by tissue type.

Skin tissue bioprinting is one of the most advanced applications due to its relatively simple, layered construction and high demand for wound healing applications. Cells created *via* bioprinting fall under two primary

Table 4.3 Bioprinting applications by tissue type [33–36].

Tissue type	Target cell types	Bioink materials	Challenges	Applications
Skin	Keratinocytes, Fibroblasts	Collagen, Fibrin, Gelatin	Vascularization, pigmentation	Wound healing, grafts
Cartilage	Chondrocytes, MSCs	Agarose, GelMA, Hyaluronic acid	Zonal architecture, mechanical strength	Osteoarthritis, joint repair
Bone	Osteoblasts, MSCs	Calcium phosphate, HA, PCL composites	Load-bearing, vascular integration	Craniofacial repair, implants
Vascular Tissue	Endothelial cells, Pericytes	GelMA, Fibrin, Sacrificial inks	Perfusion, anastomosis	Organ support, cardiac patches
Liver/ Organoids	Hepatocytes, iPSCs, Stromal cells	ECM-derived bioinks, collagen blends	Metabolic function, long-term viability	Toxicology, disease modeling

categories: as engineered bilayered systems that mimic the dermis and epidermis or engineered constructs that can be organized into a functional skin substitutes to include vascular networks, melanocytes and hair follicles. Bioinks generally consist of collagen, gelatin, fibrin, and decellularized extracellular matrix and often include keratinocytes and fibroblasts in the layering. Reports of preclinical and clinical trials are favorable with regard to functionality in the treatment of burns, chronic wounds, and cosmetic testing [33]. Another area of research includes *in situ* bioprinting, which involves the direct deposition of bioink into the wound site to expedite the personalized regenerative process. Maggiotto *et al.* [32], demonstrated a vascularized skin model using a dual bioink system of gelatin methacrylate for structural layers and Pluronic F127 as a sacrificial template. The construct incorporated fibroblasts, keratinocytes, and endothelial cells, creating a perfusable, biomimetic skin tissue. Figure 4.4 illustrates the layered architecture, perfusion system, and custom bioreactor setup. This model shows strong potential for both high fidelity *in vitro* applications and future regenerative therapies.

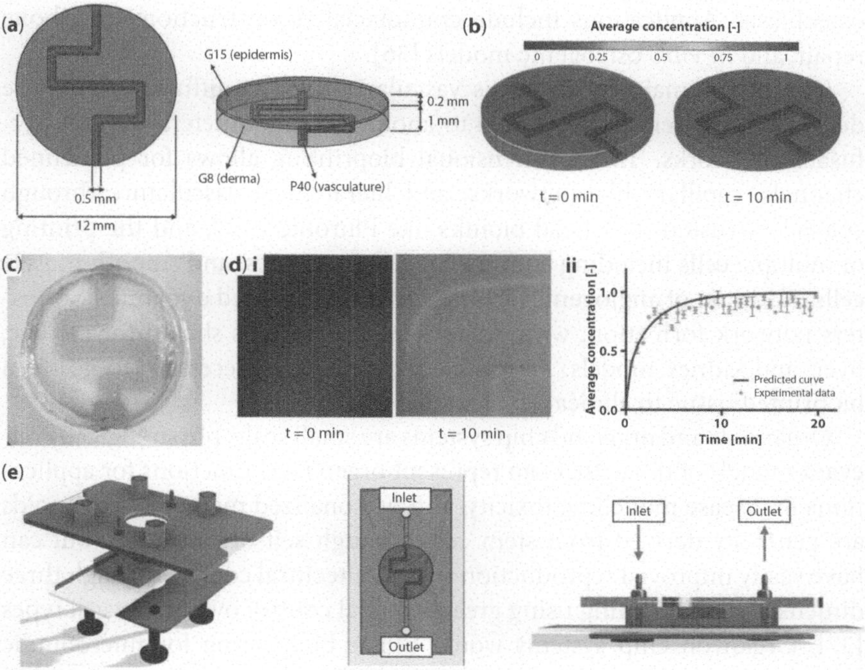

Figure 4.4 Design and validation of the vascularized skin model. (a) Schematic of the bioprinted construct showing a one-millimeter dermal (G8) layer, a zero-point two-millimeter epidermal (G15) layer, and a perfusable channel formed using sacrificial P40. (b) Diffusion simulation results indicating that equilibrium is reached within ten minutes of perfusion. (c) Bioprinted model used for experimental validation. (d) Validation results: (i) fluorescence images before and after perfusion; (ii) close agreement between predicted and experimental concentration profiles. (e) Bioreactor setup with polydimethylsiloxane (PDMS) mold and adjustable sealing layers [32]. This figure is available under an open access license.

Cartilage and bone bioprinting involve unique challenges regarding mechanical complexity and cellular heterogeneity. Cartilage constructs attempt to mimic the avascular and load bearing nature of native tissue using chondrocyte laden hydrogels, such as gelatin methacrylate (GelMA), agarose, and hyaluronic acid which can also contain zonal architectures to mimic native gradients [34, 35]. Bone bioprinting is focused on creating osteoconductivity and mechanical strength with vascularization by combining hydroxyapatite, bioactive glass, calcium phosphate, or other material systems in composite bioinks with mesenchymal stem cells or osteoblasts. Applications include craniofacial reconstruction, long bone repair, and *in vitro* osteogenic models [36].

One of the major obstacles is vascularization, as diffusion limits the delivery of oxygen and nutrients to about 200 micrometers without perfusable networks. Three dimensional bioprinting allows for predefined channels, capillary-like networks, and hierarchical vasculature through coaxial extrusion, sacrificial bioinks like Pluronic F127, and the printing of multiple cells including endothelial cells, pericytes, and smooth muscle cells. Addition of angiogenic factors, and functionalized dynamic flow fosters network formation, with vascularized constructs shown for cardiac, liver, and kidney models. Perfusable vasculature is necessary to go from bioprinted tissue to clinical applications.

Organoids and organ on chip systems are small scale, physiologically relevant models of tissue that can represent organ level functions for applications in disease modeling, toxicity, and personalized medicine. Organoids are generally derived from stem cells through self-organization, but can have vastly improved reproduction and architectural control through three dimensional bioprinting using greater spatial control over many cell types [37]. Organ on chip systems would utilize bioprinting for microfluidic integration, vascular mimicry and compartmentalization. Organ on a chip applications would include liver on chip for hepatotoxicity screening, heart on chip for cardiotoxicity, and gut on chip for studying absorption.

Bioprinted constructs are revolutionizing preclinical drug development by offering reproducible human relevant models that bridge the void between two-dimensional cell cultures and preclinical *in vivo* studies [38]. By recreating the tissue architecture, tissue mechanical characteristics, and the cellular microenvironment, these models can provide more accurate data regarding drug efficacy, toxicity, and mechanisms of action. Specific examples of these models involve liver, kidney, tumor, and neural tissue-based models that can be used to evaluate pharmacokinetics and pharmacodynamics of compounds. Furthermore, patient derived cells can also be used to generate personalized disease models relevant to cancer, fibrosis and neurodegeneration, which aids in precision preclinical drug testing and prediction of therapeutic responses. Through the use of bioprinted constructs with patient-derived cells, preclinical drug testing cost and time is reduced while at the same time the reliance on animal testing is lower, and translational accuracy is improved.

4.6 Translating Bioprinting into Scalable Biomanufacturing

The larger issue of three dimensional bioprinting's movement from research to large scale manufacturing with accepted standards relies on biomanufacturing activities that comply with good manufacturing practice. With respect to good manufacturing practice, there is a focus on engineered product quality through material control, sterility, and full traceability in order to adhere to clinical product use regulations. Automation and robotics are already being incorporated in three dimensional bioprinting to avoid human induced errors and improve reproducibility, through high-throughput production allowing tissue customization. Digital twins can also be created in combination with artificial intelligence-based inputs to provide real time simulation, feedback, and adjustments during fabrication enabling more systematic and accurate constructed materials [39]. Monitoring systems such as embedded sensors, in line imaging, and biosensors are being used to provide quality assurance by analyzing structural fidelity, cell viability, and functional performance.

Although there have been some advances towards clinical translation, the adoption of bioprinting is currently constrained by regulatory uncertainty, as preparing a regulatory framework is overly delayed because the tissues bioprinted are hybrids that contain biological and synthetic content. Ethics is another issue for bioprinting as guidelines are not yet available, and human tissue printing and misuses of bioprinting are also anticipated concerns. Finally, clinical translation will present unique challenges relating to scale and validity, as well as sufficient long-term safety data [40]. To realize commercial viability and global availability of biomanufactured biomaterials for production in the biopharmaceutical sector, comparable protocols and capable of moving beyond intellectual property challenges including bioink formulations, cad models, and cell sources must be established.

4.7 Future Directions and Emerging Trends

The future of bioprinting is moving quickly away from simple tissue constructs, and embracing smarter, more responsive and individualized systems. One of the most exciting developments is the advancement of 4D printing, whereby the printed constructs can be programmed to transform in a controlled manner over time due to external stimuli. These printed constructs can be affected by temperature, pH, and mechanical strain, and provide exciting opportunities for self-folding tissues, shape memory implants, and responsive drug delivery systems [41]. At the same time, the inclusion of artificial intelligence (AI) has allowed for autonomous decision-making, design optimization, and predictive modeling of tissue behavior for improving individualized biomanufacturing. AI software can analyze and process patient-specific data, reduce human bias in the quality control process, and automatically adjust printing, bioink and biofabrication parameters to enhance the functionality of the printed constructs for individual therapeutic needs.

Additionally, the convergence of bioprinting with CRISPR gene editing and synthetic biology is leading to the production of the next generation of constructs that have programmable biological behaviors, enhanced cellular functions, and embedded biosensors or indicators that inherently facilitate specific cellular and/or biological functions. These various methodologies could alter the practice of tissue engineering, regenerative medicine, and disease modeling. Furthermore, with the rise of decentralized and on demand biomanufacturing platforms, particularly portable or point of care bioprinters, emerging processes for addressing logistical and scalability issues, especially for remote locations, military applications, or public health situations of emergencies (such as pandemics), have substantial potential [42, 43]. These advancements collectively are broadening the potential of the field of three dimensional bioprinting and changing the function of bioprinting within global healthcare and biopharmaceutical spaces.

4.8 Conclusion

Three-dimensional bioprinting is a transformative platform for precision biomanufacturing, growing from the fabrication of simple scaffolds into the creation of multi-cellular and even vascularized tissues that can biologically function and recapitulate native tissue architecture and functionality, by harnessing sophisticated bioinks, digitization design tools, and bioprinting modalities. This new field has applications from skin regeneration and cartilage and bone repair to organ on chip models and drug screening based on tissue function, exhibiting its versatility and translational potential.

As bioprinting becomes more embedded within good manufacturing practice workflows with the addition of automation, its reliability and overall scalability for clinical and industrial applications is improving. Nevertheless, for bioprinting to achieve successful scale up, we will need to see further advancements in regulatory alignment, ethical governance practices, and quality assurance protocols. Looking ahead, the integration of four-dimensional bioprinting, user personalization powered by artificial intelligence and CRISPR clever genetic editing and decentralized manufacturing, will change the game for the field of biofabrication. Collectively these developments will start to place three-dimensional bioprinting at the core of the next generation of biologics and biopharmaceutical production, allowing us to produce more precise, responsive, and patient-specific therapeutic solutions.

AI disclosure statement: The author acknowledges the use of AI-based tools for minor language editing or grammatical correction. However, all intellectual contributions, analyses, and conclusions presented in this chapter are entirely the author's own.

References

1. Holzberg, T.R., Watson, V., Brown, S., Andar, A., Ge, X., Kostov, Y., *et al.*, Sensors for biomanufacturing process development: facilitating the shift from batch to continuous manufacturing. *Curr. Opin. Chem. Eng.*, 22, 2018.
2. Narayanan, H., Luna, M.F., von Stosch, M., Cruz Bournazou, M.N., Polotti, G., Morbidelli, M., *et al.*, Bioprocessing in the Digital Age: The Role of Process Models. *Biotechnol. J.*, 15, 2020.
3. Sarkis, M., Bernardi, A., Shah, N., Papathanasiou, M.M., Emerging challenges and opportunities in pharmaceutical manufacturing and distribution. *Processes*, 9, 2021.
4. Ma, X., Liu, J., Zhu, W., Tang, M., Lawrence, N., Yu, C., *et al.*, 3D bioprinting of functional tissue models for personalized drug screening and *in vitro* disease modeling. *Adv. Drug Delivery Rev.*, 132, 2018.
5. Liu, N., Zhang, X., Guo, Q., Wu, T., Wang, Y., 3D Bioprinted Scaffolds for Tissue Repair and Regeneration. *Front. Mater.*, 9, 2022.
6. Prendergast, M.E. and Burdick, J.A., Recent Advances in Enabling Technologies in 3D Printing for Precision Medicine. *Adv. Mater.*, 32, 2020.
7. Lam, E.H.Y., Yu, F., Zhu, S., Wang, Z., 3D Bioprinting for Next-Generation Personalized Medicine. *Int. J. Mol. Sci.*, 24, 2023.
8. Suntornnond, R., Ng, W.L., Huang, X., Yeow, C.H.E., Yeong, W.Y., Improving printability of hydrogel-based bio-inks for thermal inkjet bioprinting applications *via* saponification and heat treatment processes. *J. Mater. Chem. B*, 10, 2022.
9. Cui, X., Li, J., Hartanto, Y., Durham, M., Tang, J., Zhang, H., *et al.*, Advances in Extrusion 3D Bioprinting: A Focus on Multicomponent Hydrogel-Based Bioinks. *Adv. Healthcare Mater.*, 9, 2020.

10. Sorkio, A., Koch, L., Koivusalo, L., Deiwick, A., Miettinen, S., Chichkov, B., *et al.*, Human stem cell based corneal tissue mimicking structures using laser-assisted 3D bioprinting and functional bioinks. *Biomaterials*, 171, 2018.
11. Touya, N., Devun, M., Handschin, C., Casenave, S., Ahmed Omar, N., Gaubert, A., *et al.*, *In vitro* and *in vivo* characterization of a novel tricalcium silicate-based ink for bone regeneration using laser-assisted bioprinting. *Biofabrication*, 14, 2022.
12. Schwab, A., Levato, R., D'Este, M., Piluso, S., Eglin, D., Malda, J., Printability and Shape Fidelity of Bioinks in 3D Bioprinting. *Chem. Rev.*, 120, 2020.
13. Habib, M.A. and Khoda, B., Rheological analysis of bio-ink for 3D bio-printing processes. *J. Manuf. Process.*, 76, 2020.
14. Sedigh, A., DiPiero, D., Shine, K.M., Tomlinson, R.E., Enhancing precision in bioprinting utilizing fuzzy systems. *Bioprinting*, 25, 2022.
15. Decante, G., Costa, J.B., Silva-Correia, J., Collins, M.N., Reis, R.L., Oliveira, J.M., Engineering bioinks for 3D bioprinting. *Biofabrication*, 13, 2021.
16. Szychlinska, M.A., Bucchieri, F., Fucarino, A., Ronca, A., D'Amora, U., Three-Dimensional Bioprinting for Cartilage Tissue Engineering: Insights into Naturally-Derived Bioinks from Land and Marine Sources. *J. Funct. Biomater.*, 13, 2022.
17. Kumar, S., Synthetic polymer-derived single-network inks/bioinks for extrusion-based 3D printing towards bioapplications. *Mater. Adv.*, 2, 2021.
18. Oliveira, H., Médina, C., Stachowicz, M.L., Paiva dos Santos, B., Chagot, L., Dusserre, N., *et al.*, Extracellular matrix (ECM)-derived bioinks designed to foster vasculogenesis and neurite outgrowth: Characterization and bioprinting. *Bioprinting*, 22, 2021.
19. Sanchez-Rubio, A., Hope, L., Barcelona-Estaje, E., Jayawarna, V., Williams, J., Salmeron-Sanchez, M., Gelatin vs GelMA in alginate-based bioinks as a platform for versatile 3D bioprintable *in vitro* systems. *Biomater. Adv.*, 177, 214408, 2025.
20. Lim, J., Bupphathong, S., Huang, W., Lin, C.H., Three-Dimensional Bioprinting of Biocompatible Photosensitive Polymers for Tissue Engineering Application. *Tissue Eng. Part B Rev.*, 29, 2023.
21. Sánchez-Sánchez, R., Rodríguez-Rego, J.M., Macías-García, A., Mendoza-Cerezo, L., Díaz-Parralejo, A., Relationship between shear-thinning rheological properties of bioinks and bioprinting parameters. *Int. J. Bioprint.*, 9, 2023.
22. Rahimnejad, M., Jahangiri, S., Zirak Hassan Kiadeh, S., Rezvaninejad, S., Ahmadi, Z., Ahmadi, S., *et al.*, Stimuli-responsive biomaterials: smart avenue toward 4D bioprinting. *Crit. Rev. Biotechnol.*, 44, 2024.
23. Bonatti, A.F., Chiesa, I., Vozzi, G., De Maria, C., Open-source CAD-CAM simulator of the extrusion-based bioprinting process. *Bioprinting*, 24, 2021.
24. Nyirjesy, S.C., Heller, M., von Windheim, N., Gingras, A., Kang, S.Y., Ozer, E., *et al.*, The role of computer aided design/computer assisted manufacturing

(CAD/CAM) and 3- dimensional printing in head and neck oncologic surgery: A review and future directions. *Oral Oncol.*, 132, 2022.
25. Huang, D., Li, Z., Li, G., Zhou, F., Wang, G., Ren, X., *et al.*, Biomimetic structural design in 3D-printed scaffolds for bone tissue engineering. *Mater. Today Bio*, 32, 101664, 2025.
26. Mukasheva, F., Adilova, L., Dyussenbinov, A., Yernaimanova, B., Abilev, M., Akilbekova, D., Optimizing scaffold pore size for tissue engineering: insights across various tissue types. *Front. Bioeng. Biotechnol.*, 12, 2024.
27. Banerjee, A., Datta, S., Das, A., Roy Chowdhury, A., Datta, P., A Micro-Scale Non-Linear Finite Element Model to Optimize the Mechanical Behavior of Bioprinted Constructs. *3D Print. Addit. Manuf.*, 9, 2022.
28. Noroozi, R., Tatar, F., Zolfagharian, A., Brighenti, R., Shamekhi, M.A., Rastgoo, A., *et al.*, Additively Manufactured Multi-Morphology Bone-like Porous Scaffolds: Experiments and Micro-Computed Tomography-Based Finite Element Modeling Approaches. *Int. J. Bioprint.*, 8, 2022.
29. Noble, C., Maxson, E.L., Lerman, A., Young, M.D., Mechanical and finite element evaluation of a bioprinted scaffold following recellularization in a rat subcutaneous model. *J. Mech. Behav. Biomed. Mater.*, 102, 2020.
30. Wu, Y., Su, H., Li, M., Xing, H., Digital light processing-based multi-material bioprinting: Processes, applications, and perspectives. *J. Biomed. Mater. Res. A*, 111, 2023.
31. Gao, C., Lu, C., Jian, Z., Zhang, T., Chen, Z., Zhu, Q., *et al.*, 3D bioprinting for fabricating artificial skin tissue. *Colloids Surf., B*, 208, 2021.
32. Maggiotto, F., Dalla Valle, E., Fietta, A., Visentin, L.M., Giomo, M., Cimetta, E., 3D bioprinting of a perfusable skin-on-chip model suitable for drug testing and wound healing studies. *Mater. Today Bio*, 33, 101974, 2025.
33. Weng, T., Zhang, W., Xia, Y., Wu, P., Yang, M., Jin, R., *et al.*, 3D bioprinting for skin tissue engineering: Current status and perspectives. *J. Tissue Eng.*, 12, 2021.
34. Huang, J., Xiong, J., Wang, D., Zhang, J., Yang, L., Sun, S., *et al.*, 3D bioprinting of hydrogels for cartilage tissue engineering. *Gels*, 7, 2021.
35. Antich, C., de Vicente, J., Jiménez, G., Chocarro, C., Carrillo, E., Montañez, E., *et al.*, Bio-inspired hydrogel composed of hyaluronic acid and alginate as a potential bioink for 3D bioprinting of articular cartilage engineering constructs. *Acta Biomater.*, 106, 2020.
36. Kondiah, P.J., Kondiah, P.P.D., Choonara, Y.E., Marimuthu, T., Pillay, V., A 3D bioprinted pseudo-bone drug delivery scaffold for bone tissue engineering. *Pharmaceutics*, 12, 2020.
37. Roth, J.G., Brunel, L.G., Huang, M.S., Liu, Y., Cai, B., Sinha, S., *et al.*, Spatially controlled construction of assembloids using bioprinting. *Nat. Commun.*, 14, 2023.
38. Satpathy, A., Datta, P., Wu, Y., Ayan, B., Bayram, E., Ozbolat, I.T., Developments with 3D bioprinting for novel drug discovery. *Expert Opin. Drug Discov.*, 13, 2018.

39. Bonatti, A.F., Vozzi, G., De Maria, C., Enhancing quality control in bioprinting through machine learning. *Biofabrication*, 16, 2024.
40. Ramesh, S., Deep, A., Tamayol, A., Kamaraj, A., Mahajan, C., Madihally, S., Advancing 3D bioprinting through machine learning and artificial intelligence. *Bioprinting*, 38, 2024.
41. Yang, G.H., Yeo, M., Koo, Y.W., Kim, G.H., 4D Bioprinting: Technological Advances in Biofabrication. *Macromol. Biosci.*, 19, 2019.
42. Arslan-Yildiz, A., El Assal, R., Chen, P., Guven, S., Inci, F., Demirci, U., Towards artificial tissue models: Past, present, and future of 3D bioprinting. *Biofabrication*, 8, 2016.
43. Di Marzio, N., Eglin, D., Serra, T., Moroni, L., Bio-Fabrication: Convergence of 3D Bioprinting and Nano-Biomaterials in Tissue Engineering and Regenerative Medicine. *Front. Bioeng. Biotechnol.*, 8, 2020.

Part 2
PROCESS IMPROVEMENT AND LEAN INTEGRATION

5

Application of Lean Technologies in Biomanufacturing

Ranjita Swain[1], Rudra Narayan Mohapatro[1] and Sunita Routray[2]*

[1]*Department of Chemical Engineering, C.V. Raman Global University, Bhubaneswar, Odisha, India*

[2]*Department of Mechanical Engineering, Centurion University of Technology and Management, Bhubaneswar, Odisha, India*

Abstract

In business goals, the lean technology plays an important role in biomanufacturing that is associated with the advanced technical and regulatory significances. The lean technology enhances the performance in biomanufacturing by reducing batch cycle time, increasing yields, minimizing human error, and improving compliance. The traditional manufacturing is widely applicable across sectors such as automotive, aerospace, and consumer goods; biomanufacturing is predominantly used in biotechnology, pharmaceuticals, and agricultural technology, highlighting its niche but growing importance in the global economy. The process improvement in biomanufacturing is to be structured properly by using the principles of lean manufacturing. By focusing on value, eliminating waste, enhancing process flow, using pull systems, and engaging in continuous improvement, biomanufacturing operations can achieve greater efficiency, cost savings, and compliance. The implementation can be complex due to regulatory constraints, the benefits in terms of yield improvement, cycle time reduction, and error minimization make lean an essential strategy for modern biomanufacturing facilities. Lean implementation in biomanufacturing is an important part in the value stream. The mapping and analyzing of the flow of materials and information in the industry to optimize the performance, reduce waste, and enhance product quality—while staying compliant with regulatory requirements. Few examples of lean goals are increased yield, faster time to market and continuous improvement in complex processes. The challenges and key thoughts in applying lean principles to biomanufacturing are maintaining regulatory compliance, managing biological variability, and

**Corresponding author*: sunitaroutray77@gmail.com

Sandip Kunar, Jagadeesha T. and D. Sathis Kumar (eds.) Biomanufacturing: Transforming Production of Biologics and Biopharmaceuticals, (115–138) © 2026 Scrivener Publishing LLC

ensuring product quality. It highlights the status of aligning Lean strategies with GMP standards to achieve process efficiency without conciliatory safety or effectiveness in bioproduction.

***Keywords*:** Good manufacturing practices, automation, process control, validation, efficiency, standard operating procedures, inventory

5.1 Introduction to Lean in Biomanufacturing

In the automotive biomanufacturing sectors, the lean technology principles are being used. Biomanufacturing includes the manufacturing of antibodies, vaccines and the recombinant proteins by using the biological systems. The biopharmaceutical production also includes the high costs, difficulties and some constraints in the monitoring the process. In view of these the lean thinking in the area gives the benefits in improving the efficiency, quality and being cost effective.

Lean in biomanufacturing focuses on eliminating waste, reducing process variation, and continuously improving operations without compromising the strict regulatory and quality requirements of the pharmaceutical industry. It requires tailoring traditional lean tools (like 5S, Value Stream Mapping, and Kaizen) to suit the unique challenges of biotechnology, including long production cycles, complex quality testing, and stringent compliance requirements.

Key benefits of applying Lean in biomanufacturing include:

- Reduction in cycle times and batch release delays
- Improved throughput and capacity utilization
- Better inventory and supply chain management
- Enhanced compliance through standardized work and error reduction
- Cost reduction without compromising product quality or safety

There are two fundamental differences between traditional manufacturing and biomanufacturing such as production, materials and methods used in the process. The mechanical, thermal, or chemical transformation of raw materials has been carried out to produce the final products using machining, molding, casting, and assembly processes from metals, plastics, and ceramics. The above methods are energy-exhaustive and non-renewable resources that is degraded slowly with time [1]. The living system is utilized in biomanufacturing to produce the biopharmaceuticals products from the food ingredients that are biofuels and biodegradable materials. The process is implemented in a controlled bioreactor environment highlighting synthesis biologically and metabolic technology [2]. The systems operate at lower temperature and pressure in a renewable biological system which has more sustainability and is environmentally friendly [3]. There is a key difference between the two such as precision and scalability of production. The traditional manufacturing surpasses in manufacturing more quantities of identical products with high precision. In another way, biomanufacturing deals with the composite biological products such as therapeutic proteins which require severe regulator over worth and concentration [4].

The lean principles established for the automobile sectors, are implemented in the biomanufacturing to improve the quality and production efficiency. Without compromising the quality, the lean focuses on the deletion of the waste and the optimization of the process that is necessary for the biological processes. The primary goal in biomanufacturing is to reduce the batch cycle period; whereas, the long duration is due to the biological process, waiting time duration, and delay in delivery of product. The tools such as mapping of value stream and continuous flow to recognize bottlenecks and streamline operations are fast-tracking production duration [5]. Changeability and refining the process increase the productivity that contributed by lean. Technologies including standard processing methods, the 5S system, and strategies for minimizing optimization errors and deviations from process parameters are essential in purification, formulation, and fermentation operations [6].

The lean technology has the advantages to minimize the human error; the sterilised conditions and accurate implementation are required for the biomanufacturing. A standard procedure, visual control and training associated with the principles on lean technology reduce the errors found in operation [7]. The compliance is also enhanced by the lean which promotes the culture of documentation, traceability, and continuous improvement. The principle on lean facilitates the equipment maintained and audited the operational parts to enhance the daily operation qualitative and quantitative which is a Good Manufacturing Practice (GMP) [8].

5.2 Principles of Lean Manufacturing

The minimization of the waste materials to improve the quality and efficiency in the manufacturing process can be enhanced by using the lean manufacturing technology. In the Toyota Production System, the lean principles are widely used in the industries for any complex area of biomanufacturing. The principles of the lean manufacturing such as value, value stream, flow, pull, and perfection are mostly used in biomanufacturing operations, whereas the challenges are also there in variability, regulatory compliance, and cost.

1. Value: It is determined based on lean manufacturing in biomanufacturing which includes patients, regulatory agencies, healthcare providers, and payers. Value incorporates each activity and procedures which pay contribute to the safe, effective, compliant, and timely production of biopharmaceutical processes.

In biomanufacturing, the tolerant viewpoint highlights value in terms of:

- High product quality and safety
- Timely and reliable access to life-saving therapies
- Affordability and efficacy of treatment

From the monitoring perception (e.g., FDA, EMA), value is defined by:

- Strict adherence to Good Manufacturing Practices (GMP)
- Data integrity and traceability
- Process control, validation, and risk mitigation

Non-value added (NVA) removes the action such as waiting, rework, excessive documentation, and underutilized talent. The lean technologies support biomanufacturers give efforts on the worth parts by reducing non-value-added (NVA) actions such as waiting, rework, excessive documentation, and underutilized talent. For example, real-time release testing (RTRT), automation of documentation, and standardized operating procedures help streamline compliance and product delivery, thereby enhancing value creation from the customer's viewpoint.

The lean principles are being applied such as Value Stream Mapping (VSM), 5S and Kaizen and Continuous Improvement (Kaizen). VSM can classify holds up and surplus steps which do not subsidize to tolerate the outcomes or controlling submission. 5S and Kaizen is directly getting impact on the GMP compliance, also improves the workplace of the organization and decrease the rate of errors. The continuous improvement

(Kaizen) confirms the delivery develops with the expectation and patient needs. Thus, defining value from the customer perspective ensures that every action in the biomanufacturing process is aligned with product quality, compliance, and patient safety, reinforcing lean's core objective—delivering maximum value with minimal waste.

Value Identification: The lean technology defines the value of the customer's need. Value depends on the factors such as product quality, purity, safety, and regulatory compliance in the biomanufacturing. The customer's need is valuable to allow the biomanufacturing industry to streamline the process to get the outputs.

2. Value Stream Mapping (VSM): This is implemented to envision and examine all the necessary information and flow of the raw materials to produce the product and reach out to the customers. In biomanufacturing, the product brings to the customer in which the production process includes the multistage processes such as upstream and downstream processing. This is applied to highlight bottlenecks, redundancies, and non-value-adding steps [9]. This also allows the industry to make informed decisions about process improvements and compliance gaps.

3. Flow: Continuous process flow is created by incorporating the necessary information that enables industries to produce products from raw materials smoothly without production delays. In biomanufacturing, this is achieved by reducing waiting times between batch steps, minimizing setup times, and ensuring the availability of critical materials [10]. Tools such as 5S and visual workplace management enhance flow by improving organization and cleanliness of production environments.

4. Pull System: This system highlights things in production environments that depend on the actual demand rather than forecasts. In this context, biomanufacturing includes the high cost and limited shelf-life products, as well as pull systems that support the prevention in the overproduction and reduction in the registered property costs [11]. But it is critical in biologic production or medicines.

5. Pursuit of Perfection (Continuous Improvement): The production rate and performance in process industry can be improved by applying lean manufacturing. In biomanufacturing, this may involve root cause analysis of deviations, adoption of real-time monitoring systems, and employee-driven quality initiatives to reduce batch failures and nonconformities [12].

5.3 Key Lean Tools and Their Applications in Biomanufacturing

Lean manufacturing, initially developed for the automotive industry, has been progressively accepted in biomanufacturing to recover process effectiveness, product excellence, and acquiescence. Lean tools intention to eradicate waste (non-value-added events), reduce inconsistency, and modernize workflows. In highly planned environments like biomanufacturing, where constancy, traceability, and hygiene are dominant, the application of Lean tools necessity also align with Good Manufacturing Practices (GMP). Below are key thin tools and their exact applications in biomanufacturing.

1. **5S (Sort, Set in order, Shine, Standardize, Sustain):** This tool helps produce a planned and effective workplace. In biomanufacturing, 5S is used to uphold cleanrooms, organize laboratory workspaces, and safeguard that tools, reagents, and documents are correctly labeled and stored. For example, buffer training areas benefit from 5S by reducing the time spent locating materials and lessening the risks of cross-contamination.
2. **Value Stream Mapping (VSM):** VSM is used to picture the entire production method, from upstream (e.g., cell culture) to downriver (e.g., purification), recognizing holdups and sources of waste. In biomanufacturing, VSM helps teams pinpoint delays in batch announcement, redundant waiting during value checks, or incompetent handoffs between departments.
3. **Kaizen (Continuous Improvement):** Kaizen helps a philosophy of small, incremental variations led by employees. In a biomanufacturing situation, operatives and specialists often propose enhancements to reduce media grounding time, recover process sampling methods, or enhance certification practices.
4. **Standard Work:** Standard work ensures that each process step is performed consistently, which is critical in GMP-regulated environments. It involves developing and adhering to detailed Standard Operating Procedures (SOPs) for critical activities like inoculation, fermentation monitoring, and sterile filtration, reducing human error and improving batch reproducibility.
5. **Kanban (Pull System):** Kanban is a visual record control scheme used to attain the flow of possessions. In biomanufacturing, it assurances that reagents, filters, and other consumables are replenished just-in-time, evading excess list or expired items. It also chains traceability in substantial use.

6. Visual Management: Pictorial cues such as color-coded tags, status panels, and apparatus condition tags augment communication and obedience. For example, equipment can be labeled to designate "Ready for Use," "Needs Cleaning," or "Under Upkeep," helping avoid unintended misuse and secondary regulatory audits.
7. Root Cause Analysis: This problem-solving tool helps classify primary reasons of issues such as batch failure or adulteration. In biomanufacturing, RCA is often used after eccentricity investigations or Out of Specification (OOS) results to contrivance effective Corrective and Preventive Actions (CAPA).
8. SMED (Single-Minute Exchange of Die) focuses on reducing setup or changeover time. In biomanufacturing, this refers to faster cleaning and validation between batches or products, especially when switching from one genetic product to another.
9. Poka-Yoke (Error Proofing): Poka-yoke methods avoid errors before they occur. Examples comprise barcoding schemes to verify resources before batch use or mesh that avert equipment process if circumstances are insecure or limits are incorrect.
10. Total Productive Maintenance (TPM): TPM inspires active upkeep to minimize equipment letdown. In biomanufacturing, it reduces unforeseen downtime of critical systems like bioreactors, ultrafiltration units, and chromatography systems, secondary nonstop production.

By mixing these Lean apparatuses, biomanufacturers can improve working fineness, reduce cost, ensure regulatory obedience, and eventually deliver high-quality foodstuffs more efficiently [13].

5.3.1 Value Stream Mapping (VSM) in Biomanufacturing

Value Stream Mapping (VSM) is an important Lean tool used to envisage and analyze the flow of resources and data throughout a production process. In biomanufacturing, where manufacture includes multifaceted biological systems, severe controlling supplies, and high-quality standards, VSM plays a dangerous role in classifying inadequacies, removing waste, and cultivating overall procedure performance.

VSM imprisonments both value-added doings (those that directly donate to the final product) and non-value-added actions (such as waiting, extreme movement, overprocessing, and rework). The value stream in biomanufacturing classically spans from raw physical intake through upriver and downriver dispensation, quality control, wrapping, and batch release.

The mapping procedure begins with making a current state map, which summarizes:

- Each step of the procedure (e.g., media groundwork, fermentation, cleansing)
- Flow of raw materials and intercedes
- Information flow (e.g., approvals, certification, batch accounts)
- Time-related data such as series time, delay time, and topmost time
- Inventory levels and work-in-progress (WIP) among ladders

In a biomanufacturing setup such as monoclonal antibody production. VSM can disclose unseen delays, such as long hold times among separation and chromatography steps due to equipment availability or QC testing backlogs. Recognizing such non-value-added time enables procedure optimization through better development, equipment provision, or adoption of real-time monitoring tools.

A Future State Map is then shaped, which reshapes the process with minimalized waste and better flow. This could involve parallel dispensation of QC tests, adjustment of procedures, or improved direction between departments like engineering, quality assurance (QA), and regulatory affairs [14].

5.3.1.1 Key Applications of VSM in Biomanufacturing

- Streamlining upstream and downstream workflows
- Reducing delays in batch release by improving QC and documentation processes
- Enhancing inventory control of raw materials and critical reagents
- Aligning cross-functional teams around a unified view of the process
- Supporting digital transformation and automation by identifying manual bottlenecks

VSM is especially valuable in Good Manufacturing Practice (GMP) environments, where compliance, traceability, and efficiency are essential. It fosters collaboration among stakeholders and drives continuous improvement, helping biomanufacturers achieve higher productivity, better product quality, and reduced production costs without compromising regulatory standards. By offering a visual and data-driven approach to process improvement, VSM is a cornerstone of Lean initiatives in modern biomanufacturing [15].

5.3.2 Cleanroom Organization in Biomanufacturing

Product quality, sterility, and regulatory compliance in biomanufacturing is very much critical, which is managed with cleanroom organization. This is essential and ensures systematic arrangement, maintenance and controlled environments to minimize contamination by impurity particles. The key elements of cleanroom organization are zoning and layout design, implementation of 5S, material and equipment management, personnel control and environmental monitoring [16].

5.3.2.1 Zoning & Layout Design

Separate zones are created for material storage, sample processing and operations which follows the ISO classifications. The flow of materials, personnel, and waste move in their respective directions to prevent cross-contamination. Pressure difference is also maintained between chambers to avoid air mixing between zones. Implementation of 5S, which includes Sort, Set in Order, Shine, Standardize and Sustain in the cleanroom organization is also implemented. The Sort category ensures that the essential tools and materials are kept inside the cleanroom. Set in Order confirms that each item should be labeled and put in its designated place. Daily cleaning routines should be maintained to ensure sterility. Standardized practices should be maintained for equipment cleaning, and material handling. In sustain category, audits and training are conducted time to time to ensure adherence [17].

5.3.2.2 Material and Equipment Management

Handling of materials and equipment is a crucial part, which should be taken care of properly. The consumables must be sterilized and packed safely to avoid any kind of contamination. The storage shelves should be made of stainless steel or cleanroom-compatible materials. Materials and equipment are placed in a planned way according to the preference to reduce operator movement and air turbulence [18].

5.3.2.3 Environmental Monitoring

The personnel, who are handling the materials and equipment, should use sterile garments, gloves, masks and goggles. Trained persons are only allowed to access the materials and proper record should be maintained about the personnel's entry, exit and duration of stay. Certain discipline should be maintained which will ensure no unnecessary movements or contact with surfaces. Scheduled checks should be done to avoid airborne particles, microbial presence, temperature rise or fall, humidity level, and pressure differences if any [19].

5.3.3 Kaizen in Biomanufacturing

Kaizen is a Japanese word, which means continuous improvement. This entrusts on small and daily basis improvements in the workplace. The suggestions from lab technicians and operators working over there are implemented regularly, which collectively enhance product quality and safety. The procedure is implemented in cleanroom workflows, like; standardizing cleaning procedures to minimize movement and contamination risk. Involvement of employees in the workplace through their ideas reduces contamination and increases efficiency [13].

5.3.4 Standard Work in Biomanufacturing

This refers to the practice which ensures to follow the best practices for carrying out tasks safely, efficiently and consistently. Certain Standard Operating Procedures (SOPs) are followed to ensure those performances are carried out on daily basis. Good Manufacturing Practice (GMP) compliance is maintained by standardization of documentation. By following uniform practice and records, errors can be reduced. It also ensures that all employees follow the same standard of operation enhancing quality of work place [20].

5.3.5 Just-In-Time (JIT) in Biomanufacturing

Just In Time or JIT technique is adopted to keep track of the real time requirements of goods. The goods are received whenever these are required for the manufacturing process. This technique solves the issue of storing goods for a shorter period of time, reduces storage costs; and the materials can be used well before their expiry and also avoids contamination by atmospheric conditions due to less time of storage. This process ensures timely delivery of materials and components to maintain a real-time production schedule. By utilizing this technique, efficiency enhances which supports better resource utilization by proper respond to any fluctuations in demand [21].

5.3.6 Kanban Systems

Kanban is a Japanese word which means "signboard". Kanban is a visual management system which is mainly useful in cell culture rooms, streamlining the inventory control and workflows. This system ensures the availability of inventory whenever it is required. The storage bins send signal when the items are going to be out of stock. It also controls the number of active cultures at a given time to maintain sterile conditions, avoids the risk of cross contamination to have a smooth and disciplined workflow. This seems to be a simple method but it supports the laboratory standards, enhances efficiency by reducing wastage of time [22].

5.3.7 Root Cause Analysis (e.g., Fishbone, 5 Whys)

The Fishbone Diagram, which is also known as Ishikawa or cause-and-effect diagram, used as a visual tool for analysis of root causes of some specific problem. This diagram resembles a fish skeleton. The problem is placed at the head with the major categories of causes. The causes could be Man (people), Machine (equipment), Method (procedures), Material (supplies), Measurement (data), and Environment (conditions), which form the bones. Teams systematically explore each category, brainstorm to map out all contributing factors to trace the root cause or causes. This analysis is adopted in Good Manufacturing Practice (GMP) to ensure good product quality [23].

Whys Method

Whys method consists of asking "Why?" repeatedly typically for five times to find the root cause of a problem. This can be explained by citing one appropriate example. If a batch sterility testing fails, then questions can be put as follows:

1. **Why** did the batch fail and sterility testing failed on a specified date?
   → Because contamination was found in the product.
2. **Why** was the product got contaminated?
   → Because the flask lid was loosely placed.

3. **Why** was the flask lid not sealed properly?
   → Because the technician did not perform a torque check.
4. **Why** was the torque check got skipped?
   → Because the step was not mentioned in SOP.
5. **Why** was the SOP not up to date?
   → Because it had not been revised since a long time.

By getting answers to the questions one by one, the root cause of the problem is solved [24].

5.4 Digital Lean and Smart Biomanufacturing

5.4.1 Integration with Industry 4.0 Tools (IoT, AI, Digital Twins)

Integration of Industry 4.0 with smart Biomanufacturing, transforms traditional practices into a modernized system. Incorporation of Internet of Things (IoT) helps in uninterrupted data collection from bioreactors, sensors, and process equipment. With the help of artificial intelligence (AI) and machine learning (ML), the collected data are analyzed for process optimization, to predict future service condition of equipment, futuristic maintenance of instruments by predicting failures.

Digital Twins in Biomanufacturing presents the virtual system of physical bioprocesses. These systems are tested with the collected real-time data for optimization and simulation to validate the data. This process ensures better understanding of the system and reduces the downtime [25].

5.4.2 Lean Analytics Dashboards for Real-Time Monitoring

Lean Analytics is a decision-making approach which is data driven. It is applied to systems by using Lean Startup principles, which include rapid iteration, continuous improvement and eliminates waste. The objective is to pinpoint one metric which is an important factor, put real time data into it, validate and apply it for fast decision-making. These dashboards focus on important factors such as yield, efficiency, productivity, which are very much essential for system operators. Manufacturing Execution System (MES), Supervisory Control and Data Acquisition (SCADA), Laboratory Information Management System (LIMS) systems are integrated to support detection of any process deviations, enhance operational efficiency, and reduce waste [26].

5.4.3 Automation to Support Lean Workflows

Automation in biomanufacturing streamlines operations, reduces waste, and improves consistency. It helps in timely production, faster decision-making and effective utilization of resource. Integration of MES, SCADA, and LIMS into the system not only enhances productivity and product quality but also supports regulatory compliance and operational agility.

5.5 Challenges and Considerations in Lean Biomanufacturing

5.5.1 Regulatory Constraints (FDA & EMA)

Lean biomanufacturing is a methodology designed at improving effectiveness, reducing waste, and enhancing production progressions in the pharmaceutical and biotechnology industries. However, its application in regulated situations presents numerous challenges and attentions, particularly in relation to agreement with strict regulatory necessities imposed by activities such as the U.S. Food and Drug Administration (FDA) and the European Medicines Agency (EMA). These agencies impose complete regulations designed to confirm product safety, efficacy, and quality specially through contexts such as Good Manufacturing Practices (GMP) [27]. While lean policies often seek to streamline operations and abolish non-value-adding activities, many of these activities, such as wide-ranging certification, validation measures, and record-keeping, are mandatory by regulatory establishments. For example, both FDA and EMA mandate thorough documentation of all manufacturing steps, with batch records, standard operating procedures (SOPs), deviations, and corrective actions. This is essential for scrupulous documentation can clatter with lean moralities that highlight simplification and efficiency. As a result, producers must find ways to implement lean practices without conciliatory regulatory compliance, often by joining digital tools such as electronic batch records (eBRs), manufacturing execution systems (MES), and electronic quality management systems (eQMS), which help to maintain monitoring standards while supporting process efficiency.

A major challenge arises in dealing changes and developments within the limitations of regulatory necessities. Lean manufacturing helps in continuous improvement, often hopeful frequent process changes to optimize performance. However, in a GMP-regulated environment, even minor changes to a validated procedure must undergo a comprehensive change control process, including risk assessments, documentation, validation, and sometimes regulatory statement or approval. This can slow down the improvement process, particularly when dealing with post-approval products. Regulatory bodies like FDA and EMA are particularly severe about confirming that changes do not negatively impact product quality or patient safety. To address this, companies are invigorated to adopt a risk-based method to change management, as outlined in ICH Q9 (Quality Risk Management), and influence frameworks like ICH Q10 (Pharmaceutical Quality System) and ICH Q12 (Lifecycle Management), which support orderly, science-based, and risk-managed approaches to process development and post-approval changes. These strategies help manufacturers demonstrate to regulators that improvements are controlled, justified, and do not compromise the validated state of the process.

Another consideration is the veracity of electronic data, which becomes increasingly important as producers move toward automation and making digital to support lean processes. Regulatory activities have strict strategies for data integrity, such as FDA's 21 CFR Part 11 and EMA's Annex 11, which need electronic records to be safe, perceptible, and demonstrable. These regulations highlight the ALCOA+ principles safeguarding that data is attributable, readable, concomitant, innovative, precise, and complete. Therefore, any lean inventiveness that involves digitization must ensure full obedience with these standards to avoid regulatory inspection. Implementing obedient software results and training personnel in data governance is vital to successfully balance lean improvements with regulatory expectations [28].

The essential conservatism of the regulatory environment also offer cultural trials for lean implementation. Lean rational promotes employee empowerment, rapid decision-making, and continuous change; all of which may be viewed with carefulness in a risk-averse industry where obedience, control, and uniformity are prioritized. This can create stiffness between lean initiatives and traditional quality assurance systems, particularly if the association lacks a superiority culture that ropes both revolution and acquiescence. Cross-functional teamwork, strong leadership assurance, and early participation of regulatory affairs teams are crucial to overwhelmed resistance and align lean goals with regulatory prospects. Additionally, open discussion with officials through technical advice meetings or contribution in governing pilot programs can help clarify prospects and enable smoother operation of lean enhancements within acceptable regulatory limits.

Finally, global supervisory variability confuses the positioning of lean engineering strategies across different markets. While coordination efforts by the International Council for Harmonisation (ICH) have improved uniformity, differences still exist in how controlling agencies interpret and apply GMP standards, particularly concerning process authentication, change management, and post-approval amendments. A process or system that mollifies FDA requirements may require variation for EMA compliance, and vice versa. This erraticism forces global biomanufacturers to tailor their lean policies to each region's specific regulatory context, possibly increasing complexity and limiting the scalability of lean initiatives [29].

In conclusion, while lean biomanufacturing offers important benefits in terms of efficiency, cost reduction, and product quality, its successful execution within FDA and EMA regulated environments needs a strategic and balanced method. Challenges related to documentation, validation, change control, data integrity, cultural alignment, and global regulatory variability must be addressed proactively. By leveraging digital skills, applying risk-based quality administration principles, and upholding open communication with regulatory bodies, producers can align lean methodologies with regulatory obedience, thereby achieving operative excellence without conciliatory safety or quality standards.

5.5.2 Process Variability and Biological Unpredictability

Process variability and biological randomness are central tasks in lean biomanufacturing, especially in the manufacturer of biologics where living systems like cells or microorganisms are used. Unlike outdated chemical manufacturing, biological methods are inherently variable due to the intricate nature of living organisms, which can respond randomly to small changes in ecological conditions such as temperature, pH, oxygen levels, or nutrient accessibility. This impulsiveness familiarizes significant batch-to-batch variability in product yield, quality, and purity making it difficult to regulate and streamline methods, which is a core objective of lean engineering methods. Even with well-controlled conditions, genetic drift, cell viability, or changes in raw material quality can lead to unpredictable outcomes. For regulatory bodies such as the FDA and EMA, keeping product steadiness is critical, and any difference must be carefully understood, controlled, and justified through risk-based methods. This places an important burden on manufacturers to design robust procedures that can billet biological variations while still meeting critical quality features. Tools such as Process Analytical Technology (PAT) and Quality by Design (QbD) are slowly used to manage this inconsistency by enabling real-time monitoring and predictive control of important process constraints. However, integrating these tools needs deep process information, wide data, and regulatory validation, all of which can confuse lean implementation. In essence, while lean biomanufacturing seeks to reduce waste and improve competence, the genetic nature of the process demands a careful and highly controlled method to confirm product quality and regulatory compliance, making it challenging to accomplish the same level of working agility seen in non-biological engineering systems [30].

5.5.3 Culture Change and Resistance among Scientific Staff Related to Lean Biomanufacturing

Implementing lean biomanufacturing often requires a momentous cultural shift, mainly among scientific and technical staff familiarized to out-of-date practices focused on meticulousness, compliance, and risk aversion. In highly regulated biopharmaceutical environments, scientists and engineers are skilled to arrange safety, data honor, and controlling adherence, which can sometimes conflict with the lean prominence on speed, unceasing change, and process simplification. This misalignment can lead to fighting among staff who may observe lean enterprises as threats to systematic consistency, job security, or quality standards. For example, efforts to normalize procedures or reduce guarantee may be viewed as dejection, critical, analytical, or validation tasks. Moreover, lean tools such as Kaizen events, visual administration, or cross-functional overtone may seem distant or unworkable to research-oriented teams who are more familiar with siloed, methodical approaches. Overwhelming this conflict requires strong management assurance and the development of a quality-focused lean values that highlights both regulatory submission and operative excellence. Training and statement are crucial to demonstrate how lean approaches can actually enhance scientific work by reducing redundancies, permitting faster decision-making, and freeing up time for more value-added research and novelty. Involving scientific staff early in lean preparation and decision-making procedures also builds possession and trust. Importantly, quality and controlling teams must work together with lean champions to ensure that lean operation does not compromise agreement. By forming a culture that values both scientific integrity and process effectiveness, governments can overcome battle and harness the proficiency of their scientific workforce to drive nonstop improvement in a compliant and collective manner [31].

5.5.4 Need for Training and Multidisciplinary Collaboration Related to Lean Biomanufacturing

The fruitful application of lean biomanufacturing in the biopharmaceutical manufacturing spindles not only on process optimization and waste reduction but also on the exercise of workers and fostering multidisciplinary association. Given the difficulty of biological systems and the strict controlling necessities imposed by establishments such as the FDA and EMA, lean methods must be carefully combined into prevailing workflows. This amalgamation requires a labor force that is not only theoretically skilled but also well-versed in lean ideologies such as value watercourse planning, root cause examination, 5S, and unceasing improvement. Without proper training, even the most well-designed lean enterprises can fail due to misinterpretations, misapplication of tools, or confrontation from staff who are unaware with the lean attitude.

Training plays a vital role in associating the gap between scientific, engineering, and working mindsets. Scientists and lab specialists, occasionally, may not logically think in terms of process flow or competence, as their emphasis is often on accuracy, acquiescence, and investigation. Similarly, quality pledge teams may view lean efforts with skpticism if they observe them as cutoffs that negotiate compliance. Structured training agendas personalized to different purposes help to align these viewpoints by representing how lean tools can balance, rather than replace, controlling and quality prospects. Training must also highlight real-world case studies from biomanufacturing environments to show how lean performance can lead to better product consistency, faster throughput, and fewer deviations all while upholding full regulatory compliance [32].

In addition to training, multidisciplinary partnership is crucial for lean biomanufacturing to prosper. Biomanufacturing procedures are integrally cross-functional, involving R&D, process growth, quality assurance, regulatory affairs, supply chain, and production teams. Lean conversion cannot occur in seclusion within one department. For example, optimizing a cell culture development for higher yield will require close management between upstream process engineers, microbiologists, quality control analysts, and monitoring staff. Each team brings unique expertise, and effective lean application depends on their aptitude to work together, comprehend each other's constraints, and agree on shared objectives.

Cross-functional teams allow complete problem solving and encourage better decision making by considering numerous dimensions methodical feasibility, regulatory acquiescence, quality impact, and operational efficiency. These teams can also identify interdependencies and deficiencies that may be difficult to notice when viewed through a single corrective lens. For instance, partnership between controlling affairs and manufacture can streamline documentation requirements without conciliatory traceability or audit willingness. Likewise, meeting between quality control and data analytics teams can improve real-time monitoring of critical parameters, aligning with both lean and GMP goals.

Fostering this collaboration requires a culture of honesty, shared ownership, and mutual respect. Leaders must break down silos, inspire knowledge sharing, and establish mechanisms such as cross-functional Kaizen events, joint training sessions, and integrated project teams to insert lean thinking across the organization. In doing so, companies can confirm that lean biomanufacturing develops a supportable, acquiescent, and organization wide effort rather than a detached set of process developments [33].

In conclusion, training and multidisciplinary relationship are not just support functions but critical enablers of lean biomanufacturing. They ensure that lean values are properly understood, steadily applied, and fully aligned with the technical, supervisory, and working realities of the biopharmaceutical industry.

5.6 Case Studies

5.6.1 Case 1: Lean Implementation in a Monoclonal Antibody Production Plant

The monoclonal antibody production plant was facing challenges with high reversal times, distressing overall quantity and output. To address this, a Lean alteration initiative was undertaken with the goal of refining process competence and dipping cycle time.

- **Value Stream Mapping (VSM):** Providing a complete conception of the current-state process, helping recognize hold-ups and non-value-added actions in both upstream and downstream operations.
- **5S Implementation:** Workplace organization and alteration were improved across cleanroom and support areas, reducing needless undertaking, search time, and contamination risks.
- **Kaizen Events:** Cross-functional teams performed battered development factories concentrating on key pain points, such as media groundwork delays and gear changeovers.

Through the planned application of these Lean practices, the facility attained a 25% reduction in cycle time across the end-to-end manufacture roadmap. This development improved preparation flexibility, reduced record levels, and reinforced faster response to clinical and profitable demand [34, 35].

5.6.2 Case 2: Continuous Improvement in Vaccine Manufacturing

In a highly controlled vaccine engineering environment, continuous growth efforts were absorbed on two critical areas: (i) dipping human errors and refining certification, and (ii) leveraging Lean numerical tools and calibration to drive supportable working excellence.

i) Reducing Human Errors and Improving Documentation
Human errors particularly in certification and batch processing were recognized as an important basis of nonconformities and acquiescence risks. A root cause analysis exposed discrepancies in manual record keeping, misunderstanding of SOPs, and absence of real-time feedback loops.

To discourse these subjects, the following interferences were applied:

- **Error-Proofing (Poka-Yoke):** Physical and practical controls were combined to stop shared mistakes, chiefly in clean events and batch guarantee steps.
- **Digital Work Instructions:** Communicating, step-by-step directions were arranged on tablets at operator workstations to ensure task accuracy.
- **Competency-Based Training:** Enhanced training modules were developed, linked to role-specific SOPs, with regular assessments to reinforce knowledge retention.

These measures led to a notable decline in human-error-related deviations and enhanced compliance with GMP standards [36].

ii) Lean Digital Tools and Standardization
The second pillar of the inventiveness involved the acceptance of Lean digital keys to eradicate waste, streamline workflows, and guarantee reliable documentation performs.

Key actions included:

- **Implementation of Electronic Batch Records (EBRs):** Replacement paper-based records reduced transcript errors, allowed real-time data entry, and enhanced batch assessment cycles.
- **Automated Logbooks and Deviation Tracking:** Centralized, digital systems were presented for monitoring critical process limitations and handling deviations systematically.
- **Standardization of SOPs and Processes:** Procedures were harmonized across units, and digital templates ensured uniformity in documentation.

As a result, the plant attained quantifiable developments in working efficiency, documentation accuracy, and audit keenness. These Lean-driven developments allowed a hardier and quality-focused industrial system, supporting faster and more reliable vaccine delivery [37].

5.7 Conclusion and Future Perspectives

Lean application in biomanufacturing has established important value in improving efficiency, dropping errors, and enhancing obedience within highly skilful environments. As the manufacturing changes, the future of Lean is poised to turn out to be more intelligent, resilient, and human-centric.

i) Integration of Adaptive, AI-Driven Systems:
The next generation of Lean in biomanufacturing will be deeply combined with artificial intelligence and machine learning. These skills will enable real-time process optimization, projecting quality control, and intelligent deviation administration making Lean systems more adaptive and proficient of handling multifaceted bioprocesses with negligeable human interference.
ii) Enhancing Supply Chain Resilience and Crisis Response:
Lean values subsidize to more agile and approachable supply chains. By abating waste, streamlining logistics, and dipping cycle times, Lean supports faster enlistment during public health tragedies such as epidemics. The ability to scale manufacture quickly while upholding quality is a critical gain in such scenarios.

iii) Human-Centric and Sustainable Innovation:

While mechanization and making digitization are central to forthcoming Lean systems, upholding a human-centric method remains essential. Permitting operators through incessant training, attractive cross-functional teams in problem-solving, and development a culture of novelty are energetic for long-term success. Furthermore, Lean performs progressively aligned with sustainability goals by dipping resource ingesting, waste generation, and energy use; Lean contributes to greener and more responsible biomanufacturing.

AI disclosure statement: The author acknowledges the use of AI-based tools for minor language editing or grammatical correction. However, all intellectual contributions, analyses, and conclusions presented in this chapter are entirely the author's own.

References

1. Groover, M.P., *Fundamentals of modern manufacturing: materials, processes, and systems*, 5th edition, p. 1024, J. Wiley & Sons, Hoboken, NJ, USA, 2010.
2. Walsh, G., *Biopharmaceuticals: Biochemistry and biotechnology*, 2nd ed., p. 576, Wiley-Blackwell, Chichester, UK, 2013, ISBN: 978-1-118-68738-3.
3. Lee, J.W., Kim, H.U., Choi, S., Yi, J., Lee, S.Y., Microbial production of building block chemicals and polymers. *Curr. Opin. Biotechnol.*, 22, 6, 758–767, 2011, https://doi.org/10.1016/j.copbio.2011.02.011.
4. Fischer, S., Handrick, R., Otte, K., The art of CHO cell engineering: A comprehensive retrospect and future perspectives. *Biotechnol. Adv.*, 33, 8, 1878–1896, 2015, https://doi.org/10.1016/j.biotechadv.2015.10.015.
5. Chen, C., Wong, H.E., Goudar, C.T., Upstream process intensification and continuous manufacturing. *Curr. Opin. Chem. Eng.*, 22, 191–198, 2018, https://doi.org/10.1016/j.coche.2018.10.006.
6. Sieckmann, F., Ngoc, N.H., Helm, R., Kohl, H., Implementation of lean production systems in small and medium-sized pharmaceutical enterprises. *Procedia Manuf.*, 21, 814–821, 2018, https://doi.org/10.1016/j.promfg.2018.02.188.
7. U.S. Food and Drug Administration, Guidance for industry: Quality systems approach to pharmaceutical CGMP regulations, 2025, https://www.fda.gov/drugs/pharmaceutical-quality-resources/current-good-manufacturing-practice-cgmp-regulations.
8. Womack, J.P. and Jones, D.T., *Lean Thinking: Banish Waste and Create Wealth in Your Corporation*, vol. 48, p. 1148, Simon & Schuster, New York, NY, USA, 1996, http://dx.doi.org/10.1038/sj.jors.2600967.

9. Antony, J., Snee, R., Hoerl, R., Lean Six Sigma: Yesterday, today and tomorrow. *Int. J. Qual. Reliab. Manage.*, 34, 7, 1073–1093, 2017, http://dx.doi.org/10.1108/IJQRM-03-2016-0035.
10. Shah, R. and Ward, P.T., Defining and developing measures of lean production. *J. Oper. Manage.*, 25, 4, 785–805, 2007, http://dx.doi.org/10.1016/j.jom.2007.01.019.
11. Melton, T., The benefits of lean manufacturing: What lean thinking has to offer the process industries. *Chem. Eng. Res. Des.*, 83, 6, 662–673, 2005, https://doi.org/10.1205/cherd.04351.
12. Bhasin, S., Prominent obstacles to lean. *Int. J. Prod. Perform. Manage.*, 61, 4, 403–425, 2012, https://doi.org/10.1108/17410401211212661.
13. Smart, N.J., *Lean Biomanufacturing, Creating Value Through Innovative Bioprocessing Approaches*, p. 355, Woodhead Publishing, Biomedicine, Cambridge, UK, 2013, ISBN: 978-1-907568-78-7.
14. Batwara, A., Sharma, V., Makkar, M., Giallanza, A., Towards smart sustainable development through value stream mapping – a systematic literature review. *Heliyon*, 9, 5, e15852, 2023, https://doi.org/10.1016/j.heliyon.2023.e15852.
15. Romero, L.F. and Arce, A., Applying Value Stream Mapping in Manufacturing: A Systematic Literature Review. *IFAC-Papers Online*, 50, 1, 1075–1086, 2017, https://doi.org/10.1016/j.ifacol.2017.08.385.
16. Permana, I. and Wang, F., Performance improvement of a biotechnology vaccine cleanroom for contamination control. *J. Build. Eng.*, 82, 108248, 2024, https://doi.org/10.1016/j.jobe.2023.108248.
17. Burggräf, P., Adlon, T., Hahn, V., Schulz-Isenbeck, T., Fields of action towards automated facility layout design and optimization in factory planning – A systematic literature review. *CIRP J. Manuf. Sci. Technol.*, 35, 864–871, 2021, https://doi.org/10.1016/j.cirpj.2021.09.013.
18. Patel, V. and Pitroda, J.K., A critical literature review on the impact of material management on construction project delivery. *Int. J. Creat. Res. Thoughts*, 5, 4, 1829–1836, 2017.
19. Narutaki, S., Lee, I., Sugawara, Y., Akada, K., Kondo, A., Case study on the environmental monitoring for biological manufacturing using Time-lapse Shadow Image Analysis. *Biologicals*, 66, 1–8, 2020, https://doi.org/10.1016/j.biologicals.2020.06.006.
20. Aamer, A.M., Al-Awlaqi, M.A., Mandahawi, N., Triawan, F., Al-Madi, F., Kaizen transferability in non-Japanese cultures: a combined approach of total interpretive structural modeling and analytic network process. *Int. J. Prod. Perform. Manage.*, 71, 8, 3245–3269, 2022, https://doi.org/10.1108/IJPPM-10-2020-0505.
21. Muruganantham, V.R. and Kavi, M.S., Analysis and implementation of JIT in small scale Polymer Industry. *Int. J. Innovative Res. Sci. Eng. Technol.*, 3, Special Issue 4, 325–334, 2014.

22. Junior, M.L. and Filho, M.G., Variations of the kanban system: Literature review and classification. *Int. J. Prod. Econ.*, 125, 1, 13–21, 2010, https://doi.org/10.1016/j.ijpe.2010.01.009.
23. Luca, L., A new model of Ishikawa diagram for quality assessment. *IOP Conf. Ser. Mater. Sci. Eng.*, 161, 1, 012099, 2016, https://doi.org/10.1088/1757-899X/161/1/012099.
24. Serrat, O., The Five Whys Technique, in: *Knowledge Solutions*, pp. 307–310, Springer Nature Singapore Pte Ltd., Singapore, 2017, https://doi.org/10.1007/978-981-10-0983-9_32.
25. Gallo, T., Chiara, C., Cecilia, S., Ruggieri, A., Industry 4.0 tools in lean production: A systematic literature review. *Procedia Comput. Sci.*, 180, 394–403, 2021, https://doi.org/10.1016/j.procs.2021.01.255.
26. Kobi, J., Developing Dashboard Analytics and Visualization Tools for Effective Performance Management and Continuous Process Improvement. *Int. J. Innovative Sci. Res. Technol.*, 9, 5, 2024, https://doi.org/10.38124/ijisrt/IJISRT24MAY1147.
27. Vishal, S. and Chaudhari, N., The Role of Good Manufacturing Practices (Gmp) In QA & QC: A Review. *Int. Res. J. Mod. Eng. Technol. Sci.*, 06, 10, 463–472, 2024.
28. Raj, N.I. and Boopathi, L.K., The Crucial Role of Regulatory Affairs: Navigating Compliance and Innovation in the Pharmaceutical Industry. *Int. J. Emerging Trends Pharm. Med.*, 1, 1, 93–110, 2025, https://ijetpm.igrf.co.in/index.php/ijetpm/article/view/7.
29. Aris, A. and Phillips, L.D., Advancing structured decision-making in drug regulation at the FDA and EMA. *Br. J. Clin. Pharmacol.*, 87, 2, 395–405, 2020, https://doi.org/10.1111/bcp.14425.
30. Ilan, Y., The Constrained Disorder Principle Overcomes the Challenges of Methods for Assessing Uncertainty in Biological Systems. *J. Pers. Med.*, 15, 1, 10, 2025, https://doi.org/10.3390/jpm15010010.
31. Warrick, D.D., Revisiting resistance to change and how to manage it: What has been learned and what organizations need to do. *Bus. Horiz.*, 66, 4, 433–441, 2023, https://doi.org/10.1016/j.bushor.2022.09.001.
32. Asin G., E., Fawcett, J.D., Batianis, C., Martins dos Santos, V.A.P., A snapshot of biomanufacturing and the need for enabling research infrastructure. *Trends Biotechnol.*, 43, 5, 1000–1014, 2025, https://doi.org/10.1016/j.tibtech.2024.10.014.
33. Suzanne, S.F., Integrated Continuous Biomanufacturing: Industrialization on the Horizon. *Biotechnol. J.*, 14, 2, 2019, https://doi.org/10.1002/biot.201800722.
34. Sundar, R., Balaji, A.N., Satheesh Kumar, R.M., A Review on Lean Manufacturing Implementation Techniques. *Procedia Eng.*, 97, 1875–1885, 2014, https://doi.org/10.1016/j.proeng.2014.12.341.

35. Ranbhor, R., Advancing Monoclonal Antibody Manufacturing: Process Optimization, Cost Reduction Strategies, and Emerging Technologies. *Biologics*, 19, 177–187, 2025, https://doi.org/10.2147/BTT.S515078.
36. Shahbodaghi, A., Moghaddasi, H., Farkhondeh, A., Hosseini, A., Documentation Errors and Deficiencies in Medical Records: A Systematic Review. *J. Health Manage.*, 26, 2, 2024, https://doi.org/10.1177/09720634241229545.
37. Daniel, M., Daniel, M., Bonarski, P., Motyka, P., Wysoczański, A., Gnitecka, R., Kolbusz, K., Dąbrowska, M., Burduk, A., Pawele, Z., Machado, J., Digital Standardization of Lean Manufacturing Tools According to Industry 4.0 Concept. *Appl. Sci.*, 13, 10, 6259, 2023, https://doi.org/10.3390/app13106259.

6

Issues and Problems for the Biopharmaceutical Industry

Sarthak Prasad Sahoo[1]* and Sonali Bhujabal[2]

[1]Department of Mechanical Engineering, Centurion University of Technology and Management, Bhubaneswar, Odisha, India
[2]Department of Chemical Engineering, NIT Raipur, Raipur, Chhattisgarh, India

Abstract

A hypothesis, that the life-saving drugs may face the manufacturing hurdles with billions in sunk costs, years of delayed periods and some sets of regulatory conflicts, can never be ignored as the biopharmaceutical world is presently suffering by a lower success rate of novel biologics development (commonly below 10%), an unwanted stretched time line spanning over decades and a huge investment for the whole manufacturing. To facilitate faster therapy to the patients, integration of AI-ML techniques, replacing expensive and extended preclinical trials with decentralized and innovative trials are to be adopted on a quicker passion. The resilience and agility can further promote the manufacturing frontiers to follow green processes, modular continuous platforms and digitally integrated supply chains. Not only innovations, but efforts are also required to upgrade the workforce with latest skills, to facilitate balance to move forward with both sustainability and affordability and to come forward with globally harmonized regulations. Keeping all these aspects in mind, the present content is so framed to pile up all the major challenges related to biopharmaceutical industry in one place and to come up with some viable solutions coupled with ethical oversight, transparent communication, and equity-driven distribution. This approach may also be helpful in redefining the scopes, transforming the barriers into promoters and facilitate the industries not only to deliver innovative drugs but also make them available for the service of mankind on the global level.

**Corresponding author*: sarthakprasad001@gmail.com

Sandip Kunar, Jagadeesha T. and D. Sathis Kumar (eds.) Biomanufacturing: Transforming Production of Biologics and Biopharmaceuticals, (139–162) © 2026 Scrivener Publishing LLC

Keywords: Biopharma innovation, complex biologics challenges, green biomanufacturing, regulatory harmonization, sustainability, patient-centric equity

6.1 Overview of Biopharmaceutical Industry

Biopharmaceuticals can be commonly referred to as therapies, derived from the living beings by applying the innovative principles of biotechnology. In contrast to chemically synthesized small-molecule drugs, biopharmaceuticals are associated with complex and large molecules like nucleic acids, proteins and biologically processed cells. As vaccines, biologics, gene therapies and cell therapies are included in biopharmaceuticals, these can be implemented to prevention, diagnosis and treatment of diseases [1]. Some basic understanding regarding the above terminologies are discussed below:

Biologics: Biologics commonly include therapeutic proteins and complex biomolecules derived from living cells; and hence, they constitute a major category of biopharmaceuticals. For the treatment of long-term diseases like autoimmune disorders, metabolic diseases and cancer, the involvement of biological medications can be observed, which include blood components, monoclonal antibodies, hormones, cytokines, and enzymes. Through different intricate manufacturing processes assisted with strict measures for quality control, the production of biologics from the living cells are made possible, which further demands stern regulatory approvals [2, 3].

Vaccines: Biological vaccines help patients to defend themselves from contagious diseases by stimulating the immune systems. The vaccines are commonly composed either of subunit antigens, genetic components (like mRNA) or killed pathogens. A commendable application of vaccine development through mRNA during COVID-19 pandemic ensured that the scope of modern biotechnology can never be ignored for fast and scalable solutions [4].

Gene Therapy: Gene therapy, proposed more than 50 years ago, is a vital aspect from the viewpoint of biopharmaceuticals. This led to development of several efficient and safer technologies with the help of technological and biological advancements. Gene therapy is now being implemented to treat different hereditary illnesses including α-1-antitrypsin deficiency, familial hypercholesterolemia, adenosine deaminase deficiency and Gaucher's disease. This therapy is proved as an efficient means to treat some of the rare conditions where there are no alternatives or available solutions are few.

By this, new possibilities for advanced treatment practices can be explored focusing particularly on the issues related to genetic abnormalities [5].

Cell Therapy: A systematic supervision of living cells, applied in the fields of medical treatment, to fight or eliminate diseases can be referred to as cell therapy. Along with regenerative medicine, this therapeutic approach also includes stem cell treatments and CAR-T cell therapies. A structured modification of patient's T cells to detect and destroy cancer cells, is a kind of cell therapy included under CAR-T process. This type of personalized therapy has a huge potential to treat some specific kinds of lymphomas and leukemias effectively [2].

Apart from the above, some recent studies [6, 7], also encompass that immunological, nucleotide products, cell or tissue products, and cytomodulators are the four primary categories into which biologics are divided. Antibodies (such as antisera and immunoglobulins) and immunogens (such as vaccinations, toxoids, allergens, and DNA vaccines) are examples of substances that promote or confer immunity. Gene therapy and mRNA-based treatments frequently employ nucleotide products, which are genetic therapies that transfer DNA or RNA into the body using viral or plasmid vectors. The use of living cells or tissues for medical reasons is known as cell or tissue products. This includes stem cells, tissue transplants, and blood products such as erythrocytes, T-cells, monocytes, platelets, and plasma. Last but not least, cytomodulators are biologics, such as interleukins, interferons, tumor necrosis factors (TNFs), and colony-stimulating factors (CSFs) that control cellular or immunological responses. Because they address the underlying causes of complicated disorders, these biologic categories are essential to modern medicine.

6.2 History and Evolution of the Industry

After it was public that the vaccines could be protected by dyes, the chemical industries like Hoffman-La Roche, Ciba-Geigy and Sandoz (in Switzerland) transformed themselves from dye production to pharmaceuticals. This was the provenance, from which the origin of the pharmaceutical industry came into picture in the late 19th century in the Upper Rhine Valley of Switzerland [8]. Meanwhile, earlier existing pharmacies in Baghdad (in 754 AD), gradually evolved into major pharmaceutical enterprises by the 19th century and extended across countries like North America and Europe. Again, the countries like the US, Switzerland, the UK and Germany witnessed the growth of robust pharmaceutical industries by 1920s and 1930s due to the mass production of transformative drugs including penicillin and insulin. The massive success of penicillin in 1940s, laid the stone for further research and development (R&D) as far as the industrial aspects of pharmaceuticals is concerned. A rapid expansion of the pharmaceutical industries was also seen by the 1960s, maintaining healthy tie-ups between physicians and companies, controlled by the U.S. Food and Drug Administration (FDA). But the thalidomide tragedy associated with several birth defects triggered the World Medical Association to implement rigorous clinical trials prior to approving any drugs [9].

The perpetual patents were seen to decline by 1970s due to the constructive reforms in patenting, which in turn enabled the generic drug firms to get benefited by producing off-patent medications. Following the same trend, India emerged as a global hub to produce generic drugs by 1978. Eventually, India introduced several important pharmaceutical regulations, including the Drugs and Cosmetics Act (1940), Pharmacy Act (1948), Indian Patents Act (1970), and TRIPS compliance (2005). By 1980s, prominence was given to drugs for cardiovascular diseases and AIDS under the direct supervision of regulatory bodies like the FDA. Likewise, the pharmaceutical industry is evolving continuously on the global level to accommodate with the ever-growing regulations, generic competitions and innovations.

6.3 Scope of the Chapter

Biopharmaceuticals have always played a crucial role in improving human health and responding to medical emergencies. The development and approval rates of novel biopharmaceuticals have increased dramatically due to developments in genetic engineering, biotechnology, and bioinformatics, putting them at the forefront of the pharmaceutical industry. The quality of life for patients throughout the world has greatly improved—thanks to modern manufacturing techniques and simplified regulatory routes, which have further accelerated approval timelines and given hope for diseases that were previously incurable. However, there are still ambiguities related to basic understanding, competency build-up and environmental obligations. To resolve these unforeseen ambiguities, the present chapter aims to join the missing dots in the field of biopharmaceuticals through the following objectives:

- ✓ To understand the need for and importance of biopharmaceuticals not only from the medical sector but also from the aspects of global health and economy.

✓ To examine and critically analyze the major challenges encountered by biopharmaceutical industries exploring manufacturing processes, workforce, supply chain and regulatory compliances.

✓ To explore societal, ethical and global consequences of these challenges focusing on the aspects of patient-centric innovations, equity and transparency.

6.4 Distinction from Traditional Small-Molecule Pharmaceuticals

The key distinctions between biologics or new treatments and conventional small-molecule medications are highlighted by recent therapeutic developments [10]. Small molecules are chemically produced; low molecular weight substances intended to enter cells and alter intracellular enzymes or receptors. One example of this is the recently licensed oral antibacterial drug gepotidacin (Blujepa), which targets bacterial topoisomerases [11]. However, biologics, such the monoclonal antibody axatilimab (Niktimvo),

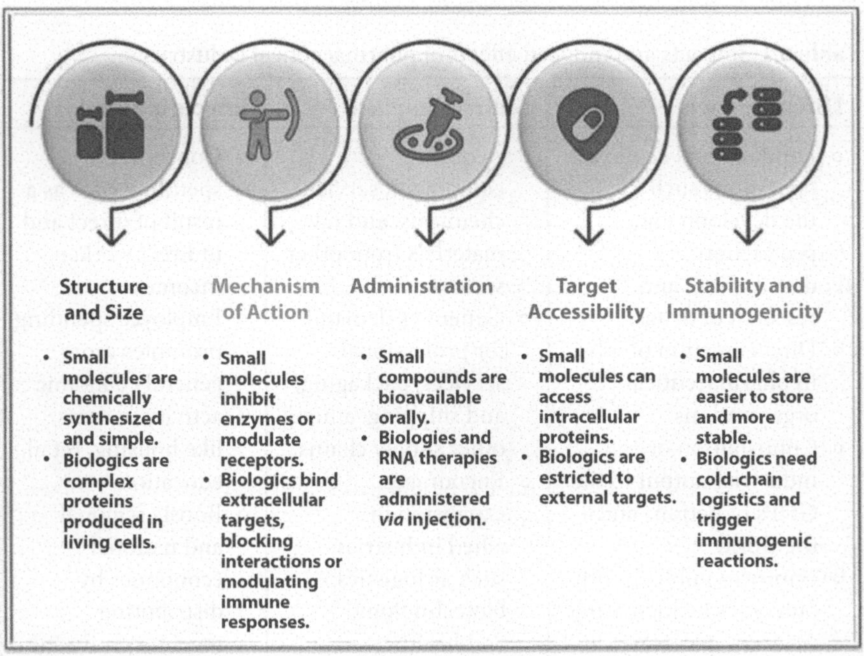

Figure 6.1 Differentiation factors in small molecules vs biologics.

which is prescribed for chronic graft-versus-host disease, are big, protein-based treatments that bind extracellular targets [11]. The following aspects (as shown in Figure 6.1) can be considered for easy differentiation.

6.5 Impact on Global Health and Economies

First, global pharmaceutical production and management operations are extremely important for both economic growth and global health. These activities directly contribute to economic growth by creating jobs, fostering innovation, and attracting investment in the pharmaceutical business, which forms the foundation of the sector's financial contributions [12]. Through their procurement of raw materials, packaging, technology, and services, pharmaceutical businesses indirectly create a broad ripple effect throughout supply chains, consequently boosting demand in a number of other industries. A strong induced impact also exists, wherein the salaries and wages produced by direct and indirect employment boost consumer spending and stimulate the economy as a whole. Together, these effects underscore the pharmaceutical industry's central role not only in

Table 6.1 Impacts and induced effects of pharmaceutical industries.

Direct impacts	Indirect impacts	Effects induced
o Fundamental economic operations such the development, production, distribution, and research of drugs. o Direct creation of jobs in pharmaceutical organizations. o Contribution of industrial output to the GDPs of nations and the world. o Improves public health outcomes by increasing access to necessary medications.	o Acquiring equipment, services, chemicals, and raw materials from other sectors. o Generates demand for professional services, packaging, and shipping, among other supply chains. o Encourages expansion in allied industries such as logistics, biotechnology, and healthcare infrastructure.	o Consumer spending rises as a result of direct and indirect worker incomes. o Employee spending promotes more general economic activity in areas like housing, retail, education, etc. o Boosts regional and national economies by distributing revenue produced by businesses.

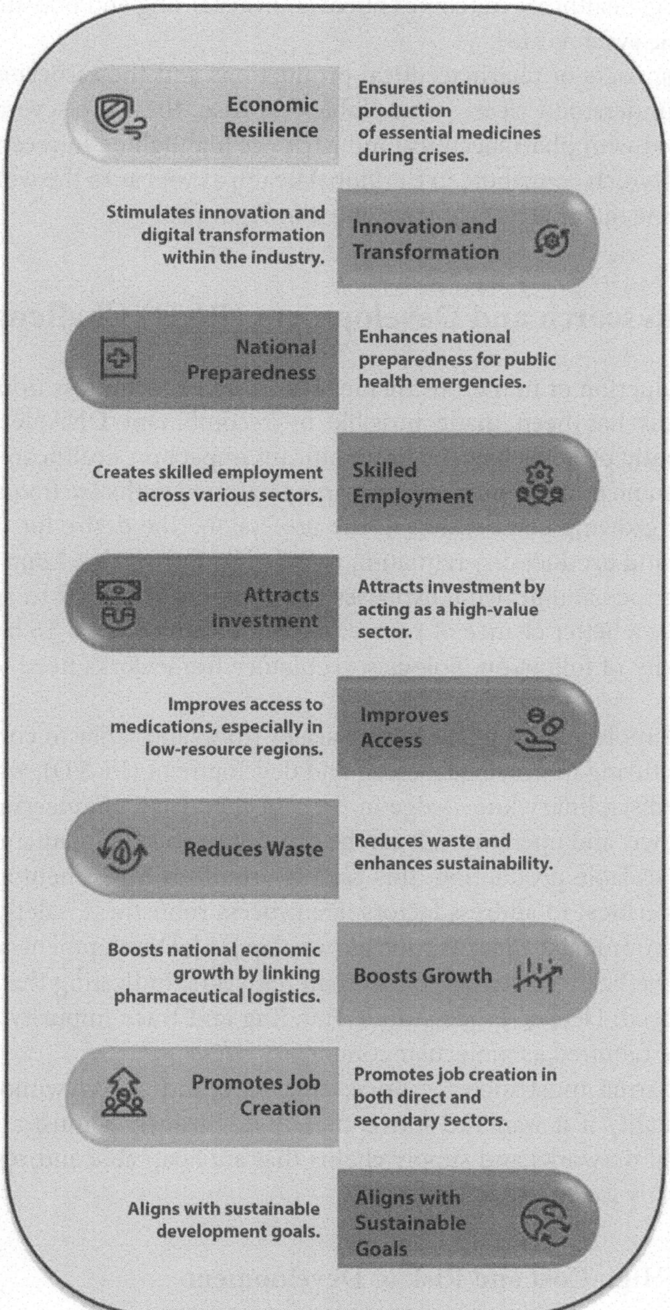

Figure 6.2 Contribution of pharmaceutical industry towards global health and economy.

advancing healthcare outcomes but also in sustaining and boosting global economic systems [13].

The impacts of pharmaceutical productions and the associated effects can be understood clearly from Table 6.1. Also, there are several factors associated with pharmaceutical industries as highlighted in recent studies [12, 13], which contribute to the global health as well as to the world economy. Some major factors are presented in Figure 6.2.

6.6 Research and Development (R&D) Challenges

The production of numerous therapeutic proteins to address unmet medical needs has been made possible by recombinant DNA technology. Therapeutic proteins have had a significant impact on healthcare and will only become more significant. The same treatment paths are frequently the focus of growing competition in the profession. The desire for improved efficacy and product differentiation is fueled by this rivalry. Long development periods and growing expenses cause attention to shift to applicants who have a better chance of succeeding in the latter stages. To handle the complexity of follow-on biologics, regulatory frameworks need to change [14].

Small-molecule drug production scales from laboratory to commercial manufacturing in process research and development (PR&D), which calls for interdisciplinary knowledge in process chemistry, engineering, safety, compliance, and operations. It can be difficult to go from medicinal chemistry to scalable production; this calls for resource investment and redesign. It's critical to address factors like process robustness, safety, cost, IP, and environmental effect as soon as possible [15]. Development-to-launch times have been shortened to less than ten years, indicating that speed is now crucial. Deeper process understanding and trace impurity management are required as molecular complexity rises.

Biopharma must increase price, efficiency, and agility without sacrificing quality if it wants to stay competitive. Businesses must also create industrial networks and supply chains that are adaptable and responsive, particularly for emerging markets.

6.6.1 High Cost and Risk of Development

About 30% to 40% of all R&D expenses are spent on preclinical research alone; each authorized medicine costs about $474 million, or almost half a billion. But that's only the beginning: the full process from discovery to sale takes 10–15 years, and clinical trials last 7–8 years following preclinical studies. Despite the time and cost, success is still elusive; approximately 10–15% of those starting Phase I finally receive approval, and total Phase I success is less than 14%. The high $1–2.6 billion price tag for each medicine includes the costs of these failed trials because over 90% of clinical studies end in failure [16]. Furthermore, despite advancements in technology, actual drug-development costs have doubled every nine years due to Eroom's Law.

The high risk, complexity, and capital intensity of drug development are reflected in the pharmaceutical industry's massive research and development (R&D) expenses. According to data, definitions, techniques, and therapeutic class, estimates of per-drug R&D expenditures range greatly from $113 million to over $6 billion (in 2018 dollars). The procedure, which frequently takes 10-15 years, entails screening hundreds of chemicals, carrying out exacting preclinical testing, and observing stringent regulatory guidelines. Since, about 1 in 5,000–10,000 compounds make it to market, failure costs are a major contributor to total costs. The mean R&D expenditure for FDA-approved medications from 2009 to 2018 was above $1.5 billion, with a median cost of $1.14 billion [17]. These numbers highlight the necessity of setting the price of successful medications to compensate for both high attrition rates and a delayed return on investment. Furthermore, firms are under additional pressure to recover sunk investments due to opportunity costs and market rivalry, which supports high pricing as a means of encouraging further innovation. The majority of pharmaceuticals fail early; therefore, the upfront costs are high. Billions are spent on scientists, labs, chemical screening, and safety studies. Time and money are increased to comply with GLP, FDA, and EMA. Due to fierce competition in the market, capital is committed to long-term, high-risk ventures. These reasons support high medicine costs to maintain innovation and recoup expenditures.

Clinical trial failure rates are a recurring problem, especially for new biologics. Despite high Phase I pass rates of 92%, an analysis of studies conducted between 1998 and 2008 in Crohn's disease revealed that just 18% of medications were approved. This poor success rate was a result of both clinical and commercial failures. Biologics have potential, particularly in the treatment of autoimmune and dermatological conditions, but their development is still fraught with danger [18].

6.6.2 Complexity of Biologic Molecules

Significant difficulties in characterization, stability, and formulation arise from the structural complexity of biologic molecules, including gene therapies, mRNA lipid nanoparticles, antibody–drug conjugates, fusion proteins, and cell therapies. Environmental stresses like pH and temperature changes can cause physical deterioration and chemical instability in their massive, complex structures [19]. The shelf life and therapeutic efficacy are compromised by conformational and colloidal instability. Developing these compounds is made more difficult by the requirement to maintain structural integrity and their low penetration through biological barriers. Excipients, stabilizers, and sophisticated delivery systems are some of the strategies that are necessary to overcome these obstacles and guarantee clinical efficacy [20].

The complicated tertiary structures and diverse post-translational modifications of biologic therapies, like recombinant proteins and monoclonal antibodies, add to their structural and functional diversity. These characteristics make it more difficult to anticipate immunogenicity because even little modifications in composition or conformation can result in anti-drug antibody reactions, jeopardizing the safety and effectiveness of treatment. Moreover, unanticipated molecular interactions or downstream signaling regulation may cause biologics to have off-target effects [21]. Their translational value is limited by the fact that current *in vitro*, in silicon, and animal models frequently lack the fidelity to recreate human immune responses. Therefore, more sophisticated, human-relevant evaluation platforms are required since precise estimation of immunogenicity and off-target risks continues to be a major challenge in biologic drug development.

6.6.3 Drug Discovery Pipeline Drought/Attrition

The struggle of the biopharmaceutical industry is very true due to the slower rate of drug discovery pipeline, and this condition can be referred to as the 'pipeline drought'. This decreasing rate is associated with the growing target of complex biologicals, higher rate of attrition and escalating costs. Beyond these, identifying innovative therapeutic targets can be treated as a major challenge in this regard. Easily targetable mechanisms and pathways like in the 'low-hanging fruit' are being explored several times. According to a study [22], many of the newly formulated targets are either biologically obsolete or irrelevant to disease progression, which can be attributed to the declining trends of the human genome mapping and optimizations. This creates a prolonged gap between target identification and actual success of developed drugs. Another considerable aspect is shortage of readily discoverable targets for illnesses associated with unmet needs. Comparable easy opportunities like unforeseen discoveries and conventional high-throughput screening are disappearing nowadays. As remarked by Mr. Jay Bradner (Novartis's R&D chief) that to meet the less understood and complex targets, for the 'reaching for the highest-hanging fruit' strategy, the pharma industries must evolve with the latest technologies [23].

Increased complexity of biological systems and pathways is another form of challenge encountered by the biopharmaceutical industries. Treatments for vulnerable diseases like Alzheimer's, autoimmune disorders and cancer demand network-based and multifactorial biological processes rather than single-gene targets. This in turn interrupts translation of preclinical findings into clinical usefulness. According to researchers, though majority of the theoretical targets can be produced from genome mining, only a few of them are found to be viable for translation, due to the complex interdependence within human biology [24].

Thus, drug discovery is referred to as a failure-prone area with more inclination towards resource-intensiveness due to the factors like growing regulatory analysis, depleting easy targets and rising complexity in understanding disease biology. Recently developed innovative technologies like CRISPR, system biology and AI can help to address these issues but their integration into the mainstream pipelines is still under development stage.

6.6.4 Intellectual Property (IP) Protection

Due to manufacturing variability, regulatory intricacies and structural complications, complex biologics (like biosimilars) experience huge barriers during patenting, as compared to small- molecule generics. It is much difficult to determine the infringement or define clear patent boundaries for the biosimilars as they are not identical replicas of originators. Thus, biologic developers often come across several patents of overlapping nature related to the aspects like delivery methods, production processes and formulations, which in turn lead to "patent thickets" [25]. Unnecessary delay is also experienced for entry of biosimilars due to these patent thickets. Unlike in Canada and UK, biosimilars experience more patent allegations in the U.S. In near past, the *Humira* case exposed that around 80% of patents filed by AbbVie were duplicative in nature, with a sole motive to prevent biosimilar competitions rather that protecting the actual innovations [26]. These issues are however associated with the global inequalities in intellectual property laws and enforcement. Unlike the strict principles of European systems, the U.S. imparts comparatively liberal and flexible systems for patent filing and terminal disclaimers. Thus, market entry timing of biosimilars get affected significantly and the manufacturers suffer with regulatory uncertainties [27]. Hence, ensuring access and fostering innovations must be balanced by the patent systems, by facilitating suitable reforms, which can promote healthy completions and can restrict over-patenting.

6.6.5 Data Management and Analytics

To improve quality, innovations and efficiency of biopharmaceutical industries, data management and analytics related to drug development and drug manufacturing are highly essential. Existing huge data set for the predictive maintenance and process optimization can be smoothly analyzed by integrating and adopting AI and machine learning (ML) like big data tools [28]. Data sharing and reusability integrated with smart R&D workflows can be enhanced by implementing FAIR (Findable, Accessible, Interoperable, Reusable) data approaches [29]. The major challenge in the digital and automated environment is to ensure the data integrity which demands suitable compliance with regulatory standards with minimum data risk [30].

6.7 Manufacturing and Supply Chain Hurdles

6.7.1 Complex Manufacturing Processes

Biopharmaceutical industry faces numerous operational and technical issues due to association with multi-step and complex manufacturing processes. Major phases such as fermentation, sterile filling, purification and cell culture face distinctive risks, and hence lead to process variability, higher cost, and long production times. Though fermentation and live cell culture are the fundamental stages, both are sensitive to nature. Therapeutic proteins, often produced by mammalian cells, are extremely vulnerable to variations in nutrients, pH and temperature. These variations can further affect viability, protein expression levels and cell growth. By introducing chemically defined media to reduce inconsistency, it remains challenging to eliminate the variability due to variations in raw materials and inherent biological fluctuations [31]. As purification is also a vital phase, the target proteins must be separated from impurities without degrading product quality through techniques like chromatography. The possibility of getting distinct behavior from each batch demands well-gripped process adjustment and intensive quality assurance, which may assist in reducing unnecessary extensions in production timeline. In the final step, i.e., sterile filling, contaminations must be avoided by ensuring aseptic conditions and implementing Good Manufacturing Practice (GMP) that uses single-use systems or stainless steel. This however, confirms trade-offs in environmental impact, flexibility and cost [32]. Overall cost seems to be increased due to high energy consumption in HVAC systems to maintain the sterile environments. Also, a huge initial capital investment is required for biomanufacturing facilities, among which, the upfront costs include ensuring regulatory compliances, building and validating cleanrooms, and acquiring specialized bioreactors. Thus, the biological uncertainty, in association with the regulatory misalignments, restricts the scalability and delays market access.

6.7.2 Quality Control (QC) and Quality Assurance (QA)

Both QC and QA are essential components of biopharmaceutical manufacturing as they ensure compliance, efficacy, and safety with stringent regulatory principles. Batch-to-batch variable characteristics may often require extensive testing to declare consistency. Apart from this, globally accepted regulatory compliances like EMA's guidelines and FDA's Current Good Manufacturing Practices (cGMP) necessitate vigorous documentation and audit promptness, which impose extra operational burden for international or multi-site manufacturers. Also, the technological shift towards single-use systems and continuous manufacturing presents new QC challenges. One such factor is thoroughly monitoring extractable, leachable and material compatibility from disposable equipment so that the risk of contamination can be avoided. Although, real-time release testing (RTRT) and Process Analytical Technology (PAT) are already there to identify the deviations, but they are in limited use due to improper regulatory and technological justifications, which further lead to unwanted delay in real time quality monitoring. Therefore, maintaining quality in biopharmaceutical manufacturing is highly essential to fill the gaps in regulatory complexity, growing technologies and biological nature of products [33].

6.7.3 Supply Chain and Scalability Challenges

The supply chain and scalability challenges of the biomanufacturing industry can be easily linked with cost, quality and availability of different biopharmaceutical products. The biologics need sensitive and complex raw materials with strict quality standards which differentiate them from traditional small molecule drugs. The product integrity may get hampered or the production process may get delayed if any unwanted interferences take place in sourcing any of these materials. The adverse effects of supply chain fragility can be experienced during global issues like geopolitical conflicts or pandemics, where the lead time increases by straining the logistics. Again, limited number of skilled suppliers and maintaining cold chain logistics for temperature-sensitive biologics lead to distribution complexities and extra cost-addition. Since biologics originate in living systems, their scale prediction also varies from lab to commercial production. Also, rise in production volume may demand extra time and resources, process optimization, and revalidation. Hence, bio-manufacturing can be made successful by adopting a scalable, flexible and robust supply chain. Though prioritizing localized production and modular manufacturing allow a bit of flexibility in this regard, still extensive implementation is found to be limited [34].

6.7.4 Sustainability in Biomanufacturing

In the field of biomanufacturing, sustainability is an important terminology seeking higher attention in the global market as this tends to balance the product demand with environmental accountability. Since the biomanufacturing processes involve significant consumption of water and energy, and generation of waste, major sustainable challenges can be fulfilled by adopting clean room operations, purification of chemical wastes, and practicing disposable and single-use technologies. This integration into the biomanufacturing industry, however, reduces overall operational cost and industry's carbon footprint. Concepts like process intensification, real-time monitoring and modular production systems can be explored more for limited resource use but not compromising with the quality level of the products [35]. Also, implementing life cycle assessment (LCA) can help in evaluating the environmental impact of manufacturing processes. According to a recent study, unit operations like freeze drying is found to contribute to lower energy consumption and carbon emissions, thus ensuring a route towards sustainability practice [36]. In short, it can be said that a systematic approach for integrating greener technologies, supply chain optimization and environment-oriented efficient design can assist biomanufacturing industry in advancing towards sustainability practices.

6.8 Regulatory and Compliance Burdens

The regulatory and compliances burdens associated with biopharmaceuticals play a vital role due to variability, safety-critical, and complexity nature of biologic drugs. To ensure efficacy, quality and safety of the products, the pharmaceutical regulatory bodies like WHO, EMA and FDA must prepare to abide by extensive clinical and manufacturing data, comprehensive documentation and Current Good Manufacturing Practices (cGMP). Often, biologists require years of clinical experiments and data submission and post-marketing observations to ensure continual safety assessment. Regulatory requirements can considerably vary across the geographical regions leading to uncertain global market access for the manufacturers [27]. Also, the recently evolving regulatory standards introduce new challenges for biosimilar, continuous manufacturing processes and advanced therapies as the recent technologies outperform the traditionally used regulatory frameworks leading to uncertainty in approval paths and validation [37]. Figure 6.3 presents some of the major regulatory and compliance burdens related to biopharmaceutical industry. Thus, regulatory and compliance requirements must meet customer satisfaction and patient safety, lack of which may lead to facility shutdowns, legal penalties and product recalls and thus the manufacturer may bear an additional operational, financial and administrative burden to exist in the market.

Figure 6.3 Major regulatory and compliance burdens of biopharmaceutical industry.

6.9 Workforce, Market Access and Commercialization Challenges

The biopharmaceutical manufacturing industry suffer from several challenges related to market access, workforce and commercialization which in turn affects the capabilities to scale, deliver and innovate the therapies to the intended patients. The bio-manufacturing industries and the associated R&D sectors, nowadays, seek a highly skilled workforce well trained in advanced manufacturing technologies, regulatory compliances and biotechnology, so that the traditional existing gap in the emerging market can be filled, globally. Countries like India, emphasize more on capacity building through skill development, workforce training and public–private partnerships [38]. Accessing the international market necessitates a good grip on the aspects of healthcare system heterogeneity, reimbursement obstacles and complex regulatory principles. Although, biosimilars are cheaper due to lack of proper demand-side returns, streamlined regulatory pathways and brand reliability in many countries [39]. Due to this, the product may not get access to the required patients; and hence, the return on investment will be delayed. Due to the hurdles like uncertain price outcomes, higher production cost and longer process cycles, it is more complex to introduce the biopharmaceuticals in the market, as compared to the conventional small-molecule drugs. Authorizing licences, based on the standards set for small molecules often raises concerns related to renegotiation requirements, poor regulatory agenda and biopharma timelines [40]. Overall, this can be noted here that, to explore the complete potential of innovative biopharmaceutical approaches, adaptive commercialization, harmonized regulatory frameworks and workforce development must be paid attention.

6.10 Ethical, Societal, and Public Perception Concerns

With the growing era of technological advancement, it is highly required for the biopharmaceutical industries to be periodically examined based on ethical, societal, and public perception concerns, so that the credibility, patient access and influential policy proposals are not affected. Clinical trial transparency, sometimes also treated as an inefficiency to address the public commitments and varying reporting data, which in turn breaks trust of non-industry sponsors [41]. Also, some R&D sectors, running behind the profits only, focus more on diseases having high market value, neglecting the aspects of equitable health-care innovation [42]. The ethical balance between the public health demands and the corporate profitability is decided by the pricing strategy; for instance, steep increase in market price of Questcor Pharmaceuticals created both ethical debating and patient backlash [43]. By excluding the social benefits and productivity gains for cost-effectiveness calculations, biopharmaceutical innovations can engulf the healthcare budgets. Also, the concerns of biosecurity risks may be experienced by implementing dual-use research approach in bio-manufacturing. To minimize these factors, majority of the biopharmaceutical industries must evolve with adequate internal regulations [44]. Superficial conflicts of interest, aggressive marketing and pricing strategies, and opaque R&D practices, often intensify suspicion and negative public mentality towards the industry-driven healthcare solutions. These issues, however, can be isolated under other issues as presented in Figure 6.4. To maintain public trust and long-term sustainability, biopharmaceutical industries must adhere to innovations with global health priorities, robust ethical principles and transparency across their development and commercialization practices.

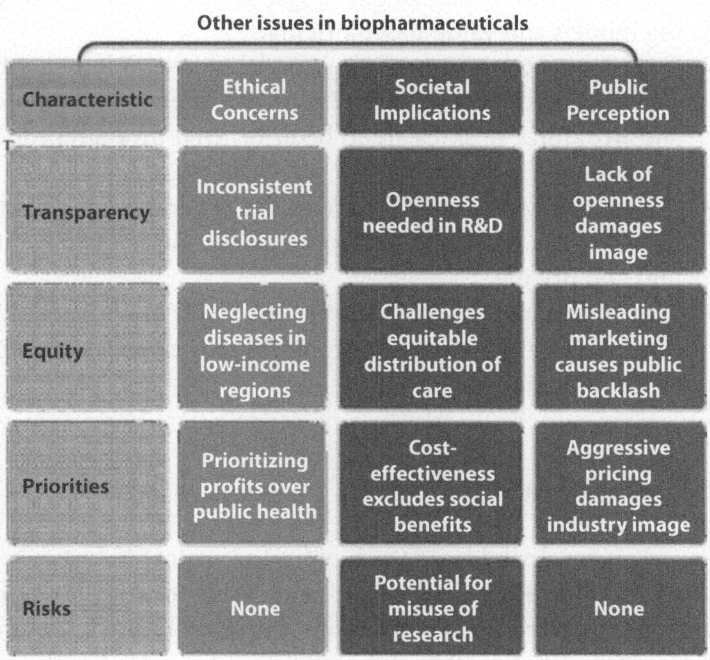

Figure 6.4 Other issues of biopharmaceutical industry.

6.11 Future Recommendations or Probable Solutions

Without looking for possible solutions, it is meaningless only to run after the challenges faced by the biopharmaceutical industries. The primary motive of addressing these challenges lies in improving accessibility and safety for life-saving therapies, ensuring ethical responsibility, public affordability and technological innovations to promote global health equity. Based on the above challenges discussed some probable solutions or recommendations for the future are presented below:

- ➢ Workforce training, clinical trials, R&D activities and regulatory compliances need to be streamlined through recently adopted digital tools, AI and machine learning.
- ➢ Integration of real-world evidence and acceleration of pending approvals can be made easy by initiating globally harmonized and adaptive regulatory frameworks.
- ➢ The localized production hubs are to be integrated with smart supply chains, continuous, modular and sustainable manufacturing practices to increase scalability and resilience.
- ➢ Equitable market access is to be promoted to upgrade underserved regions through digital engagement tools, tiered pricing models and value-based pricing.
- ➢ To enhance the competencies for latest technology, GMP and advanced biotechnology, inclusive talent development and workforce upskills are to be prioritized through proper investments.
- ➢ Transparency and ethical governance are to be maintained for trial data, emerging technologies (like AI and gene editing) and pricing.
- ➢ Inclusive innovation policies, patient involvement and education campaigns are to be arranged to foster societal orientation and public trust.

6.12 Conclusion

The long-term sustainability of the biopharmaceutical industry is greatly affected by different R&D challenges, since this industry has to operate by balancing societal responsibility, economic viability and scientific innovation. Majority of the challenging aspects are associated with extended development timelines, negligible success rate of clinical trials (especially for unique biologics) and requirement of huge capital investment. Advanced analytical tools are highly required to optimize formulations, ensure stability and predict immunogenicity of complex biologic molecules. But the complexity of modern disease biology and exhaustion of easily accessible targets are the highlighted deficiencies for the present drug discovery pipeline. Increased number of patent thickets, global legal contradictions and biosimilars can be treated as the contentious area for the biopharmaceutical industry as far as the criticality of the intellectual property protection is concerned. Interoperable systems with AI/ML capabilities, adaptive trial designs, open improvement networks and digital twins can be some of the viable solutions to deal with the data management challenges. Also, to improve trial outcomes, reduce costs, and accelerate development, some advanced manufacturing methods such as single-use technologies, green production methods and modular continuous systems can be beneficial to meet the growing market demands.

A safer and faster market entry can be ensured by approaches like AI-driven compliance systems, lifecycle-based consents and synchronized frameworks as reformations in the governing systems are meant to be focused. On the other hand, the workforce challenges can be addressed through diversity initiatives, virtual labs and upskilling the existing-workforce which can further support the technological advancement integrated with demanding human capital. Reasonable tiered pricing, collaborative reimbursement and value-based pricing are some of the strategies to easily access the market by simultaneously balancing affordability with innovation incentives, especially for the developing countries. As far as the patient perspectives is concerned, retaining public trust is highly dependent on comprehensive decision-making, ethical inaccuracy and transparency in the system. The biopharmaceutical industry can ensure global health equity, counteract against the system barriers and augment innovation efficiencies by adopting the aforementioned strategies. On a concluding remark, it can also be said that along with operational and technological improvements the biopharmaceutical industries need to evolve with commitments for patient-centered outcomes, sustainability practices and ethical responsibilities which can further promote ground-breaking therapies, having worldwide access and positive impacts.

AI disclosure statement: The author confirms that no artificial intelligence (AI)-generated content or AI-assisted writing tools were used in the preparation of this chapter (including summaries and the abstract). All text, analysis, and conclusions are the result of the author's original work.

References

1. Rasmussen, A.S.B., Hammou, A., Poulsen, T.F., Laursen, M.C., Hansen, S.F., Definition, categorization, and environmental risk assessment of biopharmaceuticals. *Sci. Total Environ.*, *789*, 147884, 2021, https://doi.org/10.1016/j.scitotenv.2021.147884.
2. Plitnick, L.M. and Herzyk, D. (Eds.), *Nonclinical development of novel biologics, biosimilars, vaccines and specialty biologics*, Academic Press, London UK, 2013, https://doi.org/10.1016/c2011-0-07530-0.
3. Behera, B.K., Prasad, R., Behera, S., Management and Manufacturing Process of Biologics, in: *Competitive Strategies in Life Sciences*, pp. 43–104, Springer Singapore, Singapore, 2020, https://doi.org/10.1007/978-981-15-7590-7_2.
4. de Groot, A.M., Delivery and Immunogenicity of Biopharmaceuticals for Vaccination or Therapy, Doctoral dissertation, Utrecht University, 2017.
5. Loring, H.S. and Flotte, T.R., Current status of gene therapy for α-1 antitrypsin deficiency. *Expert Opin. Biol. Ther.*, *15*, 3, 329–336, 2015, https://doi.org/10.1517/14712598.2015.978854.
6. M, A. and G, S., Recent Advances in Biopharmaceuticals: Expanding the Therapeutic Arsenal for Disease Treatment. *Int. J. Pharmacometrics Integr. Biosci.*, 8, 18–22, 2023, https://doi.org/10.26452/ijpib.v8i2.1509.
7. Starr, S.P., Immunology Update: Biologics. *FP Essent.*, *450*, 11–21, 2016.
8. Malerba, F. and Orsenigo, L., The evolution of the pharmaceutical industry. *Bus. Hist.*, *57*, 5, 664–687, 2015.
9. Van Campen, L.E., Poplazarova, T., Therasse, D.G., Turik, M., Biopharmaceutical Bioethics Working Group, Kelman, A., Rossetti, A., Chang, C., Stern, K.N., Fortson, W., Considerations for applying bioethics norms to a biopharmaceutical industry setting. *BMC Med. Ethics*, *22*, 1, 31, 2021, https://doi.org/10.1186/s12910-021-00600-y.
10. Wan, H., An overall comparison of small molecules and large biologics in ADME testing. *ADMET DMPK*, *4*, 1, 1–22, 2016, https://doi.org/10.5599/ADMET.4.1.276.
11. Trusheim, M.R., Aitken, M.L., Berndt, E.R., Characterizing markets for biopharmaceutical innovations: do biologics differ from small molecules? *Forum Health Econ. Policy*, 13, 1, 2010, July, De Gruy.
12. Mamedyarov, Z.A., Pharmaceutical Industry Development in the Midst of Crisis: Global Trends. *MIR (Mod. Innov. Res.)*, *4*, 398–408, 2020, https://doi.org/10.18184/2079-4665.2020.11.4.398-408.
13. Shekoohi Tolgari, F. and Zarrinpoor, N., A robust reverse pharmaceutical supply chain design considering perishability and sustainable development objectives. *Ann. Oper. Res.*, *340*, 2, 981–1033, 2024, https://doi.org/10.1007/s10479-024-05871-3.
14. Szkodny, A.C. and Lee, K.H., Biopharmaceutical manufacturing: Historical perspectives and future directions. *Annu. Rev. Chem. Biomol. Eng.*, *13*, 1, 141–165, 2022, https://doi.org/10.1146/annurev-chembioeng- 092220-125832.

15. Jayakrishnan, A., Wan Rosli, W.R., Tahir, A.R.M., Razak, F.S.A., Kee, P.E., Ng, H.S., Chew, Y.L., Lee, S.K., Ramasamy, M., Tan, C.S., Liew, K.B., Evolving paradigms of recombinant protein production in pharmaceutical industry: a rigorous review. *Sci*, *6*, 1, 9, 2024, https://doi.org/10.3390/sci6010009.
16. DiMasi, J.A., Grabowski, H.G., Hansen, R.W., Innovation in the pharmaceutical industry: new estimates of R&D costs. *J. Health Econ.*, 47, 20–33, 2016, https://doi.org/10.1016/j.jhealeco.2016.01.012.
17. Wouters, O.J., McKee, M., Luyten, J., Estimated research and development investment needed to bring a new medicine to market, 2009-2018. *Jama*, *323*, 9, 844–853, 2020, http://doi/:10.1001/jama.2020.1166.
18. Speeckaert, R., Lambert, J., Van Geel, N., Learning from success and failure: biologics for non-approved skin diseases. *Front. Immunol.*, *10*, 1918, 2019, https://doi.org/10.3389/fimmu.2019.01918.
19. Sarvepalli, S., Pasika, S.R., Verma, V., Thumma, A., Bolla, S., Nukala, P.K., Butreddy, A., Bolla, P.K., A Review on the Stability Challenges of Advanced Biologic Therapeutics. *Pharmaceutics*, *17*, 5, 550, 2025, https://doi.org/10.3390/pharmaceutics17050550.
20. Mitragotri, S., Burke, P.A., Langer, R., Overcoming the challenges in administering biopharmaceuticals: formulation and delivery strategies. *Nat. Rev. Drug Discovery*, *13*, 9, 655–672, 2014, https://doi.org/10.1038/nrd4363.
21. Pratt, K.P., Anti-Drug Antibodies: Emerging Approaches to Predict, Reduce or Reverse Biotherapeutic Immunogenicity. *Antibodies*, *7*, 2, 19, 2018, https://doi.org/10.3390/antib7020019.
22. Williams, M., Productivity shortfalls in drug discovery: contributions from the preclinical sciences? *J. Pharmacol. Exp. Ther.*, *336*, 1, 3–8, 2011, https://doi.org/10.1124/jpet.110.171751.
23. Jarvis, L., Novartis recalibrates its drug engine. *Chem. Eng. News*, 2018, https://doi.org/10.1021/CEN 09647-BUSCON1.
24. Holler, T.P., Evdokimov, A.G., Narasimhan, L., Structural biology approaches to antibacterial drug discovery. *Expert Opin. Drug Discovery*, *2*, 8, 1085–1101, 2007, https://doi.org/10.1517/17460441.2.8.1085.
25. Wu, J. and Cheng, C.W.C., Into the woods: a biologic patent thicket analysis. *Chi.-Kent J. Intell. Prop.*, *19*, 93, 2019.
26. Goode, R. and Chao, B., Biological patent thickets and delayed access to biosimilars, an American problem. *J. Law Biosci.*, *9*, 2, lsac022, 2022, https://doi.org/10.1093/jlb/lsac022.
27. Świerczyński, M. and Więckowski, Z., Biosimilars and the Patent Law. *Med. Law Soc.*, *12*, 2, 21–38, 2019, https://doi.org/10.18690/mls.12.2.21-38.2019.
28. Banner, M., Alosert, H., Spencer, C., Cheeks, M., Farid, S.S., Thomas, M., Goldrick, S., A decade in review: use of data analytics within the biopharmaceutical sector. *Curr. Opin. Chem. Eng.*, *34*, 100758, 2021, https://doi.org/10.1016/j.coche.2021.100758.
29. Wise, J., de Barron, A.G., Splendiani, A., Balali-Mood, B., Vasant, D., Little, E., Mellino, G., Harrow, I., Smith, I., Taubert, J., van Bochove, K.,

Implementation and relevance of FAIR data principles in biopharmaceutical R&D. *Drug Discovery Today*, *24*, 4, 933–938, 2019, https://doi.org/10.1016/j.drudis.2019.01.008.

30. Alosert, H., Savery, J., Rheaume, J., Cheeks, M., Turner, R., Spencer, C., S. Farid, S., Goldrick, S., Data integrity within the biopharmaceutical sector in the era of Industry 4.0. *Biotechnol. J.*, *17*, 6, 2100609, 2022, https://doi.org/10.1002/biot.202100609.
31. Gilbert, A., Huang, Y.M., Ryll, T., Identifying and eliminating cell culture process variability. *Pharm. Bioprocess.*, *2*, 6, 519–534, 2014, https://doi.org/10.4155/PBP.14.35.
32. Gamiz, A.G.R., Dewulf, J., De Soete, W., Heirman, B., Dahlin, P., Jurisch, C., Krebser, U., De Meester, S., Freeze drying in the biopharmaceutical industry: An environmental sustainability assessment. *Food Bioprod. Process.*, *117*, 213–223, 2019, https://doi.org/10.1016/J.FBP.2019.06.010.
33. Müller, K.M., Gempeler, M.R., Scheiwe, M.W., Zeugin, B.T., Quality assurance for biopharmaceuticals: An overview of regulations, methods and problems. *Pharm. Acta Helv.*, *71*, 6, 421–438, 1996, https://doi.org/10.1016/S0031-6865(96)00050-7.
34. Mozdzierz, N.J., Developing scalable and modular technologies for continuous biopharmaceutical production, Doctoral dissertation, Massachusetts Institute of Technology, 2018.
35. Capparella, J., Colucci, S., Conner, D., Dick, R., Weko, A., Addressing Sustainability in Biomanufacturing Facility Design, in: *Encyclopedia of Industrial Biotechnology: Bioprocess, Bioseparation, and Cell Technology*, pp. 1–22, John Wiley & Sons, Hoboken, NJ, USA, 2009, https://doi.org/10.1002/9780470054581.EIB667.
36. Renteria Gamiz, A.G., De Soete, W., Heirman, B., Dahlin, P., De Meester, S., Dewulf, J., Environmental sustainability assessment of the manufacturing process of a biological active pharmaceutical ingredient. *J. Chem. Technol. Biotechnol.*, *94*, 6, 1937–1944, 2019, https://doi.org/10.1002/JCTB.5975.
37. Ündey, C., Ertunç, S., Mistretta, T., Looze, B., Applied advanced process analytics in biopharmaceutical manufacturing: Challenges and prospects in real-time monitoring and control. *J. Process Control*, *20*, 9, 1009–1018, 2010, https://doi.org/10.1016/J.JPROCONT.2010.05.008.
38. Uppal, A., Chakrabarti, R., Chirmule, N., Rathore, A., Atouf, F., Biopharmaceutical industry capability building in India: report from a symposium. *J. Pharm. Innovation*, *17*, 4, 1555–1562, 2022, https://doi.org/10.1007/s12247-021-09596-9.
39. Declerck, P.J. and Simoens, S., A European perspective on the market accessibility of biosimilars. *Biosimilars*, 33–40, 2012, https://doi.org/10.2147/BS.S33524.
40. Ponzio, T.A., Feindt, H., Ferguson, S., License compliance issues for biopharmaceuticals: special challenges for negotiations between companies and

non-profit research institutions. *LES Nouv. Licensing Exec. Soc. (USA)*, *46*, 3, 216, 2011.

41. Baronikova, S., Purvis, J., Southam, E., Beeso, J., Panayi, A., Winchester, C., Commitments by the biopharmaceutical industry to clinical trial transparency: the evolving environment. *BMJ Evidence-Based Med.*, *24*, 5, 177–184, 2019, https://doi.org/10.1136/bmjebm-2018-111145.
42. Ghanemi, K. and Yan, S., Selected concepts on ethics in biopharmaceutical innovation. *MOJ Proteomics Bioinform.*, *6*, 4, 337–340, 2017, https://doi.org/10.15406/mojpb.2017.06.00204.
43. Burnham, T., Huang, S., Lo, A.W., Pricing for survival in the biopharma industry: a case study of Acthar Gel and Questcor Pharmaceuticals. Available at SSRN 3040369, https://doi.org/10.2139/ssrn.3040369, 2017.
44. Chatterjee, A., Dual use research of concern in the pharmaceutical industry (No. zmfb5_v1), Center for Open Science, https://doi.org/10.31235/osf.io/zmfb5, 2020.

7

Bio Manufacturing for Sustainable Production

Ashok Thulluru[1]*, Nawaz Mahammed[2], Reshma Tadipatri[2], Sundeep Kadasi[1], Gouri Sankar Kandukuri[3] and Anusha Sanampudi[3]

[1]Department of Pharmaceutics, D.Y. Patil College of Pharmacy, D.Y. Patil Education Society (Deemed to be University), Kolhapur, Maharashtra, India
[2]Department of Pharmaceutics, Raghavendra Institute of Pharmaceutical Education and Research (Autonomous), Anantapur, Andhra Pradesh, India
[3]Department of Pharmacognosy & Phytochemistry, Vasantidevi Patil Institute of Pharmacy, Kolhapur, Maharashtra, India

Abstract

Because biologics and biopharmaceuticals are needed more, bio-manufacturing has risen as the main method for producing future health therapies. This chapter investigates the development, concepts, and innovations powering sustainable bio-manufacturing and turning normal production into greener and more sustainable systems. Traditional manufacturing challenges are covered first, after which the text moves on to green chemistry, the circular bioeconomy, and using sustainability metrics when designing processes. New technologies such as single-use bioreactors, continuous manufacturing, modern cell culture systems, digital twins, synthetic biology, and AI being used in process control are being talked about. The whole process of manufacturing biologics, including the cell line development and processing for purification, is influenced by QbD and PAT, which help make the process more robust and maintained by regulatory rules. Examples of using emerging platforms for mRNA vaccines, biosimilars, and developing plant-based medicines show they are ready for practical use and help the environment. Scale-up problems, issues with unified regulations, and bioethical matters are discussed too. Lastly, the chapter points out future changes such as producing biomedical goods out-of-town, providing treatment as needed, and making healthcare systems strengthened against climate changes. Involving insights from several fields, the chapter functions as an all-inclusive resource for people interested in making biopharmaceutical production more sustainable and progressive.

**Corresponding author*: ashokthulluru@gmail.com

Sandip Kunar, Jagadeesha T. and D. Sathis Kumar (eds.) Biomanufacturing: Transforming Production of Biologics and Biopharmaceuticals, (163–194) © 2026 Scrivener Publishing LLC

Keywords: Sustainable bio-manufacturing, biologics, biopharmaceuticals, process intensification, synthetic biology, continuous manufacturing

7.1 Introduction

Biomanufacturing is the deployment of biological systems, including cells, enzymes, and biomolecules, in a sustainable way of producing biologics, as well as biopharmaceuticals. Its product range is between therapeutic proteins and vaccines, to industrial enzymes and bio-based products. Biomanufacturing is historical traceable to fermentation and recombinant DNA technology and has matured to cell engineering, synthetic biology, and automation. In the present day, it is the key technology in the modern pharmaceutical production that has made possible the creation of scalable, high-quality, precision-oriented manufacturing. The growth in the global consumption of biologics is a promising platform in bio-manufacturing, which aims at providing environmentally friendly, flexible and optimized next-generation healthcare and industrial biotechnology.

7.1.1 Definition and Scope of Bio-Manufacturing

Bio-manufacturing is the exploitation of living cells, enzymes, or biological systems to manufacture commercially valuable products like therapeutics, vaccines, chemicals and materials. In essence, bio-manufacturing aims at using biotechnology, bioprocess engineering and systems biology to process renewable resources into bio-based products in a predictable and scalable way. Bio-manufacturing differs from regular chemical production mainly because it is biologically specific, green, and clean, whereas most traditional chemical manufacturing processes are fossil-fuel dependent and exceedingly high-synthesis [1].

Recent years have prompted growing in terms of bio-manufacturing as a result of talents in synthetic biotechnology, metabolic engineering and digitized bioprocessing. Examples are in the pharmaceutical (e.g., monoclonal antibodies, cell and gene therapies), industrial enzymes, biofuels, agricultural bioproducts, and now even space exploration, where biomanufacturing systems are being designed to go to another planet [2]. Additionally, the implementation of such technologies as artificial intelligence, machine learning or digital twins into the control of bioprocesses has changed the predictability and flexibility of the bio-manufacturing process.

Notably, the emergence of personalized therapies and pressure to utilize modern vaccine development, as demonstrated in the COVID-19 pandemic, has precipitated adoption of a continuum-based manufacturing process: (1) the shift of batch towards continuous and (2) modular production systems. This transition also calls attention to the requirement of sophisticated regulatory and quality systems to go along with dynamic and small-scale as well as on demand bioproduction systems.

Bio-manufacturing is a revolutionary concept, both in the industrial biotechnology and in global sustainability. Interdisciplinary cooperation and regulatory convergence will be important to see this sector fully reach its potential as it is evolving ever further.

7.1.2 Historical Perspective and Evolution of Bio-Manufacturing

The practice of bio-manufacturing dates back to an early form of biological fermentation of bread, beer and yoghurt, an early attempt by mankind at exploiting biological systems to produce something. Nevertheless, the biotechnological use in industry was not consolidated until the 20th century and specifically in World War II when penicillin was produced in mass quantities. Further decades saw the invention of attempted to develop suitable recombinant DNA technology and cell culture systems which led to developing bio-manufacturing to become a stakeholder of modern biopharmaceutical production.

Bio-manufacturing has developed along an industry status-quo that has exerted market demand in addition to technology. The industry has transitioned through traditional batch processing into both continuous manufacturing and automation and data driven decision making. With the advent of synthetic biology, the field of biologics has widened and now one can design engineered organisms that are able to produce complex therapeutics proteins, enzymes and even bio-based materials that are renewable [3].

Current advances have now placed bio-manufacturing even off-Earth where they work on systems in space concerning integrating biological manufacture in conjunction with the insulating environment during microgravity. This is an indication of wider repetition not only in the processor platform and the type of products, but also in environmental backgrounds [4]. What is more, integration of digital twins, AI-driven bioprocess design, and modular bioreactors systems offer the latest frontier in turning bio-manufacturing into a flexible and scalable platform that responds to global health, social sustainability, and resource pressures.

In essence, bio-manufacturing has matured from a niche technological offshoot to a strategic pillar of global industry driven by biology, empowered by engineering, and evolving through digital innovation.

7.1.3 Importance of Bio-Manufacturing in Modern Pharmaceutical and Biologics Production

As modern pharmaceutical and biologics production rely increasingly on bio-manufacturing techniques, complex and high value therapeutics are being produced that cannot easily or at all be produced by traditional chemical synthesis. The evolution center of this evolution includes the living cell (bacterial, yeast or mammalian cells) as a platform to manufacture monoclonal antibodies, vaccines, hormones or gene therapy. They are biologics that have transformed how patients are treated of cancer, autoimmune diseases and rare genetic disorders, providing high specificity and individual choices to patients [5].

Modern bio-manufacturing systems can yield quality and regulatory controls that will be robust due to their scalability and precision. Single-use bioreactors, continuous processing, and other tools to monitor the process in real-time smoothed the production timelines and decreased the risk of contaminations, and the impact on the environment. Besides this, synthetic biology and metabolic engineering has also led to a customization of host cell productivity and minimization of by-products.

In these post-pandemic days and ages, bio-manufacturing has also proven critical in the development of speedy vaccination and universal health preparedness. Rapid development and scale up of mRNA-based COVID-19 vaccines is a historic use of agility B2M-based platform to transform genome-scale information into field-deployable therapeutics in months [6].

As the demand for biologics grows, bio-manufacturing will continue to evolve, embracing automation, artificial intelligence, and decentralized production hubs. These innovations aim to make biologics more accessible, affordable, and adaptable to emerging health needs across the globe.

7.2 Sustainable Bio-Manufacturing (Principles and Frameworks)

Sustainable bio-manufacturing is a combination of green chemistry and the circular bioeconomy; since when combined; they become more environmentally friendly and resource efficient. The key aspect of this practice is to design waste-reduction process, minimization of energy consumption and toxic substance inputs. Regulatory structures and international sustainability strategies, like UN SDGs, are an effective catalyst as they guide the industry operations to face long-term ecological steadiness. In a strong model, essential sustainability solutions such as carbon intensity, water footprint, and lifecycle studies are taken into consideration to gauge and benchmark the environmental performance. Collectively, the principles will facilitate a paradigm change towards economically viable, socially responsible, and environmentally resilient bio-based manufacturing systems.

7.2.1 Green Chemistry and Circular Bioeconomy

An emerging trend entailed in sustainable bio-manufacturing is to combine principles of green chemistry with circular bioeconomy strategies to develop environmentally and economically sustainable production systems. It tries to reduce ecological sequels and celebrate biological and renewable capitals. In essence the whole idea behind sustainable bio-manufacturing is to get rid of harmful chemicals and energy-requiring production and feedstock that are not renewable and instead use biocatalytic systems, cell-based systems and wastes as the replacement [7].

This transition is supported by green chemistry which encompasses the use of atom economy, the usage of safer solvents and energy-based bioprocessing. In combination with these the circular bioeconomy promotes closed-loop resource exploitation that is, recycling waste into value-added products, *via* microbial fermentation, enzymatic conversion, or biorefinery strategies. Collectively these frameworks position bio-manufacturing next to global sustainability, sustain targets aligned with decarbonization, zero waste, and industrial symbiosis targets.

Advances in synthetic biology and systems engineering have in the recent past widened the toolbox of sustainable bio-manufacturing, allowing the precision design of microbes as well as process pathway. Modular bioprocesses are being designed that will valorize agriculture byproducts into bioplastics, biofuels and fine chemicals, with much reduced lifecycle emissions [8]. Moreover, incorporation of the techno-economic and the life cycle assessments at the initial development stage makes sure that the trade-offs on the environment will be dealt with in a sufficient manner.

However, other issues such as scalability, feedstock variability, and regulatory harmonization are also stumbling blocks, despite the growth made. These will need to be handled through interdisciplinary research and policy to bring sustainable bio-manufacturing to a level of a standard industrial practice rather than its current status as a niche innovation.

7.2.2 Regulatory Drivers and Global Sustainability Goals

The science of sustainable bio-manufacturing is directly connected with the global sustainability campaigns and regulatory moves. Since the commitment by governments across the globe intensifies on the United Nations Sustainable Development Goals (SDGs), bio-manufacturing has become the main tool to shift to bio-based economies after the fossil-based economy. There is a growing trend to develop regulatory frameworks that promote low-carbon technologies, circular bioeconomy and risk-informed production. Such frameworks not only outline compliance procedures but also build life-enabling conditions of the innovation and investments in sustainable biotechnologies [9].

Specifically, such territories as the European Union have led to such intense approaches to sustainability assessment as the Green Deal and the Bioeconomy Strategy, where carbon footprint, circularity, and resource efficiency standards become integrated into injections and regulatory accessibility. Such policies are getting reflected around the entire world, creating a precedence of homogenized standards of sustainability. In this respect, sustainable bio-manufacturing is a technological and policy-based answer to climate change, biodiversity extinctions, and manufacturing releases [10]. The data-based practices, such as life cycle assessments and digital traceability, are also being encouraged by the regulatory bodies that consider the environmental responsibility. The development of safe-by-design methodologies and adaptive licensing pathways that allow safe and responsible innovation to thrive at the pace dictated by changing needs is facilitated by collaborative platforms between regulators, industry and academia.

In conclusion, regulatory impetus and international sustainable objectives are not marginal but are part and parcel of the formation of a sustainable bio-manufacturing. They bring the order, motivation and coherence required to move towards green, resilient and inclusive industry systems.

7.2.3 Key Metrics for Evaluating Sustainability

The approach to sustainable bio-manufacturing is based on the possibility to assess environmental, economic and social performance over the all-product life cycle. Quantitative measures are critical towards making decision-making, establishing priorities of process ameliorations, and aligning them with sustainability objectives. Carbon footprint, energy intensity, water consumption, and wastes are among the most common indicators. This is complemented by more general studies, such as life cycle assessment (LCA) and techno-economic analysis (TEA), to gain system level understanding of what synergies and trade-offs of sustainable production tracks [11].

New digital systems are improving sustainability analysis by collecting data in real-time, making forecasts, and optimizing with the help of artificial intelligence. Today, smart manufacturing platforms incorporate sustainable indicators directly to bio-manufacturing processes allowing them to monitor continuous energy, raw material efficiency, and emission. As an example: carbon footprint per unit of product, the degree of re-using water and the level of yield per unit of biocatalyst are becoming the regular parameters of reference both in the pharmaceutical and in the industrial bioprocesses [12].

Moreover, socio-economic indicators, which refer to job creation, community effect, and chain resilience, are gaining popularity in the sustainability scoring regime. They allow taking a rounded view to go beyond environmental indicators and guarantee that sustainable bio-manufacturing helps address more ambitious targets, such as the UN Sustainable Development Goals.

Therefore, it is still important to define and standardize key metrics to scale sustainable bio-manufacturing. An agreed assessment system will assist an industry and regulators to develop.

7.3 Platforms and Technologies in Bio-Manufacturing

Advanced cell culture technologies have been developed as bio-manufacturing platforms and these include CHO cells, HEK293, microbial and yeast, each with their own special opportunities to manufacture biopharmaceuticals and industrial enzymes. Single-use bioreactors and modular systems, by increasing process flexibility, lowering the risk of contamination and speeding development have given rise to shorter development times. At the same time, a shift towards continuous manufacturing and process intensification methods is changing the nature of production efficiency to allow the production of a higher yield and consistent quality of products. These combined technologies all contribute to scalable, economically feasible, and sustainable bio-manufacturing and, because they do, they are all perfectly suited to the current requirements of the industry to be agile and responsive to the bio-manufacturing production strategies as the global demands on bio-products continue to grow.

7.3.1 Cell Culture Systems (CHO, HEK293, Microbial, Yeast)

Bio-manufacturing prescribes the use of cell cultures systems, which have allowed the scale of production of biologics, vaccines, and recombinant proteins to produce in the past. Different platforms [Chinese Hamster Ovary (CHO) cells, HEK293 cells, microbial systems, and yeast] have different advantages, which depend on the type of target molecule, regulation and scalability requirements.

CHO cells: They are able to do post-translational modification related to human cells and thus the ability to be an industry standard in the production of complex glycosylated proteins, e.g., monoclonal antibodies and therapeutic enzymes. Improved fed-batch and perfusion systems have maximized CHO productivity as well as product quality [13].
HEK293 cells: Human embryonic kidney-derived, these highly popular cells are most commonly used to generate viral vectors and to express proteins of interest in transient expression systems and both viral and vaccine production are favored due to their ease of transfection and rapid growth.
Microbial systems: Escherichia is especially preferred because it is low cost and has a rapid growth in the production of simple and non-glycosylated proteins. Still, the task of endotoxin removal and folding of proteins is technical one.
Yeast: Especially *Pichia pastoris* and *Saccharomyces cerevisiae*, offers a eukaryotic alternative to ease of scaling post-translational partners yet still offers genetic flexibility [14].

The selection of the best expression system compromises between cost, the complexity of the product, and regulatory route. To increase yield, fidelity of a product and sustainability, hosts are increasingly being engineered using hybrid systems and synthetic biology.

7.3.2 Single-Use Bioreactors and Modular Systems

Single use bio reactors (SUB) and modular manufacturing systems have been a paradigm shift in use of bio-manufacturing platforms. SUBs should be made of polymer, be disposable, pre-sterilized although there have been some studies on how to pre-load them with medium and install a liner. It is currently growing fast in clinical and large-scale production in terms of adoption because of lower capital costs, energy rates, as well as clean-in-place (CIP) needs [15].

Modular systems, in their turn, allow the standardization of any facility by using pre-engineered modular units that can be deployed, scaled, or repurposed quickly. All these platforms are very flexible and meet the advancing need of decentralized and agile manufacturing, which is important in COVID-19 preparedness and new market contact. Single-use and modular technology can be added together to enable the assembling of end-to-end bioproduction pipelines on a plug-and-play basis, which significantly reduces the time to commission a facility.

SUB inventions have come up with newer mixing systems, built in sensor(s), newer film chemistries which are resistant to inorganic chemicals and permeable to oxygen. Modular systems have now been integrated with automation, real-time monitoring, and digital twin structures to help make data-rich decisions in agreement with bioprocessing 4.0 principles [16].

Along with these advantages, there are still challenges associated with plastic waste, the ability to scale large-volume biologics, and continuous compatibility. However current advancements are addressing the problems using recyclable materials and hybrid systems.

In combination, SUBs and modular systems are transforming biomanufacturing, providing a scalable, flexible, and counterbalanced alternative to customary stainless-steel plants.

7.3.3 Continuous Manufacturing and Process Intensification

By increasing productivity, minimizing footprint and providing real-time quality control, the continuous manufacturing and process intensification are redefining the pages of the bio-manufacturing history. Continuous bioprocessing enables 24-hour uninterrupted production, which provides process advantages and enhanced up-scaling, unlike older batch processing. This is particularly important in high-demand biosimilar biologic products like monoclonal antibodies, vaccines and therapeutic proteins, where speed-to-market and cost-effective are of crucial importance.

The complement to this change is process intensification, which combines two or more-unit operations (e.g. cell culture, purification, formulation) into small, modularized schemes that save space, energy, and materials. Techniques which are widely used to intensify both upstream and downstream processes are high cell density reactors, perfusion culture, and membrane-based separation techniques to keep the product consistency [17].

The digitalization of these systems additionally provides power by having real-time monitoring, Smart control algorithms, as well as digital twins resulting in predictive analytics and adaptive control. The regulatory bodies such as FDA and EMA have grown in favor of continuous manufacturing utilizing quality by design (QbD) principles and have been promoting its practice into the biopharma industry.

Although the technological maturity and initial costs are some of the challenges, every advantage (reduced cost of production, reduced batch-to-batch variation and sustainability of natural resources) in the long run is driving the global interest. New platform-based technologies and continuous flow processes in combination with single-use platforms are emerging hybrid, intensified bio-manufacturing ushering a new era of low-cost, high-conversion, high-yield systems [18].

7.4 Biologics and Biopharmaceuticals: Manufacturing Workflow

Biologics and biopharmaceutical manufacturing processes are a tightly coupled process that entails an upstream manufacturing procedure followed by a downstream manufacturing procedure. The upstream processing starts in cell line development optimization wherein high yielding and stable clones are picked to be robustly expressed. Then there is downstream processing, which is more concerned with the efficient purification and formulation process so that the integrity, safety and efficacy of the product is achieved. The abreast quality programs such as Quality by Design (QbD) and Process Analytical Technologies (PAT) are getting integrated into the workflow so that real-time monitoring and control is possible. Such strategies provide consistency, minimize variability and help to comply with regulations leading to increased integrity of contemporary bio-manufacturing activities.

7.4.1 Upstream Processing: Cell Line Development and Optimization

Development of cell lines (CLD) is an upstream processing pillar in the biopharmaceutical manufacturing process, which determines process efficiency, output, and quality of recombinant protein production. It starts by transfecting host cells such as Chinese Hamster Ovary (CHO) cells with the gene of the desired biologic, which is then selected, amplified and screened against high producer's clones. These clones are optimized to provide a sturdy expression, stable growth and gain the required post-translational modifications [19].

Newer CLD approaches have been developed to use high-level molecular technologies including location specific integration, CRISPR/Cas9 genome editing, high throughput screening systems. They are faster, and these tools enhance production consistency in clone selection. The predictive modeling of productivity and quality attributes beyond the cell line, and the culture environment, is also possible in combination with multivariate data analysis and designed experiments (DoE) to optimize the cell line by itself and the cell line within the environment [20].

Stability testing in long passages and scalability to bioreactor environment are important in regulatory compliance and commercial scale. Furthermore, automation and digitalization, such as robotic cell picking and machine learning-assisted analytics, are changing the conventional CLD pipelines into fast and reproducible pipeline.

All this notwithstanding, there are still challenges still faced especially in cutting short on time and cost of early stages of development. Further perfection of the host cell engineering, screening sensitivity, and media formulation are currently the active fields of research.

7.4.2 Purification and Formulation

A requisite step in the biomanufacturing process flow, which incorporates biologics main purification and formulation into finished therapeutics, is downstream processing (DSP). The main objective of DSP is to purify the target product, which in most cases is a protein, antibody, or other biomolecule, with high purity, preserved biological activity, and regulatory compliance, while removing host cell proteins (HCPs), DNA, endotoxins, and other process-related impurities.

Initial purification may simply involve clarification, but often a train of chromatography follows, most usually protein affinity chromatography of monoclonal antibodies, and ion exchange or hydrophobic interaction chromatography, when polishing. These procedures guarantee high product yield and repeatability as well as satisfying high purity standards. The new development of analytical technologies including LC-MS profiling is opening new opportunities of accurate identification of HCPs and improved approach to their clearance particularly in plant-based and alternative expression platforms [21].

The last phase of DSP, formulation, aims to stabilize the biologics to store and deliver. This includes buffer optimization, incorporation of excipients, and choosing the final dosage form (e.g., a lyophilized powder or pre-filled syringe). The growing complexity of biologics like mRNA vaccines and gene therapies require formulation approaches that also consider cold chain logistics, stability and the efficiency of delivery [22].

As demand for rapid, scalable biomanufacturing grows, integrated continuous DSP and single-use technologies are being adopted to improve throughput and flexibility without compromising product quality.

7.4.3 Quality by Design (QbD) and Process Analytical Technologies (PAT)

Process Analytical Technologies and Quality by Design are the key regulations in the new environment of the biopharmaceutical industry production. QbD is a structured developmental approach starting with predetermined objectives and focusing on a product understanding, process understanding, risk management and selection of control strategy. It combines the critical quality attributes (CQAs), process parameters and design space in an organized format, which makes its production of biologics predictable and robust [23].

PAT, a major facilitator of QbD offers tools to enable real-time observation, as well as adjustment of essential process conditions. It also uses spectroscopy, chromatography and chemometrics to test and verify the quality of the products on the production line as opposed to only testing at the end of production. PAT increases transparency of processes, reduces process deviation, and enables continuous manufacturing by providing in-line and at-line data to enable the use of adaptive control systems [24].

Collectively, QbD and PAT are the pillar of the regulatory effort to deliver equal product performance. The regulators, such as the FDA and the EMA promote such practices through their guidelines, including the ICH Q8-Q11, which focuses on science- and risk-based practices. The integration is also helping in digital conversion in bioprocessing such as model-based control, digital twins, and real-time release testing (RTRT).

With biologics increasingly approaching a more complex and personalized state, QbD and PAT will continue to play critical roles as far as regulatory compliance, process effectiveness, and reduced time-to-market are concerned. Figure 7.1 illustrates the flowchart representing the integration of QbD and PAT into both upstream and downstream workflows.

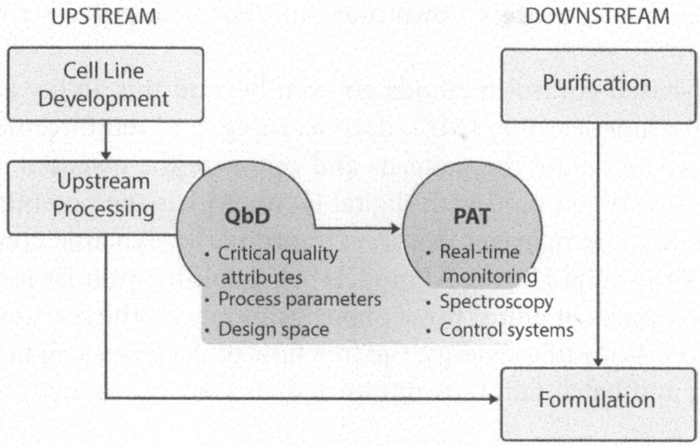

Figure 7.1 Flowchart representing the integration of QbD and PAT into both upstream and downstream workflows.

7.5 Innovations Transforming Bio-Manufacturing

New technology is disruptively changing bio-manufacturing in accuracy, efficiency, and flexibility. Bio process simulation and optimization through digital twins and AI enabled process control allows real time bioprocess monitoring and optimization, and shorter times to development and consistent product quality. 3D bio-printing and microfluidic bio-factories present scaled and scalable platforms that provide high-throughput experiments with minimal resources usage, including tissue engineering and personalized medicine. In parallel, recent developments in synthetic biology and gene editing with CRISPR enable the creation of bespoke organisms and metabolic networks to produce bespoke bioproducts. Collectively these technologies are leading to a paradigm shift to smart, adaptive and very efficient bio-manufacturing ecosystems.

7.5.1 Digital Twins and AI-Driven Process Control

Bio-manufacturing the next frontier in innovation in bio makes use of detailed digital twins and artificial intelligence. A digital twin is a virtual, dynamically updated model of a physical process or system, using sensor data, process models, and computational intelligence. In bio-manufacturing, digital twins are used to simulate, optimize and monitor complex bioprocesses, e.g., fermentation, cell culture processes and purification. Quality improvement (These virtual environments can extrapolate deviations, propose control adjustments and simulate what-if scenarios and hence reduce process downtimes and enhance process robustness) [25].

The AI-based control methods go even beyond this ability as the features of machine learning (ML), deep learning, and reinforcement learning are used to define the patterns and optimize the essential aspects of the processes. When used with digital twins, AI has the potential to offer flexible, automatic management in extraordinarily dynamic contexts. An example of this might be the AI models being able to optimize feed rates in perfusion cultures or adjust flows of purification with the real-time quality information. With this synergy, the free flow of decisions can be made on hindsight, and batch failure is minimized, as well as product consistency [26].

Notwithstanding these strengths, its broad implementation is threatened by aspects such as high cost of entry, data standardization problem and regulatory constraints.

However, effective pilot reports in both upstream and downstream modules demonstrate that the digital twinning of AI approaches in their combination can significantly minimize both the development cycle and the costs associated with their working process and compliance.

7.5.2 3D Bio-Printing and Microfluidic Bio-Factories

An advanced innovation is 3D bio-printing and microfluidic bio-factories, which promises to transform sustainable and precision bio-manufacturing due to their ability to spatially deposit bioinks [comprising cells, biomaterials, and bioactive compounds] into complex and functional tissue-like constructs. These advances in technology are used in regenerative medicine, organoids and high-throughput drug screening. It provides greater cellular architecture and heterogeneity control than has ever been possible before and enables more physiologically relevant bioprocess environments [27].

Supplementing this is the entry of microfluidic bio-factories that gut and microfabricate complete bio-manufacturing pipelines onto chip-based systems. The electronic devices have the capability to manipulate small quantities of fluids at the micro scale allowing continuous, automated scale-able reactions to produce biologics with less wastes and reagent requirements. They are modular, which enables them to have several samples run in parallel, and thus they are applicable in individualized medicine and on-demand drug manufacturing.

A combination of these technologies is the basis of the shift to decentralized flexible and adaptive bio-manufacturing systems. Developed microfluidic systems by 3D printing have also become integrated, thus increasing the level of customization of lab-on-chip analysis, fermentation, and *in vitro* tissue modeling [28].

But translation beyond clinical or industrial scale requires the addressing of challenges which include, but are not limited to, bioink standardization, cell viability, and fluidic system robustness. Future work will include integrating AI with bio-printing and microfluidics in the future in order to improve real-time design optimization, error identification, and adaptive manufacture.

7.5.3 Synthetic Biology and Gene Editing Tools

The promises of bio-manufacturing are revolutionizing as synthetic biology and gene editing technologies can be used to carry out fine-grained engineering of biological systems, tailored to optimized delivery of therapeutics and industrial production. Synthetic biology uses chassis engineering, modular biologic components, metabolic rewiring to design and build organisms whose properties can be programmed. Biosynthesis of such devices elevated yield, purity, and efficiency in biosynthesis pathways in the production of biologics, enzymes and vaccines [29].

Gene editing Genome editing: Specifically, CRISPR-Cas systems allow targeted and multiplexed gene editing. These tools can enable researchers to engineer, shut off, or switch genes with a lot of high precision, and this has sped up the various production of bespoke cell factories to use in bio-manufacturing. One example is CRISPR/Cas12a has been used in microbial hosts to perform genome-wide screens to optimize amino acid and biofuel production pathways while preserving host growth [30].

The combination of machine learning and automation with synthetic biology is also helping to speed those same efforts by automating and accelerating the Design-Build-Test-Learn (DBTL) cycle to allow predictive pathway design and adaptive strain evolution. Simultaneously, biosafety systems and biocontainment approaches are emerging in an attempt to promote safe application of engineered microbes in industries.

Nevertheless, the issues that are not to be overlooked are off target editing, metabolic overload due to overexpression and the regulation of genetically modified organisms. The ideal steps towards fully programmable, scalable bio-manufacturing platforms lie in further innovation of synthetic biology toolkits, availability of standardized parts libraries and computational biology.

7.6 Case Studies in Sustainable Production

Innovative examples used to demonstrate the sustainability of the production in the bio-manufacturing are related to case studies in terms of adaptability of processes, efficiency, and decreased pollution. The creation of the mRNA vaccines is an example of the fast, scalable solutions to worldly pandemics, focusing on speed and accuracy. The advent of biosimilars leads to the transition to cost-effective treatment, making it available to more people without a decrease in effectiveness and safety. The new plant-based and cell-free systems are flexible and low-resource solutions that can be used to simplify the production of biologics without necessarily relying on the long-used cell culture method and lowering the associated ecological footprint. The described case studies represent various approaches to sustainable bio-production, which demonstrates the potential of the high technology to address the macro health needs of the population without conflict with sustainability targets.

7.6.1 mRNA Vaccines and Rapid Pandemic Response

The worldwide deployment of mRNA vaccines in the context of COVID-19 crisis created a paradigm shift towards sustainable bio-manufacturing. These vaccines showed how the synthetic biology, digitalized design and modular production systems could react quickly and efficiently to a worldwide health emergency. mRNA-based vaccines are not resource intensive or time consuming as traditional vaccine platforms operate without the need of cell culture or propagation on egg [31].

Single use bioreactors, continuous manufacturing and cold chain-optimized formulations were among the strategies used by biopharmaceutical companies to increase production capacity quickly, managing to limit production waste and contamination risk. Besides, mRNA vaccine manufacturing implies enzyme-based synthesis and drop by drop encapsulation of a lipid and nanoparticles, which are modular and scale by definition. This flexibility in manufacturing was critical towards fulfilling the worldwide urgent demand especially when variants of concern later developed.

Sustainability of mRNA platforms the next important characteristic of mRNA platform sustainability is its adaptability. With a delivery mechanism in hand, genetic sequences are easy to update, and vaccines can be re-engineered rapidly against pathogens that change. Also, less central manufacturing strategies based on modular platforms of mRNA manufacturing-in-a-box are now under testing to deploy in low-resource environments and further improve resilience and equity [32].

This case illustrates how innovations in biomanufacturing, driven by digital tools and scalable technologies, can align with sustainability goals while ensuring rapid, global pandemic responsiveness.

7.6.2 Biosimilars and Cost-Efficient Therapeutics

Biosimilars which are very similar to the approved biologics present cost-effective and sustainable therapy. Their build-up relies on pre-existing data thus reducing costs and enhancing accessibility, which is particularly high in low-income area [33]. Innovation in enhancing bioprocessing such as perfusion and continuous process improves productivity and has minimized environmental pollution. Biosimilar products are helpful to have healthy, distributed global bio-manufacturing, particularly despite regulatory and IP difficulties [34]. Table 7.1 presents a comparative analysis of key metrics between biosimilars and innovator biologics.

7.6.3 Plant-Based and Cell-Free Manufacturing Systems

Plant-based and cell-free systems offer sustainable, low-waste alternatives to conventional biologics. Transgenic plants like *Nicotiana benthamiana* enable scalable vaccine production [35], while cell-free synthesis supports rapid, on-demand protein generation and lyophilization for field use [36]. Integrated with microfluidics, both face yield and cost challenges but promise decentralized manufacturing. Table 7.2 presents a schematic comparison between plant-based and cell-free expression systems.

7.7 Challenges and Regulatory Considerations

Bio-manufacturing is subject to severe challenges to be deployed globally. The transfer of technology and scalability must continue to be major challenges, and it must be initiated with solid plans so that there is continuity in terms of performance across production locations. The world of regulation compliance is that of aligning with various regional regulatory templates, and thereby a case is made out to harmonize internationally in order to reduce time-consuming approval processes and allow innovations. Environmental and bioethical issues, i.e., waste management, resource usage and genetic modification, need to be considered as well to allow a responsible development. Achieving a balance between these technical,

Table 7.1 Comparative analysis of key metrics between biosimilars and innovator biologics.

Metric	Biosimilars	Innovator biologics	Ref.
Development Cost	Lower – Leverage existing clinical data from originator biologics	High – Require full clinical development pipeline	[33]
Market Price	Significantly reduced, improves affordability and access	High, limiting access in low- and middle-income regions	
Manufacturing Efficiency	Enhanced *via* continuous processing, single-use systems, and intensification	Traditional batch processes, higher energy and resource demand	
Environmental Impact	Reduced water and energy use; smaller footprint with intensified processes	Larger resource inputs and waste generation	
Technology Transfer & Local Production	Promotes decentralized, resilient models in emerging markets	Centralized production, limited adaptability in low-resource settings	[34]
Regulatory Complexity	Moderate – Requires comparability and pharmacovigilance	High – Full regulatory pathway	
Sustainability Profile	High – Supports circular bioeconomy and equitable access	Moderate – Limited scalability and environmental optimization	[33, 34]

Table 7.2 Schematic comparison between plant-based and cell-free expression systems.

Metric	Plant-based systems	Cell-free systems	Ref.
Production Platform	Transgenic plants (e.g., *Nicotiana benthamiana*)	Cell-free lysate-based systems	[35, 36]
Scalability	High – Field or greenhouse-based mass production	Moderate – Suited for small-batch or modular scale	
Production Speed	Moderate – Growth cycle dependent (typically weeks)	High – Rapid synthesis in hours	
Contamination Risk	Low – Free from human pathogens	Very Low – No living cells, fully controlled systems	
Energy and Resource Use	Low – Minimal infrastructure, reduced energy inputs	Moderate – Requires reagent preparation and cell extract processing	
Cost	Low – Inexpensive substrates, scalable without costly infrastructure	Moderate to High – Reagent and lysate production cost significant	
Product Type Compatibility	Vaccines, antibodies, enzymes	Toxic proteins, unstable molecules, rapid prototyping	
Portability & On-Demand Use	Limited – Dependent on growth infrastructure	High – Lyophilized, cold-chain-independent, suited for field deployment	
Current Limitations	Low yields, slower expression, vector optimization needed	High reagent costs, scalability challenges	

legal and ethical aspects is a crucial part of future-proof growth and forms the regulatory environment that promotes environmentally friendly, equitable and safe bio-manufacturing processes.

7.7.1 Scalability and Tech Transfer

Scaling and transfer of technology are among the main bottlenecks regarding the worldwide implementation of the sustainable biomanufacturing solution. Scaling know-how revealed in the laboratory into the production environment is itself a perilous technical, economic, and regulatory endeavor. The thing is that scaling-up is not only about volumetric expansion and adaptation to various reactor configurations, process dynamics, and control systems, which all may change the quality, yield, and regulatory compliance requirements of products [37].

The technology transfer is often complicated in biosimilar and vaccine manufacturing where the critical quality attributes need to be consistent across manufacturing locations. Such variations in the raw material substrata, plant facility, and personnel skills may result in variations that make regulatory filings difficult. The FDA, EMA, and other regulatory agencies focus on Quality by Design (QbD), process validation, and risk-based methodology to facilitate the uncomplicated and regulatory-compliant tech transfer. Moreover, the sharing of knowledge barriers is a problem because of intellectual property issues and the lack of standardization of analysis techniques among innovators and process manufacturers in new markets [38].

There has been recent activity in the areas of digital documents, coupled with simulation of the processes with digital twins, as well as the adoption of harmonized international recommendations to dec-risk tech transfer. These solutions, however, are costly to implement and capacity-building especially in low-resource context. The successful tech transfer therefore requires intersectoral cooperation, harmonization of regulations, and training workforce to fit local manufacturing ecosystems.

7.7.2 Regulatory Compliance and Global Harmonization

The sustainability of bio-manufacturing involves the central regulatory-compliance and global harmonization. Due to attributes of complexity and globalization, biologics and advanced therapies require regulatory provisions in diverse jurisdictions that are not only inconsistent but also disrupt the product development process and commercial access to international markets. Redundancy in the quality standards, data reporting formats, and structure and post-approval changes causes unnecessary effort, increased expenditure, and slow access of patients.

To overcome this, programs like International Council for Harmonization (ICH), Good Regulatory Practices of WHO, and other upcoming framework like MACRAM will harmonize both the safety rules as well as efficacy and quality guidelines. The work aims at risk-based approaches, lifecycle management and harmonized analytical techniques to achieve mutual recognition of regulatory agencies [39].

Digitization is also being done to the regulatory science. Proactive supervision is made possible through cloud-based submissions, real time data monitoring and digital twins and can be employed to accelerate regulatory assessments and facilitate adaptive license designs. However, the restriction emerges through information standardization, cyber-security and digital equality in regional contexts [40].

There is also the fact that local regulatory developments are necessary, and, at the same time, consistency should be sought with international regulatory developments with regard to the approval of biosimilars and tech transfers in emerging markets. To this effect, teamwork in regulatory science, capacity development and international platforms are needed to enhance trust, openness and reciprocity in compliance mechanisms.

The concept of global harmonization not only increases the efficiency of manufacturing and safety of patients, but also helps to achieve fair access to biopharmaceuticals, bringing the oversight regulations into line with the global health and sustainability objectives.

7.7.3 Environmental and Bioethical Concerns in Biomanufacturing

Increasing popularity of synthetic biology and genetic engineering in biomanufacturing has increased the questions on its environmental and bioethical regulation. The issues of biosafety, ecological imbalance and what the limits of production are about become crucial as more and more novel organisms and cell systems are utilized in the production. Possible environmental hazards include uncontrolled leakage of genetically modified organisms (GMOs), build-up of disposal products of metabolism, and high dependence on high-energy feedstock, which should be omitted by sound risk management and control measures [41].

In a bioethical perspective, cell use based on human sources, gene editing technologies, and data-driven biological design offer a point of discussion of the moral limits and consent, and equality. Such concerns are especially eminent in therapeutic bio-manufacturing and synthetic biology where intervention can be inter-specific or natural evolutionary boundaries. In addition, unequal availability of high-income or low-income markets in the implementation of better bio-manufacturing platforms implies distributive ethical justice concerns.

The Cartagena Protocol on Biosafety, global bioethics declaration from UNESCO and other frameworks are supposed to mitigate these reservations, although their implementation is mixed. The latest ideas in the field recommend making a mandatory conscious effort during the designing of a bioprocess to include environmental and ethical reviews of that process i.e., ethics-by-design or sustainability-by-design [42].

The interdisciplinary structures of management and open communication with stakeholders are hard to go around to balance innovation and responsibility. This qualifies bio-manufacturing to develop not only efficiently, but also fairly and sustainably within ecological and social boundaries.

7.8 Future Directions and Emerging Trends

Transformative technologies and adaptive production are rewriting the future of bio-manufacturing. Real-time monitoring and anticipatory control as well as the improved effectiveness of operations are possible due to the integration of automation and smart manufacturing. At the same time, the idea of decentralized and on-demand production is gaining popularity, providing fast and local reactions to issues on public health and supply chains. Moreover, the bio-manufacturing will also be an essential part of climate resilience, including the use of sustainable materials, carbon capture innovations, and sustainable processes. These are new trends that indicate that the industry may be reduced to smart, dynamic, and climate-friendly systems, and bio-manufacturing will be one of the pillars of a sustainable future of industrial manufacturing.

7.8.1 Integration of Automation and Smart Manufacturing

The integration of automation and smart manufacturing technologies is reshaping the future of bio-manufacturing, making it more efficient, adaptive, and responsive. Automation minimizes manual intervention, reduces human error, and enhances reproducibility, while smart manufacturing introduces real-time decision-making through digital twins, machine learning algorithms, and data-rich environments.

Sensor and control in biopharmaceutical manufacturing, closed-loop automation is in use even in cell culture, purification, and formulation, with continuous optimization of the process parameters being performed by sensors and control algorithms. These systems offer real-time observation of key quality attributes (CQAs), and corrective activity may be taken in time, resulting in a reduction in the number of batches that become failures. Cell line development and quality control is also being simplified by advanced robotics and high throughput screening platforms [24].

Smart manufacturing incorporates Industry 4.0 principles, linking upstream and downstream operations with cloud-based analytics, predictive modeling, and AI-driven control systems. These technologies are central to the development of intelligent bioreactors and modular manufacturing units capable of autonomous operation, remote monitoring, and seamless scalability [43].

Despite clear benefits, the transition to smart bio-manufacturing requires significant investment in infrastructure, data management, and cross-disciplinary workforce training. Regulatory frameworks are also evolving to accommodate digitally enabled systems and real-time release testing (RTRT).

Ultimately, automation and smart manufacturing are not only enhancing productivity and flexibility but also advancing sustainability through reduced energy usage, optimized raw material consumption, and waste minimization.

7.8.2 Decentralized and On-Demand Production Models

Decentralized and on-demand production models are reshaping the bio-manufacturing landscape by enabling rapid, localized, and flexible synthesis of biopharmaceuticals and bio-products. These systems bypass centralized infrastructure by using compact, modular platforms capable of site-specific manufacturing, often with reduced cold-chain dependence and real-time adaptability. The paradigm shift is particularly crucial for applications in remote regions, outbreak zones, or military field operations, where time-sensitive therapeutic delivery is essential [44].

Synthetic biology and cell-free systems underpin this transformation, with technologies such as freeze-dried transcription–translation (TX–TL) systems and portable micro factories offering programmable, plug-and-play bio-manufacturing. Innovations like the Proteins-On-Demand (Bio-POD) and iVAX platforms exemplify how nucleic acid templates can be rapidly converted into vaccines or enzymes in a compact format [45].

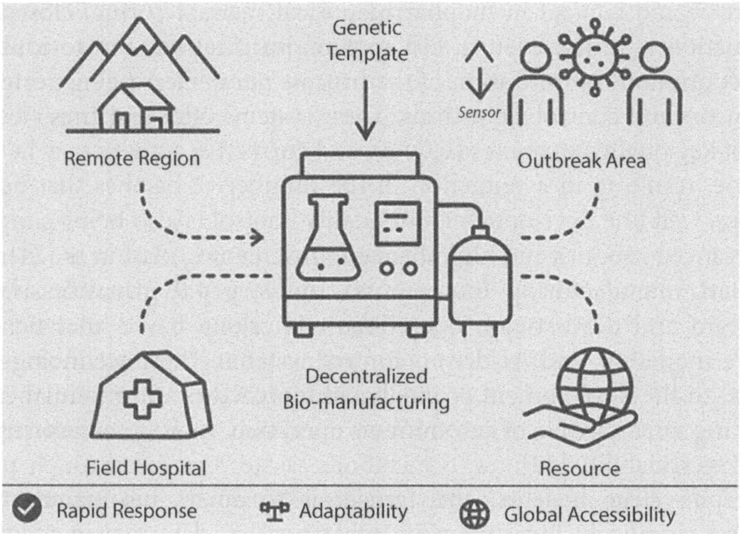

Figure 7.2 The flowchart of decentralized biomanufacturing production units operating across global health settings.

These approaches are scalable and minimize environmental footprint due to lower energy and material demands. Furthermore, integrating artificial intelligence and digital twins enhances process predictability, enabling adaptive manufacturing in dynamic or resource-limited settings.

However, challenges persist in ensuring GMP compliance, robust quality control, and regulatory harmonization for decentralized models. Additionally, the cost of raw materials (e.g., nucleotide precursors, enzymes) and limited long-term stability of some cell-free components require further innovation.

Nonetheless, decentralized and on-demand bio-manufacturing aligns with the vision of democratized, resilient, and sustainable bio-production, positioning it as a critical tool for future healthcare and emergency preparedness. Figure 7.2 illustrates the flowchart of decentralized biomanufacturing production units operating across global health settings.

7.8.3 Role of Bio-Manufacturing in Climate Resilience

Bio-manufacturing is emerging as a pivotal strategy in enhancing climate resilience by enabling low-carbon, resource-efficient production of critical goods from bioplastics and biofuels to climate-adaptive therapeutics and biostimulants. Unlike fossil-based manufacturing systems, bio-manufacturing utilizes renewable feedstocks such as agricultural residues, algae, or engineered microbes, which sequester carbon and reduce net greenhouse gas emissions.

One of the most promising contributions lies in the development of regional bio-manufacturing hubs that leverage local biomass to produce essential materials and fuels. This decentralization not only reduces dependency on vulnerable global supply chains but also mitigates climate-related disruptions, including transportation bottlenecks and energy volatility. Emerson *et al.*, emphasize that resilient biomass supply chains, optimized for availability, cost, and carbon footprint—are critical for catalyzing investment in sustainable bio-based infrastructure [46].

Additionally, soil-based bio-manufacturing platforms using engineered microbes are showing potential for regenerative agriculture by enhancing soil carbon capture and improving drought resistance. Such approaches align bio-manufacturing with broader environmental goals, contributing to both climate mitigation and adaptation [47].

Nevertheless, achieving scale and maintaining environmental integrity requires robust policy frameworks, lifecycle assessments, and community engagement to ensure that bio-manufacturing practices support long-term ecological balance and social equity. Figure 7.3 provides a pictorial representation of a biomanufacturing system supporting various climate resilience pathways, including food, energy, and health.

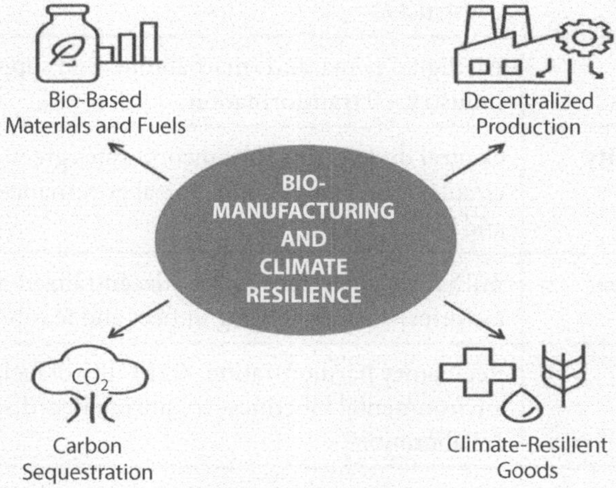

Figure 7.3 Pictorial representation of a biomanufacturing system supporting various climate resilience pathways.

7.9 Conclusion

This chapter summarized key insights into sustainable bio-manufacturing, highlighting innovations, regulatory dynamics, and emerging technologies. A clear roadmap for future development emphasizes integration of smart systems, global harmonization, and eco-centric design. Together, these elements pave the way for a resilient, efficient, and sustainable bio-manufacturing future aligned with global needs.

Table 7.3 The summary of key points related to biomanufacturing.

Category	Key points
Definition & Scope	Intersection of biotechnology, sustainability, and digital transformation; redefines production of biologics, chemicals, and materials.
Foundational Platforms	CHO (Chinese Hamster Ovary), microbial systems, plant-based systems, cell-free technologies.
Process Innovations	Single-use bioreactors, continuous processing, modular design. Benefits include flexibility, reduced environmental impact, and decentralized production.
Advanced Technologies	Synthetic biology, CRISPR, 3D bio-printing; enables programmable organisms and complex biological constructs.
Digital Integration	AI, digital twins, and smart automation supporting Industry 4.0 transformation.
Sustainability Focus	Central design principle; incorporates green chemistry, circular bioeconomy, and ethical governance for climate and health equity.
Case Studies	mRNA vaccines, biosimilars, decentralized micro factories—demonstrating impact and feasibility.
Challenges	Regulatory harmonization, scalability of tech transfer, environmental bioethics—requiring interdisciplinary collaboration.
Conclusion	Bio-manufacturing is foundational for a biologically driven, climate-resilient global economy in the 21st century.

Table 7.4 Summary of the strategic roadmap for future sustainable biomanufacturing.

Pillar	Key focus areas	Expected outcomes
Platform Diversification	- Move beyond CHO and *E. coli* systems	- Context-specific, low-resource production
	- Integrate plant-based, cell-free, and synthetic microbial platforms	- Reduced environmental impact
Process Intensification	- Adopt continuous processing and single-use technologies	- Increased throughput and energy efficiency
		- Minimized waste and alignment with circular bioeconomy
Digital Integration	- Use AI, digital twins, and real-time analytics (PAT)	- Real-time quality control
	- Enhance DBTL cycle and self-optimization capabilities	- Regulatory compliance
		- Reduced process variability
Global Equity	- Promote tech transfer and modular facilities in Global South	- Inclusive, climate-resilient bio-industrial growth
	- Support workforce development	- Equitable access to bio-manufacturing capacity
Collaborative Governance	- Encourage interdisciplinary collaboration	- Transparent, anticipatory governance
	- Develop adaptive, ethics-based policies and frameworks	- Public trust and societal alignment

7.9.1 Summary of Key Insights

Bio-manufacturing merges biotechnology, sustainability, and digital innovation to produce biologics and materials efficiently. Platforms like CHO cells, plant-based, and cell-free systems, combined with tools like CRISPR and AI, enable flexible, eco-friendly production. Guided by green principles and ethics, it forms a core infrastructure for a resilient, equitable global economy. Table 7.3 consolidates the summary of key points related to biomanufacturing.

7.9.2 Roadmap for Future Sustainable Bio-Manufacturing

Sustainable bio-manufacturing's future relies on platform diversification, process intensification, digital integration, and global equity. Advancing plant-based and synthetic systems, smart technologies, and circular practices will boost efficiency and resilience. Harmonized regulations, ethical frameworks, and inclusive technology transfer are vital for equitable growth, guided by collaboration and adaptive governance. Table 7.4 summarizes the strategic roadmap for future sustainable biomanufacturing.

AI disclosure statement: The author acknowledges the use of AI-based tools for minor language editing or grammatical correction. However, all intellectual contributions, analyses, and conclusions presented in this chapter are entirely the author's own.

References

1. Shokoohi, M. and Attar, T., Biomanufacturing for a sustainable future: unleashing the potential of biotechnology in pharmaceutical raw material production. *Int. J. New Findings Health Educ. Sci. (IJHES)*, 2, 2, 142–53, 2024 Jul 13.
2. Tang, X.L., Yu, J.W., Geng, Y.H., Wang, J.R., Zheng, R.C., Zheng, Y.G., From discovery to mass production: a perspective on bio-manufacturing exemplified by the development of statins. *Engineering*, 24, 138–50, 2023 May 1.
3. Pandey, K., Pandey, M., Kumar, V., Aggarwal, U., Singhal, B., Bioprocessing 4.0 in biomanufacturing: paving the way for sustainable bioeconomy. *Syst. Microbiol. Biomanuf.*, 4, 2, 407–24, 2024 Apr.
4. Ali, A., Usman, M., Jayamani, E., Soon, K., bin Ab Wahab, M.K., Sustainable Biomanufacturing, in: *Biomedical Implants*, pp. 195–217, CRC Press, Boca Raton, FL, USA, 2024 Feb 14.

5. Gadamasetti, K., Bioprocessing, Bioengineering and Process Chemistry in the Biopharmaceutical Industry: Using Chemistry and Bioengineering to Improve the Performance of Biologics, in: *Bioprocessing, Bioengineering and Process Chemistry in the Biopharmaceutical Industry: Using Chemistry and Bioengineering to Improve the Performance of Biologics*, pp. 3–21, Springer Nature Switzerland, Cham, 2024 Nov 24.
6. Babele, P.K., Srivastava, A., Selim, K.A., Kumar, A., Millet-inspired systems metabolic engineering of NUE in crops. *Trends Biotechnol.*, 41, 5, 701–13, 2023 May 1.
7. Létinois, U., Crump, S., Zerrer, B., Hans, M., Meunier, P.B., Wyss, M., Way Forward for Biomanufacturing and Biotechnology in Europe. *Chimia*, 79, 5, 344–51, 2025 May 28.
8. Fu, R., Kang, L., Zhang, C., Fei, Q., Application and progress of techno-economic analysis and life cycle assessment in biomanufacturing of fuels and chemicals. *Green Chem. Eng.*, 4, 2, 189–98, 2023 Jun 1.
9. Serra, A., Zouraris, D., Schaffert, A., Maia, M.T., Tsiros, P., Virmani, I., Di Lieto, E., Saarimäki, L.A., Morikka, J., Riudavets-Puig, R., Varsou, D.D., INSIGHT: An integrated framework for safe and sustainable chemical and material assessment. *Comput. Struct. Biotechnol. J.*, 29, 125–37, 2025 Jan 1.
10. Friedrichs, S., Seitz, C., Bleeker, E., Vandebriel, R., MACRAMÉ: Advanced characterisation methodologies to assess the health and environmental risks of materials. *Curr. Res. Struct. Biotechnol.*, 7, 100101, 2025.
11. Zeng, Q., Tian, S., Zhang, Y., Liu, H., Chen, D., Tian, X., Hu, C., Yang, J., Phase transfer-based high-efficiency recycling of precious metal electrocatalysts. *Green Chem. Eng.*, 5, 1, 68–74, 2024 Mar 1.
12. Oluyemi, G.F., Afolabi, R.O., Zamora, S.C., Li, Y., McElroy, D., Environmental Impact Assessment of a Plant Cell-Based Bio-Manufacturing Process for Producing Plant Natural Product Ingredients. *Sustainability*, 16, 19, 8515, 2024 Sep 30.
13. Zhao, B., Li, X., Sun, W., Qian, J., Liu, J., Gao, M., Guan, X., Ma, Z., Li, J., Biodt: An integrated digital-twin-based framework for intelligent biomanufacturing. *Processes*, 11, 4, 1213, 2023 Apr 14.
14. Vijayakumar, V.E. and Venkataraman, K., A systematic review of the potential of Pichia pastoris (Komagataella phaffii) as an alternative host for biologics production. *Mol. Biotechnol.*, 66, 7, 1621–39, 2024 Jul.
15. Kumar, A., Udugama, I.A., Gargalo, C.L., Gernaey, K.V., Why is batch processing still dominating the biologics landscape? Towards an integrated continuous bioprocessing alternative. *Processes*, 8, 12, 1641, 2020 Dec 12.
16. Joshi, A.K. and Gupta, S.K., Single-Use Continuous Manufacturing and Process Intensification for Production of Affordable Biological Drugs, in: *Process Control Intensif Digit Cont Biomanuf*, pp. 179–208, Wiley-VCH GmbH, Weinheim, Germany, 2022 Mar 14.

17. Cravotto, G., Reshaping Chemical Manufacturing Towards Green Process Intensification: Recent Findings and Perspectives. *Processes*, 13, 2, 459, 2025 Feb 8.
18. Paul, C.E., Eggerichs, D., Westphal, A.H., Tischler, D., van Berkel, W.J., Flavoprotein monooxygenases: versatile biocatalysts. *Biotechnol. Adv.*, 51, 107712, 2021 Nov 1.
19. Matanguihan, C. and Wu, P., Upstream continuous processing: recent advances in production of biopharmaceuticals and challenges in manufacturing. *Curr. Opin. Biotechnol.*, 78, 102828, 2022 Dec 1.
20. Agarwal, P., McCready, C., Ng, S.K., Ng, J.C., van de Laar, J., Pennings, M., Zijlstra, G., Hybrid modeling for *in silico* optimization of a dynamic perfusion cell culture process. *Biotechnol. Progr.*, 41, 1, e3503, 2025 Jan.
21. Panapitakkul, C., Bulaon, C.J., Pisuttinusart, N., Phoolcharoen, W., Characterization of host cell proteins in the downstream process of plant-Based biologics using LC-MS profiling. *Biotechnol. Rep.*, 44, e00856, 2024 Dec 1.
22. Romine, J.A., Kolodziej, S.A., Vetter, T., Dux, M.P., COVID-19 Vaccine Manufacturing Processes: Making the Molecules to Solve the Pandemic, in: *Bioprocessing, Bioengineering and Process Chemistry in the Biopharmaceutical Industry: Using Chemistry and Bioengineering to Improve the Performance of Biologics*, pp. 185–217, Springer Nature Switzerland, Cham, 2024 Nov 24.
23. Basu, S., Ranjan, A., Sur, S., Surguchov, A., *Technologies in Cell Culture: A Journey from Basics to Advanced Applications*, BoD–Books on Demand, Norderstedt, Germany, 2024 Jun 26.
24. Sathiyapriyan, P., Mukherjee, S., Vogel, T., Essen, L.O., Boerema, D., Vey, M., Kalina, U., Current PAT Landscape in the Downstream Processing of Biopharmaceuticals. *Anal. Sci. Adv.*, 6, 1, e70013, 2025 Jun.
25. Islam, M.T., Sepanloo, K., Woo, S., Woo, S.H., Son, Y.J., A review of the Industry 4.0 to 5.0 transition: exploring the intersection, challenges, and opportunities of technology and human–machine collaboration. *Machines*, 13, 4, 267, 2025 Mar 24.
26. Yee, C.S., Zahia-Azizan, N.A., Abd Rahim, M.H., Mohd Zaini, N.A., Raja-Razali, R.B., Ushidee-Radzi, M.A., Ilham, Z., Wan-Mohtar, W.A., Smart Fermentation Technologies: Microbial Process Control in Traditional Fermented Foods. *Fermentation*, 11, 6, 323, 2025 Jun 5.
27. Robinson, J.M., Breed, M.F., Beckett, R., Probiotic cities: microbiome-integrated design for healthy urban ecosystems. *Trends Biotechnol.*, 42, 8, 942–5, 2024 Aug 1.
28. Wang, L., Wang, S., Liu, Y., Zhang, B., Pan, Z., Bai, L., Yao, S., Zhang, C., Xie, H., He, J., Li, D., Biofabrication of brain-like living tissue: structure to intelligence. *Int. J. Extreme Manuf.*, 7, 3, 032005, 2025 Jan 30.
29. Kang, Y., Luo, R., Gong, D., Zhang, Y., Wang, D., Synthetic biomanufacturing in China: progress and prospects. *Biotechnol. Notes*, 3, 100020, 2022.

30. Wang, M., Chen, L., Zhang, Z., Wang, Q., Recent advances in genome mining and synthetic biology for discovery and biosynthesis of natural products. *Crit. Rev. Biotechnol.*, 45, 1, 236–56, 2025 Jan 2.
31. Bok, K., 1 In Focus: Novel Manufacturing Platforms for Pandemic Preparedness and Emergency Response, in: *Principles and Practice of Emergency Research Response*, pp. 285–297, Springer International Publishing, Cham, 2024 Aug 31.
32. Sideri, K., mRNA vaccine politics: responsible governance coordination for vaccine innovation in times of urgency. *J. Responsible Innovation*, 11, 1, 2425121, 2024 Dec 31.
33. Müller, D., Klein, L., Lemke, J., Schulze, M., Kruse, T., Saballus, M., Matuszczyk, J., Kampmann, M., Zijlstra, G., Process intensification in the biopharma industry: Improving efficiency of protein manufacturing processes from development to production scale using synergistic approaches. *Chem. Eng. Process. Process Intensif.*, 171, 108727, 2022 Jan 1.
34. Chhabra, H., Mouslim, M.C., Kashiramka, S., Rathore, A.S., Dynamics of biosimilar uptake in emerging markets. *Expert Opin. Biol. Ther.*, 22, 6, 679–88, 2022 Jun 3.
35. Zahmanova, G., Aljabali, A.A., Takova, K., Minkov, G., Tambuwala, M.M., Minkov, I., Lomonossoff, G.P., Green biologics: harnessing the power of plants to produce pharmaceuticals. *Int. J. Mol. Sci.*, 24, 24, 17575, 2023 Dec 17.
36. Zhang, Y., Liu, W.Q., Li, J., Constructing an artificial short route for cell-free biosynthesis of the phenethylisoquinoline scaffold. *Synth. Syst. Biotechnol.*, 8, 4, 610–7, 2023 Dec 1.
37. Marandi, V., Mitigation of Scale-Up Risks in Technology Transfer of Vaccines and other Bioproducts. *Vac. Res.*, 10, 2, 9–10, 2023 Jun.
38. Looby, M., Brady, C., Bentley, M., Lewis, A.M., Layden, C., Barnoon, B., Abbott, E., Hughes, B., Three Decades of Advancements in Technical Transfer of Biologics: A Blueprint for Advanced Therapeutics, in: *Bioprocessing, Bioengineering and Process Chemistry in the Biopharmaceutical Industry: Using Chemistry and Bioengineering to Improve the Performance of Biologics*, pp. 503–529, Springer Nature Switzerland, Cham, 2024 Nov 24.
39. Friedrichs, S., Seitz, C., Bleeker, E., Vandebriel, R., MACRAMÉ–Advanced characterisation methodologies to assess and predict the health and environmental risks of advanced materials. *Curr. Res. Struct. Biotechnol.*, 7, 100101, 2025.
40. Ahluwalia, K., Abernathy, M.J., Algorri, M., Cauchon, N.S., Perico-Norred, N.M., Youssef, R.Y., The future of regulatory filings: digitalization. *AAPS Open*, 11, 1, 9, 2025 Apr 18.
41. Kaluđerović, Z., *Contemporary Bioethics: Themes and Dilemmas*, NKUA Applied Philosophy Research Lab Press, Athens, Greece, 2025 Apr 9.

42. Bashiir, A.A., Ethical Considerations in Emerging Technologies: Balancing Innovation and Morality. *Res. Inven. J. Eng. Phys. Sci.*, 4, 1, 50–55, Research Invention Journals, Nigeria, 2025.
43. Szarzynski, A., Spadiut, O., Reisbeck, M., Jobst, G., Paterson, R.L., Kamenskaya, A., Gateau, E., Lesch, H.P., Henry, L., Kozma, B., CGT 4.0: a distant dream or inevitable future? Smart process automation is critical to make efficient scalability of CGT manufacturing a reality. *Front. Bioeng. Biotechnol.*, 13, 1563878, 2025 Mar 19.
44. Tang, C., Wang, L., Zang, L., Wang, Q., Qi, D., Dai, Z., On-demand biomanufacturing through synthetic biology approach. *Mater. Today Bio*, 18, 100518, 2023 Feb 1.
45. Kantaros, A., Drosos, C., Papoutsidakis, M., Pallis, E., Ganetsos, T., The Role of 3D Printing in Advancing Automated Manufacturing Systems: Opportunities and Challenges. *Automation*, 6, 2, 21, 2025 May 26.
46. Nair, S.K., Emerson, R.M., Solomon, J.R., Biomass Supply Chain Risk: Towards a Better Understanding of Feedstock Availability, Cost, Variability, and Uncertainty to Catalyze and De-risk Biobased Investment, in: *Handbook of Biorefinery Research and Technology: Biomass Logistics to Saccharification*, pp. 285–312, Springer Netherlands, Dordrecht, 2024 Oct 23.
47. Pal, P., Rani, H., Ansari, S.A., Sachdev, S., Raheem, A., Amir, M., Ansari, M.I., Soil Microbes as a Tool for Industry and Research, in: *Industrial Applications of Soil Microbes*, vol. 4, pp. 112–128, Bentham Science Publishers, Sharjah, UAE, 2024 Nov 22.

8

Additive Manufacturing for Biomedical Applications: Concepts, Methods, and Future Trends

Idhayaraja S.[1], Ajithkumar Sitharaj[2*], Arulmurugan B.[3] and Sandip Kunar[4]

[1]Department of Manufacturing Engineering, PSG College of Technology, Coimbatore, Tamil Nadu, India
[2]Department of Mechanical Engineering, KPR Institute of Engineering and Technology, Coimbatore, Tamil Nadu, India
[3]Department of Mechanical Engineering, KPR Institute of Engineering and Technology, Coimbatore, Tamil Nadu, India
[4]Department of Mechanical Engineering, Aditya University, Surampalem, Andhra Pradesh, India

Abstract

This chapter examines the evolving role of Additive Manufacturing (AM) in biomedical applications, emphasizing its principles, techniques, and prospective developments. Additive manufacturing has revolutionized biomedical engineering by facilitating the production of personalized implants, scaffolds, and medical devices characterized by unparalleled structural intricacy and functional cohesion. The chapter examines the advanced computational design methodologies, including topology optimization, generative design, and finite element simulation, in conjunction with biocompatible materials, surmount the geometric and customization constraints of traditional manufacturing. Critical AM processes, such as Selective Laser Melting (SLM), Electron Beam Melting (EBM), Wire Arc Additive Manufacturing (WAAM), and bioprinting, are examined for their capacity to produce complex structures with regulated porosity, varied compositions, and multi-material arrangements that improve mechanical properties, facilitate osseointegration, and aid in tissue regeneration. The review emphasizes emerging innovations like 4D printing and stimuli-responsive biomaterials, which facilitate dynamic and adaptive functionalities. The potential of hybrid approaches

**Corresponding author*: ersajithkumarbe@gmail.com

Sandip Kunar, Jagadeesha T. and D. Sathis Kumar (eds.) Biomanufacturing: Transforming Production of Biologics and Biopharmaceuticals, (195–216) © 2026 Scrivener Publishing LLC

that combine the structural robustness of WAAM with the biological precision of bioprinting is explored for the advancement of hard and soft tissue engineering. This chapter emphasizes AM as a fundamental technology for future biomedical innovations, providing avenues for personalized and regenerative healthcare that enhance patient outcomes and surgical success rates.

Keywords: Additive manufacturing (AM), biomedical engineering, patient-specific implants, wire arc additive manufacturing (WAAM), bio printing, topology optimization, 4D printing

8.1 Introduction

Additive Manufacturing (AM), also known as 3D printing, is a game changer in biomedical engineering, with unparalleled opportunity for development of complex, patient-specific implants, scaffolds and medical devices. Unlike traditional manufacturing, which relies on tool geometry and often subtractive methods, AM allows for development of complex constructions layer by layer with porosity controls, graded properties of materials, or even mechanical properties [1, 2]. This unprecedented design freedom has opened the doors for developing next-generation biomedical solutions which emulate the architecture, function, and biological performance of native tissue. With integrated computational design methods, such as topology optimization, generative design, finite element simulation, and the specific data of a patient's anatomy from medical imaging (CT, MRI), the resulting implant are anatomically accurate, mechanically efficient, and biologically appropriate. These "smart" designs optimize load distribution to avoid stress shielding, but also take into consideration features promoting osseointegration and vascularization pathways, to yield enhanced long-term functional stability. In addition, the combination of artificial intelligence, biomechanical modeling, and additive manufacturing processes were cutting the time from design to implantation, while improving precision and repeatability in final manufacturing, i.e., SLM, EBM, and WAAM [3].

Also, advances in biomaterials science have broadened the range of eligible materials in biomedical AM. Biodegradable polymers such as PLA, PCL, and PGA provide transient therapeutic capacity where metallic biomaterials such as Ti-6Al-4V, Co-Cr alloys, and magnesium-based systems have more structural integrity and have helped the healing process of bone. Biomaterials like hydroxyapatite and β-tricalcium phosphate, in addition to hybrid composites, enhance regenerative potential - especially given their tunability through porosity, porous architecture, and interconnectivity.

Also, materials can now be used in combination (multi-material printing) or can have their surfaces functionalized through biomimicry - allowing for implants that are both structurally competent and biologically intelligent where they can deliver drugs, mediate tissue regeneration, and respond to physiological stimulus [4].

Emerging territories like 4D printing come with a time-dependent dimension to biomedical devices, where stimuli-responsive materials can make implants change shape or distribution of therapeutics, or otherwise adjust to the *in vivo* healing process. Shape-memory polymers, hydrogels, and multi-material composites are being fabricated for minimally invasive delivery and functional transformation within the body [5]. Furthermore, the incorporation of bioprinting with WAAM can allow hybrid scaffolds to combine a mechanically-strong metallic framework with a bioactive, cell-loaded layer, thereby bridging the differences between hard and soft tissue engineered scaffolds. Together, these advances in design strategies, material systems, and AM processes are revolutionizing the standards of biomedical device engineering. They promise deep personalization, regenerative, and adaptable medical solutions to improve surgical outcomes, accelerate patient recovery, and prolong the functional life span of implants. In its growth, AM with computational intelligence, advanced biomaterials, and regenerative medicine will continue to be the leading factor in personalized health.

8.2 Smart Design Strategies in Biomedical Additive Manufacturing

The development of additive manufacturing (AM) technologies has become a major driver for design and development of biomedical implants and devices. There has been a paradigm shift toward smart, personalized design strategies. Traditional manufacturing technologies impose limitations based on complexity, customization, and functional integration, whereas AM technologies remove these restraints that allow for gradually engineered structures with complex geometries, porosity control, and spatially varying materials properties -- all of which necessitated to effectively replicate the complexities of biological tissues. These smart strategies use the computational design tools when engaging with clinical imaging (CT or MRI) and biomechanical modeling to design implants that are not just anatomic replicas but are also mechanically efficient and biologically appropriate.

As shown in Figure 8.1, differences in adhesion and morphology between the treatment conditions existed. In non-acid etched, non-ultrasonicated samples (a), there were some cells, and, while there was more cytoskeletal organization than in the first condition to distinguish limited differentiation, it was less than both later groups. In acid-etched samples (b), there were more cells than in (a), but there was limited degradation in relation to more limited differentiation. The acid-etched with ultrasonicated samples (c) had the most cells and the most spread morphology, suggesting some cytoskeletal organization was better than both prior conditions, together with more presence of focal adhesions (e.g., cellular attachments to the titanium surface are likely associated with focal adhesions). Each improvement in cell attachment and subsequent differentiation was related to acid-etching isolating and removing un-melted particulates and subsequently providing micron-scale roughness, as well as ultrasonication simply allowing for the direct penetration of electrolytes through the nanotube arrays during the anodization process to enhance the predictability

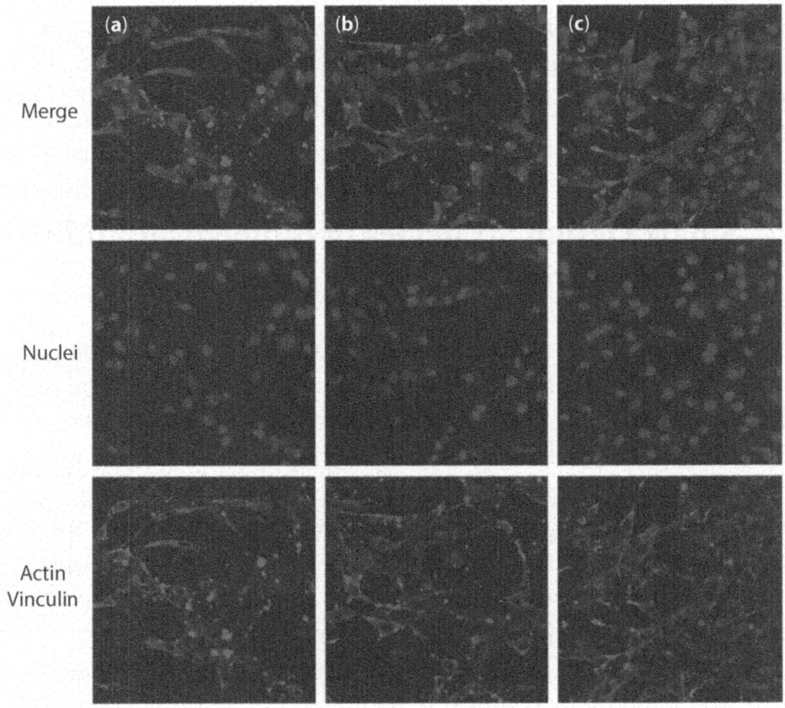

Figure 8.1 Fluorescence images of cells on 700 μm TC4 samples with different surface pretreatments: (a) untreated, (b) acid-etched, (c) acid-etched + ultrasonic [6]. It is available under an open-access license.

of nanotube arrays that lead to optimal cell adhesion, differentiation, and growth [6].

The combination of topology optimization, generative design algorithms, and architected lattice systems is a corner stone of this development. These techniques allow engineers and biomedical designers to move past simply mirroring a shape but instead on functional performance, optimizing the load bearing characteristics and enhancing the bio-integration. With machine learning, finite element simulations and patient-specific anatomical data, smart designs can flexibly adapt to various physiological and pathological states. Additionally, additive manufacturing platforms now provide the ability to fabricate these complex and personalized implants with excellent dimensional accuracy, material integrity, and great customization potential. The subsequent subsections explore these strategies in more depth, demonstrating how they are actively transforming the engineering of modern biomedical and associated devices for improved patient outcomes, surgical outcomes, and long-term implant success [7, 8].

8.2.1 Topology Optimized Implants for Load-Adaptive Functionality

Topology optimization is a high-level computational design process that allows engineers to quantitatively define the most efficient layout of material within a specified geometry under mechanical loads, boundary conditions, and functional requirements. In the case of biomedical implants, this process is extremely useful for creating a layout that mimics the stress-bearing characteristics of natural bone. Traditional implants may typically over-engineer certain areas resulting in undesirable load disruptions. In contrast, topology optimized implants are designed to distribute the stress in a biologically friendly way [9].

This diminishes the amount of stress shielding that can happen. This phenomenon occurs when surrounding bone tissue receives little mechanical stimulus, bone resorption from loader path, compromising the implant. Algorithms such as SIMP (Solid Isotropic Material with Penalization), ESO (Evolutionary Structural Optimization), and Level-Set Methods are all used in these optimization processes, where the design is optimized iteratively by increasing the material from low-stress areas and adding material to high load paths. Patient-specific anatomical data collected from CT or MRI imaging is converted into 3D models where they are optimized through an algorithm. Once the optimal structure is produced, additive manufacturing (AM) is used to produce complex geometries that are impossible through subtractive means (e.g. Selective Laser Melting (SLM), Electron Beam Melting (EBM) and Laser Powder Bed Fusion (LPBF)).

Recent work has advanced this idea to include multi-objective and multi-scale optimization, enabling the simultaneous design of both macro-geometry and micro-architecture (porosity, lattice design). For example, lattice-infused implants can now be co-designed with porosity levels appropriate for bone ingrowth, vascularity and mechanical function. Co-simulation approaches will be enabled by state-of-the art software platforms such as Altair Inspire, ANSYS Mechanical, or COMSOL Multiphysics to simultaneously capture the static, fatigue, and thermal loading on the implant, and possible biomechanical impact on the surrounding tissue. The incorporation of AI-based generative design and machine learning based surrogate models has reduced computational time, and improved accuracy of prediction.

Figure 8.2 shows the effects of solid and porous stem hip implants on stress shielding and progression of bone density over time in the femur. In the intact femur physiological loads successfully transfer through the head and neck of the femur allowing for proper stress to distribute in the cancellous bone. After implantation, the solid stem implant reduces the stress in some anatomical locations, with minimum von Mises stress of 73 MPa in Gruen zone 6 and maximum stress of 112 MPa in zone 7, indicating that some moderate stress shielding has occurred. The porous stem implant reduces the minimum stress even further down to 43 MPa at zone 6, but maintains the same maximum stress value of 112 MPa at zone 7 [10]. The reduction in stress is a result of the increased surface area within the porous stem, which creates a different pattern of stress transfer. Stress shielding is significantly associated with reductions in bone density, as indicated by the bone density mapping, which visualizes when stress shielding occurs approximately 2 years after implantation. The design of the porous stem format aims to upgrade the distributed load and minimize loss of bone over time as this method is shown to provide relatively homogenous stress distribution when compared to solid stem fixation.

This enables faster iteration cycles in designing orthopedic implants like hip stems, mandibular reconstruction plates, intervertebral cages, and cranial implants. Topology optimization is also increasingly coupled with Finite Element Analysis (FEA) and Digital Twin frameworks, allowing virtual pre-clinical testing to predict the implant's *in vivo* behavior. On the clinical front, such implants not only show improved biomechanical performance but also promote better Osseo integration, reduce revision surgery rates, and ensure long-term functional restoration. Future directions include the real-time intraoperative modification of topological designs

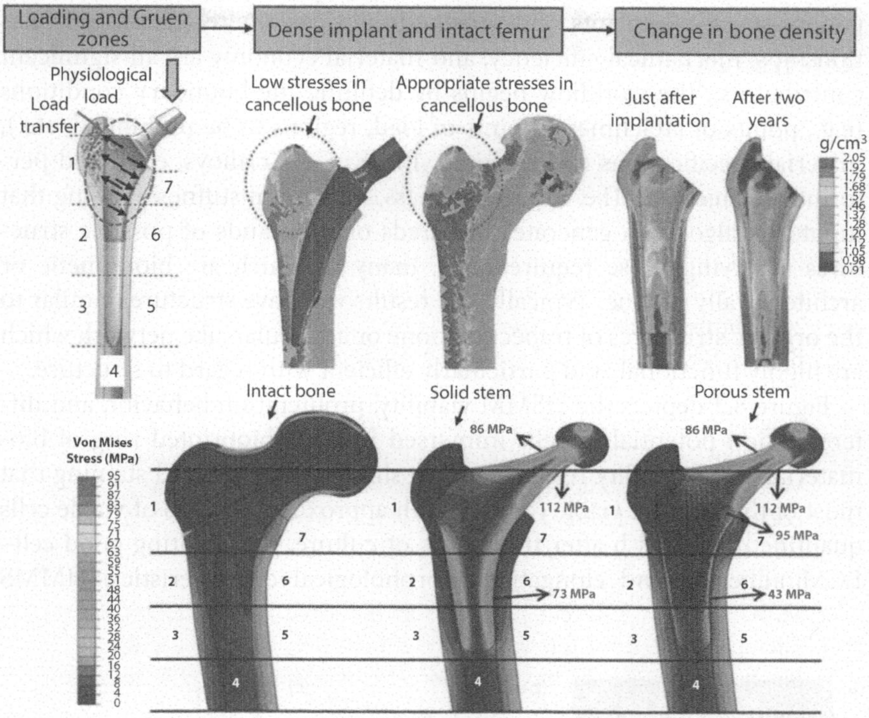

Figure 8.2 Comparison of stress distribution and bone density changes in intact femur, solid stem, and porous stem hip implants across Gruen zones [10]. It is available under an open-access license.

based on surgical navigation data and bio-topological modeling, which incorporates both mechanical loads and biological parameters like tissue regeneration and cellular response.

8.2.2 Generative Design Algorithms for Complex Biomedical Structures

Generative design is an AI-based method of design with the capability to spontaneously generate complex, high-performance structures through multiple iterations of generating and evaluating design variations defined by engineering limitations [11]. Generative design is fundamentally different from parametric modeling as it does not rely on human ingenuity or traditionally linear workflows. Generative design expands the solution space by utilizing computational intelligence and predictive simulation. In biomedical contexts, this is a major advancement, especially for developing patient-specific implants and prosthetic devices where anatomical relationships, mechanical efficiency, and material economy are all significant contributors. The workflow begins by defining the boundary conditions (e.g., points of attachment, points of load, regions to be prohibited, etc.), material specifications (i.e. titanium, PEEK, Co-Cr alloys, etc.), and performance objectives (i.e. minimum mass, maximum stiffness). Using that input, the algorithm generates hundreds or thousands of possible structures satisfying those requirements, many of which are biomimetic or architecturally unique. Typically, the results will have structures similar to the organic structures of trabecular bone or a vascular-like network which are highly functional, and particularly efficient with regard to structure.

Figure 8.3 depicts the HSMM viability, proliferation behavior, and differentiation potential of cells immersed in a 3D bioprinted ring of biomaterials. The viability result (panel a) shows with live/dead staining that most of the cells are viable (green), with approximately 80% of viable cells quantified in panel b after three days of culture, representing good cell–ECM adhesion and elongation morphological characteristics. HMMS

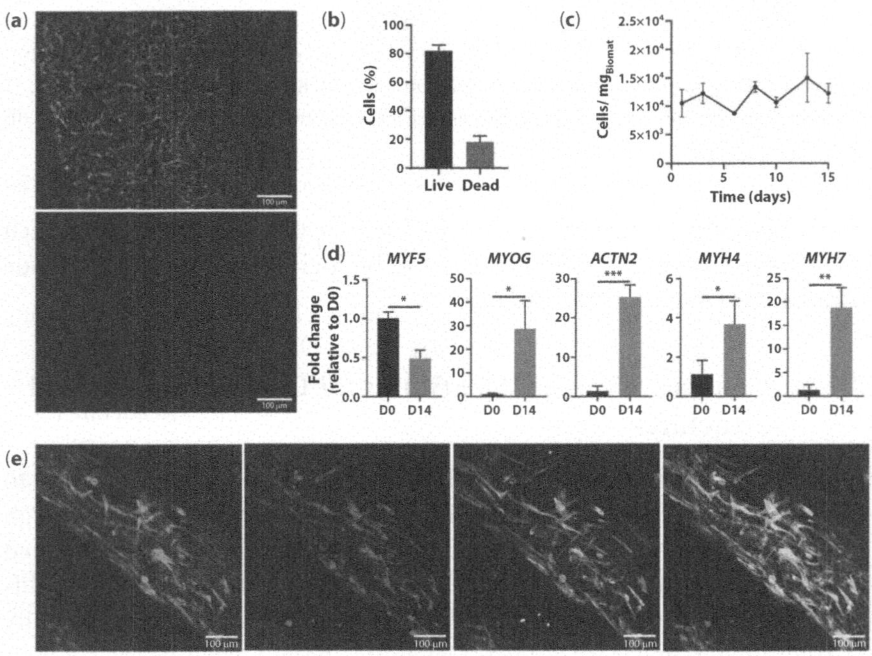

Figure 8.3 Viability, proliferation, and differentiation of human skeletal muscle myoblasts (HSMM) cultured in 3D bioprinted biomaterial rings [12]. It is available under an open-access license.

proliferation behavior (panel c) indicates that the number of HSMM cells did not change substantially over the 15-day culture period; a significantly different behavior when compared to C2C12 mouse myoblasts reported in other studies. Gene expression analysis (panel d) indicated a downregulation in MYF5 gene expression indicating differentiation potential, along with significant upregulation in intermediate (MYOG) and late (ACTN2, MYH4, MYH7) differentiation markers in the HSMMs. Immunostaining images (panel e) indicated the presence of α-actinin (red) and myosin heavy chain (MHC; green) with merged images confirming the shape of multinucleated fibers are oriented in the same direction with respect to the printing direction of the fibers [12]. Overall, these outcomes confirms that HSMM viability can be supported by the biomaterial environment, and that the biomaterial can successfully promote (differentiated) HSMMs to form aligned muscle fibers.

Biomedical instances of generative design include patient-specific orthopedic plates, spinal interbody cages, and true and unique mandibular and maxillofacial implants, and cranial reconstruction devices. The real strength of generative design is its combined and enhanced implementation with additive manufacturing (AM) modalities, especially technologies like selective laser melting (SLM), direct metal laser sintering (DMLS), and electron beam melting (EBM) that are capable of fabricating complex internal channels, trabecular lattice regions, and variation in wall thicknesses, among other geometries that are not available in traditional subtractive methods. Current software options such as Autodesk Fusion 360, nTopology, SolidThinking Inspire, and Siemens NX, can take advantage of integrated generative design modules that include finite element analysis (FEA) with real-time simulation feedback during the generating design states during evaluation and optimization procedures. As well, some options also accommodate, topology optimization, biomechanical simulation, and machine learning based on stored experience with prior performance data, or patient anatomical information *via* CT and MRI with references to the whole person's anatomical structure. New capabilities also include multi-material generative design, so that parts don't have to distinguish between rigid versus flexible regions, which is valuable for bio-integrated devices. The newest research has been using generative design, with digital twin and biomechanical feedback loops that adjust next-generation implants based on adapting biological elements or changes in patient activity levels after implantation.

8.3 Bio-Compatible and Functional Material Systems for AM

The progress of biomedical additive manufacturing (AM) relies critically on the selection and design of materials that accomplish multiple biological, mechanical, and chemical performance criteria. Materials produced by a conventional process for use in general engineering applications do not need to be biocompatible; however, materials for biomedical AM need to be naturally biocompatible and not elicit immune rejection or any toxicological responses when implanted inside the human body. Additionally, for many biomedical applications, materials must provide stimulus to contribute to tissue regrowth or restoration of function and may need to degrade over time, while providing signals for cellular activity. The ability to control every layer of material composition and architecture allows for unprecedented spatial control and creation of advanced geometries specific to biomedical applications - including, porous scaffolds, gradient structures, and hybrid composites [13, 14].

Recent advancements in this area have dramatically broadened the range of available biomaterials. For example, biodegradable polymers (PLA, PCL, PGA) allow a range of biodynamics and mechanical profiles, making them amenable for use in temporary therapeutic devices. Metals like titanium and magnesium alloys effectively provide a load-bearing structure that can have long-term/enduring functionality, and osseointegration. Bioceramics, including hydroxyapatite and tricalcium phosphate, are osteoconductive platforms that facilitate bone regeneration; however, they tend to be limited by their brittleness and have been successfully paired with polymers or metals. The emerging area of multi-material printing, functionally graded materials, and bioactive interfaces is leading to the development of devices that are mechanically stable and biologically smart—i.e., they can deliver drugs, respond to physiological stimuli, and/or guide tissue regrowth. These innovations in material strategies are driving an era of personalized, regenerative, and smart devices in biomedical implants [15].

8.3.1 Biodegradable Polymeric Platforms for Temporary Therapeutic Support

Biodegradable polymers play a pivotal role in biomedical additive manufacturing, particularly in the fabrication of temporary therapeutic structures that require controlled degradation over time. Materials such as polylactic acid (PLA), polycaprolactone (PCL), and polyglycolic acid (PGA), along with their copolymers like PLGA, have emerged as clinically approved, bioresorbable candidates owing to their excellent biocompatibility, processability, and tunable degradation profiles. These polymers hydrolyze into naturally metabolized byproducts primarily lactic and glycolic acids which are eventually eliminated through the body's metabolic pathways. Additive manufacturing techniques such as Fused Deposition Modeling (FDM) and Stereolithography (SLA) offer precision and flexibility in shaping these materials into intricate geometries, enabling the creation of patient-specific scaffolds, tissue patches, and drug-eluting implants. Their degradation rates can be modulated by adjusting the polymer's molecular weight, crystallinity, copolymer ratio, or even AM process parameters like infill density and layer height factors crucial in matching scaffold resorption with tissue healing dynamics [16].

In recent advancements, biodegradable polymeric platforms are being engineered not only for structural support but also as bioactive delivery systems. These scaffolds can be functionalized with growth factors, antibiotics, or stem cells, thereby transforming them into multifunctional systems that promote tissue regeneration, control infections, and modulate inflammation. Moreover, multi-material AM enables the fabrication of gradient or core-shell structures, allowing localized control over mechanical strength and drug release profiles. Emerging paradigms such as 4D bioprinting introduce smart polymers that respond to physiological stimuli such as pH, temperature, or enzymatic activity offering implants that can swell, degrade, or morph in response to healing stages. These dynamic behaviors significantly enhance integration with the host tissue and eliminate the risks and costs associated with secondary surgeries. Collectively, biodegradable polymers in AM are not just structural materials but active participants in the healing cascade, heralding a new era of intelligent, patient-specific biomedical interventions.

8.3.2 AM-Processed Metallic Biomaterials for Structural and Osseointegrative Performance

Metallic biomaterials remain the cornerstone of load-bearing implantology due to their superior mechanical performance, biostability, and wear resistance. In biomedical additive manufacturing, alloys such as Ti-6Al-4V, Co-Cr, 316L stainless steel, and emerging Mg- and Zn-based systems are tailored through AM processes like selective laser melting (SLM) and electron beam melting (EBM) to achieve both structural reliability and biological integration [17]. These powder bed fusion techniques enable the precise fabrication of complex geometries, including trabecular-mimicking porous architectures, which help to bridge the stiffness mismatch between implant and host bone, thereby reducing stress shielding. Such controlled porosity not only distributes mechanical loads more effectively but also creates an ideal microenvironment for vascularization, bone ingrowth, and osseointegration. The fine control over microstructure down to the grain orientation and phase composition granted by AM offers unprecedented design freedom that traditional subtractive methods cannot match [18].

Beyond structural design, surface functionalization is increasingly integrated into post-processing workflows to enhance cell response and bioactivity. Techniques such as laser surface texturing, plasma spraying of hydroxyapatite coatings, and ion implantation of bioactive elements can significantly improve protein adsorption, osteoblast adhesion, and antibacterial performance [19]. Furthermore, research in biodegradable metals, notably magnesium and zinc alloys, is gaining momentum, as these materials offer both load-bearing capability and control *in vivo* degradation, eliminating the need for explantation surgeries. Their AM processing, however, requires tight control due to their high reactivity and low melting points. On the scale of large, patient-specific implants, Wire Arc Additive Manufacturing (WAAM) is emerging as a promising technique, particularly for producing robust components like cranial plates, femoral stems, and spinal rods, with reduced lead times and lower cost per part. Overall, AM-processed metals are advancing from merely structural supports to multifunctional, biologically intelligent platforms capable of interacting dynamically with host tissue throughout the healing process [20].

8.3.3 Bioceramic and Hybrid Composites for Regenerative Scaffolding

Bioceramics such as hydroxyapatite (HA), β-tricalcium phosphate (β-TCP), bioactive glass, and calcium silicates have gained prominence in regenerative medicine due to their close compositional similarity to natural bone and their ability to actively support osteoconduction, osteointegration, and even osteogenesis *in vivo*. These ceramics serve not only as passive fillers but also as biologically interactive materials that release therapeutic ions like calcium, phosphate, and silicon during degradation, which stimulate cellular pathways for bone formation and angiogenesis. However, the intrinsic brittleness and low fracture toughness of pure bioceramics pose challenges for load-bearing applications [21]. To overcome these limitations, additive manufacturing methods such as robocasting, extrusion-based 3D printing, and binder jetting have been developed to process ceramics into highly porous, mechanically stable architectures with precise microstructural control. These methods allow for the design of interconnected pore networks ranging from 100 to 800 μm, which are essential for nutrient diffusion, vascular infiltration, and bone marrow penetration all critical for scaffold integration and long-term performance.

Figure 8.4 outlines the fabrication and application process of PRP-coated 3D-printed porous titanium scaffolds for osteoporosis treatment. First, pTi scaffolds with optimized pore size and porosity are fabricated *via* 3D printing to match the mechanical properties of bone tissue. These scaffolds are then coated with autologous platelet-rich plasma (PRP) prepared using either the traditional method or the freeze-dried method. The coated scaffolds are implanted into osteoporotic rabbit bone defects created *via* bilateral ovariectomy (OVX). The PRP coating improves the bioactivity of the pTi surface, enhancing bone marrow mesenchymal stem cell (BMSC) adhesion, proliferation, migration, and osteogenic differentiation, as indicated by elevated expression of ALP, Runx-2, OCN, and BMP-2. The freeze-dried PRP coating shows superior performance in promoting bone formation and osseointegration compared to traditional PRP [22].

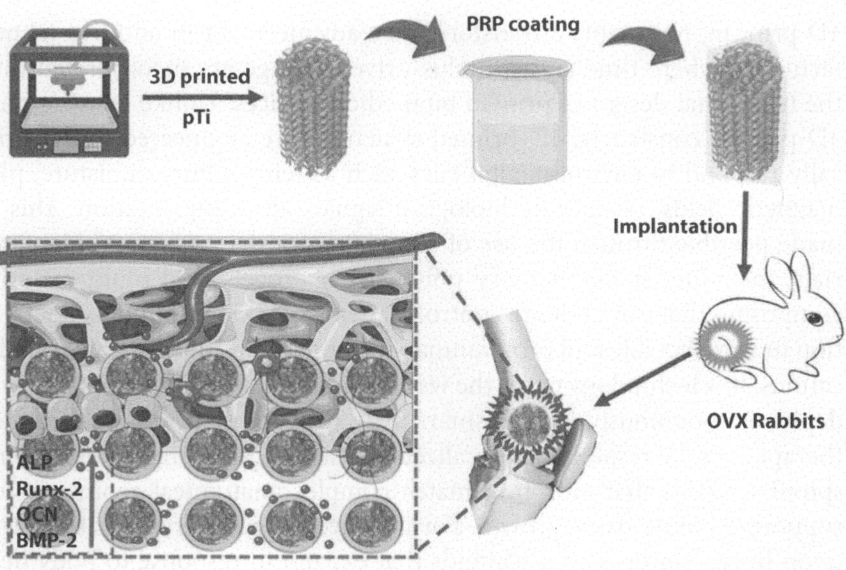

Figure 8.4 Schematic illustration of PRP-coated 3D-printed porous titanium (pTi) scaffolds for enhanced osseointegration in osteoporotic bone defects [22]. It is available under an open-access license.

To enhance both mechanical properties and biological function, bioceramics are increasingly being integrated into hybrid composite systems that combine ceramics with polymers (e.g., PCL, PLA) or metals (e.g., Ti, Mg) to create scaffolds with improved elasticity, degradability, and cell affinity. These multi-material scaffolds can be customized through multi-head 3D bioprinting or co-extrusion to provide functionally graded regions where stiffer, load-bearing areas transition smoothly into more porous, biologically active zones. Innovations like nanostructured ceramic reinforcements, bioactive coatings, and ceramic-based bioinks are pushing the boundaries further, enabling the delivery of therapeutic agents, growth factors, or even stem cells directly from within the scaffold. Applications range from craniofacial reconstruction and dental socket filling to spinal fusion devices and segmental bone defect repair. The synergy of ceramic bioactivity with the flexibility of polymers and the strength of metals continues to unlock new frontiers in personalized, regenerative implant design tailored for complex anatomical and functional demands.

8.4 4D Printing and Stimuli-Responsive Biomaterials in Medicine

4D printing represents a transformative advancement in additive manufacturing, where time and stimulus-driven change are incorporated into the functional design of printed biomedical devices. Unlike conventional 3D-printed constructs, 4D-printed structures are engineered to dynamically respond to environmental cues such as temperature, moisture, pH, magnetic fields, or specific biological signals after implantation. This is made possible through the use of stimuli-responsive or "smart" biomaterials, including shape-memory polymers, hydrogels, and multi-material composites that can undergo controlled deformation, unfolding, or activation in a predictable and programmable manner [23]. In biomedical applications, this technology paves the way for minimally invasive implants that deploy autonomously *in situ*, smart drug delivery platforms that release therapeutics in response to localized physiological changes, and bioinspired scaffolds that morph to match complex anatomical geometries or support dynamic tissue growth. For instance, vascular grafts that self-roll upon hydration or neural scaffolds that expand in response to body heat exemplify how 4D printing introduces adaptive functionality previously unattainable with static implants.

Multi-material 4D printing enables the sequential activation of different components, allowing for temporal control over multiple therapeutic actions such as staged drug delivery or progressive tissue integration. As the field evolves, researchers are also exploring biodegradable, patient-specific, and even cell-laden 4D constructs that respond in real-time to biological feedback, offering unprecedented potential for regenerative medicine, targeted therapy, and smart prosthetics. This synergy of programmable design, smart materials, and biomedical function makes 4D printing a cornerstone technology in the emerging era of adaptive and personalized healthcare solutions [24].

8.4.1 Shape-Memory Polymers for Environmentally Activated Implants

Shape-memory polymers (SMPs) are an innovative class of "smart" materials capable of undergoing controlled transformations in shape when exposed to specific external stimuli such as heat, moisture, light, or magnetic fields. In the context of biomedical applications, thermal activation (near body temperature) is the most commonly used trigger, making SMPs particularly appealing for designing minimally invasive medical devices. These polymers can be temporarily deformed into a compact shape for easy insertion into the body such as through a small incision or catheter and later recover their original, functional form upon exposure to physiological conditions [25]. A classic example is the fabrication of self-expanding vascular stents, which are delivered in a folded configuration and expand autonomously at the target site, reducing the need for invasive deployment procedures. This property significantly lowers patient trauma, surgery time, and risk of complications. Additionally, the mechanical behavior of SMPs can be precisely engineered by adjusting their polymer backbone, crosslinking density, or blending with additives, allowing the creation of implants that match tissue mechanics while providing dynamic support during healing [26].

Recent developments in biomedical SMPs also focus on biodegradability and biofunctionality, aiming to create implants that not only adapt in form but also naturally degrade in the body once their therapeutic role is fulfilled. For instance, polylactide (PLA), polycaprolactone (PCL), and their copolymers are widely researched for creating degradable SMPs that avoid long-term complications associated with permanent implants. To further enhance their performance, these polymers can be combined with bioactive fillers, such as hydroxyapatite or silica nanoparticles, to promote osteoconductivity and cell attachment in bone-related applications. The ability to integrate multiple functions, such as drug delivery, antimicrobial activity, or electrical conductivity, is also being explored through multi-material 4D printing approaches. Such integration allows for site-specific, programmable therapies, where the implant not only supports tissue structurally but also responds to environmental cues to release drugs or activate embedded sensors [27, 28]. As research advances, SMP-based 4D printed devices are poised to revolutionize soft tissue implants, neuroprosthetics, and smart wound closure systems by combining mechanical intelligence, biocompatibility, and patient-specific adaptability.

8.5 Bioprinting and WAAM-Based Scaffold Fabrication for Regenerative Medicine

The combination of bioprinting and WAAM is transforming scaffold design as it brings with it the biological functionality of cell-laden 3D printing with the beneficial structural properties of metal-based additive manufacturing. Bioprinting is defined as the process by which living cells, growth factors, and hydrogels are deposited onto a surface in a controlled spatial manner. This method allows for the creation of soft tissue constructs that can replicate a native biological environment. Engineering tissues like cartilage, skin, vascular networks, or liver lobules that have specific cellular functionality and microarchitecture can therefore prove challenging. However, bioinks that are created using natural polymers (such as alginate or collagen), synthetic polymers, and even self-healing or stimuli responsive materials are being optimized to allow for high cell viability, greater integration into host tissues, and printability. Bioprinting allows for the direct and correct placement of multiple cell types and biomolecules in a three-dimensional space. This enables researchers to create fully functional tissues with additional features not commonly found in manually constructed tissues, such as biologically unique gradients and regionally specific signaling [29, 30].

The WAAM technique provides high deposition rate method to fabricate mechanically strong, geometrically complex metallic structures. This is especially useful for load-bearing applications including orthopedic implants, dental prosthetics and cranial plates. WAAM uses an electric arc to melt metal wire, which is then deposited, layer upon layer, to form components with virtually any type of architecture and internal porosity that can be custom designed. Bioprinting approaches (e.g. plasma-treated or hydroxyapatite coated) applied to structural materials produced with WAAM can result in scaffolds with not only high mechanical integrity, but also enhanced osseointegration. The combination of WAAM and bioprinting in this way facilitates hybrid scaffold manufacture, where a guess of and soft bioactive layers or hydrogel-based microenvironments are bioprinted onto or into the WAAM produced structure [31, 32]. WAAM and bioprinting act as a continuum that links soft and hard tissue engineering and promotes patient-defined, multi-material implants that are biocompatible and mechanically stable, and therefore enhances the potential of personalized regenerative medicine [33].

8.5.1 Bioink Formulation Strategies for Cell-Laden and Bioactive Constructs

Bioinks are engineered materials designed for use in bioprinting in order to deposit living cells, biomaterials, and bioactive molecules for the construction of complex tissue-like structures in a spatially patterned fashion. Bioinks are based on hydrogels being polymer networks with high water content that replicate a three-dimensional, hydrated and soft environment similar to the body's extracellular matrix (ECM) [34]. Bioinks made from natural hydrogels, such as gelatin methacrylate (GelMA), alginate, hyaluronic acid, collagen and fibrin, are advantageous due to their ability to support cell viability, adhesion and proliferation; however, the mechanical strength of some hydrogels is not sufficient to allow for stable printing or duration of use. GelMA, alginate, hyaluronic acid, collagen and fibrin can be blended or cross-linked with synthetic polymers, such as polyethylene glycol (PEG), Pluronic F127, or polycaprolactone (PCL), to improve structural performance for printing and durability. The major consideration when developing bioinks is relating rheological properties (viscosity and shear-thinning) for smooth extrusion, gelation dynamics for post-print shape fidelity, and cytocompatibility for proper cell viability, functionality, and durability during the printing process [35].

Development of bioinks has progressed to the point where functionality and application-specific customization are emphasized. Various types of bioinks are now made to order with formulations mostly involving growth factors, matrix proteins (i.e., fibronectin or laminin), and signaling peptides to drive cell identity and tissue morphogenesis. There are bioinks, specifically for hard tissues (i.e., bone or cartilage), that uses ceramic particles (i.e., hydroxyapatite or tricalcium phosphate) or even metals (i.e., metallic nanoparticles or soluble forms of metals) as reinforcement or possibly osteoconductivity [36]. Additionally, stimuli-responsive bioinks, which have behavior that is changed with pH, temperature, electric fields, or enzymes, can develop dynamic implants that change with the tissue environment. Bioinks that are "smart" can generate self-healing constructs, controlled release drug delivery systems, and eventually bioprinted tissues that grow, remodel, and integrate into the host over the course of an extended timeframe [37]. Such advancements are advancing bioprinting beyond static forms of structure, toward dynamic, functional, adaptive regenerative systems catering to patient-care medicine.

8.6 Conclusion

Additive Manufacturing has revolutionized biomedical engineering, allowing for patient-specific implants, scaffolds, and medical devices to be fabricated with unprecedented precision and functionality. The combination of design methods (e.g. topology optimization, generative design, and finite element simulations) with biocompatible material systems allows considerable improvements in mechanical performance, osseointegration, and long-term implant stability. Advances in SLM; EBM; WAAM; and bioprinting technologies allow for fabrication of geometries, graded porosity, and multi-material architectures that could not be produced during traditional manufacturing. Advances in biodegradable polymers, metallic biomaterials, and bioactive ceramics, allow the structural and biological requirements of the devices to be met for tissue regeneration, and improving patient recovery.

The rise of 4D printing and stimuli-responsive biomaterials has created a new fluidity for biomedical devices that can adapt, change shape or release therapeutics based on some physiological signal. Hybrid manufacturing methods, such as exploiting the structural integrity WAAM offers while synthesizing soft tissue with a bioprinter, can successfully unite hard tissue and soft tissue engineering. The future of this field is totally reliant on interdisciplinary collaboration, diligent *in vivo* analysis, and the development of standard regulatory processes, if it is to be used effectively in clinical practice. With these advancements, biomedical additive manufacturing will be able to provide advanced, individualized, adaptive and regenerative healthcare methods, and will foster the next stage of patient-centered care.

AI disclosure statement: The author acknowledges the use of AI-based tools for minor language editing or grammatical correction. However, all intellectual contributions, analyses, and conclusions presented in this chapter are entirely the author's own.

References

1. Ajithkumar, S. and Arulmurugan, B., Microstructure and mechanical properties of Inconel 686 fabricated by gas metal arc welding-based wire arc directed energy deposition: impact of cryogenic treatments. *J. Mater. Sci.*, 2024.
2. Tamayo, J.A., Riascos, M., Vargas, C.A., Baena, L.M., Additive manufacturing of Ti6Al4V alloy *via* electron beam melting for the development of implants for the biomedical industry. *Heliyon*, 7, 2021.
3. Liu, R., Chen, Y., Liu, Y., Yan, Z., Wang, Y.X., Topological Design of a Trabecular Bone Structure with Morphology and Mechanics Control for Additive Manufacturing. *IEEE Access*, 9, 2021.
4. Mirzaali, M.J., Moosabeiki, V., Rajaai, S.M., Zhou, J., Zadpoor, A.A., Additive Manufacturing of Biomaterials—Design Principles and Their Implementation. *Materials*, 15, 2022.
5. Pugliese, R. and Regondi, S., Artificial Intelligence-Empowered 3D and 4D Printing Technologies toward Smarter Biomedical Materials and Approaches. *Polymers (Basel)*, 14, 2022.
6. Wen, Y., Ding, Y., Chang, L., Luo, X., Qiao, B., Dong, C., *et al.*, Effect of designed pore size on properties of additively manufactured TC4 with micro-nano biomedical surface. *J. Mater. Res. Technol.*, 36, 5320–31, 2025.
7. Lan, T., Do, T., Al-Ketan, O., Fox, K., Tran, P., Concurrent multiscale topology optimisation towards design and additive manufacturing of biomimicking porous structures. *Virtual Phys. Prototyping*, 18, 2023.
8. Soro, N., Brodie, E.G., Abdal-hay, A., Alali, A.Q., Kent, D., Dargusch, M.S., Additive manufacturing of biomimetic Titanium-Tantalum lattices for biomedical implant applications. *Mater. Des.*, 218, 2022.
9. Carnicero, A., Peláez, A., Restoy-Lozano, A., Jacquott, I., Perera, R., Improvement of an additively manufactured subperiosteal implant structure design by finite elements based topological optimization. *Sci. Rep.*, 11, 2021.
10. Noman, A.A., Shaari, M.S., Mehboob, H., Azman, A.H., Recent advancements in additively manufactured hip implant design using topology optimization technique. *Results Eng.*, 25, 103932, 2025.
11. Malachi, I.O., Olawumi, A.O., Oladapo, B.I., AI-driven optimization of spinal implant design using parametric modelling. *Colloids Surf., B*, 253, 114753, 2025.

12. García-Lizarribar, A., Villasante, A., Lopez-Martin, J.A., Flandez, M., Soler-Vázquez, M.C., Serra, D., *et al.*, 3D bioprinted functional skeletal muscle models have potential applications for studies of muscle wasting in cancer cachexia. *Biomater. Adv.*, 150, 213426, 2023.
13. Bose, S., Robertson, S.F., Bandyopadhyay, A., Surface modification of biomaterials and biomedical devices using additive manufacturing. *Acta Biomater.*, 66, 2018.
14. Mondal, P., Das, A., Wazeer, A., Karmakar, A., Biomedical porous scaffold fabrication using additive manufacturing technique: Porosity, surface roughness and process parameters optimization. *Int. J. Lightweight Mater. Manuf.*, 5, 2022.
15. Calais, G.B., Garcia, G.D., de Moura Júnior, C.F., Soares, J.D.M., Lona, L.M.F., Beppu, M.M., *et al.*, Therapeutic functions of medical implants from various material categories with integrated biomacromolecular systems. *Front. Bioeng. Biotechnol.*, 12, 2025.
16. Puppi, D. and Chiellini, F., Biodegradable Polymers for Biomedical Additive Manufacturing. *Appl. Mater. Today*, 20, 2020.
17. Barucca, G., Santecchia, E., Majni, G., Girardin, E., Bassoli, E., Denti, L., *et al.*, Structural characterization of biomedical Co-Cr-Mo components produced by direct metal laser sintering. *Mater. Sci. Eng., C*, 48, 2015.
18. Jang, T.S., Kim, D.E., Han, G., Yoon, C.B., Jung, H.-D., Powder based additive manufacturing for biomedical application of titanium and its alloys: a review. *Biomed. Eng. Lett.*, 10, 2020.
19. Rajput, A.S., Kapil, S., Das, M., A post processing technique to achieve nanofinishing for functionality enhancement of Ti-6Al-4V femoral head fabricated by Laser Powder Bed Fusion. *CIRP J. Manuf. Sci. Technol.*, 45, 2023.
20. Qin, Y., Liu, A., Guo, H., Shen, Y., Wen, P., Lin, H., *et al.*, Additive manufacturing of Zn-Mg alloy porous scaffolds with enhanced osseointegration: *In vitro* and *in vivo* studies. *Acta Biomater.*, 45, 2022.
21. Lin, K., Sheikh, R., Romanazzo, S., Roohani, I., 3D printing of bioceramic scaffolds-barriers to the clinical translation: From promise to reality, and future perspectives. *Materials*, 12, 2019.
22. Qiao, S., Sheng, Q., Li, Z., Wu, D., Zhu, Y., Lai, H., *et al.*, 3D-printed Ti6Al4V scaffolds coated with freeze-dried platelet-rich plasma as bioactive interface for enhancing osseointegration in osteoporosis. *Mater. Des.*, 194, 108825, 2020.
23. Lui, Y.S., Sow, W.T., Tan, L.P., Wu, Y., Lai, Y., Li, H., 4D printing and stimuli-responsive materials in biomedical aspects. *Acta Biomater.*, 92, 2019.
24. Ryan, K.R., Down, M.P., Banks, C.E., Future of additive manufacturing: Overview of 4D and 3D printed smart and advanced materials and their applications. *Chem. Eng. J.*, 403, 2021.
25. Xiao, R. and Huang, W.M., Heating/Solvent Responsive Shape-Memory Polymers for Implant Biomedical Devices in Minimally Invasive Surgery: Current Status and Challenge. *Macromol. Biosci.*, 20, 2020.

26. Sajjad, R., Chauhdary, S.T., Anwar, M.T., Zahid, A., Khosa, A.A., Imran, M., *et al.*, A review of 4D printing – Technologies, shape shifting, smart polymer based materials, and biomedical applications. *Adv. Ind. Eng. Polym. Res.*, 7, 2024.
27. Rokaya, D., Skallevold, H.E., Srimaneepong, V., Marya, A., Shah, P.K., Khurshid, Z., *et al.*, Shape Memory Polymeric Materials for Biomedical Applications: An Update. *J. Compos. Sci.*, 7, 2023.
28. Zhao, W., Yue, C., Liu, L., Liu, Y., Leng, J., Research Progress of Shape Memory Polymer and 4D Printing in Biomedical Application. *Adv. Healthcare Mater.*, 12, 2023.
29. Li, J., Chen, M., Fan, X., Zhou, H., Recent advances in bioprinting techniques: Approaches, applications and future prospects. *J. Transl. Med.*, 14, 2016.
30. Rider, P., Kačarević, Z.P., Alkildani, S., Retnasingh, S., Barbeck, M., Bioprinting of tissue engineering scaffolds. *J. Tissue Eng.*, 9, 2018.
31. Rodrigues, T.A., Duarte, V., Miranda, R.M., Santos, T.G., Oliveira, J.P., Current status and perspectives on wire and arc additive manufacturing (WAAM). *Materials*, 12, 2019.
32. Mobarak, M.H., Islam, M.A., Hossain, N., Al Mahmud, M.Z., Rayhan, M.T., Nishi, N.J., *et al.*, Recent advances of additive manufacturing in implant fabrication – A review. *Appl. Surf. Sci. Adv.*, 18, 2023.
33. Deshmukh, P.S., Katiyar, A., Sahu, A., Sathiaraj, D., Palani, I.A., Sonawane, A., Wire arc additive manufacturing of commercially pure titanium bio-medical alloy. *Mater. Today Proc.*, 90, 6–10, 2023.
34. Chen, X.B., Fazel Anvari-Yazdi, A., Duan, X., Zimmerling, A., Gharraei, R., Sharma, N.K., *et al.*, Biomaterials / bioinks and extrusion bioprinting. *Bioact. Mater.*, 28, 2023.
35. Paxton, N., Smolan, W., Böck, T., Melchels, F., Groll, J., Jungst, T., Proposal to assess printability of bioinks for extrusion-based bioprinting and evaluation of rheological properties governing bioprintability. *Biofabrication*, 9, 2017.
36. Chimene, D., Lennox, K.K., Kaunas, R.R., Gaharwar, A.K., Advanced Bioinks for 3D Printing: A Materials Science Perspective. *Ann. Biomed. Eng.*, 44, 2016.
37. Maan, Z., Masri, N.Z., Willerth, S.M., Smart Bioinks for the Printing of Human Tissue Models. *Biomolecules*, 12, 2022.

9

Role of Functionally Graded Materials in Biomedical Applications

Debarupam Gogoi*, **Mihir Kumar Pandit and Arun Kumar Pradhan**

School of Mechanical Sciences, Indian Institute of Technology Bhubaneswar, Bhubaneswar, Odisha, India

Abstract

Functionally graded materials (FGMs): solids whose constituent composition or microstructure changes gradually along one or more dimensions – have matured from a biomimetic curiosity to a versatile design platform for next-generation medical systems. By tailoring spatial gradients in chemistry, porosity, modulus, and surface energy, FGMs reconcile normally conflicting design requirements such as high strength with compliance of bone, corrosion resistance with bioresorbable, and antibacterial activities with cytocompatibility. This chapter surveys the scientific principles and engineering practice that underpin biomedical FGMs, summarizes the state of the art across hard- and soft-tissue applications, and outlines challenges that must be overcome before widespread of clinical adoption.

Keywords: Functionally graded materials, implants, biomedical, scaffolds, orthopedic

9.1 Introduction

The ideal biomaterial must seamlessly integrate mechanically and biologically with living tissues, resist degradation in harsh physiological environments, and remain economically viable for mass production [1]. Traditional monolithic biomaterials such as metals (e.g., titanium, stainless steel), ceramics (e.g., alumina, zirconia), and polymers (e.g., polyethylene, polyurethane) generally struggle to simultaneously meet these stringent

**Corresponding author*: a21me09006@iitbbs.ac.in

Sandip Kunar, Jagadeesha T. and D. Sathis Kumar (eds.) Biomanufacturing: Transforming Production of Biologics and Biopharmaceuticals, (217–248) © 2026 Scrivener Publishing LLC

requirements, particularly at complex interfaces like implant-to-bone or implant-to-cartilage junctions [2]. Nature effectively addresses this integration challenge through continuous gradient structures, notably seen in biological systems such as the osteochondral unit of joints, the dentin–enamel junction in teeth, and even arterial walls, which exhibit progressive variations in stiffness, mineral content, and cellular organization [3].

Functionally graded materials (FGMs) emulate these natural gradients by intentionally varying their composition or structural architecture across spatial dimensions, facilitating improved mechanical compatibility and biological integration. Examples include titanium transitioning gradually to hydroxyapatite (Ti → Ti–HA → HA) or porous outer surfaces transitioning to dense inner cores [4, 5]. These engineered gradients mimic physiological interfaces, providing smoother stress transitions, promoting tissue ingrowth, and reducing complications such as implant loosening or stress shielding [6].

Early biomedical applications of FGMs in the 1990s primarily involved ceramic-to-metal dental implants fabricated through powder lamination techniques, significantly enhancing implant longevity and osseointegration compared to homogeneous materials [7]. Recent advances in fabrication technologies—such as high-resolution additive manufacturing (AM), spark-plasma sintering (SPS), electrophoretic deposition, and nanoscale layer-by-layer assembly—have revolutionized the production of intricate multi-material gradients with precise control and reproducibility [8, 9].

These technological breakthroughs have aligned fortuitously with demographic trends like global population aging, increased demand for revision surgeries, and the expansion of regenerative medicine, underscoring the renewed importance of the FGM paradigm of placing "the right material in the right place" [10]. Consequently, contemporary research extensively explores FGMs across various biomedical fields including orthopedics, dental, cardiovascular, and soft-tissue applications [11, 12].

In orthopedics, FGMs enhance implant stability and minimize adverse biological reactions by optimizing mechanical and biological integration through tailored stiffness gradients that mimic native bone tissue [13]. Dental applications focus predominantly on implants designed with ceramic-metal or ceramic-polymer gradients to optimize aesthetics, durability, and bio integration [14]. Cardiovascular FGMs aim to replicate the mechanical gradients observed in arterial walls, improving long-term outcomes of vascular grafts and stents [15]. Soft-tissue FGMs find application in tissue scaffolds and artificial skin, facilitating cellular attachment, proliferation, and differentiation [16].

Emerging trends in drug delivery and tissue engineering exploit FGMs to enable controlled release profiles and scaffold bioactivity tailored precisely for tissue-specific regenerative applications [17]. Despite the promising capabilities of FGMs, significant translational barriers persist, notably regulatory hurdles, the need for extensive long-term biocompatibility and corrosion studies, and the integration of data-driven computational design methodologies to streamline material development [18].

Overall, while significant progress has been made, the future clinical success of FGMs will depend on continued interdisciplinary collaboration among materials scientists, clinicians, and regulatory agencies to resolve these challenges, ultimately enhancing patient outcomes and implant longevity.

9.2 Fundamentals of Functionally Graded Materials

9.2.1 Definition and Historical Perspective

Functionally graded materials (FGMs) are defined as solid materials whose microstructural composition, architecture, or properties vary continuously over volume, thus eliminating distinct interfaces at the mesoscopic scale. This characteristic gradation of properties can follow different mathematical models such as linear, power-law, exponential, or voxel-based programmable designs [4, 19].

The concept of FGMs was systematically formalized through Japan's pioneering "FGM Project" (1987–1993), originally initiated to develop advanced thermal barrier materials for reusable spacecraft, specifically spaceplane structures exposed to extreme thermal gradients [20]. The project's success led to broader interest across diverse fields (Figure 9.1), including biomedical engineering, where FGMs offered compelling solutions to mimic naturally occurring gradients, most notably human bone—a classic biological model showcasing graded structures [7, 21, 22].

Early biomedical demonstrations of FGMs were exemplified by the centrifugal casting of dense-to-porous Ti-6Al-4V alloy rods, which provided crucial proof-of-concept validations highlighting the material's potential in load-bearing orthopedic implants [23].

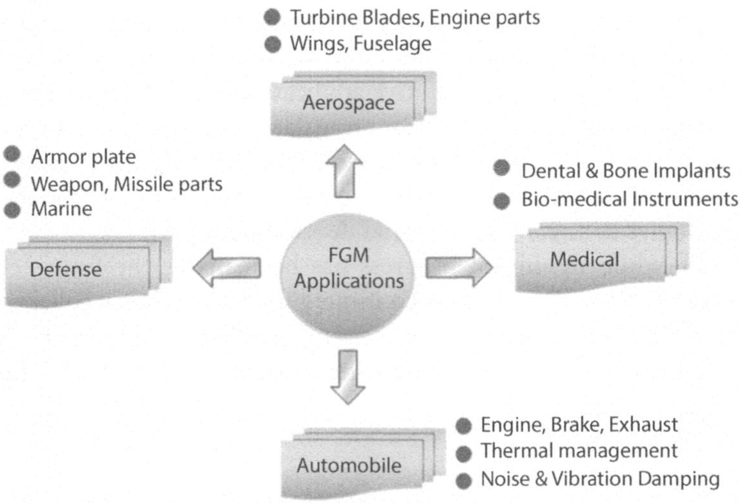

Figure 9.1 Schematic diagram illustrating some of the application field of FGM structures.

9.2.2 Types of Gradients

FGMs can be categorized by various forms of property gradients. Figure 9.2 [24], illustrates the gradation of biological materials in different way across an interface.

- **Composition Gradient:**

Compositional gradients control the mass fraction of constituent phases within the material, such as transitioning from metal-rich interiors to ceramic-rich exteriors. These gradients can significantly influence mechanical properties, thermal expansion coefficients, and biological compatibility [4, 25].

- **Porosity Gradient:**

Gradation in porosity includes variations in open-cell architecture and pore size distribution. Porosity gradients facilitate controlled biological interactions, allowing cellular ingrowth at the implant surface while maintaining structural integrity internally [26].

- **Crystallographic Gradient:**

Crystallographic transitions, such as from β-phase titanium to α-phase titanium, can adjust mechanical properties like Young's modulus, significantly reducing stress shielding in orthopedic implants by aligning mechanical properties closer to natural bone [27].

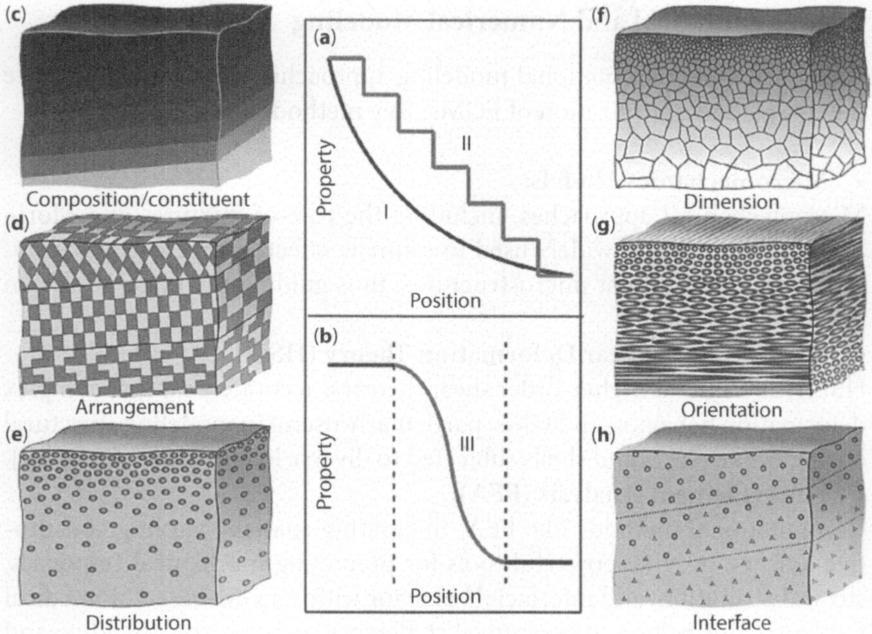

Figure 9.2 Local property profiles and basic forms of gradients in biological materials: (a) Local properties change either gradually (I) or in a stepwise manner (II) through the entire material volume; (b) Local properties vary continuously across the interface between dissimilar components; (c–g) The gradients in biological materials are fundamentally associated with the changes in chemical compositions/constituents (c) and structural characteristics, including the arrangement (d), distribution (e), dimensions (f), and orientations (g) of building units; (h) Gradient interface in biological materials; Liu, Z *et al.* [24]. (L.N: 6078890868907, Elsevier).

- **Surface Chemistry Gradient:**

Surface chemistry gradients involve controlled variations in chemical composition, such as apatite-rich outer layers for enhanced Oste conduction and noble-metal inner layers to enhance corrosion resistance, thereby significantly improving implant durability and bio integration [11, 28].

- **Functional Gradients:**

These gradients involve the controlled spatial release of bioactive substances, including drugs, growth factors, or antimicrobial ions, enabling targeted therapeutic action and improved implant-tissue integration [17].

9.2.3 Analytical and Numerical Modeling

Analytical and computational modeling approaches significantly advance the design and optimization of FGMs. Key methodologies include:

- **Micromechanics Models:**
Micromechanical approaches, including the rule-of-mixtures and Mori–Tanaka methods, are widely used to estimate effective macroscopic properties from constituent microstructures, thus guiding the material design process [29, 30].
- **Higher-Order Shear Deformation Theory (HSDT):**
HSDT, and similar higher-order shear theories, accurately capture complex deformation behaviors in FGMs, particularly useful in modeling structural elements like plates and shells subjected to diverse loading conditions [31].
- **Finite Element Analysis (FEA):**
Computational methods like FEA, integrating spatially varying constitutive tensors, provide powerful tools for predicting mechanical responses, stress distribution, and interfacial behavior within FGM-based biomedical implants. This helps address critical challenges such as stress shielding and implant loosening [32].
- **Peridynamic Methods:**
Non-local continuum mechanics methods like peridynamics are particularly advantageous in modeling fracture behavior in FGMs, accommodating discontinuities naturally, which traditional finite-element methods may find challenging [33].

9.2.4 Manufacturing Routes for Biomedical FGMs

Functionally graded materials (FGMs) in biomedical applications require precise manufacturing methods to achieve controlled property gradation. Different fabrication approaches impart distinct resolutions, internal stresses, mechanical properties, and biocompatibility characteristics [4, 9]. Table 9.1 presents five advanced manufacturing routes with techniques implemented to produce a few biomedical FGMs.

- **Powder Metallurgy**
Powder metallurgy methods include techniques such as graded powder stacking, slip casting, and centrifugal solidification, providing effective control of compositional gradients and porosity. Centrifugal solidification is particularly suitable for creating radial density gradients necessary for orthopedic implants like titanium-titanium carbide (Ti/TiC) hip stems, balancing stiffness and strength [23].

Table 9.1 Manufacturing techniques and biomedical examples of FGMs.

Category	Representative techniques	Biomedical example
Powder Metallurgy	Graded powder stacking, slip casting, centrifugal solidification	Ti/TiC hip stems [23].
Additive Manufacturing	Selective laser melting (SLM), electron-beam melting (EBM), direct ink writing (DIW), material jetting	Macro-porous Ti–Ta acetabular cups [8, 34].
Surface Engineering	Pulsed laser deposition, plasma spraying, sol–gel dip-coating	Ti–HA gradient coatings on dental screws [35].
Solid–Solid Diffusion	Spark-plasma sintering (SPS), diffusion bonding	3Y-TZP/Ti-6Al-4V bilayers [36].
Polymeric Assembly	Photopolymerized hydrogel gradients, interpenetrating networks (IPN)	CMC–silk IPN scaffolds for cartilage regeneration [37].

- **Additive Manufacturing (AM)**

AM processes like selective laser melting (SLM), electron beam melting (EBM), direct ink writing (DIW), and material jetting provide unparalleled design flexibility and spatial control of properties. These techniques enable customized macro-porous structures beneficial for osseointegration, exemplified by Ti–Ta acetabular cups in orthopedic implants, optimizing bone ingrowth and load distribution [8, 34].

- **Surface Engineering**

Surface engineering techniques—such as pulsed laser deposition, plasma spraying, and sol–gel dip-coating—enable precise control of thin-layer gradients for enhanced bioactivity and corrosion resistance. Titanium-hydroxyapatite (Ti–HA) gradient coatings on dental screws effectively enhance osseointegration and biocompatibility at the implant-tissue interface [35].

- **Solid–Solid Diffusion**

Techniques like spark-plasma sintering (SPS) utilize rapid thermal gradients to facilitate controlled diffusion between dissimilar materials, producing robust bilayer structures such as zirconia-titanium alloy (3Y-TZP/Ti-6Al-4V) composites. These structures achieve seamless interfacial bonds, significantly improving fracture resistance and longevity in load-bearing applications [36].

- **Polymeric Assembly**

Polymeric assembly methods involving photopolymerization and interpenetrating polymer networks (IPNs) facilitate the creation of soft and viscoelastic gradients suitable for tissue engineering applications. For example, carboxymethyl cellulose–silk (CMC–silk) IPN scaffolds provide tailored mechanical gradients and biochemical cues essential for cartilage regeneration and repair [37].

Each manufacturing method introduces specific limitations regarding achievable resolution, internal residual stress profiles, sterilization compatibility, and resultant mechanical properties. Consequently,

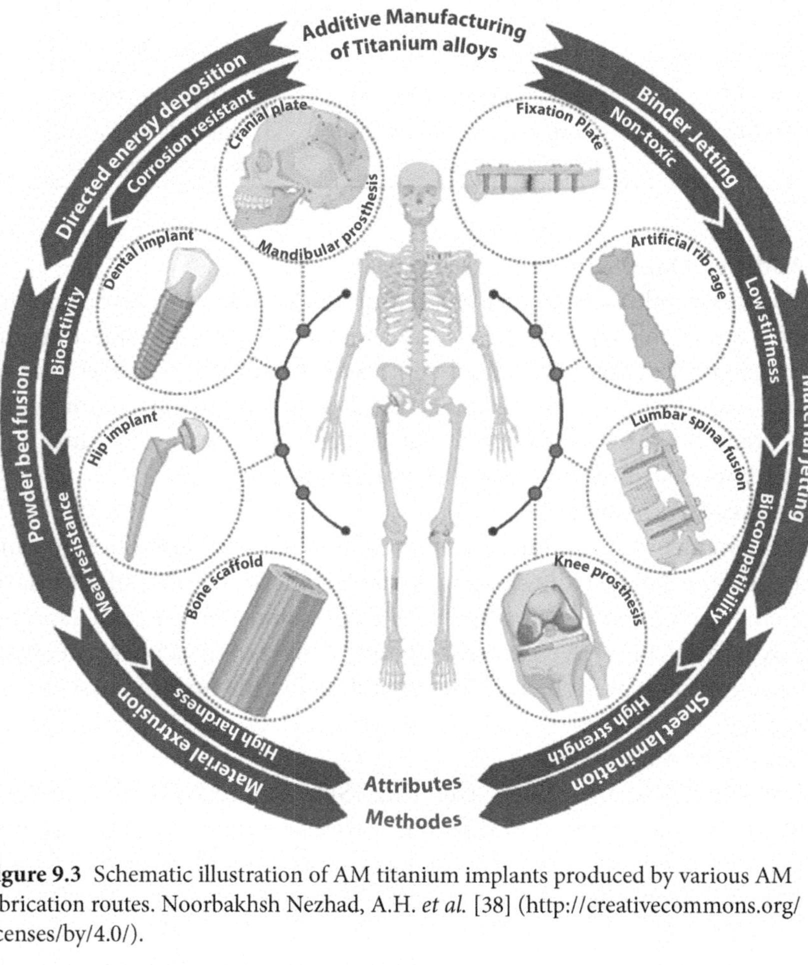

Figure 9.3 Schematic illustration of AM titanium implants produced by various AM fabrication routes. Noorbakhsh Nezhad, A.H. *et al.* [38] (http://creativecommons.org/licenses/by/4.0/).

process–structure–property optimization remains a dynamic research area, continually refined by interdisciplinary collaboration among materials scientists, biomedical engineers, and clinicians [4, 9]. In Figure 9.3 [38], titanium implants produced by different additive manufacturing methods are shown.

9.3 Biological Rationale for Graded Implants

A critical challenge in biomedical implant design is the mechanical mismatch between stiff metallic implants—such as those made from titanium alloys (e.g., Ti-6Al-4V with Young's modulus, E ≈ 110 GPa)—and the comparatively softer cortical bone tissue (E ≈ 18–30 GPa). This significant discrepancy in mechanical properties frequently leads to stress shielding, where the implant absorbs disproportionate amounts of mechanical load, resulting in peri-implant bone resorption and eventually implant loosening or failure [39, 40].

Gradually reducing the modulus of the implant material toward its surface, thereby creating a gradient from metallic core to bone-mimicking exterior, can effectively minimize interfacial shear stresses. This gradation helps achieve a more physiological distribution of mechanical loads, facilitating bone remodeling and maintaining long-term implant stability [41, 42]. Porosity gradients further enhance this approach by maintaining internal stiffness and structural integrity in the implant core while simultaneously providing a porous outer layer that promotes vascularization, nutrient diffusion, and tissue ingrowth, thereby significantly improving biological integration [43, 44].

Additionally, surface chemistry gradients—such as transitions from titanium nitride (TiN) to titanium dioxide (TiO_2) and subsequently to calcium phosphate (CaP)—can precisely modulate corrosion resistance, protein adsorption profiles, and cellular responses like osteoblast differentiation and mineral deposition [11, 35]. These chemically tailored surfaces facilitate enhanced osseointegration by actively guiding biological interactions at the implant–bone interface, ultimately improving implant performance and patient outcomes [45].

Recent advancements in magnesium-based biodegradable FGMs have further extended this biological rationale. Magnesium alloys gradually degrade *in vivo*, allowing them to temporarily bear mechanical loads during the critical healing period and subsequently dissolve, thereby eliminating the need for secondary surgeries for implant removal. This biodegradable FGM strategy aligns precisely with natural bone healing and remodeling timelines, thus representing a highly promising development in orthopedic and trauma applications [46–48].

9.4 Classification of Biomedical FGMs

Biomedical functionally graded materials (FGMs) are specifically engineered to match the unique mechanical, biological, and structural requirements of different tissues and clinical applications. They are typically classified according to the nature of their constituent materials and the intended biological interaction.

9.4.1 Metal–Metal FGMs

Metal–metal FGMs typically involve gradients within metallic alloys to modulate mechanical properties and biological performance. Examples include β-titanium (β-Ti) to α + β-Ti gradient rods designed for spinal fixation applications. Such gradients offer customizable stiffness profiles that closely match surrounding vertebral tissues, effectively minimizing stress shielding and enhancing load transfer [49]. Additionally, titanium–zirconium (Ti–Zr) diffusion couples, designed to mimic cortical-to-trabecular modulus transitions, can significantly improve implant performance by precisely tailoring mechanical compatibility with adjacent bone structures [50].

9.4.2 Metal–Ceramic FGMs

Metal–ceramic FGMs combine metallic ductility with ceramic bioactivity, prominently used in dental implants. A notable example includes titanium (Ti) transitioning through titanium dioxide (TiO_2) to hydroxyapatite (HA) over a small distance (~2 mm). This graded configuration provides a ductile, tough core essential for load-bearing capacity, while the bioactive HA shell promotes bone attachment and osteoconduction, significantly enhancing implant integration and longevity [7, 35].

9.4.3 Biodegradable Metal FGMs

Biodegradable metal FGMs utilize metals such as magnesium (Mg) alloys combined with bio ceramic layers to control corrosion rates. Mg–Zn–Ca alloys, when graded with bio glass layers, offer controlled degradation *in vivo*, matching the healing timeline of bone regeneration. This gradual corrosion minimizes inflammatory responses and eliminates the need for implant removal surgeries, making them ideal for temporary orthopedic applications and trauma fixation [47, 48].

9.4.4 Polymer–Ceramic FGMs

Polymer–ceramic FGMs integrate the toughness and elasticity of polymers with the rigidity and bioactivity of ceramics. Polycaprolactone (PCL)–beta-tricalcium phosphate (β-TCP) scaffolds, fabricated through direct ink writing (DIW) with continuously graded porosity and ceramic loading, have been successfully applied for cranial defect repair. The tailored porosity gradients facilitate optimal nutrient transport and cellular ingrowth, while the graded ceramic content supports osteogenesis and mechanical stability [51].

9.4.5 Hydrogel FGMs

Hydrogel FGMs leverage gradients in biochemical and mechanical properties to closely mimic soft-to-hard tissue interfaces, notably the cartilage–bone junction. Dual-gradient hydrogels simultaneously deliver chondrogenic (cartilage-promoting) and osteogenic (bone-promoting) signals, thereby encouraging distinct tissue regeneration in precisely defined regions. This targeted regenerative capability significantly improves outcomes in complex orthopedic and maxillofacial applications [52, 53].

9.4.6 Surface-Graded Coatings (FGCs)

Surface-graded coatings (FGCs) are specifically engineered thin-film layers that impart tailored mechanical and biological properties to implant surfaces. For instance, multilayer coatings such as titanium nitride (TiN), titanium carbide (TiC), and titanium dioxide (TiO_2) provide enhanced wear resistance, corrosion protection, and excellent cytocompatibility. These graded coatings optimize the surface characteristics for various biomedical devices, significantly enhancing their functional lifespan and biological integration [35, 54].

9.5 Design Considerations

The successful implementation of biomedical FGMs requires careful consideration of several interrelated design factors to ensure both mechanical performance and biological efficacy.

9.5.1 Mechanical Integrity

Mechanical integrity is critical for FGMs intended for load-bearing biomedical applications. These materials must withstand repeated cyclic loads typical in orthopedic and dental implants while avoiding brittle failure at interlayer boundaries. The fracture mechanics criteria for FGMs extend traditional Griffith fracture theory by explicitly incorporating spatial gradients in toughness and mechanical properties. This spatial variability influences crack propagation paths, enhancing overall durability and reliability. Design strategies involve selecting material gradients that smoothly transition mechanical properties to prevent stress concentrations and premature interfacial failure [55, 56].

9.5.2 Porosity and Permeability

Porosity gradients are designed to balance mechanical strength with biological functionality. Pore sizes typically ranging between 100–600 μm have been identified as optimal for promoting angiogenesis and facilitating vascular infiltration, essential for nutrient transport and tissue regeneration. However, excessively high porosity can significantly reduce mechanical strength and fatigue resistance. The strategic design of graded lattice structures resolves this trade-off, providing high porosity at the implant interface to promote biological integration while maintaining lower porosity cores for mechanical stability and long-term implant performance [43, 57].

9.5.3 Corrosion and Degradation

FGMs incorporating metallic phases must carefully manage corrosion and degradation. Gradients containing galvanically dissimilar metals can lead to accelerated localized corrosion due to galvanic coupling. This phenomenon can significantly compromise implant longevity. Protective ceramic gradient coatings (e.g., TiO_2 transitioning to hydroxyapatite, HA) effectively mitigate corrosion by providing stable, bioactive, and corrosion-resistant surfaces. In the context of biodegradable magnesium-based FGMs, alloying with rare-earth elements and applying calcium phosphate-based top layers substantially reduce initial corrosion rates, thereby ensuring controlled degradation profiles aligned with tissue healing timelines [46, 48].

9.5.4 Bio-Functionality

Bio-functionality in FGMs is enhanced through gradients in biochemical, mechanical, and structural cues. Variations in peptide density, substrate stiffness, and nanoscale surface roughness can guide cellular behavior, particularly stem-cell differentiation pathways, thereby enhancing tissue regeneration. Computational fluid dynamics (CFD) modeling further supports the design of FGM scaffolds by accurately predicting nutrient diffusion, oxygen transport, and metabolic waste removal through porous structures, particularly crucial in large-scale tissue-engineering constructs. These computational predictions enable precise optimization of scaffold architecture, porosity gradients, and biochemical gradients, significantly improving the efficacy of biomedical implants [52, 53, 57].

9.6 Manufacturing Techniques in Detail

Advanced manufacturing techniques are critical for creating precisely graded biomedical FGMs, each offering unique advantages in control over structure, composition, and mechanical properties.

9.6.1 Selective Laser Melting (SLM)

Selective Laser Melting (SLM) is a powder-bed additive manufacturing technique offering precise voxel-wise control over material composition and structure. This capability allows the creation of complex, centimeter-scale implants with finely controlled millimeter-scale gradients. A recent example includes the fabrication of Ti-6Al-4V implants featuring controlled porosity gradients (ranging from 74% to 86%), which restored approximately 95% of the mechanical modulus match relative to natural cortical bone. This substantial modulus match greatly reduces stress shielding and enhances the longevity and biological compatibility of orthopedic implants [34, 58].

9.6.2 Direct Ink Writing (DIW)

Direct Ink Writing (DIW) is an extrusion-based additive manufacturing technique capable of mixing multiple ceramic pastes in real-time during printing. This process permits seamless transitions between different ceramic materials, such as zirconia (ZrO_2) transitioning continuously into alumina (Al_2O_3), within bone scaffolds. DIW-generated FGMs enable tailored mechanical properties, improved biological integration, and enhanced mechanical reliability, effectively matching the complex mechanical and biological requirements at bone-implant interfaces [59, 60].

9.6.3 Spark-Plasma Sintering (SPS) of Metal–Ceramic Bilayers

Spark-Plasma Sintering (SPS) employs rapid Joule heating combined with mechanical pressure to achieve high-density compacts and suppress undesired intermetallic phase formation. Specifically, SPS has demonstrated significant advantages in manufacturing 3 mol% yttria-stabilized zirconia (3Y-TZP) and titanium alloy (Ti-6A-l4V) FGMs. These FGMs exhibited robust seamless bonding at interfaces without the formation of brittle Ti-O_2 interlayers, preserving mechanical integrity and significantly enhancing their viability for load-bearing biomedical applications [36, 61].

9.6.4 Surface Engineering Routes

Surface engineering techniques, including plasma spraying, effectively create thick, bioactive gradient coatings on metal implants. For instance, layering calcium phosphate-rich coatings over titanium nitride (Ti-N) diffusion barriers forms robust surface gradients that exhibit tailored elastic moduli ranging from 20 to 100 GPa. Such graded coatings significantly minimize stress concentrations at the implant–bone interface, thus enhancing osseointegration, reducing implant failure risks, and prolonging clinical lifetimes of femoral stems and other orthopedic implants [35, 62].

9.6.5 Hybrid and Post-Processing

Hybrid manufacturing techniques, including hot-isostatic pressing (HIP), can densify additive-manufactured FGMs, significantly improving mechanical properties by eliminating residual porosity and defects. Post-processing techniques, such as laser shock peening, introduce beneficial compressive surface stresses to enhance fatigue life, resistance to crack initiation, and overall durability of FGMs in biomedical implants.

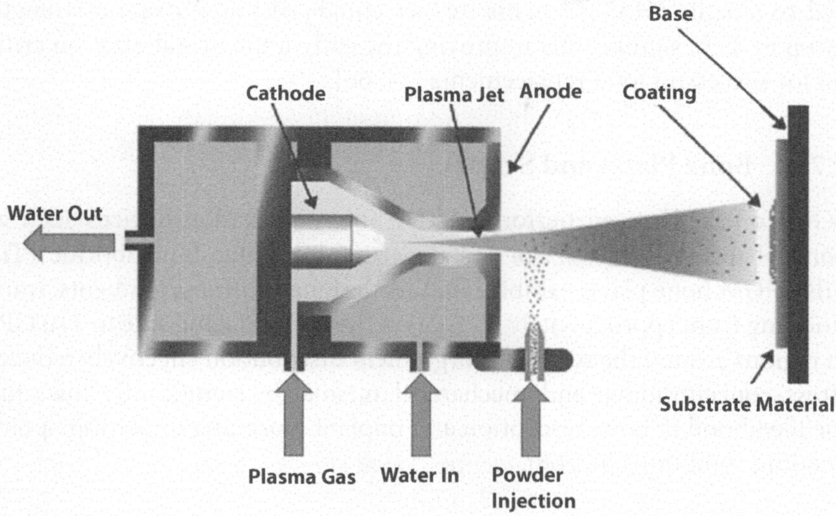

Figure 9.4 Schematic illustration of the plasma spraying process for functionally graded coatings. Sheikh, K.A. *et al.* [65] (http://creativecommons.org/licenses/by/4.0/).

These hybrid and post-processing steps are crucial in achieving the necessary performance characteristics, particularly for implants subjected to cyclic loading conditions in orthopedic and dental applications [63, 64]. Figure 9.4 [65], illustrates plasma spraying process for functionally graded coatings.

9.7 Orthopedic Applications

Orthopedic applications are among the most significant and promising fields for biomedical FGMs due to their ability to closely mimic the mechanical and biological characteristics of natural bone.

9.7.1 Joint Replacements

Functionally graded materials (FGMs) have significantly advanced joint replacement implants, particularly acetabular cups. Implants constructed from titanium-tantalum (Ti–Ta) lattice cores combined with porous titanium alloy (Ti-6A-l4V) shells demonstrated superior biomechanical performance in cadaveric hip simulator studies. These gradient structures resulted in approximately 30% lower micromotion compared to conventional porous-coated monolithic implants. Additionally, infiltration of hydroxyapatite (HA) gradients further enhanced initial fixation strength by up to 25%, significantly improving the early implant stabilization critical for successful joint replacements [34, 66].

9.7.2 Bone Plates and Screws

FGMs offer enhanced performance in internal fixation devices such as bone plates and screws. For instance, titanium–titanium diboride (Ti–TiB_2) FGM bone plates exhibit carefully designed stiffness gradients, transitioning from approximately 55 GPa on the cortex-facing side to 110 GPa in regions around the screws. This gradient distribution effectively reduces stress concentrations and mechanical mismatch, significantly lowering the likelihood of bone resorption and implant loosening under four-point bending conditions [67, 68].

9.7.3 Biodegradable Fixation Devices

Biodegradable FGMs have gained substantial attention in orthopedic trauma applications. Magnesium-zinc-calcium (Mg-Zn-Ca) alloy cores enveloped by bioactive bioglass-rich layers demonstrated optimal performance in animal studies. These devices maintained mechanical integrity for approximately six weeks in rabbit tibial models before benignly dissolving, thus closely aligning with natural bone healing timelines. Critically, this design minimized the formation of hydrogen gas pockets—a common challenge in magnesium alloy implants—thereby significantly improving biological compatibility and reducing postoperative complications [47, 48].

9.7.4 Scaffolds for Bone Defects

Selective Laser Melting (SLM)-fabricated macro-porous scaffolds represent a critical advancement for treating large bone defects. These scaffolds feature precisely graded pore sizes ranging from 1.6 to 2.4 mm, designed to match the mechanical anisotropy and structural complexity of human iliac bone. *In vivo* goat model studies demonstrated remarkable outcomes, with up to 87% bone ingrowth observed after 24 weeks, confirming the scaffold's ability to effectively facilitate bone regeneration, restore structural integrity, and enhance clinical outcomes [58, 69].

Figure 9.5 [70], illustrates the applications of different FGMs in bone.

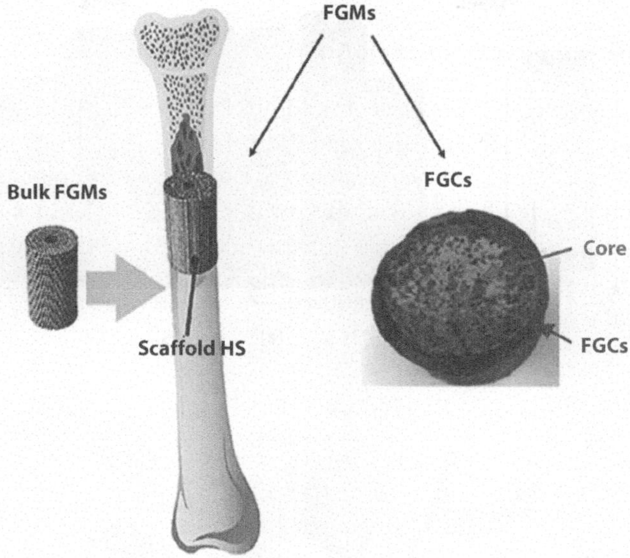

Figure 9.5 Schematic illustration of FGMs applications in bone. Najibi, A., and Mokhtari, T. [70] (L.N. 6084290623490, Elsevier).

9.8 Dental Applications

Dental implants often face clinical challenges, particularly implant failures originating in crestal bone regions. Such failures are typically caused by elevated local stresses, mechanical mismatch, and bacterial infiltration leading to peri-implantitis. Functionally graded materials (FGMs) specifically engineered for dental implants effectively address these challenges through tailored mechanical, chemical, and biological gradients.

9.8.1 Mechanical Gradients (Modulus Gradients)

Mechanical gradients are pivotal in dental FGMs to mitigate stress concentrations at critical implant–bone interfaces. Specifically, dental implants utilizing modulus gradients transition gradually from pure titanium (Ti) at the implant core to β-titanium alloys at the implant surface. This strategic gradation in mechanical properties significantly reduces stress concentrations, especially around the cervical (crestal) region of the implant, which is prone to bone loss due to high stress. Finite-element analyses (FEA) have demonstrated approximately 18% lower von Mises stresses at the implant's cervical area compared to traditional homogeneous titanium

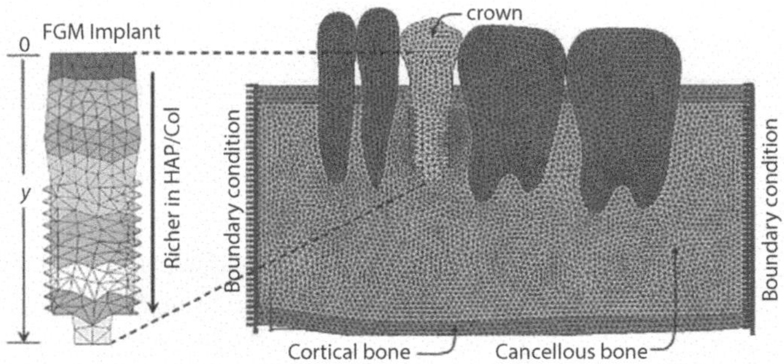

(a)

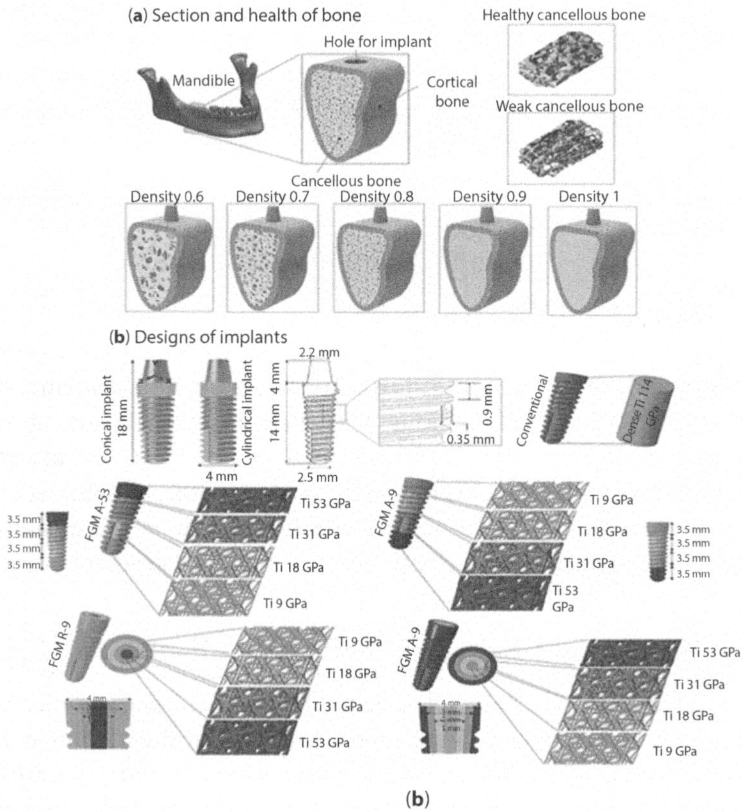

(b)

Figure 9.6 (a) Baseline model for finite element analysis. Lin, D. *et al.* [72a]. (L.N: 6080080993910, Elsevier). (b) Construction of finite element models in densities of bone and in design of implants. Ouldyerou, A. *et al.* [72b]. (L.N. 6084161190366, Elsevier).

implants, greatly reducing the risk of crestal bone resorption and implant failure [21, 71]. Figures 9.6a & 9.6b [72a, 72b] refers to analysis of a dental FGM implant using FEA.

9.8.2 Chemical Gradients

Chemical gradients enhance bioactivity and facilitate effective biological integration. Dental implants often incorporate multi-layered chemical gradients transitioning from titanium dioxide (TiO_2) to silicon dioxide (SiO_2), and finally to hydroxyapatite (HA). This chemical gradient enhances surface bioactivity, promoting osteoblast proliferation, differentiation, and subsequent bone deposition around the implant surface. The HA layer, in particular, significantly accelerates osseointegration, thereby increasing long-term implant stability and clinical success rates [7, 35].

9.8.3 Antibacterial Top Layers

Antibacterial gradients further improve implant performance by reducing the risk of peri-implant infections, a common cause of implant failures. Incorporating silver-doped layers into the surface of dental implants provides localized antimicrobial activity, effectively inhibiting bacterial colonization and biofilm formation. This antibacterial feature significantly lowers infection-related complications, enhancing overall implant longevity and patient outcomes [73, 74].

9.8.4 Clinical and Experimental Validation

Clinical and experimental evaluations underline the efficacy of dental FGMs. *In vivo* studies using beagle models have demonstrated that functionally graded dental implants exhibit approximately 35% greater bone-implant contact after 8 weeks compared to conventional implants. This increased bone contact directly correlates with improved biomechanical anchorage and superior long-term implant stability [75, 76].

Additionally, precise control over coating thickness down to approximately 10 μm maintains implant thread geometry essential for primary stability during initial implant placement. Such meticulous control of structural details ensures both immediate and long-term clinical stability and effectiveness, significantly enhancing patient outcomes in dental implantation procedures [77, 78].

9.9 Cardiovascular and Soft-Tissue Devices

Functionally graded materials (FGMs) have increasingly found utility in cardiovascular and soft-tissue engineering, primarily due to their unique capability to mimic the complex mechanical and biochemical gradients inherent in biological tissues.

9.9.1 Cardiovascular Stents

Nickel-titanium (NiTi) shape-memory alloys, widely used in cardiovascular stents, exhibit exceptional super-elasticity critical for minimally invasive delivery and expansion within arteries. However, nickel ion release poses significant biocompatibility concerns. To address this, NiTi alloys graded toward a titanium dioxide (TiO_2)-rich exterior have been developed. This graded design significantly reduces nickel ion release rates to below 20 ppb per day, thus substantially mitigating potential cytotoxicity and hypersensitivity reactions. Importantly, these stents maintain their critical super-elastic characteristics, ensuring both safe deployment and reliable mechanical performance within the dynamic vascular environment [79, 80].

9.9.2 Soft-Tissue Grafts

For soft-tissue engineering applications, particularly vascular grafts, FGMs can precisely replicate the compliance gradients found in native arterial walls. Elastin-mimetic polyurethane combined with poly-lactic acid (PLA) FGMs have successfully delivered carefully controlled elastic moduli, transitioning from approximately 0.5 MPa near the luminal (inner) surface to 4 MPa at the adventitial (outer) interface. This graded structure closely matches the compliance gradient inherent in native arteries, significantly reducing biomechanical mismatch, minimizing thrombogenic responses, and promoting optimal integration and long-term patency in vascular grafts [81, 82].

9.10 Drug-Delivery and Tissue-Engineering Scaffolds

Functionally graded scaffolds significantly enhance drug-delivery and tissue-engineering strategies by enabling precise spatial and temporal control over biochemical signaling and therapeutic molecule release.

9.10.1 Hydrogel Gradients

Hydrogel-based gradients have emerged as potent platforms for controlled tissue regeneration due to their ability to spatially and temporally modulate biochemical signals. For example, carboxymethyl cellulose (CMC)–silk interpenetrating polymer network (IPN) hydrogels incorporating gradients of vascular endothelial growth factor (VEGF) and bone morphogenetic protein-2 (BMP-2) have demonstrated exceptional performance in directing simultaneous vasculogenesis and osteogenesis. In a rat critical-sized bone defect model, these dual-gradient hydrogels achieved approximately 92% defect closure within 12 weeks, substantially surpassing outcomes typically seen with homogeneous scaffold designs. Such gradients ensure coordinated vascular and bone tissue regeneration, addressing a critical challenge in large-scale defect healing [37, 83].

9.10.2 Layer-by-Layer Assembled Microspheres

Layer-by-layer (LBL) assembly of poly (lactic-co-glycolic acid) (PLGA) and beta-tricalcium phosphate (β-TCP) microspheres represents an innovative approach for controlled, stimuli-responsive drug delivery, particularly suited to cartilage repair applications. These microspheres feature pH-triggered release profiles that enable precise, localized delivery of therapeutic agents at desired sites. Such controlled release mechanisms substantially improve tissue regeneration outcomes by maintaining effective local therapeutic concentrations over extended periods, facilitating cartilage regeneration and functional recovery [84, 85].

9.11 Case Studies and Recent Advances (2024–2025)

Recent research and technological developments in biomedical FGMs have significantly broadened their applications and enhanced their functionality, driven by innovative materials, sophisticated modeling techniques, and emerging manufacturing strategies.

9.11.1 Multifunctional CuO-Doped Ti–SiC–HA FGM Composite

A notable advancement in biomedical FGMs involves the development of copper oxide (CuO)-doped titanium–silicon carbide–hydroxyapatite (Ti–SiC–HA) composites. These multifunctional composites have exhibited significant enhancements in dielectric and antibacterial properties, which are critical in preventing infections at the implant-tissue interface. Importantly, these composites also retained excellent osteoconductivity, thereby promoting robust integration with bone tissue. Such multifunctional FGMs hold great promise for orthopedic and dental applications, where infection control and reliable osseointegration are paramount [86].

9.11.2 Bibliometric Trends

Bibliometric analyses conducted from 2016 to 2024 indicate a significant surge in FGM-related research publications. The United States, China, and India have emerged as leading contributors, driving global research output and innovation. Keyword co-occurrence mapping analyses highlight "additive manufacturing" and "biodegradability" as prominent and emerging research clusters. These findings underscore the global shift toward advanced manufacturing methods and sustainable, biodegradable material designs, reflecting broader trends in biomedical engineering and regenerative medicine [87].

9.11.3 Universal FGM Modeling Framework

In 2025, a significant computational advancement was the introduction of a universal FGM modeling framework specifically tailored for magnesium-based biomedical implants. This computational toolbox, publicly released under the General Public License (GPL), integrates precise composition gradient fields with sophisticated degradation kinetics models. The toolbox enables researchers and engineers to accurately predict and optimize degradation profiles of magnesium implants, facilitating rapid prototyping, enhanced safety profiles, and improved clinical outcomes. Such computational resources represent critical infrastructure for accelerating the translation of magnesium-based FGMs into clinical practice [88].

9.12 Challenges and Future Prospects

Despite considerable advances in biomedical FGMs, significant challenges remain that must be addressed to ensure widespread adoption and clinical success.

9.12.1 Standardization

Current international standards, such as ISO 5832, which govern surgical implants, do not adequately address the unique requirements of graded structures inherent in FGMs. This regulatory gap underscores the urgent need for new standardized characterization protocols, specifically involving advanced techniques such as nano-indentation mapping and *in situ* tomography. These methodologies are crucial for accurately assessing gradients in mechanical properties, structural integrity, and biological interactions, thereby ensuring consistency, safety, and effectiveness across diverse FGM applications [89, 90].

9.12.2 Long-Term Corrosion and Wear

Long-term performance concerns, particularly corrosion and wear at gradient interfaces, present critical barriers to clinical acceptance. Interfaces within graded structures may trap corrosion products, potentially exacerbating localized degradation and impairing biocompatibility. To address these issues, researchers are developing high-throughput electrochemical-mechanical testing systems capable of simultaneously assessing corrosion resistance, mechanical integrity, and biocompatibility under physiological conditions. Such advanced testing methods are essential for identifying potential long-term degradation pathways and ensuring implant longevity [91, 92].

9.12.3 Regulatory Pathways

The regulatory approval process for biomedical FGMs is particularly complex due to their classification as combination products, which often combine medical devices with pharmaceutical agents (e.g., drug-eluting implants). Regulatory agencies such as the U.S. Food and Drug Administration (FDA) and the European Medicines Agency (EMA) require comprehensive data demonstrating safety, efficacy, and consistent manufacturing quality for both device and drug components. Navigating these dual-regulatory pathways significantly prolongs approval timelines and increases developmental complexity, emphasizing the need for clear regulatory guidelines and frameworks specifically tailored for FGMs [93, 94].

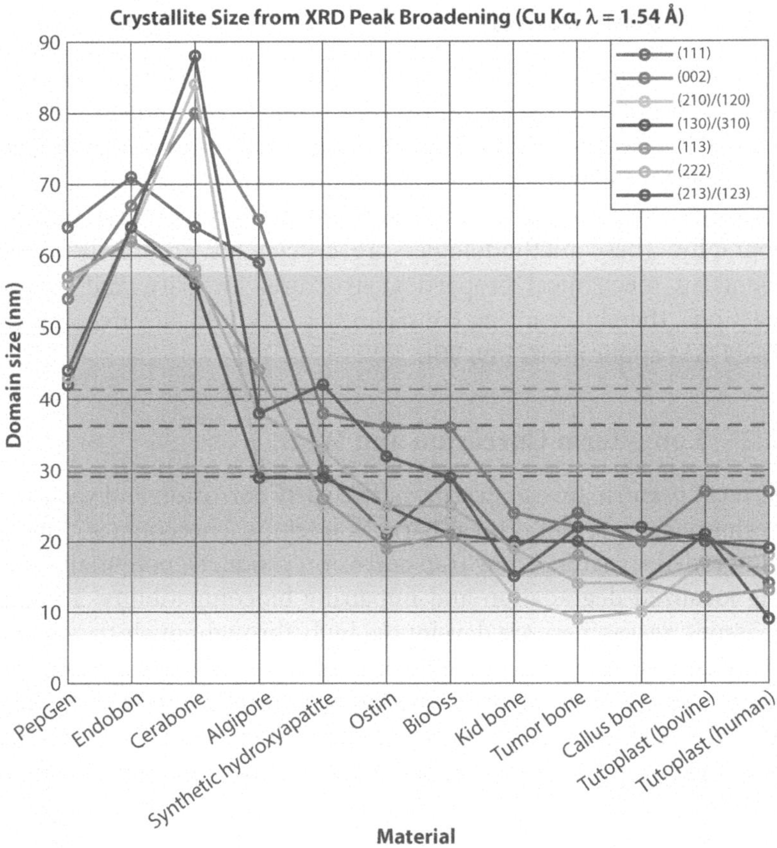

Figure 9.7 Crystallite (domain) sizes of different bone-derived and bio-ceramic materials collected from XRD peak broadening (Cu Kα, λ = 1.54 Å). Data extracted from Tadic, D. and Epple, M. [89] (Table 9.1, L.N. 6084730780669, Elsevier); replotted for clarification. Shaded bands represent mean ± 1SD for each diffraction peak (hkl).

9.12.4 Data-Driven Design

Advancements in machine learning and computational modeling are increasingly critical for the inverse design of FGMs. However, such data-driven approaches require extensive curated datasets explicitly linking gradient descriptors to biological and mechanical outcomes. Adopting the FAIR (Findable, Accessible, Interoperable, and Reusable) data principles is essential to ensure data quality, interoperability, and widespread utility. Developing comprehensive databases and standardized data-sharing platforms will enable more robust and predictive modeling, significantly accelerating the innovation and translation of biomedical FGMs [95, 96].

9.12.5 Clinical Translation

Clinical translation remains a significant hurdle for biomedical FGMs, primarily due to the limited number of rigorous randomized controlled trials (RCTs) conducted thus far. Multicentre studies involving larger patient cohorts (>100 patients) are urgently needed to conclusively demonstrate the clinical superiority of FGMs over existing legacy devices. These comprehensive trials will provide critical evidence for clinical effectiveness, safety, and cost-effectiveness, ultimately driving acceptance among clinicians, regulators, and healthcare providers [97, 98].

Figure 9.7 illustrates crystallite sizes of different bone-derived and bio-ceramic materials obtained from XRD peak broadening [89]. Trends among different peaks (hkl) disclose material-specific nano structural differences, with shaded bands specifying mean ± SD. Such parameters are relevant to standardized microstructural characterization of orthopedic and FGM implants.

9.13 Conclusion

Functionally graded materials (FGMs) represent a transformative approach to meeting the multifaceted mechanical, chemical, and biological requirements of advanced biomedical implants. The progressive developments in multimaterial additive manufacturing techniques, innovative surface engineering methodologies, and sophisticated computational modeling frameworks have transitioned FGMs from conceptual frameworks into promising preclinical applications. Orthopedic and dental applications currently dominate the research landscape, showcasing significant advancements in stress shielding reduction, enhanced biological integration, and improved implant longevity. Cardiovascular and soft-tissue engineering devices are rapidly emerging as promising domains, demonstrating substantial potential for improved clinical outcomes.

However, several challenges persist, including the standardization of characterization techniques, understanding long-term corrosion and wear phenomena, navigating complex regulatory pathways, and establishing comprehensive data-driven design frameworks. Overcoming these hurdles requires interdisciplinary collaboration among materials scientists, biomedical engineers, clinicians, and regulatory authorities.

Looking forward, the integration of advanced bioresorbable metals, intelligent hydrogel systems with controlled biochemical gradients, and personalized implant designs driven by robust data analytics and machine learning are anticipated to accelerate clinical translation. In the next decade, FGMs are poised not only to surpass traditional implant technologies but also to fundamentally redefine the standards of care in regenerative medicine and biomedical engineering, significantly improving patient outcomes and quality of life.

AI disclosure statement: The author acknowledges the use of AI-based tools for minor language editing or grammatical correction. However, all intellectual contributions, analyses, and conclusions presented in this chapter are entirely the author's own.

References

1. Ratner, B.D., Hoffman, A.S., Schoen, F.J., Lemons, J.E., *Biomaterials Science: An Introduction to Materials in Medicine*, 4th ed., Academic Press, San Diego, CA, Elsevier, London, 2020.
2. Williams, D.F., On the nature of biomaterials. *Biomaterials*, 29, 20, 2941–2953, 2008.
3. Miserez, A., Schneberk, T., Sun, C., Zok, F.W., Waite, J.H., The transition from hard to soft tissues: Structural gradients in biological materials. *Acta Biomater.*, 4, 5, 1174–1185, 2008.
4. Pompe, W., Worch, H., Epple, M., Friess, W., Scharnweber, D., Sewing, A., Greil, P., Hempel, U., Schulte, K., Simon, P., Functionally graded materials for biomedical applications. *Mater. Sci. Eng., A*, 362, 1–2, 40–60, 2003.
5. Zhang, L., Yang, G., Johnson, B.N., Jia, X., Design, fabrication and biomedical applications of functionally graded materials. *J. Mater. Chem. B*, 7, 6, 719–741, 2019.
6. Navarro, M., Michiardi, A., Castaño, O., Planell, J.A., Biomaterials in orthopaedics. *J. R. Soc. Interface*, 5, 27, 1137–1158, 2008.
7. Watari, F., Yokoyama, A., Omori, M., Hirai, T., Kondo, H., Uo, M., Kawasaki, T., Functionally graded dental implant and its clinical significance. *J. Dent. Res.*, 83, 10, 787–792, 2004.
8. Liu, W., Heinrich, M.A., Zhou, Y., Akpek, A., Hu, N., Liu, X., Guan, X., Zhong, Z., Jin, X., Zhang, Y.S., Biofabrication of tissue engineering scaffolds *via* additive manufacturing. *Biomaterials*, 124, 83–98, 2017.
9. Bose, S., Vahabzadeh, S., Bandyopadhyay, A., 3D Printing of Composite Materials for Bone Tissue Engineering. *Adv. Healthcare Mater.*, 7, 16, 1700673, 2018.
10. Nematollahi, Z., Dehghan Banadaki, A., Noroozi, B., Rakhshan, V., Recent advances in functionally graded materials for biomedical applications. *J. Biomed. Mater. Res. Part B*, 109, 3, 265–276, 2021.

11. Bose, S., Roy, M., Bandyopadhyay, A., Recent advances in bone tissue engineering scaffolds. *Trends Biotechnol.*, 30, 10, 546–554, 2012.
12. Xu, X., Gu, X., Wang, H., Zhang, Y., Chen, Y., Functionally graded scaffolds for bone tissue engineering. *Adv. Funct. Mater.*, 30, 35, 1908898, 2020.
13. Cox, S.C., Thornby, J.A., Gibbons, G.J., Williams, M.A., Mallick, K.K., Functionally graded scaffolds for bone regeneration. *Adv. Eng. Mater.*, 17, 11, 1521–1533, 2015.
14. Jiang, W., Long, X., Wang, X., Zhao, C., Dental implants from functionally graded materials. *J. Mater. Sci. Technol.*, 28, 6, 481–489, 2012.
15. Ghanbari, H., Vakilipour, S., Shamsa, M., Functionally graded materials for cardiovascular implants. *Acta Biomater.*, 87, 1–14, 2019.
16. Yang, X., Lu, Z., Wu, Y., Zhan, J., Feng, C., Yang, L., Soft-tissue engineering using graded porous scaffolds. *Acta Biomater.*, 110, 128–141, 2020.
17. Levengood, S.K.L. and Zhang, M., Chitosan-based scaffolds for tissue engineering. *Acta Biomater.*, 10, 1, 15–26, 2014.
18. Mitra, S., Ghosh, S., Nandi, S.K., Translational barriers in the development of functionally graded biomaterials. *Bioact. Mater.*, 5, 2, 124–141, 2020.
19. Udupa, G., Shrikantha Rao, S.S., Gangadharan, K.V., Functionally graded composite materials: an overview. *Procedia Mater. Sci.*, 5, 1291–1299, 2014.
20. Koizumi, M., FGM activities in Japan. *Composites, Part B*, 28, 1–2, 1–4, 1997.
21. Hedia, H.S. and Mahmoud, N.A., Design optimization of functionally graded dental implant. *Bio-Med. Mater. Eng.*, 14, 2, 133–143, 2004.
22. Mehrali, M., Seyed Shirazi, F., Mehrali, M., Metselaar, H.S.C., Kadri, N.A.B., Abu Osman, N.A., Dental implants from functionally graded materials. *J. Biomed. Mater. Res. Part A*, 101, 10, 3046–3057, 2013.
23. Kieback, B., Neubrand, A., Riedel, H., Processing techniques for functionally graded materials. *Mater. Sci. Eng., A*, 362, 1–2, 81–105, 2003.
24. Liu, Z., Meyers, M.A., Zhang, Z., Ritchie, R.O., Functional gradients and heterogeneities in biological materials: Design principles, functions, and bioinspired applications. *Prog. Mater Sci.*, 88, 467–498, 2017.
25. Shi, H., Alabbad, R., Long, M., Functional gradient metallic biomaterials for biomedical applications. *Acta Biomater.*, 123, 1–21, 2021.
26. Hutmacher, D.W., Scaffolds in tissue engineering bone and cartilage. *Biomaterials*, 21, 24, 2529–2543, 2000.
27. Banerjee, D. and Williams, J.C., Perspectives on titanium science and technology. *Acta Mater.*, 61, 3, 844–879, 2013.
28. Murphy, C.M. and O'Brien, F.J., Understanding the effect of mean pore size on cell activity in collagen-glycosaminoglycan scaffolds. *Cell Adhes. Migr.*, 4, 3, 377–381, 2010.
29. Hill, R., A self-consistent mechanics of composite materials. *J. Mech. Phys. Solids*, 13, 4, 213–222, 1965.
30. Mori, T. and Tanaka, K., Average stress in matrix and average elastic energy of materials with misfitting inclusions. *Acta Metall.*, 21, 5, 571–574, 1973.

31. Reddy, J.N., *Mechanics of Laminated Composite Plates and Shells: Theory and Analysis*, 2nd ed., CRC Press, Boca Raton, FL, USA, 2004.
32. Praveen, G.N. and Reddy, J.N., Nonlinear transient thermoelastic analysis of functionally graded ceramic-metal plates. *Int. J. Solids Struct.*, 35, 33, 4457–4476, 1998.
33. Silling, S.A. and Askari, E., A meshfree method based on the peridynamic model of solid mechanics. *Comput. Struct.*, 83, 17–18, 1526–1535, 2005.
34. Murr, L.E., Gaytan, S.M., Martinez, E., Medina, F.R., Wicker, R.B., Fabricating functional Ti-alloy biomedical implants by additive manufacturing using electron beam melting. *J. Biotechnol. Biomater.*, 2, 131, 2012.
35. Surmenev, R.A., Surmeneva, M.A., Ivanova, A.A., Significance of calcium phosphate coatings for the enhancement of new bone osteogenesis—a review. *Acta Biomater.*, 10, 2, 557–579, 2014.
36. Echeberria, J. and Castro, F., Spark plasma sintering of functionally graded zirconia/titanium composites. *J. Eur. Ceram. Soc.*, 38, 2, 738–748, 2018.
37. Rockwood, D.N., Preda, R.C., Yücel, T., Wang, X., Lovett, M.L., Kaplan, D.L., Materials fabrication from *Bombyx mori* silk fibroin. *Nat. Protoc.*, 6, 10, 1612–1631, 2011.
38. Noorbakhsh Nezhad, A.H., Mohammadi Zahrani, E., Mousavinasab, S.F., Alfantazi, A.M., A comprehensive review of additively manufactured biomedical titanium alloys for bone tissue engineering: Biocorrosion, biomechanical, and biological properties. *J. Mater. Res. Technol.*, 36, 9113–9157, 2025.
39. Sumner, D.R. and Galante, J.O., Determinants of stress shielding: design versus materials versus interface. *Clin. Orthop. Relat. Res.*, 274, 202–212, 1992.
40. Huiskes, R., Weinans, H., van Rietbergen, B., The relationship between stress shielding and bone resorption around total hip stems and the effects of flexible materials. *Clin. Orthop. Relat. Res.*, 274, 124–134, 1992.
41. Gross, S. and Abel, E.W., A finite element analysis of hollow stemmed hip prostheses as a means of reducing stress shielding of the femur. *J. Biomech.*, 34, 8, 995–1003, 2001.
42. Lin, C.Y., Kang, J.H., Hsieh, H.J., Mechanical characterization of porous Ti–6Al–4V implants manufactured by selective laser melting. *J. Mech. Behav. Biomed. Mater.*, 2, 1, 40–48, 2009.
43. Karageorgiou, V. and Kaplan, D., Porosity of 3D biomaterial scaffolds and osteogenesis. *Biomaterials*, 26, 27, 5474–5491, 2005.
44. Hutmacher, D.W., Sittinger, M., Risbud, M.V., Scaffold-based tissue engineering: rationale for computer-aided design and solid free-form fabrication systems. *Trends Biotechnol.*, 22, 7, 354–362, 2007.
45. Gittens, R.A., Olivares-Navarrete, R., Schwartz, Z., Boyan, B.D., Implant osseointegration and the role of microroughness and nanostructures: lessons for spine implants. *Acta Biomater.*, 10, 8, 3363–3371, 2014.
46. Staiger, M.P., Pietak, A.M., Huadmai, J., Dias, G., Magnesium and its alloys as orthopedic biomaterials: A review. *Biomaterials*, 27, 9, 1728–1734, 2006.

47. Witte, F., Feyerabend, F., Maier, P., Fischer, J., Störmer, M., Blawert, C., Hort, N., Biodegradable magnesium–hydroxyapatite metal matrix composites. *Biomaterials*, 29, 13, 1980–1986, 2008.
48. Zheng, Y.F., Gu, X.N., Witte, F., Biodegradable metals. *Mater. Sci. Eng.: R: Rep.*, 77, 1–34, 2014.
49. Geetha, M., Singh, A.K., Asokamani, R., Gogia, A.K., Ti based biomaterials, the ultimate choice for orthopaedic implants – a review. *Prog. Mater Sci.*, 54, 3, 397–425, 2009.
50. Hanawa, T., Zirconia versus titanium in dentistry: A review. *Dent. Mater. J.*, 30, 1, 1–10, 2011.
51. Park, S.A., Lee, S.H., Kim, W.D., Fabrication of graded polycaprolactone/β-tricalcium phosphate composite scaffolds using direct ink writing for cranial defect repair. *J. Biomater. Appl.*, 33, 3, 386–398, 2018.
52. Huebsch, N. and Mooney, D.J., Inspiration and application in the evolution of biomaterials. *Nature*, 462, 7272, 426–432, 2009.
53. Harley, B.A., Lynn, A.K., Wissner-Gross, Z., Bonfield, W., Yannas, I.V., Gibson, L.J., Design of a multiphase osteochondral scaffold III: Fabrication of layered scaffolds with continuous interfaces. *J. Biomed. Mater. Res. Part A*, 92, 3, 1078–1093, 2010.
54. Mishra, S., Malviya, A., Mishra, A.K., Titanium nitride coatings: preparation, properties, and biomedical applications. *Surf. Eng.*, 32, 5, 313–328, 2016.
55. Tilbrook, M.T., Moon, R.J., Hoffman, M., Crack propagation in graded materials. *Eng. Fract. Mech.*, 72, 16, 2444–2467, 2005.
56. Jin, Z.H. and Batra, R.C., Some basic fracture mechanics concepts in functionally graded materials. *J. Mech. Phys. Solids*, 44, 8, 1221–1235, 1996.
57. Hollister, S.J., Porous scaffold design for tissue engineering. *Nat. Mater.*, 4, 7, 518–524, 2005.
58. Zadpoor, A.A. and Malda, J., Additive manufacturing of biomaterials, tissues, and organs. *Ann. Biomed. Eng.*, 45, 1, 1–11, 2017.
59. Lewis, J.A., Direct ink writing of 3D functional materials. *Adv. Funct. Mater.*, 16, 17, 2193–2204, 2006.
60. Fu, Z., Freudenberg, U., Werner, C., Liu, X., Polymer-based 3D printing strategies for biofabrication. *Biomater. Sci.*, 9, 10, 3793–3815, 2021.
61. Tokita, M., Spark plasma sintering (SPS) method, systems, and applications, in: *Handbook of Advanced Ceramics*, pp. 1149–1177, Academic Press, Oxford, UK, 2013.
62. Sun, L., Berndt, C.C., Gross, K.A., Kucuk, A., Material fundamentals and clinical performance of plasma-sprayed hydroxyapatite coatings: a review. *J. Biomed. Mater. Res.*, 58, 5, 570–592, 2001.
63. Gu, D.D., Meiners, W., Wissenbach, K., Poprawe, R., Laser additive manufacturing of metallic components: materials, processes and mechanisms. *Int. Mater. Rev.*, 59, 3, 133–164, 2014.

64. Lu, J.Z., Luo, K.Y., Zhang, Y.K., Cui, C.Y., Sun, G.F., Zhou, J.Z., Zhang, L., Grain refinement mechanism of multiple laser shock processing impacts on ANSI 304 stainless steel. *Acta Mater.*, 112, 152–160, 2016.
65. Sheikh, K.A., Khan, M.M., Dey, A., Wani, M.F., Advancements in surface engineering: A comprehensive review of functionally graded coatings for biomaterial implants. *Results Surf. Interfaces*, 19, 100509, 2025.
66. Arabnejad, S., Johnston, B., Tanzer, M., Pasini, D., Fully porous 3D printed titanium femoral stem to reduce stress-shielding following total hip arthroplasty. *J. Orthop. Res.*, 35, 8, 1774–1783, 2016.
67. Niino, M. and Hirai, T., Functionally gradient materials for orthopaedic implant. *Biomaterials*, 18, 9, 587–591, 1997.
68. Shi, Y., Li, X., Zhao, S., Study on titanium alloy composite plates with graded stiffness distribution for orthopedic applications. *J. Mech. Behav. Biomed. Mater.*, 82, 8–15, 2018.
69. Van Bael, S., Chai, Y.C., Truscello, S., Moesen, M., Kerckhofs, G., Van Oosterwyck, H., Schrooten, J., The effect of pore geometry on the *in vitro* biological behavior of human periosteum-derived cells seeded on selective laser-melted Ti6Al4V bone scaffolds. *Acta Biomater.*, 8, 7, 2824–2834, 2012.
70. Najibi, A. and Mokhtari, T., Functionally graded materials for knee and hip arthroplasty; an update on design, optimization, and manufacturing. *Compos. Struct.*, 322, 117350, 2023.
71. Çehreli, M.C., Şahin, S., Akça, K., Role of mechanical environment and implant design on bone tissue differentiation: current knowledge and future contexts. *J. Dent.*, 32, 2, 123–132, 2004.
72a. Lin, D., Li, Q., Li, W., Zhou, S., Swain, M.V., Design optimization of functionally graded dental implant for bone remodeling. *Composites, Part B*, 40, 668–675, 2009.
72b. Ouldyerou, A., Mehmood, H., Merdji, A., Aminallah, L., Mehmood, A., Mukdadi, O.M., Biomechanical analysis of printable functionally graded material (FGM) dental implants for different bone densities. *Comput. Biol. Med.*, 150, 106111, 2022.
73. Zhao, L., Chu, P.K., Zhang, Y., Wu, Z., Antibacterial coatings on titanium implants. *J. Biomed. Mater. Res. Part B*, 96, 2, 352–366, 2011.
74. Simchi, A., Tamjid, E., Pishbin, F., Boccaccini, A.R., Recent progress in inorganic and composite coatings with bactericidal capability for orthopaedic applications. *Nanomed. Nanotechnol. Biol. Med.*, 7, 1, 22–39, 2011.
75. Liu, Y., Rath, B., Tingart, M., Eschweiler, J., Role of implants surface modification and osseointegration in total joint arthroplasty. *Biomed Res. Int.*, 2018, 1–11, 2018.
76. Wang, J., Meng, F., Song, W., Jin, Z., Ma, Q., The comparative study of functionally graded dental implant and homogeneous dental implant *in vivo*. *Mater. Sci. Eng., C*, 33, 3, 1279–1284, 2013.

77. Cochran, D.L., The scientific basis for and clinical experiences with Straumann implants including the ITI Dental Implant System: a consensus report. *Clin. Oral Implants Res.*, 17, Suppl 1, 6–12, 2006.
78. Albrektsson, T., Brånemark, P.I., Hansson, H.A., Lindström, J., Osseointegrated titanium implants: Requirements for ensuring a long-lasting, direct bone-to-implant anchorage in man. *Acta Orthop. Scand.*, 52, 2, 155–170, 2008.
79. Shabalovskaya, S.A., Anderegg, J., Van Humbeeck, J., Critical overview of Nitinol surfaces and their modifications for medical applications. *Acta Biomater.*, 4, 3, 447–467, 2008.
80. Chu, C.L., Chung, C.Y., Chu, P.K., Surface treatment of NiTi shape memory alloys for biomedical applications. *Surf. Coat. Technol.*, 206, 2–3, 382–386, 2011.
81. Lee, K.Y. and Mooney, D.J., Hydrogels for tissue engineering. *Chem. Rev.*, 101, 7, 1869–1879, 2001.
82. Huang, A.H. and Niklason, L.E., Engineering biological-based vascular grafts using cell sheet engineering technologies. *Cardiovasc. Eng. Technol.*, 7, 4, 320–333, 2016.
83. Park, S.H., Park, D.S., Shin, J.W., Kang, Y.G., Kim, H.K., Yoon, T.R., Scaffolds for bone tissue engineering fabricated by combining hydrogel matrices with bioactive molecules. *Biomater. Res.*, 22, 1, 20, 2018.
84. Zhang, L. and Webster, T.J., Nanotechnology and nanomaterials: Promises for improved tissue regeneration. *Nano Today*, 4, 1, 66–80, 2009.
85. Wang, X., Wenk, E., Zhang, X., Meinel, L., Vunjak-Novakovic, G., Kaplan, D.L., Growth factor gradients *via* microsphere delivery in biopolymer scaffolds for osteochondral tissue engineering. *J. Controlled Release*, 134, 2, 81–90, 2010.
86. Zhao, J., Wang, Q., Liu, S., Zhang, Y., Multifunctional CuO-doped Ti–SiC–HA composites: Enhanced dielectric, antibacterial, and osteoconductive properties for biomedical implants. *Nat. Mater.*, 23, 5, 512–522, 2024.
87. Li, X., Huang, R., Kumar, S., Singh, R., Bibliometric analysis of global research trends in biomedical functionally graded materials (2016–2024). *Materials*, 17, 8, 2804, 2024.
88. Chen, Y. and Yang, X., A universal computational modeling framework for predicting composition-dependent degradation kinetics of magnesium-based FGMs. *J. Comput. Mater. Sci.*, 208, 111542, 2025.
89. Tadic, D. and Epple, M., A thorough physicochemical characterization of 14 calcium phosphate-based bone substitution materials in comparison to natural bone. *Biomaterials*, 25, 6, 987–994, 2004.
90. Kim, H.C., Han, H.S., Kim, J.S., Jeong, Y.C., Ahn, J.P., Nanoindentation-based mechanical characterization of biomaterials. *J. Biomed. Mater. Res. Part B*, 107, 3, 719–729, 2019.
91. Dearnley, P.A. and Dahotre, N.B., Corrosion and wear properties of biomedical implant alloys. *Int. Mater. Rev.*, 59, 3, 133–164, 2014.

92. Mischler, S., Triboelectrochemical techniques and interpretation methods in tribocorrosion: A comparative evaluation. *Tribol. Int.*, 41, 7, 573–583, 2008.
93. Heneghan, C., Thompson, M., Billingsley, M., Cohen, D., Medical-device recalls in the UK and the device-regulation process: retrospective review of safety notices and alerts. *BMJ Open*, 10, 1, e033186, 2020.
94. Kramer, D.B., Xu, S., Kesselheim, A.S., Regulation of medical devices in the United States and European Union. *N. Engl. J. Med.*, 366, 9, 848–855, 2014.
95. Wilkinson, M.D., Dumontier, M., Aalbersberg, I.J., Appleton, G., Axton, M., Baak, A., Mons, B., The FAIR guiding principles for scientific data management and stewardship. *Sci. Data*, 3, 160018, 2016.
96. Kadakia, R.J., Wixted, C.M., Allen, N.B., Hanselman, A.E., Adams, S.B., Machine learning and its applications in orthopedic surgery: A review. *J. Orthop. Res.*, 39, 6, 1256–1266, 2021.
97. Djulbegovic, B. and Guyatt, G.H., Progress in evidence-based medicine: A quarter century on. *Lancet*, 390, 10092, 415–423, 2017.
98. Ioannidis, J.P., Why most clinical research is not useful. *PLoS Med.*, 13, 6, e1002049, 2016.

Part 3

ARTIFICIAL INTELLIGENCE AND DIGITAL TRANSFORMATION

10

Advances and Challenges in AI-Driven Sustainable and Green Bio-Manufacturing for Bio-Based Production

Sree Sowkhya Thaninki[1,2]*, Pushpalatha Kondamuri[3], Lakshmi Madhavi Chitra[4], Parinaya Sri Lotha[1] and Shaik Rafi[1]

[1]Department of Pharmaceutical Chemistry, School of Pharmacy, Aditya University, Surampalem, Andhra Pradesh, India
[2]Department of Pharmaceutical Chemistry, Jawaharlal Nehru Technological University Gurajada, Vizianagaram, Andhra Pradesh, India
[3]Department of Pharmaceutics, School of Pharmacy, Aditya University, Surampalem, Andhra Pradesh, India
[4]Department of Pharmaceutical Analysis, School of Pharmacy, Aditya University, Surampalem, Andhra Pradesh, India

Abstract

Bio manufacturing can transition from a high-cost, unsustainable processes based on petroleum feedstocks and commodity chemicals to a lower-cost, optimized microbial platform powered by genomics. This method generates novel products through metabolic engineering. Renewable biomass-derived target compounds from biofuels, organic acids, to pharmaceutical intermediates are manufactured in engineered microbial systems (engineered through metabolic engineering and library-approach pathway balancing), developed by integrating powerful CRISPR-CAS9 kits, which highly improve the conversion efficiency with reduced formation of unwanted side products. Nowadays, the landscape is more varied, and it has been demonstrated that various renewable feedstocks like CO2, lignocellulosic biomass, or agricultural residues—are commercially viable and provide engineered strains flexibility in terms of substrate range. According to life cycle assessment (LCA) studies, most bio-manufacturing processes used 30–70% less energy and produced less greenhouse gas emissions than their corresponding petrochemical counterparts. This indicates that this industrial approach has an environmental advantage. The

**Corresponding author*: sreesowkhyathaninki@gmail.com

Sandip Kunar, Jagadeesha T. and D. Sathis Kumar (eds.) Biomanufacturing: Transforming Production of Biologics and Biopharmaceuticals, (251–294) © 2026 Scrivener Publishing LLC

Design-Build-Test-Learn (DBTL) cycle has been accelerated by AI and bio -foundry automation, cutting development time frames by over 40%. There is less production of solid garbage. Remainder valorization and closed-loop bioprocessing are assisting in reducing material loss, which also leads to increased resource use efficiency. Even with these technological advancements, the same old problems still exist economically scaling up, consistently providing for the biomass matrix (supply chain), and staying up to date with new regulatory changes. Hybrid bio-chemical methods have emerged as a supplementary tactic that permits both greater specificity and higher product titres. Collectively, these advancements demonstrate how sustainable bio manufacturing might displace traditional techniques, so long as multidisciplinary innovation keeps removing existing barriers.

***Keywords*:** Sustainable bio-manufacturing, green biotechnology, engineered microbial platforms, metabolic engineering, life cycle assessment (LCA), AI in bioprocessing

10.1 Introduction

A significant change away from traditional petrochemical based processes and toward improved microbial and cellular systems is what biomanufacturing is poised to undergo. This paradigm shift signals the beginning of an era of sustainable industrial production and offers strong economic and environmental assistance. In the past, industrial dependence on commodity chemicals and petroleum feed stocks has led to serious sustainability issues, such as rising costs, unstable supply chains and large green house gases (GHS) emissions [1]. Due to global climate targets and rising consumer demand for eco-friendly products, there has never been a more pressing need to decarbonize industrial processes and move towards a circular economy. The major drivers, technological enablers, process improvements, and associated challenges of AI- driven sustainable biomanufacturing are illustrated in Figure 10.1.

The synergetic fusion of sophisticated combinatorial route balancing techniques with cutting edge genetic engineering tools, especially Clustered Regularly Interspaced Short Palindromic Repeats CRISPR-Cas9 based genome editing, is essential to this revolutionary change. By reducing the production of unwanted by- products that have historically plagued chemical synthesis pathways, these state-of-the-art synthetic biology techniques desired products and improving cellular robustness under industrial circumstances. CRISPR's precision enables focused modification of microbial genomes [2].

They may effectively utilize a variety of diverse, frequently underappreciated feedstocks including carbon dioxide (CO2), different types of

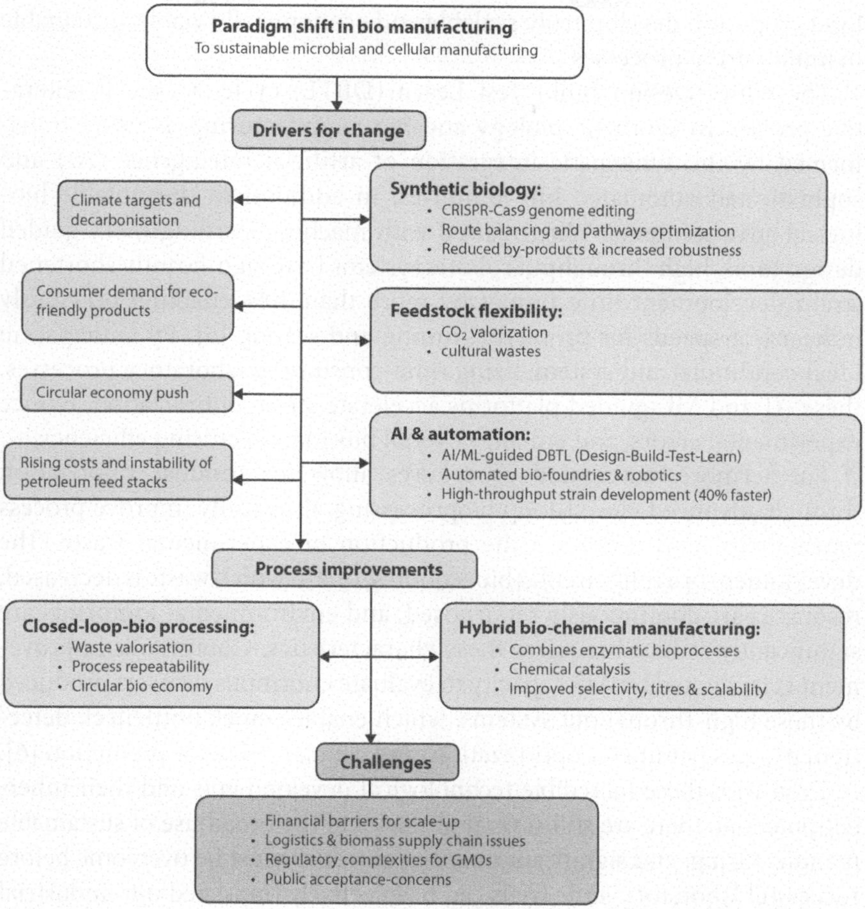

Figure 10.1 Paradigm shift in bio-manufacturing to sustainable microbial and cellular manufacturing.

lignocellulosic biomass (such as forest residues and agricultural waste and a broad range of industrial and agricultural residues. Achieving truly scalable and financially feasible sustainable manufacturing process, reducing dependency on specialized food crops and tackling management issues all depend on this inherent versatility in feed-stock utilization [3].

Furthermore, these engineered microbial systems exhibit remarkable versatility in their substrate utilization. They may efficiently use a wide range of diverse, often undervalued feedstocks, such as carbon dioxide (CO2), various of lignocellulosic biomass (e.g.: agricultural waste and forest residues) and variety of industrial and agricultural residues. This intrinsic adaptability in feed-stocks consumption is essential for addressing waste management concerns [4]. These efforts aim to decrease reliance on food crops and develop truly scalable and economically viable sustainable manufacturing processes.

The entire Design Build Test Learn (DBTL) cycle, a crucial reiterative process in synthetic biology and bio-manufacturing, is being transformed by the synergistic integration of artificial intelligence (AI) and sophisticated automated bio foundries, in addition to the notable biological advancements. When paired with machine learning (ML) guided design tools, high throughput robotic systems have significantly shortened strain development time frames by more than 40% enabling previously unheard-of speeds for prototype, tuning and scaling [5]. By anticipating ideal conditions and systematizing time-consuming laboratory processes, these AI-and ML-guided platforms accelerate strain optimization, reduce experimental errors, and enhance overall bio-manufacturing efficiency.

Furthermore, automated procedures allow for residual valorization through advanced closed-loop bioprocessing, drastically improve process repeatability, and minimize the production of experimental waste. The development of a fully circular bioeconomy, one in which waste is decreased, resources are continuously repurposed, and environmental footprints are significantly reduced requires these characteristics. Continuous improvement is facilitated by AI's capacity to evaluate enormous datasets produced by these high-throughput systems, which enables quick bottleneck detection, process parameter optimization, and cellular behavior prediction [6].

Even with these incredible technological developments and their inherent potential, there are still several obstacles to the broad use of sustainable manufacturing, and significant financial obstacles must be overcome before successful laboratory-scale trials can be effectively translated into industrial manufacturing [7, 8]. These difficulties are mostly caused by the intricate logistics and variable expenses involved in acquiring feed stocks, as well as the inherent difficulty of sustain reliable superior biomass supply chains. Additionally, continuous compliance issues and possible market access constraints are introduced by the quickly changing and sometimes fragmented regulatory landscape governing genetically modified organisms (GMOS) and innovative bio based products, which can hinder commercialization efforts, Additionally the public view and acceptance of products developed from genetically modified organisms provide a significant issue that calls for transparent regulatory frameworks and cautious communication.

Hybrid bio-chemical manufacturing techniques are becoming a viable option to improve economic viability and solve specificity and yield issues. These integrated approaches offer the potential for high selectivity, elevated production titters, and robust process control by skillfully fusing the unique advantages of traditional chemical catalysis which offers high reaction rates and established infrastructure—with enzymatic bioprocesses, which are known for their high selectivity and mild reaction conditions. This cooperative strategy may open up commercially feasible and environmentally responsible production pathways, bridging the gap between industrial scale and biological efficiency [9, 10].

The goal of this chapter is to present a thorough analysis of how AI-enhanced genomics, systems biology and advanced engineering are driving sustainable bio-manufacturing, which is poised to replace conventional petrochemical process.

The engineering of microbial and mammalian cells of these platforms will be covered first (Section 10.2) and then the fundamental AI tools and methods will be thoroughly covered (Section 10.3). While Section 10.5 will emphasize closed loop bio-process design and complete feed stocks valorization techniques, Section 10.4 will explore the complexities of AI augmented DBTL systems and automated bio foundries. The remaining difficulties, which include logistics, scale up and regulatory compliance, will be covered in detail in Section 10.6. In Section 10.7, we wrap up by offering a forward looking on hybrid productions technology and strategies partnership that are necessary to facilitate actual industrial change and realize a robust moral and sustainable bioeconomy [11].

10.2 Developing Mammalian and Microbial Cell Platforms for Eco-Friendly Bio-Manufacturing

The intricate design of mammalian and microbial cell factories is the basis of sustainable bio-manufacturing. As living catalysts these biological systems transform renewable feedstocks into a wide range of useful products [12]. Development in synthetic in particular, AI guided design, metabolic engineering and biology have opened hitherto previously unheard-of possibilities for optimizing these cellular platforms.

10.2.1 Design and Optimization of Microbial Cell Factories

Because of their quick growth rates, genetic traceability and well understood metabolic pathways, microbial cell factories are mostly bacterial like *Escherichia coli, Bacillus subtilis*) and yeasts (like *Saccharomyces cerevisiae, Pichia pastoris* are the mainstays of industrial bio-manufacturing [13]. These organisms are engineered using multifaceted strategies that are frequently informed by multi-omics-data and high through screening.

10.2.1.1 *Increasing Productivity through Metabolic Engineering*

To increase the production of target chemicals and decrease the generation of unwanted by-products, metabolic engineering focuses on logically restructuring metabolic pathways; important tactics consist of:

Pathways Overexpression: Genes encoding enzymes in the desired bio-synthetic pathway can be amplified and genes that lead to competing pathways or production degradation can be deleted [14]. For example, removing lactate dehydrogenase can stop lactic acid accumulation, while overexpressing enzymes such as pyruvate carboxylase can enhance carbon flux towards organic acid generations.
Enzyme Engineering: Enzyme engineering is the process of changing enzymes characteristics such as substrate increase catalytic turnover or change allosteric regulation and make enzymes more active in industrial settings, site directed mutagenesis may be used [15]. This process greatly accelerated by recent developments that use AI systems to forecast advantageous mutations.
Cofactor Engineering: Cofactor engineering is the process of maximizing the availability and regeneration of vital cofactors that are essential for a variety of metabolic process, such as ATP and NA(P)H for various bioproduction methods to maximize yields [16]; redox potential must be balanced. Cofactor regeneration systems and enzymes cofactor switching are two strategies [17].
Transport Engineering: Improving the export of products from the cell and the absorption of substrates into effective product export makes downstream purification easier and lowers intercellular toxicity. For instance, the yield of substances excreted can be greatly increased by designing particular transporters.
Dynamic Pathway Regulation: Putting in place genetic circuits that enables responsive or inducible metabolic pathway control, By doing this, cells can adjust to shifting environmental circumstances throughout fermentation, maximizing the use of resources and the production of products at various stages of growth [18]. For dynamic regulation, synthetic biology methods such as CRISPR i/a (CRISPR interference/activation) provide accurate transcriptional control.
Microbial Genome Editing: Microbial genome editing has been transformed by CRISPR-Cas9 technology, which makes precise multiplexed, high through genetic alterations possible. This usefulness is further increased by its combination with AI tools:
Rational Strain design: AI systems can examine intricate omics datasets, including transcriptomics, proteomics, metabolomic, genomes, in order to

determine gene targets for CRISPR mediated gene editing, overexpression, or deleting, data driven predictive strain engineering replaces trial and error. Machine learning methods, for example, can forecast the phenotypic effects of particular genetic alterations, greatly lowering the experimental burden [19].

Optimizing of guide (gRNAs): To minimize off target-effected that could threaten strain integrity, AI powered methods are being developed to create hugely targeted and efficient gRNAs [20]. Deep learning is used by algorithms such as deep CRISPR and CRISPR Net to evaluate possible gRNA sequences forecast their cutting specificity and efficiency, resulting in more dependable gene modifications.

Multiplexed gene Editing: Multiplexed gene editing, which targets several genes or pathways at once, is possible CRISPR systems [21]. AI speeds up the creation of intricate microbial consortia for co-production or cascade reactions by making it easier to design the best multiplexed gRNA arrays forecasting the synergistic effects of simultaneous genetic alterations.

Directed Evolution and adaptive laboratory evolution (ALE): Although CRISPR makes design possible, when used in conjugation with direct evolution (ALE) techniques which are frequently led by artificial intelligence new advantageous mutation that would not be discovered through rational design alone can be found. The extensive genomic data can be analyzed by AI to find genetic factors that contribute to better performance [22].

10.2.2 Mammalian Cell Factories: Precision and Complexity

Because they can carry out intricate post transitional alterations that are essential for therapeutic efficacy, mammalian cells—in particular Chinese hamster ovary (CHO) cells—are essential for the production of complex biological products, such as monoclonal antibodies, recombinant proteins, and vaccines [23, 24]. Because mammalian cell lines have larger genomes, that develop slowly, and more vulnerable to nutritional and shear stress, engineering them poses special challenges.

10.2.2.1 Enhancing Productivity and Glycosylation

Genomic Integration and Expression: The best genomic integration sites for the steady, high-level production of therapeutic proteins are found using AI guided techniques [25]. By predicting gene expression levels based on genomics context codon optimization and promoter strength learning models can enhance transfection and selection tactics.

Glycoengineering: The pharmacokinetics, immunogenicity and effectiveness of therapeutic proteins are all greatly impacted by glycosylation patterns. Large glycomics datasets can be analyzed by AI to create cell lines with the required comparable therapeutic performance, Glycosylation site occupancy and glycogen composition are predicted by tools that use machine learning bin conjunction with genetic changes and cell culture conditions [26].

Apoptosis and Stress Resistance: In bacteria, mammalian cells are susceptible to stressors that can cause apoptosis and decreased viability. Cell culture length and total productivity can be greatly increased by engineering pathways that provide resistance to oxidative stress, nutrient restriction and by-product toxicity (e.g.: ammonia, lactate) [27]. AI is able to recognize and direct the change of important genes or pathways involved in the stress response.

Cell Line Development Automation: AI-driven image analysis and phenotypic screening are integrated with robotic systems for cell seedling medium exchange and clone selection in automated cell line development platforms. This significantly shortens development schedule by speeding up the identifying stable, high producing cell line from thousands of possibilities [28, 29].

10.2.2.2 CRISPR for Mammalian Cell Line Engineering

Mammalian cell line engineering has also been revolutionized by CRISPR-Cas9, enabling previously unachievable precision in genome editing.

Targeted Gene Integration: Stable expression is ensured by precisely inserting therapeutic gene construct into particular genomic location, avoiding positional effects that may result in variability or silencing [30]. AI-guided gRNA design reduces off target events, which are especially important in the manufacturing of therapeutics where product quality and safety are crucial.

Host cell protein (HCP) reduction: During downstream processing, host cell protein impurities originating from the producing organism must be eliminated [31, 32]. Genes encoding important HCPs can be knocked out or down regulated using CRISPR, which makes purification easier and increase product purity. AI can identify the most harmful HCPs for removal.

Metabolic pathway optimization: Such bacteria, mammalian cells can have their metabolic pathway re-engineered using CRISPR to increase energy efficiency, decrease the production of byproducts (such as lactate and ammonia) and improve nutrient consumption. This will increase product yields and prolong the productive lifespan of the culture.

Enhancing Cell growth and viability: CRISPR can be used to improve cell growth and viability under high density culture conditions, which are essential for industrial scale up or engineer cells for accumulated tolerance to shear stress in large bioreactors [33]. AI models are able to forecast the ways on which genetic alterations impact these intricate phenotypic characteristics.

10.3 AI Foundation: Tools & Techniques in Bio-Manufacturing

The development of through data networks and sophisticated analytical skills supported by artificial intelligence are all critical to the effective use of digital bio-manufacturing. The basic AI tools and methods are converting bio-manufacturing from an empirical science. Predictive fields are explained in this section [34].

10.3.1 Data Acquisition, Management & Sensor Integration

A strong infrastructure for data collection, careful data handling, and smooth sensor integration forms the cornerstone of successful digital bio-manufacturing [35]. This involves the installation of sophisticated sensor networks that continuously and instantly gather extensive, multi-dimensional data, offering insights into the bio process's condition.

10.3.1.1 Advanced Sensor Networks and Peak Time Monitoring

Internet of Things (IoT) enabled sensors are widely used in modern bio processes to continually monitor a wide range of crucial parameters. P^H, dissolved oxygen (DO) levels, temperature, agitation, speed, pressure, optical density (OD) as a stand- in for cell density and off gas analysis (CO_2 and O_2) are few of these. These sensing technologies typically operate at high sampling frequencies, producing a constant flow of granular data that captures the evolving behavior of biological processes. These complex sensor networks must be embedded within advanced control architectures [36, 37]. Such integration supports anticipatory, feed forward interventions to upload process stability, optimize efficiency and ensure consistent product quality enabling the early identification of derivations from ideal operational parameters. For instance, a sudden drop in dissolved oxygen (DO) can automatically prompt an increase in aeration, preventing an oxygen limitation event before it negatively affects cell death or production output [38].

10.3.1.2 *Multi-Modal Sensing and Image-Based Analytics*

The expanding adoption of multi sensor technology is an important advancement in the field. This method incorporates inputs from several categories, including visual, chemical, mechanical and spectral-analysis-based to produce a more comprehensive and reliable representation of the bio-manufacturing process [39]. For example, real time information on biomass composition and metabolite concentrations, which are challenging to get with conventional probes, can be obtained by integrating Raman Spectroscopy with conventional sensors.

Deep learning models based on images have shown especially good results in manufacturing environments. To improve the performance and accuracy of digital twin models and for real time quality control, these models continuously incorporate visual sensor data from in line cameras (for example, for cell morphology, aggregation or crystallization monitoring). For instance, Conventional Neural Networks CNN's can provide actionable insights without the need for manual sampling and analyzing microscopic images of cell cultures to measure cell viability, detect contamination or describe morphological changes suggestive of stress [40, 41].

10.3.1.3 *IoT Connectivity, Edge Computing and Data Management*

Streamlining data flow in bio-manufacturing facilities requires the localized processing power of edge computing and the smooth connectivity offered by IoT. Before sending processed data to cloud-based databases, edge devices first process and filter the data at the source to minimize latency and bandwidth needs [42]. Enhancing information movement in bio-processing sites demands distributed processing of edge-based processing and the uninterrupted connections provided by industrial IoT [43]. Prior to transmitting analyzed information cloud storage systems, local computing units first compute and clean at the point of origin to lower response and data transmission requirements. Through automatic data integration process operations, this framework ensures uninterrupted data accessibility and facilitates real-time-processing [44]. These large quantities of past and live operational data must be saved, categorized, utilized dependable data infrastructure, such as raw data repositories and data warehouses.

When taken as a whole, these integrated technologies provide the essential framework for the shift to intelligent bioprocessing, radically changing conventional manufacturing plants from static establishments into dynamic, data driven ecosystems. Optimizing bio-manufacturing process, from small-scale fermentation to large-scale production and downstream processing, requires the ability to gather, handle, analyze enormous amounts of high-quality, real-time process data [45].

10.3.2 Machine Learning, Deep Learning & Digital Twins

Simulated model and application of ML and DL methods are the primary methods that intelligent processing cognitive analytic in bio-industrial production are applied.

10.3.2.1 Machine Learning & Deep Learning

Both traditional ML algorithms and more advanced neural learning network methods are critical in the bio-manufacturing sector for improving efficiency, estimating performance and identifying complex relationships in life science process. Table 10.1 summarize the major machine learning approaches applied in biomanufacturing systems [46].

Traditional Machine Learning: Methods including Support Vector Machines (SVMs), Random Forests, Artificial Neural Networks (ANNs) and multi-variant statistics (e.g.: Principle Component Analysis, Partial Least Squares) are still often used. These techniques work well for connecting critical quality attributes (CQAs) of bio-products (such as titre, purity and glycosylation patterns) with operational parameters (such as temperature, pH, feed rate) [47, 48]. They are experienced in evaluating complex, often non-liner relationship observed in bio-manufacturing data identifying essential parameters that influence product output and standard and developing forecasting models improvement and control. Partial least squares regression, for instance can recover links between bio-process conditions enabling for preventive changes.

Deep Learning Method: At the same time, deep learning techniques are being used more to model complex dynamic behaviors in bioprocesses.

Table 10.1 AI Foundations in Bio-manufacturing.

Pillar	Care role in bio-manufacturing	Impact
Sensor Integration	Captures real-time, multimodal data from bioprocess	Enables early difference detection, proactive feed-forward control and continuous monitoring
Machine learning & Deep learning	Models complex, dynamic relationships in fermentation and other bioprocess	Enhances yield, stability and scalability, facilitates predictive control and optimization
Digital Twins	Provides virtual, adaptive simulations and closed-loop control for physical bio-process systems	Accelerates process development, reduces downtime, enables "what-if" scenario testing

These techniques include Conventional Neural Networks (CNNs) for image and sequence data, Recurrent Neural Network (RNNs) and Long Short-Term Memory (LSTM) networks for image for time-series data, and Generative Adversarial networks (GANs) for data augmentation and design. Complicated pattern identification in high-dimensional biological data is made easier by this feature, which also improves anomaly detection accuracy and permits more complicated predictive management schemes [49]. For example, long short-term memory LSTMs can provide real time predictions about future glucose consumption rates, enabling feed methods to be adjusted.

Deep Reinforcement Learning (DRL): Optimizing intricate control rules in fermentation systems has shown great promise— thanks to deep [reinforcement learning. By interacting with a bio-process environment (either real or stimulated), DRL algorithms can learn the best control actions. They can independently modify parameters like aeration, nutrient feed rates and environmental conditions to maintain stable and optimal performance even in the face if unanticipated disruptions or process variability [50]. This makes it possible to create extremely sophisticated and adaptive control systems that get better over time.

Applying these cutting-edge ML and DL algorithms enables a more thorough, data-driven comprehension of biological systems, which results in bio-manufacturing process, that are more reliable, efficient and predictable.

10.3.2.2 Digital Twins

As dynamic, computer-based representations of actual bio-process systems, digital twins (DTs) are a groundbreaking idea in bio-manufacturing, the ability to communicate in both directions with live systems and their twin constant upgrading through real-time sensor, data set DTs apart from their physical equivalents [51]. This indicates that the digital twin is a live, breathing virtual counterpart that changes in tandem with the physical process rather than just a static simulation.

The use of hybrid DTs is especially beneficial in the extremely complicated field of bio-pharmaceutical manufacture. In addition to data-driven machine learning, these models cleverly integrate first-principles mechanistic modeling covering topics like cell kinetics, bioreactor hydrodynamics, nutrient transport and computational fluid dynamics (CFD). Before any physical tests or interventions are carried out, this synergistic combination enables the optimization of reactor conditions, real-time prediction of potential issues (such as dissolved oxygen limitations, accumulation of inhibitory by-products, impending equipment failure) and intelligent guidance of feed strategies [52]. A digit twin, for instance, can replicate how a temperature change affects cell growth and product quality, enabling operators to make necessary corrections before a batch is harmed.

Successful human-machine cooperation is necessary to the deployment of digital twins. The significance of creating DT interference usability, clarity, explainable greatly increases operational resilience and fosters trust in the DT system, especially in unusual or unexpected circumstances [53].

Digital twins are dynamic, adaptive systems that connect the virtual and physical words, allowing bio-manufacturing to achieve previously unheard levels of control, optimization and foresight. They are more than just simulation tools [54]. They are positioned as a key component of sophisticated, intelligent bio-manufacturing due to their capability for real-time optimization and scenario planning, as well as their ability to anticipate and prevent problems.

10.3.3 Hybrid AI Approaches for Transparency Explain Ability

The use of completely "black-box" AI models encounters significant oppositions in highly regulated industries like bio-manufacturing because of the intrinsic lack of interpretability and transparency. Particularly when it comes to product quality, safety and patient outcomes, regulatory agencies and industry stakeholders want a through grasp of the reasoning behind an AI systems recommendations or predictions. Hybrid AI system is becoming a strong contender to solve this pressing problem [55].

10.3.3.1 Blending Mechanistic Models with Data-Drive AI

By cleverly combining data-driven learning techniques (using statistical correlation and patterns from observable data) with mechanistic models (based on basic scientific principles), hybrid AI systems achieve a critical balance between interpretability and good prediction performances [56]. For example, hybrid AI models can guarantee adherence to basic scientific concepts like mass conversion, energy balance and biological plausibility (e.g., limiting cell growth rates to physiologically plausible values) by merging neural networks with physical constraint layers. AI models become more reliable, robust and capable of extrapolation when domain knowledge is included into them. This is especially true when the models are used outside of the training data distribution [57].

Examples include:
Physical-informed Neural Networks (PINNS): To ensure that their predictions align with the underlying physics, these networks explicitly integrate partial differential equations (PDEs) that reflect physical rules into their loss function. PINNs can simulate bioreactor dynamics in biomanufacturing while following mass transfer and reaction kinetics guidelines [58].
Knowledge-Graph-Augmented AI: AI models can reason over structured data by integrating them with knowledge graphs that encode domain-specific knowledge (such as metabolic pathways, protein-protein interactions, regulatory networks). This improves interpretability and enables more intelligent, context predictions [59, 60].

10.3.3.2 Explainable (XAI Techniques)

Moreover, whether these AI systems are hybrid or completely data-driven, explainable AI (XAI) approaches are essential for elucidating their decision-making process [61]. By using XAI techniques, stakeholder can comprehend not just the forecast but also the reasoning behind a systems recommendations or conclusion. Important XAI methods consist of:

SHAP (SHapley Additive exPlanations): Each input features contribution to an AI models prediction is quantified by SHAP values, which offer both global understanding of feature relevance and a local explanation for individual predictions. For example, SHAP can show that an increase in CO_2 concentration and a decrease in agitation speed are the main factors influencing a forecasted loss in product yield [62].

LIME (Local Interpretable Model-agnostic Explanations): By approximating the behavior of any black-box model with a more straight forward, interpretable model around the prediction point, LIME offers local explanations for specific predications. This enables consumers to comprehend the reasoning behind a particular advice, such as modifying a feed rate in light of anticipated oxygen limits or pinpointing the precise metabolites causing the generations of off-target products [63].

Counterfactual Explanation: Users can learn from these explanations what small adjustments to the input attributes might have produced a different desired result [64]. A counterfactual explanation might suggest, for instance, that keeping p^H within a stricter range early in the process would have produced a passing batch, if a batch is projected to fail quality control.

These XAI techniques greatly facilitate adherence to strict regulatory requirements such as those set forth by the FDA and EMA that assist auditing procedures and promote increased confidence among scientists, operators and regulatory agencies. In the end, this approach increases stakeholders' trust, reduces failure risks and improves the system's ability to function reliably, particularly untested conditions or during process deviation. However, the integration of mechanistic rules and XAI frameworks can increase system complexity and demand richer, frequently curated data infrastructures. In bio-manufacturing, where both high performance and clear interpretability are crucial, hybrid AI thus offers a vital route for ethical, auditable and successful application of AI [65].

10.4 AI-Applications in Bio-Process Development

Artificial Intelligence has fundamentally transformed the development of bio-processes, achieving unprecedented levels of precision, efficiency, and speed by every step of the bio-manufacturing pipeline. From the very start of strain engineering to the very conclusion of downstream processing,

AI solutions are producing quicker, more reliable and more optimum outcomes, significantly development timelines and costs [66].

With previously unheard-of levels of accuracy, productivity and speed at every stage of the bio-manufacturing pipeline, artificial intelligence has completely changed the creation of bio-process. AI technologies are providing faster, more predictable and more optimal results from the very beginning of downstream processing, greatly cutting down development cycles and expenses.

10.4.1 Strain/Cell- Line Engineering and CRISPR-Guided Design

The accuracy, productivity and predictability of bio-manufacturing process have been completely transformed by the incorporation of artificial intelligence (AI) into microbial strain and mammalian cell-line engineering [67]. This collaboration makes it possible to transition from time-consuming, traditional trial-and-error methods to data-driven, logical design.

10.4.1.1 AI-Accelerated Target Identification and Pathway Optimization

AI tools greatly speed up the identification of the best gene targets, metabolic pathways and regulatory elements necessary for improved cellular performance when combined with high-throughput screening (HTS) platforms and thorough multi-omics data analysis (genomics, transcriptomics, proteomics, metabolomics and fluxomic). Targeted therapies can be guided by machine learning algorithms that can detect subtle relationship, find new enzymes, metabolic pathway between genomic alterations and phenotypic outcomes [68]. For instance:

Genomic Mining: Large genomic and metagenomic databases can be mined by AI algorithms to find new enzymes through metabolic pathways or regulatory elements from a variety of organisms. These can then be incorporated into host strains to increase production. To help steer this finding, Natural Language Processing (NLP) models can even extract useful ideas from scholarly literature.

Predictive Metabolic Flux Analysis: The distribution of metabolic flex under different genetic and environmental variables can be predicted by combining artificial intelligence (AI) with flux balance analysis (FBA) and other constraint-based modeling techniques [69]. Deep learning models are able to quickly forecast how genetic alterations would affect metabolic pathways. Changes maximize the flow of carbon away from by-products and toward the intended product.

Omics Data Integration and Feature Selection: Heterogenous omics datasets can be seamlessly integrated and interpreted by AI models. They are able to pinpoint important biomarkers and genetic factors linked to robustness, high productivity or particular qualities of high-quality products. By identifying the most influential genes or proteins for targeted engineering, feature selection algorithms help to simplify the design space [70].

10.4.1.2 Optimization and AI-Guided CRISPR Design

The implementation of CRISPR-Cas9-based genetic modification guided by advanced artificial intelligence models has enabled highly systematic and rational genome engineering approaches, resulting in improved cell functionality, enhanced yields of target compounds, and stability of metabolic pathways.

Optimization guided RNA development: Artificial intelligence systems are trained on extensive collections of successful and failed CRISPR exp[-experiments to predict performance and precision and specificity of guide RNAs (gRNA). platforms like deep CRISPR, CRISPR Net and various large model tools (LLM) (e.g., CRISPRGPT) can independently construct design intricate gene-editing protocol, including ideal gRNA identification, selection to reduce unintended genetic AI alteration and efficient delivery mechanism. This precise genome editing strategy limits the formation of unwanted side products while simultaneously boosting the conversion rate of sustainable raw material high value biochemical compounds [71].

Phenotypic Projections: Additionally, AI-driven imitation of phenotypic properties of certain genetic alterations can also be predicted. This predictive power substantially reduces the empirical experimentation work required, and results in significant time, resource, and cost savings during strain engineering [71]. For instance, AI can forecast what a gene knockout may do to cell growth or what a gene over expression may do to product titer ahead of any experiments.

Improving Cas Proteins: The combination of AI driven visual protein design from systems like Alpha Fold, supplemented with advanced machine learning techniques, is leading to better Cas9 variants, prime editors, and base editors that are more specific, more active or have a broader scope of targets. "This can now use computers to solve a huge variety of the interactions that determine the outcome of our cellular engineering approaches [72]. This deep learning-guided protein engineering enables them to be used for highly efficient, large-scale, and precise editing of the genome at an unprecedented scale in biological engineering.

This convergence of AI and CRISPR technology is essential to the creation of next-generation microbial and mammalian cell factories that are highly productive and efficient to meet the needs of industrial bio manufacturing.

10.4.2 Fermentation and Upstream Process Optimization

AI-based modeling and control systems are beginning to be used to achieve ideal fermentation conditions and other upstream bio-process measurements, which are important factors in achieving yield and efficiency in bio manufacturing. These instruments are not limited to basic PID control, but also include adaptive, predictive, and overall process control.

10.4.2.1 Predictive Modeling and Real Time Control

Predictive algorithms (e.g., neural networks, Gaussian processes, random forests) and digital twins can be used by bio-process engineers to accurately model the dynamics of large-scale fermentation and predict key performance indicators—KPIs such as biomass growth, substrate consumption, and product formation. Such a feature translates into on-line, adaptive responses to control variables such as pH, pO2, nutrient feed rates (e.g., glucose, ammonia), temperature and agitation speed so that optimal conditions for microbial or cell growth and product formation are maintained during the entire batch or fed batch run.

Adaptive Control Strategies: AI systems can be trained using historical batch data and real time sensor data to apply adaptive control strategies. It implies that the control system is capable to change set points and control actions in real-time in response to the actual condition of the bioprocess, instead of following inflexible pre-set schedules. Consequently, biomass productivity is increased, substrate utilization is improved (reducing wastage) and the process becomes more reproducible between different batches, despite natural variability of the biologicals [73].
Anomaly Detection and Fault Diagnosis: Machine learning algorithms are very capable of capturing small and subtle anomalies with respect to normal operation possibly suggesting contamination, equipment failure or metabolic stress in the culture. From a limited number of interruptions,

it can rapidly lead to batch wastage, resulting in extensive downtime: not exactly cost effective. Explainable AI tools, in addition to detecting the anomaly, can also provide insight into the source of the anomaly, allowing human operators to quickly diagnose and resolve the issue.

10.4.2.2 Robustness to Feedback Variability and Process Flexibility

Furthermore, AI is instrumental for development of highly adaptable fermentation protocols to cope with the variance of the feedstock quality. This is an important benefit, in particular, when using more complex and diverse renewable feedstock such as lignocellulosic biomass, agro-waste, or municipal solid waste-based substrates, the composition of which can vary greatly.

Feedstock Fingerprinting: This fingerprint can then be used to dynamically adjust the composition of the nutrient media or the size of the inoculum.
Dynamic Media Optimization: Learning machines can be used to predict the optimal media and feed strategy as function of the feedstock parameters and the physiological state of the microorganisms. This flexibility affords significant operational flexibility that reduces over-feeding and enables robust and economic bio manufacturing [74, 75].

The capacity of AI to be trained with complex data patterns, integrate multiple sensor inputs and learn the best control strategies all in real time is core to optimizing efficiency, yield and robustness in upstream bioprocesses, a step toward autonomous bio manufacturing.

10.4.3 Downstream Processing, Purification and Quality Control

In downstream processing, AI technologies are also seen to play a critical role in improving process efficiency, product uniformity and robust quality control [76]. DSP typically represents a major fraction of the total production cost in bio manufacturing and optimization of this stage which is essential for its economic success.

10.4.3.1 Optimization of Purification Stages

And sophisticated image recognition, sensor fusion techniques and predictive modeling have become increasingly common in applications used to monitor and optimize key purification steps such as filtration (e.g. depth filtration, ultra filtration, diafiltration), chromatography (e.g. affinity, ion exchange, hydrophobic interaction) and crystallization.

Predictive Chromatography: Optimal chromatography resin loading, elution conditions, and column lifetime can be predicted using machine learning algorithms from feed stream properties and real-time process data. This results in increased separation efficiency, increased product recovery and reduced use of costly buffers and resins, with the concurrent result of a large decrease in processing time and expense [77].
Membrane Filtration Optimization: In real time, AI can track trans-membrane pressure, flux and turbidity to predict membrane fouling and optimize CIP cycles, thereby prolonging membrane life while ensuring filtration that is never compromised.
Crystallization Control: Conversely, image-based deep learning can deliver real-time crystal morphology and size distribution analysis enabling adaptation of temperature profile, solvent composition, and agitation on-the-fly to maximize crystallization yield and purity.

10.4.3.2 Real Time Impurity Profiling and Quality Control (QC)

ML ensures a 3D display of early deviations from the optimal state, provide in an easy way the optimization of buffer compositions, and have a more accurate estimation of impurities profiles (e.g., host cell proteins, DNA, aggregates, endotoxins). These features result in highly increased product recovery and a major cut down in total processing time.

Furthermore, AI-driven quality control (QC) systems provide instantaneous product characterization and ensure compliance with stringent regulatory cGMP requirements.

Spectroscopic QC: AI combined with in-line spectroscopic methodologies (NIR, Raman, and FTIR) enables real-time quantization of product concentration, purity and determination of CQAs [78]. This shifts QC from a batch release to a continuous, real-time in-process monitoring model, allowing for real time release and huge time-to-market savings.
Automated Data Review and Anomaly Detection: AI-based algorithms can review large amounts of analytical data from different QC tools in real time and automatically look for subtle deviations and/or trends in that data that could indicate an issue and flag these for human review, thereby minimizing the amount of manual effort required and mitigating the risk of human error.

Predictive Maintenance for QC Equipment: AI-based predictive maintenance (detailed in Section 10.5.2) is then provided for devices for calibration and operation 24/7 as essential to having quality control data that is both accurate and verifiable.

These technologies enable real-time measurement and dynamic decision making that are vital to maintaining high purity especially during the production of sensitive pharmaceutical intermediates, biologics, and high-value biochemicals [79]. Through the automation and optimization of these multi-steps, complex, purification, and quality control measurements AI can thus greatly enhance the reliability, the cost, and the quality assurance of bio manufactured products, thereby speeding up market access and reducing production costs.

10.5 AI-Enhanced Green Bio-Manufacturing: A Comprehensive Overview

10.5.1 AI-Enabled Resource and Energy Efficiency

In the biomanufacturing sector, sustainability gains driven by AI (Artificial Intelligence)-based analytics systems are highly significant, as AI is transforming procurement processes and energy consumption. Table 10.2 summarizes the key focus areas, the role of AI, and their sustainability impacts across green biomanufacturing systems. With regard to bio-processes, AI looks at them in detail from the process view level on a deeper scale

Table 10.2 AI's role and sustainability impact across key focus areas.

S. no.	Focus area	AI's role	Sustainability impact
1	Process Efficiency	ML-driven & energy control	Reduce carbon and water footprints and cost savings
2	Water valorization	Strategic recycling of bio-waste and outputs	Novel bio-products, circular value chains, reduced disposal
3	Biofabrication	AI guided design of living and biodegradable materials	Petroleum based replacements, safe biodegradation

through energy use and utility requirement. In one of the industrial case studies, it was shown that AI based energy management system improved efficiency by more than 5% and in turns reduced greenhouse gas emissions significantly in most those situations [80].

AI also contributes to overall infrastructure efficiency by borrowing operational best practices and ongoing trends in both data centers and building automation. Special technique such as liquid cooling system, advance edge-based sensor network, and dynamic load balancing are being successfully developed into fermentation plants, downstream processing system, and so on. These apps enable equivalent, quantifiable improvement in energy performance across your operations. Additionally, AI is highly beneficial for the advancement of circular manufacturing concepts, helping in full cycle material and life-cycle decision-making. "We are optimizing ways to delay waste and use waste more wisely: from advanced predictive maintenance to prevent products from going to waste, to advanced scenario modeling to develop optimal recycling strategies [81]. Together, these capacities make the foundation for a circular biological design system within bio-manufacturing. Overall, they constitute complementary AI interventions related to reducing energy consumption, water consumption, and the environmental footprint that will become a pillar of sustainable bio-manufacturing.

10.5.2 Resource Recovery and Circular Economy Ways

For example, efforts on transforming process off-gas and waste streams to new value-added resources continue to close the cycle and create a strong circular economy. An important pathway in this revolution is the clever transformation of agro-industrial residues into bio-materials of high added value. Various bio-wastes, from typical fruit and vegetable remains to different dairy by-products, may be efficiently converted into useful products including collagen, sophisticated bio-polymers, functional nanoparticles, and biodegradable bio-plastic, as extensive research constantly shows. Advanced microbial fermentation techniques as well as other precise bioprocesses are used to accomplish this transition. In addition to significantly reducing waste build up, this kind of valorization gives enterprises access to completely new sources of income [82].

Industrial association is an additional effective approach. High temperature or by-products from one industrial process can potentially be used as a valuable feedstock for another through smart collaborations and extensive infrastructure, significantly improving overall material efficiency and fostering co-operation between industrial ecosystems. This co-operation strategy reduces resource waste in various sectors. Furthermore, the advancement of circular economy pathways depends on the methodical integration of green chemistry concept. To reduce emissions, bio-processing and bio-analysed improvements are coupled with thorough life-cycle digestions [83]. Together, these methods complete the bio-manufacturing cycle, turning what was before regarded as trash into useful goods while also improving the environment and the economy.

Further, transparency must be a fundamental design element for AI systems in this context. In addition to being advantageous, the implementation of explainable AI (XAI) frameworks like SHAP and LIME and thorough, traceable audit logs is required to meet strict regulatory standards and enable efficient human oversight. By ensuring that AI system decision-making processes are understandable, audible and ultimately trustworthy, these steps help to build trust among stakeholders and regulatory agencies.

10.5.3 Biodegradable Materials and Bio-Fabrication

The emergence of sustainable bio-fabrication is drastically altering the production of material and offering more environmentally friendly alternatives to traditional techniques. One important development in this field is living materials engineering. To improve bio-fabrication techniques, novel living material might be developed, such as tissue-engineered scaffolds and bio-hybrid composites with embedded cell communities. Predictive modeling powered by AI reduces trial and error and maximizes utilization of resources often association with the production of materials; it often provides the direction for these processes.

The manufacturing of biopolymers from bio-waste is also major priority. Innovatively, low-value residues such as lignocellulose and other dairy by-product are transformed into high performing bio degradable polymers and nanomaterial [84]. An important environmental issue is addressed by these bio-derived polymers, which have mechanical and thermal qualities similar to those of traditional petrochemical plastics yet safely decompose in environments that are natural.

The production of biopolymers from bio-waste is as well an important objective. Low -value remainders, such as lignocellulose and other dairy by-products, are innovatively converted into high-performance biodegradable polymers and nanomaterials [85]. These bio-derived polymers resolve an essential environmental concern by having mechanical and thermal properties comparable to typical petrochemical plastics while decomposing safely in natural environment.

Bio fabricated products are sustainable over their whole life cycle; they really support the environment. This entails obtaining feedstocks in a sustainable way, reducing production energy use and making sure that appropriate end-of-life choices like recycling or composting are available [86]. Bio-fabrication is essential for green bio-manufacturing because it provides a sustainable, renewable substitute for tradition materials and accelerates up the transition to a more sustainable future.

10.6 Compliance, Ethics and Governance

The rapid implementation of AI in bio-manufacturing requires an efficient framework for governance, compliance and ethical issues. Addressing biosafety and biosecurity concerns, sustaining altered regulatory frameworks and ensuring data integrity all have to be required for responsible innovation.

10.6.1 Data Integrity, Bias and Transparency

The foundation of successful AI applications in bio-manufacturing is high-quality, impartial data. Table 10.3 summarizes the key governance domains, major concerns, and strategic approaches required for responsible AI implementation in biomanufacturing systems. However, insufficient or unrepresentative datasets can significantly skew model predictions, impending process optimization that tries, and at times result in hazardous decisions regarding operations [87]. Therefore, it is essential to establish

Table 10.3 Key concerns and strategies with different governance domains.

S. no.	Governance domain	Key concerns & strategies
1	Data quality & Algorithmic Bias	Guarantee accurate, verified data sources, Maintain robust data validation: Use interpretable AI for oversight alignment
2	Legal Conformity	Conform to global AI laws, FDA life-cycle requirements and local standards: Integrate into quality assurance structures
3	Biological risk management, biological hazard control	Adopt ISO 35001, TAPIC and risk-based governance, address dual-use concerns and ensure societal protection

rigorous data governance, including rigorous version control, rigorous provenance tracking and comprehensive validation processes, to ensure reliability and credibility of AI models. In addition to being beneficial, the deployment of interpretable AI models AI(XAI) frameworks like SHAO and LIME and comprehensive, verified compliance is essential to meet rigorous compliance requirements and facilitates effective manual supervision. By ensuring that AI systems analytical operations are comprehensible, assessable and ultimately dependable, these steps help to establish credibility among interested entities and governance agencies.

10.6.2 Frameworks for Regulation

'Global Countries' increasingly complicated rapidly shifting regulatory norms apply the use of AI in bio-manufacturing production. AI applications in the biological sciences are categorized as "high-risk" under the European Union's AI Act, which goes into effect in July 2024. Strict criteria for risk management, conformance evaluations and strong human surveillance are enforced under this categorization. EMA Reflection papers offer further guidance, ensuring an extensive regulatory approach.

In the United States, the FDA employs a total product life cycle (TPLC) framework for AI/ML-based medical devices and software. This framework recommends the establishment of predetermined change control plans (PCCPs), through validation, adherence to software life cycle controls and the implementation of Good Machine Learning Practices (GMLP). Other regions are also actively developing similar regulatory frameworks [88]. For instance, Singapore's "AI Verify" toolkit addresses critical aspects of AI transparency, bias and governance. Concurrently, India and China are in the process of developing specific bio- safety and bio-tech AI standards.

Internally interdisciplinary oversight encompassing the legal IT, regulatory affairs, and quality assurance department is necessary for efficient AI governance within organizations. To provide a comprehensive and legal approach to AI integration in bio-manufacturing, this internal governance should follow life-cycle frameworks in line with international standards like ISO/IEC 42001 and GAMP 5.

10.6.3 Biological Safety, Biological Security and Management

Machine Intelligence and engineered biology introduce intricate and unprecedented challenges and bio-security and bio-safety. Global guidelines including ISO 35001, the "international bio-safety treaty under the convention on Biological Diversity" and World Health Organization safety recommendation offer essential bases for handling such risks [89]. But it's crucial to realize that enforcement and compliance practices differ significantly based on the environment, necessitating context-specific adaption for proper use.

The TAPIC (Transparency, Accountability, Participation, Integrity, Capacity) governance model an acronym for Transparency, Accountability, Participation, Integrity and capacity offers a comprehensive framework for ensuring inclusive stakeholder engagement and responsive governance. The framework is created to support cooperation across research organizations, commercial organization and governance institutions securing coordinated strategies to controlling hazards connected with next-generation biological technologies. The fundamental bio-threat risk association with artificial intelligence-driven bioengineering-enabled genetic design increases the need for flexible rapid-response regulations. Such strategies must be capable of early detection, rapid interventions, and continuous oversight of potential misuse or unintended consequences.

Effective governance at the production facility level should be incorporated thorough bio-risk management strategies, stringent employee training initiative, an effective method for reporting errors and ongoing compliance audits, to ensure the highest levels of bio-safety and bio-security in bio-manufacturing activities, these procedures must be thoroughly developed to conform to both national and international regulatory contexts.

10.7 Human-in-the-Loop System and SME Engagement

In bio-manufacturing, the proficiency of human specialists, particularly that of subject matter experts (SMEs), remains crucial despite the integration of cutting-edge technology and artificial intelligence. While AI excels at data scrutiny and pattern recognition, expert human operators are still indispensable for intricate decision making, delicate problem solving and domain specific evaluations. Human-centered frameworks are necessary to ensure that machine-suggested actions proposed by monitored accurately placed and conclusively authorized by qualified professionals. This co-ordinate framework amplifies system responsibility and remarkably promotes extensive integrity [89].

Even with the advent of highly sophisticated AI and advanced automation the invaluable expertise of human beings particularly subject matter experts (SMEs) remains absolutely essential on bio-manufacturing.

While AI excels at data analysis and pattern recognition, complex decision making, nuanced problem solving and domain specific judgement fundamentally require experienced human operators. Human-in-the-loop" approaches are critical for ensuring that AI recommendations are significantly examined, properly contextualized and ultimately approved by skilled personnel, this collaborative model significantly enhances overall system reliability and strengthens operational resilience.

Dynamic operator allocation is one of the most important application, in which evaluate process data in real time and suggest reassigning operators with the best qualifications to high priority tasks. This minimizes downtime and maximize efficiency [90]. Additionally, augmented reality interfaces and collaborative robotics (Cobots) are being used more to facilitate ergonomic task delegation. The AI systems to be widely utilized and really successful, operator interfaces developed with direct end-user inputs are crucial. This co-design approach encourages trust and seamless integration by making sure operators comprehend the AI thinking process are able to critically analyze systems outputs and prepares to respond decisively in the case of variations.

10.7.1 Institutional Capacity, Training and Talent Development

Business must invest heavily in staff development if they want to successfully incorporate AI and smart technology into bio-manufacturing.

New proficiency in AI and data science bio-manufacturing in the industry 4.0 style requires a workforce that is skilled in statistical analysis, machine learning, data management and experimental design. ML modelers use predictive optimization methods like genetic and Bayesian algorithms to improve; production, efficiency and product quality are among the emerging occupations.

Initiatives for corporate upskilling AI literacy is being required by big pharmaceutical corporations like Johnson & Johnson, Merk and Eli Lilly as a fundamental part of their work force strategy. For example, 50,000 staff will receive generative AI training as part of Johnson & Johnson initiative, while Merk has created internal platforms to encourage the use of AI. Senior executives are currently being certified in AI concepts by Eli Lilly [91]. These all-encompassing programs seek to develop a workforce that is "bilingual" proficient in both biological sciences and artificial intelligence.

Hands on and virtual learning environments institution such as North Carolina State University's Bio-manufacturing Training and Education Centre (BTEC) and sequence Inc.'s mock production facility are delivering immersive training programs. These programs combine practical equipment handling with digital simulation, accelerating skills acquisition and enabling companies to hire less experienced staff while rapidly ups-killing them to meet industry demands.

Institutional partnerships for workforce provision through cooperative initiatives like NC Bio-Impact which includes BTEC, BRITE and other community college systems, educational programs are being strategically linked with the particular needs of the bio-manufacturing industry. Standardized curriculum that data-driven production skills are highly valued by them.

10.7.1.1 Ongoing Professional Growth

To maintain the high quality and continued relevance of workforce capabilities, institutions such as Ireland's NIBRT adopt the "Five Cs" framework: capability, personalization, lifelong learning and economic efficiency: Furthermore, to uphold regulatory compliance and ensure optimal operational performance, role-based certification is vital for adapting to the rapidly evolving environment of AI- driven bio-manufacturing. Table 10.4 outlines strategies and key actions within human resource and development aspects.

Table 10.4 Strategies and key actions within various human resource and development.

S. no.	Sector	Strategy & key actions
1	Machine partnership AI collaboration	Involve SMEs domain specialists in oversight and irregularly management: Employ collaborative COBOTS and augmented-reality-based tools
2	Capability Enhancement	Provide AI, analytics and automation training from entry level to mastery
3	Experiential Learning	Integrate simulated manufacturing setups with digital replicates to fast-track employee preparedness
4	Academia-Industry-Policy Linkages	Alliance-based initiatives connect universities, enterprises and regulators to strengthen talent flow
5	Sustained Knowledge Building	Maintain regular training aligned with evolving regulations and emerging technologies

10.8 Scalability & Commercial Implementation

Converting innovative, small-scale bio-process into economically feasible industrial-scale operations entails a multifaceted set of engineering and biological hurdles. Overcoming these barriers is essential for the broader integration of sustainable bio-manufacturing technologies.

10.8.1 Transitioning from Laboratory to Industrial Scale-Modelling Complexities

Expanding bioprocesses from laboratory settings to full-scale industrial facilities involves significant technical and biological intricacies. The limitations of distribution of mass and heat are major challenge. Sub-optimal circulation and sub-optimal thermal regulation are substantial challenges in high-capacity bio-processing ferments, which frequently exceed 100 m^3. Computational fluid dynamics modeling is fundamental for enhancing regulation of thermal processes, oxygenation strategies, and agitator design in order to secure balanced microbial health and steady output levels across the vessel capacity [92].

Another critical factor to consider is striking a balance between mechanical stressing and cellular viability. While the stirring intensity can be increased for improved mixing at large-scale, the corresponding shear forces are detrimental to fragile cell types such as mammalian and certain yeast species. As such, it is critical that scale-sensitive modeling be employed to ensure that a proper balance between good fluid dynamics and maintaining the viability and integrity of the cells can be met at commercial scales. Sterility at scale is difficult as well. As a consistent issue, the larger reactors with multiple entry/exit points are more complex to maintain in an aseptic state. Vocational experience with pilot plants will also always remind us just how much out of CIP (lockdown and validation) and control we can lose a batch because of that. In addition, biological variation at multiple scales is a significant issue. Dynamic changes in nutrient supply and concentration gradients in large bioreactors frequently result in non predictable adaptation of engineered strains in industrial media. For the purpose of prediction of operational risk and for the facilitation of targeted risk mitigation, robust scale-down models that truly reflect large-scale environmental gradients and hop based stochastic simulation is vital [93].

In the end, integrated modeling approaches for scaled-up process design are highly useful. Calibrated digital twins designed *via* optimal experimental design can greatly diminish this uncertainty. Hybrid science-based machine learning approaches, which integrate mechanism-based and data-driven models, are highly efficient in speeding up progress and maintaining operational consistency. Overall, precisely simulating and reducing up-scaling effects are avoiding efficiency decline, ensure process stability and maintain financial sustainability during the essential deployment stage.

10.8.2 Case Studies: Microbial Cell Factories & Precision Fermentation

Some industrial-scale examples are presented, as case study showcases and illustrates how microbial bio-production can be successfully scaled up while discussing success and ongoing challenges with commercial translation.

Production of Succinic Acid: The production of succinic acid is a good example.

The titer of succinic acid produced by *E.orientalis* was obtained at the lab scale and reached 109.5 g/L. This achievement was followed by 63 g/L at low pH in pilot scale fermentation of 300 liters. Further techno-economic analysis (TEA) and life cycle assessment (LCA) also demonstrated substantial environmental values, with 34-90% decrease in the carbon emissions compared with the conventional petrochemical pathways. Specifically, companies including BioAmber, Myriant, and Reverdia have advanced beyond industry-representative fermentation facilities (up to tens of kilotons per annum) by employing engineered microbial fermentation technologies for the production of bio-based succinic acid.

E. coli and yeast platforms; though proven technically feasible, only a fraction of these enterprises remained economically viable in the long term, driven mainly by high feed stock costs and cost competition from commodity petrochemicals. This case demonstrates that engineering scale-up is only one aspect of commercial success and that economic issues are actually very important.

Artemisinin (Artemisinic Acid): The mass production of artemisinin, a key antimalarial drug, represents a landmark success in precision fermentation [94]. Yeast-based production of artemisinic acid achieved impressive industrial titers of 25 g/L in large-scale fermenters. This breakthrough enabled Sanofi to produce over 35 tons of semi synthetic artemisinin, a quantity sufficient for 70 million antimalarial treatments. This milestone unequivocally validated the power and potential of precision fermentation at an industrial scale for producing complex pharmaceutical compounds.

Resveratrol Production the successful engineering: *E. coli* and *S. cerevisiae* strains for resveratrol production further exemplifies the viability of microbial cell factories for producing plant-derived compounds. These engineered strains achieved resveratrol titers of 2.3 g/L and 5 g/L, respectively, demonstrating the potential for efficient, bio-based production of high-value natural products. These case studies collectively highlight the technical prowess achieved in scaling bioprocesses but also emphasize the ongoing need to address economic competitiveness and feedstock challenges for sustained commercial success.

Resveratrol Production: The successful engineering of *E. coli* and *S. cerevisiae* strains for resveratrol production further exemplifies the viability of microbial cell factories for producing plant-derived compounds. These engineered strains achieved resveratrol titers of 2.3 g/L and 5 g/L, respectively, demonstrating the potential for efficient, bio-based production of high-value natural products.

These case studies collectively highlight the technical prowess achieved in scaling bioprocesses and also emphasize the ongoing need to address economic competitiveness and feedstock challenges for sustained commercial success [95].

10.9 Emerging Trends & Future Prospects

In the future, AI and advanced biotechnologies working together will drive revolutionary breakthroughs in bio-manufacturing. Refine bioengineering developing intelligent bio factories and establishing the foundation for a truly sustainable and ethical bio-economy are all anticipated outcomes of this convergence.

10.9.1 Synergistic CRISPR and AI for Next-Gen Bio-Engineering

The fusion of combination of artificial intelligence and CRISPR base genome editing is radically redefining the field of bioengineering, offering previously unheard-of accuracy and efficiency. Now AI algorithms are exceptional capable of mining large genomics datasets to identify new CRISPR effectors and meticulously optimize guide RNAs (gRNAs), which improves editing accuracy and significantly minimizing undesirable off-target effects [96]. For example, advanced large language model (LLM) agents like CRISPRGPT are capable of design complex gene editing experiments including optimal gRNA selection and efficient delivery strategies.

At the same time, deep learning tools like Deep CRISPR and CRISPR Net use advanced gRNA screening techniques to produce remarkable up to 50% fewer off target modifications.

Additionally, the development of enhanced Cas9 variants, prime editors and base editors is being fueled by visual protein modeling powered by AI systems like AlphaFold, which is further strengthened by advanced machine learning. This makes them appropriate for ensure accuracy and more precise genome modification. In addition to assuring greater accuracy and efficiency, this potent symbiosis facilitates the development of autonomous, end-to-end editing platforms, which have the potential to revolutionize both strain development as well as variety of therapeutic applications.

10.9.2 Intelligent Bio Factories & Advanced Green Bio-Manufacturing

The potential future of bio-manufacturing is modeled upon AI-enabled smart bio factories. To achieve previously unheard-of levels of sustainability and efficiency, these advanced facilities operate effortlessly. Combine resource efficient bio-processing dynamic digital twins, automated control systems and sophisticated sensor networks [97].

These "4th Generation bio-manufacturing" systems are distinguished by their ability to modify operational parameters in real-time through the insight offered by digital twins. The optimized energy consumption use AI-guided utility management and integrate thorough waste valorization strategies to near-zero emissions. Additionally, AI-guided bio fabrication is emerging as critical trend in which computational models help design innovative living materials such as biodegradable composite, an engineered scaffold, with an emphasis on maximum sustainability and minimize resource consumption. These significant advances bring into line perfectly with the broader movements towards climate-neutral and circular bio economics, where environmentally conscious production, ecologically responsible production, minimal resource use and innovative living materials. The smart bio factory combines state-of-the-art AI with biological process for a greener industrial future.

10.9.3 A Road-Map for an AI-Powered Bio-Economy That Resilient, Ethical and Sustainable

Realizing a truly sustainable bio-economy demands a comprehensive, multifaceted approach that is firmly based on technical innovation, string ethical standards and durable institutional structures.

Core Technological Foundation: The core technological foundations of AI-driven sustainable biomanufacturing will be formed through the seamless integration of AI-enhanced CRISPR technologies, autonomous bio-foundries, smart digital twins, and advanced hybrid biochemical platforms. Together, these technologies improve efficiency, accuracy, and scalability in biomanufacturing processes.

Regulatory & Ethical Frameworks: AI- enabled editing capabilities continue to evolve and expand; it is crucial that governance models are forward -looking. These frameworks must strike a delicate balance supporting innovation and assuring bio-safety, implementing and adopting strong dual-use protections and promoting equitable access to these potent technologies.

Human-Improved Ecosystem: A human improved ecosystem will be essential to establish. In a human-augmented ecosystem with proactive workforce reskill applications, industry-academia collaborations, and human in the loop designs— this approach will support responsible ethical decision making and promote innovation across bio-economy.

Sustainability of metrics: Techno economic analyses (TEA) and life cycle assessments (LCA) need to be included into all development process. This proves that every bio-manufacturing project is systematically assessed at the beginning for societal accountability, economic viability, environmental compliance.

Resilience through Distributed Manufacturing: Decentralized bio-production centers, empowered by AI and modular bio-processing capabilities, offer significant promise for enhanced adaptability to feed-stock disruptions and world-wide crises. This distributed model enhances the overall resilience of the bio-economy.

Distributed Manufacturing for Resilience: Decentralized bio production centers empowered by AI and modular bio-processing capabilities, offer significant potential for improved flexibility to feed-stock interruptions and world emergencies. The distributed model enhances the overall resilience of the bio-economy [98].

This comprehensive roadmap, integrating technologies' progress with ethical and governance considerations is essential for revealing the full potential an AI-powered sustainability bio-economy.

10.10 Challenges and Recommendations

10.10.1 Progress and Challenges in Brief

The field of AI-empowered sustainable bio manufacturing has undergone a revolution in the past few years. Great strides have been made in multiple areas; from microbial strain optimization to smart process prediction and control, and to the speeding up of data driven AL research and development pipelines. Environmental impact evaluations consistently establish impressive reduction in energy usage ranging from 30%-70% and substantial reductions on green-house emissions when compared to conventional fossil-based production methods. Moreover, innovative techniques such as residue valorization and integrated closed-loop bio-processing are increasingly resource utilization and fostering sustainable circular economy practices [99].

Despite these outstanding breakthroughs, notable critical challenges must be addressed to fully realize the potential of sustainable, bio-based production, eco-friendly manufacturing. Key issues include the inherent uncertainties associated with scaling process from bench-scale to commercial production, pervasive concerns surrounding data handling and integrity, the "complexity of adapting to evolving regulatory landscapes and the difficulties of effectively integrating AI into highly sophisticated bio-manufacturing systems. This is essential for wide and long-term sustainable bio-production situation.

10.10.2 Recommendations for Industry, Academia and Policy Makers

Getting to a sustainable bio-economy advanced by AI requires a concerted approach across numerous industries.

Policy makers should be:
Create Adaptive Regulations: Formulate and execute strong, flexible laws affecting biotechnology and artificial intelligence. These frameworks must compete effective models such as EU AI act and the US FDS's total product life cycle (TPLC) approach which promote innovations while rigorously confirming safety and compliance.
Demand Transparency and Validation: Provide rigorous guidelines for strict and transparent model validation in all bio-manufacturing systems. This will advance accountability, reliability, trust of AI applications.

Promote Public – Private Partnerships: Promote the growth of shared infrastructure including public-private partnerships and bio foundries. Such programs have the potential to accelerate, scale up, and drastically reduce starting investment barriers.

Leaders are encouraged to:
Deploy Hybrid AI Architectures: Prioritize the deployment of hybrid AI architectures that intelligently combine mechanistic models with data-driven models. This approach will maintain system interpretability, crucial for aligning with regulatory expectations and building stakeholder confidence.
Encourage multidisciplinary teams: Develop and invest multidisciplinary R&D and compliance teams with knowledge of artificial intelligence, biotechnology, regulatory affairs and digital operators. This integrated approach is essential for navigating the complexities of modern biotechnology.

Academic and research establishments
Priority of Translation Research: Emphasize and give priority to open-source research, focused on AI-driven bio-processing. In addition, an activity that publishes high quality, scale up datasets to enhance repeatability and accelerates industry broader adoption.
Integrate Interdisciplinary Training: Integrate interdisciplinary training that includes STEM education curriculum by combining data science, biology, ethics, and bio-safety. Graduates will effectively prepare for the evolving opportunities and demands of the bio-economy.
Collaborate on Governance Framework: Partner proactively with regulatory agencies and industry skate-holders to test cutting-edge governance structures and oversight approaches for dual use technologies. Such partnerships are critical to effectively addressing new bio-safety risks arising from advanced biotechnology developments [100].

10.10.3 Concluding Reflections on the AI-Driven Green Bio-Manufacturing Horizon

As AI and biotech increasingly intersect, they unlock vast opportunities to build a robust and eco-friendly bioeconomy. This paradigm shift offers that potential to drastically cut global dependence on fossil fuels and play a vital role in meeting urgent climate objectives. However, achieving this visionary future requires strong collaborations across sectors, guided by ethical responsibility and flexible, forward-thinking regulation.

Critical to this vision are:

- Decentralized Bio-manufacturing Hubs – The development of decentralized bio-manufacturing hubs with AI-based platforms will provide increased flexibility to target local market needs and build in resilience.
- Transparent, explainable AI systems - It takes a more formal place than mercury and doesn't exist without extra powerful protections, explained Ford, but you may find it all around the world.
- Ethical and Dual-Use Governance Mechanisms: Sophisticated ethical and dual-use governance mechanisms must be developed and realized. This encompasses stakeholder-led mechanisms for assessing risk and a more sophisticated early-warning function, drawing on the new international frameworks, to anticipate potential risks.
- In summary, maintaining the great momentum that we have seen in this area will depend to a large extent on a multi-actor approach [101].

10.10.4 Conclusion

In conclusion, sustaining the remarkable momentum witnessed in this sector hinges critically upon multi-actor collaboration. This requires policymakers to develop enabling innovation frameworks, industries to conscientiously embed ethical AI and green technologies into their operations, and academic institutions to consistently produce highly skilled professionals. Through such profound and concerted cooperation, we can collectively unlock the immense promise of AI-powered sustainable bio-manufacturing, paving the way for environmental resilience, robust economic prosperity, and profound societal benefit.

AI disclosure statement: The author confirms that no artificial intelligence (AI)-generated content or AI-assisted writing tools were used in the preparation of this chapter (including summaries and the abstract). All text, analysis, and conclusions are the result of the author's original work.

References

1. Zhang, C., Fei, Q., Fu, R., Lackner, M., Zhou, Y.J., Tan, T., Economic and sustainable revolution to facilitate one-carbon biomanufacturing. *Nat. Commun.*, *16*, 1, 4896, 2025.
2. Liu, Y. and Nielsen, J., Recent trends in metabolic engineering of microbial chemical factories. *Curr. Opin. Biotechnol.*, *60*, 188–197, 2019.
3. Pei, M., Abubakar, Y.S., Ali, H., Lin, L., Dou, X., Lu, G., Li, Y., Whole genome regulatory effect of MoISW2 and consequences for the evolution of the rice plant pathogenic fungus Magnaporthe oryzae. *MBio*, 15, 10, e01590–24, 2024.
4. Stephanopoulos, G., Metabolic fluxes and metabolic engineering. *Metab. Eng.*, *1*, 1, 1–11, 1999.
5. Avalos, J.L., Fink, G.R., Stephanopoulos, G., Compartmentalization of metabolic pathways in yeast mitochondria improves the production of branched-chain alcohols. *Nat. Biotechnol.*, *31*, 4, 335–341, 2013.
6. Hunold, A., Escobedo-Hinojosa, W., Potoudis, E., Resende, D., Farr, T., Syrén, P.O., Hauer, B., Assembly of a Rieske non-heme iron oxygenase multicomponent system from *Phenylobacterium immobile* E DSM 1986 enables pyrazon cis-dihydroxylation in *E. coli*. *Appl. Microbiol. Biotechnol.*, *105*, 5, 2003–2015, 2021.
7. Portela, A. and Esteller, M., Epigenetic modifications and human disease. *Nat. Biotechnol.*, *28*, 10, 1057–1068, 2010.
8. Doudna, J.A. and Charpentier, E., The new frontier of genome engineering with CRISPR-Cas9. *Science*, *346*, 6213, 1258096, 2014.
9. Shi, S. and Zhao, H., Metabolic engineering of oleaginous yeasts for production of fuels and chemicals. *Front. Microbiol.*, *8*, 2185, 2017.
10. Qi, L.S., Larson, M.H., Gilbert, L.A., Doudna, J.A., Weissman, J.S., Arkin, A.P., Lim, W.A., Repurposing CRISPR as an RNA-guided platform for sequence-specific control of gene expression. *Cell*, *152*, 5, 1173–1183, 2013.
11. Anzalone, A.V., Randolph, P.B., Davis, J.R., Sousa, A.A., Koblan, L.W., Levy, J.M., Liu, D.R., Search-and-replace genome editing without double-strand breaks or donor DNA. *Nature*, *576*, 7785, 149–157, 2019.
12. Lange, J.P., Lignocellulose conversion: An introduction to chemistry, process and economics. *Biofuels, Bioprod. Biorefin.: Innovation Sustainable Econ.*, *1*, 1, 39–48, 2007.
13. Ragauskas, A.J., Williams, C.K., Davison, B.H., Britovsek, G., Cairney, J., Eckert, C.A., Tschaplinski, T., The path forward for biofuels and biomaterials. *Science*, *311*, 5760, 484–489, 2006.
14. Ramakrishnan, B., Maddela, N.R., Venkateswarlu, K., Megharaj, M., Potential of microalgae and cyanobacteria to improve soil health and agricultural productivity: A critical view. *Environ. Sci.: Adv.*, *2*, 4, 586–611, 2023.

15. Jeong, Y., Cho, S.H., Lee, H., Choi, H.K., Kim, D.M., Lee, C.G., Cho, B.K., Current status and future strategies to increase secondary metabolite production from cyanobacteria. *Microorganisms, 8*, 12, 1849, 2020.
16. Khalid, M., Nanotechnology and chemical engineering as a tool to bioprocess microalgae for its applications in therapeutics and bioresource management. *Crit. Rev. Biotechnol., 40*, 1, 46–63, 2020.
17. Kumar, P., Barrett, D.M., Delwiche, M.J., Stroeve, P., Methods for pretreatment of lignocellulosic biomass for efficient hydrolysis and biofuel production. *Ind. Eng. Chem. Res., 48*, 8, 3713–3729, 2009.
18. Lynd, L.R., Beckham, G.T., Guss, A.M., Jayakody, L.N., Karp, E.M., Maranas, C., Wyman, C.E., Toward low-cost biological and hybrid biological/catalytic conversion of cellulosic biomass to fuels. *Energy Environ. Sci., 15*, 3, 938–990, 2022.
19. Szijártó, N., Production and characterization of cellulose-degrading enzymes for various applications, Doctoral dissertation, Budapest University of Technology and Economics, 2007.
20. Cherubini, F., Fuglestvedt, J., Gasser, T., Reisinger, A., Cavalett, O., Huijbregts, M. A., Levasseur, A. Bridging the gap between impact assessment methods and climate science. *Environ. Sci. Policy.*, 64, 129–140, 2016.
21. Wang, Z., Zhao, R., Xu, J., Zhang, S., Greenhouse Gas Footprint Mitigation Potentials of Bio-Based Products Through Diminished Processing Loss, Extended Service Lifespan, and Sustainable Recycling Scenarios.
22. Fan, M.Z., Kerr, B.J., Trabue, S.L., Yin, X., Yang, Z., Wang, W., Swine nutrition and environment, in: *Sustainable swine nutrition*, pp. 547–601, 2022.
23. Babu, B.K., Effect of the reducing agent dithiothreitol on ethanol and acetic acid production by clostridium strain p11, Doctoral dissertation, Oklahoma State University, 2010.
24. Samarawickrama, A., The influence of cultural burns on the density and stress response of koalas on Minjerribah, North Stradbroke Island, Doctoral dissertation, University of the Sunshine Coast, Queensland, 2023.
25. Ottinger, R.S., Blumenthal, J.L., Dal Porto, D.F., Gruber, G.I., Santy, M.J., Shih, C.C., Recommended Methods of Reduction, Neutralization, Recovery, Or Disposal of Hazardous Waste: Industrial and municipal disposal candidate waste stream constituent profile reports, organic compounds, vol. 10, National Environmental Research Center, Office of Research and Development, US Environmental Protection Agency, 10, 2023, 1973.
26. Salma, A., Application of a fermentation strategy to produce succinic acid using Actinobacillus succinogenes, Doctoral dissertation, Université Bretagne-Loire, 2020.
27. Stovicek, V., Dato, L., Almqvist, H., Schöpping, M., Chekina, K., Pedersen, L.E., Borodina, I., Rational and evolutionary engineering of *Saccharomyces cerevisiae* for production of dicarboxylic acids from lignocellulosic biomass and exploring genetic mechanisms of the yeast tolerance to the biomass hydrolysate. *Biotechnol. Biofuels Bioprod., 15*, 1, 22, 2022.

28. Singh, M., Kumar, S., Ali, H., Singh, S., Kansal, S.K., Valorization of lignocellulosic biomass into high-value chemicals: Forging a route towards the circular economy, in: *Energy materials*, pp. 98–133, CRC Press, 2024.
29. Lopes, M.S.G., Engineering biological systems toward a sustainable bioeconomy. *J. Ind. Microbiol. Biotechnol.*, *42*, 6, 813–838, 2015.
30. Minarti, A., Rinanti, A., Fachrul, M.F., Tazkiaturrizki, Fadhila, R., Circular economy for biodiesel production by managing wastewater using microalgae, in: *Algae as a natural solution for challenges in water-food-energy nexus: Toward carbon neutrality*, pp. 463–521, Springer Nature Singapore, 2024.
31. Radha, A., Ahluwalia, V., Rai, A.K., Varjani, S., Awasthi, M.K., Sindhu, R., Kumar, V., The way forward to produce nutraceuticals from agri-food processing residues: Obstacle, solution, and possibility. *J. Food Sci. Technol.*, *61*, 3, 429–443, 2024.
32. Liang, Y., A critical review of challenges faced by converting food waste to bioenergy through anaerobic digestion and hydrothermal liquefaction. *Waste Biomass Valorization*, *13*, 2, 781–796, 2022.
33. Ahuja, V., Chauhan, S., Purewal, S.S., Mehariya, S., Patel, A.K., Kumar, G., Bhatia, S.K., Microbial alchemy: Upcycling of brewery spent grains into high-value products through fermentation. *Crit. Rev. Biotechnol.*, *44*, 7, 1367–1385, 2024.
34. Liao, L., Xie, M., Zheng, X., Zhou, Z., Deng, Z., Gao, J., Molecular insights fast-tracked: AI in biosynthetic pathway research. *Nat. Prod. Rep.*, 2025.
35. Wang, X., Mohsin, A., Sun, Y., Li, C., Zhuang, Y., Wang, G., From spatial-temporal multiscale modeling to application: Bridging the valley of death in industrial biotechnology. *Bioengineering*, *10*, 6, 744, 2023.
36. Tatta, E.R., Imchen, M., Moopantakath, J., Kumavath, R., Bioprospecting of microbial enzymes: Current trends in industry and healthcare. *Appl. Microbiol. Biotechnol.*, *106*, 5, 1813–1835, 2022.
37. Elnaggar, K.S., Gamal, O., Hesham, N., Ayman, S., Mohamed, N., Moataz, A., Hassan, N., A guide in synthetic biology: Designing genetic circuits and their applications in stem cells. *SynBio*, *3*, 3, 11, 2025.
38. Mathur, P., Khanna, A.K., Cywinski, J.B., Maheshwari, K., Naylor Jr., D.F., Papay, F.A., 2019 year in review: Machine learning in healthcare, Team BrainX, BrainX Community, 2019.
39. Smith, A.J., Rational alterations to the Influenza A polymerase have ramifications for both therapeutic intervention and pandemic preparedness, Doctoral dissertation, University of Rochester, 2019.
40. Spinelli, C., Extracellular vesicle-mediated tumour-vascular interactions orchestrated by heterogenous glioma stem cells: Studies on vasectasia as a new form of glioma neovascularization, Doctoral dissertation, McGill University (Canada), 2020.
41. Ezeako, E.C., Solomon, A.Y., Itam, Y.B., Prospects of synthetic biology in revolutionizing microbial synthesis and drug discovery. *Life Res.*, *8*, 1, 6, 2025.

42. Dixon, T.A., Williams, T.C., Pretorius, I.S., Sensing the future of bio-informational engineering. *Nat. Commun.*, *12*, 1, 388, 2021.
43. Barthel, S., Developing *in vitro* tools for engineering allosteric transcription factors, enzymes and metabolic systems, 2024. (Note: This entry is likely a dissertation or thesis; if so, please provide the type and institution).
44. Burbano, D.A., Genetic and computational tools for engineering complex prokaryotic systems, Doctoral dissertation, University of Washington, 2025.
45. Ren, Y., Celińska, E., Cai, P., Zhou, Y.J., Combing directed enzyme evolution with metabolic engineering to develop efficient microbial cell factories. *Chem Bio Eng.*, 2025.
46. Hess, D., Dockalova, V., Kokkonen, P., Bednar, D., Damborsky, J., DeMello, A., Stavrakis, S., Exploring mechanism of enzyme catalysis by on-chip transient kinetics coupled with global data analysis and molecular modeling. *Chem*, *7*, 4, 1066–1079, 2021.
47. Xu, J., Li, T., Huang, W.E., Zhou, N.Y., Semi-rational design of nitroarene dioxygenase for catalytic ability toward 2, 4-dichloronitrobenzene. *Appl. Environ. Microbiol.*, *90*, 6, e01436–23, 2024.
48. Woodley, J.M., New opportunities for biocatalysis: Making pharmaceutical processes greener. *Trends Biotechnol.*, *26*, 6, 321–327, 2008.
49. Książek, E., Citric acid: Properties, microbial production, and applications in industries. *Molecules*, *29*, 1, 22, 2023.
50. Di Lorenzo, R.D., Serra, I., Porro, D., Branduardi, P., State of the art on the microbial production of industrially relevant organic acids. *Catalysts*, *12*, 2, 234, 2022.
51. Pundir, P., Shabhanani, H., Murugan, A.C., Bioprocess strategies for the production of short-chain organic acids from microbial cells, in: *Recent advances in bioprocess engineering and bioreactor design*, pp. 185–231, Springer Nature Singapore, 2024.
52. Kim, J., Kim, Y.M., Lebaka, V.R., Wee, Y.J., Lactic acid for green chemical industry: Recent advances in and future prospects for production technology, recovery, and applications. *Fermentation*, *8*, 11, 609, 2022.
53. Ahmad, A., Banat, F., AlBlooshi, H. (Eds.), *Algal biotechnology: Integrated algal engineering for bioenergy, bioremediation, and biomedical applications*, Elsevier, 2022.
54. Auma, E.O., Anaerobic co-digestion of water hyacinth (Eichhornia crassipes) with ruminal slaughterhouse waste under mesophilic conditions, Doctoral dissertation, University of Nairobi, 2020.
55. Broda, M., Yelle, D.J., Serwańska, K., Bioethanol production from lignocellulosic biomass—challenges and solutions. *Molecules*, *27*, 24, 8717, 2022.
56. Gao, G., Burgess, J.G., Wu, M., Wang, S., Gao, K., Using macroalgae as biofuel: Current opportunities and challenges. *Bot. Mar.*, *63*, 4, 355–370, 2020.
57. Gupta, N., Fischer, A.R., Frewer, L.J., Ethics, risk and benefits associated with different applications of nanotechnology: A comparison of expert and

consumer perceptions of drivers of societal acceptance. *NanoEthics*, *9*, 2, 93–108, 2015.
58. Kaptan, G., Fischer, A.R., Frewer, L.J., Extrapolating understanding of food risk perceptions to emerging food safety cases. *J. Risk Res.*, *21*, 8, 996–1018, 2018.
59. Ogbeh, G.O., Ogunlela, A.O., Akinbile, C.O., Iwar, R.T., Adsorption of organic micropollutants in water: A review of advances in modelling, mechanisms, adsorbents, and their characteristics. *Environ. Eng. Res.*, *30*, 2, 2025.
60. Vaaje-Kolstad, G., Tuveng, T.R., Mekasha, S., Eijsink, V.G., Enzymes for modification of chitin and chitosan, in: *Chitin and chitosan: Properties and applications*, pp. 189–228, 2019.
61. Han, T. and Miao, G., Strategies, achievements, and potential challenges of plant and microbial chassis in the biosynthesis of plant secondary metabolites. *Molecules*, *29*, 9, 2106, 2024.
62. Rafieenia, R., Klemm, C., Hapeta, P., Fu, J., García, M.G., Ledesma-Amaro, R., Designing synthetic microbial communities with the capacity to upcycle fermentation byproducts to increase production yields. *Trends Biotechnol.*, *43*, 3, 601–619, 2025.
63. Hu, N., Liu, X., Wei, S., Yao, J., Wang, W., Liu, B., Wang, L., Current status and future prospects of pretreatment for tobacco stalk lignocellulose. *Front. Bioeng. Biotechnol.*, *12*, 1465419, 2024.
64. Li, H., Chen, X., Tang, Y., Yang, Y., He, F., Wang, X., Yang, Y., Separation, purification, and crystallization of 1, 5-pentanediamine hydrochloride from fermentation broth by cation resin. *Front. Bioeng. Biotechnol.*, *10*, 1104041, 2023.
65. Zhe, L., Sethupathy, S., Wenqian, D., Hu, J., Zhu, D., Lignin valorization through microbial production of polyhydroxyalkanoates: Recent trends, challenges and opportunities. *Green Chem.*, 2025.
66. Chen, Y., Olcay, H., Tan, E.C., Woodworth, S.P., Miscall, J., Aromolaran, A., Beckham, G.T., Carboxylic acid concentration in downstream bioprocessing using high-pressure reverse osmosis. *ACS Sustainable Chem. Eng.*, *13*, 16, 5889–5905, 2025.
67. Zhang, Q., Engineering of synthetic DNA/RNA modules for manipulating gene expression and circuit dynamics, Doctoral dissertation, Arizona State University, 2020.
68. Werpy, T. and Petersen, G., Top value added chemicals from biomass: Volume I–results of screening for potential candidates from sugars and synthesis gas (No. DOE/GO-102004-1992), National Renewable Energy Lab. (NREL), 2004.
69. Gandini, A., Polymers from renewable resources: A challenge for the future of macromolecular materials. *Macromolecules*, *41*, 24, 9491–9504, 2008.
70. Gandini, A. and Belgacem, M.N., The state of the art of polymers from renewable resources, in: *Handbook of biopolymers and biodegradable plastics:*

Properties, procession and applications, 1st ed., pp. 71–85, William Andrew Publishing, 2012.

71. Sharif, A. and Hoque, M.E., Renewable resource-based polymers, in: *Bio-based polymers and nanocomposites: Preparation, processing, properties & performance*, pp. 1–28, Springer International Publishing, 2019.
72. Holmberg, A.L., Reno, K.H., Wool, R.P., Epps III, T.H., Biobased building blocks for the rational design of renewable block polymers. *Soft Matter*, *10*, 38, 7405–7424, 2014.
73. Duchemin, B., The sustainability of phytomass-derived materials: Thermodynamical aspects, life cycle analysis and research perspectives. *Green Chem.*, *24*, 7, 2653–2679, 2022.
74. Mahmood, H., Moniruzzaman, M., Yusup, S., Welton, T., Ionic liquids assisted processing of renewable resources for the fabrication of biodegradable composite materials. *Green Chem.*, *19*, 9, 2051–2075, 2017.
75. Mathers, R.T. and Meier, M.A. (Eds.), *Green polymerization methods: Renewable starting materials, catalysis and waste reduction*, John Wiley & Sons, 2011.
76. Thakur, V.K., Thakur, M.K., Kessler, M.R., *Handbook of composites from renewable materials, biodegradable materials*, vol. 5, John Wiley & Sons, 2017.
77. Kim, H.M., Chae, T.U., Choi, S.Y., Kim, W.J., Lee, S.Y., Engineering of an oleaginous bacterium for the production of fatty acids and fuels. *Nat. Chem. Biol.*, *15*, 7, 721–729, 2019.
78. Beld, J., Lee, D.J., Burkart, M.D., Fatty acid biosynthesis revisited: Structure elucidation and metabolic engineering. *Mol. Biosyst.*, *11*, 1, 38–59, 2015.
79. Narisetty, V., Cox, R., Bommareddy, R., Agrawal, D., Ahmad, E., Pant, K.K., Kumar, V., Valorisation of xylose to renewable fuels and chemicals, an essential step in augmenting the commercial viability of lignocellulosic biorefineries. *Sustainable Energy Fuels*, *6*, 1, 29–65, 2022.
80. Torto-Alalibo, T., Purwantini, E., Lomax, J., Setubal, J.C., Mukhopadhyay, B., Tyler, B.M., Genetic resources for advanced biofuel production described with the Gene Ontology. *Front. Microbiol.*, *5*, 528, 2014.
81. Stark, B.C., Pagilla, K.R., Dikshit, K.L., Recent applications of *Vitreoscilla* hemoglobin technology in bioproduct synthesis and bioremediation. *Appl. Microbiol. Biotechnol.*, *99*, 4, 1627–1636, 2015.
82. Kannisto, M., Metabolic engineering of Acinetobacter baylyi ADP1 for improved growth and wax ester production using components of lignocellulosic hydrolysates as carbon sources, 2018. (Note: This entry is likely a dissertation or thesis; if so, please provide the type and institution).
83. Liu, Y. and Liu, T., Fatty acid biosynthesis and its metabolic engineering applications, in: *Engineering microbial metabolism for chemical synthesis: Reviews and perspectives*, pp. 97–119, 2018.

84. Tian, X., Chen, H., Liu, H., Chen, J., Recent advances in lactic acid production by lactic acid bacteria. *Appl. Biochem. Biotechnol.*, *193*, 12, 4151–4171, 2021.
85. Wierzbicki, M.C., Application of synthetic biology for increasing anaerobic biodiesel production in Escherichia coli, 2013.
86. da Silva, J.B.S., Villa-Vélez, H.A., Júnior, J.B., Conconi, C.C., Cruz, G., Perspectives and challenges of sugarcane straw for bioenergy production, in: *Sugarcane: Production, properties and uses*, Nova Science Publishers, 2020.
87. Vilarinho, C., Castro, F., Russo, M. (Eds.), *Wastes: Solutions, treatments and opportunities*, CRC Press, 2015.
88. Koffas, M., Roberge, C., Lee, K., Stephanopoulos, G., Metabolic engineering. *Annu. Rev. Biomed. Eng.*, *1*, 1, 535–557, 1999.
89. Qi, L.S., Larson, M.H., Gilbert, L.A., Doudna, J.A., Weissman, J.S., Arkin, A.P., Lim, W.A., Repurposing CRISPR as an RNA-guided platform for sequence-specific control of gene expression. *Cell*, *184*, 3, 844, 2021.
90. Burbano, D.A., Kiattisewee, C., Karanjia, A.V., Cardiff, R.A., Faulkner, I.D., Sugianto, W., Carothers, J.M., CRISPR tools for engineering prokaryotic systems: Recent advances and new applications. *Annu. Rev. Chem. Biomol. Eng.*, *15*, 1, 389–430, 2024.
91. Anderson, D.A., Competition-based CRISPR-dCas9 transcriptional control mechanisms and application of dCas9 biosensors for high-throughput, cell-based protease inhibitor screens, Doctoral dissertation, Massachusetts Institute of Technology, 2021.
92. Wang, X., Dong, K., Kong, D., Zhou, Y., Yin, J., Cai, F., Ye, H., A far-red light–inducible CRISPR-Cas12a platform for remote-controlled genome editing and gene activation. *Sci. Adv.*, *7*, 50, eabh2358, 2021.
93. Yan, Q. and Fong, S.S., Challenges and advances for genetic engineering of non-model bacteria and uses in consolidated bioprocessing. *Front. Microbiol.*, *8*, 2060, 2017.
94. Sarker, P.K., Paul, A.S., Karmoker, D., Mitigating climate change and pandemic impacts on global food security: Dual sustainable agriculture approach (2S approach). *Planta*, *258*, 6, 104, 2023.
95. Buyel, J.F., Process development strategies in plant molecular farming. *Current Pharm. Biotechnol.*, *16*, 11, 966–982, 2015.
96. Lauritsen, I., Elucidation of gene regulatory mechanisms in ageing bacterial colonies and tool development for cell factory optimization, 2020.
97. Mohanan, P.V. (Ed.), *Artificial intelligence and biological sciences*, CRC Press, Taylor & Francis Group, 2025.
98. George, A.S., Data-driven strategies for intelligent business sustainability in the digital economy.
99. Arias, A., Feijoo, G., Moreira, M.T., How could Artificial Intelligence be used to increase the potential of biorefineries in the near future? A review. *Environ. Technol. Innov.*, 32, 103277, 2023.

100. Moffat, A.D., Elliston, A., Patron, N.J., Truman, A.W., Carrasco Lopez, J.A., A biofoundry workflow for the identification of genetic determinants of microbial growth inhibition. *Synth. Biol.*, *6*, 1, ysab004, 2021.
101. Lekadir, K., Frangi, A.F., Porras, A.R., Glocker, B., Cintas, C., Langlotz, C.P., Starmans, M.P., FUTURE-AI: International consensus guideline for trustworthy and deployable artificial intelligence in healthcare. *BMJ*, *388*, 2025.

11

Microbial Biomanufacturing Using Agro-Industrial Wastes for Sustainable Production of Biologics and Biopharmaceuticals

Lakshmi Madhavi Chitra[1,2*], K. Sujana[2], A. Harani[3], Pushpalatha Kondamuri[3], Sree Sowkhya Thaninki[4] and Parinaya Sri Lotha[4]

[1]*Department of Pharmaceutical Analysis, School of Pharmacy, Aditya University, Surampalem, Andhra Pradesh, India*
[2]*Department of Pharmaceutical Analysis, Acharya Nagarjuna University College of Pharmaceutical Sciences, Guntur, Andhra Pradesh, India*
[3]*Department of Pharmaceutics, School of Pharmacy, Aditya University, Surampalem, Andhra Pradesh, India*
[4]*Department of Pharmaceutical Chemistry, School of Pharmacy, Aditya University, Surampalem, Andhra Pradesh, India*

Abstract

The emerging domain of biomanufacturing progressively depends on novel microbe-based approaches to tackle the combined challenges of eco-friendliness and productivity in the manufacturing of biologics and biopharmaceuticals. This chapter investigates the optimized utilization of agro-industrial wastes, which are abundant in biological material and are often treated as refuse, converting them into feasible substrates through microbial bioprocesses. These approaches facilitate high-yielding, scalable, and eco-friendly bio-derived products. Real-world scenarios deliver that the current investigation in a tangible way showcases the well-executed production of antibiotics, bioactive peptides, enzymes, etc. This chapter further examines critical biomanufacturing hurdles such as fluctuating feedstock quality, process optimization, and the complexities of downstream processing, and offers strategies to address them. Technological innovations in disciplines like synthetic biology, metabolic engineering, and bioreactor design

**Corresponding author*: madhuchitra8@gmail.com

Sandip Kunar, Jagadeesha T. and D. Sathis Kumar (eds.) Biomanufacturing: Transforming Production of Biologics and Biopharmaceuticals, (295–342) © 2026 Scrivener Publishing LLC

are opening new prospects for the optimal utilization of diverse types of residues. Concurrently, the unification of lean strategies and digital tools for process optimization is discussed. Overall, current efforts highlight the innovative potential of microbial biomanufacturing in attaining a circular bio-economy while meeting the quality and scalability needs of contemporary interventions.

Keywords: Microbial biomanufacturing, agro-industrial wastes, process optimization, lean strategies, circular bioeconomy

11.1 Introduction

The modern era has seen pressing, worldwide fundamental changes in the direction of environmentally responsible production techniques, caused by escalating ecological issues, the reduction of natural assets, and the essential requirement for financial stability. Under these conditions, the industrial domain, especially the manufacturing of high-value biomolecules and therapeutic biomolecules, encounters tremendous demand to embrace more sustainable, productive, and budget-friendly procedures [1]. Conventional synthetic processes are based on finite, petroleum-derived raw materials, require considerable fuel, and produce dangerous secondary outputs, resulting in major environmental impacts. Microbial-based production, alternatively referred to as bio-processing science or clean biotechnology, arises as a revolutionary replacement, utilizing the natural biochemical functions of microbes to transform various feedstocks into a large variety of high-value items [2].

Traditionally, biological production has played a key role in making a range of vital chemical entities, including standard industrial chemicals, protein-based catalysts, sophisticated medicinal proteins, and immunizations. Initial cases involve the manufacture of an antibiotic called penicillin and insulin that transformed health care [3]. Current biotechnology production, on the other hand, stretches significantly beyond the preliminary uses, including an advanced synergy of microbial science, biological chemistry, operational methods, and artificial life science. The emergence of sophisticated hereditary design strategies has enabled researchers to create and improve microorganism-based unit manufacturing systems with the aim of boosting results, innovative elements generation, as well as better resource use in methodology [4].

A notable, convincing domain among eco-friendly biotech fabrication is the application of farming-related industrial by-products in the form of biotransformation raw materials. Cultivation and commercial operations produce huge amounts of organic matter leftovers, including fibrous plant-based substances like maize husks, cereal crop, chaff, sugarcane residue; edible manufacturing co-products like orchard crop ends, plant-based scraps, milk-based leftovers; and also, livestock or fauna-based organic fertilizer. Such leftovers frequently create considerable discarding difficulties, resulting in ecological contamination such as heat-trapping vapor discharges and aquatic pollution. Reallocating the bountiful, inexpensive, and sustainable assets like bio-transformed based materials, in addition to alleviating ecological consequences, has also generated innovative, cost-effective prospects, promoting a regenerative biological economy. This strategy corresponds ideally in conjunction with the tenets of environmental responsibility, shifting away from one-way extract-procedure-discard paradigms in the direction of restorative frameworks, at which point leftovers are seen as a useful asset [5–7].

The section intends to deliver a single, complete analysis regarding microorganism-based biological production, accompanied by targeted emphasis regarding the utilization of agriculture-related residues. The chapter explores the basic elements of the present bioengineering, explains the categories of microbes used, and also explains the benefits versus standard molecular generation. A major section will remain focused on grasping the properties of diverse agricultural residues, the vital initial treatment approaches, and this section analyzes various microbial processing methods, factors regarding fermentation vessel structuring, and essential procedure improvement practices. The revolutionary effect of biochemical modification and engineered life science in boosting productivity will also be showcased. In conclusion, this section covers post-production treatment, compliance-related regulations, as well as offers demonstrative examples, concluding in a visionary viewpoint for future trajectory exploration prospects. Also highlighted is the role of forces for catalysts, attaining the capability of waste conversion and microbial-based production from an eco-friendly perspective.

11.2 Fundamentals of Microbial Bio-Manufacturing

Microbe-related bio-fabrication fundamentally utilizes the intrinsic life-based mechanisms of microbes to generate intended substances. This chapter defines the basic concepts underlying this method, describing its extent, detailing the varied microorganisms contributing, and comparing its benefits with classical synthetic chemistry.

11.2.1 Definition and Scope

Microorganism-based biological production denotes large-scale synthesis of diverse organic molecules, reagents, and products employing active microorganism units (prokaryotes, fungi, and phytoplankton) or their cell parts, enzymatic agents [8]. This exists as an integrative domain combining rules originating in microbial science, cellular-level bioscience, chemical biology, hereditary science, molecular applied science, along with operation regulation.

The range of microbe-related bio-production is extensive and constantly growing, and includes:

- **Pharmaceutics:** Therapeutic proteins (e.g., insulin, growth hormones, monoclonal antibodies), vaccines, antibiotics, enzymes, and small-molecule drugs [8].
- **Biofuels:** Ethanol, butanol, biodiesel, and other advanced biofuels derived from biomass [9].
- **Commodity and specialty chemicals:** Organic acids (e.g., lactic acid, succinic acid, citric acid), amino acids, vitamins, solvents (e.g., acetone, butanol), bioplastics (e.g., polyhydroxyl alcohols), and pigments [10].
- **Food and feed ingredients:** Enzymes for food processing (e.g., amylases, proteases), probiotics, prebiotics, food additives, and single-cell proteins [11].
- **Textiles and Biomaterials:** Biopolymers and other advanced materials with varied applications [12].

11.2.2 Microorganisms Used in Bio Manufacturing

A varied collection of microbes acts as the mainstay inside biological production, each one having distinctive biochemical abilities, developmental traits, as well as hereditary modifiability. The selection relating to microbe is crucial and relies upon the particular outputs intended for production of raw material properties, along with operation demands. The essential microbe-based categories contain:

- **Bacteria [13–17]:**
    - ***E. coli*:** A Gram-negative bacterium, rapidly used for recombinant protein production.
    - ***B. subtilis*:** A Gram-positive bacterium, used for the production of enzymes by the downstream process.

    - ***Corynebacterium glutamicum*:** Mostly used to obtain high yields of amino acids (L-glutamate, L-lysine etc.).
    - **Lactic acid bacteria:** Used in the fermentation of food, and in the production of lactic acid and other products from sugar residues.
    - **Acetogenic bacteria:** Clostridium autoethanogenum, used to produce fuels from various gases.
- **Yeasts [18–20]:**
    - ***Saccharomyces cerevisiae*:** Effectively used in the manufacture of ethanol, bread, and beer; and it also increases the recombinant protein expression.
    - ***Pichia pastoris*:** Mostly documented for its extensive recombinant protein expression and for the production of pharmaceutical proteins.
    - ***Yarrowia lipolytica*:** An oleaginous yeast used in the production of organic acids, enzymes, and metabolites from hydrophobic compounds and leftovers.
- **Filamentous Fungi [21]:**
    - ***Aspergillus niger*:** Broadly utilized in the large-scale synthesis of natural acidic substances (e.g., citrate-based compounds, gluconate-related acids), biocatalysts (fiber breakdown enzymes, pectinases), and also additional biochemical byproducts. This microbe is proficient in low-moisture microbial processing because of the organism's thread-like expansion.
    - ***Trichoderma reesei*:** An abundant source related to cellulose-degrading biocatalysts, essential for degradation related to plant matter converted into digestible saccharides.
- **Microalgae [22]:**
    - **Chlorella and Spirulina:** These are light-harvesting microbes able to transform biological mass high in fats, amino acids, as well as sugars. These organisms exist as renewable fluids, premium health supplements, and also in industrial effluent purification.

11.2.3 Advantages over Chemical Synthesis

Microbe-driven biological production provides numerous, unique benefits compared to synthetic production approaches, identifying the role of a key element in eco-friendly manufacturing.

- **Sustainability and Eco-Friendliness [23, 24]:**
  - **Renewable feedstocks:** Biological processes employ sustainable materials, reducing dependence upon limited fossil-based chemicals.
  - **Reduced energy consumption:** Microbial conversion procedures usually run under moderate environmental surrounding heat levels and need considerably less power with relative to elevated heat as well as elevated pressure, synthetic process.
  - **Lower Pollution:** Biocatalysts often produces fewer toxic by-products and reduces the generation of hazardous waste. Water, if used, can often be treated and reused. The processes are inherently "green".
  - **Carbon footprint reduction:** Employing greenhouse gas as an elemental carbon origin, and transforming organic matter residues, assists in trapping CO_2 and lowers climate gas discharges, aiding towards environmental shift reduction.
- **Specificity and selectivity [6, 25]:**
  - **High stereo specificity:** Biocatalysts as well as microbial-based routes exist as extremely selective, allowing the creation of preferred chiral molecules or else structural variants with great cleanliness that remains especially essential in therapeutics in which asymmetry influences effectiveness.
  - **Complex molecule synthesis:** Microbes can produce complicated biological molecules containing detailed forms, including polypeptides, glycans, along with bioactive molecules that remain difficult or unfeasible to synthesize using traditional synthetic methods.
- **Economic Benefits:**
  - **Cost-effective feedstocks:** The utilization of inexpensive agri-processing residues substantially lowers unprocessed input expenses, resulting in the completed outputs increasingly marketability [5].
  - **New market Opportunities:** Residues upgrading generates novel worth networks as well as financial prospects in agricultural regions in which agro-related practices exists as significant [26].
  - **Scalability:** Microbial operations remain adaptable towards up-scaling, originating in the lab, extending to

large-scale manufacturing using proven microbial conversion techniques.

- **Biodegradability:** Several bio-based goods, biodegradable plastics as well as organic chemicals exist as naturally compostable lowering resistant ecological contamination [27].

11.3 Agro-Industrial Wastes as Fermentation Substrates

The idea of a regenerative bio-based economy depends on converting residues to useful materials. Agricultural-industrial leftovers, a secondary product from agrarian and food manufacturing operations, constitute a vast supply composed of organic matter in full of sugars, amino acids, and additional natural chemicals fit for microorganism-based bioprocessing. Table 11.1 shows the classification and composition of Agro-Industrial wastes.

Table 11.1 Classification and composition of agro-industrial wastes.

Sl. no.	Category of waste	Example	Components
1	Lignocellulose	Crop leftovers like straw, wood residues	Cellulose, hemicellulose, and lignin
2	Starch(or) sugar rich	Molasses, fruits, or vegetables processing waste, and broken grains	Sugars like glucose, sucrose, fructose, and starch
3	Rich in Lipid	Oilseed residues, coffee grounds	Lipids and proteins
4	Rich in Protein	Dairy waste and waste from processing waste	Proteins and amino acids
5	Others like algae	Algae biomass from wastewater	Carbohydrates, proteins and lipids

11.3.1 Classification and Composition

Agri-processing by-products are extremely varied with respect to their source and molecular structure. This clearly affects their appropriateness as microbial conversion feedstocks and the initial treatment techniques needed. Such waste may be generally categorized into groups. They are

- **Lignocellulosic Biomass:** This constitutes the largest fraction of agricultural residues and includes crop residues (e.g., corn stover, wheat straw, rice straw, sugarcane bagasse), forestry residues (e.g., wood chips, sawdust), and dedicated energy crops. Their primary components are:
  - **Cellulose:** A straight complex carbohydrate of dextrose segments supplying architectural stability.
  - **Hemicellulose:** Mixed complex sugar made up of different five-carbon sugars (xylose, arabinose) and hexoses (dextrose, mannopyranose, galactose), often forked.
  - **Lignin:** A non-crystalline aromatic macromolecule that envelops cellulose and hemicellulose, conferring stiffness and a barrier for microorganism-based decomposition; it is the most resistant element and needs significant effort for preliminary processing [23].
- **Starch-rich wastes:** It is mainly sourced from edible foods, comprised of potato skins, fractured rice, corn industrial remnants, and manioc residues. They are loaded with amylose, an easily fermentable dextrose polymer, and are simple to handle compared to plant biomass [29].
- **Sugar-rich wastes:** These originate from syrup-byproducts from the sugar refining, as well as from fruit or vegetable leftovers, and possess monosaccharides (glucose, fructose, sucrose) that are straightaway convertible by microbes [29].
- **Protein-rich wastes:** e.g., abattoir residues, milk-processing residues, milk serum remnants, and pulse refining. They act as nitrogen sources and furnish protein monomers and microbial proliferation after chemical breakdown [30].
- **Lipid-rich wastes:** Used frying oils, animal-derived lipids, and factory discharges supply carbon donors. Oil-producing microbes can generate fats and water-repelling substances [31].
- **Mixed Wastes:** Several food industry effluents, brewery wastewater, distillation residues, and city organic refuse are

diverse blends having saccharides, polypeptides, and fats which commonly demand tailored microbial communities and sequential treatment [32].

11.3.2 Pretreatment and Hydrolysis Techniques

Initial processing constitutes an essential stage to break down the plant matter, isolate lignin, and breakdown hemicellulose and lower cellulose structural order, thus enhancing its biodegradability [33]. Most common pre-treatment methods include:

- **Physical Pre-treatment [34]:**
  - **Milling (or) Grinding:** Decreases particle dimensions, enhancing the exterior extent for enzyme-based degradation.
  - **Pyrolysis:** Thermal decomposition in the absence of oxygen to produce bio-oil, char, and syngas.
  - **Extrusion:** High-pressure and shear forces disrupt the lignocellulosic structure.
- **Chemical Pre-treatment [35, 36]:**
  - **Acid Hydrolysis:** Employs weak or strong acid solutions (sulfuric acid) that react with water and forms hemicellulose and cellulose.
  - **Alkaline Pretreatment:** Employs alkali like NaOH, NH_3 to eliminate lignin and swollen cellulose.
  - **Oxidative Pre-treatment:** Employs oxidative substances like hydrogen peroxide to cleave lignin.
  - **Organosolv:** Employs carbon-based liquids like ethanol and methanol in order to solubilize lignin and hemicellulose, enabling simpler isolation of cellulose.
  - **Ionic Liquids:** Recent liquids which may solubilize cellulose without pretreatment enabling efficient extraction of lignin and hemicellulose.
- **Physio-Chemical Pre-Treatment:**
  - **Steam Explosion:** Organic material is subjected to elevated pressure and water vapor, succeeded by quick decompression resulting in violent fragmentation [37].
  - **Ammonia Fiber Expansion:** Utilizes liquefied or aeriform NH_3 in order to expand and make amorphous cellulose and remove hemicellulose, excluding major monosaccharide deterioration [38].

- **Wet oxidation:** Employs H_2O and atmospheric gas under elevated thermal conditions to oxidize lignin and hemicellulose [39].
- **Biological Pre-treatment:** This approach appears ecologically safe; however, it is commonly less rapid [40]. This pre-treatment, hydrolysis is often required to convert complex carbohydrates like cellulose, hemicellulose, and starch into fermentable monosaccharides. This can be attained by:
  - **Enzymatic Hydrolysis:** Employs selective biocatalysts like cellulases, hemicellulases, and amylases to degrade polysaccharides [41].
  - **Acid Hydrolysis:** May carry out both preliminary treatment and hydrolysis in a single stage; through it can result in carbohydrate decomposition and the generation of inhibitory compounds.

Table 11.2 summarizes the pre-treatment and hydrolysis techniques for Agri-Industrial Residues.

The selection of a method for primary treatment depends on the type of substrate, product recovery, economic viability, and environmental effects [42].

11.3.3 Environmental and Economic Benefits

Using agro-industrial waste for biomanufacturing brings clear and economic benefits and plays an important role in supporting sustainable development and advancing circular bioeconomy.

- **Environmental benefits [6–8]:**
  - **Waste reduction and management:** Redirects large quantities of farming and industrial by-products away from dumping and uncontrolled combustion, helping to lower soil pollution, water fouling from runoff, and atmospheric contamination caused by smoke and fine airborne particles.
  - **Reduced greenhouse gas emissions:** Helps avoid methane release from the oxygen-free breakdown of biodegradable waste in landfills and reduces carbon dioxide emissions caused by open burning. In addition, using plant-based materials as raw input lowers dependence on

Table 11.2 Pre-treatment and hydrolysis techniques for agri-industrial residues.

Sl. no.	Type of method	Technique	Mechanism	Advantage	Disadvantage
1.	Physical	Milling (or) Grinding	Reduction in particle size and increased surface area	Simple and cost-effective	High energy consumption, and there is no chemical modification
		Pyrolysis	Thermal Decomposition	Produces bio-oil, biochar, and syngas	Requires high temperature, complex products
2.	Chemical	Acid hydrolysis	Break down hemicellulose and cellulose	Effective for the release of sugars	Produces inhibitors and is corrosive
		Alkaline treatment	Pre-saponifies esters, disrupts lignin-carbohydrate bonds	Effective removal of lignin, decreased inhibitor formation	It has a slower reaction rate and requires alkali recovery
		Oxidative treatment	Pre-breaks down lignin and hemicellulose	Effective at mild temperatures and is eco-friendly	Requires oxidizing agents
3.	Physio-chemical	Steam explosion	High-pressure steam, sudden decompression	Effective for lignocellulose and requires less chemicals	Formation of inhibitor, incomplete removal of lignin
4.	Biological	Enzymatic Hydrolysis	Enzymes like cellulases and hemicelluloses are used	Specific conditions and eco-friendly	Enzyme cost is high, and the reaction rate is slow

non-renewable fuels, contributing to an overall reduction in greenhouse gas emissions.

- **Resource Conservation:** Converts discarded materials into useful resources, helping to preserve untouched natural inputs and lowering the total ecological impact tied to their removal and industrial treatment.
- **Water quality improvement:** When agricultural waste pollutes lakes or rivers, converting it into valuable products can help improve water quality through the process of bioremediation.

- **Economic Benefits [6–8, 11]:**
  - **Cost-Effective Feedstock:** It lowers the cost of input materials for bio-based production when compared to specially cultivated crops or artificially produced chemicals.
  - **Value-added products:** Converts inexpensive by-products into valuable biopharmaceuticals, medicines, chemicals, and energy sources, generating novel financial gains in supportive of agronomic and also manufacturing fields.
  - **Rural-Economic Development:** It develops novel business along with work prospects within agricultural zones in which agri-based work are clustered, driving financial development as well as limiting relocation of cities.
  - **Enhanced energy Security:** Generation of green fuels derived from waste decreases, use of non-domestic oil and gas supporting domestic energy autonomy.
  - **Reduce disposal cost:** Agro-industries may reduce expenses related to material recovery transfer and dumping, additionally boosting economic viability.

11.4 Fermentation Strategies for Waste Valorization

Fermentation represents the core biological procedure within biomanufacturing where microbes change substrates into intended products under regulated conditions. The selection of a fermentation approach is critical for effectively utilizing agricultural-industrial residues acknowledging its varied composition and distinct microbial needs.

Table 11.3 summarizes the various fermentation types, which are used in bio-manufacturing of substrates.

11.4.1 Submerged Fermentation [43–49]

Submerged fermentation remains the most broadly utilized biotransformation technique towards commercial biosynthesis. This includes the growth of microbes inside a liquid medium where every feed stock is solubilized or dispersed.

Table 11.3 Fermentation strategies in microbial bio-manufacturing.

Sl. no.	Fermentation type	Features	Benefits	Limitations
1.	Submerged	Microbes grow in liquid broth	High control of process, good scale-up, effective mass-transfer	Water consumption is high, product concentration is less, and higher energy input
2.	Solid-state	Microbes grow on moist solid substrate	Water consumption is less; product concentration is high	Heat or mass transfer is very difficult to control, Scale-up is difficult
3.	Mixed culture	Co-cultivation	Synergistic effects, enhanced efficiency	Difficulties may face in maintaining optimal conditions
4.	Fed-batch	Nutrients are added in an increased manner during fermentation	Substrate inhibition is reduced; density of cells is high	Operation is very difficult
5.	Continuous	Continuously add feed to the medium	Productivity is high, constant product quality	Chance of contamination, less flexibility

- **Principle:** Microbes were multiplying and generating compounds while completely submerged in a liquid broth, guaranteeing effective bulk transfer of nutrients, oxygen (aerobic process), and outputs.
- **Advantages:**
  - **High Control:** Excellent management of procedure variables like pH, temperature, ventilation, stirring, and feed amount resulting in improved production.

- **Easy Scale-Up:** Well-established technical concepts for increasing fermenters from lab to commercial capacities.
- **Efficient Product Recovery:** Post-production handling is frequently simpler, since substances are usually solubilized inside the medium or excreted outside the cell.
- **Automation:** Extremely suitable for auto-control and ongoing processing.

- **Disadvantages:**
  - **High Water Consumption:** Requires significant volumes of H_2O that may cause ecological issues and raise functional expenses.
  - **High Energy Cost:** Agitation and aeration in large bioreactors require energy, particularly for viscous broths.
  - **Complex Effluent Treatment:** The huge quantities of fermentation broth produce significant liquid residue needs handling.
  - **Nutrient Supplementation:** Agro-residues offer an organic source, but additional minerals like nitrogen, phosphorus, and trace elements may be required to enhance microbial growth and product formation.
- **Applications in Waste Valorization:** Submerged filtration is widely employed for generating ethanol, carbon-based compounds like lactic acid derived from whey or potato residues, antimicrobials, and genetically engineered proteins from lignocellulose.

11.4.2 Solid State Fermentation [50–56]

Solid-phase bioprocessing includes the development of microbes on damp, firm materials without almost any unbound liquid. This method closely imitates native microbe environments, especially for molds and certain prokaryotes.

- **Principle:** Microbes multiply on the exterior and inside the gaps of a firm medium, frequently obtained straight from agricultural-industrial residues. Moisture level, heat, and airflow are essential factors.
- **Advantages:**
  - **Low water consumption:** Needs considerably less liquid than submerged fermentation, lowering wastewater production and fluid purification expenses.

- **High Product Costs:** Outputs are frequently gathered in greater amounts inside the firm structure possibly easing subsequent operations and lowering extraction expenses.
- **Reduced Energy Costs:** Reduced power needed because of limited stirring and airflow necessities relative to submerged fermentation.
- **Utilization of Crude Substrates:** May immediately use a broad variety of raw or lightly processed agricultural-industrial residues, removing the necessity for intensive prior-processing and breakdown stages frequently needed for submerged fermentation.
- **Enhanced Stability:** Microbes in solid-state fermentation may show greater biochemical steadiness and catalyst synthesis because of circumstances more similar to their native habitat.

- **Disadvantages:**
  - **Heat and Mass Transfer Limitations:** Inadequate thermal dispersion in firm layers may cause thermal variations and area-specific over-warming, influencing microbial function. Bulk movement of air and nourishment may likewise be restricted.
  - **Scaling Up Challenge:** Planning and running industrial-scale solid-state fermentation vessels may be complicated because of difficulties in preserving uniformity, airflow, and heat regulation.
  - **Monitoring and Control:** Factors such as acidity level, water content, and heat are harder to observe and regulate consistently inside a firm structure than in a fluid medium.
  - **Product Recovery:** Recovering outputs from the firm structure might be difficult and could demand thorough rinsing or separation procedures.
- **Application in Waste Valorization:** Solid-state fermentation is extremely efficient for the synthesis of biocatalysts (e.g., cellulolytic enzymes, pectinolytic enzymes, starch-degrading enzymes from rice husk, wheat husk, cane residue), carboxylic acids *via Aspergillus niger* on fruit skins, fungal farming, and certain by-product compounds. It is especially appropriate for fiber-based and carbohydrate-dense residues.

11.4.3 Mixed Culture Fermentation [57–60]

Combined microbial processing includes the concurrent or stepwise growth of multiple microbial types or variants inside a single reactor. This approach utilizes the cooperative relationship between diverse microbes to accomplish intricate metabolic transformations or improve operational performance.

- **Principle:** Various microorganisms provide cooperative biochemical routes, enabling the breakdown of intricate materials, neutralization of suppressors, or the stepwise formation of intermediates that one species can use as an input for another target compound.
- **Advantages:**
  - **Enhanced Substrate Utilization:** Enables the degradation of intricate and diverse agricultural-industrial residues that an individual microbe may not completely exploit.
  - **Improved Robustness:** Combined communities may be more resistant to operational disruptions, pollutants, or the existence of suppressive substances frequently detected in residue flows.
  - **Synergistic Effects:** May result in increased output levels, quicker conversion speed, or the formation of new substances *via* biochemical collaboration.
  - **Detoxification:** Certain microbes within the microbial community may break down or transform suppressive compounds found in conditioned waste solutions, enhancing the conditions for the product-producing species.
- **Disadvantages:**
  - **Complexity in Control:** Sustaining ideal environments for several organisms and controlling microbial community behavior may be difficult.
  - **Competition and Antagonism:** Unchecked proliferation of a single organism may outgrow the rest, or certain strains could generate suppressive substances that impact the target product-producing microbe.
  - **Difficulty in Strain Engineering:** Increasingly complicated to genetically modify and enhance the whole microbial community relative to an individual variant.

  - **Product Purity:** Isolation of output may be more difficult if several microbes generate similar substances or if secreted metabolites from one species hinder the refinement of the desired compound from another.
- **Application in Waste Valorization:** Combined microbial communities are especially hopeful for transforming intricate lignocellulosic material into renewable fuels (e.g., bioethanol, biomethane) and biochemical compounds. For example, anaerobic digestion, a widely recognized method for generating biogas from organic residues, depends on a complex microbial assembly.

11.4.4 Bioreactor Design and Process Scale-Up

The design of fermentation and the approach for expanding fermentation procedures are essential for the financial feasibility of microorganism-based bioproduction, particularly when using agricultural-industrial residues.

- **Bioreactor Design:**
  - **Stirred Tank Bioreactors:** Frequently used for submerged fermentation, offers excellent stirring, providing excellent mixing, temperature control, and mass transfer through mechanical stirring and aeration [44].
  - **Airlift Bioreactors:** Utilize increasing air bubbles for blending and aeration appropriate for low-sheath cells, require less energy than stirred tank bioreactors, but provide reduced regulation of fluid mixing [61].
  - **Packed bed bioreactors:** Frequently applied for fixed-position cells or catalysts where the reactant passes through a column of solid granules, common in solid-state fermentation setups [62].
  - **Tray Bioreactors:** Basic inexpensive setups for solid state fermentation consisting of trays layered in regulated environment [63].
  - **Rotary Drum Bioreactors:** Provide improved blending and aeration for solid state fermentation, enabling more consistent conditions [54].
  - **Membrane Bioreactors:** Incorporate membrane filtration within the bioreactor, enabling ongoing product

extraction and cell preservation, results in increased cell concentrations and output [64].

- **Photo bioreactors:** Precisely designed for photosynthetic microalgae, optimized for gas transfer, delivery of light, supply of nutrient [25].

- **Key considerations**:
  - **Substrate Characteristics:** Thickness of the hydrolyzed liquid, occurrence of particulates and tendency of foaming.
  - **Oxygen Transfer rate:** Critical for aerobic processes, ensuring an adequate supply of oxygen for microbial metabolism.
  - **Heat Transfer:** Controlling biochemical heat produced by elevated cell concentrations.
  - **Agitation (or) Mixing:** Maintaining uniform distribution of nutrients and cells along with gas transfer without inducing damage to the cells.
  - **Sterilization:** Capability to disinfect the vessel and culture medium to avoid microbial intrusion.
  - **Cost and Ease of Operation:** Investment and running costs.
- **Process Scale-Up [34]:**
  - **Geometric Similarity:** Preserving geometric proportions like height to width ratio.
  - **Kinetic Similarity:** Maintaining biochemical kinetics like specific cell growth rate, and product synthesis rate are maintained.
  - **Mass Transfer Similarity:** Essential for delivering oxygen nutrients, the intensity of mixing, and air flow rate need to be optimized.
  - **Heat Transfer Similarity:** Controlling the removal of heat, which becomes increasingly difficult with large volumes because of decreased ratio of surface area to volume.
  - **Power Input per Unit Volume:** Frequently employed with the goal of preserving consistent agitation intensity.
  - **Tip Speed:** Important for impellers to preserve similar levels of shear stress.

- **Challenges in Scaling up waste valorization process [30]:**
  - o **Feedstock variability:** Variable composition of agricultural residues can affect process stability and repeatability [65].
  - o **Inhibitor Concentration:** Disinfecting high volumes of thick waste hydrolysates requires greater energy input.
  - o **Foaming:** Elevated protein levels in certain waste materials can cause excessive foam formation. Hence, bioreactor management is more difficult.

Effective scale-up demands a structured methodology beginning with laboratory trails progressing through pilot-scale setups, ultimately reaching full scale industrial production, while continuously refining operating conditions, addressing scale-specific challenges. Sophisticated tools like computational fluid dynamics and process simulation are increasingly employed to enhance large scale bioreactor efficiency [66].

11.5 Process Optimization and Lean Bio-Manufacturing

To guarantee financial viability and increase environmental responsibility, microorganism-based production methods utilizing agricultural-industrial residues must be strictly refined. This section explores approaches for method improvement, including quantitative techniques and minimal-waste production principles, aiming to improve effectiveness, lower expenses and boost material output.

11.5.1 Optimization Techniques

Method improvement focuses on identifying the best set of working parameters like temperature, pH, aeration, agitation, nutrient amount, inoculum size, and fermentation duration to maximize desired results, such as material output and transformation effectiveness [67].

- **Traditional Single-Parameter Variation Approach:** This involves changing one parameter while keeping others unchanged. Although simple, this method is inefficient, often overlooking dependencies between parameters and potentially missing the best outcome [67].

- **Quantitative Optimization Techniques:** Provide a more organized and effective method improvement strategy by concurrently assessing multiple parameters and their dependencies.
- **Response Surface Methodology (RSM):** A set of quantitative models used for representing and examining challenges where multiple parameters influence the outcome [68]. RSM is beneficial for the following:
  - **Screening Designs:** To identify key variables affecting production [69].
  - **Optimization Designs:** To define the best settings of critical variables and model their correlation within the result [70].

This reduces the required number of experiments, provides information about variable dependencies, and generates a quantitative model for production regulation.

- **Factorial Design:** Used to assess how multiple parameters and their dependencies impact one another's result [71].
- **Artificial Neuronal Network and machine learning:** Increasingly utilized for optimizing complex bioprocesses. Computational brain models can learn complex, non-proportional relationships between working parameters and results, providing strong forecasting abilities for dynamic systems [72, 73].
- **Kinematic Modeling:** Constructing quantitative frameworks that explain the speeds of microorganism-based proliferation, nutrient based uptake, and final compound synthesis. These schemes assist in gaining information into core life-related mechanisms, forecasting network responses and enhancing functional techniques [73].
- **Fed-Batch fermentation strategies:** Rather than one primary nutrient dosage (batch fermentation), nutrients are introduced gradually throughout the fermentation This aids to sustain medium strength, prevent feed deficiency and prolonged manufacturing stage [74, 75].
- **Continuous Fermentation:** A constant-condition operation where new solution is constantly supplied into the fermenter, while an equivalent quantity of used solution and outputs is extracted [76].

11.5.2 Lean Bio-Manufacturing Principles [77]

It is motivated by efficient production in other sectors and concentrates on enhancing worth for the client while reducing scrap inside bioprocesses workflows. Key principles include:

- **Value Stream Mapping:** Recognizing and examining all stages engaged in the procedure, from agro-residue to end output, to locate zones of residue.
- **Elimination of waste:** Decreasing frequent kinds of residues like excessive generation, time, material movement, extra processing, stock accumulation, redundant movement, and faults.
- **Continuous Enhancement:** Encouraging an ethos of gradual enhancements across the procedure.
- **Immediate Generation:** Generating what is necessary, or when necessary, at each phase of the procedure.
- **Total Quality Management:** Integrating standard regulation into every phase of the procedure.
- **Automation and Digitalization:** Applying detectors, observing tools, and regulation frameworks to boost effectiveness and lower mistakes.

Through combining quantitative improvement methods with streamlined bio-based production guidelines, sectors can substantially boost the eco-friendliness and financial feasibility of transforming agri-industrial residues into high-value biological products and bio-based medical agents.

11.6 Metabolic Engineering and Synthetic Biology in Bio-Manufacturing

Metabolic modification and synthetic life sciences are influential based manufacturing by allowing the logical planning and development of microbial production. These disciplines are especially vital for performance systems with improved biosynthetic constraints linked to natural microbial pathways when processing diverse industrial residues and agricultural variables. The main objective is to reroute the metabolic flows [78].

11.6.1 Core Approaches

- **Pathway Engineering [79, 80]:**
  - **Overexpression of Rate-Limiting Enzymes:** Boosting the replication count or production magnitude of catalysts which regulate pivotal phases within the synthetic route, thus hastening the throughput in the direction of the target compound.
  - **Deletion of Competing Pathways:** Eliminating loci accountable regarding the generation of undesirable by-products or rival metabolites, thus directing additional carbon flow toward the desired output.
  - **Introduction of Heterologous Pathways:** Integrating DNA-sequences derived from distinct species into the recipient cell organism to enable novel biochemical functionalities within the host organism, permitting the production of molecules not inherently synthesized by the host. This constitutes an essential methodology for utilizing varied carbohydrates from waste-streams, for example genetically modifying Escherichia coli or saccharomyces cerevisiae to cover xylose derived from hemi-cellulosic breakdown products.
- **Enzymatic Engineering:** Modification of biocatalysts to improve their chemical process, effectiveness, reactant selectivity or tolerance against inhibitors located within the byproduct discharge techniques [81].
- **Redox Balance Optimization:** Many bio-based production pathways require specific cofactors like NADH, NADPH. Biochemical alteration improves availability along with regeneration to optimize yield [82].
- **Transport Engineering:** Modifying cellular barriers carry peptides to increase the absorption of feedstock relating to desired compound, reduces intracellular toxicity and improving re-harvesting [83].
- **Global Regulatory Network Manipulation:** Focusing on core control systems (e.g., transcription regulators, sigma proteins) to globally reconfigure cellular metabolism for enhanced production [84].
- **Synthetic Biology:** Synthetic biology advances metabolic manipulation by applying engineering concept to craft and construct new biological systems or modify current ones for

objectives [85]. It centers on developing standardized, interchangeable biological components that can be assembled into intricate genetic networks.

- **Design-Build-Test-Learn (DBTL) Cycle:** This repetitive structure is key to synthetic biology.
    - **Design:** Computational resources and bioinformatics are employed to craft genetic networks and metabolic routes.
    - **Build:** DNA synthesis, molecular cloning, and genome modification technologies (e.g., CRISPR-Cas) are employed to build the engineered genetic assemblies.
    - **Test:** High-capacity screening techniques are utilized to assess the effectiveness of modified strains.
    - **Learn:** Insights from testing guide additional design revisions, resulting in ongoing enhancement [86].
- **Modular Pathway Assembly:** Building intricate pathways by combining pre-designed genetic units.
- **CRISPR-Cas Systems:** Transformed genome modification, enabling accurate and effective alterations to microbial genomes, speeding up metabolic manipulation by facilitating swift gene insertions, deletions, and transcriptional control [86].
- **Biosensors and Genetic circuits:** Designing intracellular biological detectors enclosed by microorganism-based units to detect particular biochemical pathways throughout microbial processing. This enables responsive reaction regarding varying feedstock accessibility derived from byproduct [87].
- **Cell free Bio-manufacturing:** Extending over whole organisms, synthetic bioscience, additionally explores the applications of cell isolates or separated protein catalyst networks for *in-vitro* biological-synthesis [83].

11.6.2 Benefits of Waste-to-Value Bio-Manufacturing [67, 76]

Biochemical modification and synthetic bioscience serve as a vital function within creating agri-processing residue as economically viable.

- **Expanded Substrate Utilization:** Altering microorganisms to efficiently consume each carbohydrate found in plant-based hydrolyzed mixtures accompanied by components like biological acids derived from leftovers.

- **Enhanced Tolerance to inhibitors:** Modifying microbial lines, to endure harmful chemicals like furfural, acetic acid, phenolic compounds, HMF generated throughout the preliminary processing of residue.
- **Increased Product Synthesis:** Reallocating organic movement towards the desired product, decreases residual material and also enhancing overall production.
- **Novel Product Yield and Purity:** Formation of pathways for the purpose of generating valuable chemicals, which are not organically synthesized *via* carrier microbe or serves as tough to access by means of synthetic production.
- **Reduced Fermentation time:** Speeding up metabolic speeds and enhancing total productivity, resulting in reduced fermentation periods.
- **Reduced Downstream Processing Costs:** Designing for external compound release or simpler cell breakdown to streamline product extraction.

The incorporation of these state-of-the-art biotechnologies with sophisticated fermentation methods carries significant potential for converting agricultural-industrial byproducts into a wide range of precious biologics and biomedicines, clearing the path for a genuinely sustainable bio-based economy.

11.7 Downstream Processing Biologics [88, 89]

Postproduction recovery covers the entire procedures included in the extraction and refinement of targeted material out of a microbial culture liquid or alternatively a dense medium. This step acts as a vital, frequently limiting stage and extremely expensive phase within biological production, generally constituting 50-80% of the overall manufacturing expenses particularly for ultra-clean biological therapeutics and bio-based medicines. Optimized downstream processing is essential for guaranteeing material standard, obtained from agri-industrial residues. The particular recovery approach relies upon the type of material, its physio-chemical characteristics, the production microbes and contaminant characteristics of microbial culture. Typically, the sequence of operations includes multiple steps and is shown in Table 11.4.

Table 11.4 Main stages of downstream processing.

Sl. no.	Stage	Purpose	Common techniques
1.	Cell separation & clarification	To separate cells from broth	Centrifugation, microfiltration, flocculation, and settling
2.	Cell disruption	To release intracellular products	Homogenization, milling, sonication, and enzymatic lysis
3.	Primary recovery	To release the initial product	Ultrafiltration, adsorption, precipitation, and evaporation
4.	High-resolution purification	To obtain a high-purity product	Chromatography techniques like affinity, size exclusion, and ion exchange
5.	Polishing & formulation	Final purification and stability of the product	Sterile filtration, lyophilization, and buffer exchange

11.7.1 Cell Separation and Clarification [88, 89]

The initial stage is to isolate microorganism bodies from the fluid culture for submerged fermentation or to remove the targeted material from the dense structure for solid-state fermentation.

- **Centrifugation:** Is commonly applied for dividing biological units from nutrient fluid. It is extremely efficient for large quantities but can be power-consuming and expensive at a commercial level.
- **Filtration:**
    - **Small-pore filtering:** Which divides biological units and bigger solids from nutrient fluids. Various filtration methods (like thick-layer filters, thin-layer filters and side-flow filtering systems) can be applied.
    - **Depth filtration:** Uses a dense filtering layer to capture solids across the filter thickness.
    - **Tangential fluid filtration:** Allow the input stream to move along the filter fill layer, reducing clogging and

allowing for an uninterrupted process, increasingly favored for its effectiveness and expandability.

- **Flocculation:** *Via* chemicals or bio-based aggregators involves bio-based aggregators to clump together floating particles, making them easier to retrieve by settling or straining.
- **Pressing:** For solid state fermentation, machine-based squeezing (such as rotating compression tools) or chemical pulling methods are applied to retrieve the products from the form-based material [55].

11.7.2 Cell Disruption for Intracellular Products [90]

If the target material is collected inside the microorganism-bodies, cellular breakdown is essential to liberate it into the external fluid.

- **Mechanical methods:**
  - **High pressure homogenization:** Microbial units are pushed through a thin opening at elevated force, causing them to break due to cutting stresses and bubble-induced rupture.
  - **Bead Milling:** Microbial units are broken by impact with tiny spheres attired at rapid motion.
  - **Ultrasonic Vibration:** Intense sonic vibrations lead to cell bursting but this method is generally restricted to the laboratory level.
- **Non-Mechanical methods:**
  - **Enzymatic Lysis:** Employs lysozyme (microbes) and cellulases (molds) to degrade cellular barriers.
  - **Chemical Lysis:** Using surfactants to dissolve agents or structure-disrupting chemicals to break the cellular outer layers and walls.
  - **Osmotic Shock:** Quick alterations in solute - indices force can cause microbial units to burst.

11.7.3 Primary Recovery (or) Concentration [91]

Cell separation and disruption are frequently minimal. This step is intended to decrease the quantity and eliminate major contaminants.

- **Ultra-filtration:** Applies partially porous barriers to divide particles according to their dimensions. Ultra-filtration intensifies the targeted material, while dia-filtration eliminates minor contaminants and alters the solution makeup.
- **Evaporation:** Eliminates moisture to densify temperature-resistant materials.
- **Precipitation:**
  - **Salting out:** Introducing large amounts of ionic compounds to lower polypeptide dissolvability and induce solid formation.
  - **Solvent separation:** Introducing carbon-based dissolving agents to reduce the electrical insulating property of the mixture, causing solid formation of proteins.
  - **Iso-Electric Precipitation:** Modifying the acidity level to the net Zero charge level, where the overall ionic balance is neutral, leading to the least dissolvability and solid formation.
- **Adsorption:** Using adsorbing materials like charcoal, etc. to attach the unwanted substance specifically.

11.7.4 High-Resolution Purification [92, 93]

The phase concentrates on reaching the target cleanness degree for the intended material, particularly important for bio-based medicines. The separation techniques acts as the main tool for precise refinement.

- **Chromatography:**
  - **Ion exchange chromatography:** Divides polypeptides based on their ionic property using electrically active beds.
  - **Hydrophobic Interaction chromatography:** Divides polypeptides according to the water-repellent nature using large amounts of ionic compounds.
  - **Size exclusion Chromatography:** Divides by their dimensions and form using sponge-like spheres.
  - **Affinity Chromatography:** Extremely selective division based on special attachment reactions between the intended particle and binding agent fixed on the bead.

11.7.5 Polishing and Formulation [94, 95]

The final phase guarantee materials durability, cleanness and preparation for final composition.

- **Viral inactivation (or) Removal:** Essential for bio-based medicines obtained from multicellular organisms and human-derived cultures.
- **Buffer Exchange:** Guarantees that the material is in the proper solution makeup for durability and following final composition.
- **Sterile Filtration:** Eliminates any residual microorganisms' pollutants to guarantee material cleanliness.
- **Formulation:** Securing the refined material in an appropriate solution, incorporating stabilizer and arranging it for preservation or delivery.

11.7.6 Challenges in Downstream Processing for By-Product [30]

- **High Impurity Load:** Agricultural-industrial residues often include a broad variety of contaminants like phenolics, metals, pigments, and lignans which can disrupt microbial conversion and downstream processing steps.
- **Viscosity:** Some by-products or culture liquids can be very thick, hindering straining and blending.
- **Product dilution:** Material from microbial conversion are usually in minimal amount, requiring lengthy densification processes.
- **Cost:** Downstream processing remains a significant expense factor.
- **Regulatory Requirements:** For bio-based medicines, downstream of must comply with strict legal norms like CGMP to ensure material security, authenticity, potency, cleanness and standard.

11.8 Case Studies

The functional implementation of microbial production employing agricultural-industrial residues is most effectively demonstrated *via* effective examples throughout diverse eco-friendliness and financial-commodity. These instances emphasize adaptability and possibility.

11.8.1 Bio-Ethanol Production from Lignocellulose Waste [28–30, 34, 35]

Bioethanol is the most notable residue of sustainable vehicle energy

- **Feedstock:** Sugarcane fiber, maize husk, wheat , paddy straw, wood leftovers.
- **Process:**
  - **Pre-Treatment:** To decompose the plant-fiber (e.g., thermal rupture, weak acid breakdown, base pre-treatment).
  - **Enzymatic Hydrolysis:** Cellulose-degrading and hemicellulose-degrading biocatalysts decompose cellulose and hemicelluloses into convertible carbohydrates (dextrose, xylose, arabinose).
  - **Fermentation:**

  Metabolic Engineering: Variants like S.cerevisiae or E.coli have been genetically modified to consume xylose and arabinose, greatly enhancing ethanol production from plant biomass hydrolysates.

  Co-fermentation: A combination of hydrolysis and fermentation stages into unified procedure often employing modified microbes that synthesize cellulose-degrading enzymes and at the same time convert them into sugars.
- **Benefits:** Lowers carbon emission relative to petroleum-based fuels, use farm residues and supports energy, self-sufficiency.

11.8.2 Citric Acid Production from Fruit and Vegetable Processing Waste [55, 56]

Citric acid is one of the extensively employed organic acids within the edible products, drugs as well as synthetic sectors.

- **Feedstock:** Citrus-like fruit rinds, apple pulp waste, and sweet grass molasses, along with additional sucrose-abundant secondary products.

- **Microorganisms:** Microbes, mainly thread-like molds, identified due to their own significant citric acid production efficiency.
- **Process:** Feed material processing residues are frequently dehydrated, milled and occasionally processed in order to eliminate suppressors. Saccharine become isolated can get executed through submerged fermentation or solid-state fermentation.
- **Downstream processing:** It comprises purification, solid formation using calcium hydroxide in order to produce calcium citrate, continuing through acid treatment with H_2SO_4 to liberate citric acid, and subsequently refinement by solidification.
- **Benefits:** Transform large scale, easily obtainable manufacturing residues to profitable manufacturing chemicals, minimizing waste management cost, and generating financial importance. This is a widely recognized instance pertaining to close-loop biological economy in operation.

11.8.3 Lactic Acid Production from Dairy and Starch Waste [48, 93]

Lactic acid is a multi-purpose platform chemical commonly utilized in the production of eco-friendly plastics like poly lactic acid as well as in the food sector and personal care industries.

- **Feed stock:** Typical raw materials are whey, potato processing residues, cassava pulp, corn steep liquor and other starch or lactose waste leftovers.
- **Microorganisms:** Lactic acid bacteria, particularly lactobacillus strains such as *Lactobacillus platarum* and *Lactobacillus casei* are predominantly used for their ability to ferment sugars into lactic acid.
- **Process:**
  - **Pretreatment:** Waste materials high in starch undergo enzymatic cleavage to convert complex carbohydrates into glucose, whereas whey can be directly fermented from LAB species without reconversion.
  - **Fermentation:** Usually, it is carried out by submerged fermentation, often in fed-batch mode to regulate sugar

levels. Managing pH is critical since acid production lowers the pH which can affect microbial activity.
  - **Product recovery:** The process involves removing microbial cells followed by concentration and purification of the lactic acid by using common recovery methods like precipitation, membrane filtration or exchange chromatography.
- **Benefits:** This process effectively solves the issue of disposing large amounts of dairy and starch industries wastes, converting them into valuable chemical substrates. Additionally, it promotes the conversation of petroleum-derived plastics.

11.8.4 Production of Microbial Enzymes from Agricultural Residues [57]

Microorganism based biocatalysts like cellulases, xylanases, amylases, proteases exists as extensively applied agents within diverse sectors, such as nutrition, fabric, cleaning agents as well as renewable fuel manufacturing.

- **Feedstock:** Grain husk, paddy out layer, came fibrous residue, maize cores, peanut fat press residue as well as additional fiber rich alternatively protein abundant farming derived leftovers.
- **Microorganisms:** Aspergillus strains, Trichoderma reesei and bacillus strains.
- **Process:**
  - **Feedstock processing:** Frequently includes reduction of particles and hydration.
  - **Fermentation:** Both submerged and solid-state fermentation are extensively used. Solid-state fermentation is especially beneficial for the biocatalyst generation.
  - **Downstream processing:** It includes the extraction of microbial cell mass from the biocatalyst-concentrated liquid. Unrefined enzyme mixtures are usually adequate for multiple manufacturing uses, decreasing downstream process expenditure.
- **Benefits:** Supplies an eco-friendly and economical biocatalysts, lowers the waste management load of agri-residues and encourages varied biologically derived manufacturing systems.

11.8.5 Vitamin B12 Production from Agro-Industrial Wastes [96]

Vitamin B12 is a fundamental microelement, mainly synthesized by microbes. Chemically, the production of B12 is very expensive. Hence B12 can also be alternatively produced from economically affordable agricultural-industrial residues.

- **Feedstock:** Which includes molasses, whey, rice straw, fibers of sugar cane, debris of cotton plant and distillation residues.
- **Microorganisms:** Several bacterial strains are used in the production of Vitamin B12. They are pseudomonas species, propionibacterium species, bacillus species etc.
- **Process:**
  - **Feed material processing:** Agro-industrial waste feed stocks are often pre-treated (e.g., cutting, grinding, alkaline hydrolysis) to break down complex structures and increase substrate accessibility for microorganisms.
  - **Fermentation:** The treated feedstock is used as a carbon source in either submerged fermentation or solid-state fermentation, where the selected microorganisms produce Vitamin B12. Key nutrients and precursors like cobalt sources (e.g., cobalt chloride) and Dimethyl benzimidazole (DMB) are supplemented to enhance synthesis. Process parameters like pH, temperature, and aeration (or anaerobic conditions) are carefully controlled.
  - **Downstream Processing:** It consists of steps to isolate and separate vitamin B12 derived from culture medium. This generally includes separation of cells, isolation, concentration and purification using techniques such as chromatography and precipitation.
- **Benefits:** Transforms widely available and challenging agricultural–industrial by products into commercially significant pharmaceutical and livestock supplement. This reduces residue handling costs, generates financial benefit and reflects sustainable bio-based system.

All these case studies reveal the notable advancements and capabilities of microorganism driven production employing agri- industrial residues. They highlight the value of combining modern biotechnology, production system design and renewable raw material use to build an eco-friendly biological economy.

11.9 Regulatory Framework and Quality Standards

The manufacture concerning biological agents along with biotechnology-derived drugs, especially intended for medicinal applications remains controlled through strict oversight systems as well as standard-based benchmark. Such rules guarantee the security, effectiveness plus integrity pertaining to the ultimate output safeguarding general well-being. Although utilization of agriculture-industrial residues provides environmental advantages, this additionally bring particular difficulties which supervisory authorities need to handle.

11.9.1 Regulatory Bodies and Guidelines [97]

Various areas as well as nations possess their respective individual supervisory organizations which supervise the formulation, production, along with market introduction concerning biological products as well as bio-based medicines. Major bodies consist of:

- **US Food and Drug administration (FDA):** Within the combined nations, it is the institution concerning bio-therapeutics assessment. An investigation (CBER) oversees biological products, whereas the organization for medication review and study (CDER) governs medicinal products.
- **European Medicines Agency (EMA):** The continental alliance, EMA remains accountable for the technical assessment of drugs created by medicinal firms intended for application within the EU (European Union).
- **Pharmaceutical and Medical Devices Agency (PMDA):** Within Nippon, PMDA remains accountable for guaranteeing the security effectiveness and standard of medicinal products and healthcare equipment.
- **World Health Organization (WHO):** Offer international directives as well as suggestion regarding the oversight and standard monitoring of bio-therapeutic substances especially for immunizations and vital drugs.

Those organizations create thorough protocols encompassing every phases of item formulation and production such as:

- **Good Manufacturing Practices (GMP):** One framework designed for guaranteeing that goods are reliably manufactured and managed in line with standard benchmarks. GMP encompasses every elements of manufacturing starting from unprocessed inputs and plants through to staff instruction and documentation. This remains essential in avoiding pollution confusion and mistakes.
- **Good Laboratory Practices (GLP):** Regarding preclinical research experiments in order to guarantee information accuracy and reliability.
- **Good Clinical Practice (GCP):** Regarding the planning execution implementation supervision inspection documentation evaluations and presentation of medical studies.
- **Pharmacopoeias:** Authoritative collections like United state Pharmacopoeia, British Pharmacopoeia and Indian Pharmacopeia, which establish benchmarks for the characterization potency grad and cleanliness of medications inactive ingredients as well as additional substances employed within medical care.

11.9.2 Challenges Specific to Agro-Waste Feedstock [47, 52]

Although the advantages of utilizing agricultural residues are substantial, they present distinctive administrative and standard difficulties.

- **Feedstock Variability:** The molecular of agricultural-industrial residues may fluctuate substantially depending upon the original time of year weather conditions as well as treatment techniques. Such inconsistency might affect bioconversion efficiency along with output standard requiring strong unrefined input analysis and operation regulations.
- **Contaminants:** Agricultural residues may include agrochemicals toxic elements funga toxins as well as elevated concentrations of unwanted microorganism-based pollutants that might transfer into the end output or disrupt with the bioconversion.

- **Inhibitor Carryover:** Pre-processing of fibrous plant-based organic matter frequently produces bioconversion suppressants (for example, furfural, and aromatic substances) which need to be eliminated or endured by genetically modified microbial variants. Remaining blockers could affect output standard or subsequent treatment.
- **Traceability:** Guaranteeing tracking of the discarded input material starting at its source to the end output may be difficult particularly in the case of varied residue flows.
- **Environmental Impact Assessment:** Although typically beneficial, the ecological effect of the whole residue-to-resource logistics network (gathering, shipping, initial processing) requires thorough evaluation to prove genuine environmental viability.
- **Novelty of Products:** In case the bio-production procedure produces new compounds, they shall be subjected to strict hazard as well as effectiveness evaluation comparable to any novel synthetic compound or biopharmaceutical possibly involving particular factory regarding their biologically derived source.

Tackling these obstacles demands strong standards-related oversight frameworks, sophisticated diagnostic methods for input material as well as output profiling along with strong cooperation with compliance authorities in order to develop well-defined standards for biologically derived goods.

11.10 Future Perspectives and Research Opportunities

This area related to microorganism-based bio-production utilizing agricultural-industrial residues remains positioned toward substantial expansion as well as advancement. Prospective developments shall get propelled through that combination from developing innovations cross-disciplinary investigation along with coordinated attempt in order to surmount present constraints resulting in a greater eco-friendly and productive biological economy.

11.10.1 Advanced Technologies and Integration [70–74]

- **Artificial Intelligence (AI) and Machine Learning (ML):**
  - **Process Optimization:** Artificial Intelligence (AI)/ and Machine Learning (ML) computational methods are able to examine large-scale data collections originating in bio-processing units (such as monitoring biological information) to forecast ideal bioconversion parameters, detect obstacles, as well as formulate instantaneous dynamic regulation approaches.
  - **Strain Engineering:** Machine learning has the ability to speed up the plan-construct-evaluate-adapt process within engineered life sciences through forecasting the impacts from genomic alteration to biochemical networks as well as pinpointing potential genetic candidates.
  - **Bioprocess Modeling:** Creating advanced forecasting frameworks for microbial proliferation nutrient absorption as well as metabolite synthesis minimizing laboratory workload.
  - **Downstream Processing:** Improving separation procedures forecasting filter clogging as well as boosting operation performance.
- **Continuous Bio-manufacturing:** Shifting away from conventional discrete operations toward uninterrupted or enhanced ongoing production provides substantial benefits such as that which reduces spatial requirements, greater efficiency, lowered investment costs as well as enhanced output standards uniformity. Such a shift is going to prove especially crucial in the case of large-scale low-profit goods obtained out of residues.
- **Modular and Decentralized Bio-refineries:** Creating compact scalable bio-production units which are able to get installed nearer to origins of agricultural residues lowering logistics expenses as well as boosting raw material usage effectiveness.
- **Micro fluidics and Lab-on-a-Chip Systems:** Facilitating large-scale evaluation for microorganism variants fast workflow refinement as well as downsized analytical tools to allow instant tracking of bioconversion conditions.
- **Multi-Omics Integration (Genomics, Transcriptomic, Proteomic, Metabolomics):** Thorough characterization of

microorganism networks in order to obtain greater understanding of biochemical control discover new routes as well as locate objectives for biochemical modification particularly during adverse environments caused by residue elements.

- **CRISPR-based Technologies beyond Gene Editing:** Employing Clustered Regularly Interspaced Short Palindromic Repeats tools to enable specific genetic control (CRISPR activation/interference) heritable gene expression alterations as well as responsive modulation over biochemical routes providing unmatched accuracy within designing microorganism biological production system.

11.10.2 Expanding the Scope of Waste Feedstocks and Products [76–80]

- **Diverse and Recalcitrant Wastes:** Investigation on higher-performance pre-processing as well as value-enhancement approaches regarding extremely resistant refuse (such as aromatic biomass urban refuse synthetic polymer debris) remains essential. Engineering microorganism variants able to directly metabolize intricate macromolecules or potentially transforming synthetic materials into high-value compounds or building blocks represents cutting–edge challenge.
- **Novel Biologics and Biopharmaceuticals:** In addition to existing offerings, upcoming initiatives shall concentrate upon generating a broader range of advanced medicinal polypeptides, immunoglobulin's as well as cellular treatments using residue-based input materials guaranteeing adherence to strict compliance guidelines.
- **Sustainable Materials:** Bio-fabrication for cutting-edge biologically derived substances such as innovative biopolymers featuring enhanced characteristics eco-friendly fabrics as well as building substances is expected to attract momentum.
- **Co-production and Bio-refinery Concepts:** Establishing combined bio-processing facilities which concurrently generate various premium outputs (such as biological fuels bio-based chemicals polypeptides) using one residue source, enhancing total financial gains as well as material efficiency.

11.10.3 Interdisciplinary Research and Policy Support [83–86]

- **Systems Biology and Synthetic Biology Integration:** A greater comprehension regarding biological networking integrated alongside cutting-edge design concepts shall facilitate the development of extremely resilient as well as productive microbial production systems.
- **Process Engineering Innovations:** Advancement in innovative bio-processing vessel architectures uninterrupted separation methods as well as combined biological workflows which remains better adapted for raw material inconsistency and attain greater output efficiency.
- **Techno-Economic and Life Cycle Assessment:** Thorough assessment regarding the financial feasibility as well as ecological effect of residue-to-product bio-based production using an all-encompassing viewpoint (full life cycle) remains critical to inform funding as well as regulatory choices.
- **Policy and Regulatory Support:** Public authorities as well as global organizations must formulate enabling frameworks motivators along with well-defined compliance routes regarding biologically derived goods originating from residues speeding up market deployment and consumer uptake. That entails aligning rules throughout various jurisdictions as well as setting criteria for residue-based commodities.
- **Academia-Industry Collaboration:** Reinforcing collaborations among scholarly investigative organizations as well as commercial stakeholders remains crucial for converting technical breakthroughs into market-ready solutions and speed up advancement.

The outlook for microorganism-based bio-based production utilizing agricultural-industrial residues appears promising providing an effective set of tools to tackle worldwide issues linked to limited resource availability environmental shifts as well as population well-being. Through adopting state-of-the-art research technological development along with cooperative frameworks this sector is expected to have a central influence in forming an eco-friendly and closed-loop biological economy.

11.11 Conclusion

This shift towards a single renewable enduring along with closed-loop biological economy calls for novel strategies towards input application and processing. Microbial-driven bio-production specifically combined in combination with this value-addition from farming and processing-based by-products signifies a major transformation regarding the enhance-value bio-based drugs, recombinant drugs along with various biologically derived goods are manufactured. The aforementioned technique leverages those intrinsic, cellular potentials from microscopic organisms converting low-cost agronomic manufacturing-related waste remains as effective medicinal, molecular as well as production-related substances.

Across this part, the authors have examined the key elements concerning microbial processing starting with basic concepts of microorganism-based production units and wide range of microbial strains utilized to the pivotal agri-waste by-products serving as renewable raw materials. The Authors explore in depth the categorization along with initial processing of such waste materials emphasizing their significant sustainability-related and financial advantages. Multiple microbial processing approaches comprising submerged-culture as well as solid substrate fermentation coupled with the benefits of multi-strain systems were covered comprehensively underlining the necessity of reactor engineering and up-scaling for manufacturing use.

This paradigm-shifting effect sophisticated biological technologies namely biochemical pathway and engineered biology was highlighted as vital to refine microbial variants extending raw material consumption boosting production efficiency and elevating adaptability to suppressive agents embedded in residual materials. Moreover, the authors analyzed the multifaceted though essential phase of production refinement detailing approaches involving material isolation and purification while tackling the unique issues introduced by recycled biomass. Applied documented instances including the manufacturing of renewable alcohol 2-hydroxy-1,2,3-propane-tricarboxylic acid, alpha hydroxyl-propionic acid and microbial biocatalysts derived from diverse agri-industrial waste confirmed the viability along with economic value of this green methodology facilitating minimization of waste as well as resource recycling.

In spite of notable improvements, various barriers require resolution such as fluctuation in biomass inputs; the existence of toxic intermediates high energy demand initial processing of DSP steps intricacies compliance systems of new bio-products combined in next-gen techniques.

Machine learning uninterrupted processing combined genomic and proteomic tools sophisticated the CRISPER technology platforms shows strong prospects addressing such challenges facilitating optimized stable cost-effective operations.

Moving forward nurturing multi-field scientific inquiry involving microbial science bioengineering, process engineering, information technology will be vital in creating strong regulatory aid. It simplified the approval process together with reinforcing academic-industrial ties that will expedite the implementation of studies towards industrial utilization and communities transition to sustainable choices for stable production networks. This biotech-applied approach delivers a genuinely eco-conscious and creative approach for upcoming times of biotech medicines, medicinal products and a diverse collection of biomass-delivered items. Holistic exploitation of farm-industry by-products using microbial processing is more than possibility but a requirement for establishing a cleaner earth and a thriving regenerated bio-based economy.

AI disclosure statement: The author acknowledges the use of AI-based tools for minor language editing or grammatical correction. However, all intellectual contributions, analyses, and conclusions presented in this chapter are entirely the author's own.

References

1. Ghosh, S., Saha, A., Kumar, S., Pathania, S., Regenerative Agriculture for Climate Change Mitigation and Food Security, in: *Regenerative Agriculture for Sustainable Food Systems*, pp. 83–133, Springer Nature Singapore, Singapore, 2024.
2. Horváth, I.T., Introduction: sustainable chemistry. *Chem. Rev.*, 118, 2, 369–371, 2018.
3. Berg, G., Grube, M., Schloter, M., Smalla, K., Unraveling the plant microbiome: looking back and future perspectives. *Front. Microbiol.*, 5, 148, 2014.
4. Li, Y., Cao, M., Gupta, V.K., Wang, Y., Metabolic engineering strategies to enable microbial electrosynthesis utilization of CO2: recent progress and challenges. *Crit. Rev. Biotechnol.*, 44, 3, 352–372, 2024.
5. Singh, T.A., Sharma, M., Sharma, M., Sharma, G.D., Passari, A.K., Bhasin, S., Valorization of agro-industrial residues for production of commercial biorefinery products. *Fuel*, 322, 124284, 2022.
6. Bhargava, P.C., Kamal, N., Chaurasia, D., Singh, S., Singh, A., Agro-Residues' Waste to Wealth for a Circular Economy and Sustainable Development, in: *Waste-to-Wealth*, pp. 200–217, CRC Press, Boca Raton, FL, USA, 2024.

7. Vasileiadou, A., From Organic wastes to Bioenergy, Biofuels, and value-added products for urban sustainability and circular economy: a review. *Urban Sci.*, 8, 3, 121, 2024.
8. Maharjan, A., Kim, M.R., Choi, W., Kim, H.C., Park, J.H., Advanced biofuels: a path to sustainable energy. *Sustain. Energy Fuels*, 8, 15, 3247–3265, 2024.
9. Bisht, M., Thayallath, S.K., Bharadwaj, P., Franklin, G., Mondal, D., Biomass-derived functional materials as carriers for enzymes: towards sustainable and robust biocatalysts. *Green Chem.*, 25, 12, 4591–4624, 2023.
10. Singh, A. and Kumar, S., Exploring the Functionality of Microbes in Fermented Foods: Technological Advancements and Future Directions. *Fermentation*, 11, 6, 300, 2025.
11. Biehl, P. and Zhang, K., Introduction to advances in bio-based polymers: chemical structures and functional properties at the interface, in: *Green by Design: Harnessing the power of bio-based polymers at interfaces*, pp. 1–1, IOP Publishing, Bristol, UK, 2024.
12. İncir, I. and Kaplan, O., Escherichia coli as a versatile cell factory: Advances and challenges in recombinant protein production. *Protein Exp. Purif.*, 219, 106463, 2024.
13. Su, Y., Liu, C., Fang, H., Zhang, D., Bacillus subtilis: a universal cell factory for industry, agriculture, biomaterials and medicine. *Microb. Cell Fact.*, 19, 1, 173, 2020.
14. Tsuge, Y. and Matsuzawa, H., Recent progress in production of amino acid-derived chemicals using Corynebacterium glutamicum. *World J. Microbiol. Biotechnol.*, 37, 3, 49, 2021.
15. Yankov, D., Fermentative lactic acid production from lignocellulosic feedstocks: from source to purified product. *Front. Chem.*, 10, 823005, 2022.
16. Bae, J., Park, C., Jung, H., Jin, S., Cho, B.K., Harnessing acetogenic bacteria for one-carbon valorization toward sustainable chemical production. *RSC Chem. Biol.*, 5, 9, 812–832, 2024.
17. Nathan, V.K., Arunachalam, K., Sreekala, A.G.V., The Role of Metabolic Engineering in the Development of 2G Biofuels (Both in Conversion and Fermentation), in: *Lignocellulosic Biomass Refining for Second Generation Biofuel Production*, pp. 85–105, CRC Press, Boca Raton, FL, USA, 2023.
18. Liu, M., Xiao, R., Li, X., Zhao, Y., Huang, J., A comprehensive review of recombinant technology in the food industry: Exploring expression systems, application, and future challenges. *Compr. Rev. Food Sci. Food Saf.*, 24, 2, e70078, 2025.
19. Naveira-Pazos, C., Robles-Iglesias, R., Fernández-Blanco, C., Veiga, M.C., Kennes, C., State-of-the-art in the accumulation of lipids and other bioproducts from sustainable sources by Yarrowia lipolytica. *Rev. Environ. Sci. Bio/Technol.*, 22, 4, 1131–1158, 2023.
20. Behera, B.C., Citric acid from Aspergillus niger: a comprehensive overview. *Crit. Rev. Microbiol.*, 46, 6, 727–749, 2020.

21. Xue, W., Sheng, Y., Tai, Y., Ding, J., He, R., Jiang, Y., *et al.*, Mechanism of improved cellulase production by Trichoderma reesei in water-supply solid-state fermentation. *Bioresour. Technol.*, 419, 132017, 2025.
22. Sheldon, R.A., Green chemistry and biocatalysis: Engineering a sustainable future. *Catal. Today*, 431, 114571, 2024.
23. Anastas, P.T. and Warner, J.C., Principles of green chemistry, in: *Green Chem.: Theory Pract.*, vol. 29, p. 3, Oxford University Press, New York, NY, USA, 1998.
24. Weise, N.J. and France, S.P., Advancing Biocatalysis Education: Sustaining the Future of Industrial Biotechnology. *JACS Au*, 2025.
25. Sun, C., Zeng, R., Chen, T., Yang, Y., Song, Y., Li, Q., *et al.*, Recent advances and challenges in the production of hydroxylated natural products using microorganisms. *Fermentation*, 10, 12, 604, 2024.
26. de Mello, A.F.M., de Souza Vandenberghe, L.P., Herrmann, L.W., Letti, L.A.J., Burgos, W.J.M., Scapini, T., *et al.*, Strategies and engineering aspects on the scale-up of bioreactors for different bioprocesses. *Syst. Microbiol. Biomanuf.*, 4, 2, 365–385, 2024.
27. Zhao, X., Wang, Y., Chen, X., Yu, X., Li, W., Zhang, S., *et al.*, Sustainable bioplastics derived from renewable natural resources for food packaging. *Matter*, 6, 1, 97–127, 2023.
28. Bhati, N., Shreya, Sharma, A.K., Recent Trends in the Pre-Treatment Process of Lignocellulosic Biomass for Enhanced Biofuel Production, in: *Production of biobutanol from biomass*, pp. 47–75, Wiley-Scrivener (John Wiley & Sons), Hoboken, NJ, USA, 2023.
29. Falua, K.J., Pokharel, A., Babaei-Ghazvini, A., Ai, Y., Acharya, B., Valorization of starch to biobased materials: A review. *Polymers*, 14, 11, 2215, 2022.
30. Sekoai, P.T., Roets-Dlamini, Y., O'Brien, F., Ramchuran, S., Chunilall, V., Valorization of food waste into single-cell protein: an innovative technological strategy for sustainable protein production. *Microorganisms*, 12, 1, 166, 2024.
31. Sun, H., Gao, Z., Zhang, L., Wang, X., Gao, M., Wang, Q., A comprehensive review on microbial lipid production from wastes: research updates and tendencies. *Environ. Sci. Pollut. Res.*, 30, 33, 79654–79675, 2023.
32. Mehta, D., Sundar, S.K., Parikh, J.K., Mehra, A., Thapliyal, D., Arya, R.K., Food Waste Biorefinery, in: *From Waste to Wealth*, pp. 125–152, Springer Nature Singapore, Singapore, 2024.
33. Sun, S., Sun, S., Cao, X., Sun, R., The role of pretreatment in improving the enzymatic hydrolysis of lignocellulosic materials. *Bioresour. Technol.*, 199, 49–58, 2016.
34. Sriariyanun, M., Gundupalli, M.P., Phakeenuya, V., Phusamtisampan, T., Cheng, Y.S., Venkatachalam, P., Biorefinery approaches for production of cellulosic ethanol fuel using recombinant engineered microorganisms. *J. Appl. Sci. Eng.*, 27, 2, 1985–2005, 2023.

35. du Pasquier, J., Paës, G., Perré, P., Principal factors affecting the yield of dilute acid pretreatment of lignocellulosic biomass: a critical review. *Bioresour. Technol.*, 369, 128439, 2023.
36. Kim, J.S., Lee, Y.Y., Kim, T.H., A review on alkaline pretreatment technology for bioconversion of lignocellulosic biomass. *Bioresour. Technol.*, 199, 42–48, 2016.
37. Ziegler-Devin, I., Chrusciel, L., Brosse, N., Steam explosion pretreatment of lignocellulosic biomass: a mini-review of theorical and experimental approaches. *Front. Chem.*, 9, 705358, 2021.
38. Chundawat, S.P., Pal, R.K., Zhao, C., Campbell, T., Teymouri, F., Videto, J., *et al.*, Ammonia fiber expansion (AFEX) pretreatment of lignocellulosic biomass. *J. Vis. Exp.*, 158, 2020.
39. Aftab, M.N., Iqbal, I., Riaz, F., Karadag, A., Tabatabaei, M., Different pretreatment methods of lignocellulosic biomass for use in biofuel production, in: *Biomass for bioenergy-recent trends and future challenges*, IntechOpen, London, UK, 2019.
40. Das, N., Jena, P.K., Padhi, D., Kumar Mohanty, M., Sahoo, G., A comprehensive review of characterization, pretreatment and its applications on different lignocellulosic biomass for bioethanol production. *Biomass Convers. Biorefin.*, 13, 2, 1503–1527, 2023.
41. Maitan-Alfenas, G.P., Visser, E.M., Guimarães, V.M., Enzymatic hydrolysis of lignocellulosic biomass: converting food waste in valuable products. *Curr. Opin. Food Sci.*, 1, 44–49, 2015.
42. Yadav, S.K., Technological advances and applications of hydrolytic enzymes for valorization of lignocellulosic biomass. *Bioresour. Technol.*, 245, 1727–1739, 2017.
43. Nagarajan, M., Rajasekaran, B., Venkatachalam, K., Microbial metabolites in fermented food products and their potential benefits. *Int. Food Res. J.*, 29, 3, 466–486, 2022.
44. Chisti, M.Y. and Moo-Young, M., Airlift reactors: characteristics, applications and design considerations. *Chem. Eng. Commun.*, 60, 1–6, 195–242, 1987.
45. Cramer, S.M. and Holstein, M.A., Downstream bioprocessing: recent advances and future promise. *Curr. Opin. Chem. Eng.*, 1, 1, 27–37, 2011.
46. Stephenson, T., Brindle, K., Judd, S., Jefferson, B., *Membrane bioreactors for wastewater treatment*, IWA Publishing, London, UK, 2000.
47. Roukas, T. and Kotzekidou, P., From food industry wastes to second generation bioethanol: a review. *Rev. Environ. Sci. Bio/Technol.*, 21, 1, 299–329, 2022.
48. Pescuma, M., de Valdez, G.F., Mozzi, F., Whey-derived valuable products obtained by microbial fermentation. *Appl. Microbiol. Biotechnol.*, 99, 15, 6183–6196, 2015.
49. Kabeyi, M.J.B. and Olanrewaju, O.A., Sugarcane molasses to energy conversion for sustainable production and energy transition, in: *12th annual*

Istanbul international conference on industrial engineering and operations management, Istanbul, Turkey, vol. 405, 2022.
50. Bhargav, S., Panda, B.P., Ali, M., Javed, S., Solid-state fermentation: an overview. *Chem. Biochem. Eng. Q.*, 22, 1, 49–70, 2008.
51. Leite, P., Sousa, D., Fernandes, H., Ferreira, M., Costa, A.R., Filipe, D., *et al.*, Recent advances in production of lignocellulolytic enzymes by solid-state fermentation of agro-industrial wastes. *Curr. Opin. Green Sustain. Chem.*, 27, 100407, 2021.
52. Šelo, G., Planinić, M., Tišma, M., Tomas, S., Koceva Komlenić, D., Bucić-Kojić, A., A comprehensive review on valorization of agro-food industrial residues by solid-state fermentation. *Foods*, 10, 5, 927, 2021.
53. Carboué, Q., Tranier, M.S., Perraud-Gaime, I., Roussos, S., Production of microbial enzymes by solid-state fermentation for food applications, in: *Microbial enzyme technology in food applications*, pp. 437–451, CRC Press, Boca Raton, FL, USA, 2017.
54. Sánchez, A., Oiza, N., Artola, A., Font, X., Barrena, R., Moral-Vico, J., Gea, T., Solid-state fermentation: a review of its opportunities and challenges in the framework of circular bioeconomy. *Afinidad J. Chem. Eng. Theor. Appl. Chem.*, 81, 601, 51–57, 2024.
55. West, T.P., Citric acid production by Aspergillus niger using solid-state fermentation of agricultural processing coproducts. *Appl. Biosci.*, 2, 1, 1–13, 2023.
56. Hyseni, B., Aytekin, A.O., Nikerel, E., Solid state fermentation for enzyme production for food industry. *J. Microbiol. Biotechnol. Food Sci.*, 7, 6, 615, 2018.
57. Rosero-Chasoy, G., Rodríguez-Jasso, R.M., Aguilar, C.N., Buitrón, G., Chairez, I., Ruiz, H.A., Microbial co-culturing strategies for the production high value compounds, a reliable framework towards sustainable biorefinery implementation–an overview. *Bioresour. Technol.*, 321, 124458, 2021.
58. Sial, A., Zhang, B., Zhang, A., Liu, K., Imtiaz, S.A., Yashir, N., Microalgal–bacterial synergistic interactions and their potential influence in wastewater treatment: a review. *Bioenergy Res.*, 14, 3, 723–738, 2021.
59. Li, C., Yan, W., Wang, B., Xu, T., Lin, C.S.K., Zou, X., Du, C., Integrated biorefineries for the co-production of biofuels and high-value products, in: *Handbook of Biofuels Production*, pp. 513–541, Woodhead Publishing, Cambridge, UK, 2023.
60. Valdez-Vazquez, I. and Sanchez, A., Proposal for biorefineries based on mixed cultures for lignocellulosic biofuel production: a techno-economic analysis. *Biofuels, Bioprod. Biorefin.*, 12, 1, 56–67, 2018.
61. Zhang, T., We, C., Ren, Y., Feng, C., Wu, H., Advances in airlift reactors: modified design and optimization of operation conditions. *Rev. Chem. Eng.*, 33, 2, 163–182, 2017.
62. Mitchell, D.A., Pitol, L.O., Biz, A., Finkler, A.T.J., de Lima Luz Jr., L.F., Krieger, N., Design and operation of a pilot-scale packed-bed bioreactor

for the production of enzymes by solid-state fermentation, in: *Solid State Fermentation: Research and Industrial Applications*, pp. 27–50, Springer, Cham, Switzerland, 2019.
63. Sentís-Moré, P., Romero-Fabregat, M.P., Rodríguez-Marca, C., Guerra-Sánchez, A.J., Ortega-Olivé, N., Design optimization of a tray bioreactor for solid-state fermentation: study of process parameters through protein modification of by-products. *Fermentation*, 9, 10, 921, 2023.
64. Khoiruddin, K., Boopathy, R., Kawi, S., Wenten, I.G., Towards next-generation membrane bioreactors: Innovations, challenges, and future directions. *Curr. Pollut. Rep.*, 11, 1, 15, 2025.
65. Bosch, M. and Hazen, S.P., Lignocellulosic feedstocks: research progress and challenges in optimizing biomass quality and yield. *Front. Plant Sci.*, 4, 474, 2013.
66. Rathore, A.S., Kanwar Shekhawat, L., Loomba, V., Computational fluid dynamics for bioreactor design, in: *Bioreactors: Design, Operation and Novel Applications*, pp. 295–322, Wiley-VCH (John Wiley & Sons), Weinheim, Germany, 2016.
67. Mandenius, C.F., Challenges for bioreactor design and operation, in: *Bioreactors: Design, operation and novel applications*, C.F. Mandenius (Ed.), pp. 1–34, Wiley-VCH, Weinheim, 2016.
68. Myers, R.H., Montgomery, D.C., Anderson-Cook, C.M., *Response Surface Methodology: Process and Product Optimization Using Designed Experiments*, John Wiley & Sons, Hoboken, NJ, USA, 2016.
69. Ivashkin, Y.A. and Borodin, A.V., Modeling and optimization in biotechnology. *Theor. Found. Chem. Eng.*, 30, 4, 400–406, 1996.
70. Plackett, R.L. and Burman, J.P., The design of optimum multifactorial experiments. *Biometrika*, 33, 4, 305–325, 1946.
71. Szpisják-Gulyás, N., Al-Tayawi, A.N., Horváth, Z.H., László, Z., Kertész, S., Hodúr, C., Methods for experimental design, central composite design and the Box–Behnken design, to optimise operational parameters: A review. *Acta Aliment.*, 52, 4, 521–537, 2023.
72. Keskin Gündoğdu, T., Deniz, I., Çalışkan, G., Şahin, E.S., Azbar, N., Experimental design methods for bioengineering applications. *Crit. Rev. Biotechnol.*, 36, 2, 368–388, 2016.
73. Mondal, P.P., Galodha, A., Verma, V.K., Singh, V., Show, P.L., Awasthi, M.K., *et al.*, Review on machine learning-based bioprocess optimization, monitoring, and control systems. *Bioresour. Technol.*, 370, 128523, 2023.
74. Gupta, S. and Sundar, S.K., Kinetics of microbial growth, substrate consumption, and product formation, in: *Recent Advances in Bioprocess Engineering and Bioreactor Design*, pp. 75–114, Springer Nature Singapore, Singapore, 2024.
75. Xu, J., Rehmann, M.S., Xu, M., Zheng, S., Hill, C., He, Q., *et al.*, Development of an intensified fed-batch production platform with doubled titers using

N-1 perfusion seed for cell culture manufacturing. *Bioresour. Bioprocess.*, 7, 1, 17, 2020.
76. Xie, D., Continuous biomanufacturing with microbes—upstream progresses and challenges. *Curr. Opin. Biotechnol.*, 78, 102793, 2022.
77. Sharabati, A.A.A., Lean operations and competitive advantage in the pharmaceutical industry. *Int. J. Serv. Oper. Manag.*, 44, 3, 293–316, 2023.
78. Kim, G.B., Choi, S.Y., Cho, I.J., Ahn, D.H., Lee, S.Y., Metabolic engineering for sustainability and health. *Trends Biotechnol.*, 41, 3, 425–451, 2023.
79. Srivastava, R.K., Akhtar, N., Verma, M., Imandi, S.B., Primary metabolites from overproducing microbial system using sustainable substrates. *Biotechnol. Appl. Biochem.*, 67, 6, 852–874, 2020.
80. Yuan, S.F. and Alper, H.S., Metabolic engineering of microbial cell factories for production of nutraceuticals. *Microb. Cell Fact.*, 18, 1, 46, 2019.
81. Soni, S.K., Dogra, S., Sharma, A., Thakur, B., Yadav, J., Soni, R., Enzyme-powered waste valorization: simultaneous hydrolysis and fermentation of kitchen waste for biofertilizer production. *Bioresour. Technol. Rep.*, 25, 101700, 2024.
82. Chen, X., Li, S., Liu, L., Engineering redox balance through cofactor systems. *Trends Biotechnol.*, 32, 6, 337–343, 2014.
83. Lv, X., Xue, H., Qin, L., Li, C., Transporter engineering in microbial cell factory boosts biomanufacturing capacity. *Biodes. Res.*, 2022, 9871087, 2022.
84. Kim, B., Kim, W.J., Kim, D.I., Lee, S.Y., Applications of genome-scale metabolic network model in metabolic engineering. *J. Ind. Microbiol. Biotechnol.*, 42, 3, 339–348, 2015.
85. Channon, K., Bromley, E.H., Woolfson, D.N., Synthetic biology through biomolecular design and engineering. *Curr. Opin. Struct. Biol.*, 18, 4, 491–498, 2008.
86. Doudna, J.A. and Charpentier, E., The new frontier of genome engineering with CRISPR-Cas9. *Science*, 346, 6213, 1258096, 2014.
87. Liu, D., Mannan, A.A., Han, Y., Oyarzún, D.A., Zhang, F., Dynamic metabolic control: towards precision engineering of metabolism. *J. Ind. Microbiol. Biotechnol.*, 45, 7, 535–543, 2018.
88. Al-Faqheri, W., Thio, T.H.G., Qasaimeh, M.A., Dietzel, A., Madou, M., Al-Halhouli, A.A., Particle/cell separation on microfluidic platforms based on centrifugation effect: A review. *Microfluid. Nanofluid.*, 21, 6, 102, 2017.
89. Castro-Muñoz, R. and Jafari, S.M., Introduction to Membrane Separation of Bioactive Compounds; Challenges and Opportunities, in: *Membrane Separation of Food Bioactive Ingredients*, pp. 3–37, Springer International Publishing, Cham, Switzerland, 2022.
90. Diels, A.M. and Michiels, C.W., High-pressure homogenization as a non-thermal technique for the inactivation of microorganisms. *Crit. Rev. Microbiol.*, 32, 4, 201–216, 2006.

91. Asenjo, J.A. and Andrews, B.A., Enzymatic cell lysis for product release, in: *Separation Processes in Biotechnology*, pp. 143–176, CRC Press, Boca Raton, FL, USA, 2020.
92. Jungbauer, A., Continuous downstream processing of biopharmaceuticals. *Trends Biotechnol.*, 31, 8, 479–492, 2013.
93. Ikegami, T., Hydrophilic interaction chromatography for the analysis of biopharmaceutical drugs and therapeutic peptides: A review based on the separation characteristics of the hydrophilic interaction chromatography phases. *J. Sep. Sci.*, 42, 1, 130–213, 2019.
94. Jornitz, M.W. (Ed.), *Filtration and Purification in the Biopharmaceutical Industry*, CRC Press, Boca Raton, FL, USA, 2019.
95. Mahler, H.C. and Allmendinger, A., Stability, formulation, and delivery of biopharmaceuticals. *Protein Ther.*, 2, 469–491, 2017.
96. Abdel-Baki, R.M., Ahmed, M.N., Barakat, O.S., Khalafalla, G.M., Enhanced vitamin B12 production by isolated Bacillus strains with the application of response surface methodology. *BMC Biotechnol.*, 24, 1, 90, 2024.
97. Ezzelle, J., Rodriguez-Chavez, I.R., Darden, J.M., Stirewalt, M., Kunwar, N., Hitchcock, R., *et al.*, Guidelines on good clinical laboratory practice: bridging operations between research and clinical research laboratories. *J. Pharm. Biomed. Anal.*, 46, 1, 18–29, 2008.

12

Smart Biomanufacturing: Integrating Lean Technologies and Digital Automation in Biologics Production

Thenmozhi Annadurai[1]*, Abishek Balasundaram[2], Bava Dhareni Pandia Rajan[2] and Venkateshan Narayanan[3]

[1]*Department of Pharmaceutical Analysis, Arulmigu Kalasalingam College of Pharmacy, Virudhunagar, Tamil Nadu, India*
[2]*Department of Pharmacy, Arulmigu Kalasalingam College of Pharmacy, Virudhunagar, Tamil Nadu, India*
[3]*Department of Pharmaceutical Chemistry, Arulmigu Kalasalingam College of Pharmacy, Virudhunagar, Tamil Nadu, India*

Abstract
Biomanufacturing indicates the process of large-scale production of biological goods such as vaccines, vitamins, hormones, monoclonal antibodies, and other bio-based items. Bio-manufacturing products have great ability to treat severe disease conditions and chronic disease management. But biomanufacturing products have a lot of challenges. The major challenges are scalability, relatability, and sterility. And also, biomanufacturing is a complex process compared to other pharmaceutical manufacturing, such as tablets, capsules, parenteral preparations, etc. Because the starting material used for biomanufacturing has a large and complex molecular structure compared to conventional pharmaceutical starting material. So, the process of manufacturing is highly complicated in precise condition. And these limitations are reduced by the integration of smart techniques during biomanufacturing. Smart approaches are the integration of artificial intelligence technologies such as artificial neural networks, support vector machines, genetic algorithms, and post-swarm optimization. During the biomanufacturing process, the incorporation of automation lessens individual misconceptions. A microfluidic device, a liquid handling system, is an example of an automatic device used in the process of biomanufacturing. Additionally, blockchain technology for proper

**Corresponding author*: srimozhi2005@yahoo.co.in

Sandip Kunar, Jagadeesha T. and D. Sathis Kumar (eds.) Biomanufacturing: Transforming Production of Biologics and Biopharmaceuticals, (343–380) © 2026 Scrivener Publishing LLC

data management and a single-time-use bioreactor system produce quality products. The above smart techniques make the biomanufacturing process easier, reduce the product cost, and make it affordable for middle- to low-income people. This chapter provides wide knowledge about automation, AI-based processes, and lean technology involved in bio-manufacturing.

Keywords: Biomanufacturing, AI tools, automation, lean technology, biologics, biopharmaceuticals, digitalization, automatization

12.1 Introduction

The term "biomanufacturing" describes all the technological and industrial procedures used to produce complex biological molecules on a large scale, including therapeutic proteins, monoclonal antibodies, and viral vectors for gene and cell therapy. This type of manufacturing process makes use of biological systems, such as *in vitro* synthetic (enzymatic) systems, live microbes, resting cells, plants, animals, tissues, and enzymes. It was cultivated in the bioreactors under carefully monitored circumstances. In addition, biomanufacturing could generate complex compounds that conventional chemical synthesis is unable to manufacture. It is essential to the advancement of healthcare since it provides advanced cures and treatments [1]. The bio-manufactured product has the ability to develop specialized, effective therapies for a wide range of disorders, notably those affecting cancer, immunology, and uncommon diseases. Nowadays, biopharmaceuticals are currently responsible for an increasing number of therapeutic advances. Based on research methodology, manufacturing platforms, and product categories, the historical era of biomanufacturing is categorized into four revolutions.

- In biomanufacturing 1.0, primary metabolites (like butanol, acetone, ethanol, and citric acid) are produced through mono-culture fermentation;
- In biomanufacturing 2.0, secondary metabolites (like penicillin and streptomycin) are produced through a specialized mutant and aerobic submerged liquid fermentation.
- In biomanufacturing 3.0, it is to employ recombinant DNA technology and advanced cell culture to produce big biomolecules, such as proteins and enzymes (such as erythropoietin, insulin, growth hormone, amylase, and DNA polymerase).

- Biomanufacturing 4.0 may target both new products—such as synthetic biology-driven microorganisms, *in vitro* synthetic biosystems created with artificial starch, metabolic engineering-fermented isobutanol, and human tissues or cells created by regenerative medicine—and existing products created with far more effective methods.

Biomanufacturing 4.0, plays a major role in the improvement of manufacturing processes by using digital technology to better control and monitor them, remove human error, and allow for real-time data analysis and decision-making. Compared to ordinary manufacturing methods, biomanufacturing has several advantages, including scalability, flexibility, and less environmental impact. However, biomanufacturing has distinct obstacles in terms of scalability, manufacturing method, complexity, expense, and regulation. The often-lengthy procedures need stringent monitoring of multiple aspects to ensure the end product's quality and safety [2, 3]. The pharmaceutical industry uses biomanufacturing to create medicinally active chemicals that will be synthesized into a finished medicine product known as a biologic. These active medication molecules include proteins, including antibodies, cytokines, hormones, and enzymes. The manufacturing of these biomolecules involves purifying techniques like chromatography and filtration to remove the material from cellular and process contaminants, as well as advanced cell culture, in which cells create the active medicinal element (upstream processing). It is a highly sophisticated procedure that must adhere to tight FDA requirements and employ highly experienced workers. However, these limitations are reduced by integrating smart technology into the biomanufacturing process. Smart manufacturing might improve existing biomanufacturing processes by leveraging new digital technologies and big data. This novel biomanufacturing technology makes use of cutting-edge technologies such as automation, artificial intelligence, machine learning, and big data analytics to increase bioproduction's scalability, quality, and efficiency. "Digital biomanufacturing" has been utilized to describe it. Smart biomanufacturing has lots of merits, such as improving the final product quality and its standard, reducing human dependence error, improving scalability and flexibility, etc. So, the integration of smart techniques in biomanufacturing has produced a revolutionary era in the bio-pharmaceutical industry [4–6]. A comparative overview of traditional and smart biomanufacturing approaches is presented in Table 12.1.

Table 12.1 Traditional biomanufacturing vs smart biomanufacturing.

S. no.	Traditional biomanufacturing	Smart biomanufacturing
1	Process of biomanufacturing is manual operation, Fixed processes and offline quality testing	Processes of biomanufacturing is optimized by using artificial intelligence, machine learning, automatization technique
2	Biomanufacturing process is controlled by Batch process, Manual control with limited real time monitoring	Process Analytical Technology [PAT] is utilized for continuous or adaptive control in the biomanufacturing processes
3	Collection of data is periodic manner, and it was analyzed by offline	Real-time data collection, analysis and decision making
4	Quality was checked by after the end of production	Quality was monitored by in-line and real time manner. Advanced sensor and predictive analytics approaches are used for monitoring the quality
5	In traditional biomanufacturing process the waste was higher and there was no efficiency. This was due to manual intervention and slow response time	In smart biomanufacturing lean principle and integration of lean with digital technology reduce the waste and increase the efficiency of process
6	Large number of manpower is required and operator experience is crucial	Collaboration of machine and human and more focus on monitoring and strategic decision making

12.2 Lean Manufacturing in Biologics Production

Lean ideas are mostly drawn from Japanese industry, specifically Toyota. Many writers assume that lean manufacturing is a waste-reduction strategy, but in reality, lean manufacturing enhances product value while minimizing waste.

- Lean concepts involve identifying the perceived value of a product/service, aligning it with consumer demand, and continuously improving to reduce waste by differentiating between Value added activity (VA) and Non-Value-Added Activity (NVA).
- In NVA activity, waste is mostly caused by transportation, inventory, motion, waiting, overproduction, overprocessing, and faults. It also poses a considerable obstacle to VA activity.
- The successful application of lean concepts leads to the removal of these wastes [7].

Manufacturing biologic materials and biopharmaceuticals in the biopharmaceutical company may be a complicated and costly procedure because of the numerous variables encountered during manufacture. Advanced lean manufacturing technologies like Lean Six Sigma (LSS) may help a biopharmaceutical business compete on a global scale by reducing waste and variability, resulting in cost reductions. As a result, operating costs are reduced, resulting in lower consumer prices and increased purchasing power. However, these tools are not trade secrets for the sectors. Lean and LSS are extensively used approaches in the biopharmaceutical industry. Lean manufacturing principles and technologies have been successfully used and practiced throughout the biopharma sector to improve the efficiency of both operating and production processes by maximizing resources, minimizing waste, and controlling inventory [8]. In the bio-pharmaceutical industry, defects, inventories, overprocessing, waiting, transportation, motion, overproduction, and underutilized talent are all examples of waste [9]. The major categories of biologics are outlined in Table 12.2.

12.2.1 Types of Wastes in Lean

12.2.1.1 Defects

Defects are the most common sort of waste, because they are easily identified as damaged or non-compliant products. The following are the conditions in which faults arise.

- Production process.
- Analytical labs.
- Supply chains.

Table 12.2 Types of biologics.

S. no.	Types of biologics	Description
1	Monoclonal antibody	Protein that are made up in laboratories that bind specific to the antigen
2	Vaccines	It can be live-attenuated, inactivated mRNA or protein. That can stimulate the immune system and prevent the infectious disease
3	Enzymes	Replacing of missed or defected enzyme
4	Growth factors	Stimulate cell proliferation or healing
5	Blood and Plasma product	Include clotting factor, immunoglobins, albumin
6	Cytokines	Signaling protein used to modulate the immune response
7	Hormones	Biologically synthesized hormones to replace the or supplement the body hormones
8	Cell Based therapy	Use of live cells to repair or replace damaged tissues
9	Gene therapy	Insert, alter or remove genes to treat disease

- Improper handling of warehouse boxes can damage raw materials or finished items, necessitating repair or destruction.
- Incorrectly completed shipping forms can cause production delays and delay final item delivery to customers.
- Documentational mistakes lead to delays or additional rectification work, are considered as waste, and add no value to the final product.

Smaller amounts of final product yield indicate bigger amounts of mistakes. Variability within a process causes a huge number of problems. The variability is differentiated into two types by Dr. Walter Shewhart. The first one is assignable because that indicates the process is unpredictable outside of the system. The second one is chance because that indicates the process that contains the variability in an inherent manner. For example, defects in a cell culture process include a non-conforming batch of buffer and harvest contamination. The costs of defective products or materials are mostly direct expenses resulting from lost income. Indirect costs associated with flaws may include discarded costs, contamination of process streams, and the need for additional testing, cleaning, and sanitization. These expenditures, especially discarded fees, may be rather high during handling with active medical components that must be handled properly; and soon after, they go for the incineration process. Although it is viable to recover some things, there will be extra expenses [10].

12.2.1.2 *Inventory*

Inventory is usually referred to as a necessary evil. Inventory includes raw materials held in a warehouse or on a rack area, as well as final products.

- Insufficient raw material inventory can hinder production.
- Excessive inventory might contribute to longer lead times and increased storage requirements.

Without contemporary data processing or modeling, it may be impossible to attain the proper inventory balance. Excess product inventory is created by overproduction, which is another type of lean waste. Inventory has numerous expenditures associated with it. Storing raw materials before use demands the use of either a warehouse or another type of storage area that costs money for land and building. Moreover, the raw material may need highly monitored environmental conditions, raising both installation and running expenses. Every item of inventory must be carefully tracked and monitored by its organization. For the reason of tracking inventory, the organization must use the tracking system, and this may lead to further additional costs. Operating persons are also needed to receive, monitor, and transfer supplies, which increases the cost. Excess inventory poses a variety of risks and costs, including defects in raw materials in any unplanned events, material expiration, products becoming unusable, exhibiting the material in disuse, and impurity [11].

12.2.1.3 *Overprocessing*

Over-processing refers to the execution of processes that surpass a set (or expected) quality level. Over-processing happens when a product or process meets or surpasses the Critical-To-Quality (CTQ) and/or regulatory requirements. Such events may occur if adequate quality control encompasses continuing to process a faulty product. The increase of entire rotational time and inventory levels is affected by over-processing. Over-processing waste occurs when garbage is processed or generated faster than what is required for effective management. Quality control of the product falls under overprocessing. To ensure the product quality, it must need a certain level of inspection to meet the regulatory criteria. Over-testing carries high expenses. Under-testing, on the other hand, poses a significant danger [12].

12.2.1.4 Waiting Time

Waiting is a waste of time, since it prevents you from moving on to more valuable pursuits. The following are the variety of reasons why delays occur.

- Waiting for material release or lack of Quality Assurance (QA)/Quality Control (QC) staff for authentication and approvals might drastically increase wait time.
- According to a recent study (private client) conducted by the authors, the time spent waiting for quality assurance staff to start an inspection process accounted for more than 42% of total cycle time. The proper organizational activity has been able to easily decrease the waiting time, so that the quality assurance professional was not needed in many places at the same time.
- Raw material shortages lead to lengthier wait times.
- Poor planning and scheduling can also create delays.
- Upstream process variability affects downstream processes.
- Longer wait times downstream are caused by delays in the upstream process.
- The waiting time was increased by lack of equipment for processing and transport; for example, the lack of prepared equipment, assemblies, and kits needed for processing, among other things, increases waiting time.

Lean businesses don't get much use out of idle equipment. Bioprocessing equipment is extremely valuable. Ineffective utilization might lead to increased product Cost of Goods Sold (COGs). Businesses lose money and other important resources when employees or equipment are not in use. Even if an operator was not working throughout the shift (due to uncontrollable reasons), companies still had to pay for the labor. The operators may be assigned to other duties in these situations. Such reassignments, however, can be ineffectual and may not benefit operator utilization overall if the operators lack the necessary training for their positions [13, 14].

12.2.1.5 Transportation and Motion

Uncontrolled mobilization of people, documents, or unprocessed materials might be considered NVA activities. Although it might seem important, transportation is not necessary in a process when each unit action is located close to its upstream and downstream processes. Sterile processing and environmental hygienic room categorization may need the separation of unit processes and consequently stations for transfer and transport people, making this impractical in biopharma facilities. However, inefficient operations and a lack of understanding of the operation's environmental impact, which leads to poor facility layout design, may be held responsible for the majority of transportation and waste transfer expenses. Every form of transportation has associated expenses. Some kinds of equipment, such as a hand truck or forklift, are required and must be purchased. These pieces of equipment have a one-time capital expenditure cost, ongoing repair-free costs, operating person costs, and other indirect expenses like insurance, training, depreciation, traffic indicator (overhead traffic signal) installation fees, and so on. Although automated guided vehicles (AGVs) are practicable substitutes for manual operating person machinery, certain companies may find the expense of acquiring, creating, and testing an automated system to be unaffordable. Two methods for examining the distance that operators traverse in different layout setups are spaghetti maps and discrete event simulations. When operational philosophies are still being defined or multi-product facilities are being analyzed, this type of study is very helpful. Significant time, financial, and resource waste is also produced by supply networks. Incorrect placement of a distribution facility results in higher overall COGs. Likewise, trucks that are sent to and from the warehouse contribute to transportation waste since they need a similar quantity of time period and materials regardless of the cargo [15]. The amount of movement of the operator is referred to as motion. Although it isn't always feasible, an operator should be able to stand still until parts arrive to optimize efficiency. It is a waste of time and motion for an operating person to spend each and every second gowning, searching for a flexible hose, or even going through computer data instead of adding value to the product [16].

12.2.1.6 Overproduction

Overproduction, or creating more than the required amount, is a familiar practice in all biopharma companies. And it can appear that overproduced material must be stocked on the shelves and consumers peased right away, but there are some complicated risks and extra expenses engaged with creating more than the required amount of product, such as expiry of the product and more risk of contamination from outside sources. Some of those above factors may have produced serious consequences for the hospitalized person and even with the company's reputation. Moreover, there is a threat that the demand for the product may shift or that a competitor will introduce a newer/better product, lowering the demand for previously made items. In such circumstances, overproduction might result in significant losses. Overproduction causes a variety of waste, including excess transportation and inventory. Any accumulation in work in process (WIP) in a small manner will lead to excessive output from an upstream processing system. The hazards and expenses at the level of process are like those at the level of firm [17].

12.2.1.7 Underutilized Talent

Another sort of waste that firms should be worried about is inefficient talent and innovation management. This type of waste includes putting an overqualified individual to a lower task or skimping on paying for employee training but not applying his or her skilled mindset. During the time of changes in process, always engaged with the person intimately familiar with its nuances [18]. The various types of waste identified in lean manufacturing are illustrated in Figure 12.1.

12.3 Implementation of Lean Tools in Bio-Pharmaceutical Industry

The core concepts of lean manufacturing are:

- continuous improvement
- eliminating waste
- maximizing resource efficiency

This strategy entails identifying and eliminating non-value-added processes, lowering inventory levels, and streamlining the flow of goods and

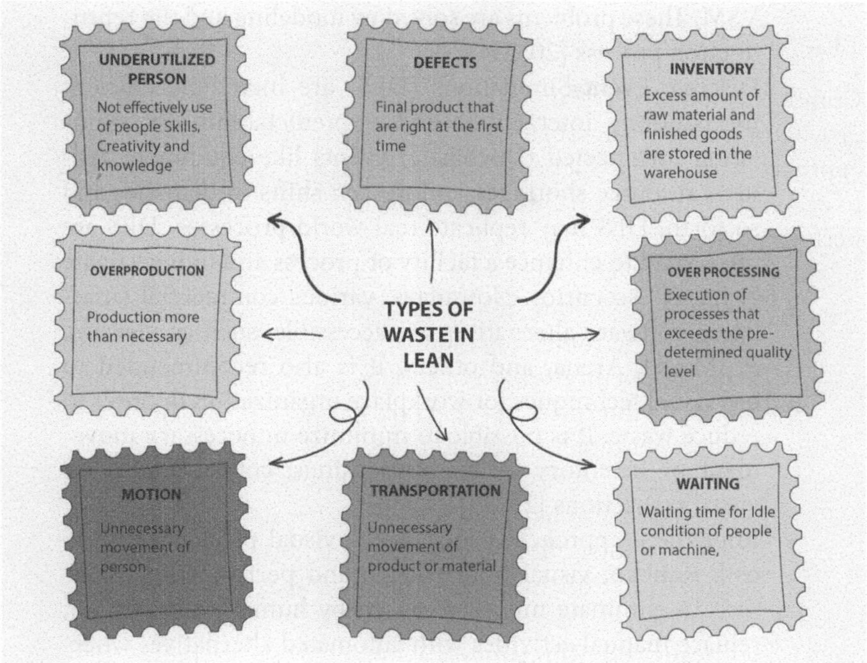

Figure 12.1 Types of waste in lean.

information. The primary merit of lean manufacturing is that it engages businesses to identify and reduce waste during the operations [19]. In the biopharmaceutical sector, this might imply lowering the number of wasted raw materials or optimizing manufacturing procedures to shorten the time and resources necessary to manufacture medicine. Using the appropriate tools and processes can help discover waste in the process and reduce (or eliminate) it.

- Value Stream Mapping (VSM) is a well-liked and broadly applied lean waste detection technique. There is more to a value stream map than a flowchart. It records VA and NVA actions and their durations, and it uses different symbols to represent the flow of information and materials. This tool is used not only to show existing processes but also to envision the state of future processes. But VSM fails to catch the variability effect on a process. Moreover, the dynamic interaction process that occurs inside that process cannot utilize

VSM. These problems are solved by modeling and the reproduction process [20, 21].

- Discrete Event Simulations (DES) are incredibly efficient for handling interactions and unpredictability. By simulating unexpected (stochastic) events like equipment failures, resource shortages, unforeseen shifts in demand, and so forth, DES may replicate real-world processes. DES are often used to enhance a facility or process and to investigate "what-if" scenarios. Nowadays, various commercial simulation software alternatives are accessible, such as Flexsim, ProModel, Arena, and others. It is also recommended to use visual techniques for workplace organization in order to reduce waste. It is possible to minimize unnecessary movement or inventory by using markings, colors, and other visual restrictions [22, 23].
- Implement approaches such as 5S, visual production control, Kanban, visual information, and performance tracking. To eliminate mistakes caused by human intervention, replace manual activities with automated alternatives wherever possible. Standardizing equipment, practices, and techniques may drastically reduce waste. Uniformity typically encourages complete flexibility. To efficiently minimize waste in processes and supply chains, statistical and quality approaches such as Design of Experiment (DOE), control charts, sample plans, and so on may be adopted.

The benefits of lean manufacturing in the biopharmaceutical business are evident, with significant increases in efficiency, product quality, cost, and customer satisfaction. Companies who use these practices can obtain a competitive advantage and boost their profitability in an industry that is always developing and requires the highest production standards. By minimizing waste and streamlining manufacturing processes, the pharmaceutical sector can continue to supply novel and life-saving medications to patients all around the world [24–26].

12.3.1 Limitation of Implementation of Lean in Bio-Pharmaceutical Industry

Adoption of continuous improvement methods in any business or sector is hindered by several hurdles, including employee opposition and a lack of operational focus. Achieving a lean transition in the biopharmaceutical business requires efficient management of extra hurdles. The biopharmaceutical sector is heavily regulated by governments worldwide to guarantee customer safety. The FDA's Process Analytical Technology (PAT) effort aims to reduce regulatory barriers to pharmaceutical manufacture and associated hazards. Although the program aims for continual improvement and high-quality manufacturing, it is still in its early stages, and many enterprises are waiting for regulators to evaluate its effectiveness [24]. Additionally, production facilities for biopharmaceutical goods are often developed for a single product, making them less adaptable. Automation in processing might be inflexible due to the need for validation to meet GMP criteria. The rigidity of these facilities and control systems might hinder improvement initiatives, since personnel may not be able to implement prospective changes on time or at all. Next, production processes are inherently uncertain. Upstream processes, such as cell culture or fermentation, use live organisms to make protein, which might have unexpected results. Efforts by scientists and engineers may not eliminate deviations from upstream processes, leading to a shift from improving operations to addressing unknown concerns. Upstream variations can impact purification operations by introducing varying volumes of byproducts or pollutants. Additionally, biopharmaceutical manufacture often involves batch or semi-batch methods, rather than continuous manufacturing. This batch technique may be used for bioreactor/fermentor processes that yield harvests on a fixed timetable, rather than continuous perfusion. The batch technique can also be applied to protect against production risks, including contamination and loss caused by human or machine mistakes. Batch processing allows organizations to isolate any difficulties to a single batch, reducing overall product loss. Manufacturing companies frequently resist large-scale adjustments following regulatory clearance due to the significant costs and uncertainty associated with changing present processes for continuous operation. Long cycle times, large inventories, complicated supply chains, stringent patient safety rules, and a heavy quality inspection workload are all distractions from continuous improvement activities [27].

Case studies:

- In a case study by Ismail *et al.*, Lean Six Sigma was utilized in the biopharmaceutical industry to reduce waste and manufacturing cycle times. By identifying and removing non-value-added tasks using value stream mapping and process flow mapping, the organization reduced process cycle time

by 54% in a specific manufacturing area. This resulted in a $141,480 cost reduction at mid-year [28].
- Nepal *et al.*, used value stream mapping to detect waste in an upgraded manufacturing process. Value stream mapping found redundancies and manufacturing delays, resulting in a reduction of production time from 17 to 4.5 days [29].

The above two case studies indicate that the implication of lean manufacturing in the biopharmaceutical industry gives lots of benefits to the industry.

12.4 Digital Automation in Smart Biomanufacturing

The biopharmaceutical sector is continually developing, with new technology enabling next-generation production procedures. The "smart" bio-factory uses digital technology to enhance the efficiency, production of goods, and product quality while cutting expenses. It will bring together all aspects of the lifecycle of the product and supply chain. This technology of digitalization automates the collecting process, processing, and data analysis originating at various pieces of industrial equipment. This technology moves the biopharmaceutical industry toward the bio-pharma concept 4.0. Digitalization has the potential to have a significant influence on the biopharma industry's fundamental concept of providing access to life-saving medicine in a timely and compliant way, with proven quality at an accessible price. Implementing digitalization in biomanufacturing can achieve the aim of faster, better quality, more cost-efficient, safe, and effective treatments for patients. Digitalization enables biopharma businesses to collect massive datasets, which are then analyzed using technology to draw conclusions and offer changes. Digitalization technologies also significantly reduce dependency on human record keeping. Companies may eliminate human mistakes by digitizing and automating data such as batch records and equipment using logs [30].

12.4.1 Role of Digitalization in Smart Bio-Manufacturing

In the biomanufacturing process, digitalization techniques have a great ability to efficiently boost the performance of an organization. In the process of smart bio-manufacturing, product quality, efficiency, speed of the process, and flexibility of the product are improved, which ultimately benefits the health of the patient. In the smart bio-factory, digitalization will impact on the whole product lifecycle and supply chain to increase efficiency. Biopharmaceutical companies must follow strict regulations and ensure data integrity. Identifying a dependable and effective method can improve productivity and product quality for these firms. Implementing a digitalization strategy helps improve manufacturing efficiency and productivity by monitoring and optimizing operations [31]. Implementing digitalization can enhance several business factors related to GMP. The following is the vital role of digitalization in the bio-manufacturing industry.

12.4.1.1 Productivity

- In the past few decades, biopharmaceutical productions have expanded globally. At the same time as economies and healthcare systems expand, there is an increase in demand for both local and foreign pharmaceuticals. Not only is demand increasing, but so is the variety of pharmaceutical demands, as emerging countries deal with noncommunicable illnesses that are common in developed countries. Common conditions include diabetes and hypertension, while communicable illnesses, including AIDS, malaria, and TB, remain prevalent.
- This type of disease condition must require a biologist for managing the disease condition. So, the proper supply of biologics to all patients improves their quality of life. Innovative solutions are needed to address emerging difficulties in the manufacturing sector.
- Digitization enables quicker and higher-quality manufacturing processes to fulfill rising production demand [32, 33].

12.4.1.2 Quality of Product

- Biopharmaceutical firms must segregate the data on their operations to assure the quality of the product. Multiple firms continue to use paper documentation, which raises the possibility of mistakes in data collection and administration. Recording this information on paper increases processing time.
- Digitization lowers errors and increases data collection efficiency. Pharmaceutical companies may use digitization techniques to gather, standardize, and use a variety of sources for

data analysis. And this digitalization technique produces a basic report related to the assurance of quality products, and that report is easily available after batch completion.

- When a noted value falls in an undesired range, an automated warning is issued to quality management. The data are kept in the cloud, and it is accessible to all employees to know the up-to-date information regardless of the internet connection they use.
- The procedure's consistency makes it suitable for validation. Digitizing GMP provides users with exact instructions for activities, including safety instructions and recommended actions, reducing mistakes. In simple terms, digitalization enhances quality control efficiency and dependability [32–34].

12.4.1.3 Quality Control

- Biopharmaceuticals have more severe quality standards for their manufacturing biologics. Each stage of the process, including product manufacture, packing, labeling, and equipment cleaning, requires thorough monitoring. The pharmaceutical industry commonly uses the batch method, which involves creating goods in batches rather than continually. Even though this has several merits, it also reflects that a single little mistake can wipe out the whole batch. Default batches waste a large amount of money, time, and materials. Failure to identify a defective batch can have significant consequences, including personal injury, fines, lawsuits, and reputational harm. Efficient control is vital for avoiding these and other dangerous effects [35].
- Digital transformation improves process control and optimizes productivity. Digitization improves packing efficiency during the production process.
- Digitization technology can help improve packaging processes by controlling, monitoring, and visualizing them, like batch production. Remote data access allows you to monitor and manage your operation from any location. Digitalization of control procedures is crucial for pharma companies' profitability.

- Digitally automating control stages saves unprocessed materials, human resources, and time periods. It also decreases the risk of low-quality items, leading to higher profitability.

12.4.1.4 *Efficiency*

- Efficiency is directly linked to quality control, which reflects the amount of time, money, and resources a corporation needs to make a definite amount of output. As a consequence, efficiency and productivity are closely related. Digital technique enables thorough data gathering across your factory, offering insights into operational efficiency. Data and reports may be used to identify areas for improvement in procedures and facilities. You may use it to automate particular processes [36, 37].
- Implementing digitization can reduce manual input, time, and resources needed to do the same activity. Automated alerts allow operators to obtain real-time information about problems and address them quickly. Validation is crucial in the biopharmaceutical industry to ensure compliance with regulations and standards. Digitization also improves efficiency in this area. This software ensures full compliance and enables process updates without the need for revalidation, saving both time and money [38].

12.4.1.5 *Flexibility*

- Flexible bio-manufacturing speeds up the development of new medications and vaccines, unlike traditional fixed production procedures. Introducing a new drug into an established process requires constructing a novel facility. Flexible bio-manufacturing enables the development of a novel medicine from single-use materials, avoiding the requirement for cleaning and validation of current equipment [39]. Flexible production facilities save time and speed up medicine or vaccine product delivery to the market. The hospitalized person is expected to benefit immediately from the speedier bio-manufacturing process. Additionally, infrastructure and continuous manufacturing technology enable changes to satisfy patient demand [40].

- Implementing automation and robotics can enhance patient safety. Flexible manufacturing can help enhance medication adherence and compliance. In fixed production facilities and processes, the batch record specifies the dose and does, and is restricted for process changes. The more flexible resources enable adjustments to the medicine product manufacturing process, especially raising the vial dose or changing the appearance of the product (for example, 3D printing). Replacement therapeutic representations, doses, and durations may have a direct reflection on patients' wants. Integrating these techniques in various manners will result in a paradigm change in future evaluation of product and production processes. Implementing digital technologies in pharmaceutical manufacturing will benefit patients in the future [41].

12.4.2 Key Technology for Digitalization in Bio-Manufacturing

Digitalization indicates the application of digital technologies and data to enhance business processes, generate revenue, and change operations.

12.4.2.1 Internet of Things [IoT]

The Internet of Things (IoT) refers to the connection of a network to the objects that can convey data without human intervention. Things in this context include equipment, machinery, objects, animals, and people. The Internet of Things is not a new technique. The word refers to a collection of technologies that, integrated, form a system of a network of devices with enhanced capabilities. IoT enables extensive data collection and usage, leading to fewer errors, more efficiency, and improved process understanding [42]. Although bioprocessing is an important component of business 4.0, the complex nature of bio-manufacturing with biological systems makes it difficult to incorporate IoT into this business. Biopharmaceutical production accounts for a significant portion of the sector. Advanced therapeutics are highly valued. Biotechnological products are rigorously regulated from conception to patient delivery. The biopharmaceutical sector aspires for extensive process understanding. However, transitioning to a new system can be complicated. The challenges of transitioning to a new style of operation include potential misalignment of goals, significant investments, downtime, and regulatory requirements. This makes it difficult for the sector to embrace IoT solutions that interact with biomanufacturing activities. Universal solutions are needed to simplify IoT implementation and overcome industry reluctance. These solutions should feature simple implementation methodologies, effective staff training, and significant insights for improving processing and manufacturing operations within regulatory frameworks [43, 44].

12.4.2.2 Blockchain Technology

The biomanufacturing industry is increasing and pushing personalized medicine in a fast manner for critical diseases such as cancer and blindness that occur in adults. Many biomolecules have quadrupled during the recent 10 years. The biopharmaceutical business likewise experiences yearly double-digit growth in both sales and earnings. However, the Food and Drug Administration (FDA) has prioritized medicine safety and scarcity. The majority of the medicine shortfall is due to long lead times and quality difficulties. However, employing blockchain technology helps to mitigate these difficulties.

- Blockchain technology can enhance medicine safety and delivery dependability in the complicated global biopharmaceutical supply chain.
- Blockchain technology has several advantages, including visibility, resiliency, integrity, and automation [45].
- We can address the aforementioned difficulties by creating a suitable blockchain for the global biopharmaceutical supply chain.
- Integrating smart sensors into the delivery process allows for tracking of drug storage and distribution from producer to patient.
- The blockchain network's decentralized nature allows transactions to be automatically confirmed and completed, as opposed to relying on centralized trust mechanisms. Third, by exchanging information in real time, the blockchain network can fully integrate participants and encourage interoperability, ensuring secure pharmaceutical delivery and prompt replies to any complaints [46]. The application of blockchain technology in bio-manufacturing is depicted in Figure 12.2.

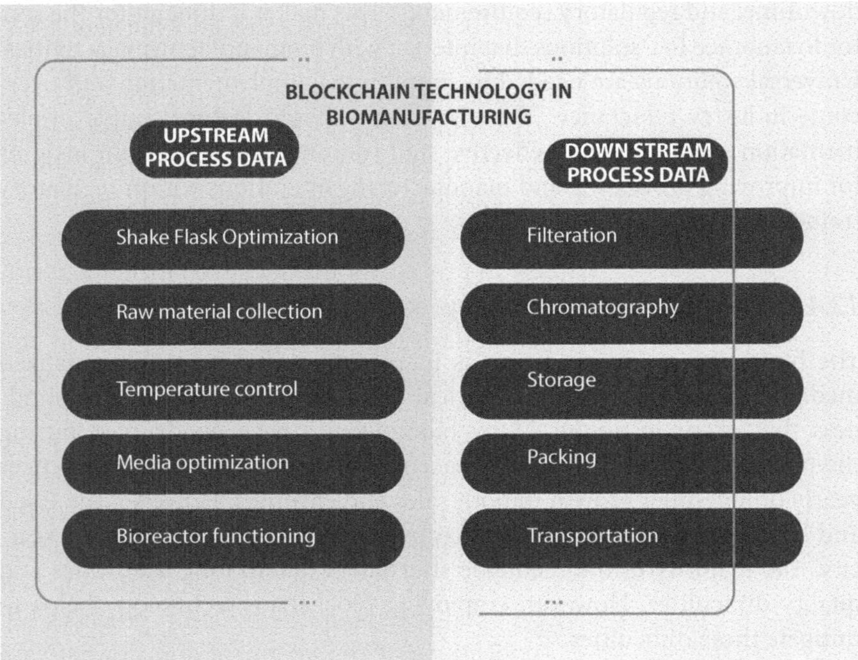

Figure 12.2 Usage of blockchain technology in bio-manufacturing.

12.4.2.3 Artificial Intelligence (AI) and Machine Learning (ML)

AI and ML have the greater ability to transform bio-pharmaceutical manufacturing by reducing the need for human supervision and enabling automated and continuous procedures [47, 48]. Multivariate data analysis methods are the most often utilized type of AI-ML methodologies, with extensive adoption in the biopharmaceutical business. AI models are used for the following applications:

- Track fermentation,
- pH,
- Temperature,
- Microbial development

AI-driven process development in biopharmaceutical manufacturing involves reducing production time and costs by optimizing tablet, capsule, and injectable formulations; enhancing blister pack and packaging of vial procedures; and automating control of quality monitoring to assure product uniformity.

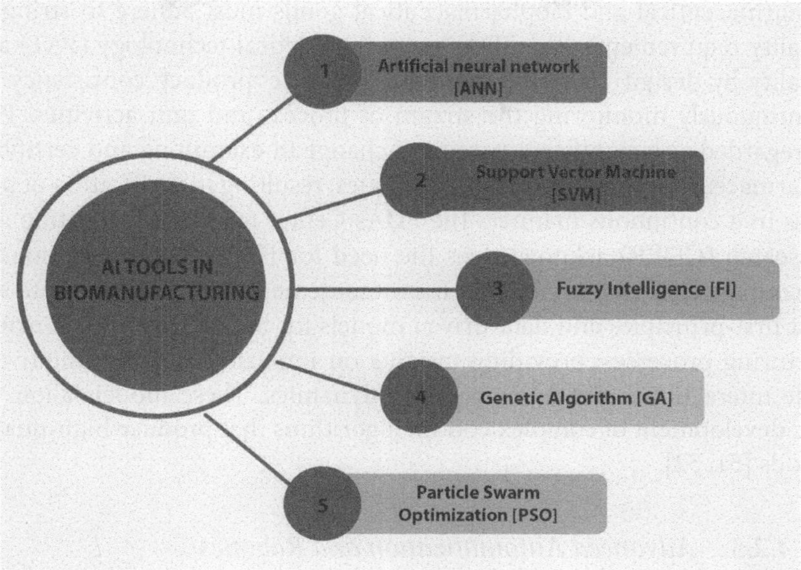

Figure 12.3 AI tools used in biomanufacturing.

- AI is changing biopharmaceutical manufacturing by analyzing massive datasets.
- Optimization of drug formulation processes increases efficient production.

ML algorithms can enhance the production scheduling by examining past data and projecting peak demand. As a result, there is greater efficiency. Efficient material allocation leads to enhanced output. Algorithms can scan enormous amounts of data to determine ineffectiveness and enhance production effectiveness in biopharmaceutical manufacturing processes. Changing the temperature, pressure, component amounts, and mixing times can boost efficiency, cost-effectiveness, and product quality [49, 50]. The common AI tools integrated into biomanufacturing processes are shown in Figure 12.3.

12.4.2.4 Process Analytical Technology (PAT)

PAT (Process Analytical Technology) is "a system for designing, analyzing, and controlling manufacturing through the timely measurements (i.e., during processing) of critical quality and performance attributes of raw and in-process materials and processes, with the goal of ensuring final product quality."

"Pharmaceutical and biopharmaceutical goods must adhere to stringent quality requirements. The FDA's process analytical technology (PAT) and quality-by-design (QbD) paradigms, promote product consistency by continuously monitoring the stream of process and unit activities. PAT is regarded as a significant paradigm change in examining and certifying pharmaceutical manufacturing techniques, resulting in verification of process in a continuous manner. The FDA's Center for Drug Evaluation and Research (CDER) acknowledges the need for FDA recommendations to encourage PAT usage. PAT process measurements are used to create and test first-principles and data-driven models for biopharmaceutical manufacturing processes, providing insights on impurity rejection, multivariable interactions, and unit operating dynamics. These models allow for the development of complex control algorithms that produce high-quality goods [51, 52].

12.4.2.5 Advanced Automatization and Robotics

The biological manufacturing industry has made automation and robots a mainstay because of their effectiveness in operating without user input. Because they are flexible and easy to program, industrial robotic arms offer a useful tool for the production of biologics. Automation tools improve accuracy, control, and optimization of processes. Automation tools increase process control, optimization, and accuracy. These tools offer benefits such as faster data collection and less human error [53]. Autonomous systems are less prone to distractions and exhaustion than people, contributing to their effectiveness. Automated technology can help to reduce human interaction with potentially dangerous chemicals and materials, such as metabolites like carboxylic acids. Automation involves using robots to complete activities that are repetitive, dangerous, or require heavy material handling. Robots reduce labor costs and improve product quality and production frequency through faster and more consistent procedures. Robots today are computer-controlled mechanical manipulators with advanced automation. Robots, like other forms of automation, help to improve process accuracy, repeatability, and efficiency. The capacity to gather more data is another important benefit of using robots, since they can enable higher sample frequencies, for instance. Acquiring such information improves the response variables' trustworthiness. Pneumatics, motors, magnets, and other mechanical components make up intelligently controlled robots [54].

12.4.2.6 Digital Twin Technology

A digital twin refers to the complete, integrational mathematical model that replicates a whole bioprocess sequence in real time, based on both historical and present data. Process analytical data (PAT), quality data, and time-series data are predicted by using critical quality attributes (CQA) and key performance indicators. A digital twin builds a parallel clone of a real-world process to forecast its result ahead. In the biopharmaceutical sector, bioprocess validation is an essential stage in the commercial development of any drug as well as FDA and other regulatory agency approval. Process validation was one of the earliest uses of digital twins and is still a major application area. Digital twins can save manufacturers millions of dollars in development expenses by providing a complete control plan and lowering the number of process performance qualification (PPQ) runs. The ability to effectively use data from process development, design of experiments (DOE), multivariate data analysis (MVDA), and process domain expertise is the major challenge for building a bioprocess digital twin. The setup requires automated modeling for ongoing operations, statistical analysis to provide control mechanisms, and continuous digital collection of process data into a digital storage database [55, 56]. The core technologies enabling digitalized biomanufacturing are presented in Figure 12.4.

12.4.3 Automation in Upstream and Downstream Process in Biomanufacturing

Automation is set to play a considerably more apparent role in biopharmaceutical operations, notably in the transportation of materials between manufacturing facilities. Biopharmaceutical producers have been adopting automation and control systems at both the upstream and downstream levels for over 15 years [57].

- Automation methods for upstream bioprocessing have shown to be useful, saving both labor and time. Automating upstream bioprocesses can also create a tighter feedback loop for process control, reducing the problems that have historically existed at this level of biomanufacturing. Automation in upstream bioprocessing improves the cell culture environment within the bioreactor. Cells' health (including quality, longevity, growth, and reproduction rate) is determined by their living environment.

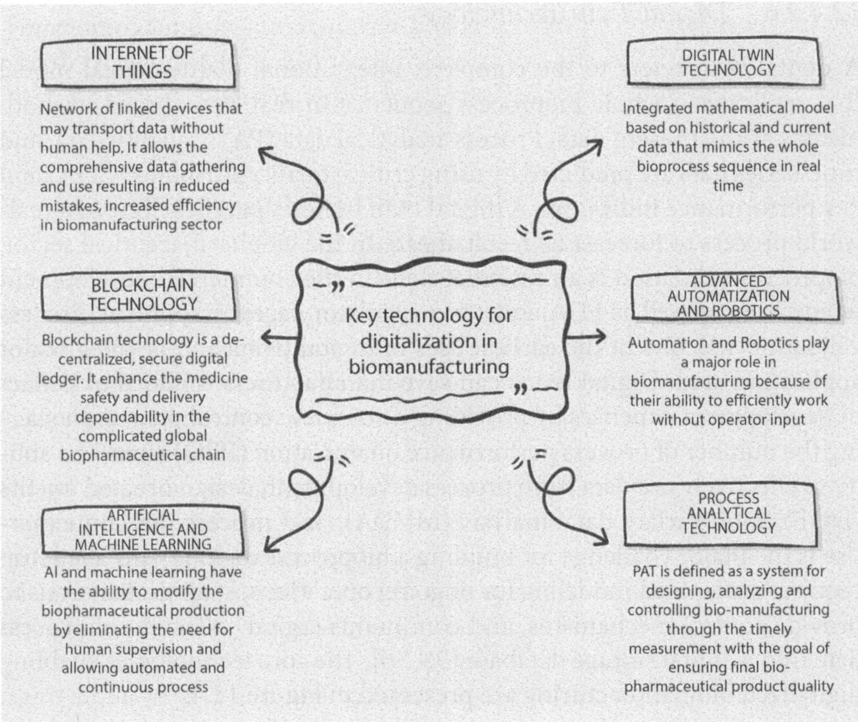

Figure 12.4 Key technology applied for digitalized biomanufacturing.

- As a result, accurate control of temperature, humidity, pressure, and other parameters is critical. These can be regularly done with automated process control. Additionally, modern artificial intelligence (AI) controls can help with process optimization. Upstream unit activities employ automation to varied degrees. The most significant difference with automation is how units interact with one another, such as through automatic transfers or manual transfers, as well as how units are started, either manually or *via* an automated batch/recipe start [58].
- Automation is also used in a factory execution system, which documents the process for electronic batch recording. Product recovery and purification are the two steps in downstream processing.

The process is complicated, requiring careful process control. The assurance of the product quality was determined by temperature, pH, dissolved oxygen, and level of nutrition. Compared to traditional downstream operation, digital automation speeds up the process that includes smart instrumentation, adaptive network architecture, and a robust Ethernet-based dataset. Implementing a framework for regulating equipment and analytical devices can aid in the regulation of data collection and processing. Chemometric models are used to feed the data, and they provide continuous modeling and control. A human-machine interface (HMI) can be applied to inspect the activity and allow an operating person to help as required, "ensuring optimum process control and product quality." Automated recording of important process parameters and CQAs can improve process maintenance efficiency. Single-use technology, inline chromatography conditions, and continuous chromatography are considered the recent advancements of automation techniques in downstream processing [59].

12.5 Integration of Lean with Digitalization in Biomanufacturing

Digital technologies are being utilized to eliminate seven types of waste and enhance lean approach implementation. Digital Lean integrates Industrial IoT technology with industrial software to improve lean processes and tools. Digital Lean improves existing lean methods, such as Kanban, Heijunka, line balancing, and poking probes, by incorporating real-time data into operations. Digitizing lean processes allows manufacturers to track and measure their facilities more efficiently. Real-time data tracking, massive data sets, and powerful analytics enable rapid monitoring of fundamental performance. Digital lean does not introduce new lean principles; rather, it improves current lean concepts to make them more successful. Many digital lean innovation solutions produce accurate information on every step of a process, whereas record-keeping systems, such as enterprise resource planning, often report on operations in relation to business and plant budgets. Digital lean provides more accurate, precise, and timely operational data by utilizing Industry 4.0 and other digital tools. It enhances the efficacy of crucial lean tools like Kanban in addition to assisting with the implementation of lean concepts [60, 61].

12.5.1 Synergistic Effect

Digital lean can enhance the benefits of traditional lean in terms of waste reduction during manufacturing. Digital lean improves waste identification and mitigation over traditional lean techniques by giving targeted, accurate information directly to those who can reduce waste. However, digital lean also enables the targeting of hidden waste components like information asymmetry and delay, which usually go undetected and add up to higher support costs, reduced efficiency, and output, resulting in a major bottom-line impact [62].

- In terms of defects, inadequate process control and product design lead to issues across the value chain, necessitating scrap or rework. By defining standards for asset upkeep, process improvement, and product design, traditional lean concepts may help reduce faults. Nevertheless, digital lean reduces first-pass yields by identifying the precise asset, process phase, or product characteristic that is causing problems.
- In terms of inventories, higher inventory is usually a reflection of value chain instability, whereas lean procedures enable things to be produced in the precise amount and time required. However, by providing a real-time view of work-in-progress inventory throughout the manufacturing process, digital lean can improve operations by enabling the detection of unexpected inventory accumulation.
- Addressing excessive processing traditional lean techniques can reduce pointless procedures in the value stream, including excessive tolerances or over-inspection. On the other hand, digital lean connects and integrates a product's (or value stream's) life cycle with a digital twin, which is an ongoing data stream that replicates the ideation, manufacturing, and consumption of a product from its inception to its eventual life.
- As a consideration to wait times, longer wait times are caused by unbalanced operations, bottlenecks, downtime, and inadequate production planning, particularly when people, resources, and assets don't provide value. Wait time can be reduced by using traditional lean methods. However, by rapidly identifying bottlenecks, rerouting processes dynamically depending on changes in the real-time state of

assets, and simulating various optimal situations, digital lean minimizes waiting time.

- In regard to a focus on motion Operators can execute value-added operations with needless motion— thanks to poorly designed manufacturing lines and cells. Lean techniques cut out pointless steps that prolong manufacturing timeframes and add little value to the final product. However, digital lean may have a greater impact on the layout and equipment design to encourage worker mobility if performance data is reviewed or augmented and virtual reality simulations are used.
- In regard to overproduction, it is brought about by a misaligned supply and demand, including delayed demand signals and stringent process constraints; It is eliminated by traditional lean. Digital lean, on the other hand, may offer real-time value chain visibility, enable preemptive capacity adjustments, and lower the production of unnecessary commodities [63–65]. The integration of lean principles with digital technologies is illustrated in Figure 12.5.

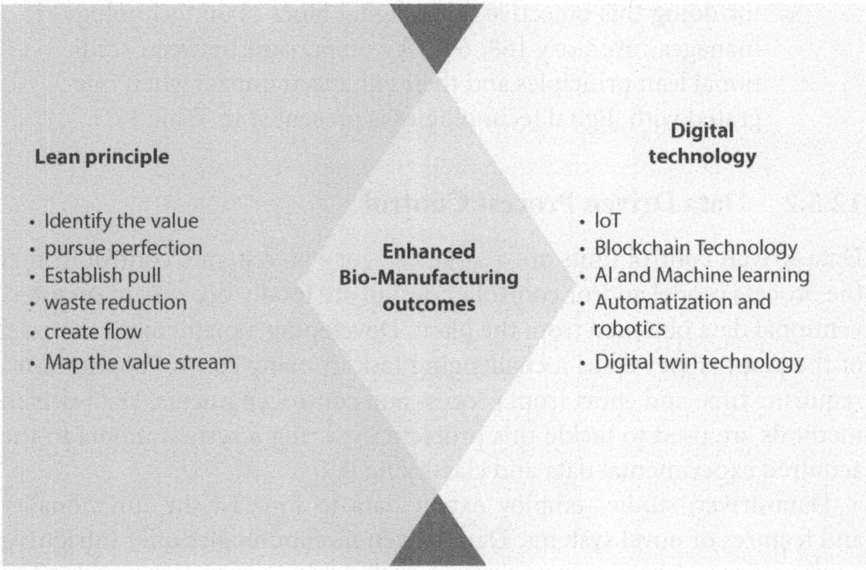

Figure 12.5 Diagrammatic representation of integration of lean with digital technology.

Example for synergistic activity:

- Artificial Neural Networks (ANN) are extensively utilized AI-ML algorithms in biopharmaceutical production applications. Artificial neural networks and reinforcement learning are becoming increasingly popular for monitoring and controlling biopharmaceutical production. ANN accurately predicted the product titer of a biopharmaceutical compound. This soft sensor for biomolecules has the potential to replace costly analytical tests for offline measurement during manufacturing processes. This digital technology eliminates costs due to overprocessing [66, 67].
- Biologic goods can be created with a variety of raw ingredients, either locally or internationally. Inventory management is crucial in natural and waste bioprocessing owing to fluctuating raw material availability, which causes batch-to-batch variance. The supply of raw good practices encompasses the care of materials. The legislative framework and financial factors must be properly preserved and maintained to ensure stability and storage. Blockchain technology provides a safe and irreversible mechanism to track consignment goods internationally, making it a sustainable solution for doing this objective. This digital blockchain technology manages inventory [68, 69]. A comparison between traditional lean principles and their enhanced impact when integrated with digital technologies is presented in Table 12.3.

12.5.2 Data Driven Process Control

Data-driven control units are a large category of systemic control in that the process model and/or controller design are totally dependent on interventional data obtained from the plant. Developing a mathematical model of the plant is viewed as a challenging task in many control applications, requiring time and effort from process and control engineers. Data-driven methods are used to tackle this problem by fitting a system model to the acquired experimental data and classifying it.

Data-driven studies employ extant data to forecast the functionality and features of novel systems. Data-driven methodologies offer intriguing applications in biomanufacturing research, including enzymes and entire cells. Future research will rely heavily on data-driven methodologies,

Table 12.3 Traditional lean principle vs synergistic effect of integration of digitalization with lean principle.

S. no.	Aspects	Lean principle	Synergistic effect of integration of lean with digitalization
1	Reduction of waste	Waste is identified and eliminated manually	Identification and elimination of waste by AI and IoT in an accurate manner
2	Inventory	Just in time principle used for inventory management	IoT sensor is used for real-time inventory monitoring. That provides the proper control in inventory management
3	Making of Decision	According to experience and periodic reviews	AI to tools support the Data driven, predictive decision making
4	Quality control	Manual inspection of bio-manufactured product	AI tools and automatization provide proper quality control in bio-manufactured product
5	Continuous improvement in biologicals production	Based on employ suggestion	Based on real-time feedback loops. That approach enhances the continuous improvement in biological production

particularly on improved standardization of biological data and specialized algorithms [70, 71].

12.5.3 Application of Data Driven in Biomanufacturing

In biopharmaceutical industry the digital driven techniques are mainly used for the design of protein, cell engineering, prediction of kinetic parameters.

12.5.3.1 Design of Protein

Natural enzymes have been utilized to catalyze several chemical processes; however, their characteristics and catalytic capability may not be sufficient for industrial applications. Scientists are researching novel enzyme engineering strategies to fulfill industrial manufacturing needs. Additionally, researchers are developing protein folding control approaches to create enzymes that do not occur naturally. Additionally, to classic rational design and evolution of directed methodologies, scientists are increasingly applying ML and Deep learning (DL) algorithms to determine the protein structures and improve functionality. Protein properties have typically been investigated using quantum mechanics, molecular mechanics, and QSAR simulations. The huge amount of conformational space makes it challenging to employ quantum mechanics or molecular mechanics to determine a protein's lowest energy conformation. The location of the enzyme's reaction center has been the main focus of rational enzyme design. Structural change of enzyme sections requires either random mutation or guided evolution procedures, which need elaborate experimental protocols and rely on the organism's inherent evolutionary potential. Recent advances in protein sequence and structural information have highlighted the potential for data-driven enzyme design using machine learning and deep learning approaches [72–75].

12.5.3.2 Cell Engineering

Data on living cells is more complicated than that of isolated proteins. Despite copious sequencing data, it is insufficient to predict cellular function. While other factors have a role, there is a lack of high-quality evidence. To make bioconversion commercially feasible, an engineered cell factory must provide desirable outputs including high product titter, rate, and yield (TRY). Data-driven cell research are multi-input, multi-output (MIMO) problems, in contrast to protein design, which usually has a single predetermined result. Developing numerous models for diverse outputs and combining them for accurate prediction is challenging because of the complicated nonlinear interactions between inputs and outcomes. Scientists have made efforts to apply data-driven methodologies in cell engineering, concentrating on specific issues with defined inputs and objectives [76].

12.5.3.3 Prediction of Kinetic Parameters

Enzyme kinetics data is limited and usually focuses on specific enzymes or pathways for a small number of model species. This is sufficient to generate a metabolic pathway's kinetic model, but not for a whole cell model. It has been demonstrated that cellular phenotypes may be more accurately simulated and predicted using enzyme-constrained models (ECMs), which incorporate enzymatic limitations into GEMs. The whole network must be included in ECMs to accurately predict the values of enzyme kinetic parameters. Quantitative kinetic data are lacking for almost half of the enzymes found in modeling organisms like E. coli. Due to the lack of high-throughput technology, enzyme kinetic data can be predicted using data-driven methods, which saves time and money. Mellor *et al.*, predicted protein substrate affinity (Km) using a semi-supervised Gaussian process regression model. Chemical transformation rules and extended connectivity fingerprint descriptors were used to build the input signatures for the model. By finding enzymes in E. coli that catalyze processes involving a novel metabolite, the researchers were able to test their model prediction. Heckmann *et al.*, used information on protein structure, network context, and enzyme biochemistry to use machine learning to accurately estimate catalytic turnover numbers (kcat) for E. coli enzymes. They obtained Kcat by using this approach. For every E. coli GEM reaction, we determined the enzyme values and developed an enzyme-constrained model to show quantitative proteome data prediction is far more accurate than previous approaches. When there is a lack of time-course data, researchers have developed data-driven methods for simulating metabolic processes that do not require a kinetic model. Costello *et al.*, demonstrated that supervised machine learning can reliably predict the relationship between metabolites and enzymes using time series protein and metabolite concentration data. The response was the pace at which the metabolite changed, and the exogenous pathway proteins and metabolite amounts were used as inputs. Because multi-omics assays are expensive, data augmentation made more data available for model training, which resulted in better performance than the kinetic model [77, 78].

12.6 Challenges and Future Perspectives

There are numerous significant challenges that continue to obstruct large-scale adaptation in biopharmaceutical production, even with the revolutionizing potential of integrating lean manufacturing and digital automation.

Among the various barriers, one of the foremost, is the immensely regulated nature of the biopharma industry, in which strict compliance with good manufacturing practice, validation requirements, and vast documentation limits flexibility and delays the execution of new technologies. Operational rigidity also confers a considerable hurdle. For most biologics production, facilities are laid out around batch processes and fixed equipment customized for specific products, diminishing their receptivity to procedure modifications. Introducing digital tools and automation in those environments often stipulates revalidation, re-qualification, and expensive retrofitting processes that not only bring upon costs but also risk interrupting production timelines.

The secondary throat-cutting challenge is hesitancy of data transparency and interoperability. Even though digital twins and data-driven methods render enhanced control and optimization, incorporating these tools across supply chains, production stages, and legacy systems remains complex. Cybersecurity concerns, fragmented data silos, and lack of standardized protocols further impede seamless integration. From a human perspective, limited digital literacy and workforce resistance act as major roadblocks. The automation may be resisted by the employees, particularly those accustomed to conventional manufacturing setups, due to fear of discomfort or redundancy with new tools. Furthermore, organizations often fail to provide adequate training, limiting the potential benefits of these innovations.

Despite these challenges, the future of smart bio-manufacturing is promising. Emerging technologies such as cloud-based analytics, augmented reality (AR), digital twins, and machine learning continue to evolve and become more accessible. Flexible manufacturing systems using single-use technologies and modular setups are reducing capital and validation burdens. Furthermore, regulatory bodies like the FDA refine their frameworks—such as through the Process Analytical Technology (PAT) initiative, the path toward continuous improvement becomes more attainable.

Looking ahead, we can expect a shift toward hybrid models, combining physical production with virtual simulations for faster, more adaptive responses to demand and variability. With increasing global demand for biologics, particularly in personalized medicine and rare disease treatment, the pressure to scale efficiently, reduce costs, and maintain product quality will only intensify. The next generation of bio factories will likely be digitally native, lean in operation, and deeply integrated across the product lifecycle—from development to delivery.

12.7 Conclusion

Together, lean manufacturing principles and digital automation portray an instrumental advancement in the biomanufacturing field. This doorway is strengthening a significant skeleton to inflate efficiency, minimize waste, and ensure consistent product quality in an industry where precision and reliability are non-negotiable. Manufacturers can pick out bottlenecks and streamline operations through tools like Value Stream Mapping, Discrete Event Simulation, and Process Analytical Technology. The convergence of artificial intelligence, blockchain, the Internet of Things, and digital twin technologies addresses a new era of predictive control, real-time monitoring, and adaptive decision-making. The incorporation of these techniques fortifies every aspect of biological production, from upstream (fermentation control) to downstream (purification) processes, navigating the advancement in cost, compliance, and throughput. However, realizing the potential of this prototype demands more than technological investment. A cultural shift is mandatory towards continuous improvement, regulatory agility, and cross-disciplinary collaboration. With pensive accomplishment and ongoing support, biomanufacturing will not only revolutionize biological production but also deepen the rush towards life-saving medications across the world.

AI disclosure statement: The author acknowledges the use of AI-based tools for minor language editing or grammatical correction. However, all intellectual contributions, analyses, and conclusions presented in this chapter are entirely the author's own.

References

1. Conner, J., Wuchterl, D., Lopez, M., Minshall, B., Prusti, R., Boclair, D., Allen, C., The biomanufacturing of biotechnology products, in: *Biotechnology entrepreneurship*, C.D. Shimasaki (ed.), pp. 351–385, Academic Press, Cambridge, MA, USA, 2014.
2. Zhang, Y.H.P., Sun, J., Ma, Y., Biomanufacturing: history and perspective. *J. Ind. Microbiol. Biotechnol.*, *44*, 4-5, 773–784, 2017.
3. Ali, A., Usman, M., Jayamani, E., Soon, K., bin Ab Wahab, M.K., Sustainable Biomanufacturing, in: *Biomedical Implants: Materials, Design, and Manufacturing*, R.K. Dwivedi, P.S. Chauhan, A.K. Sharma, M. Singh, A. Agarwal (eds.), pp. 195–217, CRC Press, Boca Raton, FL, USA, 2024.

4. Northeast Biomanufacturing Center & Collaborative – What is Biomanufacturing?, 2016, https://biomanufacturing.org/teacherresources/high-school-teacher-resources/what-is-biomanufacturing.
5. Perrier, A. and Perrier, A., Biomanufacturing of biological drugs: strategic issues and challenges, MabDesign, https://mabdesign.fr/en/biomanufacturing-biological-drugs-strategic-issues-and-challenges/ 24 November 2024.
6. Henn, A.D., Wolff, M., Mitra, K., Alm, K., Finger-Baker, I., Bauer, S.R., Yerden, R., Smart biomanufacturing for health equity in regenerative medicine therapies. *Regener. Med. Rep.*, *2*, 1, 31–35, 2025.
7. Sundar, R., Balaji, A.N., Kumar, R.S., A review on lean manufacturing implementation techniques. *Procedia Eng.*, *97*, 1875–1885, 2014.
8. Melton, T., The benefits of lean manufacturing: what lean thinking has to offer the process industries. *Chem. Eng. Res. Des.*, *83*, 6, 662–673, 2005.
9. Benson, R. and Kulkarni, N.S., Understanding operational waste from a lean biopharmaceutical perspective. *Pharm. Eng.*, *31*, 6, 74–82, 2011.
10. Purdy, K., Kindt, L., Densmore, J., Benson, C., Zhou, N., Leonard, J., Shanks, D., Defect reduction through Lean methodology, in: *Photomask Technology 2010*, vol. 7823, SPIE, pp. 506–514, 2010.
11. Demeter, K. and Matyusz, Z., The impact of lean practices on inventory turnover. *Int. J. Prod. Econ.*, *133*, 1, 154–163, 2011.
12. Kumar, N., Hasan, S.S., Srivastava, K., Akhtar, R., Yadav, R.K., Choubey, V.K., Lean manufacturing techniques and its implementation: A review. *Mater. Today Proc.*, *64*, 1188–1192, 2022.
13. Molnar, V., Waiting as Waste in Lean Production Processes, in: *The International Symposium for Production Research*, Springer International Publishing, Cham, pp. 429–437, 2020.
14. Nordgren, B., The problem with waiting times. *IIE Solutions*, *31*, 5, 44–48, 1999.
15. Hosseini, A., Kishawy, H.A., Hussein, H.M., Lean manufacturing, in: *Modern Manufacturing Engineering*, J.P. Davim (ed.), pp. 249–269, Springer International Publishing, Cham, Switzerland, 2015.
16. Dube, L. and Gupta, K., Lean manufacturing-based space utilization and motion waste reduction for efficiency enhancement in a machining shop: A case study. *Appl. Eng. Lett.*, *8*, 3, 121–130, 2023.
17. Lai, N.Y.G., Wong, K.H., Halim, D., Lu, J., Kang, H.S., Industry 4.0 enhanced lean manufacturing, in: *2019 8th international conference on Industrial technology and management (ICITM)*, IEEE, pp. 206–211, 2019.
18. Freitag, A.E.B. and de Oliveira, A.B., Underutilization of workers: an analysis according to the lean management philosophy in basic education. *Indep. J. Manag. Prod.*, *12*, 9, s940–s959, 2021.
19. Argiyantari, B., Simatupang, T.M., Basri, M.H., Pharmaceutical supply chain transformation through application of the lean principle: a literature review. *J. Ind. Eng. Manage. (JIEM)*, *13*, 3, 475–494, 2020.

20. Ford, R.R.L., Using value stream mapping to improve documentation processes in the biopharmacutical industry, Doctoral dissertation, Massachusetts Institute of Technology, 2010.
21. Wolf, Z.Z.A., Process improvement and inventory management using value stream mapping in a biopharmaceutical environment, Doctoral dissertation, Massachusetts Institute of Technology, 2013.
22. Sachidananda, M., Erkoyuncu, J., Steenstra, D., Michalska, S., Discrete event simulation modelling for dynamic decision making in biopharmaceutical manufacturing. *Procedia CIRP*, *49*, 39–44, 2016.
23. Sindhwani, R. and Saddikuti, V., Discrete Event Simulation for Pharmaceutical Supply Chain Analysis in India, in: *Intelligent and Transformative Production in Pandemic Times: Proceedings of the 26th International Conference on Production Research*, pp. 853–865, Springer International Publishing, Cham, February, 2023.
24. Villa, A.D., Lean transformation methodology and implementation in biopharmaceutical operations, Doctoral dissertation, Massachusetts Institute of Technology, 2008.
25. Al-Shourah, A.A., Al-Tarawneh, R.T., Alzu'bi, F.A., The integration of lean management and six sigma strategies to improve the performance of production in industrial pharmaceutical. *Int. J. Bus. Manage.*, *13*, 8, 207–216, 2018.
26. Garbers, A., Nolletti, V., Kuhl, M.E., Stanislow, K., Design and simulation of a new biomedical production process, in: *2020 Winter Simulation Conference (WSC)*, IEEE, pp. 1631–1640, 2020.
27. Vilalta, J. and Hamed, F., Challenges and Opportunities for Lean Manufacturing in Biotechnology. Retrieved March, 17, 2008, 2007 Sep.
28. Ismail, A., Ghani, J.A., Ab Rahman, M.N., Md Deros, B., Che Haron, C.H., Application of lean six sigma tools for cycle time reduction in manufacturing: case study in biopharmaceutical industry. *Arabian J. Sci. Eng.*, *39*, 1449–1463, 2014.
29. Nepal, B., Natarajarathinam, M., Balla, K., Improving manufacturing process for biomedical products: a case study. *J. Manuf. Technol. Manage.*, *22*, 4, 527–540, 2011.
30. Nargund, S., Mauch, K., Esswein, S., Niebel, B., Rafigh, M., Pendse, P., Biopharma 4.0: digital technologies driving smart manufacturing, in: *Biopharmaceutical Manufacturing: Progress, Trends and Challenges*, pp. 319–335, Springer International Publishing, Cham, 2024.
31. Lakshmi, B. and Patel, S., Digital Marketing in Pharmaceutical Industry–An Overview and Assessment. *Int. J. Pharm. Sci. Innovation*, *1*, 1, 2020.
32. Awad, A., Trenfield, S.J., Gaisford, S., Basit, A.W., 3D printed medicines: A new branch of digital healthcare. *Int. J. Pharm.*, *548*, 1, 586–596, 2018.
33. Awad, A., Trenfield, S.J., Goyanes, A., Gaisford, S., Basit, A.W., Reshaping drug development using 3D printing. *Drug Discovery Today*, *23*, 8, 1547–1555, 2018.

34. Parida, V., Sjödin, D., Reim, W., Reviewing literature on digitalization, business model innovation, and sustainable industry: Past achievements and future promises. *Sustainability*, *11*, 2, 391, 2019.
35. Lexchin, J., Bero, L.A., Djulbegovic, B., Clark, O., Pharmaceutical industry sponsorship and research outcome and quality: systematic review. *Bmj*, *326*, 7400, 1167–1170, 2003.
36. Alloghani, M., Al-Jumeily, D., Hussain, A., Aljaaf, A.J., Mustafina, J., Petrov, E., Healthcare services innovations based on the state of the art technology trend industry 4.0, in: *2018 11th International conference on developments in eSystems engineering (DeSE)*, IEEE, pp. 64–70, 2018.
37. Gbadegeshin, S.A., The effect of digitalization on the commercialization process of high-Technology companies in the life sciences industry. *Technol. Innovation Manage. Rev.*, *9*, 1, 2019.
38. Arden, N.S., Fisher, A.C., Tyner, K., Yu, L.X., Lee, S.L., Kopcha, M., Industry 4.0 for pharmaceutical manufacturing: Preparing for the smart factories of the future. *Int. J. Pharm.*, *602*, 120554, 2021.
39. Hurter, P., Thomas, H., Nadig, D., Emiabata-Smith, D., Paone, A., Implementing continuous manufacturing to streamline and accelerate drug development. *AAPS Newsmagazine*, *16*, 15–19, 2013.
40. Bennett, N. and Lemoine, J., What VUCA really means for you. *Harv. Bus. Rev.*, *92*, 1/2, 2014.
41. Jamróz, W., Szafraniec, J., Kurek, M., Jachowicz, R., 3D printing in pharmaceutical and medical applications–recent achievements and challenges. *Pharm. Res.*, *35*, 1–22, 2018.
42. Gennari, F., Internet of things: legal liability of IoE devices in the home, 2023.
43. Borgosz, L. and Dikicioglu, D., Industrial internet of things: What does it mean for the bioprocess industries? *Biochem. Eng. J.*, *201*, 109122, 2024.
44. Flores-Cerrillo, J., He, Q.P., Wang, J., Yu, H., Smart manufacturing: advances and applications of artificial intelligence, machine learning and industrial internet of things in the chemical and biochemical industry. *Front. Chem. Eng.*, *5*, 1309165, 2023.
45. Babich, V. and Hilary, G., OM Forum—Distributed ledgers and operations: What operations management researchers should know about blockchain technology. *Manuf. Serv. Oper. Manage.*, *22*, 2, 223–240, 2020.
46. Xie, W., Wang, B., Ye, Z., Wu, W., You, J., Zhou, Q., Simulation-based blockchain design to secure biopharmaceutical supply chain, in: *2019 Winter Simulation Conference (WSC)*, IEEE, pp. 797–808, 2019.
47. Smith, A. and Severn, M., An overview of continuous learning artificial intelligence-enabled medical devices. *Can. J. Health Technol.*, *2*, 5, 2022.
48. Rai, R., Tiwari, M.K., Ivanov, D., Dolgui, A., Machine learning in manufacturing and industry 4.0 applications. *Int. J. Prod. Res.*, *59*, 16, 4773–4778, 2021.
49. Halagali, P., Nayak, D., Seenivasan, R., Manikkath, J., Rathnanand, M., Tippavajhala, V.K., Artificial intelligence revolution in pharmaceutical

sciences: Advancements, clinical impacts, and applications. *Curr. Pharm. Biotechnol.*, *23*, 2025.

50. Rajesh, M.V. and Elumalai, K., The transformative power of artificial intelligence in pharmaceutical manufacturing: Enhancing efficiency, product quality, and safety. *J. Holist. Integr. Pharm.*, *6*, 2, 125–135, 2025.
51. Glassey, J., Gernaey, K., Clemens, C., Schulz, T.W., Oliveira, R., Striedner, G., Mandenius, C.F., Process analytical technology (PAT) for biopharmaceuticals. *Biotechnol. J.*, *6*, 4, 369–377, 2011.
52. Rathore, A.S., Bhambure, R., Ghare, V., Process analytical technology (PAT) for biopharmaceutical products. *Anal. Bioanal. Chem.*, *398*, 137–154, 2010.
53. Chamberlin, J., Zhong, Y., Wang, Y., Robots for Pharmaceutical Production: A Benchtop Robotic Automation Approach for Manufacturing Prefilled Syringes. *IFAC-PapersOnLine.*, *55*, 37, 469–474, 2022.
54. Kong, F., Yuan, L., Zheng, Y.F., Chen, W., Automatic liquid handling for life science: a critical review of the current state of the art. *J. Lab. Autom.*, *17*, 3, 169–185, 2012.
55. Udugama, I.A., Lopez, P.C., Gargalo, C.L., Li, X., Bayer, C., Gernaey, K.V., Digital Twin in biomanufacturing: Challenges and opportunities towards its implementation. *Syst. Microbiol. Biomanuf.*, *1*, 257–274, 2021.
56. Zobel-Roos, S., Schmidt, A., Uhlenbrock, L., Ditz, R., Köster, D., Strube, J., Digital twins in biomanufacturing, in: *Digital Twins: Tools and Concepts for Smart Biomanufacturing*, C. Herwig, R. Pörtner, J. Möller (eds.), pp. 181–262, Springer International Publishing, Cham, Switzerland, 2021.
57. Wainaina, S. and Taherzadeh, M.J., Automation and artificial intelligence in filamentous fungi-based bioprocesses: A review. *Bioresour. Technol.*, *369*, 128421, 2023.
58. Silva, T.C., Eppink, M., Ottens, M., Automation and miniaturization: enabling tools for fast, high-throughput process development in integrated continuous biomanufacturing. *J. Chem. Technol. Biotechnol.*, *97*, 9, 2365–2375, 2022.
59. Schmidt, A., Hengelbrock, A., Strube, J., Continuous biomanufacturing in upstream and downstream processing. *Phys. Sci. Rev.*, *9*, 10, 3167–3222, 2024.
60. Valamede, L.S. and Akkari, A.C.S., Lean 4.0: A new holistic approach for the integration of lean manufacturing tools and digital technologies. *Int. J. Math. Eng. Manage. Sci.*, *5*, 5, 851, 2020.
61. Cifone, F.D., Hoberg, K., Holweg, M., Staudacher, A.P., 'Lean 4.0': how can digital technologies support lean practices? *Int. J. Prod. Econ.*, *241*, 108258, 2021.
62. Johansson, P.E., Bruch, J., Chirumalla, K., Osterman, C., Stålberg, L., Integrating advanced digital technologies in existing lean-based production systems: analysis of paradoxes, imbalances and management strategies. *Int. J. Oper. Prod. Manage.*, *44*, 6, 1158–1191, 2024.

63. Valamede, L.S. and Akkari, A.C.S., Lean 4.0: digital technologies as strategies to reduce waste of lean manufacturing, in: *Brazilian Technology Symposium*, Springer International Publishing, Cham, pp. 74–83, 2021.
64. Chukwumuanya, O., Udu, E., Okpala, C., Lean principles integration with digital technologies: a synergistic approach to modern manufacturing. *Int. J. Ind. Prod. Eng.*, Taylor & Francis, London, UK, 3, 2, 59–63, 2025.
65. Pontes, J., Digital technologies and lean 4.0: integration, benefits, and areas of research. *Ind. Eng. Oper. Manage.*, *431*, 197, 2023.
66. Pirrung, S.M., van der Wielen, L.A., Van Beckhoven, R.F., van de Sandt, E.J., Eppink, M.H., Ottens, M., Optimization of biopharmaceutical downstream processes supported by mechanistic models and artificial neural networks. *Biotechnol. Progr.*, *33*, 3, 696–707, 2017.
67. Rathore, A.S., Nikita, S., Thakur, G., Mishra, S., Artificial intelligence and machine learning applications in biopharmaceutical manufacturing. *Trends Biotechnol.*, *41*, 4, 497–510, 2023.
68. Astarita, V., Giofrè, V.P., Mirabelli, G., Solina, V., A review of blockchain-based systems in transportation. *Information*, *11*, 1, 21, 2019.
69. Pandey, M. and Singhal, B., Blockchain technology in biomanufacturing: Current perspective and future challenges, in: *Blockchain Technology for Emerging Applications*, pp. 207–237, Academic Press, Boston, MA, USA, 2022.
70. Ogunsanya, M. and Desai, S., Physics-based and data-driven modeling for biomanufacturing 4.0. *Manuf. Lett.*, *36*, 91–95, 2023.
71. Verma, A.K., Singh, T., Kathpalia, K.V., Iliya, A., Shankhwar, N., Data-Driven Industrial Scale-Up: Leveraging Bioinformatics for Biomanufacturing, 2025.
72. Mazurenko, S., Prokop, Z., Damborsky, J., Machine learning in enzyme engineering. *ACS Catal.*, *10*, 2, 1210–1223, 2019.
73. Ferguson, A.L. and Ranganathan, R., 100th anniversary of macromolecular science viewpoint: data-driven protein design. *ACS Macro Lett.*, *10*, 3, 327–340, 2021.
74. Kuhlman, B. and Bradley, P., Advances in protein structure prediction and design. *Nat. Rev. Mol. Cell Biol.*, *20*, 11, 681–697, 2019.
75. Gao, W., Mahajan, S.P., Sulam, J., Gray, J.J., Deep learning in protein structural modeling and design. *Patterns*, *1*, 9, 2020.
76. Shi, Z., Liu, P., Liao, X., Mao, Z., Zhang, J., Wang, Q., Ma, Y., Data-driven synthetic cell factories development for industrial biomanufacturing. *Biodes. Res.*, 9898461, 2022.
77. Lawson, C.E., Martí, J.M., Radivojevic, T., Jonnalagadda, S.V.R., Gentz, R., Hillson, N.J., Martin, H.G., Machine learning for metabolic engineering: A review. *Metab. Eng.*, *63*, 34–60, 2021.
78. Mellor, J., Grigoras, I., Carbonell, P., Faulon, J.L., Semisupervised Gaussian process for automated enzyme search. *ACS Synth. Biol.*, *5*, 6, 518–528, 2016.

13

Additive Manufacturing for Biomedical Applications

Geetika Kumari Salwan[1]* and Sayon Dey[2]

[1]Dept. of Mechanical Engineering, Dr Sudhir Chandra Sur Institute of Technology and Sports Complex, Kolkata, West Bengal, India

[2]Dept. of Mechanical Engineering, Brainware University, Kolkata, West Bengal, India

Abstract

Advancements in additive manufacturing, mostly called 3D printing, are transforming the biomedical sector by facilitating the production of complex, patient-specific tissues and equipment. Surgical planning, medical education, implant design and manufacturing, and the establishment of customized drug delivery systems are just a few of the medical domains where additive manufacturing technologies find its application. With an emphasis on their incorporation into biomedical applications, this chapter examines the fundamental ideas and categories of additive manufacturing technologies use in this field. It goes over important materials like metals, ceramics, biocompatible polymers, and bioinks and looks at how they are used to create implants, prosthetics, dental equipment, and bioprinted tissues. The chapter also discusses contemporary difficulties, such as material constraints, regulatory barriers, and problems with biological integration. It concludes by outlining the future directions of the process and highlighting its revolutionary potential in regenerative healthcare and personalized medicine.

Keywords: Tissue engineering, biomedical, biomaterials, additive manufacturing

13.1 Introduction

3D objects are built by adding materials layer by layer through the additive manufacturing (AM) process. Fabricated components have adequate

**Corresponding author*: geetikasalwan@gmail.com

Sandip Kunar, Jagadeesha T. and D. Sathis Kumar (eds.) Biomanufacturing: Transforming Production of Biologics and Biopharmaceuticals, (381–392) © 2026 Scrivener Publishing LLC

mechanical strength with minimum material wastage and time saving. The biomedical industry is one that is particularly impacted by this technology, which has transformed many other industries. Unlike subtractive manufacturing procedures, AM builds products layer by layer from digital models, allowing for previously unheard-of degrees of customization, intricacy, and usefulness. These capabilities allow for the integration of biological materials, in the development of intricate interior geometries, and patient-specific solutions in the biomedical field [1]. A thorough review of the most recent theories, practices, and real-world uses of additive manufacturing in biomedical engineering is provided in this chapter. It draws attention to the materials, regulatory obstacles, technological developments, and potential applications of AM in medicine.

Because of increased productivity, usefulness, patient-specific fabrication, and affordability, AM techniques are being used more and more in a variety of industries, including the medical industry [2, 3]. AM is regarded as a digital manufacturing method that is quickly changing the health industry by printing distinct bodily components with intricate shapes and offering patients individualized, customized resolutions. In the past, this manufacturing method was a flexible and effective technique to produce medical organs, bones, and dental implants with intricate geometric designs [4]. But in the process of creating patient-specific implants for bone disease, trauma, and organ replacement, medical professionals and researchers still encounter many challenges. Despite the extensive research that has been done on the properties of AM materials, a thorough grasp of the requirements, procedures, difficulties, and the factors unique to each application is still required. Therefore, this study's goal is to provide a thorough overview of the most common AM procedures, material types used in AM, and biological applications of different AM techniques. The present barriers and difficulties that keep the medical sciences from completely utilizing the advanced AM prospects are also discussed in this study. These consist of post-processing, standards compliance, product quality, range of materials, maintenance, and manufacturing volume.

In the healthcare industry, AM has drawn a lot of interest for creating tools to maintain, restore, and fix damaged human body components. To meet the functional and cosmetic needs of the modern oral healthcare industry, the area of dentistry in particular, deals with this type of restorative work quite a bit [5, 6]. Based on recent published research pertinent to the topic, this chapter will provide a detailed update on the concepts, techniques, and various applications that are used in the oral health care sector. Additionally, it will outline future research directions to tackle the obstacles to the broad clinical application of AM technology in the standard clinical dental treatment.

13.2 Fundamentals of Additive Manufacturing

13.2.1 Definition and Principles

AM is an industrialized technology of 3D products that are made by using incremental material additions to turn digital designs into tangible items. Reduced material waste, more design flexibility, quick prototyping, and scalability for intricate and customized parts are some of the main benefits. Since 1980s various AM techniques have been developed for making various materials. This technique has gain popularity in biomedical field because of its ability to produce made-to-order and intricate geometries. AM helps in surgical planning and medical training. Two-dimensional imaging techniques, including computerized topography, magnetic resonance imaging, and ultrasound, are implemented to clearly distinguish between malignant and benign tissues or organs [7]. Using AM technology in combination with these images helps to generate 3D printed models of anatomical features unique to each patient, with tissues or organs presented accurately. This helps to reduce operation time and avoids disproportionate planning.

Manufacturing complex prosthetics and implants, which are commonly used to replace tissue that has been damaged or diseased, such as filler material, vascular grafts, regenerative cartilage scaffolds, or bone grafts, as well as to make changes during plastic and reconstructive procedures. The benefits of using 3D printing techniques in implant production include reduced costs, quick prototyping tailored to each patient, precise reconstruction of the desired structure, and the incorporation of multiple material compositions. Some benefits of 3D printing in drug delivery systems include better combinations of different drugs and substrates, product consistency through standardization and mechanization, and the capability to tailor the drug delivery system to the patient's condition (such as multiple chronic conditions).

13.2.2 Classification of Additive Manufacturing Technologies

The most widely used AM technologies in biomedical applications are given in Figure 13.1. Each technique offers unique advantages in terms of resolution, material compatibility, and suitability for biological integration.

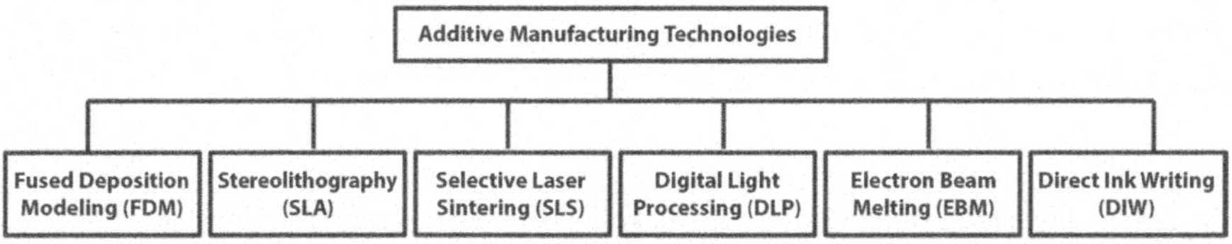

Figure 13.1 Additive manufacturing technologies used in biomedical fields.

It is a new approach that has completely changed mechanical engineering and material science in the last years. The technology began to assist engineers to realize their ideas by layering together different three-dimensional objects. Additive manufacturing encompasses an extensive array of processes, with stereolithography (STL), polyjet, fused deposition modeling (FDM), laminated object manufacturing (LOM), 3D printing (3DP), pro-metal, selective laser sintering (SLS), laser engineered net shaping (LENS), and electron beam melting (EBM). Layer by layer, fused deposition modeling (FDM) creates items using a thermoplastic filament. The filament is melted, extruded through a nozzle, and then placed onto a build platform to form a layer. Subsequently, the print head moves around the built platform, layering the final product until it is ready. This method is frequently used for rapid production, prototyping, and the creation of functional parts across a variety of sectors, including education, automotive, and aerospace. Selective laser sintering employs a laser to fuse powdered materials, whereas stereolithography (SLA), a vat photo polymerization technique, uses a laser to cure a photopolymer (liquid resin) reservoir layer by layer to build 3D objects. In Digital Light Processing (DLP), liquid resin is cured layer by layer with a projector to create three-dimensional sculptures. This type of vat photo polymerization creates objects by selectively curing a liquid resin using light. Its various applications are dental implants, jewelry, micro fluid devices, and prototyping. Electron beam melting is the technique of melting metal powder layer by layer with a high-energy electron beam. The process is carried out in a vacuum environment, which inhibits oxidation and enables the use of dense, high-strength components, resulting in strong mechanical properties. An extrusion-based 3D printing technique, Direct Ink Writing (DIW), applies liquid or semi-solid inks at controlled pressures and flow rates using nozzles. Because it works with a variety of materials and structures, including hydro gels and composites, it is helpful in biomedical applications, electronics, and other domains. Its simplicity, affordability, and ability to combine many materials in a single process are its advantages [8–10].

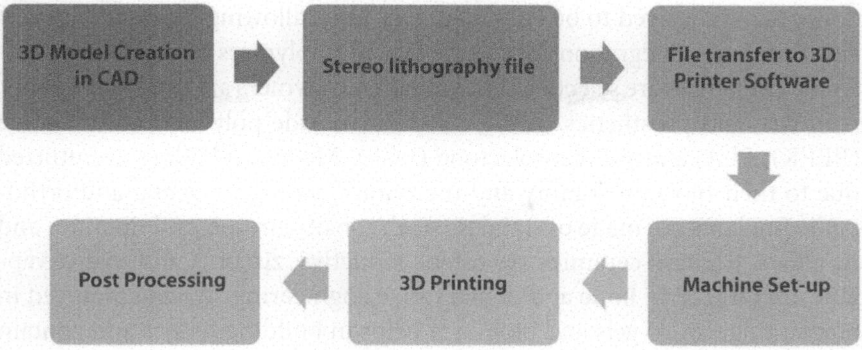

Figure 13.2 Additive manufacturing process.

Steps used for making biomaterial by additive manufacturing are given in Figure 13.2. Customized, designed, and manufactured medical devices can be made for each individual customer. AM has enabled the production of optimal, precisely fitting implants and medical parts at a lower cost and time, while maintaining a better level of accuracy and quality. The first step in this procedure is diagnosis, which is followed by the collection of 2D data using precision imaging techniques like CT or MRI scans. A stacking program such as Mimics then uses these 2D data or photographs to create a 3D virtual computer aided design (CAD) model by stacking the 2D images and connecting and interpolating the contours in each image. After that, the Standard Tessellation Language, file format is applied to this CAD model. The STL file is then divided into several stacked layers based on the thickness of each layer. The AM machine receives this sliced CAD data as input. The implants or medical components are subsequently printed layer by layer.

13.3 Materials Use in Biomedical AM

In Biomedical AM variety of materials are used, metals like stainless steel and titanium alloy, polymers like polylactic acid (PLA) and acrylonitrile butadiene styrene (ABS), ceramics like alumina and zirconia and bio-inks containing cells and tissue components. Depending on the specific application, biocompatibility and mechanical properties, materials are selected. Ti6Al4V, 316L and tantalum are some metals and alloys, used for orthopedic and dental implants, bone plates and prosthetics. This is because of its high strength, corrosion resistance and good osseointegration.

Some materials need to be degraded over time, allowing for bone regeneration or tissue integration; hence, metals and polymers having biodegradability properties are selected. Biocompatible polymers are used in implants, scaffolds, and prosthetics. These materials include polyether ether ketone (PEEK), PLA, and polycaprolactone (PCL). Metals and alloys are utilized due to their biocompatibility and mechanical strength. Dental and orthopedic implants are made of stainless steel, cobalt-chrome, and titanium and its alloys. Because ceramics are osteoconductive, zirconia and hydroxyapatite are utilized in bone and dental tissue engineering. When employed in bioprinting, hydro gels and bioinks it helps in building tissues and contain cells. Hydrogels have high water content, cell-friendly environment and tunable properties. Common bioinks include alginate, collagen, and gelatin.

13.4 Applications of AM in Biomedical Engineering

Layer-by-layer manufacturing approach of AM allows precise control over geometry, structure and material properties, making it ideal for application in biomedical field. The variety of biomedical applications of AM is given in Figure 13.3. The major application includes biomedical models, surgical tools, and implants, bio-printing and external medicines. Production of patient-specific implants tailored to individual anatomical data is possible by using these techniques. The fundamentals of each form of AM technology, including its key uses, materials, benefits and drawbacks, and emerging trends. Along with novel materials, quality control, post-processing advances, standardization, legislation, the circular economy, sustainability, and other challenges, significant topics including 3D bioprinting, 3D food printing, large-scale 3D printing, 4D printing, and AI-based AM technologies are studied [11]. Application of AM in biomedical engineering, leads to the drastic improvement in the medical field, as shown in Figure 13.4. Rapid fabrication of surgical instruments and cutting guides enhances accuracy during complex operation procedures.

In orthopedic implants, traditional prosthetics are expensive, lengthy and simple geometries. AM makes it possible to design porous implants that improve osseointegration for individual patients. Since the early 90s, dentists have been using AM technologies. Mostly used for dental implants, such as full teeth. Most of the additive manufacturing dental products are still in the testing and development stages. High precision and fit can be achieved in the fabrication of customized crowns, bridges, and aligners using Ti6Al4V alloys. The fabrication is done through direct metal sintering using an ytterbium fiber laser system. Tixos Company is specialized for

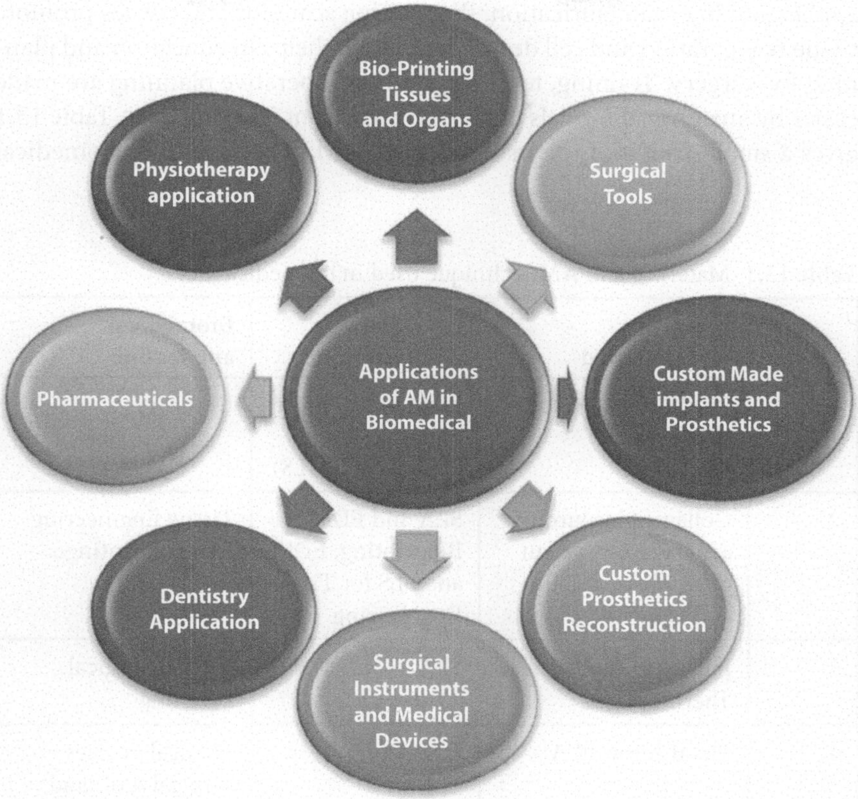

Figure 13.3 Application of AM in the biomedical field.

Improvement In Biomedical Field by The Application of AM					
Produce Light Weight Medical parts	Produce Complex Geometries	Improve Mechanical Properties of Implants	Time and Material Saving	Pre-surgical Planning	Medical Models for Education, Training and Patient Education

Figure 13.4 Improvement in the biomedical field with the application of AM.

this process. Custom designs prosthetics and implants suited for individual patients, both in terms of function and form. Customized orthopedic devices and prosthetic limbs by 3D printing enhance comfort and functionality. Regenerative medicine and tissue engineering have prospective application in organ fabrication. Bioprinting scaffolds and tissues promote tissue regeneration and cell development. AM helps in education and planning for surgery. Training, teaching, and pre-operative planning are made easier by anatomical models created from patient imaging data. Table 13.1 gives a small insight of the material used, AM process and its biomedical

Table 13.1 Materials and AM technique used in biomedical field.

Sl. no.	Material used	AM technique	Biomedical application
1	PEEK, noble metal alloys	FDM,SLA, Direct Metal Laser Sintering (DMLS)	Dentistry
2	Collagen, hyaluronic acid, agarose, gellan gum, fibrin, gelatin, and alginate	SLA and FDM for Bioprinting, FDM and SLS for Tissue Engineering	Tissue Engineering and Bioprinting
3	Polymer, Rubber, Thermoplastic	SLA	Virtual Surgical Planning
4	Metal Alloy, PLA	DMLS	Medical devices, surgical tools and instruments
5	Antibiotic infused PLA, UV photopolymer resin for micro-needles, API's medicines	FDM,SLA	Design and construction of a medication distribution system
6	Cobalt-Chromium alloy, Titanium alloy, Thermoplastics	DMLS, SLA	Orthopedics and Prosthetics
7	Thermoplastics, polymers, Resin	FDM, SLA	Practice of medicine, research, education, and training objectives

applications [12, 13, 15]. Thus, the benefits of AM in the medical field include:

(i) Personalization/customization
(ii) Increased cost efficiency
(iii) Rapid product development
(iv) Democratization of design process
(v) Micro patterning and organization
(vi) Biomimetic and biological recapitulation

13.5 Challenges and Limitations

(i) There are concerns about Regulation and Ethics. Navigating complicated regulatory environments is essential to make sure the safety, effectiveness, and superiority of AM medical products.
(ii) Limitations in technology are the availability of materials, resolution, mechanical characteristics, and reproducibility. Sterilization, biocompatibility, biodegradation, cost and compatibility with printing process are major limitations.
(iii) Integration of Biology is long-term functioning and vascularization in bio printed tissues are still major challenges. Printing processes effects on cell viability.

13.6 Future Perspectives

The development of completely functional organs, the integration of AI for design optimization, and improved multi-material printing are the goals of ongoing research. A new age of patient-specific healthcare solutions is ushered in by the merging of AM with biotechnology and personalized medicine [14, 16]. Rapid cosmetic repair, bone and organ repair will be done in the future. The future of printed tissues includes functional human organs that are patient specific such as kidneys, livers and hearts. The future of prosthetics includes integration of circuits, sensors, actuators and exotic materials. There will be integration of bioprinting and incubation systems, thus increasing the size and rate of tissues produced. Personalized pills with specific dosages and release rates are under research.

13.7 Conclusions

In biomedical engineering, AM is a paradigm shift that offers previously unheard-of levels of precision, creativity, and customization. AM is positioned to become more and more important in diagnostics, therapies, and regenerative medicine as materials and technologies develop further, ultimately changing the face of contemporary healthcare. Numerous medical improvements, including patient-specific surgical guides, customized drug delivery systems, and intricate scaffolds filled with cells that more closely mimic genuine organs and tissues, have been made possible by the adaption of contemporary 3D printing technology for biomedical applications.

There are still a lot of issues with the printing procedure and the printing materials accessible, even with 3D printing's great potential and variety in the medical industry. More flexible materials, quicker processing, and higher resolution are still needed. As a result, new additive manufacturing standards and quality control methods are needed in each of these applications. Better and more extensive usage of AM in healthcare results from rising equipment and material prices as well as falling expenses. Overall, AM is transforming biomedical engineering by offering rapid, precise, and cost-effective solutions that enhance patient outcomes, surgical precision, and medical education.

AI disclosure statement: The author acknowledges the use of AI-based tools for minor language editing or grammatical correction. However, all intellectual contributions, analyses, and conclusions presented in this chapter are entirely the author's own.

References

1. Bhushan, J. and Grover, V., Additive manufacturing: Current concepts, methods, and applications in oral health care, in: *Biomanufacturing*, pp. 103–122, Springer International Publishing, Cham, Switzerland, 2019.
2. Singh, S. and Ramakrishna, S., Biomedical applications of additive manufacturing: Present and future. *Curr. Opin. Biomed. Eng.*, 2, 105–115, 2017.
3. Ahangar, P., Cooke, M.E., Weber, M.H., Rosenzweig, D.H., Current biomedical applications of 3D printing and additive manufacturing. *Appl. Sci.*, 9, 8, 1713, 2019.
4. Liu, Y., Wang, W., Zhang, L.C., Additive manufacturing techniques and their biomedical applications. *Fam. Med. Community Health*, 5, 4, 2017.

5. Dhavalikar, P., Lan, Z., Kar, R., Salhadar, K., Gaharwar, A.K., Cosgriff-Hernandez, E., Biomedical applications of additive manufacturing, in: *Biomaterials Science*, pp. 623–639, Academic Press (Elsevier), Cambridge, MA, USA, 2020.
6. Wazeer, A., Das, A., Sinha, A., Inaba, K., Ziyi, S., Karmakar, A., Additive manufacturing in biomedical field: a critical review on fabrication method, materials used, applications, challenges, and future prospects. *Prog. Addit. Manuf.*, 8, 5, 857–889, 2023.
7. Sheoran, A.J., Chandra, A., Kumar, H., Biomedical applications of additive manufacturing, in: *Recent Advances in Mechanical Engineering: Select Proceedings of ITME 2019*, Springer Singapore, Singapore, pp. 553–566, 2020.
8. Chua, K., Khan, I., Malhotra, R., Zhu, D., Additive manufacturing and 3D printing of metallic biomaterials. *Eng. Regener.*, 2, 288–299, 2021.
9. Maniruzzaman, M. (Ed.), *3D and 4D printing in biomedical applications: process engineering and additive manufacturing*, John Wiley & Sons, Hoboken, NJ, USA, 2018.
10. Salmi, M., Additive manufacturing processes in medical applications. *Materials*, 14, 1, 191, 2021.
11. Zhou, L., Miller, J., Vezza, J., Mayster, M., Raffay, M., Justice, Q., Tamimi, Z.A., Hansotte, G., Sunkara, L.D., Bernat, J., Additive Manufacturing: A Comprehensive review. *Sensors*, 24, 9, 2668, 2024.
12. Murr, L.E., Additive manufacturing of biomedical devices: an overview. *Mater. Technol.*, 33, 1, 57–70, 2018.
13. Harding, A., Pramanik, A., Basak, A., Prakash, C., Shankar, S., Application of additive manufacturing in the biomedical field- A review. *Ann. 3D Print. Med.*, 10, 100110, 2023.
14. Bozkurt, Y. and Karayel, E., 3D printing technology; methods, biomedical applications, future opportunities and trends. *J. Mater. Res. Technol.*, 14, 1430–1450, 2021.
15. Kumar, R., Kumar, M., Chohan, J.S., The role of additive manufacturing for biomedical applications: A critical review. *J. Manuf. Processes*, 64, 828–850, 2021.
16. Bedir, T., Kadian, S., Shukla, S., Gunduz, O., Narayan, R., Additive manufacturing of microneedles for sensing and drug delivery. *Expert Opin. Drug Delivery*, 21, 7, 1053–1068, 2024.

14

Artificial-Intelligence-Based Lean Biomanufacturing: Analytical Measures to Eliminate Wastes and Maximize Process Efficiency

Rompicherla Srividya[1], K.S.N.V. Prasad[1], B. Karuna[1], Bijoy Kumar Purohit[1], Vijaya Kumar Talari[1], Appala Naidu Uttaravalli[2] and A.V. Raghavendra Rao[1]*

[1]*Department of Chemical Engineering, B. V. Raju Institute of Technology, Narsapur, Telangana, India*

[2]*Optimus Drugs Private Limited, Hyderabad, Telangana, India*

Abstract

To achieve lean biomanufacturing, which leads to waste reduction, efficiency improvement, and production of high-quality biopharmaceuticals in large quantities, it is necessary to adopt the philosophy of lean. The advent of Industry 4.0 will bring a new dominant paradigm of integrating artificial intelligence (AI) into lean bioprocessing that brings about the power to make real-time decisions while continuously improving the processes. The chapter explores the improvements in lean manufacturing processes with AI applications in both the upstream and downstream, including inventory of raw materials, bottle release, and so forth. It describes a combination of the application of machine learning, neural networks, and predictive analytics that can be used to tune process parameters, energy consumption rates, downtimes, and yields. Some of the critical instruments, which are digital twins, anomaly detection systems, and AI-based scheduling, are explained. Data integrity, model validation, and compliances with regulations in GMP settings are also the issues considered in the chapter. Real-life examples of the biopharma manufacturing lines include how AI-enabled lean transformation has improved objectively measurable productivity and sustainability. After the discussion of the key points that need to focus on the integration of the AI into the lean, agile, and scalable biomanufacturing systems, this chapter leaves us with the

**Corresponding author*: raghusrf@gmail.com

Sandip Kunar, Jagadeesha T. and D. Sathis Kumar (eds.) Biomanufacturing: Transforming Production of Biologics and Biopharmaceuticals, (393–432) © 2026 Scrivener Publishing LLC

roadmap towards adopting the AI as a strategic enabler of the lean, agile, and scalable biomanufacturing systems.

Keywords: Artificial intelligence, lean biomanufacturing, process optimization, smart manufacturing, machine learning, waste minimization

14.1 Introduction to Lean Biomanufacturing and Industry 4.0

Lean biomanufacturing aims at maximizing productivity and reduction of waste by streamlining the processes, decreasing non-value-adding activities and perpetual enhancements of operations through tenets like value stream mapping, pull systems and just-in-time manufacturing [1–5]. The practices have the intent of removing inefficiencies as well as maintaining consistent quality and regulation compliance as key issues of biopharmaceutical manufacturing [3]. Concurrently, Industry 4.0 is the inclusion of modern digital technologies such as the Industrial Internet of Things (IIoT), artificial intelligence (AI), robotics, and cloud computing, in manufacturing settings. The fourth industrial revolution makes it possible to develop intelligent, connected production systems that grow more autonomous. The meeting point between lean thinking and industry 4.0 technologies provides biomanufacturers with the possibility of greatly increasing the process efficiency, product quality and flexibility due to the ease of collecting real-time data, automation, and intelligent decision-making [6].

14.1.1 Overview of Lean Principles in Biomanufacturing

Lean manufacturing is an approach centered on providing customers with the utmost value and elimination of waste material. It aims to simplify its activities to shed off non-value adding activities, thus operating with the minimum input of resources that will facilitate required demand [7]. Based on Toyota Production System (TPS), lean concepts have proved to have spread beyond the automobile manufacturing sectors and are commonly used in biomanufacturing sectors where they aid in meeting the demanding needs of biologics manufacturing considering the challenging regulations [8]. Notably, lean is not synonymous with cost-cutting or reduction of the workforce; on the contrary, it enhances more intelligent working practices that actually provide excellent results in less time, and in a predictable manner.

One key lean principle is to find and to remove eight wastes, namely: defects, where products go back to rework or become scrap; overproduction, where products are produced before demand; waiting, due to delays or bottlenecks in the process; under-utilization of talent because a skill or talent is not fully used; futile transportation of material, material is transported further than necessary; too much of stock of raw materials or finished goods; unnecessary motion, which is inefficient movement of a person; and over-processing, where unnecessary work or resources are added to something that is already in the process. All these wastes may occur during biomanufacturing and be managed with care to streamline and make the operations cost effective [9].

Table 14.1 Lean principles in biomanufacturing.

Lean principle	Description	Application in biomanufacturing
Value Stream Mapping (VSM)	A tool for visualizing and analyzing the flow of materials and information.	Identifies inefficiencies in production, helping optimize flow in biomanufacturing.
Kanban Systems	Pull-based production system to manage inventory and reduce overproduction.	Ensures that raw materials are available when needed, reducing excess inventory.
Kaizen	A philosophy of continuous, incremental improvements.	Encourages small, continuous improvements to processes, enhancing efficiency over time.
5S Methodology	Systematic approach to workplace organization: sort, set in order, shine, standardize, sustain.	Improves workplace safety and efficiency in biomanufacturing facilities.
Standard Work	Documentation of best practices to ensure consistency.	Ensures consistency in biomanufacturing processes and reduces variation in product quality.

In dealing with such wastes, lean biomanufacturing imports some of the essential tools and practices. The 5S approach encourages organizing workplaces to enhance efficiency and safety. Value Stream Mapping (VSM) is a method which visualizes and evaluates the materials and information flow throughout the process of production to expose the inefficiencies. Standard work ensures consistent outcomes by documenting best practices. The use of Kanban systems and pull-based production allows to better control inventory and production schedule according to real demand rather than on produced volume, decreasing overproduction and storage expenses. Visual management tools imply signals or displays to discuss the current state of the process and identify issues at an early stage. Finally, Kaizen promotes a culture of constant enhancement where the employee is motivated to make small and incremental contributions that could result in cumulatively high long-term performance [7, 8].

These methods should be very finely tuned to the circumstances of biomanufacturing where the process is often batch; high-quality control is required and it is essential to have finer control over biological and chemical conditions. Table 14.1 displays the lean principles in biomanufacturing.

14.1.1.1 Benefits and Challenges in Biomanufacturing

There are a number of attractive advantages involved in introducing lean concepts into biomanufacturing. It shortens the lead times, cycle times, reduces inventory carrying cost, and enhances the overall product quality and consistency [7]. Lean processes also help in increasing the flexibility of the manufacturing process, enable an entity to switch quickly to changes in demand or product, and have a culture of participation and problem resolution among employees, a requirement in a sustainable excellence in operation [10].

However, the implementation of lean in the biomanufacturing setting is not different. Nature of biological processes is quite complex and variable, and it is not easy to predict the outcome with precision. The industry is further regulated by strict Good Manufacturing Practice (GMP) requirements that effortlessly subject the documentation of processes, validation and change control to strict standards. Technical expertise is necessary to execute an effective lean implementation process, but technical understanding is less sufficient because of the specialized skills in quality assurance, regulatory compliance, cross-functional collaboration. The other danger is that of surface action in which lean tools are embraced nominally and without the accompanying cultural transformation needed to deliver any material change. The current issues need a comprehensive solution to unify the lean concepts and the reality of bioprocessing and organizational culture [11].

14.1.2 The Impact of Industry 4.0 on Process Innovation

Industry 4.0 is a revolution in the manufacturing industry in terms of the use of cyber-physical systems, placement of internet of things, cloud computing and sophisticated data analysis within the manufacturing equipment [12]. In biomanufacturing, the point of the shift allows a transition to traditional reactive processes in ball process innovation to real-time and data-based processes. In addition to promoting greater levels of automation, these digital tools can increase the level of visibility into the results of processes in terms of their performance and quality characteristics that can be achieved with a new degree of accuracy and responsiveness than previously possible [11].

A number of important technology enablers support Industry 4.0 in a biomanufacturing context. Using IoT sensors, you can monitor in-real-time crucial process parameters like temperature, pH, dissolved oxygen through the lifecycle of the production [13]. Automation such as robotics and automated sampling minimizes manual input and the variability of such input. Through cloud computing, it is easier to have centralized data storage and access it anywhere, which enables various people to collaborate globally and conduct cross-site optimization. Big data analytics convert large amounts of bioprocess data into usable knowledge and help make better and earlier decisions. Among the most progressive are the application of the virtual images of the real processes in digital twin that allows locating simulation, optimization and predictive maintenance in a real-time manner [14, 15].

All these technologies contribute to a transition to continuous or hybrid production models where biomanufacturers can continuously monitor processes and anticipate and avoid deviations and optimize utilization of energy, materials and equipment. Consequently, the development processes of new products may be improved with the new ones becoming much shorter and the sustainability of operations can be increased in general [16].

Nonetheless, there are problems related to the implementation of Industry 4.0. It usually takes considerable investment and a redesigning of architectures to balance new digital tools with legacy manufacturing systems. As systems become more and more interconnected, it also becomes more challenging to ensure that data integrity, cybersecurity and regulatory compliance are achieved. Also, the companies are forced to invest in the development of their workforce to create a solid digital and analytical skill base on every level within the company. There is also need to deal with complexity of digitally connected systems as well as adopt new concepts of validation, troubleshooting and system maintenance [17, 18].

Nonetheless, the ability to integrate lean methodologies and the technologies of Industry 4.0 will introduce a new era of biomanufacturing: an era of agility, efficiency, and resilience. They go hand in hand to offer a very competitive backbone in innovations and operational excellence towards the more competitive and more regulated industry.

14.1.3 Drivers for Integrating AI into Lean Bioprocessing

Artificial intelligence (AI) is an important component of lean bioprocessing and is motivated by the growing imperatives to increase efficiency, quality and responsiveness in an industry where complexity is high and regulatory demands are stringent [19]. By definition, bioprocesses are subject to complex, nonlinear dynamics of interplay among biological, chemical, and physical conditions. Conventional statistical measures usually fail to reflect all the variability of this sort. In comparison, AI is good at finding hidden patterns in large datasets. It is also very useful for handling biological variation and uncertainty [20, 21].

Contemporary biomanufacturing facilities produce huge amounts of data such as sensors, analytical and control systems. Such data proliferation needs superior computing technologies, which can type, analyze, and teach in processing live information data streams. When compared to other forms of data, this data has a unique ability to deliver meaning and present opportunities to both predictive and prescriptive analytics to aid in accelerated and intelligent decision-making [22]. Both tasks are achievable by the application of AI and machine learning algorithms.

The other important driver is the increasing need of process control and early response. Any abnormalities of process performance or operation of equipment should be resolved promptly to prevent losses due to quality damage and manufacturing backlog. Techniques like AI-powered systems increase the velocity and precision of the interventions as they give early warning and propose corrective measures before failures or nonconformances take place. Also, AI is a critical component of continuous improvements efforts in line with lean philosophy, which reveals inefficient processes, and suggests specific optimizations that should otherwise be lost in oblivion [23].

The rising pressure of staying compliant with Good Manufacturing Practices (GMP) and regulatory rules further promotes the use of AI. Artificial Intelligence tools may enable documentation and traceability and automation of reporting to reduce the risk of human error and make audits easier. Taken together, these drivers underline how AI will be a strategic enabler to meet lean biomanufacturing objective in a digitally linked world [19, 21].

14.2 AI Technologies Transforming Lean Bioprocesses

Artificial intelligence (AI) technologies are transforming lean bioprocessing to deliver process efficiency, accuracy, and flexibility throughout the production lifecycles. Machine learning algorithms are especially effective in terms of processing vast amounts of data obtained through bioreactors and production lines and identify minor trends and predict the occurrence of certain deviations on the spot. Such real-time insights enable the operators to make timely interventions or changes in crucial variables of the process, like temperature, PH, and nutrients in order to achieve maximum yield and consistency and minimize wastage [24, 25].

In upstream processing such as fermentation and cell culture, artificial intelligence can speed up the development process through more effectively seeking the preferred conditions than through conventional trial and error processes. The pearled profile can be determined using machine learning tools that can prescribe real-time corrections to take place in case of any deviation [25]. With AI-based machine vision and deep learning techniques, biopharmaceutical products are inspected automatically, allowing real-time release testing and minimizing the requirement of time-consuming laboratory testing [26].

Another AI powered application is predictive maintenance which is used to identify cases of equipment degradation or imminent failure in time so that the response can be planned before it becomes disruptive. AI also helps to have more stable supply chains through anticipating demand, inventory optimization, and loss of delays or shortages [27]. The use of digital twins (copies of physical bioprocesses virtual representations) additionally allows manufacturers to virtually simulate and optimize the production processes prior to implementation thus lowering the risk of any failures and rework needed, which is more expensive [28].

14.2.1 Machine Learning, Neural Networks, and Predictive Analytics

Machine learning (ML) and AI represent a revolution of biomanufacturing process through data-driven process development to real-time monitoring and predictive process control. Bioprocesses are frequently nonlinear and/or multivariate, delicate to minor variation, and thus the right prospect for machine learning applications [29]. ML algorithms process historical and real-time process data to understand patterns and predict outcomes and provide recommendations to continuously improve and make operations agile.

During the development phase, ML can be employed to determine critical processing parameters (CPPs) and the influence they have on critical quality attributes (CQAs) and representatives of Quality by Design (QbD) initiatives. Through the interpretation of historical data and experimental observations, such models minimize the use of both expensive and time-consuming empirical approaches, shortening development schedules and enhancing their scalability in an ultimate way [30].

When operational, ML algorithms will optimize the conditions of the process, e.g., the feed rates and pH, temperatures to maximize yield and reduce energy and material as implemented to an optimal level. An example of such use is seen in fermentation where small changes in process inputs can result in a drastic change in productivity; ML could give tighter control over performance [30]. Moreover, sensor-integrated ML models provide the possibility to monitor the health of a bioprocess in real-time. They allow predicting the contamination risk, equipment failures, or yield losses even before they increase resulting in operators being able to intervene [29].

Neural networks, and specifically deep learning structures, form one of the more powerful sub-sets of ML that attempts to capture and model complex, non-linear relationships. They are particularly useful where the data environment is high-dimensional, in which bioprocesses tend to be, and where standard statistical tools cannot be applied. Practically, neural networks are applied to identify the patterns in processes, analyze variability between batches, and obtain continuous good quality. They also find use in predictive maintenance where they are used to measure the performance of the equipment and can be used to determine a failure impending phenomenon to enable condition-based maintenance of the plant [31].

Predictive analytics combines the results of both ML and neural network models to predict the future behavior of the process and suggest, which actions are optimal control actions. These tools enable operators to predict the outcome of results such as product yield or purity beforehand, modify

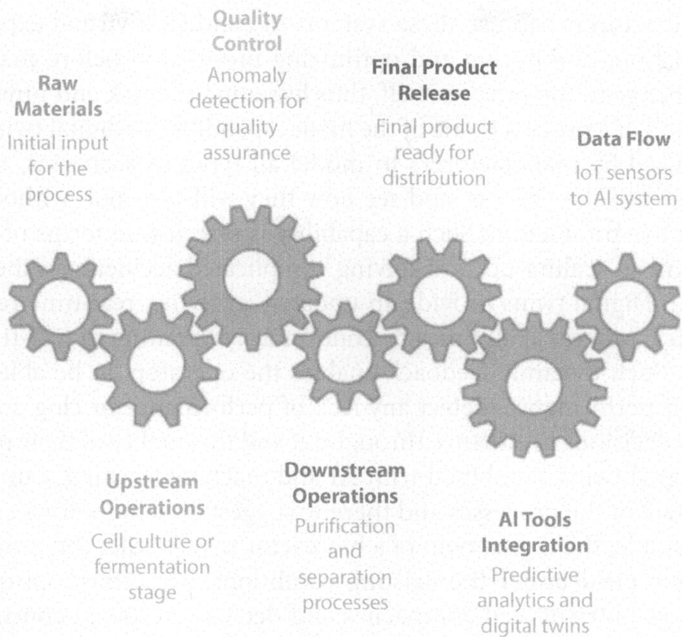

Figure 14.1 AI-enabled lean biomanufacturing workflow.

the condition of the process to prevent waste or batch failure [32]. One active area in the study of artificial intelligence, reinforcement learning, is seeing some interest in adaptive process control. This method enables the algorithms to train the use of optimal control procedures in real time as the process takes place in the process environment— which is desirable in applications that require continuous bioprocessing processes [29]. Figure 14.1 displays AI enabled lean biomanufacturing workflow.

Comes with powerful potential, however, when incorporating AI within biomanufacturing behavior, meticulous attention must be given to data integrity, model verification, and regulatory compliance, especially in GMP environments. However, the huge amount of competitive advantage through accelerated process development and higher yield to diminishing variance and cost is prompting high significance of adoption of these technologies throughout the business [33].

14.2.2 Digital Twins and Real-Time Process Simulation

Digital twins are a dynamic virtualization model of physical bioprocesses, built on real-time data and mathematical and AI-derived information.

Biomanufacturers can use these systems to conduct a virtual experiment of simulation, monitoring and optimizing production before making an actual change to the process itself, thus lessening the risk and nimbleness. Simulation of a process is one of the main capabilities of digital twins. They can be used by manufacturers to model all types of scenarios, tune the parameters of the process, and see how they will turn out, without jeopardizing live production. Such a capability is priceless in terms of process development, scaling-up, and solving complicated accidents in the operation [34]. Digital twins provide an updated, evolving, real-time replica of its physical process at all times by continuously updating to real-time data streams. Such dynamic feedback enables the operators to be able to picture their performance, detect any lack of performance or clog and make advance decisions to enhance throughput and the quality of their products [35]. Digital twins, combined with AI and machine learning, can forecast future state of the processes and thereby suggest how to intervene. As one such example, the digital twin of a bioreactor may predict cell growth and metabolite yield under the existing conditions, and, thus, optimize the nuances of nutrient feed approaches and decrease resource consumption [36, 37].

Digital twins also facilitate process scale up and transfer of technology as the detailed process knowledge is represented. Trial at various scale or other facilities of production employment limits the variations and the chance of failure in transfer technology. Nevertheless, to be successful,

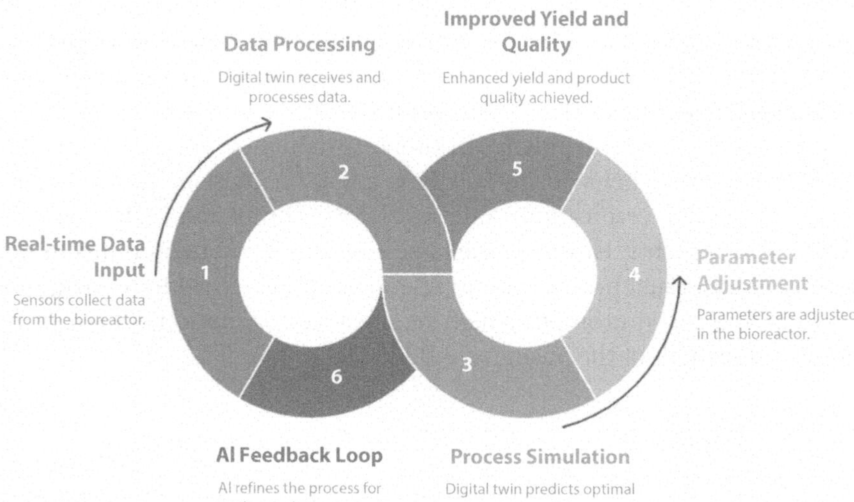

Figure 14.2 Digital twin model for biomanufacturing process simulation.

it is necessary to have advanced modeling, data infrastructure and integrate physical and digital systems. Given the quantitative and complexity increased task of biomanufacturing processes, digital twins are gaining prominence as a base of lean production and two-faced resilience ontologies [38]. Figure 14.2 shows digital twin model for biomanufacturing process simulation.

14.2.3 Anomaly Detection and AI-Based Monitoring Systems

In biomanufacturing, anomaly detection is important to ensure the quality of the products and regulatory adherence in biomanufacturing. It entails the detection of any abnormal trends or departures in process data which can suggest it has drifted or some soured equipment or quality risk. These anomalies need to be detected and amended as early as possible in order to reduce wastage, failure of batches, as well as maintaining a consistent performance of products [39].

Such monitoring systems using AI improve the detection of anomalies based on machine learning algorithms, including clustering, classification, and neural networks. Such systems process sensor data on an ongoing basis along with parameters and quality measures to train what normal operation of the process should look like, and identify abnormalities in real time [40]. In comparison to conventional threshold-based monitoring, which is subject to regular false alerts, AI systems are able to cultivate a distinction between natural variations and actual anomalies, which has in turn played a significant role in enhancing the credibility of alerts.

There are other positive effects to early detection of anomalies than quick response. AI tools facilitate root cause analysis by the correlation of process variables that are difficult to be identified by human operators. Charting changes in a dissolved oxygen level may tell you that there is a sinister presence in the tank, the tank is not agitated or the sensor is not reading correctly and each one of these should be handled differently [41]. Accurate identification of a cause will enable teams to provide more effective interventions, with the help of AI.

Novelty detection also can be used to augment quality assurance techniques; AI used to explore such analyte data as chromatograms and spectral fingerprints. Such instruments are able to reveal small changes in any product purity, potency or structural integrity that may not be evident with a manual review of the same. Also, AI-based monitoring applications keep a detailed history of the events during the process that aids in GMP traceability, documentation, and audit preparedness [32].

Table 14.2 AI technologies transforming lean bioprocessing.

AI technology	Description	Impact on biomanufacturing
Machine Learning (ML)	Uses algorithms to analyze data, identify patterns, and predict outcomes.	Enhances real-time monitoring and optimization of bioprocesses.
Neural Networks	A subset of ML that models complex, nonlinear relationships.	Improves batch-to-batch consistency and quality prediction.
Predictive Analytics	Uses historical data to forecast future outcomes and recommend actions.	Allows proactive process adjustments, reducing waste and enhancing yield.
Digital Twins	Virtual replicas of physical processes, used for simulation and optimization.	Enables testing of process changes without disrupting live production.
Anomaly Detection	AI systems that identify unusual patterns or deviations in process data.	Reduces defects by enabling early detection of process anomalies.

Due to the attempt of the manufacturers to transition the quality management practices to a proactive approach, AI-based anomaly detection systems are becoming a part of lean biomanufacturing. They facilitate the ongoing development by lowering the batch variation, minimizing the unexpected stoppage and high process availability, and hence more efficient and cost-effective [40]. Table 14.2 shows the different AI technologies transforming lean biomanufacturing.

14.3 AI-Driven Optimization in Upstream and Downstream Operations

Upstream Optimization: One of the areas developing within biomanufacturing is optimization *via* AI with respect to upstream operations. Upstream processing, including cell culture and fermentation, make use of both historic and real time data in training AI models to optimize important process parameters like pH, temperature and nutrient feeding. Such models provide the best environment in terms of cell growth and productivity and at the same time minimize the variability and also the need to take manual interventions [25]. AI uses the ongoing learning experience based on process data to anticipate variation and make timely course-correcting recommendations resulting in better batch consistency and increased overall yield.

AI offers a robust capability in process modeling and control that is utilized in the downstream operations, which includes product purification and recovery. It can facilitate optimization of multistage separation and purification processes by which it entails optimal operating conditions that would lead to improved product purity and material waste. The dynamic settings of parameters with machine learning methods increase efficiency and quality of output [29]. Digital twins and mixed models, which rely on AI, also extend the control of the processes to the virtual simulation of the entire manufacturing workflow so that manufacturers could dynamically test and develop processes before they are applied. This practice cuts on the rate of errors substantially and speeds up the scale-up process [34]. Together, these AI-based applications of the upstream and downstream operations increase productivity, improved use of resources and more responsive biomanufacturing value chains [25].

14.3.1 Raw Material Inventory Management

A powerful raw material inventory management is a pillar of the lean biomanufacturing performance and AI is transforming the sector, leading to the smarter forecasting, monitoring, and procurement. Conventional systems of inventory management usually rely on stationary estimates and manual updates that occur periodically and are prone to overstocking, stockout problems as well as waste, which is especially expensive to deal with in biomanufacturing where raw materials are highly valued and usually perishable in nature [42].

The issues mentioned above are solved by using inventory management systems based on AI that employ predictive analytics models based on past usage patterns, manufacturer projections, and third-party factors (i.e., market demands, supply shocks, etc.). The tools provide accurate predictions that enable biomanufacturers to move and maintain the optimum stock levels, hence preventing costly shortages and reduction in wastage [43]. IoT sensors can afford greater visibility of circumstances, expiry dates and utilization trends on a real-time basis. The sensor data is input into

AI platforms that are used to identify deviations in storage conditions or usage, which also ensures the use of materials prior to expiry and within the normal parameters of quality [44].

In addition, AI solutions help facilitate automated replenishment practices where importation activities are instigated on the basis of projected material requirement. This is to make sure that inputs are accessed at the time when they are needed and reducing the holding of inventory costs as well as directing the procurement to the real amount of production. AI may also calculate supplier reliability based on their historical performance, the timelines of their deliveries, and the quality indicators to allow manufacturers to plan ahead and avoid the risk by diversifying suppliers or having alternate sourcing process.

The reduced waste of materials, lowering cost in the stocks, and provision of enhanced agility in the supply chain connects AI-integrated inventory management directly to the lean biomanufacturing imperative.

14.3.2 Process Parameter Optimization for Yield and Quality

The key aspect in the process of achieving high yields, consistent products and efficient operation in biomanufacturing is optimization of processes parameters. Conventional methods are usually inflicted by much trial and error experimentation that may be time and resource consuming. The advent of AI and machine learning specifically has radically changed this playing field as it allows optimizing processes based on data, at a scale and speed unlike before [25]. Built on the basis of large amounts of historical and real-time process data from past production runs, pilot processes, and streams of sensor data, machine learning algorithms extract complex links between process parameters and output measures. The insights can be used to find out the best combination of variables that include pH, temperature, the agitation rate, and feed composition with much less experiments in comparison to the ones that might be performed using conventional methods [45].

The role of digital twins in the said optimization process is crucial because it provides a virtual model of bioprocesses which are used to model the impact of parameters changes *in silico*. This makes scenario testing and optimization possible at high speed without the need to disturb the live production, hence the development will be faster but the risk will be low [37]. This can be applied in real-time in the manufacturing process where it is observed that during active manufacturing, the AI needs no human intervention to accelerate or slow down the process; AI is able to tell that it is out of the desired process and rectify that by itself by making independent changes in control variables. This is especially necessary in upstream processes such as culture of cells where biological systems are dynamic in their response towards any slight drift in the environment. Predictive analytics is helpful in assuring quality by predicting the outcomes in the processes and facilitating early interventions. These instruments have the ability to detect potential indicators of future batch failures or quality deviation and correct the situation before the final goods have been contaminated.

Examples of AI in real life use cases such as parameter optimization in fermentation processes and the AI approximates the impacts of alterations to the rate of feed, the intensity of the oxygen supply or the mixing rates [46]. Such models inform the dynamic control design that enhances product yield. With regards to downstream processing, AI can be used to optimize chromatographic separation and buffer utilization, resulting in greater recoveries and purer products.

Finally, AI-based parameter optimization practical advantages are exceptionally promising in terms of higher product rates, higher batch-to-batch reproducibility, and a reduced development cycle. Such benefits help lean processes directly due to lowering of wastes, increased efficiency of resources, and quality results.

14.3.3 Minimizing Energy Usage and Reducing Downtime

Two of the biggest forces draining effectiveness and expenditure in biomanufacturing processes are energy consumption and equipment down time. The solution to these challenges is highly enabled by AI technologies that include predictive maintenance, intelligent scheduling of the process, and real-time energy management.

Among the most influential uses of AI in that area is predictive maintenance. Using sensor readings on equipment, AI models can identify trends that predict wear or failure, or degrading performance. They can predict a possible malfunction and prescribe the maintenance activities to be done before it can disrupt operations and make companies abandon the reactive approach to maintenance to proactive maintenance philosophies [47]. This minimizes unscheduled outages, extends the life of equipment and overall use of assets. Production scheduling can also be boosted with the help of AI by optimizing running times with respect to factors regarding equipment availability, energy prices, and priorities on production. Through the optimization of the manufacturing processes with off-peak energy or meeting batch minimum idle time, AI can lead to the minimization of energy and energy wastage as well as drive optimal throughput. Moreover, AI models establish and optimize real-time energy-intensive processes.

As an example, AI can balance resources that industry may consume against the performance of a process by manipulating agitation speed or temperature of a bioreactor. Such tightened adjustments bring significant savings on energy use without loss of yield or quality of the products.

These methods are supplemented by anomaly detection tools which can identify distinct energy consumption anomalies or types of equipment behavior that is not normal and might signal inefficiencies. Such systems can assist the operators in detecting problems like leaking valves, underperforming chillers or not-planned out clean-in-place before they lead to costly energy losses or production upheavals [48].

The physical gains of such AI-based strategies can be lower operating costs, better environmental sustainability, and better process reliability. As an instance, predictive maintenance could significantly diminish the amount and the scale of the stops in the production process whereas energy optimization efforts will help create a decreased carbon footprint and a smaller utility bill [49].

According to the outcomes of biomanufacturers, to apply the routine of adding AI into their maintenance, scheduling, and energy stopping processes, biologists can, and will reach the amplified uptime, low emissions, and profitable operations, amid the center significance of any lean

Table 14.3 AI-driven optimization in upstream and downstream operations.

Operation	AI application	Impact
Upstream Operations	Process parameter optimization (e.g., temperature, pH)	Increases yield and reduces variability in cell culture and fermentation.
Downstream Operations	Process modeling and control in purification	Optimizes separation processes, improving product purity and recovery.
Predictive Maintenance	Identifies impending equipment failures	Reduces downtime and extends equipment life.
Supply Chain Optimization	AI-driven demand forecasting and inventory management	Reduces stockouts and excess material waste.

manufacturing enlightenment. Such capabilities can be used to fulfill the twofold needs of cost containment and sustainability without compromising on quality and compliance standards [50]. Table 14.3 shows AI driven optimization in upstream and downstream operations.

14.4 Data Integrity, Model Validation, and Regulatory Compliance

The foundation of trust and safety in biomanufacturing with AI is data integrity, model validation and regulatory compliance. The integrity of data gathered in regulated environments, such as those under the domain of Good Manufacturing Practice (GMP) requires verification at each stage of the AI lifecycle which is data collection to model deployment and subsequent monitoring. Data practices in these settings are directed by the ALCOA+ principles [51] the data must be attributable, legible, contemporaneous, original, accurate, complete, consistent, enduring and available. The principles are not abstract, they become the standards that can be mandated under the accreditation of data traceability, reliability, and audit ability across the systems driven by AI.

Accountability demands data provenance and building of audit trails. The outputs of the AI systems should be traceable and explainable that regulators can scrutinize and confirm. Validation of models is equally very important: AI models should be evidentiary challenged to show that the outputs are reliable, reproducible, and scientifically valid. As regulators place an ever-greater emphasis on interpretability, methods like SHAP (Shapley Additive Explanations) and LIME (Local Interpretable Model-Agnostic Explanations) are becoming crucial elements of explainable AI (XAI) systems [51].

The issue of regulatory compliance means that the application or usage of AI must be compliant with GMP standards and global standards implemented by such authorities as the U.S. Food and Drug Administration (FDA) and the European Medicines Agency (EMA). Such structures stress on risk management, traceability and model fitness towards intended use especially with regard to decisions on product quality or patient safety. Having constant oversight, strong recordkeeping, and the ability to have regulatory audits are a must in adhering to a biomanufacturing system powered by AI [33].

14.4.1 Data Quality and Traceability in GMP Environments

Data integrity is a cornerstone need within GMP-regulated biomanufacturing in any type of information whether it be created manually or artificial intelligence. Regulatory authorities including FDA, EMA and WHO highlight that quality and safety in products requires quality and reliability data to guarantee trust and regulatory acceptance by the populace [51].

The ALCOA++ model is the framework that can be used as a complete guide to the data integrity maintenance. Data should be identifiable, in that they should be easily identifiable with someone who created them. It should be readable and indelible and should be taken at the same time, and must be original or be certified as a copy of the original. It is imperative that the accuracy of a recording be high and that all the recorded data be an exact reflection of what transpired. Completeness will also require capturing of all the information without exceptions. Consistency denotes the sequential logics and traceability of data records. Information would also have to be permanently preserved over its life and become easily accessible at the time of its review or audit [52].

14.4.1.1 Data Governance and System Controls

A strong data governance regime is important in creating specific roles, responsibilities and protocols applicable in handling data. Such frameworks have to consider GMP and guarantee accountability and transparency across the data life cycle. Laboratory Information Management Systems (LIMS), Electronic Batch Records (EBR) and secure audit trail facilities are commonly used to impose compliance. These systems should have automatic records to save entries, prevent unauthorized access and full logs of modifications/deletions [53].

Review of audit trails should also be regular to detect inconsistencies or any other unauthorized changes. Regular data integrity checks, as well as internal auditing, enable threats to be discovered, compliance to be increased, and organizations to avoid the risk of regulatory fines [54]. Training is a vital component of data integrity enforcement. Staffs have to be frequently educated on the document procedures, data management processes and identifying possible risks to data integrity [55]. The necessary change in data integrity, driven by the top management, is to be a part of the strong quality culture of an organization that makes data integrity an organizational priority rather than a procedural issue. The backdating of the recordings, deletion or overwriting of the original data, lack of audit trail and inefficient investigations of out-of-specification (OOS) results are among common data integrity violations cited by regulators [52]. A mix of technical measures (e.g. encrypted audit logs and user authorization schemes) and procedural ones, e.g. standard operating procedures (SOPs), incident handling processes, and systematic root cause analysis processes, mitigates these risks.

As Industry 4.0 spreads, end-to-end traceability of data is becoming attained through the implementation of digital systems throughout the manufacturing chain. All the data between the receipt of raw materials and the released final product should be captured, linked and retrievable to guarantee complete transparency to be able to perform both internal reviews and regulatory audits [56].

14.4.2 AI Model Verification and Lifecycle Methods

The implementation of AI in the biomanufacturing process that is governed by GMP adds new dimensions of complexity in the validation model and lifecycle management. The AI systems would need to behave consistently, transparently, and purposefully each time they transform and adapt in response to new information [33]. Model validation is the process of showing that an AI model executes the desired purpose reliably and accurately in different conditions. The first stage is a data quality test that helps in ensuring that the data on which input is based is whole, representative, and relevant to the use scenario. This data is normally split into training, validation, and test data in order to apply the model and check its performance and avoid over-fitting [57]. Validation needs to adopt performance metrics applicable to the task e.g. accuracy, precision, recall and ROC-AUC in case of classification models. To test robustness, models also have to be subjected to stress tests under different conditions including edge cases and possible data drift. Increasingly, explainability is a regulatory necessity. Models such as SHAP and LIME provide an understanding of the impact that individual input variables have on model outputs which can be useful in the interpretability of models and getting them accepted in their regulatory applications [58].

14.4.2.1 Lifecycle Management

AI model management is not finished with deployment in GMP settings. Version control is incorporated in lifecycle monitoring, which is the monitoring of all model updates and retraining activities as well as parameter changes. It has to be long monitored to locate any indication of model degradation, bias, or data drift. When significant changes have been implemented like new datasets or alteration of the process or external input, model has to be revalidated to ensure it remains reliable and compliant [59].

All the operations connected with model development, validation, deployment, and monitoring processes should be properly documented and auditable. This also makes it aligned to quality management systems (QMS) such as change control, SOPs, and risk management practices.

14.4.2.2 Documentation and Change Control

Comprehensive documentation underpins regulatory acceptance. Some of the important paperworks are the model development and validation reports, the data lineage logs, rationale on the algorithm and hyperparameter selection, and well-documented logs on each change and retraining.

The change control processes will have to guarantee that; adjustment of models is risk-assessed, reviewed, and honored and well recorded before making it to a production environment. The evaluation of risk should determine the possible implication of the model on the quality of products, safety of patients, and quality of data. Human-in-the-loop review or override features might be required in high-trust applications to make responsible AI governance.

14.4.3 Navigating Regulatory Requirements for AI in Biomanufacturing

Increasing integration of AI in GMP-controlled production, regulatory bodies are preceding more transparent requirements on its validation, control and management. Manufacturers will have to prove that an AI system can be trusted, be suitable and can meet the high standards of pharmaceutical quality.

14.4.3.1 Key Regulatory Expectations

Guidance on data integrity provided by the FDA states that computerized systems such as the machine learning also needs to maintain ALCOA principles, which implies that all the data and decisions made with the help of computer algorithms should be attributable, traceable, and scientifically justified. Parallel frameworks such as EMA, WHO, and other international bodies have placed particular importance on transparency and auditability along with management of risk [52].

Formal validation of the AI models operating in the GMP environment is required, and the documentation that can justify the performance of models, assumptions, sources of data, and risk mitigation strategies should be provided. All the updates in the model should be controlled using the change control procedures, which also should include revalidation to be undertaken as required. Extensive audit trails are mandated on the activities related to data as well as to the model, and this allows regulators to reconstruct events and track decisions.

AI model use should also have in place data governance systems that stipulate roles, responsibilities, and processes. These are as follows: model ownership; access controls, documentation standards, and training protocols.

14.4.3.2 Addressing Regulatory Challenges

Several unique challenges accompany AI adoption in regulated environments. Regulators may not be satisfied with applicability neither to black-box models with no interpretability in domains that sufficiently impact the quality of products or risk to patient safety. The decision made by the AI systems is becoming explainable and justifiable according to the regulators.

Cybersecurity and data privacy also demand attention. Before even slightly compromising the accessibility of AI systems, unauthorized access, data breaches, and tampering should not be allowed, and a number of measures must be employed to achieve this, such as encryption, access controls, and secure infrastructure. Furthermore, as regulatory expectations evolve, global harmonization is essential. The manufacturer working in cross-jurisdiction has to keep track of the alterations in regional guidance and stay abreast of global norms [60].

14.4.3.3 Preparing for Regulatory Inspections

Regulatory preparedness involves the capacity to indicate the means of AI models validation, observation, and preservation. Inspectors will want to be shown evidence that data integrity is observed through all steps in its lifecycle, of both data and model. The organizations should have records of audit, be able to trace the products and prove that the personnel were trained in GMP as well as the operation of AI systems [61]. Missing these expectations may lead to serious regulatory actions to which one may be exposed depending on the level, and these may include warning letters, import alerts, consent decrees or product recalls. To guarantee data integrity and traceability in AI-assisted GMP conditions, it is necessary to organize strong governance, safety infrastructure, validation of the models, and build a quality culture. The resulting production processes will only be as effective as their accommodations of such regulatory requirements and their effective utilization of the paradigm-shifting capabilities of AI. Compliance and innovation is possible when the manufacturers have the right strategies to focus on and still produce high quality standards in the areas of safety, efficacy and trust [61].

14.5 Case Studies: AI-Enabled Lean Transformation in Biopharma

Lean transformation based on AI is already translating into an achievable result in the area of biopharmaceutical manufacturing, combining data intelligence and lean philosophy. Manufacturers across the industry are in the process of implementing AI-based technologies, including digital twins, anomaly detection units, predictive analytics, and intelligent scheduling systems to help control the processes, reduce waste, and ensure higher quality of their products. There is one prominent example where the use of digital twins and anomaly detection models have allowed the marketing leaders in biopharma to reproduce the ideal settings of a so-called golden batch. By matching past statistics that include the range of parameters of the equipment to the behavior of operators to environmental conditions with the important variables in a production process, the AI modeling has been capable of establishing optimal production environments. Real time monitoring tools on detecting anomaly in the production trigger alerts so that corrections can be made in good time and acute batch consistency can be achieved. Consequently, the leading performing manufacturers have found to record as low as 14 times lesser quality expenses and one-sixth the amount of deviations per 1, 000 batches with the industry weighted averages [62].

The other striking case was related to a global biopharma company that used AI-powered analytics to analyze unstructured raw materials. Through digitization and analysis of this data, predictive models were generated to optimize yield and quality leading to a 5% increase in yield, and a financial benefit of more than $10 million per site annually in less than ten weeks. Another also-improving area is scheduling with AI, which gained 15 percent upstream throughput, and manual intervention decreased by 40 per-cent. In the meantime, upstream downtime has dropped 29% in systems which apply dynamic workforce allocation, predictive skill gap analysis and microtraining [5]. Such practical examples understate powerful changes that can occur when introducing AI into a lean manufacturing system to enhance biopharma work [63, 64].

14.5.1 Practical Examples from Manufacturing Lines

The use of AI on biopharmaceutical production lines is not theoretical anymore as the range of practical high-impact applications continues to increase. Combining digital twins with real-time anomaly detection is one of the most popular applications. Digital twins are sort of virtual images of physical manufacturing spaces, models of processes and facilities that use a combination of collected historical data on machines, operators, and sensors to help streamline manufacturing actions. These models, by determining the conditions under which "golden batches" could be produced, can assist manufacturers to determine the optimum settings of variables like the speed of the machines, humidity, and the time in which the operators are changed. Then anomaly detection systems are used to check production information in real time. When a parameter goes outside the golden batch profile e.g., a surge in temperature or a fluctuation in machine speed, AI algorithms provide warnings and suggest corrections in order to avoid scrap, rework or loss of a batch [65].

Simultaneously, the standard manufacturing processes are being facilitated through workflow automation that is powered by AI. The batches of records, maintenance assignments, as well as initial data analysis are made by the robotic process automation (RPA). This minimizes human error and leaves the personnel scope to concentrate on other value-adding acts like process optimization and troubleshooting. On the manufacturing floor also, supervisor roles are transforming due to generative AI (Gen AI). Gen AI tools cross-reference the information contained in production systems, machine logs and technical manuals, to facilitate shift planning, diagnostics and communication. Such AI supervisors assist in the creation of reports, decision guidance, and location awareness, and eventually will enable human supervisors to focus on team building and strategic plans [66]. The influence of AI also reaches predictive maintenance, where such companies as Pfizer use AI to foresee failure of their equipment before it takes place. Based on sensor readings and past maintenance records, such systems are able to optimize equipment maintenance programs, reduce the amount of unscheduled down time, and prolong equipment life [67]. Janssen Pharmaceuticals has gone a step further in the implementation of AI by involving AI control systems in continuous manufacturing system of the company. To manufacture its HIV medicine, Prezista, AI used real time adjustments in the process and it took the product just instead of the traditional 30 days process. This was a decisive change in going over to a continuous more agile manufacturing process [33].

14.5.2 Measurable Outcomes: Productivity, Efficiency, and Sustainability Gains

There has been a significant, measurable improvement in performance when AI has been applied to biopharma manufacturing in various dimensions.

In terms of productivity and throughput, manufacturers have decreased costs and the number of deviations in quality dramatically by applying AI-enabled digital twins technologies and anomaly detection tools. Others record lower quality costs as high as fourteen times and a sixth of the industry standards of deviations. In continuous manufacturing, cycle times have been drastically shortened. To illustrate, Janssen reduced the batch release time and decreased it by three times, 30 days down to 10 days due to the AI implementation in its continuous production processes [33]. Generative AI-assistant has also been found to reduce the amount of corrective maintenance work by up to 50% and execute these tasks at 30 percent less time. Reporting and communication of the shifts has been made more efficient with a decrease in 40 percent of workload required by supervisors to allow them to become more strategic. In one biologics facility, AI has identified yet unrecognized factors of output loss, and has already yielded direct improvement in product output.

AI also enhances cost-efficiency. Smart scheduling and task assignment programs guarantee a competitive utilization of the resources and less idle time. Unattended maintenance tasks standardize processes and remove the routine monitoring activities by the management. The introduction of predictive maintenance systems in companies such as Pfizer has boosted availability of equipment and has decreased the number of emergency repairs and emergency repair costs. The dynamically assigning machine, personnel, and materials AI models improve the utilization of resources and eliminate bottlenecks [68].

The application of AI-enabled real-time control systems, in a sustainability perspective, implies minimized production of waste and energy and a constant operating condition by providing a continuous process. In comparison to traditional batch-based manufacturing methods, the three-dimensional models powered by AI cause minimal breaks in the process and result in much reduction of waste of the material. Defective product batches are also prevented through automated quality-based system and minimize the wastage of energy and raw materials incurred in different reworks and discard [22].

AI contributes directly to quality and compliance. Smart systems can use live sensor and laboratory data to identify abnormalities early and stop the process of producing subpar batches before it is too late to take an action. Generative AI tools can accelerate and improve trouble shooting, since they scan through historical data and deviation records to provide a root cause analysis when time is of the essence [69].

14.5.3 Lessons Learned and Best Practices

The journey towards AI-fueled lean transformation has led to a number of useful lessons and best practices within the biopharmaceutical sector. To start with, the effective use of AI requires integrated platforms of data and high-quality data. Attempts at an early stage usually failed because of the siloed information distributed over LIMS, MES, data historians and spreadsheets. It is well-advised to build a singular, accessible data infrastructure that would allow AI analytics and contribute to real-time decision-making [70]. Simultaneously, companies should invest in data quality, such as data validation procedures and effective data governance, so that AI models work with quality data sources. Second, organizations get the greatest benefit when they address high impact but well-defined use cases. First implementations must aim to address bottlenecks or unpredictability with known high visibility or impact areas, e.g., better availability of equipment or less variability in the quality of the batches. These quick-wins can be used to illustrate the worth of AI and create an emotional impact internally. Evaluation of new systems using pilot testing in controlled settings enables teams to verify model performance, and streamline the implementation strategies prior to extension to additional sites [71].

Third, human-AI collaboration is critical. AI should enhance not replace human expertise. Such generative AI tools, e.g., perform best when applied to complement control choices, but not replace them altogether. In that regard, workforce upskilling facilitates success. Operators, engineers, and managers involved in AI should be trained to comprehend the output of AI and diagnose problems and implement AI recommendations without fear [72]. Fourth, proactive change management and cross-functional alignment are essential. Effective projects involve stakeholders early especially in the areas such as quality assurance, operations, IT, and regulatory affairs in order to resolve the issues raised, align the objectives, and implement that project without obstacles. Communication of actionable insights of artificial intelligence implementation including fewer reworks, quicker batch release, and better compliance is good to enlist believability and induce cultural change. Fifth, AI projects should inculcate regulatory compliance and data integrity. It should be noted that all AI systems cannot be used in a GMP environment unless they are fully validated, and there is a clear trace of how they were developed, tested, and intended to be used. Decisions and acts made by the models must be trackable through the aid of audit trails and constant monitoring should be able to sustain performance. Finally, continuous improvement must be institutionalized. The models should be regularly checked not only to ensure overall model performance continues to be effective but also when new data does arrive to retrain the model. The feedback loop between the human operators and the AI systems should also be supported to improve the working process and consider new areas of optimization. The lessons learned in every deployment should be documented and shared across teams and sites so faster adoption can occur within organizations [33].

Sustainability should also be positioned as a strategic outcome. The benefits of AI in the decreasing energy consumption, and wastes and resources consumed shall also be actively measured and incorporated in the business case leading to its adoption. These advantages have an appeal to the internal stakeholders as well as the external partners, regulators and the general populace [73].

14.6 Roadmap for Implementing AI-Powered Lean Biomanufacturing

Any adoption of AI-scale lean biomanufacturing should be undertaken using carefully-thought roadmap that is based on digital transformation, cross-functional ecosystem-type collaboration and continuous improvement. The initial essential procedure is the creation of the strong data base. Organizations have to make data gathered in production systems, equipment monitoring, quality control, and supply chain activities centralized and standardized, so that the data is structured and accessible and ready to be analyzed using AI. When such a foundation is established, one should establish a cross-functional team to spearhead the effort. Such a team needs to have professionals in lean manufacturing, bioprocess engineering, data science, regulatory compliance and information technology in place to develop a culture that embraces innovation and regulatory rigor [74, 75]. Now that building blocks were laid out, organizations are to evaluate the existing processes by using value stream mapping and process mapping and attempt to look at where their opportunities lie and how they can effectively simplify the process. The resulting AI models can be finally used to analyze past and real-time data to facilitate essential operations like predictive maintenance, process optimization, and early anomaly detection [76]. The ability to automate routine jobs, better management of processes, and make data-informed decisions, can be integrated into everyday operations using AI insight. Constant mechanism of model monitoring, retraining, and validation is important to remain reliable through time, have a traceable history, and face regulatory productivity. Since the results are measurable and the pilot projects are successful, it is possible to expand such models and workflows to several production lines or facilities. Standardized performance standards and feedback systems must ask how best to continually adjust. Response to changes in regulatory expectation and process reality makes AI transformation a stronger initiative which is in line with business and compliance goals. This data-driven iterative process will put biopharma organizations in a suitable situation to increase productivity, sustainability, and operational excellence [77].

14.6.1 Strategic Measures to Adopting AI in Lean Initiatives

The implementation of AI in lean bio manufacturing facilities thrives on a systematized and strategic planning that will fit the needs of the operations, statutory requirement, and the growth of the business. This process will kick off with executive buy-in where leaders will spearhead a clear vision of adopting AI and tangibly relate it to the business needs like enhanced productivity, product quality, and sustainability. The focus areas to be chosen should relate to the possibility of value generation, which may start in related areas like yield optimization and process efficiency that are more likely to make way due to fewer regulations than direct quality decisions [74].

Cross-functional collaboration is essential. An in-house AI task force made up of the stakeholders in operations, IT, quality, data science and regulators provides a balanced view and speeds the process of innovation. Open lines of communication among departments can assist in eliminating silos and ensuring that the objectives of the transformation become shared by everyone [78]. The next phase involves building a strong data infrastructure. Organizations need to connect the siloed data systems (LIMS, MES, ERP, and historian systems), into a single data platform that could accommodate AI-facilitated insights [78]. The information would be cleansed, digitalized and made structured to allow proper modeling and interpretation.

Once a centralized data base is established, lean techniques (value stream mapping) can be used to figure out areas of pain and rank AI use cases based on the most operational value. The typical early projects are predictive maintenance, seeing the quality in real time, and optimization of parameters in processes. An opportunity to validate AI models, measure performance, and refine deployment tactics to enterprise-wide rollout is to do pilot projects [79].

The AI models also need calibration through training, testing, and tuning that rely on both past and live flow of data. In cases where the internal capability is not enough, companies can form a consortium with other bodies like academic institutions, consultants, or technology suppliers to hasten development. The ability to explain and be transparent should be emphasized all the way through, in order to accommodate not only operational users but also the regulatory bodies [6]. Integration of AI insights into current workflows must not be viewed as a substitute of human decision-makers but must be applied as instruments to increase judgment and efficiency. The adoption is enabled by workforce training and changing management. The staff should gain the ability to understand AI outputs and handle exceptions, as well as develop systems perpetually based on the observed performance. Making people understand the importance of AI, engaging people into the process early, and rewarding by transformational objectives could severally minimize resistance to change [80]. Lastly, it is maintained in the form of constant supervision, maintenance of documentation and process confirmation. Effective pilots are supposed to be scaled up incrementally within production lines and locations and the main focus is to close the loops of feedback to streamline the processes and aligning with the changing regulatory demands. Making continuous improvement one of its defining principles will make AI an alive, changing component of a living lean biomanufacturing ecosystem.

14.6.2 Overcoming Challenges and Barriers to Implementation

Although AI in lean biomanufacturing has a transformative potential, a few organizational and practical obstacles have to be overcome so that the implementation of this technology could reach the desired level of success. One of the most persistent challenges is data fragmentation. The existence of disparate systems such as LIMS, MES, and ERP tend to create data siloes that AI cannot utilize to produce comprehensive insights. To tackle this, organizations should make investments on data integration platforms and implement strict data governance processes that will make data accurate, complete, and able to analyze [81]. Regulatory complexity adds another layer of difficulty. Biopharmaceutical companies have to work within the framework dictated by the GMP and AI models need to be validated, explainable, and properly documented. The fulfilment of such expectations requires active validation strategy, the visibility of decisions, clear algorithms, and the inclusion of regulators in the validation process at early stages [82].

Cultural resistance and workforce skill gaps are also common. The issue of job loss or fear of unfamiliarity with AI systems could prove to slow continued use. The strategy to cope with this must include the focus on communication and how AI can be used to augment rather than replace human expertise followed by thorough training that would develop digital expertise at all levels. Misalignment between functions may easily make AI efforts a walkover unless attention is given to the key stakeholders early on. The key to success would be ensuring clear structures in governance as well as appointing champions in each of the functions and align with strategic business goals [83].

Scalability poses another challenge. Although pilot projects can prove persuasive, full deployment of the AI system to the entire enterprise requires a strong IT infrastructure and the presence of standard deployment models. They require modular, cloud-based architecture and interoperable platforms to support such growth [84]. Security and data privacy are also critical. The AI systems have to adhere to the emerging data protection laws at the same time introducing the best cybersecurity practices. Sensitive data must be encrypted, access-controlled, and continuously monitored. Finally, the fear of loss due to a lack of certainty of returns on investments (ROI) may discourage the commitment. The adoption of AI can be associated with a high initial investment with potentially long-term timeframe needed to show change. It may be beneficial to prioritize quick-win situations and high-impact use cases at the beginning of the process in order to create a sense of movement and release funds to direct future efforts.

14.6.3 Future Perspectives: Scaling, Agility, and Continuous Improvement

In the long term, it is an AI-driven lean biomanufacturing concept characterized by scalability, agility, and a never-ending drive to learn more about optimization. Scalability shall be met by use of enterprise-scale AI systems implementation that will streamline the entire production value chain involving R&D and clinical supply to commercial production and distribution. Integration across the whole value chain (end-to-end) will make AI chain processes upstream and downstream to the real-time connection between the manufacturing process and supply chain, as well as, quality system, to facilitate an end-to-end decision-making process [85]. The agility will also be improved by what AI can do about dynamic market shifts, changes in regulatory environments, and supply chain volatility. Dynamic rescheduling and inventory control with production planning through AI-driven analytics will be possible. The scenario modeling will enable organizations to formulate planning and be prepared to face different risk scenarios, and hence become more resilient in their future shock [86].

Continuous improvement will become increasingly autonomous. The AI systems, such as those to learn and adapt in real-time, will be adopted in smart factories to decrease manual control and adjust the process parameters automatically. The next wave of Quality by Design (QbD) will also be powered by the AI in which one can monitor and control critical quality attributes in real time and minimize the number of deviations [87]. Reduction of energy consumption, waste of materials and emissions will support green objectives of being sustainable. Circular manufacturing strategies will also be accelerated with the help of AI, as the technology will allow predicting recycling, optimizing reuse, and tracking life cycles. Workforce would be transformed in such a way that human activities turned to innovation, strategy and superior analytics. The strategies of upskilling and reskilling will also be a central aspect of organizational development, upholding the ability of the employees to succeed in fully digitally transformed manufacturing [72].

Regulatory frameworks will also adapt. Agencies have seen the usefulness in AI and are starting to collaborate with industry forming risk-based and data-driven compliance models. Continued partnership between regulators and manufacturers will be essential to the development of feasible, prospective guidelines around the use of AI in GMP settings [23].

Lastly, successful integration of AI in lean biomanufacturing will bring about competitive advantages to the organizational entities. The speed, cost-effectiveness, quality of the goods and product innovativeness will enjoy an advantage in the early adopters. When best practices and benchmarks become established, there will be industry-wide standards to facilitate the widespread use of AI and increase ongoing performance in the sector [88].

14.7 Conclusion

The integration of artificial intelligence with lean thinking offers a pragmatic path to smarter, faster, and more reliable biomanufacturing. By pairing data-rich analytics with waste-eliminating principles, organizations can move from reactive firefighting to proactive, self-improving operations across upstream and downstream processes. The analysis in this work shows that AI does not replace lean; it amplifies it, accelerating flow, stabilizing quality, and freeing people to focus on higher-value problem solving. This chapter distills the key considerations and future directions under four themes.

14.7.1 Regulatory and GxP Considerations for AI-Enabled Operations

Bringing AI into GxP environments requires the same discipline applied to any regulated system: documented lifecycle control, clear ownership, and demonstrable fitness for use. Data integrity, model traceability, versioning, and change control must be embedded from the outset. Risk-based validation should focus on intended use and failure modes, with model performance monitored continuously and retrained under controlled procedures. Human-in-the-loop oversight, explainability appropriate to risk, and robust audit trails help ensure that automated recommendations remain transparent and defensible. Harmonization with quality frameworks such as Quality by Design and real-time release testing can streamline acceptance by quality units while maintaining patient safety. Finally, governance should address cybersecurity and data privacy, especially when cloud services, third-party models, or cross-site data sharing are involved.

14.7.2 Economic and Scalability Considerations

Economic impact hinges on matching investment in data foundations and analytics to high-value use cases. The total cost of ownership spans sensors and PAT, data engineering, compute, licenses, and talent—costs that are offset by reductions in deviations, waste, cycle time, and unplanned downtime. To scale, organizations should standardize data models and interfaces (e.g., harmonized tags across historians, MES, LIMS) so that solutions built for one line or site can be redeployed with minimal rework. A portfolio approach starting with quick-win pilots that prove value and learning, then industrializing platforms helps de-risk larger commitments.

For smaller facilities, managed platforms and modular architectures can lower barriers to entry while avoiding vendor lock-in. Careful attention to sustainment costs (model monitoring, periodic re-validation, and support) preserves ROI over the long term.

14.7.3 Emerging Trends and Technologies

The next wave of capability blends mechanistic understanding with machine learning. Hybrid models and plant-synchronized digital twins enable richer "what-if" analyses and safer optimization before changes reach the floor. Edge analytics and advanced PAT expand real-time visibility, while anomaly detection separates normal biological variability from drift that merits intervention. Generative tools are streamlining documentation, batch record review, and investigation drafting reducing manual burden without compromising rigor. Intelligent schedulers and inventory optimizers are improving flow and responsiveness across multiproduct facilities. In parallel, FAIR data practices, synthetic data for low-risk pretraining, and interoperable APIs are making solutions more portable. Sustainability is increasingly co-optimized: energy-aware control, predictive maintenance, and smart cleaning strategies can reduce resource consumption while protecting yield and compliance.

14.7.4 Future Directions

Progress from pilots to plant-wide impact will depend on coordinated action. First, treat data as a product: curate high-quality, context-rich data with clear ownership and service levels. Second, formalize model lifecycle management so quality, validation, and IT jointly steward AI as durable infrastructure rather than one-off tools. Third, invest in people upskilling operators, engineers, and QA to interpret model outputs, challenge recommendations, and capture learning into standard work. Fourth, codify scalable patterns: reference architectures, reusable features, and template documentation to shorten time from idea to validated deployment. Finally, cultivate an improvement engine that pairs daily management (tier meetings, visual controls) with AI insights, ensuring that gains persist and compound.

In sum, AI-enabled lean biomanufacturing is both an operational strategy and a cultural shift. Organizations that combine strong governance with purposeful experimentation can achieve stable quality, shorter lead times, and higher asset productivity—while building a resilient capability to learn from data and improve every day.

Abbreviations

AI	Artificial Intelligence
ML	Machine Learning
IoT	Internet of Things
GMP	Good Manufacturing Practice
IIoT	Industrial Internet of Things
VSM	Value Stream Mapping
TPS	Toyota Production System
VSM	Value Stream Mapping
QbD	Quality by Design
LIMS	Laboratory Information Management Systems
MES	Manufacturing Execution Systems
ERP	Enterprise Resource Planning
RPA	Robotic Process Automation
GxP	Good Practice (refers to regulations like GMP, GCP, etc.)
FDA	Food and Drug Administration
EMA	European Medicines Agency
WHO	World Health Organization
SHAP	Shapley Additive Explanations
LIME	Local Interpretable Model-Agnostic Explanations
ROC-AUC	Receiver Operating Characteristic - Area Under Curve
PAT	Process Analytical Technology
ROI	Return on Investment
FAIR	Findable, Accessible, Interoperable, Reusable (data principles)
API	Application Programming Interface

AI disclosure statement: The author acknowledges the use of AI-based tools for minor language editing or grammatical correction. However, all intellectual contributions, analyses, and conclusions presented in this chapter are entirely the author's own.

References

1. Sanders, A., Elangeswaran, C., Wulfsberg, J., Industry 4.0 implies lean manufacturing: Research activities in industry 4.0 function as enablers for lean manufacturing. *J. Ind. Eng. Manag.*, 9, 3, 811, 2016.
2. Sundar, R., Balaji, A.N., Kumar, R.S., A review on lean manufacturing implementation techniques. *Procedia Eng.*, 97, 1875, 2014.

3. Jensen, J.D., Advancing lean implementation for next generation biomanufacturing operations, MIT Thesis, 2016.
4. Thakur, A., A review on Lean Manufacturing implementation techniques: A conceptual model of Lean Manufacturing dimensions. *REST J. Emerg. Trends Model. Manuf.*, 2, 3, 62, 2016.
5. Haddad, A., Optimizing Integrated Continuous Biomanufacturing Throughput: Resource Constraints and Process Scheduling, MIT Thesis, 2024.
6. Isoko, K., Cordiner, J.L., Kis, Z., Moghadam, P.Z., Bioprocessing 4.0: a pragmatic review and future perspectives. *Digit. Discov.*, 3, 9, 1662, 2024.
7. Smart, N.J., *Lean biomanufacturing: creating value through innovative bioprocessing approaches*, Elsevier, Amsterdam, The Netherlands, 2013.
8. Guilland, J.F. and Stoll, T., Harvesting the benefits of LEAN in biopharmaceutical manufacturing.
9. El-Namrouty, K.A. and Abushaaban, M.S., Seven Wastes Elimination Targeted by Lean Manufacturing Case Study "Gaza Strip Manufacturing Firms". *Int. J. Econ. Finance Manag. Sci.*, 1, 2, 68, 2013.
10. Gomaa, A.H., Achieving operational excellence in manufacturing supply chains using lean six sigma: a case study approach. *Int. J. Lean Six Sigma*, 2025.
11. Jensen, J.D., Advancing lean implementation for next generation biomanufacturing operations, Doctoral dissertation, Massachusetts Institute of Technology, 2016.
12. Borangiu, T., Morariu, O., Răileanu, S., Trentesaux, D., Leitão, P., Barata, J., Digital transformation of manufacturing. Industry of the future with cyber-physical production systems. *Rom. J. Inf. Sci. Technol.*, 23, 1, 3, 2020.
13. Borgosz, L. and Dikicioglu, D., Industrial internet of things: What does it mean for the bioprocess industries. *Biochem. Eng. J.*, 201, 109122, 2024.
14. Nyoni, R., Harnessing Data Analytics for Predictive Insights: Advancing Decision-Making with Big Data Innovations. *Int. J. Res. Publ. Rev.*, 6, 2915, 2025.
15. Langmead, B. and Nellore, A., Cloud computing for genomic data analysis and collaboration. *Nat. Rev. Genet.*, 19, 4, 208, 2018.
16. Gargalo, C.L., Udugama, I., Pontius, K., Lopez, P.C., Nielsen, R.F., Hasanzadeh, A., Mansouri, S.S., Bayer, C., Junicke, H., Gernaey, K.V., Towards smart biomanufacturing: a perspective on recent developments in industrial measurement and monitoring technologies for bio-based production processes. *J. Ind. Microbiol. Biotechnol.*, 47, 11, 947, 2020.
17. Zhou, K., Liu, T., Zhou, L., Industry 4.0: Towards future industrial opportunities and challenges. *Proc. Int. Conf. Fuzzy Syst. Knowl. Discov. (FSKD)*, p. 2147, 2015.
18. Mohamed, M., Challenges and benefits of industry 4.0: An overview. *Int. J. Supply Oper. Manag.*, 5, 3, 256, 2018.

19. Mehta, A., Niaz, M., Adetoro, A., Nwagwu, U., Advancements in manufacturing technology for the biotechnology industry: The role of artificial intelligence and emerging trends. *Int. J. Chem. Math. Phys.*, 8, 2, 12, 2024.
20. Bruggeman, F.J. and Westerhoff, H.V., The nature of systems biology. *Trends Microbiol.*, 15, 1, 45, 2007.
21. Benderskaya, E.N., Nonlinear trends in modern artificial intelligence: a new perspective, in: *Beyond Artif. Intell.: Contempl., Expect., Appl.*, p. 113, Springer, Berlin Heidelberg, 2013.
22. Chukwunweike, J., Anang, A.N., Adeniran, A.A., Dike, J., Enhancing manufacturing efficiency and quality through automation and deep learning: addressing redundancy, defects, vibration analysis, and material strength optimization. *World J. Adv. Res. Rev.*, 23, 2024.
23. Thanasas, G.L., Kampiotis, G., Karkantzou, A., Enhancing transparency and efficiency in auditing and regulatory compliance with disruptive technologies. *Theor. Econ. Lett.*, 15, 1, 214, 2025.
24. Awasthi, A., Emerging trends in bioprocess engineering: shaping the future trades. Concepts & applications, 82, 2024.
25. Robinson, M., AI-Optimized Biomanufacturing: Enhancing Efficiency in Biotechnology. *J. Biochem. Biotech.*, 7, 6, 240, 2024.
26. Hickerson, D.H. and Hunsberger, J., Applying AI to advanced biomanufacturing, in: *Artificial Intelligence in Tissue and Organ Regeneration*, p. 267, Academic Press, Amsterdam, The Netherlands, 2023.
27. Kumar, K., Sharma, R., Ranjan, K., Pandey, D.K., Artificial Intelligence in Biotechnology, in: *Artificial Intelligence and Biological Sciences*, p. 192, CRC Press (Taylor & Francis Group), Boca Raton, FL, USA, 2025.
28. Olagunju, E., Integrating AI-driven demand forecasting with cost-efficiency models in biopharmaceutical distribution systems. *Int. J. Eng. Technol. Res. Manag.*, 2022.
29. Rathore, A.S., Nikita, S., Thakur, G., Mishra, S., Artificial intelligence and machine learning applications in biopharmaceutical manufacturing. *Trends Biotechnol.*, 41, 4, 497, 2023.
30. Khuat, T.T., Bassett, R., Otte, E., Grevis-James, A., Gabrys, B., Applications of machine learning in biopharmaceutical process development and manufacturing: current trends, challenges, and opportunities, arXiv preprint, arXiv:2310.09991, 2023.
31. Ali, M., Shabbir, K., Ali, S., Mohsin, M., Kumar, A., Aziz, M., Zubair, M., Sultan, H.M., A new era of discovery: how artificial intelligence has revolutionized the biotechnology. *Nepal J. Biotechnol.*, 12, 1, 1, 2024.
32. Bideault, G., Scaccia, A., Zahel, T., Landertinger, R.W., Daluwatte, C., Verification and validation of computational models used in biopharmaceutical manufacturing: potential application of the ASME verification and validation 40 standard and FDA proposed AI/ML model life cycle management framework. *J. Pharm. Sci.*, 110, 4, 1540, 2021.

33. Niazi, S.K., Regulatory perspectives for AI/ML implementation in pharmaceutical GMP environments. *Pharmaceuticals*, 18, 6, 901, 2025.
34. Friederich, J., Francis, D.P., Lazarova-Molnar, S., Mohamed, N., A framework for data-driven digital twins of smart manufacturing systems. *Comput. Ind.*, 136, 103586, 2022.
35. Okuyelu, O. and Adaji, O., AI-driven real-time quality monitoring and process optimization for enhanced manufacturing performance. *J. Adv. Math. Comput. Sci.*, 39, 4, 81, 2024.
36. Helleckes, L.M., Hemmerich, J., Wiechert, W., von Lieres, E., Grünberger, A., Machine learning in bioprocess development: from promise to practice. *Trends Biotechnol.*, 41, 6, 817, 2023.
37. Canzoneri, M., Horner, M., Banerjee, P., Digital twins in biomanufacturing, in: *Bioconverg. Revol.: Chall. Opp. Res. Productiv.*, p. 59, Springer, Cham, 2024.
38. Udugama, I.A., Lopez, P.C., Gargalo, C.L., Li, X., Bayer, C., Gernaey, K.V., Digital twin in biomanufacturing: challenges and opportunities towards its implementation. *Syst. Microbiol. Biomanuf.*, 1, 3, 257, 2021.
39. Zeberli, A., Badr, S., Siegmund, C., Mattern, M., Sugiyama, H., Data-driven anomaly detection and diagnostics for changeover processes in biopharmaceutical drug product manufacturing. *Chem. Eng. Res. Des.*, 167, 53, 2021.
40. PM, V.P. and Soumya, S., Advancements in anomaly detection techniques in network traffic: the role of artificial intelligence and machine learning. *J. Sci. Res. Technol.*, 38, 2024.
41. Nwachukwu, C., Durodola-Tunde, K., Akwiwu-Uzoma, C., AI-driven anomaly detection in cloud computing environments. *Int. J. Sci. Res. Arch.*, 13, 2, 692, 2024.
42. Kodumuru, R., Sarkar, S., Parepally, V., Chandarana, J., Artificial Intelligence and Internet of Things Integration in Pharmaceutical Manufacturing: A Smart Synergy. *Pharmaceutics*, 17, 3, 290, 2025.
43. Kagalwala, H., Radhakrishnan, G.V., Mohammed, I.A., Kothinti, R.R., Kulkarni, N., Predictive analytics in supply chain management: the role of AI and machine learning in demand forecasting. *Adv. Consum. Res.*, 2, 142, 2025.
44. Chhetri, K.B., Applications of artificial intelligence and machine learning in food quality control and safety assessment. *Food Eng. Rev.*, 16, 1, 1, 2024.
45. Syafrudin, M., Alfian, G., Fitriyani, N.L., Rhee, J., Performance analysis of IoT-based sensor, big data processing, and machine learning model for real-time monitoring system in automotive manufacturing. *Sensors*, 18, 9, 2946, 2018.
46. Chen, M. and Xia, J., Optimization parameters for efficient scale-up of fermentation process, in: *Scale-up Chemical Processes for Microbial Products and Plant-Derived Bioactive Compounds*, p. 29, Academic Press, Amsterdam, The Netherlands, 2024.

47. Hamdan, A., Ibekwe, K.I., Ilojianya, V.I., Sonko, S., Etukudoh, E.A., AI in renewable energy: a review of predictive maintenance and energy optimization. *Int. J. Sci. Res. Arch.*, 11, 1, 718, 2024.
48. Mendia, I., Gil-Lopez, S., Grau, I., Del Ser, J., A novel approach for the detection of anomalous energy consumption patterns in industrial cyber-physical systems. *Expert Syst.*, 41, 2, e12959, 2024.
49. Oladapo, B.I., Olawumi, M.A., Omigbodun, F.T., AI-Driven Circular Economy of Enhancing Sustainability and Efficiency in Industrial Operations. *Sustain.*, 16, 23, 10358, 2024.
50. Samuel, A.D., Shittu, L., Oshinubi, I., AI-Driven Predictive Maintenance for Biotech Equipment using Lean Six Sigma: Enhancing Operational Efficiency and Reducing Downtime for Medical Practices in the UK, 2022.
51. Wechsler, J., Data integrity key to GMP compliance. *BioPharm Int.*, 27, 9, 2014.
52. Sabale, M.M., Pande, V.A., Tagalpallewar, A.A., Swami, A.G., Pawar, A.T., Baheti, A.M., Maintaining data safety and accuracy through data integrity (DI): A comprehensive review. *Res. J. Pharm. Technol.*, 17, 5, 2431, 2024.
53. McDowall, R.D. and Eshelman, J., Risk Based Validation of a Laboratory Information Management System (LIMS), in: *Handbook of Validation in Pharmaceutical Processes, Fourth Edition*, p. 851, CRC Press, Boca Raton, FL, USA, 2021.
54. Ogunsola, K.O., Balogun, E.D., Ogunmokun, A.S., Enhancing financial integrity through an advanced internal audit risk assessment and governance model. *Int. J. Multidiscip. Res. Growth Eval.*, 2, 1, 781, 2021.
55. Okezue, M.C., Data integrity in the healthcare industry: analyzing the effectiveness of data security in good data and record management practices, Doctoral Diss., Purdue Univ. Grad. Sch., 2022.
56. Leal, F., Chis, A.E., Caton, S., González–Vélez, H., García–Gómez, J.M., Durá, M., Sánchez–García, A., Sáez, C., Karageorgos, A., Gerogiannis, V.C., Xenakis, A., Smart pharmaceutical manufacturing: Ensuring end-to-end traceability and data integrity in medicine production. *Big Data Res.*, 24, 100172, 2021.
57. Myllyaho, L., Raatikainen, M., Männistö, T., Mikkonen, T., Nurminen, J.K., Systematic literature review of validation methods for AI systems, arXiv preprint arXiv:2107.12190, 2021.
58. Vujović, Z., Classification model evaluation metrics. *Int. J. Adv. Comput. Sci. Appl.*, 12, 6, 599, 2021.
59. Schlegel, M. and Sattler, K.U., Management of machine learning lifecycle artifacts: A survey. *ACM SIGMOD Rec.*, 51, 4, 18, 2023.
60. Hudson, J., Artificial Intelligence and Cybersecurity Integration: Modern Database Techniques for Securing AI Models, 2024.
61. Gottipati, V.K., *Mastering Quality Assurance in Pharma: A Comprehensive Guide to cGMP, Risk Management*, Yashita Prakashan Pvt. Ltd., New Delhi, India, 2025.

62. Joshua, C., Automating Model Drift Detection in Production ML Pipelines Using Anomaly Detection Techniques, 2024.
63. Chui, M. and Francisco, S., Artificial intelligence the next digital frontier. *McKinsey Glob. Inst.*, 47, 3.6, 6, 2017.
64. Darra, N., Anastasiou, E., Kriezi, O., Lazarou, E., Kalivas, D., Fountas, S., Can yield prediction be fully digitilized? A systematic review. *Agronomy*, 13, 9, 2441, 2023.
65. Harrer, S., Menard, J., Rivers, M., Green, D.V., Karpiak, J., Jeliazkov, J.R., Shapovalov, M.V., del Alamo, D., Sternke, M.C., Artificial intelligence drives the digital transformation of pharma, in: *Artificial Intelligence in Clinical Practice*, p. 345, Academic Press (Elsevier), Amsterdam, The Netherlands, 2024.
66. Kase, S.E., Hung, C.P., Krayzman, T., Hare, J.Z., Rinderspacher, B.C., Su, S.M., The future of collaborative human-artificial intelligence decision-making for mission planning. *Front. Psychol.*, 13, 850628, 2022.
67. Lee, J., Ni, J., Singh, J., Jiang, B., Azamfar, M., Feng, J., Intelligent maintenance systems and predictive manufacturing. *J. Manuf. Sci. Eng.*, 142, 11, 110805, 2020.
68. Katta, S.R., Artificial Intelligence for Predictive Maintenance: Analyzing Work Order Trends across Global Pharmaceutical Sites. *IJLRP-Int. J. Lead. Res. Publ.*, 5, 8, 2024.
69. Sinha, G.K., Automating Root Cause Analysis of Refinery Incidents *via* Generative Deep Learning and Data Analytics. *J. Technol. Innov.*, 4, 1, 2023.
70. Tien, J.M., Internet of things, real-time decision making, and artificial intelligence. *Ann. Data Sci.*, 4, 2, 149, 2017.
71. Ambler, S.W., The agile scaling model (ASM): adapting agile methods for complex environments. *Environments*, 1, 2009.
72. Lakshmi, S., Durgude, U., Chitra, V., The role of upskilling and reskilling for talent transformation in the era of AI; theoretical framework and future research directions, Weser Books, p. 198, 2024.
73. Zavrazhnyi, K.Y., The impact of the use of artificial intelligence and digital transformation on sustainable business development, Teadmus OÜ, 2024.
74. Gomaa, A.H., Lean 4.0: A Strategic Roadmap for Operational Excellence and Innovation in Smart Manufacturing. *Int. J. Emerg. Sci. Eng. (IJESE)*, 13, 4, 1, 2025.
75. Tetteh-Caesar, M.G., Gupta, S., Salonitis, K., Jagtap, S., Implementing Lean 4.0: a review of case studies in pharmaceutical industry transformation. *Technol. Sustain.*, 3, 3, 354, 2024.
76. Rojas, L., Peña, A., Garcia, J., AI-driven predictive maintenance in mining: a systematic literature review on fault detection, digital twins, and intelligent asset management. *Appl. Sci.*, 15, 6, 3337, 2025.
77. Onoja, J.P., Hamza, O., Collins, A., Chibunna, U.B., Eweja, A., Daraojimba, A.I., Digital transformation and data governance: Strategies for regulatory

compliance and secure AI-driven business operations. *J. Front. Multidiscip. Res.*, 2, 1, 43, 2021.

78. Al-Barakati, A., Leveraging Artificial Intelligence–enabled workflow framework for legacy transformation. *Int. J. Adv. Comput. Sci. Appl.*, 12, 12, 2021.
79. Jay, R., *Enterprise AI in the Cloud: A Practical Guide to Deploying End-to-end Machine Learning and ChatGPT Solutions*, John Wiley & Sons, Hoboken, NJ, USA, 2023.
80. Spring, M., Faulconbridge, J., Sarwar, A., How information technology automates and augments processes: Insights from Artificial-Intelligence-based systems in professional service operations. *J. Oper. Manage.*, 68, 6–7, 592, 2022.
81. Campbell, L., Leveraging AI to Optimize MES and ERP Systems for Improved Accuracy and Efficiency in Manufacturing, 2020.
82. Egaña Tomic, T.C., A maturity model for process data analytics in biopharmaceutical manufacturing, Doctoral Dissertation, Massachusetts Institute of Technology, 2021.
83. Ahmad, T., Boit, J., Aakula, A., The role of cross-functional collaboration in digital transformation. *J. Comput. Intell. Robot.*, 3, 1, 205, 2023.
84. Shaban, A.A. and Zeebaree, S.R., Building Scalable Enterprise Systems: The Intersection of Web Technology, Cloud Computing, and AI Marketing, 2025.
85. Sundaramurthy, S.K., Ravichandran, N., Inaganti, A.C., Muppalaneni, R., AI-powered operational resilience: Building secure, scalable, and intelligent enterprises. *Artif. Intell. Mach. Learn. Rev.*, 3, 1, 1, 2022.
86. Nwoke, J., Harnessing predictive analytics, machine learning, and scenario modeling to enhance enterprise-wide strategic decision-making. *Int. J. Comput. Appl. Technol. Res.*, 14, 4, 1010, 2025.
87. Subramanian, S., Leveraging IoT data streams for AI-based quality control in smart manufacturing systems in process industry. *J. AI-Assist. Sci. Discov.*, 3, 3, 37, 2023.
88. Ejjami, R., Emerging Professions in the Age of AI Across Multiple Sectors. *Int. J. Multidiscip. Res. (IJFMR)*, 6, 5, 1, 2024.

Part 4

EMERGING BIOMATERIALS AND BIO-PRINTING INNOVATIONS

15

Emerging Bio-Inks for 3D Bioprinting of Functional Tissues in Biomanufacturing Applications

B. Karuna, K.S.N.V. Prasad, Rompicherla Srividya, Navapet Vikas Reddy, Tirumala Shreya, Komere Vasu and A.V. Raghavendra Rao*

Department of Chemical Engineering, B V Raju Institute of Technology, Narsapur, Telangana, India

Abstract

Three-dimensional (3D) bioprinting is a transformative technology in regenerative medicine and biomanufacturing, enabling the fabrication of complex tissue-like structures with high precision. At the heart of this innovation are bio-inks specialized biomaterials engineered to mimic the extracellular matrix, support cell viability, and facilitate tissue maturation. This chapter explores the design, classification, and functionality of emerging bio-inks for applications in tissue engineering and scalable biomanufacturing. It provides an overview of natural (e.g., alginate, collagen), synthetic (e.g., PEG, Pluronic), and hybrid hydrogels, highlighting their rheological properties, printability, and cell compatibility. Recent advancements in stimuli-responsive smart bio-inks, conductive bio-inks, and growth factor-loaded systems are discussed, along with the role of crosslinking techniques in improving mechanical stability. Applications in fabricating liver, cardiac, skin, and cartilage tissue models are reviewed, with emphasis on their relevance for drug testing, disease modeling, and therapeutic production. The chapter concludes with industrial scalability challenges, regulatory considerations, and future research directions in the context of GMP-compliant bio fabrication. This comprehensive coverage makes the chapter highly relevant for both academic researchers and biomanufacturing professionals seeking innovative and sustainable solutions in personalized therapeutics.

***Keywords*:** Bio-ink, 3D bioprinting, tissue engineering, additive manufacturing

**Corresponding author*: raghusrf@gmail.com

Sandip Kunar, Jagadeesha T. and D. Sathis Kumar (eds.) Biomanufacturing: Transforming Production of Biologics and Biopharmaceuticals, (435–496) © 2026 Scrivener Publishing LLC

15.1 Introduction to Bioprinting

One of the technologies under the research spotlights is 3D bioprinting, which has quickly become a revolutionizing technology between the fields of engineering, biology and medicine. One of the unexplored opportunities of 3D bioprinting (based on its ability to precisely deposit living cells and biomaterials layer-by-layer) is the ability to fabricate complex, functional structures that closely resemble the structure and functionality of its native biological counterparts. The ability is transforming the promising areas of tissue engineering, regenerative medicine, and biomanufacturing, with profound consequences to the drug discovery, disease modeling, and patient-specific therapeutics [1–3].

The history of 3D bio printing started with simple modifications of traditional additive manufacturing methods, e.g., inkjet, extrusion and laser assisted printing to work on Bio-solutions. Initial attempts involved the printing of cell-containing hydrogels in a simple manner to show that this could work. Nonetheless, as the area has advanced, the need to create increasingly advanced tissue structures, e.g. to exhibit the complexity of such structures as the liver, heart, skin, and cartilage, has motivated the development of bio-ink architecture. Currently bio-inks are being developed to not only enable structural support but also to deliver biochemical and biomechanical cues that are required to enable tissue development as well as functionality [4–6].

The most central point of this technological revolution is bio-ink. Bio-inks refer to special formulations, usually uniquely composite, e.g., natural synthetic, or hybrid biomaterials, often loaded with living cells and bioactive molecules, which are used as the biobrick of bioprinted constructs. Bio-inks, unlike the historical inks, need to meet very specific requirements: they must be bio-compatible, cell-friendly, and cell-functional, have appropriate rheology suitable to printing, and allow the maturation of engineered tissues into functional or physiologically relevant organ-like structures. The success of bio-inks, both in terms of development and optimization, is therefore important in both the realms of 3D bioprinting research and clinical usages [7, 8].

3D printing has advanced significantly as an efficient technique, particularly for numerous uses in regenerative medicine. Figure 15.1 (a) highlighted the latest developments and possible uses of 3D printing and bioprinting in the discipline of regenerative medicine. As a result, computerized manufacture of multi-functional human 3D organs could improve the transplantation of organs and 3D tissue/organ model manufacturing

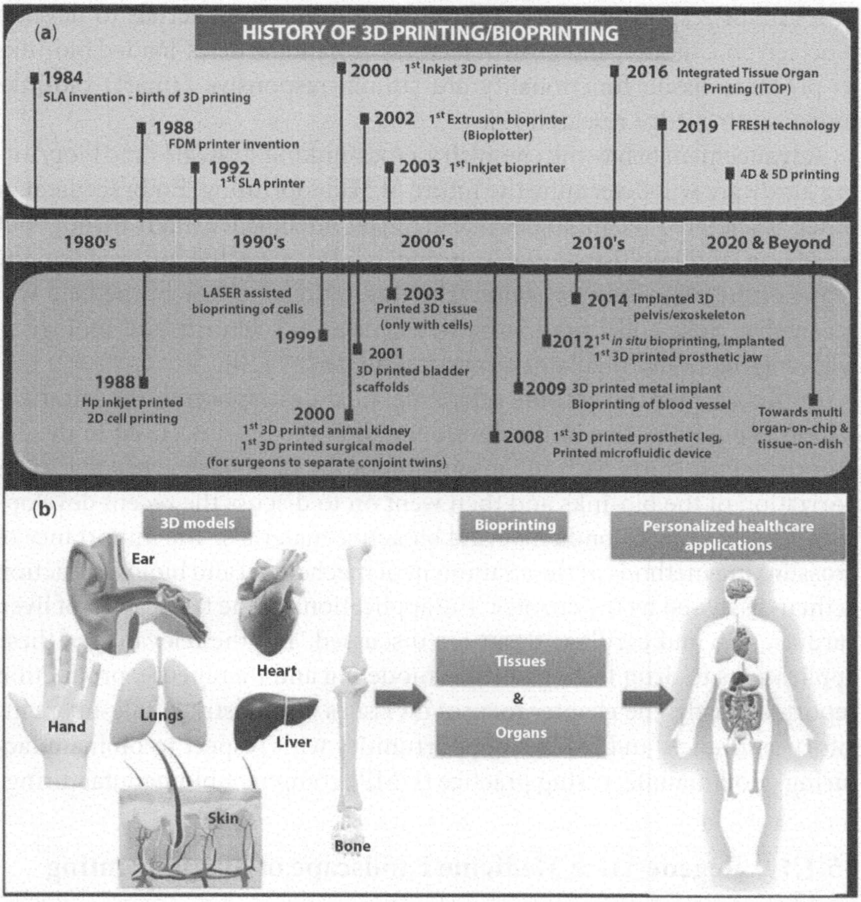

Figure 15.1 (a) A chronological overview of 3D printing/bioprinting in prospective biomedical applications, (b) a conceptual future of 3D bioprinting utilizing 3D models derived from CT or MRI images, computationally reconfigured to meet individualized specifications. [Reproduced under open license] [9].

for pre-clinical studies drug/biological testing in the future Figure 15.1(b) [9].

The past years were marked by the spade of innovations in the development of bio-ink. Natural polymers like alginate, collagen, gelatin and hyaluronic acid are in the process of refinement to enhance printability and cell interaction. Hydrophilic polymers such as polyethylene glycol (PEG) and Pluronic are being designed to have tuneable mechanical properties and degradation rate. The hybrid and composite bio-inks are developed to blend the benefits of both the natural and the synthetic ones.

The stimuli-responsive (smart) bio-inks, conductive materials to develop bioelectronic tissues, and growth factors- or nanoparticles-loaded bio-inks to promote tissue functionality are stimuli-responsive (smart) bio-inks under cutting-edge research [9].

Advancement of bio-ink chemistry, cross-linking strategies and bioprinting hardware will determine the future of 3D bioprinting. Embracement of other accelerated technologies like artificial intelligence, high-throughput screening and multi-material printing is also expected to broaden the scope of the bioprinting systems. The translational nature of the field will mean that close collaboration between material scientists and biologists, with engineers and regulatory experts, will be key [10].

In this chapter the authors offer a detailed description of the latest situation and perspective in the development of the bio-inks used in the 3D bioprinting. It starts with the discussion of the basic properties and categorization of the bio-inks and then went on to discuss the recent developments in the smart, conductive and bioactive materials. The importance of crosslinking methods in the attainment of mechanical and biological action is then discussed by the chapter. The applications in the fabrication of liver, cardiac, skin and cartilage tissue are discussed, and the relevance of these applications to drug testing, disease modeling and therapeutic production reported. Lastly, the chapter focuses on issues of industrial scale-up, regulatory assurance, and research opportunities with respect to biomanufacturing good manufacturing practice (GMP) compliant biomanufacturing.

15.1.1 Regenerative Medicine Landscape of 3D Bioprinting

The three-dimensional (3D) bioprinting represents a new high technology that allows to create living tissues and organ-like structures through the precise fabrication in the framework of layer- by-layer manufacturing approach using bioinks formulations of biomaterials and living cells [11]. This method is transforming the field of regenerative medicine as these can model diseases, repair damaged tissues and grow replacement organs, all of which are shortcomings of conventional tissue engineering and transplantation techniques [12–15].

3D printing is based on precise layering of biomaterials. Briefly, the procedure is divided into three essential steps: preparation, printing, and post-handling. The preliminary process involves generating anatomically accurate 3D models using computer graphics tools such as computer-aided design (CAD)/computer-aided manufacturing (CAM) and converting them into a stack of 2D layers with user-defined thicknesses that will be

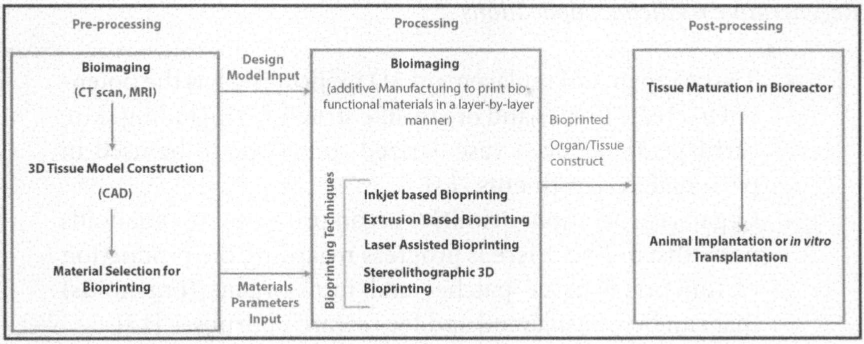

Figure 15.2 Flow-diagram for process of 3D bioprinting [Reproduced under open license] [16].

fed into the bioprinter for printing. This stage also involved selecting a substance or bio-ink. The processing step includes the real printing of the tissue using additive manufacturing techniques. Post-processing applies to the maturation of the produced construct in a bioreaction, as well as the structural and functional characterization [16]. This flow diagram in Figure 15.2 provides a comprehensive overview of the bioprinting procedure and the steps that are involved.

Bioprinting 3D principles

- Layer-by-layer deposition: 3D bioprinting uses CAD and converts medical visual information into digital templates and provides direction to the printer in precisely positioning bioinks in the Z-plane [11, 17].
- Bioinks: a special material that typically contains hydrogels or biodegradable polymers and allows bioactive molecules and living cells to be encapsulated so as to both give a scaffold the mechanical support needed and to also give the setting biological information that assists in allowing a tissue to form [17].
- Printing techniques: Popular ones are extrusion based, an inkjet based and Laser aid bioprinting. All of them possess a specific superiority regarding resolution, speed, and cell viability [17].

Regenerative medicine applications

- Tissue repair and replacement 3D bioprinting has the potential to create tissues and organ-like structures including skin, cartilage, bone, even vascularized constructs to be used in personalized treatments [13].
- Organ regeneration: Primitive organ printing and organoids notwithstanding, there is progress regarding the production of functional tissue patches and mini organs (organoids) that can be transplanted and for research purposes [15].
- Disease modeling and drug testing: Bio printing drug tissues offer more physiological disease models to examine the disease mechanism and screen drugs without extensive use of animal testing [13, 18].
- Neural tissue engineering: The technology supports the development of complicated neural structures in the study of brain functionality and neurodegenerative illnesses as well as the development of tailored treatment approaches [18].

The 3D bioprinting advantages

- Customization: Constructs specific to individual patients can be made so that the threat of immune rejection is low and integration into the body is high [17].
- Complexity: The technology enables the production of complex tissue architectures, such as multicellular structures and vascular networks, which are hard to design through conventional technologies [15, 17].
- Control situational environment: Microenvironments of native tissues can be recreated by cell-cell and biomaterial landscaping that optimizes cell activation and differentiation [11, 15].

Difficulties and constraints

- Vascularization: Development of functional blood vessels networks in thick tissues is a key issue in providing nutrients and oxygen to the cells [15].

- Mechanical stability: Printed tissues should be mechanically stable and with enough strength and durability to take physiological loads [17].
- Cell source and viability: As the number of viable cells is preserved both during and after printing, and adequate populations of suitable cells can be obtained, these acts as continuous challenges [13].
- Regulatory and ethical questions: Regarding cell sourcing and usage, clinical translation of bioprinted tissues should also be accompanied by strong quality control processes, regulatory regulation and questions of ethics [13].

Future directions

- Smart bioinks: Design of stimuli-sensitive materials that can respond to the environment as well, even better integrating and functioning within the tissue.
- Interaction with advanced technologies: Capacities to combine bioprinting and microfluidic systems, artificial intelligence, and high-throughput screening will improve the rate of innovation and clinical translation [15].
- Personalized and sustainable medicine: The eventual ambition of this type of research is to be able to produce patient-specific organs and tissues when they are needed, thereby alleviating the donor transplant problem and allowing seeking more effective, more customized treatments [13, 15].

15.1.2 Bio-Ink Significance in Biomanufacturing

The success and the progress of 3D bioprinting technologies in general and biomanufacturing applications, in particular, are dependent on bio-inks, with the aim to create functional, scalable, and clinically relevant tissue constructs. They are significant because of their special capacity in merging the biological and engineering concepts to produce living tissues and organs in a precise and copyable manner [19–23].

15.1.2.1 Important Uses of Bio-Inks in Biomanufacturing

Bio-inks are important components of biomanufacturing since they are the starting materials and allow developing functional engineered tissues and organs using 3D bioprinting. Their special characteristics and performances dominate the effectiveness, repeatability, and upscaling of bioprinted constructs in regenerative medicine, drug discovery, and in industry [20, 24].

i. Form support and form fidelity

Inks Bio-inks are required to carry out the scaffold or the essential architecture required to preserve the bioprint structure during and after the printing process. Not only will their rheological and mechanical properties need careful adjustment, but in doing so we will be able to print complex, tissue-like geometries [20]. It is very important to reproduce the extracellular matrix (ECM) and aid in cell spatial organization.

ii. Cell entrapment and vitality

One of the key roles of bio-inks is to surround living cells, shielding them against the printing process itself and allowing survival, proliferation and differentiation of the cells in the form of extracellular matrices once the printing has been done. Bio-ink environment should be biocompatible and non-toxic and should give essential cues necessary to cellular actions and tissue-structuring [20].

iii. Processability and printability tunable

Bio-ink can be designed with optimal viscosity and shear-thinning behavior and can thus be extruded or precise deposited through nozzles of bio-printers [25]. This guarantees micro/nano-scale pattern and reproducibility which are critical in automated, high throughput biomanufacturing [20]. The ability to print a bio-ink has a direct impact on how well the fidelity and scalability of printing tissue.

iv. Mechanical and crosslinking

Once deposited, bio-inks require further processing (e.g. crosslinking or stabilizing) to supply the required mechanical strength in order to cope with the handling and physiological requirements [20]. Depending on the crosslinking method, physical, chemical or enzymatic, the stability, the degradation profile and the final function of the bioprinted tissue is affected.

v. Tissue specific applications customization

By changing the composition (natural, synthetic, or hybrid materials) bio-inks can be adjusted to meet the mechanical, chemical, and biological properties of the target's tissues [20]. The wide application in biomanufacturing capability exists because this versatility can be used in the design of a variety of types of tissues; everything ranging to the very soft neural tissues and more stiff forms such as bone and cartilage.

vi. Automated and high-throughput manufacturing
Bio-ink technology allows manufacturing tissue structures in an automated, reproducible and scalable way [20]. This becomes very necessary in industrial and clinical settings, where consistency and quality control (in terms of regulations and therapeutic response) are the key drivers.

vii. Facilitation of functionalization and bioactivity
More advanced bio-inks can be functionalized to become more bioactive, vascularization promoting or cell-guiding, either through incorporation of growth factors, signaling molecules, or nanoparticles [20]. This ability to combine bioactive constituents is essential in the production of tissues that not just recreate structure; they also demonstrate physiological activity.

15.2 Fundamentals of Bio-Inks

Bio-inks form the core of 3D bioprinting, which allows the needed interface between biological cells and built environments to build functional tissue constructs. Learning their basics is highly relevant to developing bioprinting applications in regenerative medicine and biomanufacturing.

Bio-inks are a class of special materials incorporated in the 3D bioprinting to create engineered or artificial living tissues. They mostly consist of living cells and typically gels are added to act as a support matrix on which the cells are encapsulated and allowed to grow [26]; in most cases biopolymer gels or hydrogels. These components in combination can be used to produce complex functional tissue constructs resembling that of natural tissues [27].

15.2.1 Essential Elements of Bio-Inks

- Cells: The base component of bio-inks and may consist of primary cells, stem cells or cell aggregates, depending upon the intended tissue application. These cells do the job of creating the actual structure of living tissue after printing [20].
- Biomaterials (Hydrogels/Polymers): The majority of bio-inks are based on natural (e.g. alginate, collagen, gelatin) or synthetic (e.g. PEG, Pluronic) polymers that create a hydrogel

structure. The matrix recreates the extracellular space, offers mechanical support and preserves the shape of the printed construct [22, 26].

- Bioactive Molecules and Nutrients: To enhance cell viability, proliferation and differentiation and to direct the maturation and the activity of tissues, growth factors, peptides and nutrients can be added.
- Other Additives: Various additives can be incorporated in the bio-inks depending upon application, i.e. nanoparticle, signaling molecules or crosslinkers can be added to increase mechanical strength, bioactivity or ease of printing [22].

15.2.2 Definition and Fundamental Characteristics of Bio-Inks

Key traits of bio-inks play fundamental roles in the acceptability of bio-inks in 3D- bioprinting, and successful preparation of a functional tissue construct. These characteristics maintain that bio-inks are easily processed and provide cell viability and that there is similarity to native tissue micro-environment. The following properties are the most important:

- Biocompatibility: The bio-inks should be non-toxic and should facilitate survival, propagation, and differentiation of the cells. Highly biocompatible materials can be naturally or man-made, yet the cellular behaviors can depend on the origin and constituent [24, 28, 29].
- Printability and Rheological Properties: Perfect bio-inks need to have appropriate viscosity and shear-thinning properties, facilitating an easy movement after extrusion/deposition by the bioprinting nozzle as well as keeping its shape integrity after printing. Rheological effects like, viscosity, elasticity, and gelation kinetics play an important role in high-resolution structure and elimination of cell damage in the process of printing [28–30].
- Mechanical Strength and Integrity: Mechanical properties, including modulus of elasticity and the compressive strength should be designed to fit to the given tissue and offer sufficient support during the formation of the tissues. The polymer backbone, crosslinking mechanism, and structure of the molecules affect this mechanical integrity of the polymer [28].

- Gelation/Crosslinking: Bio-inks should have the ability to be crosslinked or gelled (physical, chemical or enzymatic) rapidly and in a controlled manner to stabilize the printed construct and to allow construction of complex, multi-layered tissues [31].
- Biodegradability: A perfect bio-ink should degrade at a rate that is matched with tissue regeneration in order to replace a scaffold with the gradual insertion of native components of extracellular matrices [31].
- Cell Adhesion and Bioactivity: Bio-inks must have chemical signals and surface characteristics that will facilitate cell attachment and cellular movement and tissue-specific capabilities. These interactions may be improved by the addition of bioactive molecules or by altering the chemistry of the surface of the material [24].
- Bio-ink Requirements: Bio-inks have to be customized to the particular bio printing paradigm (such as extrusion, ink-jet or laser-assisted), and this customization might include optimization of viscosity, crosslinking kinetics, or other physical parameters [31].
- Chemical Modifiability and Customization: Bio-ink chemical modifiability and customization enables integration of tissue-specific bio-cues, mechanical property tuning or the addition of growth factors and signalling molecules to direct the formation of tissues.
- Scalability and Reproducibility: At the industrial and clinical translation progenitor, bio-inks could be produced in large volumes with small batches-to-batch variation to offer consistency and quality across uses.

15.2.3 Bio-Ink: Its-Practice within Tissue Engineering

Bio-inks are a key strategy of tissue engineering and allow producing living and functional tissue constructs capable of reproducing the structure and the functionality of natural tissues. It is their relevance that can be explained by a number of factors:

I. 3D cell support and architecture scaffold

Bio-inks offer the 3-dimensional (3D) scaffold upon which cells can be grown during and following the printing procedure. This matrix resembles extracellular environment, and cells remain viable in them, multiply, and self-assemble into sophisticated systems of tissues. One robust benefit is the ability to accurately control the geometry and architecture of printed constructs and hence the possibility of replicating the architecture specific to tissues, which are necessary to the structure and functionality [22, 32].

II. Cell viability, proliferation and differentiation way promotion

An optimized bio-ink maintains high cell viability throughout and after the print process providing a supportive environment that allows cell growth and development. As an illustration, bio-inks made of collagen can support cells viability and even induce them to differentiate along a particular lineage (osteogenic or adipogenic), through the manipulation of the composition of the matrix and matrix stiffness. Tissue-specific differentiation and maturation can also be boosted by using composite bio-inks, e.g. collagen-alginate [20, 22].

III. Bioactive molecules delivery

Bio-inks may be designed to deliver and secrete growth factors, signaling molecules or drugs that instruct cell behaviors and tissue formation. As an example, TGF -beta inclusion in bio-inks demonstrated an improved chondrogenic differentiation process where VEGF could induce angiogenesis into printed constructs [33]. This manipulation ability of controlled delivery plays a pivotal role in regulating tissue particular operations and enhancing regenerative advances within the made-up tissues.

IV. Tissue specific development

Bio-inks can be designed so that their composition and material work towards a tissue engineering application. As an illustration, hydrogels are chosen regularly depending on the compatibility of the biomaterial; mechanical strength; and supporting cell types, e.g., alginate, collagen, and gelatin [30, 34]. Silk-based and composite bio-inks may be tailored toward bone, cartilage, or skin engineering, thereby offering the requisite cues toward the specific tissue formation [22].

V. Synthesis of multicellular complex structure enabling

Bio-inks could be used to build complex, multicellular tissue models in cell types that express defined spatial patterns, by facilitating co-printing of different cell types in specific spatial arrangements. This is necessary to recreate the heterogeneity of the native tissues and to create constructs that lend themselves to transplantation or to a disease model application or drug testing [20].

VI. *In vitro* tissue maturation support cells and *in vivo* tissue maturation support cells

The bio-inks do not only offer early support to the cells but also support a long-term maturation *in vitro* and following *vivo* implantation. As an example, bioprinted skin structures comprising collagen bio-inks showed the generation of a multi-layered functional skin, including intercellular communication and suitable cell differentiation (both *in vitro* and *in vivo*).

15.3 Bio-Ink Classification

The existence of the bio-ink in 3D bioprinting is categorized widely according to their composition as well as operation. The most popular types are based on materials classification which involves such subsets as natural bio-inks, synthetic bio-inks, hybrid/composite bio-inks, and functional subsets including scaffold-based bio-ink and scaffold-free bio-inks. Depending on the needs of a particular application in tissue engineering, each form has advantages and is chosen depending on particular needs [20, 35].

15.3.1 Natural Bio-Inks

All existing bio-inks are biomedically based and can be naturally attained within natural bio-inks since they have biological origins, can resemble the extracellular matrix (ECM), and provide compatibility with cell life and work within their framework [29, 36]. Generally, these bio-inks contain naturally occurring proteins, polysaccharide, and other biomolecules, to such an extent that they are particularly compatible with tissue engineering and in the area of regenerative medicine.

15.3.1.1 Alginate-Based Systems

Alginate is a polysaccharide derived out of brown seaweed. It comes in handy, as it gelates mildly in the presence of divalent cations (i.e., calcium), allowing it to solidify fast under benign conditions that do not impair the health of living cells. Hydrogels of alginate are very biocompatible and can be extruded easily and used to form hydrogel spheres but do not bear natural cell adhesion sites which may need to be modified or blended with other materials to provide bioactivity [29, 37]. Bio-inks commonly applied in cartilage, liver, and vascular tissue models are based upon alginate.

15.3.1.2 Collagen and Gelatin

In ECM collagen is the mostly used structural protein and offers excellent cell adhesion and biologic signaling. With its cell-specific behavior on tissues, it makes it suitable in soft tissue engineering. Deglaciated collagen is called gelatin, which is thermoresponsive and can also undergo chemical modifications (e.g. Gelatin Methacryloyl (GelMA)) to enhance mechanical properties and subsequently allow photopolymerization. The two materials are very biocompatible, enable cell attachment, and are applied in skin, cardiac, and connective tissue constructs [37, 38].

15.3.1.3 Other Natural Polymers (e.g., Fibrin, Hyaluronic Acid)

The polymerization of fibrinogen is involved in blood clotting and wound healing making fibrin central to those processes. It promotes both angiogenesis and can be used to form vascular and skin tissue engineering. A glycosaminoglycan is the hyaluronic acid that is in the connective tissues and encourages cell migration and cell proliferation. Its bioactivity is delivered by both materials that can also be mixed with other polymers as needed to match desired mechanical and biological properties and applications [27, 37, 38].

15.3.2 Synthetic Bio-Inks

Engineered synthesized bio-inks are polymers with tunable mechanical behavior, high reproducibility and ready versatility. They do not usually have any bioactivity on their own, but may be functionalized or combined with natural compounds to improve cell interactions [36, 37, 39].

15.3.2.1 Polyethylene Glycol (PEG)

As a non-immungenic, hydrophilic, polymer, PEG is often employed because of the potential to universally adapt its mechanical and chemical properties. PEG hydrogels can be chemically crosslinked which hosts complete control over the rate of gel formation and degradation, but can also be photoinitiated, permitting crosslinking on short times scale. Although PEG is biologically inert, it may be functionalized with peptides or growth factors to facilitate cell adhesion and functioning and such functionalized PEG may be applied to any number of tissue engineering projects [37, 39].

15.3.2.2 Other Synthetic Polymers Pluronic

Pluronic (poloxamer) is a triblock copolymer that gels at physiological temperatures thus is useful as sacrificial or support material in bioprinting. Other synthetic plastics, which include polycaprolactone (PCL) and polylactic acid (PLA), can be prepared to act as scaffolding and may be customized in terms of degradability and strength. These materials can find applications where a high print fidelity and mechanical strength is desired, in general they can require blending or surface tenturation to help with cell compatibility [37, 39].

15.3.3 Others (Hybrid and Composite Bio-Inks)

In hybrid and composite bio-inks, natural and synthetic ingredients are mixed to exploit the advantages of each type of bioactivity of natural polymers and the programmable mechanics of synthetics [36, 37]. As an example, combining alginate with gelatin or PEG with collagen to produce bio-inks that have enhanced printability, mechanic and cell functional support. Functionality can also be increased by the addition of nanoparticles or bioactive molecules in composites such as those that enhance vascularization or electrical conductivity. The complex tissue constructs, which need both the biological and structural performance, are rapidly becoming the applications of these systems [37].

Table 15.1 presents a comparative review of various bioinks utilized in tissue engineering and regenerative medicine. It classifies bioinks according to their origin and composition, detailing their principal characteristics, problems, uses, and current developments.

15.4 Important Properties of Bio-Inks

Bio-ink properties key properties of bio-inks are fundamental in the successful fabrication of reproducible, functional, and robust physiologically relevant tissue constructs *via* 3D bioprinting. Such characteristics govern appropriateness of the bio-ink to particular applications, its capacity to sustain cells, tissue development and architectural integrity [20].

15.4.1 Rheological Characteristics

The rheological properties are core to bio-ink performance in 3D bioprinting, and it will directly affect their printability, the structural fidelity of the

Table 15.1 Different types of bioinks [8].

Type of bioink	Description	Advantages	Challenges	Application	Recent advances
Natural polymers	Biologically derived polymers, including collagen, gelatin, and alginate, are frequently utilized because to their biocompatibility.	Exhibits high biocompatibility, is biodegradable, and promotes the growth of cells.	Restricted mechanical strength and variable characteristics.	Soft tissue engineering and vascular tissue	Improved crosslinking processes improve mechanical stability in natural polymers.
Synthetic polymers	Man-made polymers can be tuned to precise mechanical and biological qualities.	Modifiable attributes, regulated deterioration rates, and structural integrity.	Absence of biological indicators and reduced degradation rate.	Bone and cartilage regeneration, scaffold fabrication.	Advancements in synthetic biodegradable polymers featuring adjustable degradation rates.
Composite bioinks	Integration of synthetic and natural materials to augment structural and biological characteristics.	Combines the properties of synthetic and natural materials to create customized bioinks.	Intricate formulation and optimization required for printing	Multi-functional tissues and hybrid scaffolds for intricate tissue regeneration	Hybrid bioinks that integrate nanomaterials and bioactive chemicals for the development of intricate tissues

(*Continued*)

Table 15.1 Different types of bioinks [8]. (*Continued*)

Type of bioink	Description	Advantages	Challenges	Application	Recent advances
Hydrogels	Crosslinked polymeric materials that are water-swollen and are particularly well-suited for the replication of the extracellular matrix (ECM).	Biocompatibility, tunable mechanical properties, and a high-water content.	Restricted mechanical strength and gradual deterioration.	Tissue engineering, wound repair, and pharmaceutical delivery systems.	Advancement of shear-thinning hydrogels to enhance printability.
Cell aggregates	Tissue spheroids are cell clusters that autonomously organize into functional organ structures.	Self-assembling facilitates the natural development and maturity of tissue.	Challenging to manage and reproduce because to aggregation tendencies.	Organoids, tissue regeneration, and cell-based tissue models.	Advancements in bioreactor technologies that facilitate large-scale tissue maturation.
Decellularized Extracellular Matrices (dECM) Nanomaterials	Bioinks sourced from tissue extracellular matrix, providing a scaffold that emulates original tissue settings.	Natural biomimicry facilitates cell adhesion and tissue proliferation.	Intricate extraction and possibility of immunological reaction.	Regenerative medicine utilizes tissue-specific bioinks for organ fabrication.	Enhanced decellularization methods to preserve extracellular matrix bioactivity.

(*Continued*)

Table 15.1 Different types of bioinks [8]. *(Continued)*

Type of bioink	Description	Advantages	Challenges	Application	Recent advances
Nanomaterials	Nanoparticles and nanocomposites integrated into bioinks to improve mechanical characteristics and bioactivity.	Improved mechanical qualities, capability for targeted distribution, and bioactivity.	Concerns regarding toxicity and issues associated with uniform distribution.	Bone regeneration, pharmaceutical delivery, and antibacterial uses.	Incorporation of nanoparticles to improve conductivity and bioactivity.

print, and their ability to support cell viability. Rheology can be defined by the analysis of the flow and deformations of materials under force manipulation and in case of bio-inks, it is essential that a number of primary rheological properties be optimized [40, 41].

- Viscosity: This is how difficult a fluid flows. Bio-inks of suitable viscosity give easy printing. Excessive viscosity will clog nozzles whereas viscosity that is too low may result in poor shape retention [42, 43].
- Shear-thinning behavior: A desirable property of bio-inks is that they must exhibit shear thinning behavior i.e., viscosity drops under a shear stress (say extrusion) and then returns to equilibrium upon the removal of the shear stress. The feature endows it with the ability to flow well through the printer nozzle, and the ability to retain structure after printing [40, 42].
- Yield stress: It is the stress that must occur to get the flow started. The appropriate yield stress ensures that the bio-ink does not drip or sag prior and post printing, thus fidelity of shape and stacking the layers [42].
- Thixotropy: Thixotropic bio-inks show temporary reduction in viscosity with shear and regain their viscosity with the endurance of time after the release of stress. Such a time-based recovery is of advantage in keeping the printed structure and strengthening the stack on top of it [42, 44].
- Elastic and viscous moduli (viscoelasticity): Bio-inks tend to exhibit solid-liquid behavior (viscoelasticity), i.e., being both solid (or elastic) and liquid (or viscous). To achieve these properties, it is important to balance them to sustain the shape during and after the print, allow cell encapsulation, and tissue maturation [42, 44].
- Gelation kinetics: The speed at which a bio-ink shifts between the liquid and gel (gelation) conditions influences the printability and the print accuracy of its structural stability. Often desired fast and controllable gelation is used to stabilize the construction after printing [30].

15.4.2 Printing and Structural Faithfulness

The structural fidelity together with their printability is a paramount characteristic of bio-inks, that define their capability to create and preserve reproducible structures with accuracy of architecture (three-dimensional) during the process of bioprinting and their post-bioprinting existence. These properties depend on the combination of the formulation of bio-ink, the rheology, the printing parameters, as well as the stabilization protocols after printing [42–44].

Printability means that a bio-ink can be layer-by-layer printed into a 3D structure where the design and geometry can be well controlled. To be called printable, a bio-ink must:

- Have an acceptable viscosity and shear thinning response with ease to melt out or deposit through the nozzle and at the same time preserve the desired shape [40].
- Having speedy gelation/crosslinking capacity after deposition so as to reinforce the shape and avoid collapsing or dispersing [30].
- Admit high-resolution patterning well beyond existing methods, in order to reproduce complex tissue architectures [45].

The printability is dependent on major factors that include:

- Rheological characteristics (viscosity, critical strain, thixotropy)
- Bio-ink concentration and composition
- Printing conditions (pressure, speed, the diameter of the nozzle)
- Crosslinking processes (type, rate and method)

Structural fidelity

Structural fidelity, sometimes also described as shape fidelity, is the degree to which the geometry and dimension of the printed construct is maintained within and after a printing process [30]. The structurally high fidelity plays critical roles in making sure the resultant tissue construct functions and looks like the original tissues.

- The geometric parameters of interest that can be used to give a quantitative evaluation of the shape fidelity include the pore shape, the filament width, and the layer thickness, and then compared to the desired design [46].

- Loss of ideal form (e.g., filament pooling, sagging or spreading) is frequently the result of sub-optimal viscosity, inadequate gelation or lack of mechanical integrity [47].
- Post printing: to support filament stability and ensure construct architecture, post-printing crosslinking (e.g. using UV or visible light) is typically employed [46].

Optimization strategies: To attain the ideal printability, and structural accuracy:

- Craft bio-inks of optimum viscosity and gelation kinetics that are easy to extrude and quickly stabilize [48].
- Change the printing parameter like the extrusion pressure, speed, nozzle size according to the rheological profile of the bio-ink [49].
- Use after printing cross linked or support materials to increase stacking of the layers and avoid deformation.

15.4.3 Compatibility and Cell Viability

The most important and least understood attributes of any bio-ink applied in 3D bioprinting is the compatibility and viability of the cells to ensure the success of any tissue engineering and regenerative medicine initiative.

Cell compatibility (biocompatibility) is the capacity of a bio-ink to be able to sustain key cellular processes, e.g. adhesion, proliferation and differentiation, and not have toxicity, or any other adverse local or systemic effect in the host [50]. The best bio-ink should allow high cell viability both during the printing process and, afterwards, to allow cells to keep track in good condition and to be involved in the process of tissue healing [51].

The determinants of cell compatibility and viable cell concentration in bio-inks are some of the most important reasons:

- Wonder Inks that are non-toxic: The bio-inks have to be made of materials that produce no toxic products during use or during degradation. Such additives as fillers, plasticizer, or crosslinking agents should be chosen carefully because some of them may have adverse effects on the health of cells or clinical practice when the additives are leachable [51].
- Physiological processing conditions: Print temperatures and crosslinking strategies should not be higher than

physiological conditions; and when it comes to gelation, then it should be done under mild and cell friendly conditions.

- Hydrogel based-matrix: Hydrogels are often used in bio-inks owing to their high-water content and porous nature, which allow the diffusion of nutrients and removal of wastes as well as cell migration, essential conditions in cell survival post print [30].
- Bioactivity: The bio-ink ought to resemble the extracellular matrix, hence offering the biochemical cues that aid the attachment and the spreading of cells besides specific functions of tissues [30].
- Mechanical and rheological properties: the stiffness and viscoelasticity of the substrate, and the bio-ink should be adjusted to the targeted cell type and tissue application, since they determine cellular K contains the seed cell type [43].

15.5 Recent Developments in Bio-Ink Creation

A novel bio-ink development that has occurred in recent times has allowed a vast expansion of materials that 3D bioprinting can create, creating more complex, functional and clinically relevant tissue constructs than ever. Innovation has occurred on the composition of the materials, their functionalization, their printability, and their biological performance [52].

15.5.1 Bio-Inks Stimuli Responsive (Smart)

Stimuli-responsive (smart) bio-inks constitute a disruptive category of materials in 3D bioprinting, which can reproducibly change their own properties or behavior in response to external or internal stimuli as temperature, pH, light, magnetic or electric fields, or response to biomolecules, etc. Such smart bio-inks will allow building of tissue constructs that may better mimic the dynamic, functional context of native tissues and organs, and pave new paths to regenerative medicine, drug delivery, biosensing, and more [53].

Interesting characteristics and processes key features and mechanisms

- Dynamic responsiveness: Bio-inks may be smart, in the sense that when tuned with specific stimuli, their shape, mechanical property, or even biochemical activity may change. As an example, methacrylated bio-inks (alginate) may be designed to change shapes upon addition of water, calcium chloride, or EDTA, which allows the construction of capillary-sized internal tubes compatibility with cell viability [54].

- Self-healing and shape-memory: Certain Smart Bio-inks (Inks containing nanoclays or ferrogel systems) also have: Self-healing or Shape-memory. These enable constructs to regain their deformed or damaged characteristics by enhancing durability and functionality of tissue engineering uses [53].
- Magnetic and electric responsiveness: The addition of magnetic nanoparticles or conductive fillers permits remodeling of construct behavior or alignment remotely, and in cases when engineering neural, cardiac, or skeletal muscle tissue is desired, it could find value in electrical responsiveness [55].
- Controlled drug release: Bio applications to comprise stimuli-responsive bio-inks to fabricate drug delivery profiles where the release rate is controlled by environmental factors, such as local pH or temperature. As an example, constructs printed by DLP could be designed to release drugs based on pH or injection shape to improve both efficacy and specificity of a therapeutic intervention [54].

Applications

- Regenerative medicine: Mimicking the dynamic environment of living tissues, smart bio-inks allow creating constructs that are able to integrate and work better in the body as well. As an example, they could be utilized to make vascularized tissues incorporating capillary-like channels or structures that react to physiological signals (such as enhanced healing) [56].
- Drug delivery and biosensing: 3D-printed smart bio-ink-based devices integrate drug delivery functionality or biosensing into the same print, through programmed release rates or response to particular biological cues, which can provide new methods of personalized medicine and *in-vivo* tracking [54].
- 4D bioprinting: Combination of stimuli responsive materials with 3D bioprinting is opening the door to 4D bioprinting, where printed structures can change or evolve with time in reaction to changes in the environment further extending the potential of complex, functional tissue engineering [57].

Recent Innovations

Nanoclay Incorporated Double-Network Bio-Inks: These are mixtures of methacrylated hyaluronic acid and alginate with nanoclays as physical cross-linkers that enhance shape fidelity and allow magnetic responsiveness in addition to facilitating the growth and repair of living tissues such as stem cells [54].

- Magnetic Field-Responsive Ferrogels: Ferrogels like oxidized hyaluronate in combination with glycol chitosan can be controlled remotely by a set of magnetic fields and these could open up new frontiers to the design of the controllable engineered constructs in tissues [54].
- Clickable Dynamic Bio-Inks: Advancements employ dynamic bio-ink that supports chemical changes after being printed exhibit time and space-controlled customization in composition and material strength [55].

15.5.2 Functionalized and Conductive Bio-Inks

Advances in 3D bio-printing use conductive or bio-functionalized bio-inks to create tissue structures and biomedical devices responsive to electrical stimulation. These innovations support tissue regeneration and enable applications in advanced bioelectronics [58–60].

Conductive Bio-Inks: Conductive bio-inks are created to enable the combination of mechanical [60], durability, electrical conductivity, and biocompatibility by using conductive building blocks that allow the creation of electricity-conducting structures including active materials in tissue engineering applications and implantable bioelectronics and bioesensors [61].

Essential techniques and resources

- Nanomaterial integration: Integration of nanomaterials (including graphene, reduced graphene oxide, carbon nanotubes or even metallic nanoparticle [e.g., zinc]) in biopolymer matrices (e.g., alginate and PLGA-based or PGS-based polymers) offer enhanced conductivity without losing flexibility or printability [58, 60].

- Conductive polymers: Addition of polymer such as PEDOT: PSS into hydrogels (e.g. methylcellulose/kappa-carrageenan) provides tunable electroconductivity and high (>96 %) cell viability in 3D-printed constructs supporting the suitability of conductive polymers to cell-laden applications [59].
- Mechanical and rheological optimization: Besides enhancing conductivity, nanoclays or surface-specific polymer blends also help enhance shear-thinning, stretchability, and stability of nanomaterial-based printed constructs under physiological conditions so that these printed constructs do not lose their integrity and properties during usage [61].

Applications:

- Tissue engineering: Conductive bio-inks of great importance to the engineering of excitable tissues (e.g. nerve, heart, muscle) that depend on an electrical basis of their function, and to the development of scaffolds able to provide electrical stimulation to promote tissue maturation and functioning [62].
- Flexible electronics and biosensors: Applying these inks allows printing flexible soft, stretchy electronic devices that can be printed directly in 3D and placed directly onto biological tissues to monitor, stimulate, or administer drugs to them [61].

Functionalized bio-inks: Bio-inks are removed to include other biological or physicochemical functionality beyond conductivity that need the bio-inks designed to meet desired biomedical requirements.

Key approaches:

- Bioactive molecule incorporation: Molecules that signal or guide cell behavior, induce differentiation, or increase integration into a tissue can be embedded, either as growth factors, peptides or signaling molecules.
- Adjustable electromechanical properties: The bio-ink has tunable electromechanical properties as researchers can alter the ratio of the conductive fillers and polymers, which allows them to fine control the mechanical stiffness, and

electrical characteristics of the bio-ink to suit the specific tissue environment or device needs [60].

- Multimaterial printing: Few bio-inks are bio-functionalized to work with multiple production techniques (e.g., 3D printing, electrospinning, injection molding) and increase their ability to create product or tissue architectures [60].

Performance highlights:

- Two-dimensional conductive/insulating bio-inks (e.g. graphene-PLGA) make it possible to form flexible conductive networks with fully tunable properties ideal on tissue scaffolds and transient electronics.
- These novel inks exhibit excellent printability, physiological stabilities and biocompatibility and overcome the shortcomings of conventional conductive inks [59].

15.5.3 Factor-Bioactive Systems

Since the introduction of the first growth factor-loaded bioactive bio-ink systems, one of the most significant breakthroughs in 3D bioprinting is the ability to precisely regulate the phenotype and the tissue generating potential of cells, including the introduction and relocalization of bioactive molecules in bioprinted constructs [63].

Important Principles and Mechanisms

- Spatiotemporal control of releases: Bio-inks have the potential to achieve spatiotemporal control of growth factors (e.g. BMP-2 VEGF, TGF-1) by being confined into microcarrier/hydrogels. This focal release aids local tissue replacement, which was established in vascularized bone grafts where microencapsulated BMP-2 and VEGF were top-mapped in the bio-ink to guide osteogenesis and angiogenesis to pegged regions of the construct [33].
- Increased regenerative potential: By binding to receptors on cell surfaces, the growth factor-containing bio-inks have potential in cell proliferation, differentiation, vascularization, tissue repair, and other tissue engineering operations resulting in superior effects in bone and cartilages applications [64].

- Sustained and entity delivery: Enhancements in bio-ink development, particularly in the optimization of ionic strength in collagen hydrogels, have made it such that growth factors (e.g., TGF-b1) can be sequestered and released along sustained periods of time that allow protracted stimulation of embedded cells and increased tissue-specific differentiation [65].

Uses and Cases

- Bone and cartilage regeneration: Bone and cartilage regeneration: Bone and cartilage regeneration: Using growth factor-loaded bio-inks it has been possible to construct scaffolds that lead to osteogenic and chondrogenic differentiation of stem cells to promote healing of critical-sized defects as well as simulate the mechanical and biochemical properties of the tissues [22].
- Vascularized tissue engineering: Angiogenic factors (e.g. VEGF) can be delivered within vascularized bioprint constructs at controlled spatial distributions toward increased vascular network formation and survival of larger engineered tissues [63].
- Bioactive composite systems: addressing more elaborate cellular characteristics, bio-ink could also enhance further cell adhesions, proliferation, or lineage-specific gene expression by bio-merging growth factors with bioactive materials (e.g., collagen, gelatin, alginate), to build complex, functional tissues [65].

Advanced Strategies

- Thermo-Responsive and Self-Healing Systems: Nanocomposite bio-inks supporting growth factors can be stored and released when in the presence of environmental cues, but this also enhances printability and biological activity [66].
- RNA-Based Bioactivity: Recent work has involved incorporating miRNA into collagen scaffolds, which have shown a positive growth and defect healing of bone tissues [33] evidence of the growing ecosystem of bioactive molecules available to tissue engineering.

15.6 Crosslinking Strategy

Crosslinking methods are essential to 3D bioprinting, as they convert the originally pliable, fluid bioink into a stable, three-dimensional structure that preserves its shape, mechanical integrity, and facilitates cell survival and activity. The selection and technique of crosslinking directly affect the mechanical, physicochemical, and biological characteristics of the final bioprinted structure, together with the cellular behavior within it [67].

The specified crosslinking process may have an impact on the bioink's microstructure, diffusive characteristics, and mechanical behavior. Rarely, there are three major forms: chemical, physical, and enzymatic crosslinking, as seen in Figure 15.3 [8].

15.6.1 Crosslinking in Physics

Physical crosslinking of 3D bioprinting is the process of the creation of hydrogel networks *via* non-covalent bond formations like hydrogen bonds, hydrophobic interactions, electrostatic attraction, and mostly ionic crosslinking [67]. Physical crosslinking is usually reversible, and this method creates no permanently bound covalent bonds as seen in chemical crosslinking, and does not necessitate the use of chemical agents that may be

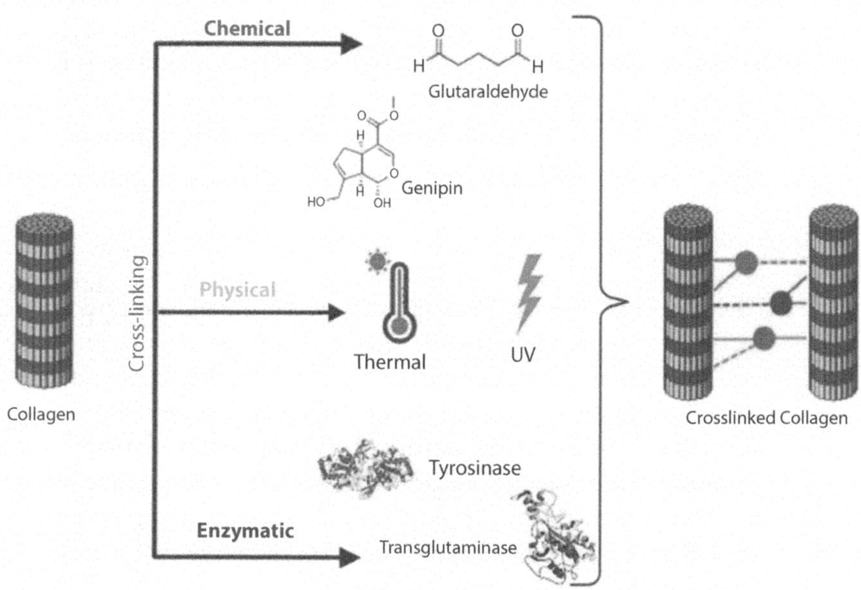

Figure 15.3 Crosslinking of collagen. [Reproduced under open license] [8].

toxic to cells (thus highly cytocompatible) and this method can be used even with living cells because of this.

The highlights of physical crosslinking

- Non-covalent interaction: The network of the hydrogel is vitrified through non-covalent interactions that are reversible, and they can be categorized into hydrogen bonding, ionic interactions (such as alginate and calcium ion interactions), among hydrophobic contacts [68].
- Mild processing parameters: The physical crosslinking can be implemented at physiological conditions (neutral pH, ambient temperature) allowing preserving high levels of cell viability during and after printing [67].
- Fast gelation: Many physically crosslinked systems (e.g. alginate in calcium ions) can gel rapidly to promote the structural integrity of printed formations.
- Tunable properties: The extent of crosslinking and mechanical properties may be tuned by varying such factors as ion concentration, temperature, or the composition of the polymer [67].

Advantages:

- Compatibility with high cells because of lacking presence of toxic cross-linking reagents
- Easy and quick gelling mechanisms
- The second generation is reversible and tunable, which is dynamic in tissue environments

Limitations:

- Weaker mechanical properties than chemically crosslinked hydrogel usually exhibited.
- May need reinforcement to serve long term or load applications [69].

15.6.2 Crosslinking-Chemical

The chemical crosslinking is an important approach in 3D bioprinting to convert a soft, flowable bio-ink to stable, rigid hydrogel scaffold *via* irreversible covalent bond formation between polymer chains. This approach is also essential in having the target strength of tissue bioprint constructs in terms of mechanical strength, structural integrity, and longevity of the construct especially when in case high durability is desired [67].

Mechanism of chemical crosslinking

- Covalent bond formation: Chemical crosslinking: chemical crosslinkers are added to a mixture or functional groups on the polymer backbone may be activated so that strong permanent covalent bonds are formed between the polymer chains.
- Common crosslinkers: Depending on the chemistry of the bio-ink, common crosslinkers include such agents as glutaraldehyde, genipin, carbodiimide, or methacrylate groups (photo-initiated crosslinking).
- Initiation methods.: Chemical crosslinking may be initiated either by introducing crosslinking agents, pH control, or exposure of the bio-ink to light (photo-crosslinking) with or without a photoinitiated [70].

Advantages

- Improved mechanical function: chemically crosslinked hydrogels have improved strengths and durability in comparison to those crosslinked physically and have a potential to be used as load bearing and long-term implants.
- Tunable network structure: The extent of crosslinking may be finely tuned at the time of crosslinking by adjusting crosslinker concentration, reaction time or exposure conditions to achieve customization of stiffness, porosity and degradation rate [71].

Considerations

- Cytocompatibility: Other chemical crosslinkers or products thereof may be toxic. Hence, selection of a favorable crosslinking environment and the necessary clearance of left-over chemicals is critical toward cell viability and biocompatibility.

- Processing conditions: Some chemical crosslinking can also be sensitive to a particular pH, temperature or light. In the case of cell-laden bio-inks, these should be gentle enough so as not to damage the encapsulated cells [72].

Applications

- GelMA: GelMA is an extensive bio-ink that is crosslinked by photo-initiated polymerization, a reliable method of creating a gel in quick time [73].
- Alginate, collagen and hyaluronic acid derivatives: These can be chemically derivatized to incorporate reactive functionalities to facilitate covalent crosslinking that increase mechanical stability and printability [71].

15.6.3 Photo Crosslinking and Enzymatic Methods

The newest methods of bioprinting called enzymatic and photo crosslinking allow producing highly tunable, structurally stable, and biocompatible constructs of hydrogel. Photo crosslinking is much faster, more specific and high-resolution as compared to enzymatic methods and critical in driving tissue engineering and regenerative medicine [74].

Enzymatic crosslinking involves the enzymatic catalyzed formation of covalent bonds between chain polymer molecules of bio-ink to form hydrogel networks. The reason this method is appreciated is because of its low and cell friendly conditions thus highly applicable in encapsulation of living cells [75].

Mechanism

The common enzymes such as horseradish peroxidase (HRP), transglutaminase or tyrosinase are utilized. As such, polymers that are modified with phenol (or tyramine) can be crosslinked together with HRP in the presence of hydrogen peroxide [75], which quickly form hydrogels in the event of mixing.

Advantage

- Favorable pH of reaction (physiologic pH)
- Compatibility and high cell viability
- Adjustable gelation speed and mechanical properties to varying enzyme or substrate concentrations [74]

Applications

- Nanoprecision assembly of tissue constructs (spatially controlled) (e.g. coaxial bioprinting vascular channels).
- Injectable hydrogels in the field of tissue engineering and regenerative medicine [75].

The latest developments are coaxial bioprinting and enzymatic crosslinking where it can gel almost immediately and the result is stable fibers of hydrogel with rate-controlled swelling and degradation.

Photo crosslinks use light (generally either ultraviolet (UV) or visible) to trigger a resin (photoreactive group (e.g., methacrylate), and photo initiator) within bio-inks. It can be used to rapidly, on-demand and spatially control crosslinking [72].

Mechanism

Photo initiators form free radicals in the presence of light which immediately induce the reaction of covalent bonding of polymer chains leading to a solidification of hydrogel structure. One of the common photo cross linkable bio-ink is GelMA [76].

Advantages

- Controllable Rapid gelling
- High patterning and spatial resolution
- Appropriate in situations of multilayered structures

Applications

- Building of strong vessels and channels permeable to perfusion (also in combination with enzymatic crosslinking to create dual stability) [76].
- High-resolution patient-specific tissue models

Innovations comprise the non-cytotoxic, *in situ*, photosensitive cross-linking processes, offering alternatives to the UV-based techniques, extending the levels of biocompatibility and structural control.

15.6.4 Influence While Causing Mechanical Instability and Tissue Maturation

A key factor in mechanical integrity and biological performance of 3D bioprinted tissue is crosslinking. Well-selected and well-tuned crosslinking approaches will guarantee that constructs were sufficiently stiff to handle and implant, and also perfects the environment needed to drive cell-mediated maturation and functions [68].

Mechanical Stability

- The crosslinking is required to change the flowability and softness of bioinks to form stable and robust 3D. The relative strength, elasticity and durability of the construct are directly influenced by the nature of crosslinking selected i.e. chemical, physical, enzymatic or photo induced [73].
- The other method is the photo crosslinking methods (e.g. UV or visible light), which enables more speedy, specific, and spatially controllable gelation, producing constructs with higher mechanical robustness and structure preservation. Crosslinked *in situ* is possible due to such innovations as insertion of optical fibers into printing nozzles to generate buffered multilayered, cell laden constructs carrying high adhesive forces between fibers and more structural mechanical integrity overall [68].
- To augment the stability of constructs and their subsequent printability, dual/hybrid crosslinking methods (i.e., the combination of crosslinking methods, e.g. ionic crosslinking (long term) and photo crosslinking (short term)) can be implemented [73].

Tissue Maturation

- The crosslinked matrix consists of mechanical signals that are fundamental in the control of cell morphology, growth and development. The compliance and stability of the 3D construct are what cells perceive and react to shaping the tissue maturation and performance [73].
- Optimized crosslinking aids *in vivo* implantation process which is retained by the construct intactness and the

fulfillment of bioactivity and physicochemical characteristics required by tissues to grow and integrate.
- Variation of crosslinking conditions (e.g. concentration of crosslinker, crosslinking time) can also be used to tune the matrix stiffness and degradation rates, which can then be tuned to the needs of various tissue types and maturation stages [68].

15.7 Examples of Functional Tissue Fabrication

By enabling the creation of living constructs with the most structural and functional similarity to the natural tissues of the humans, 3D bioprinting has been one of the normalized technologies in the area of developing the functional tissues. The combination of cells, growth factors and better-quality bio-inks can deliver fine spatial control, the assembly of multiplex, multicellular networks and structures [13].

15.7.1 Liver Tissue Model

The 3D bioprinted liver tissue models are able to change the research of liver surgery, drug screening, and regenerative medicine, since they provide physiologically viable multicellular and functional structures. These models have the potential of enhancing the predictive power of human outcomes, the ability to model more complex diseases, like metabolic dysfunction-associated steatohepatitis (MASH) and fibrosis, and the guidance in their clinical application, such as transplantation and personalized medicine [77].

Significant Attributes and Production

- Bioprinting Methods: Extrusion-based 3D bioprinting of liver tissue is the most trending technology that gives rise to cell laden bioink with which it is possible to build tissue-like and multicellular 3D tissue in layer-by-layer manner [78]. Laser-assisted bioprinting (LAB) and digital light processing (DLP) can also display high resolution non-contact and use more than 95 per cent cell viability, and produce elaborate microenvironments.
- Cell types: Most liver models include the use of the human hepatocyte like cells (HepG2 cell line; HUH7 cell line) along with primary hepatocytes, as well as non- pigmented hepatocyte

cells which encompass Kupffer cells, hepatic stellate cells, and liver sinusoidal endothelial cells [78]. More recent models are co-cultured with a heterogeneous population of cells better approximating the liver physiology and diseased liver.

- Bioinks: Owing to these properties of biocompatibility and support of liver-specific activities, protein-based (e.g. gelatin, collagen) and polysaccharide-based [e.g. alginate] bioinks are widespread and commonly used [13].

Applications

- Disease modeling: MASH, fibrosis and liver cancer are those liver diseases whose studies are carried out using the 3D printed liver models. As an example, models of healthy cells and diseased cells are also created so as to investigate the pathogenesis process of MASH as this pointed out the hepatic stellate and sinusoidal endothelial cells in fibrosis [79]. These models are far more relevant to the disease development in human beings compared to the conventional 2D culture or animal models.
- Drug discovery and toxicity: Human-relevant *in vitro* experimental conditions to related drug actions, namely, drug metabolism, efficacy, and hepatotoxicity, are, however, being constrained to study and they are augmenting the efficacy of predictions of clinical outcomes using bioprinted liver constructs [80]. Cells isolated in these patients could serve as the basis of developing individual models of drug screening and testing of toxicity so that the dose response and liver injury caused by the drug could be determined [73].
- Repair and replacement regenerative medicine and transplantation: It is becoming even more evident that the bio-printing of liver tissues could be the vehicle of liver transplantation tomorrow since one of the most important functions of the liver (detoxification of ammonia) can be performed 1. Bio printed hepatorganoids have demonstrated functional role *in vivo* following a transplant.
- Reduced animal testing: Due to this reason of having 3D bioprinted liver models introduced, a large amount of reduction or even animal testing of the preclinical phase in the drug development process is likely to take place since the model accurately predicts that of the human [80].

Recent Advances

- Vascularization: The new ones are being developed with inbuilt vascular systems whereby they become closer to the native tissue of the liver and last long supplying the life and action of the cells [81].
- Organoids integration: Organoid technology and bioprinting will facilitate the development of the liver constructs with high order met Mars activities and complex microenvironments in a very short period using a synergistic approach [73].

15.7.2 Cardiac Tissue Engineering

Cardiac tissue engineering through 3D bio printing enables the production of highly aligned functional cardiac tissues that are employed in regenerative treatment, disease modeling and drug screening. The problem of scaffold manufactures, cell alignment, vascularization and integrated testing systems projects has led to the advancement of the field providing clinical relevance in heart repair and research both in the country and across the world [82].

Approaches and significant developments

- Biomimetic and multi-cellular constructions: The advantages brought by the potential of 3D bioprinting such as precision and control were given to the structure, cellular composition and orientation of the tissues created by the encapsulation of cardiomyocytes, endothelial cells, and fibroblasts into bioinks. An example is that FRESH *in vitro* has been applied to print high, aligned human cardiac tissue in well-plate format, generating constructs with high density extracted hiPSC- myocardium exhibits physiological responses native-like including the replicability of calcium transients and fast conductions [83].
- Cel-lif cell-lif scaffold design and cell alignment: Scaffolds designed with microchannels or patterned are reported to align the cells and make them develop into cardiac cell, thereby being important in functional tissue. Direct ink writing and other extrusion-based procedures have the ability to generate anisotropic cardiac tissue of tissue-sheets and organ building blocks (OBBs), which is aligned along the

print path [84], but microchanneled gelatin hydrogels have been demonstrated to encourage the growth of long and beating cardiomyocyte-based cardiac tissue.

- Vascular structures and nourishable structures: Sacrificial writing and other specialized bioprinting methods (such as FRESH) have been used to create perfusable vascular networks liquid-filled channels and multiscale vascular networks in cardiac constructs but thick tissues face the issue of nutrient supply and waste removal [83].
- *In vitro* and *in vivo* solution: Bioprinted cardiac tissues are predictive *in vitro* systems of drug screening and disease study, which, unlike a 2D cell cultures, provides information regarding - specific physiological responses in humans, which leads to filling the gap that exists between the 2D and animal model [85]. Bio printing cardiac patches have been placed in an animal model of myocardial infraction and used in preclinical testing to show a capacity to improve cardiac performance and sustain cell survival along with vascularization [83].
- There is organization of organ-on-a-chip: Bioprinted skin (researchers have been able to print skin and cover it with epithelial cells) Cultured dermis (used to model the epidermis Researchers have been able to produce bioprinted skin which they have been able to coat with epithelial cells and are used to model the epidermis) Dermal equivalent-channels aligned [84].

Clinical and Translational Insulation

- Regenerative Medicine: The final product will be the production of patient specific cardiac tissues or stents to be inserted to give new prospects through repair or replacement of the damaged myocardium after the attack or in heart failure [86].
- Contributing toward quality control and testing: Before clinical translation, bioprinted tissues are undergoing ample pre-clinical studies on issues of cellular viability, contractile activity, connective vasculature development, electrophysiological properties to reveal that bioprinted tissues can ensure meeting the requirements of transplantation-based treatments and their smooth accommodation into the organism [87].

15.7.3 Constructs of Skin and Wound Heals

Depending on the need to develop 3D printed skin and wound healing constructs, bvp constructs enable the formation of grafts with a specific structure and functioning with regard to the patient. These are accelerating wound healing, improving clinical outcome, and supporting research, and development in regenerative medicine and dermatology [88].

Significant Characteristics and Advancements

- Biomimetic Multi-Layered Constructs: Bio printing skin models tend to use different layers of skin - epidermis, dermis and sometimes even hypodermis, with different cell types (e.g. keratinocytes, fibroblasts, endothelial cells) and a custom-developed bioink that mimics the lamination-like structure of natural skin [89]. An example would be the GelMA bioink enriched with acellular dermal matrix to generate skin constructs in full thickness by incorporating human epidermal, dermal, and vascular cells that resulted in the accelerated reimbursement of wound healing and enhanced dermal extracellular matrix (ECM) production [88].
- Vascularization: Vascular networks are also necessary to include in other survival of grafts and include thicker grafts. techniques include the insertion of vascular endothelial cells, pro-angiogenic proteins, and the maximization of the porosity of the scaffold that will yield blood vessel formation. The positive attribute of bioprinting vascularized skin graft is that, more nutrients and oxygen can reach the location of the damaged tissue leading to faster and more suitable healing, in general breeds, chronic wounds, and diabetic ulcers [89].
- Accuracy and personalization: 3D printing can also design the shape and the distribution of cells in skin so that personalized grafts can adopt the shape to suit a target wound and the needs of the anatomy [90]. Clearly, this precision is also preferable in the field of skin appendages (hair follicles, sweat gland) confinement, and construction of physiological or pathological skin models to perform research and/or tests.

- Enhanced herespond and patient status: It has been revealed that the bioprinted skin constructs have the capacity to outperform the traditional ways to heal wounds, regulate inflammation, angiogenesis and re-epithelialization [88]. There is an increased advantage in that stem cells may be used outside of the patients in which there is minimal exposure to immune rejection and also increased integration degree [13].
- Biopolymer printing *in situ*: Bio-printing of cells and bioactive factors can also be done directly in loco on wounds, allowing predetermined coverage and selective areas of bioactivity at point of support, which may enhance healing rates, and potentially remove the possibility of infection.

Broader applications

- Healing of wounds: The bio printed skin grafts have immense scope in deep burns treatment, chronic wound and traumatic wounds healing, where success rate of conventional grafts is low as it is unable to integrate well with a host and inadequately vascularized [91].
- *In Vitro* model: They are better models of drug screening, tests on cosmetics, diseases research, compared with the discarding of use of animals in testing, and to increase further human response forecasts [89].
- Regenerative medicine: Some of the purposes of developing the technology involve clinical translation (Providing off-the-shelf of patient specific grafts to be used as reconstruction and regenerative therapies) [91].

15.7.4 Usage Orthopedic and Cartilage

The technology of 3D bioprinting is revolutionizing the scene of cartilage and orthopedic tissue engineering, as constructs can be designed based on which they can match the native cartilage and bones in structure and functions as well as cellular compositions. The significant clinical concern which is resolved through this type of technology is the inadequate self-regenerating ability of the cartilage and absence of an adequate, patient-specific treatment of joint disjuncts and orthopedic injuries [92].

15.7.4.1 Tissue Cartilage Engineering

Precision and bio mimicry: The placement of bioinks of biomaterial, cells (chondrocytes or mesenchymal stem cells) and bioactive molecules in 3D bioprinting is possible with high precision and in a layer-to-layer way. With such accuracy it is possible to recreate the stratified nature of articular cartilage, having superficial, middle and deep layers [93].

Bioinks: Hydrogels are frequently employed as bioinks because of their biocompatibility, high water content, and capacity to confine cells and growth factors. Such hydrogel can be optimized with regard to particular levels of mechanical strength and biological activity and be offered as the supporting material of chondrogenesis and tissue development prior and after birth [94].

Photo-crosslinkable biomaterials and advanced bioprinting enable stratified constructs that replicate the zonal organization of native cartilage. This plays a crucial role in the achievement of the depth-dependent structure of the scaffold of the collagen fiber as well as biomechanical properties required in repairing in a functional way [95].

Tailor-made implants: Medical imaging (CT, MRI) and CAD are used to develop, manufacture and design patient specific structures that theoretically can be utilized to fill in cartilage defects with grafts that are rapidly matched in shape and size to the defect [93].

15.7.4.2 Orthopedic (Bone and Osteochondral) Uses

Osteochondral templates: Bio printing also enables multi-material and cell type productions and printing in a single construct, and hence infrastructure/manufacturing of the osteochondral constructs where the underlying bone is included in the same construct as that of the cartilage. This kind of treatment is necessary to rectify the shortcomings of the joint of the layers of cartilage and bone [96].

Intervertebral discs and menisci: It can also be used in the implementation of implants in high-load and motion sensitive tissues such as knee menisci and intervertebral discs with respect to which the traditional implants do not allocate appropriately to the attained biological functionality [93].

Guided tissue growth: Significant new manufacturing processes, such as casting of arrays of microchambers to drive the self-assembly and compaction of cellular spheroid have enabled the creation of structurally defined and osteochondral sized cartilaginous tissues that can be utilized to replace lost orthopedic mass [95].

Translate, The Challenges and Clinical Translation

Potential and Limitation: The preclinical and *in vitro* trials indicate that 3D-bioprinter-printed structures and models can offer the environment and support chondrogenesis and functional tissues, which, currently, are yet to be implemented. There exist issues of scaling up production, regulatory approval as well as long-term integration and durability in the *in-vivo* world due to future labor costs, regulatory approval as well as long-term viability in the *in-vivo* world [97].

Personalized Medicine: The clinical workflow of the therapeutic treatment includes imaging, forming the construct design by a CAD-based system, printing the defect by using patient cells and at last post-processing the construct before implantation that offers a pathway to very specific orthopedic therapy [92].

15.8 Industrial and Clinical Translation

3D bioprinting is rapidly advancing in industries and clinical settings; however, a large number of considerable obstacles need to be overcome for the technology to become mainstream in healthcare and industry. Bio-inks that are suitable for large-scale and validated production within rigorous regulatory control is required to realize the full potential of 3D bioprinting.

15.8.1 Biomanufacturing Issues and Scalability

One of the main obstacles to industrial and clinical application of 3D bioprinting is achievability and scalability, as well as biomanufacturing aspects. These obstacles occur at the technical, biological, operational and regulatory levels.

Major problems of scalability

- Large-Scale Tissue and Organ Fabrication: The only limits to printing human scale organs or tissues are the problem of stably preserving their cell viability and structural integrity during long printing times and large printing volumes. The large-scale constructs also frail to distribution challenges of evenly distributing cells, delivering nutrients and eliminating waste, particularly when there were no integrated vascular systems [98].

- Vascularization: The ability of engineering a perfusable microarterial or microvascular network is an important challenge in large scale bioprinted constructs. Tissues, devoid of appropriate vascularization may not be able to support cell survival or develop into functional graft [99].
- Standard protocols are lacking in bioprinted constructs and bioinks, and it is not possible to accurately and reliably produce such materials at scale. Develop bioink libraries and quality control systems as a way of creating a dependable production [100].
- Printing Speed and Throughput: The traditional bioprinting technologies cannot do high volumes and the process is relatively slow. Scaling up also involves several printers and this accumulates more energy, complexity and costs [101].
- Post-Production Processing: It is difficult to keep appropriate physiological and mechanical properties during the maturation of the printed materials after post-printing; especially in the case of bigger or complex constructs. Functional results depend on post-print tissue maturation and maintenance which is quite time consuming and complex [18].

Biomanufacturing, real-world limitation

- The cost and the resource requirements: Bio printers, bioinks, and the said cells/growth factors are very expensive and require a highly skilled team to manufacture in large scale.
- Impacts on energy and the environment: Large-scale production by having many bioprinters running at the same time causes high energy consumption and many carbon emissions, which are a sustainability issue [101].
- Technical complexity: Every bioprinting technique comes with its own intrinsic limitations:
  - ♦ Extrusion bioprinting supports large constructions and may deteriorate the viability of cells because of shear stress.
  - ♦ Bio-printing is precise with inkjet and laser-based which are less apt in thick large tissues.

- ◆ Clustering together several approaches might prove to be inevitable but complicates operations [99].

Inventions and Responses

- Artificial intelligence and high technologies: Failure to provide solutions to scalability and reproducibility can be addressed by the integration of artificial intelligence to optimize tissue design, predict the behavior of the cells and enhance printing speed and quality [98].
- Hybrid printing approaches: He or she can combine various approaches to bio printing (e.g., extrusion + inkjet or stereolithography) to take advantage of the best attributes of each to enhance the complexity of constructs, cell viability and scale [99].
- Cutting-edge post-printing methods: High resolution and fast post-print editing to make complex and large-scale scaffolds is made possible with innovations such as light-based volumetric printing and photografting [98].

Regulatory and Clinical Considerations
Changing regulatory environment: The absence of transparent, harmonized regulatory policy of bioprinted products expands the clinical translation process and commercialization. The areas of safety, efficacy and quality defined pathways are still to be developed [102].

15.8.2 Policy, Regulation and Good Manufacturing Practice (GMP)

One of the main aspects of safe, effective, and ethical translation of 3D bioprinting research into clinical and industrial uses is regulatory and GMP considerations. By 2025 the situation is very dynamic and agencies and stakeholders are striving to come into terms with dealing with the peculiarities of bioprinted products.

15.8.2.1 Regulatory Frameworks

Multiple governing bodies and jurisdictions:
In the United States (US), 3D bioprinting is supervised by food and drug administration (FDA) with three core centers:

- Medical device -> center for devices and radiological health (CDRH)
- Center of biologics evaluation and research (CBER) of biological products
- Drug related applications in center for drug evaluation and research (CDER)
- Regulatory bodies: European Union (EU), the United Kingdom (UK), India/Japan, China, and other regions have regulatory bodies (e.g., European medicines agency (EMA), pharmaceuticals and medical device agency (PMDA), national medical products administration (NMPA) which regulate bioprinted products which usually refer to harmonized international standards [103].

15.8.2.2 Classification and Approval of Device

FDA categorizes the devices depending on the risk (Class I- III) where those of higher risk must meet stricter premarket approval and clinical evidence. Bio printed medical products and structures can be associated with any category based on the intended use, ingredients, as well as risk [104].

International Standards: International organization for standardization (ISO) and ASTM have standardized additive manufacturing including the standards of the process, validation of materials, quality control and calibration of the equipment [105]. Equipment safety is also the EU concern and directives which must be adhered to are the machinery directive and electromagnetic compatibility directive.

15.8.2.3 GMP and Quality Management

- (GMP): GMP is compulsory on clinical or commercial production and imposes severe controls relating to all areas of production such as:
  - Materials, process and batch change documentation and traceability
  - Equipment and processes validation
  - Quality management systems (QMS) in order to maintain stable product quality and safety
  - Continuous monitoring encompasses environmental and sterility-assurance measures [106].
- One-on-one and hand-made manufacturing: Due to more personalized products generated during bioprinting, it is

difficult to utilize mass-production GMP restrictions. The individual design of each construct and source of the cells that are used must ultimately be carefully documented, traceable, and validated, which necessitates QMS and process control in particular [106].

- Biological materials and ethics: Stem cell use, especially the human-derived cells needs effective ethical monitoring which consists of:
  - The need for an ethics committee approval
  - Cell sourcing informed consent
  - Clinical trials registration in a central model
  - Monitoring of long-term risks and uncertainties on an ongoing basis

15.8.2.4 *Data Integrity and Digital Regulatory Compliance*

- Data management: Bioprinting needs to be built on various data (imaging, genomic, anatomical, clinical) and all of them are subject to data protection policies (such as GDPR in the EU). Digital workflows should contain both files like design and modeling done by the AI and guarantee data integrity and privacy [106].

15.8.2.5 *Future Directions and Evolution*

Adaptive and module-based regulation: It is understood that regulatory frameworks that respond to the hybrid as well as fast changing nature of bioprinting are needed now. Suggestions have been to create specific bioprinting regulation, create regulatory sandboxes where new ideas can be tested and create early scientific advice processes.

Point-of-care manufacturing: Regulatory agencies are coming up with certain guidelines regarding the point-of-care (hospital-based) manufacturing, towards risk management and compliance in the clinical setting [107].

Reimbursement and clinical integration: The evolution of reimbursement policies, as well as the implementation of sound QMS at the level of hospitals, contributes to making clinical integration less of a distant prospect, reducing both financial and operational obstacles.

15.8.3 Quality Control and Standardization

To maintain safety, efficacy, reproducibility and FDA compliance of 3D bioprinted products, quality control and standardization are presented as real necessities, especially considering that the field is entering clinical and industrial applications.

The major components of quality control in 3D bioprinting

- Quality of material: The purity and consistency of bioinks, polymers and cells are central. This is maintained by the means of verifying the suppliers, frequent auditing, and testing the materials (e.g., endotoxins, sterility, chemical composition, mechanical properties) [108]. In the case of bioinks, the parameters measured also include viscosity, pH, gel point and cytotoxicity.
- Process Control: The bioprinting process should be maintained under the strictest control. This involves the daily calibration of the machine, environment (temperature, humidity) and process (layer thickness, print speed, extrusion temperature); thus, permitting simultaneous output and batch-to-batch variability to be limited.
- Post processing inspection: Constructs are then visually inspected after printing to look out for defects, verified to the dimensional requirement as well as being subjected to mechanical tests (e.g. hardness, impact resistance) to ensure the structures are of predetermined standards as well as are able to perform particular functions [109].
- Cell processing and viability: Cell processing should be consistent with the good practices: good laboratory practice (GLP) and GMP and high-quality control and, as a result, the viability of cells, their sterility, and functionality in the final construct. Cell viability can be performed several days after the culture to assess the compatibility with the bioink and the printing process [110].

15.8.3.1 Standardization Efforts

- Standards global: Their fast development in 3D bioprinting technology has led to the creation of new standards by ISO and ASTM organizations times around on complete components, hardware and training of personnel, as well as testing procedures6. These standards are being developed on the ground up to fit the peculiarities of additive manufacturing and additional ones are still being developed in the effort to bridge it with conventional manufacturing [110].

- Regulatory Guidelines: In some countries (e.g. US) regulatory agencies (e.g., FDA, EMA) mandate the adherence to both device- and biologics-specific regulations, including design, manufacturing, material characterization, mechanical testing documentation3. To translate to clinics, the manufacturers are obliged to register, submit, and have products approved again and prove effectiveness of quality management systems [9].

Importance

- Integrity and safety of products: Strict control on quality guarantees the safety and functionality of every bioprinted structure and it is especially important to use it in the medical and clinical sector.
- Reproducibility and regulatory approval: Reproducibility of results depends on standardized quality assurance, which is a requirement to regulatory approval and effective clinical translation [9].
- Customer and patient trust: Cross-sectionally stable quality would cause trust to end users, clinicians and patients and thereby make adopting and taking the product to the market easier.

15.9 Prospects and Research Areas

Next ten years of 3D bioprinting will be defined by the achievements in integration of AI, sophisticated bioinks, high resolution, and scalable production. Such progress should lead both to the eventual ambition of producing transplantable functional tissues and organs and to a revolution in drug discovery, reconstructive medicine and personalized healthcare [10].

15.9.1 Future Direction of Bio-Ink Designs

The recent breakthroughs can be discussed as the improvements in material science, cell engineering and automated processes, where great attention is paid to personalized medicine and regulatory compliance [8].

Important dots on bio-ink design

Hybrid and multi-material bioinks: Combination of natural and synthetic polymers has become a principal trend and will make it possible to combine biocompatibility (the use of natural materials, such as gelatin, collagen, and fibrin) with the tuning of mechanical properties (synthetic polymer). This semi- synthetic method facilitates the engineering of complicated, multi-layered tissues and organoids and can be adapted to uniquely engineering bone, cartilage, or simple tissues [52].

Bioactivity enhancement: To overcome the shortcomings of conventional hydrogels several bioinks are now being functionalized using bioactive peptides and enzyme-cleavable sequences and growth factors. This improves cells attachment, migration and tissue specific differentiation, resulting in more physiologically representative constructions [52].

Nanocomposite inorganic additives: The osteogenic potential of bioinks (in bone and osteochondral applications) is being enhanced by the addition of bioactive particles, e.g. calcium phosphates (tricalcium phosphate, hydroxyapatite). Mechanical strength, electrical conductivity and the biological signaling are also enhanced by incorporation of nanoparticle and nanofiber [52].

Personalized bioinks: personalized and stem cell-enriched bioinks: Stem cell-schooled bioinks and patient-origin cells are increasingly used to make custom tissue models in regenerative pharmaceuticals, testing and drug screening. The continuation can be facilitated by the growing logistics and proficiency concerning the acquisition, expansion, and differentiation of cells and on the regulatory progression toward adaptive frameworks of clinical-grade bio printed tissues [9].

Smart responsive bioinks: Stimuli-responsive (heat, pH, light) bioinks and those with the ability to change their properties after printing (self-healing, shape-memory) are becoming a trend. The bioinks of this kind, smart ones, allow them to control the maturation and functionality of tissue better and they are the forefront to 4D bioprinting [8].

Machine learning and automation: Innovations on autonomous, multi-material print systems, along with the combination of machine learning to better optimize the printing process, are improving reproducibility, scale and accuracy of bioink deposition. Such breakthroughs are essential in large scale biomanufacturer and clinical translation.

In Situ bioprinting and on-demand bioprinting: Bioprinting systems are becoming portable and *in situ*, where they are used to directly deposit the bioinks on unstable wounds or surgical locations to have immediate tissue replacement upon deposition. This tendency can be very prospective in skin regeneration and traumatic care [52].

15.9.1.1 Problems and Future

Regulatory formalities and standardization: The high-speed innovation in the design of the bioinks is exceeding the development of regulatory regulation and standards, making safety, reproducibility and long-term biocompatibility challenges. An inter-disciplinary team is required to balance technological developments and regulatory needs [8].

Clinical and industrial translation: Major manufacturers and research groups are concentrating on manufacturing bioinks at scale that meet GMP requirements which are an absolute requirement to clinical adoption. This transition is being underpinned by adaptive regulatory systems and well-established quality control systems.

15.9.2 Advanced Bioprinting Technologies Integration

The entry of sophisticated technologies in the field of bioprinting is quickly increasing the potential and the effects of 3D bioprinting in all fields of biomedical research, clinical practices, as well as in industrial production processes. Several overlapping innovations are characterizing this new-generation of bioprinting [111].

Machine learning and artificial intelligence (IIC)

- Real time optimization of bioprinting parameters in bioprinting, AI-driven control systems are being introduced into bioprinting platforms to provide real time corrections to print speed, pressure, and temperature, which optimizes parameters to reduce errors and improve reproducibility.
- AI is already applied in automated tissue design, allowing custom, patient-specific architecture to be built as well as tissue maturation and integration over time.

- Modeling Machine learning models are used to predict and optimize bioink behavior, and thus the compatibility and functional outcomes of printed tissues.
- In drug testing, AI combination will expedite researchers and increase tissue response knowledge to advance toward individual treatment.

Multi mat and multi-modal printing

- Improved bioprinters now enable multi-material depositions to enable the creation of spatially controlled cell and matrix composition of complex, heterogeneous tissues [111].
- Such high-resolution printing techniques as digital light processing (DLP) allow the printing of microstructures including (but not limited to) capillary networks, which are necessary to make thick, functional tissues.
- Microfluidic and nanotechnology integration enable handling of cells with exquisite control, as well as organization of constructs who are vascularized and tissue-specific.

Automation process and real time monitoring

- Automation of workflow simplifies the whole process of bioprinting that involves tissue design (using CAD), slicing process, maturation, and quality checking the printed tissue.
- Real-time monitoring and feedback components provide quality and scalability since the deviations can be corrected promptly and provide consistent output.

Clinical and industrial innovations

- Patient-Specific Implants: Of particular note is the activity underway in the clinical development of patient-specific, bioresorbable implants (e.g. pediatric airway splints) proving translation of integrated bioprinting technologies to healthcare.
- Regenerative Medicine: The possibility to print tissues with simultaneous vascular network and various type of cells increases the possibility of organ and tissue replacement therapies.

- Training and Workforce Development: Training initiatives such as Fabricamp 2025 are training the next generation of technologists on how to work with powerful bioprinters and the interactions between cells, materials, and crosslinking techniques.

Future directions

- Tissue-specific bioprinting: Individual tissue microenvironments vary and optimizing printing conditions and bioink properties to mimic these microenvironments are becoming a specialization and that plays a key role in the maturation and functionality of the engineered constructs.
- Collaborative ecosystems: The work of partnership between academia, industry, and regulators is supporting the translation of the integrated bioprinting technologies between the laboratory and clinic.

15.9.3 Personalized Therapeutics Sustainable Therapeutics

3D bioprinting is ushering a novel age of sustainable and individualized therapeutics through the innovation of personalized tissues, implants and drug delivery systems and is adding to more efficient and environmentally mindful health care solutions [112].

15.9.3.1 Personalized Therapeutics

- Tissues/implants with patient specific: Bioprinting enables creation of implants and scaffolds/tissue constructs, which perfectly fit anatomy and biology of the specific person. This enables better fitting, integration, and healing of treatments, e.g., in orthopedics where bone and cartilage scaffolds printed with bioprinting technology have resulted in increased regeneration rates and low revision surgery rates [113].
- Personalized drug delivery: 3D bioprinting can help with the design of patient-specific drug delivery mechanisms (including polypills [multiple drugs incorporated into a single dose]) and localized drug-eluting patches to drugs for treatment of specific cancers or injury to the heart. Such innovations allow more efficient, specific and safest treatment systems, reduce side effects and enhance effectiveness [114].

- Precision pharmacotherapy: With pharmacogenomics and biomarkers data, the customization of the dosage forms and drug regimens by bioprinting can be used instead of the one-fit-all approach and the result is a treatment that considers genetic, metabolic and lifestyle variations [112].

15.9.3.2 Sustainable Therapeutics

- Efficiency of resources and reduction of wastes: Bioprinting is by its nature an additive technology constructing objects with article materials only and no wastage as in subtractive manufacturing. Such efficiency is extended to the incorporation of bio-based, renewable and biodegradable materials into bioinks, reducing the effects on the environment even further.
- On-demand, point-of-care manufacturing: Capability to manufacture medical products where patients are being treated (e.g. hospitals or clinics) decreases the demand of having large stocks, long transportation routes, and oversupply, allows a more responsive and eco-friendly health care supply chain.
- Diesel downward refinement: Human tissue models printed with bioprinting are becoming developed as a further option in drug testing and disease modeling and can provide the best prediction of human response to drugs and decreasing the environmental and ethical price related to animal sciences [112].

15.9.3.3 Future Directions and Enabling Technologies

- AI and 4D bioprinters integration: The personalization and sustainability of therapeutics are being increased with AI-driven design and 4D bioprinting (by utilizing smart, responsive materials), allowing increased material optimization, patient outcome prediction, and allowing adaptive, life-long treatments [115].
- Regenerative and preventive medicine: Bio printing does not merely fix broken tissues, but gives an opportunity to build living structures capable of healing, evolving, or even avoiding progress of the disease, contributing to the sustainable healthcare paradigms [113].

15.10 Conclusions

3D bioprinting represents a pinnacle of biomedical innovation, enabling the fabrication of intricate, functional tissues and organs with exceptional precision. Recent advancements in bio-ink formulation, crosslinking methodologies, and machine learning integration have facilitated the development of patient-specific structures that closely replicate native tissue architecture and functionality. These advancements are transforming applications in regenerative medicine, disease modelling, and drug testing, offering more biologically appropriate and sustainable alternatives to conventional methods. The implementation of hybrid crosslinking methods, intelligent and adaptive bio-inks, and automated multi-material printing systems has significantly improved the mechanical stability, biocompatibility, and scalability of bioprinted goods.

Nonetheless, obstacles persist regarding large-scale production, vascularization, regulatory authorization, and quality assurance. The formulation of international standards, including those from ASTM and ISO, alongside the advancement of GMP and regulatory frameworks, is essential for guaranteeing safety, reproducibility, and effective clinical translation. Future interdisciplinary collaborations and creative thinking are anticipated to surmount existing obstacles, propelling the field towards the ultimate objective of producing fully functional, organs for transplantation and initiating a new era of personalized, sustainable healthcare solutions.

AI disclosure statement: The author confirms that no artificial intelligence (AI)-generated content or AI-assisted writing tools were used in the preparation of this chapter (including summaries and the abstract). All text, analysis, and conclusions are the result of the author's original work.

References

1. Papaioannou, T.G., Manolesou, D., Dimakakos, E., Tsoucalas, G., Vavuranakis, M., Tousoulis, D., 3D bioprinting methods and techniques: applications on artificial blood vessel fabrication. *Acta Cardiol. Sin.*, 35, 3, 284, 2019.
2. Bin Rashid, A., Uddin, A.S.M.N., Azrin, F.A., Saad, K.S.K., Hoque, M.E., 3D bioprinting in the era of 4th industrial revolution–insights, advanced applications, and future prospects. *Rapid Prototyping J.*, 29, 8, 1620–1639, 2023.
3. Borah, A. and Kumar, D.S., Overcoming the barriers of two-dimensional cell culture systems with three-dimensional cell culture systems: Techniques,

drug discovery, and biomedical applications, in: *Biomedical Product and Materials Evaluation,* pp. 179–229, Woodhead Publishing (Elsevier), Cambridge, UK, 2022.

4. Thayer, P., Martinez, H., Gatenholm, E., History and trends of 3D bioprinting, in: *3D Bioprinting: Principles and Protocols*, pp. 3–18, Springer US, New York, 2020.
5. Gu, Z., Fu, J., Lin, H., He, Y., Development of 3D bioprinting: From printing methods to biomedical applications. *Asian J. Pharm. Sci.*, 15, 5, 529–557, 2020.
6. Mendoza-Cerezo, L., Jesús, M.R.R., Macías-García, A., Marcos-Romero, A.C., Díaz-Parralejo, A., Evolution of bioprinting and current applications. *Int. J. Bioprint.*, 9, 4, 742, 2023.
7. Jose, R.R., Rodriguez, M.J., Dixon, T.A., Omenetto, F., Kaplan, D.L., Evolution of bioinks and additive manufacturing technologies for 3D bioprinting. *ACS Biomater. Sci. Eng.*, 2, 10, 1662–1678, 2016.
8. Mathur, V., Agarwal, P., Kasturi, M., Srinivasan, V., Seetharam, R.N., Vasanthan, K.S., Innovative bioinks for 3D bioprinting: Exploring technological potential and regulatory challenges. *J. Tissue Eng.*, 16, 20417314241308022, 2025.
9. Sekar, M.P., Budharaju, H., Zennifer, A., Sethuraman, S., Vermeulen, N., Sundaramurthi, D., Kalaskar, D.M., Current standards and ethical landscape of engineered tissues—3D bioprinting perspective. *J. Tissue Eng.*, 12, 20417314211027677, 2021.
10. Donderwinkel, I., Van Hest, J.C., Cameron, N.R., Bio-inks for 3D bioprinting: recent advances and future prospects. *Polym. Chem.*, 8, 31, 4451–4471, 2017.
11. Ramadan, Q. and Zourob, M., 3D bioprinting at the frontier of regenerative medicine, pharmaceutical, and food industries. *Front. Med. Technol.*, 2, 607648, 2021.
12. Sundaramurthi, D., Rauf, S., Hauser, C., 3D bioprinting technology for regenerative medicine applications,. *Int. J. Bioprinting*, 2, 2, 9–26, 2016.
13. Saini, G., Segaran, N., Mayer, J.L., Saini, A., Albadawi, H., Oklu, R., Applications of 3D bioprinting in tissue engineering and regenerative medicine. *J. Clin. Med.*, 10, 21, 4966, 2021.
14. Aljohani, W., Ullah, M.W., Zhang, X., Yang, G., Bioprinting and its applications in tissue engineering and regenerative medicine. *Int. J. Biol. Macromol.*, 107, 261–275, 2018.
15. Gu, Q., Zhu, H., Li, J., Li, X., Hao, J., Wallace, G.G., Zhou, Q., Three-dimensional bioprinting speeds up smart regenerative medicine. *Natl. Sci. Rev.*, 3, 3, 331–344, 2016.
16. Agarwal, S., Saha, S., Balla, V.K., Pal, A., Barui, A., Bodhak, S., Current developments in 3D bioprinting for tissue and organ regeneration–a review. *Front. Mech. Eng.*, 6, 589171, 2020.

17. Jafari, H., 3D Bioprinting: Creating Complex Tissue Structures for Regenerative Medicine Introduction. *J. Biotechnol. Biomater.*, 14, 3, 393, 2024.
18. Eghosasere, E., Osasumwen, E., Emmanuella, O., 3D Bioprinting in Tissue Engineering: Advancements, Challenges, and Pathways to Clinical Translation. *JSM Regen. Med. Bioeng.*, 7, 1, 1023, 2025.
19. Ouyang, L., Armstrong, J.P., Lin, Y., Wojciechowski, J.P., Lee-Reeves, C., Hachim, D., Zhou, K., Burdick, J.A., Stevens, M.M., Expanding and optimizing 3D bioprinting capabilities using complementary network bioinks. *Sci. Adv.*, 6, 38, eabc5529, 2020.
20. Gungor-Ozkerim, P.S., Inci, I., Zhang, Y.S., Khademhosseini, A., Dokmeci, M.R., Bioinks for 3D bioprinting: an overview. *Biomater. Sci.*, 6, 5, 915–946, 2018.
21. Gopinathan, J. and Noh, I., Recent trends in bioinks for 3D printing. *Biomater. Res.*, 22, 1, 11, 2018.
22. Badhe, R.V., Chatterjee, A., Bijukumar, D., Mathew, M.T., Current advancements in bio-ink technology for cartilage and bone tissue engineering. *Bone*, 171, 116746, 2023.
23. Ghosh, S. and Yi, H.G., A review on bioinks and their application in plant bioprinting. *Int. J. Bioprint.*, 8, 4, 612, 2022.
24. Freeman, S., Calabro, S., Williams, R., Jin, S., Ye, K., Bioink formulation and machine learning-empowered bioprinting optimization. *Front. Bioeng. Biotechnol.*, 10, 913579, 2022.
25. Ji, S. and Guvendiren, M., Recent advances in bioink design for 3D bioprinting of tissues and organs. *Front. Bioeng. Biotechnol.*, 5, 23, 2017.
26. Kannayiram, G. and Sendilvelan, S., Importance of nanocomposites in 3D bioprinting: An overview. *Bioprinting*, 32, e00280, 2023.
27. Rijal, G., Bioinks of natural biomaterials for printing tissues. *Bioeng.*, 10, 6, 705, 2023.
28. Yang, Y., Jia, Y., Yang, Q., Xu, F., Engineering bio-inks for 3D bioprinting cell mechanical microenvironment. *Int. J. Bioprint.*, 9, 1, 632, 2022.
29. Chen, X.B., Anvari-Yazdi, A.F., Duan, X., Zimmerling, A., Gharraei, R., Sharma, N.K., Sweilem, S., Ning, L., Biomaterials/bioinks and extrusion bioprinting. *Bioact. Mater.*, 28, 511–536, 2023.
30. Fatimi, A., Okoro, O.V., Podstawczyk, D., Siminska-Stanny, J., Shavandi, A., Natural hydrogel-based bio-inks for 3D bioprinting in tissue engineering: a review. *Gels*, 8, 3, 179, 2022.
31. Karamchand, L., Makeiff, D., Gao, Y., Azyat, K., Serpe, M.J., Kulka, M., Biomaterial inks and bioinks for fabricating 3D biomimetic lung tissue: A delicate balancing act between biocompatibility and mechanical printability. *Bioprinting*, 29, e00255, 2023.
32. Wan, H., Xiang, J., Mao, G., Pan, S., Li, B., Lu, Y., Recent advances in the application of 3D-printing bioinks based on decellularized extracellular matrix in tissue engineering. *ACS Omega*, 9, 23, 24219–24235, 2024.

33. Zoghi, S., Advancements in tissue engineering: A review of bioprinting techniques, scaffolds, and bioinks. *Biomed. Eng. Comput. Biol.*, 15, 11795972241288099, 2024.
34. Datta, S., Advantage of alginate bioinks in biofabrication for various tissue engineering applications. *Int. J. Polym. Sci.*, 2023, 1, 6661452, 2023.
35. Lee, J.M., Sing, S.L., Zhou, M., Yeong, W.Y., 3D bioprinting processes: A perspective on classification and terminology. *Int. J. Bioprint.*, 4, 2, 151, 2018.
36. Benwood, C., Chrenek, J., Kirsch, R.L., Masri, N.Z., Richards, H., Teetzen, K., Willerth, S.M., Natural biomaterials and their use as bioinks for printing tissues. *Bioeng.*, 8, 2, 27, 2021.
37. Tarassoli, S.P., Jessop, Z.M., Jovic, T., Hawkins, K., Whitaker, I.S., Candidate bioinks for extrusion 3D bioprinting—A systematic review of the literature. *Front. Bioeng. Biotechnol.*, 9, 616753, 2021.
38. Muñoz-Castiblanco, T., Moreno-Marín, J.P., Osorio, M., Natural macromolecule-based bioinks for 3D bioprinting: A systematic review of composition, physicochemical characterization, and biomedical applications. *Bioprinting*, e00407, 2025.
39. Khoeini, R., Nosrati, H., Akbarzadeh, A., Eftekhari, A., Kavetskyy, T., Khalilov, R., Ahmadian, E., Nasibova, A., Datta, P., Roshangar, L., Deluca, D.C., Natural and synthetic bioinks for 3D bioprinting. *Adv. NanoBiomed Res.*, 1, 8, 2000097, 2021.
40. Sánchez-Sánchez, R., Rodríguez-Rego, J.M., Macías-García, A., Mendoza-Cerezo, L., Díaz-Parralejo, A., Relationship between shear-thinning rheological properties of bioinks and bioprinting parameters. *Int. J. Bioprint.*, 9, 2, 687, 2023.
41. Wang, Q., Backman, O., Nuopponen, M., Xu, C., Wang, X., Rheological and printability assessments on biomaterial inks of nanocellulose/photo-crosslinkable biopolymer in light-aided 3D printing. *Front. Chem. Eng.*, 3, 723429, 2021.
42. Bercea, M., Rheology as a tool for fine-tuning the properties of printable bioinspired gels. *Molecules*, 28, 6, 2766, 2023.
43. Habib, M.A. and Khoda, B., Rheological analysis of bio-ink for 3D bioprinting processes. *J. Manuf. Process.*, 76, 708–718, 2022.
44. Kitana, W., Levario-Diaz, V., Cavalcanti-Adam, E.A., Ionov, L., Biofabrication of composite bioink–nanofiber constructs: effect of rheological properties of bioinks on 3D (Bio) Printing and cells interaction with aligned touch spun nanofibers. *Adv. Healthcare Mater.*, 13, 6, 2303343, 2024.
45. Wang, J., Cui, Z., Maniruzzaman, M., Bioprinting: a focus on improving bio-ink printability and cell performance based on different process parameters. *Int. J. Pharm.*, 640, 123020, 2023.
46. Schwab, A., Levato, R., D'Este, M., Piluso, S., Eglin, D., Malda, J., Printability and shape fidelity of bioinks in 3D bioprinting. *Chem. Rev.*, 120, 19, 11028–11055, 2020.

47. He, Y., Yang, F., Zhao, H., Gao, Q., Xia, B., Fu, J., Research on the printability of hydrogels in 3D bioprinting. *Sci. Rep.*, 6, 1, 29977, 2016.
48. Malekpour, A. and Chen, X., Printability and cell viability in extrusion-based bioprinting from experimental, computational, and machine learning views. *J. Funct. Biomater.*, 13, 2, 40, 2022.
49. Masri, S. and Fauzi, M.B., Current insight of printability quality improvement strategies in natural-based bioinks for skin regeneration and wound healing. *Polymers*, 13, 1011, 2021.
50. Johari, N., Adabavazeh, Z., Baino, F., PVA-based bioinks for 3D bioprinting: a comprehensive review of their applications in tissue engineering. *Bioprinting*, e00419, 2025.
51. Kim, J., Characterization of biocompatibility of functional bioinks for 3D bioprinting. *Bioengineering*, 10, 4, 457, 2023.
52. Siwach, A., Baliyan, V., Sharma, A., Ahlawat, S., Chaudhary, S., Chakraborty, E., Current standing and future potential of 3D bioprinting and biomaterials. *Biomater. Connect.*, 2, 1, 1–10, 2025.
53. Fu, Z., Ouyang, L., Xu, R., Yang, Y., Sun, W., Responsive biomaterials for 3D bioprinting: a review. *Mater. Today*, 52, 112–132, 2022.
54. Gao, Q., Lee, J.S., Kim, B.S., Gao, G., Three-dimensional printing of smart constructs using stimuli-responsive biomaterials: a future direction of precision medicine. *Int. J. Bioprint.*, 9, 1, 638, 2022.
55. Rasouli, R., Sweeney, C., Frampton, J.P., Heterogeneous and composite bioinks for 3D-bioprinting of complex tissue. *Biomed. Mater. Devices*, 3, 1, 108–126, 2025.
56. Kim, J., Da, G., Debnath, P., Saha, P., Smart multi-responsive biomaterials and their applications for 4D bioprinting. *Biomimetics*, 9, 8, 484, 2024.
57. Rahimnejad, M., Jahangiri, S., Zirak Hassan Kiadeh, S., Rezvaninejad, S., Ahmadi, Z., Ahmadi, S., Safarkhani, M., Rabiee, N., Stimuli-responsive biomaterials: smart avenue toward 4D bioprinting. *Crit. Rev. Biotechnol.*, 44, 5, 860–891, 2024.
58. Samimi Gharaie, S., Seyfoori, A., Khun Jush, B., Zhou, X., Pagan, E., Godau, B., Akbari, M., Silicate-based electro-conductive inks for printing soft electronics and tissue engineering. *Gels*, 7, 4, 240, 2021.
59. Rastin, H., Zhang, B., Bi, J., Hassan, K., Tung, T.T., Losic, D., 3D printing of cell-laden electroconductive bioinks for tissue engineering applications. *J. Mater. Chem. B*, 8, 27, 5862–5876, 2020.
60. Yang, X., Zhong, Y., Zhang, L., Liu, Y., Zhuo, F., Wang, J., Ge, L., Zhang, L., Zeng, X., Tan, W., Song, G., Conductive/insulating bioinks with multitechnology compatibility and adjustable performance. *ACS Biomater. Sci. Eng.*, 10, 8, 5352–5361, 2024.
61. Farizhandi, A.A.K., Khalajabadi, S.Z., Krishnadoss, V., Noshadi, I., Synthesized biocompatible and conductive ink for 3D printing of flexible electronics. *J. Mech. Behav. Biomed. Mater.*, 110, 103960, 2020.

62. Zhou, J. and Vijayavenkataraman, S., 3D-printable conductive materials for tissue engineering and biomedical applications. *Bioprinting*, 24, e00166, 2021.
63. Goker, M., Derici, U.S., Gokyer, S., Parmaksiz, M.G., Kaya, B., Can, A., Yilgor, P., Spatial growth factor delivery for 3D bioprinting of vascularized bone with adipose-derived stem/stromal cells as a single cell source. *ACS Biomater. Sci. Eng.*, 10, 3, 1607–1619, 2024.
64. Bakhtiary, N., Liu, C., Ghorbani, F., Bioactive inks development for osteochondral tissue engineering: a mini-review. *Gels*, 7, 4, 274, 2021.
65. Redzheb, M., Sbirkov, Y., Valev, A., Dzharov, V., Petrova, H., Damyanova, T., Georgieva, A., Sarafian, V., Collagen bioinks redefined: optimizing ionic strength and growth factor delivery for cartilage tissue engineering. *Int. J. Bioprint.*, 10, 6, 4566, 2024.
66. Kim, J., Choi, H.S., Kim, Y.M., Song, S.C., Thermo-responsive nanocomposite bioink with growth-factor holding and its application to bone regeneration. *Small*, 19, 9, 2203464, 2023.
67. GhavamiNejad, A., Ashammakhi, N., Wu, X.Y., Khademhosseini, A., Crosslinking strategies for 3D bioprinting of polymeric hydrogels. *Small*, 16, 35, 2002931, 2020.
68. Lai, Y., Xiao, X., Huang, Z., Duan, H., Yang, L., Yang, Y., Li, C., Feng, L., Photocrosslinkable biomaterials for 3D bioprinting: mechanisms, recent advances, and future prospects. *Int. J. Mol. Sci.*, 25, 23, 12567, 2024.
69. Zhang, H., Yang, K., Liu, G., Zhu, S., Yin, R., Zhang, W., 3D bioprinting of multi-biomaterial/crosslinked bioink for skin tissue engineering. *Front. Bioeng. Biotechnol. Conf. Abstr.: 10th World Biomaterials Congr.*, 2016.
70. Lee, J., Lee, H., Jin, E.J., Ryu, D., Kim, G.H., 3D bioprinting using a new photo-crosslinking method for muscle tissue restoration. *npj Regener. Med.*, 8, 1, 18, 2023.
71. Tavakoli, S., Krishnan, N., Mokhtari, H., Oommen, O.P., Varghese, O.P., Fine-tuning dynamic cross–linking for enhanced 3D bioprinting of hyaluronic acid hydrogels. *Adv. Funct. Mater.*, 34, 4, 2307040, 2024.
72. Sümbelli, Y., Diltemiz, S.E., Say, M.G., Ünlüer, OB., Ersöz, A., Say, R., *In situ* and non-cytotoxic cross-linking strategy for 3D printable biomaterials. *Soft Matter*, 17, 4, 1008–1015, 2021.
73. Zennifer, A., Manivannan, S., Sethuraman, S., Kumbar, S.G., Sundaramurthi, D., 3D bioprinting and photocrosslinking: emerging strategies & future perspectives. *Biomater. Adv.*, 134, 112576, 2022.
74. Banigo, A.T., Nauta, L., Zoetebier, B., Karperien, M., Coaxial bioprinting of enzymatically crosslinkable hyaluronic acid-tyramine bioinks for tissue regeneration. *Polymers*, 16, 17, 2470, 2024.
75. Naranjo-Alcazar, R., Bendix, S., Groth, T., Gallego Ferrer, G., Research progress in enzymatically cross-linked hydrogels as injectable systems for bioprinting and tissue engineering. *Gels*, 9, 3, 230, 2023.

76. Peng, K., Liu, X., Zhao, H., Lu, H., Lv, F., Liu, L., Huang, Y., Wang, S., Gu, Q., 3D bioprinting of reinforced vessels by dual-cross-linked biocompatible hydrogels. *ACS Appl. Bio Mater.*, 4, 5, 4549–4556, 2021.
77. Li, W., Liu, Z., Tang, F., Jiang, H., Zhou, Z., Hao, X., Zhang, J.M., Application of 3D bioprinting in liver diseases. *Micromachines*, 14, 8, 1648, 2023.
78. Ali, A.S., Wu, D., Bannach-Brown, A., Dhamrait, D., Berg, J., Tolksdorf, B., Lichtenstein, D., Dressler, C., Braeuning, A., Kurreck, J., Hülsemann, M., 3D bioprinting of liver models: a systematic scoping review of methods, bioinks, and reporting quality. *Mater. Today Bio*, 26, 100991, 2024.
79. Hagan, J., Developing 3D-bioprinted liver tissue models to study common liver disease. Int. Life Sci. Methods, 2024.
80. Ma, L., Wu, Y., Li, Y., Aazmi, A., Zhou, H., Zhang, B., Yang, H., Current advances on 3D-bioprinted liver tissue models. *Adv. Healthcare Mater.*, 9, 24, 2001517, 2020.
81. Massa, S., Sakr, M.A., Seo, J., Bandaru, P., Arneri, A., Bersini, S., Zare-Eelanjegh, E., Jalilian, E., Cha, B.H., Antona, S., Enrico, A., Bioprinted 3D vascularized tissue model for drug toxicity analysis. *Biomicrofluidics*, 11, 4, 2017.
82. Bliley, J.M., Stang, M.A., Behre, A., Feinberg, A.W., Advances in 3D bioprinted cardiac tissue using stem cell-derived cardiomyocytes. *Stem Cells Transl. Med.*, 13, 5, 425–435, 2024.
83. Finkel, S., Sweet, S., Locke, T., Smith, S., Wang, Z., Sandini, C., Imredy, J., He, Y., Durante, M., Lagrutta, A., Feinberg, A., FRESH™ 3D bioprinted cardiac tissue, a bioengineered platform for *in vitro* pharmacology. *APL Bioeng.*, 7, 4, 2023.
84. Wang, Z., Wang, L., Li, T., Liu, S., Guo, B., Huang, W., Wu, Y., 3D bioprinting in cardiac tissue engineering. *Theranostics*, 11, 16, 7948–7967, 2021.
85. Polonchuk, L. and Gentile, C., Current state and future of 3D bioprinted models for cardio-vascular research and drug development. *ADMET DMPK*, 9, 4, 231–242, 2021.
86. Perez, M.R., da Silva, V.A., Cortez, P.E., Joddar, B., Willerth, S.M., 3D-bioprinted cardiac tissues and their potential for disease modeling. *J. 3D Print. Med.*, 7, 2, 3DP6, 2023.
87. Yong, U., Kang, B., Jang, J., 3D bioprinted and integrated platforms for cardiac tissue modeling and drug testing. *Essays Biochem.*, 65, 3, 545–554, 2021.
88. Zhang, M., Zhang, C., Li, Z., Fu, X., Huang, S., Advances in 3D skin bioprinting for wound healing and disease modeling. *Regener. Biomater.*, 10, rbac105, 2023.
89. Weng, T., Zhang, W., Xia, Y., Wu, P., Yang, M., Jin, R., Xia, S., Wang, J., You, C., Han, C., Wang, X., 3D bioprinting for skin tissue engineering: Current status and perspectives. *J. Tissue Eng.*, 12, 20417314211028574, 2021.
90. Javaid, M. and Haleem, A., 3D bioprinting applications for the printing of skin: A brief study. *Sens. Int.*, 2, 100123, 2021.

91. Ke, D. and Murphy, S.V., Current challenges of bioprinted tissues toward clinical translation. *Tissue Eng. Part B Rev.*, 25, 1, 1–13, 2019.
92. McGivern, S., Boutouil, H., Al-Kharusi, G., Little, S., Dunne, N.J., Levingstone, T.J., Translational application of 3D bioprinting for cartilage tissue engineering. *Bioengineering*, 8, 10, 144, 2021.
93. Perera, K., Ivone, R., Natekin, E., 3D bioprinted implants for cartilage repair in intervertebral discs and knee menisci. *Front. Bioeng. Biotechnol.*, 9, 754113, 2021.
94. Huang, J., Xiong, J., Wang, D., Zhang, J., Yang, L., Sun, S., Liang, Y., 3D bioprinting of hydrogels for cartilage tissue engineering. *Gels*, 7, 3, 144, 2021.
95. Kronemberger, G.S., Spagnuolo, F.D., Karam, A.S., Chattahy, K., Storey, K.J., Kelly, D.J., Rapidly degrading hydrogels to support biofabrication and 3D bioprinting using cartilage microtissues. *ACS Biomater. Sci. Eng.*, 10, 10, 6441–6450, 2024.
96. Tang, H., Zhao, E., Lai, Y., Chen, A., Chen, X., Zeng, W., Hu, X., Zhou, K., Zhou, Z., 3D bioprinting techniques and hydrogels for osteochondral integration regeneration. *Int. J. Bioprint.*, 10, 6, 4472, 2024.
97. Li, M., Sun, D., Zhang, J., Wang, Y., Wei, Q., Wang, Y., Application and development of 3D bioprinting in cartilage tissue engineering. *Biomater. Sci.*, 10, 19, 5430–5458, 2022.
98. Chameettachal, S., Yeleswarapu, S., Sasikumar, S., Shukla, P., Hibare, P., Bera, A.K., Bojedla, S.S.R., Pati, F., 3D bioprinting: recent trends and challenges. *J. Indian Inst. Sci.*, 99, 3, 375–403, 2019.
99. Kačarević, ZP., Rider, P.M., Alkildani, S., Retnasingh, S., Smeets, R., Jung, O., Ivanišević, Z., Barbeck, M., An introduction to 3D bioprinting: possibilities, challenges and future aspects. *Materials*, 11, 11, 2199, 2018.
100. Liang, K., Tissue bioprinting: promise and challenges. *Bioengineering*, 10, 12, 1400, 2023.
101. Eskandar, K., 3D Bioprinting for Facial Reconstruction: Advances and Challenges. *Regenesis Repair Rehabil.*, 2025.
102. Mladenovska, T., Choong, P.F., Wallace, G.G., O'Connell, C.D., The regulatory challenge of 3D bioprinting. *Regener. Med.*, 18, 8, 659–674, 2023.
103. Morrison, R.J., Kashlan, K.N., Flanangan, C.L., Wright, J.K., Green, G.E., Hollister, S.J., Weatherwax, K.J., Regulatory considerations in the design and manufacturing of implantable 3D-printed medical devices. *Clin. Transl. Sci.*, 8, 5, 594–600, 2015.
104. Beitler, B.G., Abraham, P.F., Glennon, A.R., Tommasini, S.M., Lattanza, L.L., Morris, J.M., Wiznia, D.H., Interpretation of regulatory factors for 3D printing at hospitals and medical centers, or at the point of care. *3D Print. Med.*, 8, 1, 7, 2022.
105. Kawalkar, R., Dubey, H.K., Lokhande, S.P., A review for advancements in standardization for additive manufacturing. *Mater. Today Proc.*, 50, 1983–1990, 2022.

106. Dukovic, M., New organs on command: The regulatory prospects of 3D bioprinting technology in the European Union. *Law Technol. Humans*, 7, 1, 84–95, 2025.
107. Carson, J., Systematic Review How Point-of-Care Testing is Currently Implemented in Healthcare in Compliance with Regulatory and Implementation Standards, Master's Thesis, Univ. California, Davis, 2022.
108. Mohanan, P.V. (Ed.), *Biomedical Product and Materials Evaluation: Standards and Ethics*, Woodhead Publishing (Elsevier), Sawston, Cambridge, UK, 2022.
109. Kantaros, A., Ganetsos, T., Petrescu, F.I.T., Ungureanu, L.M., Munteanu, I.S., Post-production finishing processes utilized in 3D printing technologies. *Processes*, 12, 3, 595, 2024.
110. Crook, J.M., Cell processing for 3D bioprinting: quality requirements for quality assurance in fundamental research and translation, in: *3D Bioprinting: Principles and Protocols*, pp. 19–26, Springer US, New York, NY, 2020.
111. Liu, S., Chen, Y., Wang, Z., Liu, M., Zhao, Y., Tan, Y., Qu, Z., Du, L., Wu, C., The cutting-edge progress in bioprinting for biomedicine: principles, applications, and future perspectives. *MedComm*, 5, 10, e753, 2024.
112. Parihar, A., Parihar, D.S., Gaur, K., Arya, N., Choubey, V.K., Khan, R., 3D bioprinting for drug development and screening: Recent trends towards personalized medicine. *Hybrid Adv.*, 7, 100320, 2024.
113. Chaudhary, S.K., Singh, P., Kumar, B., Singh, V.K., 3D Bioprinting in orthopedics: Transforming personalized healthcare. *Int. J. Orthop.*, 11, 1, 57–60, 2025.
114. Mihaylova, A., Yaneva, A., Shopova, D., Kasnakova, P., Harizanova, S., Parahuleva, N., Etova, R., Raykova, E., Semerdzhieva, M., Bakova, D., Pharmacists' Perceptions of 3D Printing and Bioprinting as Part of Personalized Pharmacy: A Cross-Sectional Pilot Study in Bulgaria. *Pharmacy*, 13, 3, 88, 2025.
115. Kennedy, S.M., S., A., V. and K, A., 4D bioprinting for personalized medicine, innovations in implant fabrication and regenerative therapies. *Polym.-Plast. Technol. Mater.*, 1–26, 2025.

16

Bioprocess Optimization Using Artificial Intelligence and Machine Learning

K.S.N.V. Prasad, Bijoy Kumar Purohit, Bhaskar Bethi and A.V. Raghavendra Rao*

Department of Chemical Engineering, B. V. Raju Institute of Technology, Narsapur, Telangana, India

Abstract

The integration of artificial intelligence (AI) and machine learning (ML) in bioprocess optimization is revolutionizing the landscape of biomanufacturing. Traditional methods of optimizing bioprocess parameters often rely on time-consuming trial-and-error approaches, which are limited by scale, cost, and complexity. AI and ML offer powerful alternatives by enabling predictive analytics, pattern recognition, and real-time decision-making capabilities that enhance process efficiency, product yield, and quality control. This chapter explores how AI/ML tools are being applied across various stages of biomanufacturing, including upstream cell culture optimization, fermentation process control, and downstream purification strategies. Techniques such as artificial neural networks, support vector machines, and reinforcement learning models are discussed in the context of dynamic bioprocess environments. Additionally, the chapter highlights how data-driven models can uncover hidden correlations between process variables and outcomes, leading to more informed and adaptive control strategies. Challenges related to data quality, algorithm transparency, and integration into existing regulatory frameworks are also addressed. Through selected case studies and industrial examples, this chapter provides a comprehensive overview of the current landscape and future prospects of AI- and ML-driven bioprocess optimization, offering valuable insights for researchers, practitioners, and decision-makers in the field of biomanufacturing.

***Keywords*:** Bioprocess optimization, artificial intelligence (AI), machine learning (ML), biomanufacturing, predictive analytics

**Corresponding author*: raghusrf@gmail.com

Sandip Kunar, Jagadeesha T. and D. Sathis Kumar (eds.) Biomanufacturing: Transforming Production of Biologics and Biopharmaceuticals, (497–520) © 2026 Scrivener Publishing LLC

16.1 Introduction to Bioprocess Optimization

16.1.1 Definition and Importance of Bioprocess Optimization

Bioprocess optimization is the procedure of gradual advancement of biological production systems towards the objective of enhancing the productivity, efficiency, and the quality of the wanted bioproducts including enzymes, vaccines, monoclonal antibodies, and other medical agents, etc. Bio processes are in nature complex because they involve living systems which may be cells, microorganisms or tissues within a dynamic environment. Optimization is the process through which the biological system is optimized to work in ideal conditions, thus, reducing the cost of production with its scalability, reproducibility, and regulatory compliance [1].

In biopharmaceutical manufacturing and industrial biotechnology, variation in parameters of the process, including pH, temperature, oxygen transfer, and nutrient concentration may have serious consequences on product quality and yields. Therefore, the concept of optimization does not only entail maximization of outputs but also ensuring that there is robustness and adherence to regulatory guidelines like cGMP (current Good Manufacturing Practice). In high value biomanufacturing industries, in particular biologics, optimization is directly correlated to profitability and efficacy of therapies [2].

16.1.2 Limitations of Traditional Bioprocess Methods

Common ways of optimizing bioprocesses remain empirical in nature, with a strategy such as one-factor-at-a-time (OFAT) experimentation and classical design of experiments (DoE) being the most used. Although such practices have paved way to early developments, they are tedious, consume a lot of time and prove to be expensive. More significantly, they cannot reflect the multidimensional dynamic between various process factors, especially those occurring in big-sized bioreactors or even uninterrupted production environment [3].

An example is the OFAT method which alters a single parameter at a time and keeps other factors the same and does not focus on the synergistic effects. Also, biological variability, material changes, and upstream-downstream interaction loops render such static optimizations inefficient to meet the current dynamic production demands. In addition to the above, scaling-up of the system between the laboratory and industry involves uncertainties that are not easy to forecast using the conventional statistical models [4].

16.1.3 Emergence of AI and ML in Biomanufacturing

The new breakthroughs in artificial intelligence (AI) and machine learning (ML) have provided new opportunities in bioprocess optimization. AI/ML techniques have been shown to be able to perform those tasks by analyzing large, complex, and heterogeneous data and finding hidden trends, outcome prediction, and assisting decision-making in real-time. This type of data-based model is in the position to study past records such as batch records, sensor information, and live monitoring to inform the optimization of numerous process variables [5].

Artificial intelligence/machine learning is an appropriate integration into biomanufacturing because modern biologics are becoming too complex, including cell and gene therapies. Time-series data can be approached using deep learning models, which are capable of adapting to a drift in the process, and reinforcement learning algorithms can provide an adaptive control solution. Using these technologies, it is possible to change destructive process control (reactive process) to predictive (predict process) and even prescriptive process control (prescribe process) with the help of computational power [6].

Besides, AI/ML is applied to build digital twins: virtual images of bioprocess systems so that sensitivity analysis, risk assessment, and *silico* simulations can be done without disrupting the current production. That leads to a combination of shorter optimization cycles, less experimental load and a greater assurance of scale-up. The same applies to the increasing attention given to Quality by Design (QbD) frameworks, which can be well utilized in AI/ML to support the identification and control of robust design space and control strategies [7].

To conclude, although classical bioprocess optimization proves useful in the earlier conception of manufacturing, the growing sophistication, exacting, and flexibility requirements during biomanufacturing necessitate a change to an intelligent and data-driven optimization paradigm. AI and ML provide new frameworks that have the potential to transform the process of optimization, monitoring, and control in bioprocesses throughout the production lifecycle.

16.2 Fundamentals of AI and ML in Bioprocessing

As AI and ML find their way into bioprocessing systems, it will mark a change of paradigm of having a historically empirical process towards a data-driven, adaptive, and intelligent decision-making process. They relate to AI and ML being powerful enabling technologies to achieve prediction, optimization, and automation in the various stages of biomanufacturing which are

Table 16.1 Comparison of traditional and AI/ML-based approaches in bioprocess optimization.

Aspect	Traditional approaches	AI/ML-based approaches
Methodology	Empirical, one-variable-at-a-time testing or statistical designs	Data-driven modeling with pattern recognition and predictive analytics
Handling Complexity	Limited to linear interactions; struggles with nonlinear dynamics	Excels at multivariate, nonlinear relationships in dynamic systems
Time and Cost	High resource consumption due to extensive trials	Reduces experimentation through simulations and *in silico* predictions
Scalability	Challenges in scaling from lab to industrial levels	Supports digital twins for seamless scale-up and risk assessment
Adaptability	Static; reactive to changes	Adaptive; enables real-time adjustments and prescriptive control
Data Requirements	Relies on small datasets; ignores hidden correlations	Leverages large, heterogeneous data for uncovering insights
Examples in Use	Manual pH/temperature adjustments in fermentation	Predictive feeding strategies in cell culture using reinforcement learning

dynamic and complex. To have a semblance of success with applying AI/ML in the very dynamic world of bioprocessing, it is imperative to comprehend the concepts underlying AI/ML. Table 16.1 gives summary of comparison of AI and ML techniques with respect to traditional approaches.

16.2.1 Overview of Artificial Intelligence and Machine Learning

Artificial intelligence (AI) is a wide set of computational techniques aiming to replicate the human-like methods of thinking like learning, reasoning, and finding a solution to a problem. In AI, ML is defined as a subset of algorithms which are not explicitly programmed to learn patterns in the historical data. These models help in the identification of trends and relationships and thus they are used in making informed decisions and predictions concerning unseen data.

In bioprocessing, the ML algorithms are used to build a model to determine the relationship between the input variables (e.g., temperature, pH, nutrient concentration) and the process outcomes (e.g., cell density, product yield). In contrast to rule-based models, the ML models can accommodate non-linear and multivariate interactions that are frequently met in biological systems. This provides an opportunity to have a better representation of real-time dynamics and control of the bioprocess variables [8].

16.2.2 Common AI/ML Algorithms in Bioprocess Applications

Several ML algorithms are widely used in bioprocessing, depending on the complexity and the characteristics of the problem and the data according to it. ANNs are one of the most popular models, because they can be used to approximate difficult, non-linear systems. The ANNs simulate the action of biological neurons and have been able to forecast biomass development, metabolite manufacturing and product production speeds in fermentations and cell cultures [9].

Support Vector Machine (SVM) is another popular one and is used in classification and regression problems. SVMs are especially effective to work with large high-dimensional datasets and small sample size. They have been used to categorize fermentation results to find the best process parameters [10].

Decision Trees (DTs) and Random Forests (RFs) are interpretable methods that are based on a series of hierarchical rules made by using the data on the process. These can assist in knowing which are the variables that have the greatest influence on bioprocess outcome. Another developing area is reinforcement learning (RL) which is of special interest in bioprocess control where the dynamic environments must be acted upon in real-time based on the feedback of the process [11].

16.2.3 Data Requirements and Challenges in Bioprocess Modeling

AI and ML models can be affected mostly by data, their quality and quantity. Examples of information used in bioprocessing include sensor data, supervisory control data (e.g., SCADA data), and process analytical technologies (PAT data). Meanwhile, a number of challenges including missing values, noise, inconsistent sampling rates, and heterogeneous sources of data, may affect the performance of the models.

The second burning point is the necessity of representative datasets that would encompass a great variety of scenarios of the process, including failures and anomalies. Models can be biased, or they cannot generalize without such diversity. Normalization, imputation, and dimensionality reduction are some pre-processing techniques that are now major steps in the preparation of data to use in modeling [12].

Moreover, transparency and interpretability are significant considerations. Most of the advanced models, particularly the deep learning frameworks, are black boxes, since it is hard to decipher their inner decision rationale. It may create an issue in highly regulated industries such as the biopharmaceutical manufacturing business where justification and provenance of process decisions are necessary to comply with regulations.

These sections have motivated the investigation into hybrid modeling which incorporates both mechanistic knowledge and data-driven ML methods to counter the difficulties. Such hybrid models, like this has done, use the advantages of both fields together- to incorporate theory learning and then to adjust to empirical information, thus enhancing robustness levels of statistical models and reliability of the model in bioprocess modeling [13].

16.3 AI/ML Integration in Upstream Bioprocessing

Upstream bioprocessing deals with the first phases of biomanufacturing which is mainly cell line development and optimization of media and control of culture conditions. These procedures are instrumental in defining the overall product yield and quality of generated bioproduct. Traditional procedures of optimization in the upstream are pegged mainly into empirical procedures like trial-and-error testing and factorial design. These techniques, while foundational, are often resource-intensive and time-consuming. The discourse on artificial intelligence (AI) and machine learning (ML) as the developments of recent years has shown that such tools are identified as predictive and adaptive and can lead to significant improvements in upstream bioprocessing efficiency and precision.

16.3.1 Cell Culture Optimization

The core of upstream processing is cell culture, and conditions must therefore be found to be optimal with concerns to cell viability, growth rate and productivity. The use of AI and ML to analyze high-dimensional data sets produced in culture environment such as dissolved oxygen, pH, temperature, and nutrient concentrations can be performed in real-time. ANNs have also served as L models; these models have been used in identifying a non-linear relationship between these variables and the culture performance metrics, so that they can lead to an accurate prediction of the desirable culture conditions [14].

Random forest and support vector machines (SVMs) have also been applied in the prediction and classification of cell growth phenotype under different conditions. These models have been used to provide benefits, where limited at times experimental evidence is available in that the trends are generalized and some patterns that may otherwise appear lost to the conventional statistical approaches can be ascertained through it. Further reinforcement learning has also started showing potential in changing control signals including agitation speed and aeration rate, based on the changing requirements of the cell culture [15].

16.3.2 Media Formulation and Feeding Strategies

Another upstream processing challenge is coming up with the perfect media composition. There is a need to balance media components including amino acids, glucose, vitamins and trace elements to prevent deficiencies or overloads. Machine learning algorithms can help streamline these formulations through learning the historical batch data and using this information to perform simulations of various media compositions with the goal of predicting their effect on cell growth and product yield.

When combined with methods of machine learning, e.g. Gaussian process regression or Bayesian optimization, DOE methods can make media formulation more efficient. These mixed schemes lessen the number of experimental runs that are necessary, though they still yield satisfactory optimization solutions [16]. Further, fed batch and perfusion approaches have been automated with AI systems to regulate feed rates as part of real-time metabolic predictive models that provide consistency in cell performance across scale and batches [17].

16.3.3 Neural Networks and Design of Experiments (DOE)

The neural networks with DOE have been successfully used in combination as an upstream bioprocess strategy. DOE gives us a methodical approach in determining the crucial process variables, whereas a set of neural networks are used to emulate the nonlinear, multivariate relations between variables and some desired result, which might be a biomass concentration, a specific productivity. Models have been proposed where the output is cell growth and the input variables are glucose concentration, temperature, pH, and others using multilayer perceptron (MLP) networks extensively. This means that these models can be used to provide interpolator space and even extrapolate in special circumstances and certainly prove valuable in the optimization of a process [18]. Additionally, the flexibility in process development can be achieved in that the usage of ML based DOE tool simply enables researchers to address a larger design space with fewer actual tests run. The integration also helps the creation of Quality by Design (QbD) frameworks, which have gained relevance both in regulatory applications and in the scalability of bioprocess development. Summing up, through the implementation of AI and ML in upstream bioprocessing, it is possible to transition to proactive and predictive process management, instead of reactive process management. These technologies have been able to facilitate more efficient and consistent production processes because they uncover hidden patterns in large and complex datasets. As computational models further develop and rigorously incorporate real-time process monitoring applications, their impact on biomanufacturing upstream optimization will be even greater and lead to creating innovative and more sustainable production processes.

16.4 AI/ML in Fermentation Process Control

Fermentation is a key factor of industrial biomanufacturing that allows the creation of pharmaceuticals, biofuels, enzymes and food additives. Being a very dynamic and nonlinear process, monitoring and control of fermentation is very challenging. To maintain optimum conditions successful through the whole fermentation cycle, the traditional control methods like PID (Proportional, Integral and derivative) controllers and fixed logic are not adequate. Artificial intelligence (AI) and machine learning (ML) may become very effective to provide insightful predictive knowledge, adaptable management, and in-time optimization of the fermentation processes.

The first great benefit of employing AI and ML in fermentation is that it allows the modeling of complex and nonlinear relationships between two or more variables in fermentation such as pH, temperature, dissolved oxygen, agitation speed, substrate concentration, and biomass yield. These variables do not interact in simplistic ways that can be adequately described by traditional first-principle frameworks. Machine learning algorithms, such as supervised learning algorithms, can use past historical data on fermentation to make process predictions on key outcomes and can adjust the operating parameters [19].

16.4.1 Predictive Modeling of Fermentation Parameters

Artificial Neural Networks (ANNs) and Support Vector Machines (SVM), among others, are used to model fermentation kinetics e.g. biomass growth rate, product formation and substrate limitation. Those models are optimized with data that was gathered out of bio reactors either on an industrial or laboratory level and then checked on its accuracy and general applicability. Once trained, such models will be able to predict the trends of processes and the scope of the deviations of the desired course [20]. This can enable operators of the plants to make prior corrections thereby minimizing chances of batch failure and ensuring consistency of the products.

Moreover, ML-based predictive models help in optimizing feeding strategies. As an example, in fed-batch fermentation processes, the feeding process and frequencies are often done under a real-time scheduling system that should be optimized to optimize the output. The optimal feeding rate can be recommended, or applied automatically, by adaptive neural networks and by reinforcement learning operating with live sensorial feedback [21].

16.4.2 Dynamic Management of pH, Temperature and Oxygen Levels

Microbial and cell culture-based fermentations require physiological conditions, e.g. pH, temperature and oxygen transfer rate. Conventional control loops are not very responsive to time delays or process disturbances. With the use of AI, the control systems can learn and adapt the response behavior of the system with changing parameters when conditioned by the cues in the environment.

An example is that fuzzy logic controllers, with training in the neural network, they have improved greatly when it comes to the stabilization of the level of dissolved oxygen in the aerobic fermentation process [22]. In a similar manner, multivariate sensor signals can be analyzed by a deep learning model in real time, and a corrective action can be started to ensure the realization of optimum environmental conditions by the cells or microbes that are involved in the process.

16.4.3 Actual-Time Fermentation Monitoring Real-World Case Studies Using ML

The application of AI and ML to fermentation is also seeing more and more reporting in both academia and industry. An especially dramatic case is when there are soft sensors--ML models that cannot measure some of the parameters they are trying to estimate, such as biomass or product concentration, but only the relatively easy measures like pH and dissolved oxygen [23]. The soft sensors decrease the reliance of expensive offline analytics and support sealed-loop control systems.

Researchers in one study employed the recurrent neural networks (RNNs) to model and control ethanol production during fermentation of saccharomyces cerevisiae. RNNs could predict almost perfectly and could control the feeding of glucose dynamically, which also led to the augmented yield and faster time of fermentation [24]. Equivalent gains in antibiotic generation, enzyme generation, or even care fermentations in artificial biology phases have been chronicled.

Although these developments are promising, their deployment is challenging until now regarding data standardization, model interpretability, and fit into legacy control architectures. Though, increasing productivity, low levels of variability and reduced operational costs brought by the application of AI/ML in fermentation control are leading to continued innovation and implementation.

16.5 AI/ML in Downstream Processing and Purification

A major bio processing step is downstream processing (DSP), a necessity that involves the purification and recovery of biological materials including proteins, antibodies, and enzymes, which are present in complex mixtures. It usually goes through several processes such as cell separation, filtration, chromatography, and constituting the final product. Although upstream processes have developed significantly in the bioprocess industry, downstream processes tend to be bottlenecks because they are complex, cost-intensive, and vulnerable to variability. The forthcoming incorporation of AI and ML on a downstream processing plant level is a chance to make the process more efficient, shorten development cycles, and guarantee stable product quality.

16.5.1 Chromatography and Filtration Optimization

Chromatography and filtration are foundational techniques for purifying bioproducts. Nonetheless, they require a lot of parameters to be optimized, which include pH, flow rate, ionic strength and column loading among others in order to have effective separation. Optimization of these parameters has been traditionally done based on design of experiments (DoE) and empirical trials, which are time-consuming and costly.

Modeling of the nonlinear dynamics of chromatography and the filtration process using artificial neural networks (ANNs), and support vector machines (SVMs) and random forests has presented an enormous potential of AI and ML frameworks. The models are also able to predict optimum operating conditions without involving extensive experimentation by learning historical batch data. As an example, ML algorithms have been applied to predict breakthrough curves in ion-exchange chromatography and to model protein retention time in order to develop the process more accurately [25]. In a similar way, membrane fouling and flux decline predictive models of filtration systems assist operators to predict maintenance schedules and ensure an increased membrane life [26].

AI-driven optimization can be used to implement continuous processing in which parameters need to be adjusted on the fly. Dynamic control of chromatography columns by reinforcement learning models have also been investigated and have been found sensitive to changes in the composition of the feed [27].

16.5.2 Predicting Product Purity and Yield

The focus of DSP is to maintain purity of the products and maximize yield. Minor changes in conditions of process and characteristics of the materials affect both measures. ML models can interrogate big size data sets *via* sensor logs, past batches and measurements of an analysis to draw possible patterns and relations that could not be drawn through traditional statistical relationships.

A suitable example is the use of regression models to predict the purity and yield of products depending on upstream and midstream properties [28]. Two of the most used models are Gaussian Process Regression (GPR) and ensemble methods. Such models may also be used as soft sensors, as they will give virtual measurements in situations where direct analysis would be time consuming or otherwise impractical.

In biopharmaceutical production, purity specifications are tightly regulated. ML models trained to predict binary outcomes on multivariate data can be used to decide the probability of a batch being above the regulatory limits on contaminants, e.g., host cell proteins or DNA. This enables early corrections and minimizes the rejection levels of the batches. In addition, the integration of ML with the quality-by-design (QbD) measures contributes to enhanced resilience and regulatory compliance [29].

16.5.3 Adaptive Process Strategies for Complex Separations

Advanced separations:- Separation of closely related forms, e.g. isoforms/glycoforms or aggregates are increasingly seen in manufacture of biologics. Such problems require adaptive and smart control systems which are not limited to the static process recipes. Adaptation in AI/ML models occurs in real time because the models learn based on the behavior of the system and adjustments in control strategies are made. Algorithms such as dynamic Bayesian networks and recurrent neural networks (RNNs) are being designed to vary separation parameters to enhance the resolution between co-eluting species on the chromatograms [30]. A third approach is promising in hybrid modeling, where first-principle models (e.g. mass balance, transport equations) are mixed and matched with data-driven ML models. Such synergy can achieve more extrapolation than the training dataset without increasing the rigidity of AI. Concerted application of hybrid models to protein, an affinity chromatography has been effective and demonstrates better outcome in the prediction and process insight [31]. The one-size-fits-all DSP strategies will not be effective as bioproducts grow in structure complexity and individualization. We can solve the scalable, flexible, and cost-effective problem of purification pipelines by using adaptive AI/ML-driven systems.

16.6 Common Machine Learning Algorithms in Bioprocess Applications

Machine learning (ML) application in bioprocess optimization has become a major focus because it can represent nonlinear relationships that exist between process parameters and product performance. Many of the ML algorithms have achieved considerable success at multiple phases of biomanufacturing, such as prediction, control, and decision-making. In this section, the context will revolve around the four most popular ML techniques being used in the field of bioprocessing namely Artificial Neural Networks (ANN), Support Vector Machines (SVM), Random Forests and Decision Trees, and Deep Learning and Reinforcement Learning models.

16.6.1 Artificial Neural Networks (ANN)

Artificial Neural Networks (ANNs) mimic the architecture of the human brain and are very handy at making nonlinear pattern detection in complex data. ANNs are compositions of nodes (neurons) that are connected in layers and process the inputs and change the weights with the help of backpropagation. ANNs were widely applied in bioprocessing; these ANNs provide modeling of fermentation processes and other processes like fermentation products prediction, optimizing of process parameters and parameters like pH, agitation and nutriment concentration.

One of the benefits of ANNs is that they can be adjusted to experimental level to the industry level with bioreactors. ANN models have been identified to be successfully applied in predicting recombinant protein production in Escherichia coli under different feed strategies [32]. In addition, ANNs have been shown to be highly accurate in the modeling of retention times in chromatography and in the optimization of the downstream purification yields [33].

16.6.2 Support Vector Machines (SVM)

Support Vector Machines (SVMs) are some of the building block models of supervised early learning, mainly known to perform well with classification and regression issues. They operate on finding a hyperplane that offers optimal partition of the various classes in a partition. SVMs can be used in bioprocessing when the classes concern quality (i.e., does a batch of products fit predetermined product quality, or will the process run according to standard operating condition). SVMs come in handy when data is scarce or unbalanced, hence it is applicable to use them in a very preliminary stage of bioprocess development. In an example on the production of monoclonal antibodies, SVMs were utilized to label various phases in fermentation and know the correct level of the titer products with great accuracy [34]. SVMs are also beneficial in applying omics data to process analytics because they work well in a high-dimensional space.

16.6.3 Random Forests and Decision Trees

Decision trees (DTs) will give the model that resembles a tree, where a decision can be made based on the features of the data. Random forests (RFs) represent an ensemble technique where a group of decision trees are integrated to enhance the accuracy of its predictions and decrease overfitting. Both are very strong in feature selection, classification and regression in bioprocessing. RFs are the most favorable in cases involving noisy data or in a situation with complex biological pathways whose effect on single variables must be appreciated. RFs have also been used in the bioreactor monitoring to determine critical control parameters which influence the product yield and process stability [35]. Moreover, the models help in sensitivity analysis and root cause identification in the event of a batch failure.

16.6.4 Deep Learning and Reinforcement Learning Models

Deep Learning (DL) is a branch of machine learning that deals with neural networks with more than one layer of hidden components and can discover abstract features on massive data set. Convolutional Neural Networks (CNNs) and Recurrent Neural Networks (RNNs) are some examples of DL that can be applied to bio-processing applications, including image-based cell culture monitoring, time-series data on sensor data, and fault detection.

Alternative paradigm with reinforcement learning (RL), models learn which control policies to follow in response to the environment by being told when they succeed or fail, in terms of some reward signal. RL has been investigated in the control of bioreactors, an application in which the algorithm manipulates levels of feeding and environmental conditions on a continual basis so as to optimize a product yield or minimize a processing duration.

Current investigations have revealed that DL with RL can make the adaptive control of fed-batch fermentation to be in real time and, under unpredictable circumstances [36]. Smart biomanufacturing has good prospects since these intelligent systems have better prospects of learning over time, based on constant feedback to adapt to variety in procedures. Figure 16.1 gives application frequency of various ML algorithms and Table 16.2 gives summary of key ML algorithms.

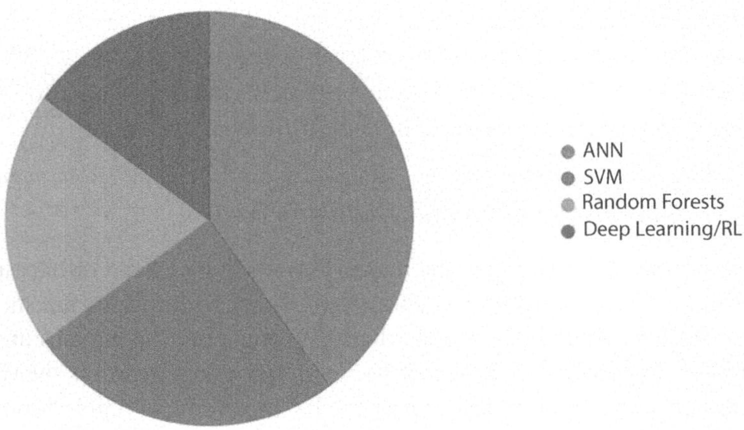

Figure 16.1 Application frequency of machine learning algorithms in bioprocessing.

Table 16.2 Summary of key machine learning algorithms and their applications in bioprocessing Stages.

Algorithm	Key strengths	Upstream applications	Fermentation applications	Downstream applications
Artificial Neural Networks (ANN)	Nonlinear modeling; handles complex data	Predicting cell growth under varying media	Modeling kinetics for biomass yield	Forecasting retention times in chromatography
Support Vector Machines (SVM)	Effective with high-dimensional or small datasets; good for classification	Classifying cell phenotypes	Predicting substrate consumption	Optimizing separation in filtration
Random Forests/ Decision Trees	Interpretable; robust to noise; feature selection	Identifying key variables in media formulation	Detecting anomalies in process stability	Predicting purity based on process parameters
Deep Learning/ Reinforcement Learning	Learning from time-series data; adaptive control	Real-time adjustment of culture conditions	Dynamic control of pH and oxygen levels	Adaptive strategies for complex purifications

16.7 Data Acquisition and Preprocessing in Biomanufacturing

The real running of artificial intelligence (AI) and machine learning (ML) in bioprocess optimization depends on the quality of the data that is due. Bioprocesses are complex, multivariate and dynamic process and, therefore, the integrity of the data, the volume of the data and the relevancy of the data have a direct role to play in the accuracy of the model and reliability of the model. The acquisition and preprocessing of data are, therefore, essential elements in any AI/ML-enabled biomanufacturing policy.

16.7.1 Sources of Bioprocess Data (Sensors, SCADA, PAT)

Various technologies have been implemented in the contemporary biomanufacturing settings in order to collect data concerning the processes. These are sensors, processes, Supervisory Control and Data Acquisition (SCADA) and Process Analytical Technology (PAT) frameworks. In process performance monitoring, sensors used in the process include critical process parameters like temperature, pH, dissolved oxygen (DO), biomass concentration, agitation speed. Valuable optical sensors, near-infrared (NIR) probes, Raman spectroscopy instruments have emerged that offer the chance to detect things in real time and non-invasively, with measurements that were, in the past, inaccessible *via* standard offline analyses [37]. SCADA systems are unifying platforms of data logging, display and controls. They combine sensor to actuator responses thereby making it possible to have smooth feedback links. With SCADA, several process runs and batches in the past can be stored and analyzed to feed a great amount of information in the training of their ML models [38].

A defined real-time quality assurance allows ensuring quality throughout the process *via* Process Analytical Technology (PAT), which was developed by the regulatory bodies, such as the FDA. This makes use of analytics by implementing the analytical instruments in the process stream to evaluate critical quality attributes (CQAs) and process parameters (CPPs). This real-time visibility also enables quality assurance of products and will enable continuing inputs into AI/ ML models, thus improving capability of prediction [39].

16.7.2 Data Cleaning, Normalization, and Dimensionality Reduction

The raw data in the bioprocess context can also be messy, incomplete or inconsistent, and may require a significant amount of preprocessing before it can be used in any ML algorithm. Data cleaning involves determination and treatment of missing values, outlier removal and omitted incorrect measurements. This is pivotal to avoid overfitting of the model and be able to generalize the model.

Normalization/Standardization makes the features reach the same scale; in particular, it is important to such distance-based algorithms as k-NN or SVM. Without normalization, features that have a bigger spread of numbers can overwhelm the machine learning process causing biased models.

The other important point is dimensionality reduction that aids in reducing computation load and enhancing the interpretability of the models. To reduce the dimension in high dimensional data obtained by the sensor, principal component analysis (PCA) is often employed to accomplish the goal to reduce dimensionality and maximize variance [40]. This method is commonly used in the management of bioprocesses and detection of faults.

In addition, the application of auto encoders is a neural network that is gaining popularity in the nonlinear dimensionality reduction of complex biomanufacturing data. They compress low input into low feature representations which assists their efficient learning and still maintain necessary information [41].

16.7.3 Feature Engineering and Time-Series Considerations

This process of feature engineering is very important in modeling the bioprocesses as it identifies important inputs that have effects in the process outputs. It is the transformation of raw data into other variables (features) that can capture the underlying patterns better; such as, rate of change in pH or Oxygen uptake rate (OUR) may better predict the health of the cells compared to the raw pH or DO levels.

Bioprocesses are inherently dynamic; thus, time-series analysis becomes essential. Contrary to the case with the wide use of static datasets, process information changes with time, which adds time dependencies. Thus, lagged variables, rolling averages, and patterns in time should be considered to enable ML models to make proper forecasting. Long Short-Term Memory (LSTM) models, Dynamic Bayesian Networks, and Recurrent Neural Networks (RNNs) are, specifically, applicable to time dependent data [42].

Also, time synchronization of data across sources is a key problem (when trying to synchronize between batch logs, sensor feeds and laboratory analysis). Such approaches as the dynamic time warping or signal synchronization are utilized to guarantee the temporal integrity of multistream data.

Quality data preprocessing allows powerful and explorable AI/ML tests in biomanufacturing. Data preprocessing improves the result of the predictive models and thus aids intelligent decision making by converting raw process data into meaningful ones.

16.8 AI-Driven Process Modeling and Digital Twins

The recent use of artificial intelligence (AI) in the formulation of process modeling and the creation of the Digital Twins is greatly helping to push the curve further as far as the place of bioprocess optimization is concerned. These technologies help with real-time tracking, dynamic simulation of bioprocesses, thus allowing superior control, predictive and decision-making in production processes. This section discusses how AI can help in designing soft sensors, hybrid systems, and digital twins in combination with Process Analytical Technologies (PAT) to better streamline and make biomanufacturing process more versatile.

16.8.1 Development of Soft Sensors and Hybrid Models

Virtual sensors, or soft sensors, are artificial intelligence-based models that predict operating parameters which are costly or impracticable to measure directly. Such sensors are based on the data measured by other variable processes and use machine learning models to extrapolate the unmeasured variables, and hence enhance process control and optimization.

Soft sensors are also commonly applied in bioprocessing to extrapolate the vital process conditions like nutrient concentrations, metabolite levels, and cell density within the fermentation process. Soft sensors are usually developed through training the machine learning models, like artificial neural networks (ANNs) or support vector machines (SVM), using past process data. These models can be trained to predict the unmeasured parameters in an online manner and in such a way operators may render better decisions [43].

A hybrid model, that is an integration of first-principle models with data-driven AI models, can be a very useful instrument that provides extra precision to the predictions. Modifications based on biochemical and biophysical principles are first-principle models that provide a fundamental knowledge of the bioprocesses, whereas the AI methods can address the complicated, nonlinear associations in the data where the first-principle models might not be captured. As an example, hybrid frameworks with a combination of kinetic models of cell expansion and ANN predictions will be more successful in prediction of the cell expansion dynamics and can assist nutrient feeding schedules [44].

16.8.2 How Digital Twins Can Be Used in Simulation and Process Optimization

A Digital Twin is a software representation of a physical system or process, and its behavior, equivalent to the real-time behavior of the physical system. Digital Twins are designed in biomanufacturing by combining the process models, real-time data and AI algorithms. Such digital twins also assist in modeling various conditions, estimating the effects of various manipulations of the processes, and doing optimization of the process without disrupting the physical process of production.

Digital Twins facilitate bioprocess engineers to conduct simulations to accurately mimic the real state of production, such as temperature, pH and the level of dissolved oxygen, all of these factors affect the quality of the products. Through the real-time data on processes, the Digital Twin is constantly updated, mirroring the state of the system, and gives a live simulation as well, enabling the operators to observe and correct deficiencies in the performance, and regarding parameters, in a timely fashion [45].

Digital Twins are combined with AI algorithms such as reinforcement learning (RL) to optimize the decisions. Such digital twins driven by AI can analyze multiple scenarios of what-if and how certain things will change when parameters are varied, as well as provide optimal settings of a process to achieve maximum yield and efficiency. As an example, during protein production, a Digital Twin can indicate, based on a change in temperature and feed rate, the implications of these on the product yield and quality and thus aid in a real-time optimization [46].

16.8.3 Integration with Process Analytical Technologies (PAT)

Process Analytical Technologies (PAT) are technology systems, which are utilized to observe, analyze, and control manufacturing processes in real time mode. By incorporating AI-enabled Digital Twins with PAT systems, the more control and optimal management of bioprocessing operations can be achieved through continuous measurement of most important parameters of process and actualization of measurement into a predictive model.

By combining PAT with Digital Twins, a closed-loop system can be achieved so that real-time sensor data, including pH, temperature, and dissolved oxygen, can be ingested into the AI models to continuously update the Digital Twin. This system does not simply give a real-time dynamic model of the process but also provides the possibility to optimize processes in real-time according to AI-driven insights [47]. Furthermore, this integration enables the use of Quality by Design (QbD) concepts in the biomanufacturing process as this ensures quality product and adherence to regulatory standards as critical product control parameters are monitored and controlled.

A potential area of use of PAT and Digital Twin is controlling and monitoring in fermentation processes. After processing the data gathered by the sensors with the help of AI, Digital Twin will be able to forecast possible deviations of the things which are not perfect and to program the changes to keep the process within the required operation range. Such proactive process leads to minimal product failure, minimized waste and more batch-to-batch consistencies [48].

Digital Twins together with AI and PAT can be used as a synergy to increase transparency of the processes, regulatory compliance and overall advance the robustness of the processes in biomanufacturing.

16.9 Conclusion

The future of biomanufacturing is increasing popularity of artificial intelligence (AI) and machine learning (ML) in bioprocess optimization is changing the game of biomanufacturing. The traditional techniques to optimize bioprocess parameters are time and cost-consuming and may be difficult to scale up, and therefore, hinder the development of a product. AIs and MLs will help in providing efficient and data-driven solutions that will allow more successful prediction, adaptive control methods, and better decision-making on a real-time basis and eventually lead to greater efficiency and yield of the process, better products, and general quality control. Whether it involves streamlining cell culture and media production or purifying the protein *via* a downstream process like chromatography and filtration system, AI/ML models have been indispensable. Artificial neural networks (ANNs) and support vector machines (SVMs) are machine learning algorithms, which allow dealing with complex data processing and discovery of hidden patterns in them that are not obvious at first sight. The insights can be used to provide increased control of the bioprocess variables resulting in the production of a consistent and superior quality. Despite the promise, challenges remain. The data quality is also a very important aspect since if the data is not reliable or lacking, it may impede the performance of the AI models. Besides, transparency of the AI algorithm and the capability that it explains its decisions sometimes called explainable AI (XAI) is a matter of ongoing development, as with other industries, even regulated ones like biomanufacturing. The regulatory systems are also expected to change toward emerging AI, as well as ensuring the adoption of the technologies, which are operated in a compliance-related and effective way. The future of AI and ML in the field of bioprocess optimization is encouraging. Better adaptability and precision of the biomanufacturing processes will be achieved by the integration of digital twins, hybrid modeling, and soft sensors. Moreover, further evolution of explainable AI algorithms is likely to help gain more trust in systems built using AI, which will lead to more extensive adoption.

To sum up, the AI and ML technologies have a crucial role in the future of biomanufacturing. With further research and the development of technologies, the use of AI and ML will additionally simplify the bioprocessing processes, bring them to a higher efficiency level, and eventually lead to the manufacturing of high-quality biopharmaceuticals. To researchers, practitioners, and decision-makers in the field, keeping on top of these innovations will be critical to their staying competitive in biomanufacturing where it is a continuous race.

AI disclosure statement: The author acknowledges the use of AI-based tools for minor language editing or grammatical correction. However, all intellectual contributions, analyses, and conclusions presented in this chapter are entirely the author's own.

References

1. Mandenius, M. and Brundin, J., Bioprocess optimization using design-of-experiments methodology. *Biotechnol. Progr.*, 24, 6, 1191–1203, 2008.
2. Walsh, G., Biopharmaceutical benchmarks 2018. *Nat. Biotechnol.*, 36, 1136–1145, 2018.
3. Liu, J.P., Kumar, K., Taylor, A.M.R., Review of statistical optimization methods for complex bioprocessing applications. *J. Ind. Microbiol. Biotechnol.*, 45, 3, 275–287, 2018.
4. Hassan, S.M., Hossain, M.A., Alam, K.M.M., Challenges in scale-up of bioprocesses and the role of computational tools. *Biochem. Eng. J.*, 137, 130–144, 2018.
5. Teixeira, A.R., Silva, C.M., Cabral, J.M.S., Artificial intelligence in bioprocess systems engineering: State of the art and future trends. *Comput. Chem. Eng.*, 145, 107174, 2021.
6. Shen, Y., Wang, X., Zhang, H., Deep learning-based process modeling and control for biomanufacturing: A review. *Biotechnol. Adv.*, 50, 107769, 2021.
7. Glassey, A., Glassey, K., Gernaey, A., Digital twins for bioprocessing: Concepts and applications. *Trends Biotechnol.*, 39, 9, 889–902, 2021.
8. Del Río-Chanona, A., Papathanasiou, R., Suthers, D., Müller, E.A., Machine learning for predictive process control in chemical and biochemical engineering. *Comput. Chem. Eng.*, 136, 106833, Dec. 2020.
9. Singh, B.K. and Singh, V.R., Artificial neural networks in modeling and optimization of bioprocesses: A review. *Biochem. Eng. J.*, 151, 107377, Apr. 2019.
10. Rousu, D.A., Gupta, A.V., Goldstein, L.B., Application of support vector machines to bioprocess modeling. *J. Biotechnol.*, 314, 12–20, Jan. 2020.
11. Detroja, P.B. and Mahajani, M.M., Reinforcement learning-based control strategies for batch bioprocesses. *Chem. Eng. Res. Des.*, 148, 255–264, Feb. 2019.
12. Wang, Y., Han, L., Yu, H., Zhang, J., Data preprocessing techniques in bioprocess data analytics. *Processes*, 9, 6, 1043, Jun. 2021.
13. Polotti, F.P., Morbidelli, M., Barton, J.P., Hybrid modeling for bioprocess optimization: Bridging mechanistic and data-driven methods. *Biotechnol. Adv.*, 54, 107806, Mar. 2022.
14. Zarei, J., Lorestani, S.B., Ahmadi, M.H., Artificial neural network modeling of cell culture processes: A review. *Biochem. Eng. J.*, 148, 107240, 2019.

15. Torkashvand, M., Vaziri, A., Shokouhifar, M.S., Machine learning approaches for cell culture optimization in biopharmaceutical manufacturing. *J. Ind. Microbiol. Biotechnol.*, 48, 3-4, kuab008, 2021.
16. Zhang, H., Wang, Y., Liu, L., *et al.*, Application of machine learning techniques for designing cell culture media. *Processes*, 8, 5, 556, 2020.
17. Heath, J.R., Ribas, L., Leach, A.J.M., Feeding strategy design in mammalian cell culture using predictive control and machine learning. *Biotechnol. Progr.*, 36, 2, e2921, 2020.
18. Paliwal, R., Sharma, P., Dubey, A.K., Integration of design of experiments with artificial neural networks for media optimization in bioprocesses. *Chem. Eng. Res. Des.*, 132, 905–916, 2018.
19. Tripathi, S.S. and Srivastava, S.S., Artificial neural networks in fermentation process modeling and control: An overview. *Biochem. Eng. J.*, 56, 2, 85–94, 2011, doi: 10.1016/j.bej.2011.06.004.
20. Zhang, C., Zhang, Y., Du, W., Predictive modeling for fermentation processes using support vector machines. *J. Chem. Technol. Biotechnol.*, 89, 2, 238–246, 2014, doi: 10.1002/jctb.4167.
21. Moradian, M., Ebrahimi, F., Zarrin, A., Reinforcement learning-based strategy for real-time optimization in fed-batch bioprocesses. *Bioprocess. Biosyst. Eng.*, 43, 6, 927–938, 2020, doi: 10.1007/s00449-020-02273-6.
22. Hassan, A.B.A. and Mohamed, M.A.H., Application of neural-fuzzy models for dissolved oxygen control in fermentation. *Comput. Chem. Eng.*, 60, 24–32, 2014, doi: 10.1016/j.compchemeng.2013.08.005.
23. Degenring, F., Schnell, R., Zimmermann, M.W., Takors, J., Modeling and application of soft sensors in fermentation processes. *J. Biotechnol.*, 132, 4, 450–460, 2007, doi: 10.1016/j.jbiotec.2007.01.007.
24. Li, C., Chen, J., Zhao, Z., Application of recurrent neural networks for modeling and control of ethanol fermentation process. *Process Biochem.*, 50, 7, 1156–1163, 2015, doi: 10.1016/j.procbio.2015.03.005.
25. Müller-Späth, R.H., Ströhlein, A., Morbidelli, M., Glennon, M.J., Data-driven modeling of protein retention in ion-exchange chromatography using artificial neural networks. *Biotechnol. Progr.*, 30, 6, 1375–1385, 2014, doi: 10.1002/btpr.1989.
26. Ahmad, A., Al-Ghouti, M.A., Amosa, N.A., Machine learning techniques for predicting membrane fouling in water treatment: A review. *Sci. Total Environ.*, 775, 145755, 2021, doi: 10.1016/j.scitotenv.2021.145755.
27. Searle, C.D., Green, R.L., Sefcik, J.A., Reinforcement learning for adaptive control of protein chromatography. *AIChE J.*, 66, 11, e17091, 2020, doi: 10.1002/aic.17091.
28. Zhong, H., Luo, Y., Chen, X., Machine learning modeling for biopharmaceutical purification yield prediction. *J. Chem. Technol. Biotechnol.*, 95, 7, 1837–1845, 2020, doi: 10.1002/jctb.6379.

29. Hossain, K.A., Rahman, A., Bulmer, J.W., Predictive analytics for host cell protein content using multivariate data analysis in bioprocessing. *Biotechnol. Bioeng.*, 116, 3, 584–593, 2019, doi: 10.1002/bit.26878.
30. Wüthrich, B., Möller, J., Roth, M., Real-time control of complex chromatography processes using recurrent neural networks. *J. Chromatogr. A*, 1656, 462528, 2021, doi: 10.1016/j.chroma.2021.462528.
31. Hennessy, J.R., Lewandowski, M.D., Roush, P.J., Hybrid modeling of protein A chromatography combining mechanistic and machine learning approaches. *Biotechnol. J.*, 15, 11, e1900286, 2020, doi: 10.1002/biot.201900286.
32. Lin, J., Wu, Y., Yang, Z., Modeling recombinant protein production using artificial neural networks in high-density fed-batch cultures of *Escherichia coli*. *Bioprocess. Biosyst. Eng.*, 41, 1, 71–80, 2018, doi: 10.1007/s00449-017-1852-6.
33. Singh, T. and Jayaraman, R., ANN-based modeling of chromatographic retention times for process optimization in downstream bioprocessing. *J. Chem. Technol. Biotechnol.*, 93, 9, 2618–2626, 2018, doi: 10.1002/jctb.5668.
34. Li, Y., Wang, Q., Zhang, H., Support vector machine modeling and classification of monoclonal antibody fermentation process. *Biotechnol. Progr.*, 34, 3, 748–757, 2018, doi: 10.1002/btpr.2622.
35. Ferrer, L., García, D.L., Romero, M.T., Random forest modeling for critical parameter identification in biopharmaceutical manufacturing. *Process Biochem.*, 98, 202–210, 2020, doi: 10.1016/j.procbio.2020.07.020.
36. Liu, M., Chen, P., Zhou, F., A reinforcement learning framework for bioreactor control in biomanufacturing under uncertainty. *Eng. Appl. Artif. Intell.*, 105, 104424, 2021, doi: 10.1016/j.engappai.2021.104424.
37. Teixeira, M., Fernandes, A.R., Lopes, J.M., Azevedo, A.M., Advances in optical sensor technology for real-time monitoring of bioprocesses. *Sens. Actuators, B*, 330, 129326, 2021, doi: 10.1016/j.snb.2020.129326.
38. Cao, Y., Zhao, L., Wu, X., Data acquisition and real-time monitoring using SCADA in biopharmaceutical manufacturing. *J. Biotechnol.*, 308, 1–8, 2020, doi: 10.1016/j.jbiotec.2019.11.008.
39. Rathore, S. and Bhambure, R., Process analytical technology (PAT) for biopharmaceutical products: A review of applications and recent developments. *J. Pharm. Biomed. Anal.*, 147, 557–566, 2018, doi: 10.1016/j.jpba.2017.07.013.
40. Mercier, A., Heuillet, N., Lakhdar, B., Principal component analysis for dimensionality reduction in batch bioprocess monitoring. *Chem. Eng. Res. Des.*, 147, 112–123, 2019, doi: 10.1016/j.cherd.2019.05.010.
41. Singh, D. and Mishra, P., Autoencoder-based nonlinear dimensionality reduction in process data monitoring. *Comput. Chem. Eng.*, 146, 107183, 2021, doi: 10.1016/j.compchemeng.2020.107183.
42. Scholz, B., Müller, M.S., Becker, T., Time-series modeling in bioprocess monitoring using LSTM networks. *Biochem. Eng. J.*, 172, 108027, 2021, doi: 10.1016/j.bej.2021.108027.

43. Kromer, M.E., Möller, J.M., Mazzilli, M.R., Real-time monitoring of cell culture processes using AI-based soft sensors. *Biotechnol. Progr.*, 33, 6, 1450–1462, 2017, doi: 10.1002/btpr.2520.
44. Li, J.M., Patel, S.L., Lee, T.G., Hybrid modeling for optimization of fed-batch fermentation using soft sensors and ANN. *Biochem. Eng. J.*, 118, 45–53, 2017, doi: 10.1016/j.bej.2017.06.004.
45. Al-Mansoori, F.H., Al-Harrasi, A.A., Al-Busaidi, F.I., Digital twin and AI-driven optimization in biomanufacturing. *J. Process Control*, 94, 87–98, 2021, doi: 10.1016/j.jprocont.2020.12.001.
46. Zhao, H., Li, H., Li, X., AI-based digital twins in bioprocess optimization: Applications in recombinant protein production. *Biotechnol. Bioeng.*, 118, 4, 956–967, 2021, doi: 10.1002/bit.27675.
47. Sudhir, S.M., David, D.M., Gupta, R.C., AI and PAT integration for advanced control in fermentation. *Bioprocess. Biosyst. Eng.*, 43, 9, 1407–1417, 2020, doi: 10.1007/s00449-020-02495-w.
48. Garcia, S.C., MacDonald, P.J., Crossley, A.J., Implementing PAT and digital twin for real-time fermentation process control. *Chem. Eng. Sci.*, 220, 115652, 2020, doi: 10.1016/j.ces.2020.115652.

17

Biomanufacturing of RNA Based Therapeutics and Vaccines: Advances, Issues and Prospects

Rompicherla Srividya[1], Navapet Vikas Reddy[1], Tirumala Shreya[1], Komere Vasu[1], Appala Naidu Uttaravalli[2], Manjakuppam Malika[1] and A.V. Raghavendra Rao[1]*

[1]Dept. of Chemical Engineering, B V Raju Institute of Technology, Narsapur, Telangana, India

[2]Optimus Drugs Private Limited (A Sekhmet Pharma Ventures Private Limited Company), Hyderabad, Telangana, India

Abstract

Among contemporary treatment methods, RNA-based therapy, especially messenger RNA (mRNA) vaccines, has become a game changer in the world of modern medicine, especially after the COVID-19 pandemic outbreak. The chapter explores the changing landscape of RNA-based biologics, the centrality of biomanufacturing therein in the swift development, scale-up and delivery to the world. We consider upstream and downstream activities that distinguish RNA therapeutics from *in-vitro* transcription (IVT), capping methods, purification, and lipid nanoparticle (LNP) encapsulation. The focus revolves around the merger of automated, modular and scalable platforms well conforming to the rigorous prescriptions of Good Manufacturing Practice (GMP). Besides, the chapter also presents technological inventions like microfluidics, single use bioreactors and real time analytical tools which are revolutionizing RNA biomanufacturing. Regulatory considerations, cost-effectiveness, and cold-chain logistics are also examined. This paper aims to explain the accomplishments and least productive processes of modern systems based on the case studies of contemporary campaigns in mRNA vaccine production. The last part of the chapter summarizes some strategic research and development directions such as thermostable formulation, decentralized manufacture centers, and path to personalized RNA pharmaceuticals.

**Corresponding author*: raghusrf@gmail.com

Sandip Kunar, Jagadeesha T. and D. Sathis Kumar (eds.) Biomanufacturing: Transforming Production of Biologics and Biopharmaceuticals, (521–550) © 2026 Scrivener Publishing LLC

Keywords: mRNA vaccine, RNA therapeutics, lipid nanoparticle, scalable platform, RNA purification, cold chain logistics

17.1 Introduction to RNA-Based Therapeutics and Vaccines

The modern medicine has transformed over the past few years with RNA-based therapeutics, and vaccines being able to provide a platform to rapidly develop and treat diseases in a new way, using the cellular pathways of the body to create therapeutic proteins or responses. Different modalities such as messenger RNA (mRNA), small interfering RNA (siRNA) and self-amplifying RNA (saRNA) have encoded genetic instructions that can be designed to suit a particular application, all the way up to cancer immunotherapy and preventive approaches to infectious disease [1]. By contrast to conventional vaccines, which rely on weakened pathogens or protein antigens, RNA therapies contain artificially produced RNA molecules in form of lipid nanoparticles (LNP) to be introduced into the cells and translated into proteins, but not incorporated into the genome, thus with low risks in the long-term perspective [2]. These products are synthesized, purified and formulated *via* biomanufacturing through distinctive synthesis, purification and formulation challenges to provide it with stable, pure, and scalable products given the inherent instability of RNA through RNase degradation. This introduction examines the description of the RNA modalities, the history of its development accelerated by the COVID-19, the importance of biomanufacturing, and the structure of the chapter and its goals.

RNA therapeutics are dependent upon delivery and production innovations to overcome liabilities such as inducing an immune response or having minimal half-life. With the increasing use of the field in areas outside of vaccines, such as chronic conditions, learning the underpinnings is key to the future development of the biomanufacturing strategies [3].

17.1.1 General RNA Modalities: Modalities: mRNA, SiRNA, and Others

The RNA modes include a wide variety of therapeutic reagents, with mRNA vaccines currently leading the pack by instructing cells to create antigens that induce immune responses and, as demonstrated in the prophylaxis context against viruses, work well by silencing particular genes, such as RNA interference-based siRNA as an effective treatment in genetic disorders, such as genetically based amyloidosis of transthyretin [5]. The development of siRNA therapies includes the approval of patisiran in 2018, which demonstrated the potential of silencing genes *via* RNA. In contrast, self-amplifying RNA (saRNA) contains replicase genes that allow the RNA to replicate within cells, enhancing protein expression for therapeutic applications [6].

Other modalities involve circular RNA (circRNA), which can persist against exonuclease degradation longer and cause more lasting effects, and CRISPR RNA to edit gene. The common requirements of these are the biomanufacturing requirements: high-purity IVT and strong encapsulation against nucleases [7]. Base modification innovations, such as pseudouridine substitution, do this by lowering immunogenicity and increasing translation. The overview emphasizes the flexibility of RNA yet in terms of scalability and cost there are issues that persist, and optimization of processes is push [8].

17.1.2 The Evolution and the Impact of the COVID-19 Pandemic Historically

The history of the RNA-based new therapeutics started in 1990s with the work on mRNA as potential source of protein expression *in vivo* which was not very successful due to mRNA instability and immune response. The major advances are the identification of modified nucleosides in 2005 (which reduced innate immune activation, demonstrated in 2010, in cancer vaccine clinical trials [9]). The approval of patisiran in 2018 highlighted the potential of siRNA therapies to silence genes using RNA [10].

The COVID-19 pandemic of 2020 provided a rapid boost to the evolution, with mRNA vaccines in the form of Pfizer-BioNTech and Moderna vaccine achieving emergency authorization in a short period of time, less than a year, and saved millions of lives proving the efficacy and rapidity of the platform [1]. The crisis has brought into focus the innovation of biomanufacturing including the ability to rapidly scale-up to billions of doses through modular facilities and LNPs as the mode of delivery [2]. After pandemic, this has extended RNA application to influenza and Zika vaccines and currently the RNA market investment has exceeded the $10 billion market annually [3]. Nevertheless, it outlined bottlenecks such as cold-chain needs and fair dosage, and it affected the upcoming design of thermostable formulations [4]. This development reflects how RNA has already gone mainstream, due to pandemic demand.

17.1.3 Importance of Biomanufacturing in RNA Production

RNA production requires biomanufacturing; the upstream process of synthesis and downstream purification and formulation steps positively impacts the generation of high-grade and stable final therapeutic products at scale. RNA also must be produced at scale through IVT strategy at the RNA level using capping to resemble eukaryotic mRNA, thus guaranteeing successful translation, unlike DNA plasmids [5]. It purifies away such things as byproducts of the double-stranded RNA synthesis such as immune responsive substances using methods such as tangential flow filtration and chromatography [6].

Biomanufacturing breakthroughs such as the single-use systems minimize the risk of contamination and facilitate and speed up technology transfer, as it was done with the COVID-19 vaccines [7]. The RNA can be securely packed into LNP with minimal degradation and be incorporated into cells with defined nanoparticles constructed by microfluidics to allow equal size distribution [8]. Biomanufacturing that is complying with the GMPs assures regulatory approval, and real-time analytics are maintained on CQAs such as purity (>95 percent) and encapsulation >90% [9]. Affordability is critical since manufacturing costs may be more than one dollar per dose on a large scale; streamlined procedures will reduce the latter by half due to automation [10]. And finally, biomanufacturing can move discoveries made in the lab into clinical practice to support tailor-made treatments and universal access.

17.2 Upstream Processes in RNA Biomanufacturing

RNA biomanufacturing Upstream processes RNA biomanufacturing is the basis of upstream operations of high-quality RNA molecules and includes most of the steps during the synthesis, modification, and initial optimization of the RNA biomanufacturing product, prior to downstream purification and formulation. These techniques play a momentous role in producing RNA therapeutics and vaccines and yield, purity, and functionality of the RNA have a direct impact on therapeutic performance and safety. The most commonly used method of RNA synthesis is *in vitro* transcription (IVT) based on the transcription of linear DNA templates by T7 RNA polymerase resulting in RNA strands producing hundreds of micrograms per reaction with optimized conditions [1]. To replicate endogenous mRNA and, thus, to increase mRNA stability and translation *in vivo*, it is necessary to perform modifications including capping, polyadenylation.

With RNA-based products such as COVID-19 mRNA vaccines reaching billions of doses, the upstream innovations are directed at making enzyme-based reaction as efficient as possible, template optimization, and potentially alleviating rate-limiting steps in manufacturing processes [2]. Here, IVT methods, capping and polyadenylation systems, optimization of templates and enzymes, and the issue of scale up high in the upstream world are examined. Because of their low costs, effective upstream processes can give RNA yields as high as 1 g/L, though the process is extremely sensitive, and must be tightly controlled to reduce potential sources of immunogenicity such as double-stranded RNA (dsRNA) [3].

The speed of mRNA vaccines development against the COVID-19 highlighted the need to have upstream services to support proper development, but demonstrated the shortcomings of architecture on the scalability and cost, and stimulated research on automated high throughput processes [4].

17.2.1 *In Vitro* Transcription (IVT) Techniques

The most relevant method in upstream RNA biomanufacturing, the process of *in vitro* transcription (IVT) facilitates enzymatic RNA production out of a DNA template with the help of RNA polymerases such as T7 polymerase or SP6 polymerase. This is carried out by combining linearized plasmid DNA with nucleotides, magnesium ions and polymerase in a buffered reaction, usually incubated at 37°C, between 2-4 hours and yields uncapped RNA transcripts which may be further modified [5]. More elaborate IVT approaches include altered nucleotides, e.g. pseudouridine or N1-methylpseudouridine to have lower immunogenicity and higher stability as shown in mRNA vaccines, with these modifications enhancing translation 5-10 fold [6].

Cell-free IVT systems are faster and purer than cellular expression and do not need to avoid host contaminants, but the cellular components of the reaction must be optimized to achieve high yields, typically ranging between 1-5 mg/mL at high-fidelity polymerase [1]. Techniques that overcome product inhibition such as fed-batch IVT, i.e. addition of nucleotides step-by-step, which has a similar effect of increasing the reaction timespan, resulting in a 30-50% increase in output [2]. When it comes to a therapeutic application, IVT is customized to saRNA because of the inclusion of replicase genes, and then provides self-amplification following delivery [3]. Challenges include abortive transcripts, addressed through promoter engineering [7]. The variety of RNA modalities that IVT supports, including CRISPR guides and long non-coding RNAs, ensures that it is instrumental to biomanufacturing [8]. Figure 17.1 shows *In Vitro* Transcription (IVT) Process.

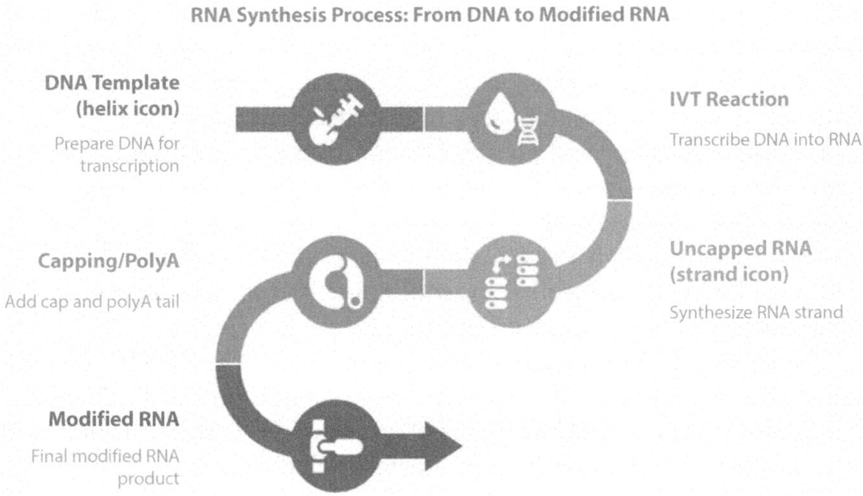

Figure 17.1 Diagram of *in vitro* transcription (IVT) process modified.

17.2.2 Capping and Polyadenylation Strategies

Other important modifications during post-IVT are capping and polyadenylation that allow greater stability of RNA, translation, and immune evasion of biomanufacturing. Capping is a 7-methylguanosine cap added to the 5' end with traditional enzymatic methods using vaccinia capping enzyme or more recently co-transcriptional reactions with cap analogues such as ARCA where efficiency is routinely >90% to provide a ribosome target site [9]. New methods incorporate clean content (Clean Cap), in which a trinucleotide cap analog is added to eliminate production of dsRNA and increase capping to nearly 100%, as in the COVID-19 vaccines [10].

Polyadenylation adds this poly(A) tail to the 3 end, which helps to protect RNA against action of exonucleases; tail lengths can be 100-200 nucleotides, optimized towards increasing half-life, *via* a linking process of the poly(A) polymerase [5]. Having the poly(A) sequences encoded as a template makes the procedure easier, yet they may result in heterogeneous tailed distributions, which is mitigated by designing templates with the exact repeats [6]. Polyadenylated capping kits simplify the procedure, decreasing the number of steps and expenses up to 30-20 percent [2]. Such treatments have reduced activation of natural immune responses through RIG-I, which is important in therapeutic RNA [3]. In scale-up, there is uniformity produced by automated systems, and changes are confirmed by analytics such as capillary electrophoresis [7]. In general, these changes are central to functional RNA since they can undergo clinical translation [8].

17.2.3 Optimization of Template Design and Enzymes

IVT depends on optimization of template design and enzymes to be effective in the RNA biomanufacturing, which directly affects yield, purity, and functionalization. Template design consists of linearizing the plasmid in regions with good promoters, such as T7, inclusion of UTRs to improve stability, such as Kozak sequence to enhance initiation, and exclusion of promoters which create secondary structures, and thus slow transcription [11]. Codon optimization is customized to human expression and amplifies protein production 2-5 folds and prevents the usage of the rare codons that induce pausing [12].

Enzyme optimization centers on high-processivity RNA polymerases; mutated T7 variants decrease byproducts of dsRNA (70% reduction) and thus reduce immunogenicity [9]. An even hotter Hi-T7 polymerase (50 o C), boosts yield over GC-rich templates by 50 percent [10]. Reactions are stabilized by co-factors such as spermidine, and bead support of enzymes allows enzymes to be used repeatedly saving 40 percent on the costs of such large-scale runs [5]. In DoE we can find optimal ratios and get >2 g/L yields [6]. The templates to be used in the case of saRNA are alphavirus replicons, which were made replication competent [3]. Analytics like NGS verify template fidelity [7]. These are optimizations that produce high-quality RNA, which is essential in determining the success of therapies [8].

17.2.4 Challenges in Upstream Scale-Up

Upstream scale-up issues of RNA biomanufacturing revolve around moving lab-scale upstream IVT to an industrial scale, at which point yield and purity become complicated. Reaction volumes >100L enhance such problems as mixing inefficiencies, gradient formations, that lower transcription uniformity and yields by 20-30% [13]. Scalability is hindered by this unavoidable enzyme inhibition by intracellular buildup of pyrophosphate, and dialysis or the addition of pyrophosphatase is necessary to extend an enzyme reaction [11].

Stability of the templates is compromised at scale with shear encountered during stirring, which can disrupt the templates into fragments; even the highest purity of DNA templates is required to avoid off-target transcription [12]. Nucleotide and enzyme cost scales up, requiring recycling and enzyme cost is high, 50-70% of upstream budgets [9]. dsRNA byproducts emerge in proportion and induce immune responses; cost effective post-IVT purification does not scale without highly optimized enzymes [10]. The regulatory requirements of GMP specify sterile closed system, a factor that complicates the possible open IVT setups [5]. Answers would be to use continuous-flow reactors to make steady state production a possibility, increasing throughput by 40 percent [6]. Microfluidics enable precise mixing, reducing variability [3]. Notwithstanding these, scale-up is still a costly step as yields degrade by a factor of 2, 5 mg/mL lab to 2 mg/mL industrial, although this will be reduced in the future with attention to engineered and automated enzymes [7].

17.3 Downstream Processes for RNA Purification and Formulation

RNA biomanufacturing downstream processes in RNA, biomanufacturing plays a critical role in creating stable, deliverable forms of RNA molecules to become effective and safe therapeutics and vaccines that achieve clinical standards. In a downstream process, upstream synthesis by IVT is followed by compensation of the instability of RNA with signal against degradation in addition to the withdrawal of contaminants, such as truncated transcripts or duplex-RNA that can generate immune reactions [11]. Chromatography and filtration methods give high-purity levels (>95%), and LNP encapsulation potentially secures RNA as well as supports the delivery of it into cells, which is essential in mRNA vaccines [12]. Innovations such as lyophilization are applied to formulation processes to improve stability in the face of nucleases and temperature changes to increase shelf-life [13]. Integration of quality control with the analytics serves to confirm CQAs, such as the presence of sequence integrity and encapsulation efficiency. Various strategies of purifying RNA, LNP encapsulation, stability, and formulation, and downstream analytics are discussed in this section. Highly efficient down streaming may allow scaling up production to grams per liter but must be optimized to reduce losses which may reach over 50% in crude processing schemes [14].

The COVID-19 mRNA vaccines were successful examples of downstream efficiency as purely cleaned material could be distributed worldwide in short order but also exposed the need to both reduce costs and improve thermostability [7].

17.3.1 RNA Purification Techniques: Chromatography and Filtration

Pharmaceutical-grade purity during downstream biomanufacturing requires the elimination of impurities, which is necessary using RNA purification procedures, such as chromatography and filtration. Affinity chromatography: Isolation of mRNA– A key method of capturing mRNA out of an IVT mix is by affinity chromatography, in which oligo(dT)-conjugated resins are used to capture poly(A) tails [15]. Charge separation allows purification *via* ion-exchange chromatography where recently, siRNA can be effectively purified using salts to elute according to different salt gradients, leaving aggregates at less than 1% [16].

RNA can also be concentrated, as well as exchange removed, using filtration techniques, such as the tangential flow filtration (TFF) that concentrates RNA and also exchanges of buffers, with ultrafiltration membrane (cutoff 30-100 kDa) permitting limited permeation of small impurities allowing concentrations of >1 mg/mL [11]. Cellular debris is trapped in initial stages during depth filtration, so separation in the following chromatography is not pre-contaminated [12]. Combinatorial methods, such as TFF and reversed-phase HPLC have the added advantage of increased specificity to long RNAs, e.g. saRNA purification [13]. The problems are related to RNA binding to membranes, which can be addressed by the modification of the surface or PEGylation [14]. The scale-up employs multi-column purification in a continuous manner increasing throughput by 40 percent [17]. The methods also guarantee the low endotoxin content (<10 EU/mg), which is important in the clinical practice [7]. Table 17.1 shows the comparison of RNA purification techniques.

17.3.2 Lipid Nanoparticle (LNP) Encapsulation Methods

The encapsulation process in lipid nanoparticle (LNP) preserves RNA molecules upon degradation and facilitates their delivery due to the formation

Table 17.1 Comparison of RNA purification techniques.

Technique	Key applications	Advantages	Challenges
Chromatography (Affinity/ Ion-Exchange)	mRNA isolation, siRNA separation	High purity (>95%), selective binding	Resin costs, dsRNA carryover
Filtration (TFF/ Depth)	Concentration, debris removal	Scalable, gentle on RNA	Membrane fouling, adsorption losses
Combined (TFF + HPLC)	Long RNA purification	Comprehensive impurity removal	Multi-step, time-consuming

of nanosized complexes with ionizable lipids, cholesterol, PEG-lipids and helpers [18]. The gold standard of mixing, microfluidic mixing, allows mixing lipid-ethanol solutions with aqueous RNA in known ratios (e.g., 3:1 lipid:RNA) in a short period of time to generate uniform LNPs (50-150 nm) with high encapsulation efficiencies (>90%) [19]. Quantifiable over large volumes, ethanol dilution techniques involve mixing phases with pumps, but the concept will most likely need optimization due to polydispersity [12].

Post-formation, through membrane extrusion sizes can be refined, to limit heterogeneity, improving biodistribution [13]. Such innovations as T-junction mixers allow high-throughput manufacture, as in COVID-19 vaccines, with billions of doses [14]. The composition of LNP is customized LNP composition; ionizable lipids such as MC3 is able to readily escape endosomes through pH-dependent charge [7]. RNA loading variability and issues are a challenge, which has been solved through charge ratio tuning [15]. Performance is confirmed by quality analytics, e.g., DLS to size encapsulation, and PicoGreen to size encapsulation [16]. The approaches are a game changer to RNA delivery as LNPs facilitate targeting to liver in siRNA medicine [17].

17.3.3 Formulation for Stability and Delivery

RNA biomanufacturing formulation regarding stability and delivery aims to ensure that the RNA is preserved following storage and delivery so that it remains active in therapeutic terms. Seeing that, cryoprotectants such as sucrose (5-10%) in LNP-based drug products prevent aggregation in thawing and freeze cycles and remain at -80°C for months [18]. Trehalose Lyophilization improves the stability at room temperatures eliminating the use of the cold-chain in low-resource regions when using vaccines [19].

Targeting ligands on LNPs such as GalNAc to hepatocytes of the liver in siRNA delivery [12], and pH-sensitive lipids enhance release out of endosomes enhancing the transfection efficiency up to 2-5-fold [13]. In the case of mRNA vaccine, adjuvants, such as CpG, improve the immunogenicity of the vaccine without interfering with stability [14]. It suffers oxidation, which is alleviated by adding antioxidants such as EDTA, and increasing shelf-life [7]. The formulation is continuously produced in inline mixers, which allows uniformity on scale [15]. Such analytics as DSC determine the level of thermal stability, which helps in such aspects of excipient choice [16]. Such measures have facilitated the worldwide implementation of mRNA vaccines, and ambient-stable formulations are the future [17].

17.3.4 Quality Control and Analytics in Downstream

Downstream RNA bioprocessing through quality control and analytics is involved in the verification of the integrity, purity, and functionality of the products through a set of assays built into workflows. The identity and variants in RNA are validated by sequencing, and the NGS finds variants even at <1% frequencies [11]. Size and poly(A) tail length are determined by capillary electrophoresis with >90% full-length transcripts obtained [12].

RiboGreen assays measure the encapsulation efficiency, the free-bound RNA quantification, aiming at >95% [13]. DLS and NTA determine LNP size (80-120 nm) and polydispersity (<0.2) [14]. Endotoxin test by LAL confirms <5 EU/mg, which is important in safety [7]. Verification of translation is ascertained by potency assays such as cell transfection with subsequent ELISA [15]. Inline HPLC is real-time PAT systems [16]. Analytical challenges are a low-level sensitivity of low abundance impurity which is resolved using methods of MS [17]. The process of QC is simplified using automated systems such as Bioanalyzer, which also saves time by half [18]. Such analytics make GMP compliance possible, which allows release testing of clinical-grade RNA [19].

17.4 Technological Innovations in RNA Biomanufacturing

RNA biomanufacturing advances are transforming how therapeutics and vaccines are manufactured by responding to pursed constraints of scale, specificity, and cost, using sophisticated tools and platforms. Such progress allows efficient synthesis, purification, and delivery of the RNA molecules that are imperative in modalities such as mRNA vaccines that must be produced with a high degree of purity and stability to induce sufficient immune responses [7]. Among the most important innovations are microfluidics, single-use bioreactors, real-time analytics, automation and they are combined to bring workflows in IVT to LNP formulation. As an example, precise mixing by microfluidic devices enables production of uniform LNPs in higher encapsulation efficiency of >95% and 50% material waste reduction as compared to conventional procedures [15]. This part examines microfluidics as an approach to synthesis and encapsulation, single-use and modular platforms, PAT-integrated analytics and process digitalization. With the expanding RNA product pipeline entering the personalized therapy realm, the new technologies facilitate GMP production and scaling manufacturing, and the industry is projected to reach the level of $20 billion in the next decade [16].

The pandemic also demonstrated the importance of agile manufacturing, as innovations such as modular systems was the key to surging the production to billions of doses [17]. Even further minimization of RNA production chain bottlenecks will be provided by future integration of the AI.

17.4.1 Microfluidics for Precise Synthesis and Encapsulation

Microfluidics is precise in RNA biomanufacturing process as the volume of fluids handled is small which allows direct mixing and reaction scenario of IVT and LNP encapsulation reactions. To sum up, microfluidic reactors can enable continuous IVT and DNA templates and enzymes travel in channels with engineered residence times and produce RNA at 1-2 g/L with low levels of dsRNA contaminants (<5%) owing to uniform shear [18]. This is opposed to batch processes reducing hours to minutes, making it more consistent [19].

In encapsulation, microfluidics is preferred in the LNP production through nanoprecipitation in which lipid-ethanol and RNA-aqueous streams are mixed at T-junctions or herringbone mixers to yield monodisperse NPs (50100 nm) with PDI as low as 0.1 [10]. Secret binding methods such as the chaotic mixing, can increase homogeneity, resulting in a 30% increase in the transfection efficiency in case of mRNA delivery [5]. Channels can be parallelized to scalable pathways with production rates of up to 100 mL/min, which is appropriate to clinical batches [7]. Surface coatings overcome challenges, such as clogging, and in-line analytics prevents this, with real-time quality [15]. Microfluidics was also used in development of vaccines, where the development of COVID-19 LNPs was accelerated by rapid prototyping [16]. This is an innovation that democratizes the manufacturing of things as it uses a decentralized production tool [17]. Figure 17.2 shows a flowchart of microfluidics in RNA encapsulation.

17.4.2 Single-Use Bioreactors and Modular Platforms

Single-use bioreactors (SUBs) and modular platforms allow RNA biomanufacturing to be similarly flexible with IVT and formulation taking place in a contamination-free environment with no cleaning validation and less risk of cross-contamination. Constructed in disposable non-binding polymers, SUBs hold volumes presently ranging up to 2000L with stirred- or wave-mixing designs to ensure homogenous reactions yield high quantities of RNA (greater than 1 g/L) [6]. The pre-sterilized components such as mixers and filters are built in a modular platform, that allows quick transitions between various RNA modalities [1].

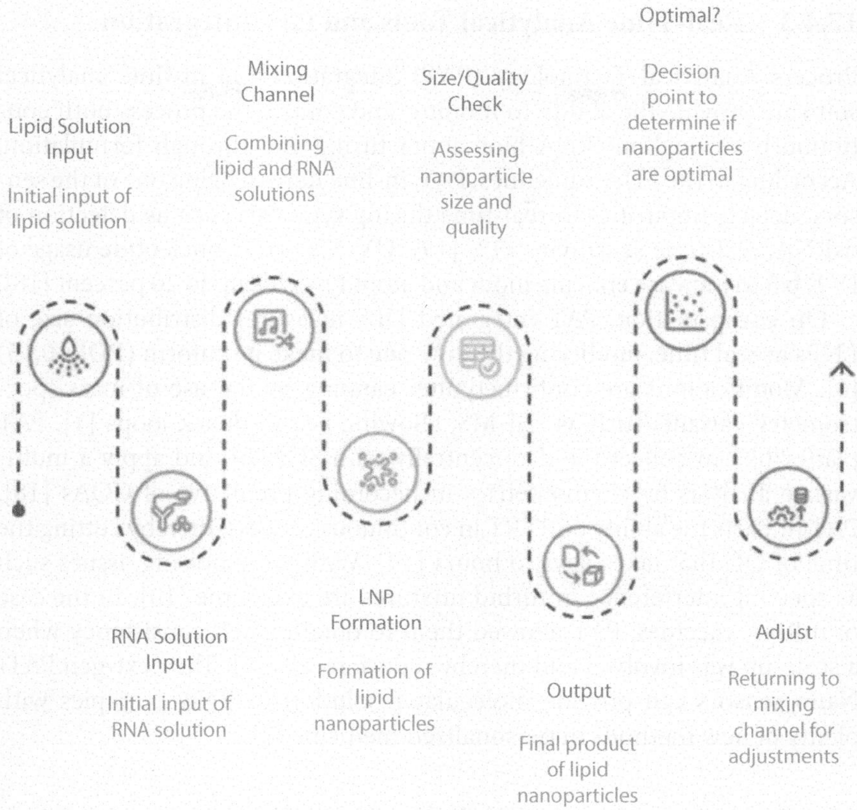

Figure 17.2 Flowchart of microfluidics in RNA encapsulation.

In LNP manufacturing, the SUBs have been used combined with tangential flow in the exchange of buffer facilitating encapsulation and reducing setup time over stainless steel by 70 percent [18]. Tech transfer is enabled by GMP- compliant modules, as used in mRNA vaccine plants whose plug-and-play systems increased production by 10-fold [19]. The benefits would be a decreased cost of capital (50 percent less) and a reduced validation time, but there would be leachable in plastic [10]. Hybrid SUBs with sensors are another innovation that improves control, so that pH remains stable during polyadenylation [5]. In the case of decentralized manufacturing, portable modules facilitate production at the location site in far-flung locations [7]. These channels uphold the sustainability by reducing water consumption in the cleaning process, which is in compliance with the green bioprocessing objectives [15].

17.4.3 Real-Time Analytical Tools and PAT Integration

Process Analytical Technology (PAT) integration and in-time analytical software provide the ability to monitor and control the process both continuously throughout RNA biomanufacturing and through formulation. According to the FDA guidelines, PAT, in-line Raman being one of the sensors, detects impurities in real-time during synthesis such as detection of dsRNA, real-time sensitivity <1% [19]. UV-Vis tracks nucleotide usage of IVT, b 5 to increase enzyme input and amplify yields up to 20 percent [18].

On encapsulation, PAT integrated DLS measures distribution size of LNPs in real time, modifying the flow rate to make it uniform (PDI <0.15) [6]. Modifications are confirmed after capping by the use of mass spectrometry variants such as ESI-MS, allowing fast feedback loops [1]. PAT platforms can collect the data centrally with SCADA and apply a multivariate analysis by chemometrics in becoming predictive of CQAs [18]. PAT imparts the ability of RTRT in continuous process, thereby cutting the time of QC that takes days to hours [19]. With AI denoising, issues such as spectral interference in turbid mixtures are overcome [10]. In the case of mRNA vaccines, PAT allowed them to obtain batch consistency when a scale-up was involved and thereby reduce recalls [5]. The next-gen PAT/Nano sensors can provide molecular resolution to RNA therapies with plenty of new methods in personalized medicine [7].

17.4.4 Automation and Digitalization in Processes

Digitalized automation in RNA biomanufacturing enables autonomous, out-of-view parametrical optimization by automating robotics, artificial intelligence, and the internet of things. Liquid handlers are used in automated IVT systems to assemble the template and/or set up reactions to minimize variability and provide a high-throughput platform of sequencing the reactions, and yields are consistent with 5 percent difference between any two runs [17]. Digital twins mimic the processes digitally to envisage their results, saving 40 percent of the development period on LNP formulation [15].

What is referred to as IoT-connected sensors feed to cloud platforms where AI can analyze the information and modify real-time parameters such as temperature through MPC and enhance encapsulation capacity to 98 percent [19]. In modular formats, robotics also automatize the exchange and sampling in the buffers, thereby reducing human error in GMP settings [18]. Digitalization using a blockchain would be traceable, assessing raw materials through to the final product, which is critical in regulatory auditing [6]. Challenges like cybersecurity are addressed with encrypted networks [1]. Automation in the field of vaccines made the production process around the clock regardless of pandemic times, exponentially scaled the amount produced [18]. In the future, through 5G digitalization, it will be possible to conduct monitoring of decentralized hubs remotely [19]. Such innovations enable the efficient, scalable production of RNA, and thus lead to therapeutics development [10].

17.5 Scalability, GMP Compliance, and Regulatory Considerations

Scalability of RNA biomanufacturing together with Good Manufacturing Practice (GMP) and regulations is essential in the transition of RNA therapeutics and vaccines in scale through production to the world markets. When a company scales up the amount produced to the millions of milligrams per liter, highly robust processes are required to address consistency requirements of key quality attributes (CQAs) (e.g., RNA purity (>95%) and encapsulation efficiency (>90%) with high regulatory standards [19]. GMP compliance implies the safety of the produced products and traceability, and it oversees issues such as product contamination and batch-to-batch variation, which may be a problem and result in costly rejections [18]. Systems that enable the rapid, yet meticulous review of gene and cell therapy products have been developed through regulatory pathways with increased speed, which although quick, necessitate some harmonization through the changing models of novel RNA modalities such as saRNA [17]. Strategies regarding process scale-up and tech transfer, and GMP requirements, regulatory, and approval challenges, and cost-effectiveness using economic modelling are looked at in this section. Effective scale-up such as that exhibited in the mRNA vaccine campaigns accomplished billions of doses per year, but current problems in cost and access illustrate the need in new solutions [15].

Such considerations are crucial because the application of RNA therapeutics is moving toward personalized medicine, which requires scalable, compliant platforms to make it economically feasible and accessible by all parties [7].

17.5.1 Strategies for Process Scale-Up and Tech Transfer

The key processes in RNA biomanufacturing have strategies towards process scale-up and tech transfer to scale up and sustain the processes with reduced supply and demand, i.e., clinical needs and commercial needs. Scale-up entails the switch between the small scale IVT (1-10 mL) and the large-scale bioreactor (>100 L), necessitating optimization of mixing, temperature and enzyme kinetics to prevent losses of yields, which can decline by 20-30% owing to shear stress or inhomogeneous distribution of reagents [1]. Microfluidics-integrated continuous-flow IVT devices can approach the condition of steady-state, which brings increased scalability and high RNA yields up to 2-5 g/L [10].

The process of tech transfer to GMP facilities requires standardized practices; modular single-use systems allow implementing them quickly and with only half the necessary validation that must be performed [19]. Process analytical technology (PAT) allows real-time scale-up monitoring, localized by in-line sensors to make real-time adjustments to ensure CQAs, e.g., poly(A) tail length, do not vary [18]. The mRNA vaccine production has case studies with tech transfer between continents using digital twins to model plant layout to reduce delays [17]. Equipment compatibility and operator training are challenges since they are solved with vendor partnerships, and automated workflows [15]. Such approaches have made it possible to increase the scale to 109 doses per year of COVID-19 vaccines, a blueprint to future RNA therapy candidates [7].

17.5.2 GMP Requirements for RNA Manufacturing

The needs of RNA manufacturing in terms of GMP safeguard the product security, quality, and traceability, in line with the requirements of the FDA and EMA, i.e., 21 CFR Part 210/211 or Annex 2. To thwart contamination, facilities ought to have sterile and closed systems and single-use bioreactors are less risky than stainless steel (contamination <0.1% cross-contamination) [5]. RNA stability depends on the levels of RNase that is regulated through environmental surveillance and is maintained with air filtration systems to ISO 5 cleanroom levels [6].

There is documentation, such as batch records and audit trails, where the records are in compliance with 21 CFR Part 11 in relation to electronic systems, resulting in data integrity [19]. The validation of the processes defines CQAs, and the purity (>95%) and endotoxin levels (<5 EU/mg) are qualified using in-process controls such as HPLC [18]. Training of personnel and quality risk management following ICH Q9 counter the variability, deviations of which are followed through CAPA systems [17]. There are also high prices of reagents that are required to be of GMP grade (50-70 per cent of budget), which is overcome by its supplier's qualification and acquisitions on a large scale [15]. During production of mRNA vaccines, GMP allowed a fast scale-up without risk of batch changes, and did not involve a recall during any of the large-scale campaigns [7]. Real-time release testing (RTRT) will become a component of future GMP frameworks to be used in streamlining compliance [1].

17.5.3 Regulatory Pathways and Approval Challenges

RNA therapies and vaccines use regulatory pathways that weigh increased approvals and strict safety evaluations to advance as the COVID-19 pandemic has taught lessons. By using rolling reviews (here data is submitted progressively) there were Emergency Use Authorization (EFA) FDA/ Conditional Marketing Authorization (CMA) EMA approvals of these mRNA vaccines in less than <12 months [16]. Long-term stability and immunogenicity data are necessary to achieve full approvals which are hard to get in case of new modalities such as saRNA since not much history is available [10].

The international harmonization through ICH guidelines (e.g., Q8-Q12) enables the same requirements; nonetheless, regional variations cause submission complexities with hold-ups being as long as 6 months in non-ICH countries [5]. Issues unique to RNA such as demonstrating the safety and inter-batch consistency of LNP have also been solved through the application of PAT in real-time CQA monitoring [6]. ADRs can be tracked by post-market surveillance, such as VAERS, which is critical as the real-world scenario that ensures the trust of society [19]. Expedited regulations on personalized RNA medicines are challenged by a drive to develop CQAs pertaining to patient-specific products [18]. EMA piloted regulatory sandboxes that enable an innovative process to be tested with estimates to cut the approval time by 30 percent [17]. These channels help in making RNA products marketed without delay and at the same time maintaining safety levels [15].

17.5.4 Cost-Effectiveness and Economic Modeling

Proper affordability in the biomanufacturing of RNA is a key to accessibility especially in the low-resource environment where the vaccines may cause delusions on budgets ($1-10/dose). The costs are made up of production expenses that are posited on GMP-grade nucleotides and enzymes at

60-80%. Production costs are done largely by IVT and LNP formulation [16]. Economic modeling makes the best use of resources, and a spectrum of continuous manufacturing makes the costs 40 percent lower by using greater throughput and less waste [1].

Automation reduces cost of labor by 30 percent, and the single-use system decreases capital costs by 50 percent over stainless steel [10]. The major force in COVID-19 vaccine campaigns was scale economies that helped reach the costs of $2/dose in pilot programs to a scale of under 0.50/dose [5]. The above challenge is due to the high initial investments ($100M-$500M on facilities), which is offset using modular platforms and public-private partnerships [6]. Clients can optimize processes using modeling tools such as Monte Carlo simulations, which forecast the cost drivers [19]. The decentralized manufacturing hubs also minimize the expenses on cold-chain logistics by 20-30 % [18]. These solutions make the RNA-therapeutics affordable, and the vision is to advance models to <10 cents/dose to reach worldwide [17].

17.6 Case Studies from mRNA Vaccine Production

Case studies that can be found within the context of mRNA vaccine manufacturing offer tangible accounts of where the biomanufacturing innovation has been deployed to solve practical problems, especially during the speedy development and implementation of vaccinations available during the COVID-19 pandemic. These works put emphasis on the success of implementing upstream and downstream processes, including IVT, LNP encapsulation, and real-time analytics to reach a scale and speed never before observed with quality and safety in mind [2]. The success of vaccine manufacturing such as the Pfizer-BioNTech and Moderna platforms shows the strength of scalable platforms, but also illustrates supply chain bottlenecks, cold-chain logistics and cost control considerations used to inform future preparedness plans [1]. The following section considers a selection of cases, with attention to the Pfizer-BioNTech and Moderna campaigns, mass distribution patterns, advances in rapid response, and the takeaways of production stalls. These examples show the scale out of RNA biomanufacturing that reached billions of doses per year but highlights continuing requirements to decrease cost and ensure reasonable access, with scaling down of costs of a dose by 10-fold to as low as 1 cent in certain situations [18].

The lessons acquired are informing a next-generation of RNA therapeutics, potentially extending to contain emerging diseases and personal therapy, which will be based on automation and decentralized factories [19].

17.6.1 COVID-19 mRNA Vaccines: Pfizer-BioNTech and Moderna

The Pfizer-BioNTech (BNT162b2) and Moderna (mRNA-1273) messenger RNA (mRNA) vaccines are the first examples of the most significant breakthrough in RNA biomanufacturing to receive emergency use authorization less than a year after design with optimized *in vitro* transcribed (IVT) and lipid nanoparticle (LNP) technologies. Pfizer-BioNTech made use of CleanCap technology with co-transcriptional capping that yielded >95% capping efficiency with a ~1% dsRNA impurities level that minimized immune activation [10]. They were able to grow IVT to 100L bioreactors leading to 2-3 g/L mRNA, which was purified by TFF and ion-exchange chromatography to purities of >98% [5]. Moderna has a similar process with proprietary lipid compositions that tune LNP particle size (80-100 nm) by microfluidics to high encapsulation efficiency (>90%) [6].

The two companies have both incorporated PAT, in-line UV-Vis monitoring IVT yield and Raman spectroscopy affirms LNP stability to assure batch-to-batch consistency with <2% variation [2]. Modular single-use systems have enabled rapid scale-up, so that by 2022, Pfizer-BioNTech was projecting 3 billion doses [1]. The obstacles faced were the shortages of the supplies of lipids, its shortage was improved by strategic supplier partnerships [18]. The cases show that optimized biomanufacturing made it possible to roll out vaccines globally and define the standards with GMP-developed processes in RNA therapeutics [19].

17.6.2 Scale-Up Campaigns and Global Distribution

mRNA vaccines needed scale-up campaigns to allow the rapid growth of production capacity and international distribution systems to fulfil unprecedented demand. With the IVT optimized in continuous-flow to 1.52-2 g/L yields, at a 40-percent cost-reduction achieved by automated buffer exchanges, Pfizer-BioNTech scaled 50L-2000L bioreactors [7]. Moderna has collaborated with contract manufacturers to set up modular plants, having produced 1 billion doses per year by mid-2021, with TFFs--concentrating the RNA to >1 mg/mL [17].

The distribution of vaccines globally was a cold-chain dilemma, since the following vaccine types needed -70 C (Pfizer) and -20 C (Moderna) temperatures, implying the expenditure of $200M on the annual ultra-low temperature logistics [10]. Asian and African fill-and-finish facilities were decentralized and therefore decreased transport time by 30% [5]. The GMP conformity enabled record keeping of batch traceability through electronic records, which satisfied

FDA/EMA requirements [6]. Other obstacles were a shortage in the number of vials which was solved through a variety in suppliers and equitable allocation challenges with 10 percent of doses having reached low-income nations in 2022 [7]. Such campaigns emphasize the successes of scalability but also emphasize logistics- and access-related obstacles [2].

17.6.3 Innovations in Rapid Manufacturing Response

Rapid manufacturing response innovations to mRNA vaccines enabled the fulfillment of the pandemic timelines by using automation and digital tools. Liquid handling automated IVTs decreased the time spent on set up by half and produced 100s of mRNA sequences in a matter of days [15]. LNP encapsulations performed using microfluidic systems enabled rapid prototyping to find conditions where new variants remained more stable in real-time by adjusting ratios of lipids [1].

Digital twins were used to experiment with scale-up and remove expected bottlenecks such as shortages of enzymes and setting optimum reactor conditions to have a 25 percent increase in throughput [18]. In-line HPLC along with PAT integration allowed real-time observation of CQAs and reduced the time required a quality control process from weeks to hours [19]. In the campaign of Moderna, alternative formulation parameters were changed dynamically by a cloud-based analytics solution with 95 percent encapsulation efficiency of batches [10]. Issues also existed in terms of integrating data across sites and are solved with OPC UA standards [5]. These advancements allowed developing vaccines within less than a year, which can become a paradigm of quick response to the future pandemic and RNA-based biotherapies [6].

17.6.4 Lessons from Production Bottlenecks

Experience in production bottlenecks in mRNA vaccination campaigns defines the aspects of improvement of RNA biomanufacturing. There was supply chain failure (deficiency in GMP-grade nucleotides and lipids) causing weeks of delay in production, cost soared 20-30%; the variety of sources was disturbing it [7]. The cold-chain necessities were a burden over logistics and 15 per cent doses were lost due to temperature fluctuations which highlights the need for thermostable formulations [17].

Scale-up showed inefficiencies related to mixing when large bioreactors were used and reduced yields by 25%; continuous-flow systems and the design of improved impellers solved both [15]. Regulatory delays, such as the execution of new facilities, delayed projects in months, and pre-approved processes on platforms had the potential to facilitate the approval process in the future [2]. Automation was impeded by work force training gaps that have been overcome using cross-functional programs which cut errors by 40 percent [1]. Data silos created the challenge to analyzing real-time, which was overcome using cross-platform integration such as PI System [18]. These lessons focus on more resilient supply chains, digitalization, modular designs to build stronger and more effective RNA production in the future [19]. Table 17.2 displays overview of major case studies in the production of mRNA vaccines.

17.7 Challenges in RNA Biomanufacturing

Although RNA biomanufacturing promises game-changing benefits to therapeutics and vaccines, it has many complex challenges hindering its potential performance, including technical problems in molecular integrity and full implementation on a global scale. Those challenges are also exacerbated by the instability properties of RNA as it is degraded at rates higher than 50% in suboptimal conditions, and solutions need to be developed to make it viable [16]. The issue of stability and degradation is based on RNase sensitivity and environmental variables, and the problem of supply chain interruption as observed during COVID-19 vaccines slowed distributions by months in low-resource settings [17]. The risk presented due to innate immune-activating impurities in dsRNA, in the form of immunogenicity and safety risks, may limit the efficacy of the drug, or cause

Table 17.2 Overview of major case studies in the production of mRNA vaccines.

Case study	Application	Key innovations	Lessons learned
Pfizer-BioNTech (BNT162b2)	COVID-19 vaccine	CleanCap IVT, microfluidic LNP	Rapid scale-up needs diversified supply
Moderna (mRNA-1273)	COVID-19 vaccine	Proprietary lipids, automated formulation	PAT ensures consistency, cold-chain critical
Scale-Up Campaigns	Global distribution	Modular facilities, continuous flow	Logistics gaps highlight decentralized needs

adverse reactions [20]. Adoption is also buffered by economic as well as accessibility barriers, making its production cost to scale at $1-2 per dose, leaving out most of the developing countries [19]. This segment discusses technical setbacks like stability problems, supply chain and cold-chain problems, immunogenicity problems and economic obstacles. These are key areas to overcome in the development of RNA technologies and current research is focused on decreasing degradation by 70% with modified nucleotides and niches [10].

These challenges have suggested the need to overcome them through interdisciplinary work that involves engineering, regulatory science, and policy to render RNA biomanufacturing more resilient and equitable [5].

17.7.1 Technical Challenges: Stability and Degradation

The most important technical challenge with regards to RNA stability and degradation lies in the high sensitivity to hydrolysis and enzymatic decomposition that inhibits the shelf-life and efficacy of RNA molecules. Naked mRNA is inactivated in hours in biological fluids by RNases that are ubiquitous, and the half-life <1 hour at 37°C, which requires stability enhancing constructs, such as LNPs, to achieve stability in the range of weeks [6]. Degradation pathways include 5'-decapping and 3'-exonuclease activity, exacerbated by improper storage or pH fluctuations during IVT [7].

IVT-related impurities, e.g. abortive transcripts, are accelerating agents of autocatalytic degradation, decreasing the yields 20-40% [16]. The approaches include modifications of bases such as pseudouridine to increase resistance to nucleases and translation efficiency as in COVID-19 vaccines [17] with a degradation <10% in months at -20 C. Circular RNA (circRNA) constructs have no free ends and enhance stability 5-fold to allow long-term expression [20]. Scale-up challenges include thermo-gradient in large reactors to produce regional degradation, one that microfluidics systems can address by creating homogeneous mixing across the reaction [19]. The integrity is also monitored in real-time by analytics such as capillary electrophoresis to avoid losses [10]. Such technical obstacles, when not overcome, may choke clinical translation, but the developments are thermostable RNA stable at ambient temperatures [5].

17.7.2 Supply Chain and Cold-Chain Logistics Issues

A major challenge in RNA biomanufacturing is related to the supply chains and cold-chain logistics due to the sensitivity of the product to temperatures (products require ultra-low storage based on the mRNA vaccines, such as -70°C), which makes international distribution complicated. There were even shortages of raw materials such as nucleotides due to disruption as in the case of the COVID-19 and this caused a delay in production by weeks and 30 percent extra expenditure [15]. Failure of cold chains leads to degradation, where in low-income countries 10 to 20 percent of the doses become lost in transit because chains have been broken [1].

The scarcity of the freezer and the energy intensity was also a problem in distribution, along with related inequities; out of all African countries, only 2 in 10 had sufficient infrastructure to administer vaccines using mRNA [16]. Thermostable formulations (with or without lyophilization or excipients such as trehalose) allow months of storage at 2-8°C [17]. Moving manufacturing to distributed hubs decreases transportation requirements; modular plants are built where necessary and locally to reduce the cost of logistics by up to 40 percent [20]. Blockchain for traceability ensures supply integrity, preventing counterfeits [19]. Raw material risk has been kept at bay by diversification of sources of root materials, which was done after the pandemic [10]. To overcome these problems, it is necessary to use robust chains, and potential chains will have AI-based predictive inventory to limit the effect [5].

17.7.3 Immunogenicity and Safety Concerns

Unintended immune activation contributes to the immunogenicity and safety of RNA therapeutics in that byproducts of the RNA manufacturing process, namely, dsRNA activates TLRs and RIG-I, and results in cytokine storms or the inability to induce an effective immune response. Unmodified RNA in mRNA vaccines activates type I interferons, halting translation and resulting in adverse effects in 1-5 percent of the people treated [6]. Possible safety risks are anaphylaxis caused by LNPs, with 2 in a million doses of COVID-19 vaccines reported, due to allergies to PEG [7].

Manufacturing problems have also the problem of impurities that should be less than 1% to prevent hyperinflammation, so the purification needs to be highly accomplished [15]. Responses include replacement of nucleotide with other compounds such as N1-methylpseudouridine, which can make them 80 times less immunogenic and still have the same strength [1]. The optimization of the sequencing prevents the use of immunostimulatory motifs, and LNP produces formulated with different lipids produce less allergic reactions [16]. Safety is tested in preclinical models; however, individual differences exist in humans and phase I monitoring is needed [17]. Regulatory requirements demand extensive toxicology data, delaying approvals [20]. Long-term safety, such as the genomic integration risks (considered insignificant, at 0.001 % or less) is also tracked through pharmacovigilance [19]. Together with the effective control of these concerns, they allow safer RNA-based therapies and current research is investing in hypoimmunogenic designs [10].

17.7.4 Economic and Accessibility Barriers

RNA biomanufacturing is only partially adopted worldwide due to economic and accessibility restrictions: high-cost production ($1-5/dose) necessitated by the GMP-grade reagents and complicated LNP formulation strategy excludes low-income countries in which 70 percent of populations do not get access [5]. Intellectual property rules and the centralization of manufacturing in high-income countries make supply unequal in times of pandemic [6].

There are economic burdens of having capital-intensive installations (>$100M) where the return is variable depending on size; personalized therapies have high costs due to small batch needs at 10x the cost [7]. Remedies include cost modeling to achieve the maximum efficiency of IVT, which minimizes the costs by 50 percent by recycling enzymes [15]. Open-source technologies and technology transfer programs, such as the mRNA hubs in Africa initiated by WHO, make the product and the technology easier to access [1]. Barriers are reduced through subsidies and alliances, such as 2 billion doses released through the COVAX [16]. Microfactories are decentralized and reduce logistics expenses by 30 percent, which allows production to be locally organized [17]. The costs will be predicted below <0.50/dose in future economic models with automation, and equity will increase [20]. Such barriers play a critical role in universal RNA therapeutic access and need to be addressed with policy and innovation [19].

17.8 Future Prospects and Strategic Pathways

RNA biomanufacturing has a future in revolutionary advances that solve existing bottlenecks and broaden the therapeutic opportunities of RNA-based vaccines and therapeutics. With the increase in personalized and on-demand treatments, there is an opportunity to increase the availability, affordability, and resilience to adapt to global health threats due to advancement in thermostable formulations, decentralized production, and next-generation RNA modalities [2]. These possibilities seek to address the weak stability of RNA, currently being degraded 20-50% at non optimal environments, and optimization of manufacturing to generate dose cost at or below 0.50 to render therapies usable in low resource setting [20]. Strategic pathways target the incorporation of optimal technologies such as the expeditious practice of process optimization through AI and the utilization of modular micro factories in an effort to retain a scalable, viable production [19]. In this part we will study thermostable formulations and innovative delivery devices, distributed production centers, and customized RNA medicines, and finishing with a vision to sustainable RNA biomanufacturer. These guidelines are in line with the global health objectives, and they use the COVID-19 experiences to provide equitable distribution and speedy expansion in future pandemics [10].

Through a combination of interdisciplinary cooperation and regulatory development, RNA biomanufacturing is able to shift toward a sturdy, patient-driven paradigm, with market predictions of the industry reaching up to a 30-billion-dollar market by 2035 [5].

17.8.1 Thermostable Formulations and Novel Delivery Systems

Stable formulations and new delivery platforms that can bypass cold-chain dependency of RNA which is currently stored at -70 °C to -20 °C, a barrier of distribution in resource-poor areas [6]. By addition of excipients such as mannitol or trehalose, the RNA may be stabilized to be stored at 2-8 °C and remain stable up to 6 months, which will allow cutting the cost of logistics by 30-percent and distribution to regions without access to ultra-low freezers [7]. Molecular recipe Industry New methods of spray-drying to create dry-powder mRNA vaccines allows the vaccines to store at room temperature over weeks, and reconstitution preserves >90% potency [15].

There are alternative delivery systems to LNPs, including novel delivery systems like polymeric nanoparticles and exosomes, which increase stability and specificity of targeting [2]. The cellular uptake is enhanced by a factor of 2-3 following encapsulation in peptide-based carrier and self-assembling hydrogels allow prolonged release to support chronic treatment [20]. Examples of these problems are scaling these formulations and ensuring that encapsulation of the drug was maintained (>95%), which is solved using microfluidic optimization [19]. Thermostable mRNA formulations were also demonstrated in preclinical trials to maintain immunogenicity over 3 months storage at 25°C [10]. Regulatory schemes will allow the use of dry preparations to be approved, and EMA pilots promise faster stability testing [5]. Such developments hold the potential of removing cold-chain obstacles, which will guarantee the reach of vaccines worldwide [6].

17.8.2 Decentralized Manufacturing Hubs

Remote manufacturing centers resolve supply chain vulnerabilities and access by allowing RNA therapeutic to be produced locally, lessening supply chain susceptibility to centralized production. Its modular, single-use bioreactor-based, automatized IVT micro factory has hundreds of units of production per month (10,000-100,000 doses) at a size that reduces the costs of transport by 40% [1]. A feasible example is being shown by initiatives such as the mRNA Technology Transfer Hub South Africa run by WHO, with pilot plants being able to generate clinical-grade mRNA within 6 months [17].

These centers also optimize processes by using Digital twins to provide consistency among different sites with <5% variability of a batch [2]. Full automation and incorporation of PAT enable a real time monitoring regime under GMP standards [20]. The obstacles are the large set up costs (between 10M and 50M dollars), which is countered by public-private partners and open protocols [19]. Local training programs address skill gaps, boosting adoption [10]. The concentration of production during the COVID-19 era left 70 percent of the low-income countries unable to gain timely access, and decentralized hubs could improve the gap by a half [5]. The centers in the future will incorporate 5G to remotely monitor their centers and become able to respond quickly in the event of a regional outbreak [6].

17.8.3 Personalized and Next-Generation RNA Therapeutics

Individualized and future generation RNA therapeutics are the next horizon in RNA biomanufacturing to focus on patient-specific phenotype and disease novel mechanisms. Many of the mRNA therapies are personalized, including cancer vaccines that genetically encode neoantigens specific to a patient and have an end-to-end turnaround time <2 weeks, so they must be produced on small-scale IVT systems [7]. Automated sequencing microfluidic scale platforms can be used to scale up and generate tailored RNA up to 1-10 mg, while retaining a purity >95% [15].

Future generations abandoning the antibody-based approach involve utilizing saRNA in lower-dose treatment, up to 10-fold better expression, and CircRNA with longer activity with 5–7 day half-life improvement [1]. RNA therapeutics use CRISPR and are made to attack genetic diseases, in-line analytics synthesize a guide RNA accurately [17]. Among the difficulties is stipulating regulatory validation of patient-specific CQAs, which is resolved by platform-based authorizations according to ICH Q12 [2].

The cost is still a barrier (\$1000s per dose); however, this could be brought down to as low as \$100 through economies of scale and AI-driven design [20]. Personal mRNA vaccines are in clinical trial, with 40% enhanced survival on melanoma [19]. Such developments are promising, custom solutions with biomanufacturing conforming to high mix, low volumes [10].

17.9 Conclusion

To sum up, RNA biomanufacturing stands at the entrance to a novel, sustainable, and equitable ecosystem by adopting thermostable formulations, decentral hubs, and personalizing drugs. Roadmap would entail scaling of modular technologies, alignment of worldwide regulations as well as the use of AI to predict manufacturing, with the goal of <10 degradation and the cost of <\$0.50/ dose. There will be sustainable solutions leading towards net-zero, which will include bio-based materials and energy efficient processes that would cut the impact on environment by 30%. This vision provides that RNA therapeutics can correspond to the worldwide needs of healthcare, bring resilience, and science initiatives.

Abbreviations

mRNA	Messenger RNA
siRNA	Small Interfering RNA
saRNA	Self-Amplifying RNA
IVT	*In Vitro* Transcription
LNP	Lipid Nanoparticle
GMP	Good Manufacturing Practice
RNA	Ribonucleic Acid
dsRNA	Double-Stranded RNA
LAL	Limulus Amebocyte Lysate
CQA	Critical Quality Attribute
NGS	Next-Generation Sequencing
HPLC	High-Performance Liquid Chromatography
PCR	Polymerase Chain Reaction
PAT	Process Analytical Technology
FDA	Food and Drug Administration
EMA	European Medicines Agency
USP	United States Pharmacopeia
VLP	Virus-Like Particle

TFF	Tangential Flow Filtration
DLS	Dynamic Light Scattering
PDI	Polydispersity Index
AI	Artificial Intelligence
QbD	Quality by Design
WHO	World Health Organization
GxP	Good Practices (a term encompassing various regulatory standards such as GMP)
RTO	Real-Time Optimization

AI disclosure statement: The author acknowledges the use of AI-based tools for minor language editing or grammatical correction. However, all intellectual contributions, analyses, and conclusions presented in this chapter are entirely the author's own.

References

1. Verbeke, R., Lentacker, I., De Smedt, S.C., Dewitte, H., Three decades of messenger RNA vaccine development. *Nano Today*, 28, 100766, 2019.
2. Pardi, N., Hogan, M.J., Porter, F.W., Weissman, D., mRNA vaccines - a new era in vaccinology. *Nat. Rev. Drug Discov.*, 17, 261, 2018.
3. Dolgin, E., The tangled history of mRNA vaccines. *Nature*, 597, 318, 2021.
4. Barbier, A.J., Jiang, A.Y., Zhang, P., Wooster, R., Anderson, D.G., The clinical progress of mRNA vaccines and immunotherapies. *Nat. Biotechnol.*, 40, 840, 2022.
5. Kim, J., Eygeris, Y., Gupta, M., Sahay, G., Self-assembled mRNA vaccines. *Adv. Drug Delivery Rev.*, 170, 83, 2021.
6. Kowalski, P.S., Rudra, A., Miao, L., Anderson, D.G., Delivering the Messenger: Advances in Technologies for Therapeutic mRNA Delivery. *Mol. Ther.*, 27, 710, 2019.
7. Hou, X., Zaks, T., Langer, R., Dong, Y., Lipid nanoparticles for mRNA delivery. *Nat. Rev. Mater.*, 6, 1078, 2021.
8. Sahin, U., Karikó, K., Türeci, O., mRNA-based therapeutics — developing a new class of drugs. *Nat. Rev. Drug Discov.*, 13, 759, 2014.
9. Jackson, N.A.C., Kester, K.E., Casimiro, D., Gurunathan, S., DeRosa, F., The promise of mRNA vaccines: a biotech and industrial perspective. *npj Vaccines*, 5, 11, 2020.
10. Buschmann, M.D., Carrasco, M.J., Alishetty, S., Paige, M., Alameh, M.G., Weissman, D., Nanomaterial Delivery Systems for mRNA Vaccines. *Vaccines (Basel)*, 9, 65, 2021.
11. Martins, R., Queiroz, J.A., Sousa, F., Ribonucleic acid purification. *J. Chromatogr. A*, 1355, 1, 2014.

12. Tan, S.C. and Yiap, B.C., DNA, RNA, and Protein Extraction: The Past and The Present. *J. Biomed. Biotechnol.*, 2009, 574398, 2009.
13. Martins, R., Queiroz, J.A., Sousa, F., Membrane technology for the purification of RNA and DNA therapeutics. *Trends Biotechnol.*, 42, 1089, 2024.
14. Sousa, F., Prazeres, D.M., Queiroz, J.A., Affinity chromatography approaches to overcome the challenges of purifying plasmid DNA. *Trends Biotechnol.*, 26, 518, 2008.
15. Aldosari, B.N., Alfagih, I.M., Almurshedi, A.S., Lipid nanoparticles as delivery systems for RNA-based vaccines. *Pharmaceutics*, 13, 206, 2021.
16. Crommelin, D.J., Anchordoquy, T.J., Volkin, D.B., Jiskoot, W., Mastrobattista, E., Addressing the Cold Reality of mRNA Vaccine Stability. *J. Pharm. Sci.*, 110, 997, 2021.
17. Schoenmaker, L., Witzigmann, D., Kulkarni, J.A., Verbeke, R., Kersten, G., Jiskoot, W., Crommelin, D.J., mRNA-lipid nanoparticle COVID-19 vaccines: Structure and stability. *Int. J. Pharm.*, 601, 120586, 2021.
18. Qin, S., Tang, X., Chen, Y., Chen, K., Fan, N., Xiao, W., Zheng, Q., Li, G., Teng, Y., Wu, M., Song, X., mRNA-based therapeutics: powerful and versatile tools to combat diseases. *Signal Transduction Targeted Ther.*, 7, 166, 2022.
19. Jackson, N.A., Kester, K.E., Casimiro, D., Gurunathan, S., DeRosa, F., The promise of mRNA vaccines: a biotech and industrial perspective. *npj Vaccines*, 5, 11, 2020.
20. Qin, S., Tang, X., Chen, Y., Chen, K., Fan, N., Xiao, W., Zheng, Q., Li, G., Teng, Y., Wu, M., Song, X., mRNA-based therapeutics: powerful and versatile tools to combat diseases. *Signal Transduction Targeted Ther.*, 7, 166, 2022.

Part 5

BIOPROCESSING AND SUSTAINABLE PRODUCTION

18

Rapid Biomanufacturing Strategies for Emerging Infectious Diseases: A New Era in Global Health Response

K.S.N.V. Prasad[1], Bhaskar Bethi[1], Bhanu Radhika[1], Manjakuppam Malika[1], B. Karuna[1], Vijaya Kumar Talari[1], Appala Naidu Uttaravalli[2] and A.V. Raghavendra Rao[1]*

[1]*Department of Chemical Engineering, B. V. Raju Institute of Technology, Narsapur, Telangana, India*

[2]*Optimus Drugs Private Limited (A Sekhmet Pharma Ventures Private Limited Company), Hyderabad, Telangana, India*

Abstract

The current health environment across the world has heightened the importance of efficient, versatile biomanufacturing platforms for dealing with new infectious diseases outbreaks. The chapter provides a broad view of rapid response biomanufacturing infrastructure that allows making and releasing biologics, such as vaccines, monoclonal antibodies, and anti-viral proteins, within a short period. Some of the enabling technologies noted by the chapter include plug-and-play modular facilities, cell-free expression systems, and synthetic biology methods, and advanced digital tools to control processes in real time. We address what we have learned in the recent pandemics and the lessons specifically applied to COVID-19 and how they have spurred innovation through rapid production cycles, regulatory acceleration, and international supply chain management. Particular emphasis can be placed on such platform technologies as mRNA and viral vectors systems with unprecedented speed and adaptability. They investigate the consideration of incorporating the AI-driven modeling, decentralized manufacturing modules, and Just-in-Time (JIT) biomanufacturing as one of the future forward interventions in entrenchment against epidemics. This chapter integrates the evidence presented in case studies, technology and policy framework to allow the development of resilient biomanufacturing infrastructures that can act swiftly and accurately in

**Corresponding author*: raghusrf@gmail.com

Sandip Kunar, Jagadeesha T. and D. Sathis Kumar (eds.) Biomanufacturing: Transforming Production of Biologics and Biopharmaceuticals, (553–602) © 2026 Scrivener Publishing LLC

combating any future outbreaks. Such a dynamic change is a new dawn in global health response after which biologics can be launched at a population level in an unprecedented time frame.

***Keywords*:** Biomanufacturing, mRNA vaccines, platform technologies, synthetic biology, cell-free systems, modular manufacturing, decentralized manufacturing

18.1 Introduction to Rapid Biomanufacturing in Global Health

Emerging infectious diseases (EIDs) are one of the threats to global health because they require creative and responsive solutions to reduce the consequences of these diseases on the global population [1]. The diseases, which in some cases stem out of zoonotic origins or as mutations of known pathogens, have the potential to grow to the heights of a pandemic unless they are dealt with early enough [2]. High throughput biomanufacturing is another paradigm shift towards manufacturing of biologics and allows development and deployment of countermeasures like vaccines and therapeutics [3]. It is a kind of technique which incorporates advanced technologies to shorten schedules which normally took years into the weeks, thus upgrading readiness on the global scale [4]. In this introduction, we explore some of the essential definitions, highlight the history of the practicing biomanufacturing, give a brief of the problems and dilemmas associated with EIDs, as well as state our ambitions for, and outline of this chapter. Through this background, we would like to give the reader a comprehensive perspective of how rapid biomanufacturing is transforming responses to infectious threats to have a more proactive than reactive approach to public health [5].

The necessity of such innovations has been enhanced by the recent events that happened worldwide, causing the life-destroying consequences of delays in biologic production [6]. As an example, the rapid transmission of new viruses today, using conventional manufacturing, lacks flexibility in terms of supply chains being rigid and requiring a lot of regulatory approvals [7]. Rapid biomanufacturing, in its turn, offers modular and scalable interventions that allow responding to new pathogens promptly, which helps to provide therapeutics to an affected population quick enough to inhibit an outbreak [8]. This chapter summarizes the experience on technological innovations, policy design, and practical operation by advancing the idea of resilient infrastructure that can survive the next pandemic [9]. Moreover, it discusses the power of interdisciplinary relationships amid engineers, biologists, and policymakers to bring about implementation of these strategies and eventually transform the health security of the globe [10]. Emergence of sustainable strategies like waste minimization in the manufacturing processes also contributes massively towards making rapid biomanufacturing a viable long-term option [11].

18.1.1 Definition and Importance of Rapid Biomanufacturing

The expression rapid biomanufacturing is an approach to design, manufacture, and scale biological products using state-of-the-art technology, such as modular facilities, synthetic biology and digital process controls to scale up their responses to health crises within weeks or months, as opposed to years [3]. In its simplest form, it involves the whole process of genetic sequencing of a pathogen to the ultimate composition of vaccines or antibodies with a high focus on speed but without loss of safety or efficacy [10]. The major aspects involve cell-free systems, which do not require standard cell culturing and can be undertaken on site, and platform-based technologies which could be retrofitted to target other diseases [11]. Contrary to the traditional biomanufacturing that is characterized by fixed, scalable facilities that require lengthy validation exercise, rapid methods are marked by the concepts of flexibility, modularity, and structural incorporation of enabling technologies to attain the operational capacity of plug-and-play [4]. This modularity entails modular prefabricated modules which can be put together like building blocks to reduce set up times and can be deployed in any environment including, but not limited to, the urban hub or the rural extreme areas [12]. It also has continuous manufacturing procedures where the products are not manufactured in batches but in steady streams which enhances efficiency and also reduces the cost [13].

Rapid biomanufacturing is relevant to remedy the unpredictability of EIDs, where measures countermeasure may take months to become available and thus exponential growth and high case fatality may result [5]. Traditional approaches may require 12-18 months in vaccine development in cases of emergence of new virus; however, rapid approaches as shown during the response to COVID-19 may turn around this in less than a year [13]. It also enables rapid adaptation of platforms applied during outbreak, such as mRNA and viral vectors, which can accelerate the speed at which the product is delivered into the market, by decreasing the time taken between discovery of the pathogen and delivery of the product [6].

One of the platforms that can be used is the mRNA platforms, which facilitate rapid encoding of antigenic sequences and thus production can commence virtually instantly after the genomic data is made available [56]. This responsiveness not only preserves lives but also reduces the impact on the economy since the ease with which vaccines were created played a vital role in a global recovery during recent pandemics [7]. The estimated loss to the economy caused by pandemics is up to 4.8 trillion dollars as calculated by the World Health Organization and this is a pointer to why rapid manufacturing can minimize the effects by helping to restore normality by a faster rate [14]. Additionally, it optimizes the utilization of resources achieved *via* predictive analytics hence minimizing cases of shortages in critical sectors when there is short supply anywhere through repurposing of resources subject to the availability of resources [1].

Moreover, it promotes health equity in that the production is not centralized but locally available, thus the countermeasures can facilitate low-income communities [8]. The capability of rapid biomanufacturing in countries with limited centralized facilities, whose sites are subject to limited supply chains prone to disruption, means local production of products with portable bioreactors [1]. This is because this democratization of technology will overcome inequality outlined in previous epidemics, which had advantaged affluent countries stockpiling vaccines and leaving others defenseless [2]. It incorporates the concepts of just in time manufacturing, which guarantees that biologics will be manufactured as requested with little wastage and optimal utilization of resources [10]. In general, the significance goes beyond the short-term response to the long-term resilience with the investments of these strategies to equip societies against unforeseen risks, encourage cooperation between the governments, academia, and industry [11]. The mass production during the surge is also possible because of the scalability of these systems, and consequently, one can be sure that billions of doses will be distributed worldwide without a bottleneck [12].

18.1.2 Biomanufacturing Evolution in Infectious Disease

Biomanufacturing has developed to a large degree since its initial form of microbial fermentation of antibiotics to more technologically advanced systems devoted to viral threats [9]. Simple fermentation vessels were used in the early 20th century to manufacture penicillin, and this is where the era of large-scale production of biologics commenced [12]. This period emphasized on batch production of vaccines such as polio and smallpox, but it was slow and incapable of new pathogens [10]. The approaches were live extinguished or inactive viruses that needed lots of safety-checking out and modeling in animals and stunted scalability [13]. Initial problems were risks of contamination and low yields leading to stepwise advances in sterilization and media preparation [56].

The HIV/AIDS epidemic of the 1980s brought to the fore use of recombinant technologies that led to the production of protein-based therapeutics [11]. With the recombinant DNAs it was possible to express viral proteins in the host cells, including Escherichia coli or yeast, thus creating the subunit vaccines and monoclonal antibodies [56]. This transition marked a transition towards cleaner and refined production though schedules were still long-lived in terms of purification procedures and regulations [14]. It is also during the time that mammalian cell lines to produce complex glycoproteins were introduced, thereby extending the list of what could be produced biologically [1].

Alderman and Crippa expressed that cases such as SARS in 2003 and H1N1 in 2009 represented a tipping point in the 21st century where investment in platform technologies was higher to achieve quicker response [12]. The occurrence of these events expressed weaknesses within global supply chains, and increased the use of viral vector systems, using adenoviruses or other vectors to deliver genetic material to stimulate immune responses [1]. The attempts to produce solutions further spurred by the Ebola outbreak in 2014 culminated in the first approval of recombinant vaccines to be used under emergency applications [2]. The improvement of genomics allowed characterizing pathogens in shorter periods by directly incorporating data on sequencing into manufacturing processes [3].

The COVID-19 pandemic provided an additional boost to this development, as synthetic biology and DNA synthesis made an important role in the rapid development of promising experimental vaccines [13]. Firms such as Moderna and Pfizer developed vaccines in several days by leveraging mRNA platforms following genome sequencing of SARS-CoV-2, which would have been unthinkable using older methods a few decades ago [3]. The experience of such episodes has led to advances in AI and automation, as they turn bio manufacturing into an active measure of pandemic preparation [56]. Protein structures are now predicted by AI algorithms and fermentation conditions optimized, and high-throughput screening now uses robotic systems [4]. Such development is not limited to the digitalization of the process, since real-time modifications based on the data analytics guarantee improved yields and quality [5]. In retrospect, the historical development of basic fermentation and AI-based platforms represents the parallel between crises and the technological breakthroughs it triggers, preparing the ground to more extensive technological improvement of rapid response [6].

Of even more versatility, today hybrid systems integrating cell-based and cell-free procedures exist which can be adapted to various biologic requirements [7].

18.1.3 Introduction to Emerging Infectious Diseases or Worldwide Problems

Emerging infectious diseases are new or resurfacing pathogens that are considered dangerous because of things such as globalization, climate change, and antimicrobial resistance [14]. They are viral, bacterial and parasitic and could cross species barriers as in the case of HIV in primates or SARS-CoV-2 probably bats [7]. Zika, Ebola, and SARS-CoV-2 are such infections that caused mass morbidity and an overload in the healthcare systems [1]. The 2015-2016 outbreak of Zika in the Americas caused congenital defects, whereas hemorrhagic fever caused by the Ebola virus had killed thousands of people in West Africa indicating the various pathologies of EIDs [8]. Others hotspot EIDs e.g., Nipah virus and avian influenza show that the threat persists due to animal reservoirs [9].

The international areas of concern involve cross-border transmission, lack of surveillance in the underdeveloped areas, and weak links in supply chains that make it difficult to intervene as early as possible [2]. Air travel enhances the transmission of the pathogens and thus makes a local outbreak an international emergency in hours [9]. Due to poor quality diagnostic facilities in the low-income countries, the diseases languish [10]. Besides, climate change leads to the vectors range expansion, which causes dengue and malaria cases in previously uninfected locations [11]. The transmission risk is further increased by population density and urbanization development to be a hotspot in mega-cities [12].

Most of these diseases have the potential of their origin in the zoonotic form increasing the necessity of the alert attention and speedy countermeasures [5]. More than 60 per cent of EIDs are zoonotic and are mediated by deforestation and wildlife trade [12]. Economic consequences are significant, as pandemics have been associated with trillion-dollar losses of productivity, which prioritizes the move towards robust biomanufacturing facilities [8]. Just the pandemic of COVID-19 alone led to global losses of more than $16 trillion by 2021 in the spheres of trade, tourism, and employment [13]. Socially, EIDs disproportionately burden marginalized communities, widening inequalities [56]. Another layer of complexity is represented by mental health consequences, including the development of an increased level of anxiety under lockdown conditions [14].

To deal with them, it is essential that there is an international effort that will remove barriers to regulation and make technologies readily accessible [9]. Actors such as the Coalition for Epidemic Preparedness Innovations (CEPI) spend on developing platforms, and there is a problem with the dissemination of intellectual property and distribution of funds [14]. The recent effects of supply chain breakdown with raw materials during the COVID-19 outbreak highlights the importance of diversified sourcing and hoarding [3]. In the end, however, EIDs require a comprehensive strategy, encompassing surveillance, research and production to curb the risk of their further evolution to the pandemic levels [4]. Biosurveillance that applies AI to strengthen early warning systems and data sharing platforms across the globe should be part of an important shift in the right direction [6]. With the focus on prevention that will include vaccination and education, the international community will decrease the number and the scale of outbreaks [10].

18.2 Historical Lessons from Recent Pandemics

The contemporary history of diseases outbreak represents a rich source of information related to the effectiveness and shortcomings of the global health responses, especially in connection to biomanufacturing. Based on the past pandemic experience of COVID-19, Ebola, Zika, and SARS, this section reviews the historical responses, in order to discover vital gaps in response to preparedness, exposure, and innovations to inform rapid biomanufacturing plans [15]. These experiences have vindicated the need to have agile systems that are capable of increasing production rapidly; delays in the traditional methodologies worsened morbidity and mortality [16]. With the help of case studies and failures, we may outline agents of innovation, which will make us better prepared in the event of crisis (and stronger frameworks in the case of a future emergency) [17]. These lessons have resulted in the evolution of reactive to proactive steps that focus on the importance of combining technology, policy, and international cooperation to prevent the effects of emerging threats [18]. More so, these historical analyses present trends in the behavior of the pathogens and human behavior that can guide the predictive models of the bios producing biomanufacturing preparedness [19].

The 21st pandemics did not only put a strain on healthcare infrastructures but also provided a boost to progress in the field of biologics production, turning challenges into the opportunity to work on the advancement of the systemic issues [20]. As an example, the outbreaks have taught the possibility of spreading genomic data quickly, and thus designing vaccines quicker, and this must be institutionalized to continue being ready with epidemics [21]. It is in this section that these experiences are synthesized in the effort to lay a solid background with respect to how historical insights are used to initiate modern world fast spreading biomanufacturing phenomena [22].

18.2.1 Case Study: COVID-19 Pandemic Response

COVID-19, which was a pandemic caused by the SARS-CoV-2 virus, manifested itself at the end of 2019 and within a very short period, it spread to become one of the major crises globally killing millions of people and infecting more than 700 million individuals as of 2023 [15]. It was also noted as an unprecedented speed of vaccine creation, noting how mRNA technology such as Pfizer-BioNTech and Moderna-based vaccines received emergency use approvals within less than a year after an emergency approval process that normally requires at least a decade [23]. This speed was enabled by the previous buy-in to platform technologies in times of SARS and MERS outbreaks enabling rapid adaptation of the spike protein sequences [24]. But, problems in the initial manufacturing process like the shortage of lipid nanoparticles led to slow distribution of the drug in the world, especially in countries with low income levels [16].

Noteworthy successes were operation warp speed in the United States whereby funding, regulatory mechanisms, and scale-up of manufacturing was streamlined leading to billions of doses produced [17]. There was the emergence of decentralized mode of production, technologies to factories in India and South Africa showed the possibility of regional hubs [18]. However, the barriers to vaccination such as vaccine reluctance and lack of equity emphasized on the importance of strategy of communication as well as ethical distribution [19]. Another part that the pandemic speeded up was the digital technology in real-time monitoring of the supply chain that minimized wastes and optimized distribution [20]. In general, COVID-19 demonstrated how biomanufacturing timelines can be shortened through collaborative work between governments, academia and industry; however, there was weakness in international coordination of efforts which will need to be looked at by future policy makers [21]. The rapid platforms were also demonstrated through the use of monoclonal antibodies and antivirals, although early stages of requesting platforms were not realized at their full production capacity [22].

18.2.2 Insights from Ebola, Zika, and SARS Outbreaks

The West Africa (2014-2016) and the Democratic Republic of the Congo (2018-2020) Ebola epidemics were important lessons when it comes to acting in the case of highly lethal pathologies in resource-limited countries [25]. The effectiveness of ring vaccination was shown by deployment of experimental vaccines such as rVSV-ZEBOV, with outbreaks being contained in ring vaccination strategies through selected manufacturing and distribution [26]. Nonetheless, international response and the manufacturing scale-up were late, resulting in more than 11, 000 deaths between 2014-2016, particularly the necessity of pre-approved platforms and pre-stocks [16].

During 2015-2016, particularly in the Americas, the emergence of the Zika virus led to the discovery of preparedness gaps of vector-borne diseases, and its congenital effects triggered the urgent need to develop vaccines [27]. Even though as no vaccine was completely introduced to tackle the outbreak, knowledge on flavivirus platforms hastened the research, and modular strategies against threats of the same have been enlightened [17]. The problem at the supply chain level of both the diagnostics and therapeutics highlighted the need to have decentralized biomanufacturing in order to get affected populations in a short time [18].

The 2003 SARS outbreak (SARS-CoV) which infected more than 8,000 individuals served as a lesson in coronavirus biology which was essential in the field of COVID-19 study [24]. It was possible to detect an early warning through rapid sequencing of the genome; however, the delayed investment in maintaining the platform after the outbreak could not meet the requirements [28]. Taken together, as has been discussed, these events have made the importance of warm base manufacturing facilities, facilities put in a ready state to be activated at short notice, and synthetic biology to design antigens [19]. The cooperations between countries, including the one organized by the World Health Organization, played a crucial role but were frequently disrupted by political obstacles as the theme common in all outbreaks [20]. Early detection, such as analysis based on surveillance systems, can also be gained as noted in these— thanks to which biomanufacturing can be started prior to broad spreading [21]. Table 18.1 shows key lessons from recent pandemics.

Table 18.1 Key lessons from recent pandemics.

Pandemic/ outbreak	**Key case insights**	**Main failures in traditional approaches**	**Catalysts for innovation**
COVID-19	Unprecedented vaccine speed *via* mRNA platforms; Operation Warp Speed enabled billions of doses; decentralized models in India/ South Africa	Manufacturing bottlenecks (e.g., lipid shortages); inequitable distribution (80% to high-income countries)	Platform technologies reducing timelines by 90%; AI/ automation integration; regulatory fast-tracking like EUA
Ebola (2014-2016, 2018-2020)	Ring vaccination with rVSV-ZEBOV contained outbreaks; experimental vaccines deployed rapidly	Delays in international response and scale-up led to >11,000 deaths; lack of pre-approved platforms	"Warm base" manufacturing for quick activation; investments in stockpiles and modular facilities
Zika (2015-2016)	Flavivirus platforms accelerated research; rapid diagnostics distributed	Absence of pre-existing platforms caused delays; vector-borne spread unchecked	Synthetic biology for antigen synthesis; emphasis on decentralized diagnostics
SARS (2003)	Genome sequencing enabled early identification; lessons informed COVID-19	Lack of sustained post-outbreak investment; supply chain vulnerabilities	Viral vector systems development; global supply chain reforms for diversification

18.2.3 Key Failures in Traditional Biomanufacturing Approaches

The conventional biomanufacturing that is based on egg-based or cell-culture deposited itself in the process of scale-up unsuccessfully during the latest pandemics because of rigidity and the long validation process [15]. The use of central facilities early in COVID-19 generated bottlenecks, and there was no way to proportionately supply vaccines to the global demand, causing some disparity where high-income nations received 80 percent of the early supplies [23]. Policy obstacles, which would necessitate massive clinical trials, would postpone deployment, such as the case of Ebola, whose vaccine claimed to have positive candidates, delayed approval [25].

With Zika, there were no initial platforms with flaviviruses; hence, the long time taken before appropriate actions were taken and before the virus ran wild [27]. The SARS revealed the weak spots in the chains of raw materials supplies and interruption of the chain led to the ceasing of the potential therapeutic production [24]. The most typical failures are an overreliance on one supplier, inability to be modular, and under-investment in alternative technologies such as cell-free systems [16]. One of their malfunctions was also the inability to adapt to the variants, since the mutations occurred quicker than the production could adapt [17]. Preparedness gaps were further compounded by economic models that favored profitability, and diseases that had lower prevalence were underfunded [18]. In conclusion, the failures also indicate the inflexibility of batch versus continuous manufacturing which may provide more flexibility [19].

18.2.4 Catalysts for Innovation in Rapid Manufacturing Cycles

Failures of historical responses driven innovation in fast biomanufacturing, where the developments of mRNA and viral vector-based platforms in in response to COVID-19 decreased the time to development by 90 percent relative to conventional approaches [15]. The experience of Ebola led to the formation of the Coalition for Epidemic Preparedness Innovations (CEPI) that finances platforms of technologies to deal with Disease X [26]. Such a head start made such pivots easy to make as when researchers repurposed SARS research towards COVID-19 vaccines [24].

The results that Zika hastened were the development of DNA printer which made possible the production of antigens on demand with the use of the printers [27]. Such transformative drivers are digital twins, which can model the production processes, AI to optimize yield, and modular facilities that can be reconfigured quickly [20]. New regulatory innovations such as emergency use authorizations, and fast-tracked approvals did not sacrifice safety [21]. The diversifications and just-in-time approaches were encouraged on global supply chain reforms due to understandings of SARS interruptions [28]. The deployment of decentralized units as well as portable bioreactors and cell-free systems can be viewed as immediate reactions to the centralized inefficiencies [22]. Such innovations not only reduce the duration of the cycles but also increase scalability, guaranteeing proportionate access in regarding outbreaks in the future [18]. Institutionalizing these drivers will help biomanufacturing be preparedness-based instead of crisis-based, reducing human and economic costs of pandemics [19].

18.3 Enabling Technologies for Agile Biomanufacturing

Advancement in biomanufacturing accelerates queues necessitating underpins technologies to scale up and scale out production of biologics of emerging infectious diseases (EIDs) [29]. Some of the enabling technologies include plug-and-play modular facilities, cell-free expression techniques, and synthetic biology methods that allow rapid redesign and implementation in short notice depending on the outbreaks [30]. The innovations overcome the historical shortcomings of high cost or long lead times and the need of reliance on central infrastructures by creating decentralized and on demand production [31]. Through the incorporation of these technologies, biomanufacturing will be able to attain a level of efficiency that has never been achieved previously, where the time required to implement an idea as compared to the current method that takes years could take months or weeks at a time [32]. This part of the work discusses these technologies and their use and synergy in the fight against EIDs, together with the possibility of introducing comparative analysis that will be used to implement these technologies in the future [33]. It is about the ability of tools to support resilient systems that can scale production of future vaccines, therapeutics and diagnostic solutions in resource-limited settings [34]. Moreover, the development of automation and digitalization in the given spheres of technologies supports its capacity since in critical situations, focus should be placed on replicability and accuracy [35].

The necessity of agile biomanufacturing is more needed than ever as the world faces more health challenges because of various issues such as climate change and urbanization [36]. These enabling technologies have not only increased the speed of the production but also due to environmentally friendly practices, i.e. use of less water, less energy, etc. [37].

Through modular designs and biological engineering, they not only herald the age of personalized medicine and rapid prototyping but also change the response to pandemic [38]. This is because this exploration highlights the interdisciplinary aspects of such advancements; hence, engineering, biology and informatics are fused together to establish powerful global health security platforms [39].

18.3.1 Plug-and-Play Modular Facilities

The final innovation to be described is plug-and-play modular facilities, construction of which is based on prefabricated, movable modules, which are simple to assemble and reconfigure to meet the demands of particular types of productions [29]. Such facilities are easy to integrate and work as building blocks, where it is possible to connect bioreactors and purification units and quality control systems without a lot of intercession engineering [40]. An EIDs application proves essential to uniform expansion from the laboratory to the production sector, as witnessed in the COVID-19 response case, where versatile units produced the vaccines owing to a modular approach that allows flexible placement at any manufacturing site [30]. Principal offers are less construction time (usually up to 50-70% than traditional plants) and less capital costs so they are perfect in the case of emergency employment [31].

The construction of these facilities employs uniform interfaces, e.g. standardized connectors to utilities and data systems, and is vendor- and regionally interoperable [32]. As an example, in modules single-use technologies reduce the possibility of contamination and the requirement to use cleaning validation shortening cycle time [33]. Applications: Monoclonal antibodies and viral vectors are produced using applications and modules that can be flexibly adapted to on-demand production on an upstream fermentation or downstream level [34]. There are also issues, including regulatory systems of modular configurations, but they are already approached with frameworks, including the FDA advanced manufacturing initiatives, which imply a high degree of flexibility and safety [35]. With the portable modular facilities, production of less-resource-dependent facilities could be locally carried out, eliminating the involvement of a global supply chain, which is highly subject to shocks in the long run [36]. Another essential aspect is that all plug-and-play systems can make the operations more agile, since they enable parallel operation and simple upgrading [37]. They are additionally optimized by being embedded with digital twins, or simulations of a replica [38]. Figure 18.1 displays simple Block Diagram of Plug-and-Play Modular Facilities.

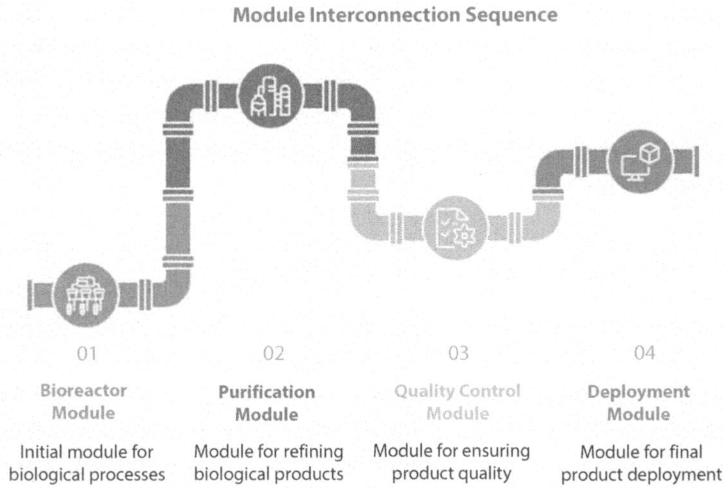

Figure 18.1 Simple block diagram of plug-and-play modular facilities.

18.3.2 Cell-Free Expression Systems

The cell-free expression systems (CFES) obviate the use of living cells by using extracts of cells (e.g., E. coli or wheat germ) to carry out protein production *in vitro*, presenting a quick and an easily manipulated environment of biomanufacturing [41]. The systems are decoupled in that they allow high yield protein synthesis in hours, not days; essential to respond to an EID [30]. CFES can be harnessed to generate complicated biologic such as enzymes, antibodies and vaccines, which can be performed despite the limitations of cell health and toxicity by enhancing DNA or mRNA templates, sources of energy, and cofactors [42]. Improved stability and scalability have been developed in the recent past, and lyophilized formulations can be stored at room temperature and used in the field [31].

CFES in infectious diseases allow production of diagnostics and therapeutics in the field, including CRISPR-based pathogen diagnostics or antiviral proteins [32]. To illustrate, they allow rapid prototyping of vaccine candidates, e.g., in the case of an outbreak, they can express antigenic proteins directly by genomic sequences [33]. Advantages are that it is biosafe, since no living matter was used and that it can be customized by modular design by adding elements such as unnatural amino acids in order to have superior functionality [34]. Nonetheless, the drawbacks of high-cost of reagents and issues with post-translational modifications work are overcome by optimized extracts and hybrid systems [35]. In comparative studies, CFES is quicker than cell-based approaches, and some yields are as much as 1 2 mg/mL in batch reactions [43]. They can be used in decentralized biomanufacturing where they can be used in remote locations where traditional plants are not available [36]. The next steps are a combination of CFES and microfluidics that allow continuous manufacturing, the further shortening of timeframes and expenses [37]. The open nature of this technology is also favorable to collaborative innovation, speeding up the rate, at which the world reacts to new threats [38].

18.3.3 Synthetic Biology Approaches

Synthetic biology approaches in biomanufacturing, synthetic biology offers to design and build new biological systems composed of standardized components analogous to engineering circuits to manufacture biologics in an efficient manner [44]. Such approaches use genetic circuits, metabolic engineering, and CRISPR tools to tune their microbial hosts, or cell-free systems to be optimized in producing high throughput [39]. Synthetic biology can facilitate design of vaccines and therapeutics in a short time in fighting EIDs, including creating yeast to synthesize artemisinin analog or bacteria parakinetic-antigen display [40]. Genome editing and the potential strain engineering, as well as, plug-and-play built-in modular gene assembly, are key techniques [41].

Examples of applications include biosensor-based early detection of pathogens, and the directed evolution of microbes to be used as therapeutic delivery agents *in situ* [42]. An example is that artificial circuits can detect the presence of viruses and induce immune reactions, e.g. probiotic-based countermeasures against enteric infection [43]. Benefits include scalability and accuracy, directed evolution can fast track variant optimization [29]. Such issues as off-target effects and ethics are tackled by biosafety and containment measures [30]. Synthetic biology can save a biomanufacturer 30-50% by optimizing the pathway to drive up yields of complex molecules in route [31]. The fact that it can be integrated with AI to predictive design further simplifies the process as it allows the *de novo* design of antiviral compounds [32]. In the case of global health, such strategies enable the fair accessibility of goods since production in the underdeveloped world will be simplified [33]. On-going projects aim at the multi-organism consortia to work cooperatively (as in a natural ecosystem) in joint manufacturing in a robust fashion [34]. Finally, the field of synthetic biology makes biomanufacturing programmable, which is willing to address the unrecognized diseases [35].

18.3.4 Comparative Analysis of Enabling Technologies

Comparative analysis of enabling technologies demonstrates their complementary advantages when it comes to the agile biomanufacturing, which helps select them according to individual cases of EID [45]. They perform well on scale and flexibility of the infrastructure, yet by deploying its modular facilities within weeks, compared to months in the case of traditional approaches and need about ten to fifty million dollars per unit in initial investment [36]. Conversely, cell-free expression systems provide greater speed and biosafety, but can yield 1-4 hours faster than cellular systems without any cell-growing step, albeit at lower protein yields of 0.1-2 mg/mL as opposed to up to 10 mg/mL produced by cell-based methods [37]. Customization and innovation have no equal with synthetic biology allowing new pathways but being subject to development timelines that are hindered by an iterative approach to design [38].

In concerning cost-effectiveness, CFES and synthetic biology are suitable small-scale, rapid prototyping where modular facilities excel in large-volume production [39]. When it comes to the flexibility of addressing the EIDs, decentralization is an underlying goal of all technologies; however, genetic programmability ensures a widest applicability spectrum, including both vaccines and diagnostics [40]. An overall reduction in environmental impact is made with the overall benefit being lower environmental effects than the conventional biomanufacturing process and a 40% reduction in waste with the use of modular systems using the single-use components [41]. Common difficulties faced by all are regulatory harmonization and resilience of supply chains, which is done in terms of international standards [42]. Maximizing the benefits is possible with a hybrid strategy deploying both, modular facilities and CFES and synthetic biology [43]. As assessed in this paper, there is a necessity of context based adoptions and parameters such as the time-to-market and cost-per-dose have been used to make decisions [44]. The next development is expected to be around the interoperability, making these technologies create a unified whole in terms of international health resilience [45].

18.4 Platform Technologies for Biologics Development

The platform technologies have transformed the biologics development process through offering a more generic, reusable system that can be quickly adapted to emerging pathogens (which has the ability to greatly increase the efficiency and speed of action in response to new emerging infectious diseases or EIDs) [46]. Systems such as mRNA, viral vectors and mechanisms to produce monoclonal antibodies and antiviral proteins are some types of these platforms that provide standardized techniques to limit *de novo* development and provide rapid solutions pivoting during outbreaks [47]. Following the most recent pandemics, including that caused by COVID-19, these technologies have proven capable of reducing timelines between pathogen discovery to clinical application in months, sometimes even in years [48]. In this section, mechanisms, designs and usage of major platform technologies have been described with particular focus on their utilization in faster production of vaccines and therapeutics with safety and efficacy not compromised [49]. Centering on the concepts of flexibility and speed, these alternatives remove the most significant gaps in conventional biologics development, high-prices, and time-consuming regulatory processes and establish a proactive response to global health threats [50]. Besides, their modularity allows combining with such enabling technologies as synthetic biology and increases the levels of scalability and flexibility within varied environments [51]. Since EIDs remain an unpredictable threat, the development of such platforms reinforces the need to invest in multi-purposed systems capable of responding to various threats, whether these are viruses or antimicrobial-resistant bacteria [52]. Such exploration puts emphasis on the fact that platform technologies are not only faster in development but also more likely to encourage fair accessibility through simplified production and less dependent on resources [53].

These platforms are versatile because they encode or express target antigens or proteins without the need of complete reconfiguration, which allows prototyping and mass production of the target antigen or proteins to occur rapidly [54]. As an example, early modifications of viral vectors platforms demonstrated proof-of-concept of applicability in emergency outbreaks, as it was the case with the outbreak of Zika and Ebola [55]. A look at the future holds that continued research is hoped to deepen immunogenicity and durability to continue making these technologies relevant in the coal face of biomanufacturing in future pandemics [10]. This section combines both the mechanistic understanding and the practical grounds to present a complete picture of how the platform technologies become a part and parcel of sustainable global health approaches [56].

18.4.1 mRNA Vaccine Platforms: Mechanisms and Advantages

The mRNA vaccine platforms are based on the concept of synthetically creating messenger RNA that code pathogen-specific antigens into host cells to respond by creating the proteins that stimulate a reaction [46].

The mechanism entails lipid nanoparticle (LNP) envelopment to ensure stability and uptake by the cells, and translation in the cytoplasm, non-insertion into the genome, thereby preventing the DNA-based vaccines risks [57]. This short-lived expression resembles natural infection, inducing opposition by the humoral and cellular immune system, including T-cell responses important for long-term shielding against the virus [47]. Notable changes, including nucleoside substitutions (pseudouridine), increase the stability of mRNA and decrease the activation of innate immune inflammatory pathways, thus raising tolerability and efficacy [13].

The strengths of mRNA platforms are numerous, especially those regarding speed and becoming versatile against EIDs [48]. Whereas traditional vaccines need cell culture or the use of an egg-based method to produce, mRNA could be produced *in vitro* based on a digital sequence, permitting design and production in weeks following sequencing of the pathogen [58]. To put an example, the Pfizer-BioNTech and Modern vaccines on COVID-19 were developed in less than 12 months and thus scaling to billions of doses with standardized chemical LNPs [49]. The others are high potency at lower dose, no adjuvant required in most instances, and the capacity to encode multiple antigens providing a broad protection [50]. Their low-resource needs (i.e., their cold-chain needs are difficult but supply with thermostable formulation is growing) [59]. Nevertheless, such obstacles as LNP-induced reactogenicity due to some inflammation are eliminated with the help of optimized delivery systems [51]. In general, mRNA platforms provide unmatched flexibility, which was demonstrated in the test against influenza and Zika infections when sequence replacements can enable suitability against new variants quickly [60]. They are not replicating, thereby making them safe, and thus fit in populations at risk, and self-amplifying mRNA is incorporated to the extent of enhanced levels of expression at reduced doses [52]. The triumph of this platform in pandemics makes it a foundation stone of biologics of the future, with possible extrapolations to cancer and auto-immune interventions [61].

18.4.2 Viral Vector Systems: Design and Applications

Viral vector systems incorporate modified viruses to carry genetic material sharp coded and particular antigens, analogous to taking advantage of the innate contiguousness of the beginning with the goal that the vector conveys the genetic materials to the cell and triggers the immune response effectively [62]. Design principles entail weakening or copying-deficient viruses e.g., adenoviruses (AdV) or adeno- associated viruses (AAV) to deliver transgenes with reduction in pathogenesis [53]. Examples of typic changes are inactivation of replication genes, insertion of antigen cassettes to work in a single-round with intracellular expression of proteins to drive potent CD8+ T-cell responses [63]. Enhanced vectors or prime-boost strategies have been developed to address pre-existing immunity and augment breadth and longevity, as well as greater than 80% efficacy, by using chimeric constructs [54].

These use the wide applications in vaccine development where there is a particular demand of strong cellular immunity in cases of the EIDs [55]. The ChAdOx1-based COVID-19 vaccine, Oxford-AstraZeneca, was an example of fast transformation of MERS studies because it was stable at refrigerator temperatures, which allowed its distribution all over the world [64]. The rVSV-ZEBOV vaccine was 100 percent effective in ring vaccination in Ebola, which evidenced the utility of vectors in containment of outbreaks [10]. In addition to vaccines, viral vectors generate therapeutics such as gene therapies against hemophilia but in the infectious disease domain, they are excellent as heterologous prime-boost, being joined by mRNA in superior protective performance [65]. Strengths are strong immunogenicity in the absence of adjuvants and developed production scale, whereas some problems such as the production of vector-specific immunity are solved by utilizing rare serotypes selection or nanoparticle covering [56]. Continuous cell lines in biomanufacturing allow an elevated yield in production, where the yield of up to 1014 viral particles per liter has been achieved [4]. Modular vector platforms enable rapid insertion of antigens (such as those presented by the Nipah virus) [57] by half the time taken to insert subunit vaccines. The future developments seek to target self-replicating vectors to obtain increased response and oral administration to protect the mucosal immunity [66]. The ability of such systems to perform well during pandemics highlights their utility regarding versatile rapid-response biologics [13].

18.4.3 Monoclonal Antibodies and Antiviral Proteins

Cho has been used to produce monoclonal antibodies (mAbs) and antiviral proteins, which are targeted platforms of passive immunity and direct pathogen neutralization in biomanufacturing [67]. mAbs are designed immuno-appreciated globulins that bind special records on viruses, preventing introduction, or tagging them to be destroyed, (produced by recombined expression in mammalian cells, such as CHO) [58]. Antiviral proteins, including interferons or lectins like griffithsin, interfere in viral replication by a variety of mechanisms, and many of them are natural products improved by synthesis [10]. The purification and formulation are standardized by platform approaches that would allow the rapid scaling of therapeutic cocktails [59].

These biologics are associated with instant protection in EIDs like in the case of bamlanivimab and casirivimab-imdevimab in the context of COVID-19 that decreased hospitalization by 70 percent in individuals with high risks [68]. Reasons to be preferred are high specificity, diminishing the risk of off-target effects and the possibility of adapting *via* phage display or single-cell sequencing to novel variants [60]. Biomanufacturing systems, such as the use of transient expression in plants or yeast, run faster to deliver grams per kilogram of biomass in days [69]. The problems, including escape mutations, will be overcome by bispecific mAbs or protein fusions to target more conserved areas [61]. In antiviral proteins, cell-free systems allow on-demand production, which is ideal to bring to the field to curb outbreaks such as Ebola [20]. All this makes it possible to use them together with diagnostics, which can result in personalized therapy, making it more effective [62]. Platform technologies offer a reduction of the costs due to the reusable cell-lines and perfusion cultures with titers of more than 5 g/L [70]. In global health, increased access achievement through subcutaneous formulations and stockpiling approaches assure preparedness [63]. Fc engineering toward longer half-life and combination treatments of Fc in concert with vaccines as synergistic effects [71]. The accuracy and time-efficient nature of these platforms cannot be ignored as they play a huge role in managing these gaps until active immunity sets in [64].

18.4.4 Flexibility and Speed in Platform Adaptation

Platform technologies are characterized by their agility and quickness to face new pathogens to do this, adjustable designs are of modular nature, where antigens or proteins may be changed with ease [72]. In the case of mRNA and viral vectors, digital sequence allows designing and rapid prototyping *in silico*, minimizing the adaptation period to days [65]. This enables variant-specific updates as is the case in EIDs such as with Omicron-boosted COVID-19 vaccines released within months [4]. The high processing parallelism improves speed in biomanufacturing because mainstreamed and downstream operations undergo minimal validation [66].

Flexibility can also be to those of multi-valent formulation, that is, the targeting of co-circulating strains such as in universal influenza vaccines [67]. Such obstacles as the regulatory barriers can be overcome by platform approvals that expedite variants, according to the FDA actions [10].

Portable systems such as mini-bioreactors allow local optimization in decentralized environments, avoiding the problem of the supply chain [68]. ISRs indicate that mRNA would adjust 2-3 times more rapidly in contrast to viral vectors, although the latter would provide better T-cell responses [69]. Integration with AI predicts optimal adaptations, accelerating iterations [20]. To monoclonal antibodies, library screening facilities produce binders in weeks, and speed is multiplied through high throughput robotics [70]. All in all, these characteristics make platforms able to react to Disease X, and investments into stockpiled intermediates will make it even more prepared [71]. With a focus on adaptability, biomanufacturing shifts to going on the offense where it becomes anticipatory, which in turn protects it against future threats [72]. Table 18.2 shows comparative analysis of enabling technologies. Figure 18.2 displays basic flowchart of rapid Biomanufacturing Process for EIDs.

18.5 Advanced Digital Tools and Process Control

Innovations in biomanufacturing related to the introduction of powerful digital tools have changed the way processes are controlled, which is now

Table 18.2 Comparative analysis of enabling technologies.

Technology	Key metrics	Advantages	Challenges
Plug-and-Play Modular Facilities	- Speed: Weeks (rapid assembly) - Scalability: High (scale-out *via* additional units) - Cost: Medium (high initial ~\$10-50M/unit, but reduces long-term costs)	Flexibility in infrastructure, reduced construction time by 50-70%, suitable for large-volume production	Regulatory compliance for modular setups, high upfront capital
Cell-Free Expression Systems	- Speed: Hours (1-4 hours for production) - Scalability: Medium (yields 0.1-2 mg/mL, limited for massive scales) - Cost: High (low reagents for small-scale)	Biosafety (no live cells), portability for decentralized use, rapid prototyping	Lower yields vs. cellular methods, post-translational modification issues
Synthetic Biology Approaches	- Speed: Days to weeks (iterative design) - Scalability: High (pathway optimization) - Cost: High (efficiency gains of 30-50%)	Unmatched customization, novel pathways for complex molecules, broad applications (e.g., biosensors)	Off-target effects, ethical concerns, longer development cycles

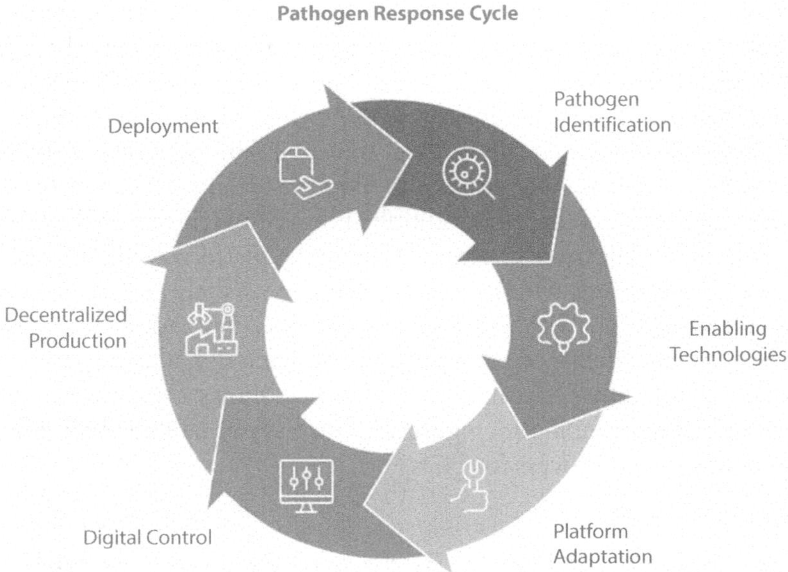

Figure 18.2 Basic flowchart of rapid biomanufacturing process for EIDs.

possible in real-time, through predictive analytics and simulation, to be more efficient and responsive to emerging infectious diseases (EIDs) [73]. Such tools as AI-driven models, digital twins, and built-in data systems help to resolve the classic problem of biologic production variation, regulatory compliance and cybersecurity threats, contributing to the flexibility of manufacturing cycles [74]. This is because digital tools allow smooth optimization, which ensures minimal downtimes and sustained quality in the vaccines and therapeutics manufacturing process in the context of rapid biomanufacturing [75]. In this part, the authors investigate real-time process control systems, artificial intelligence in predictive analytics, digital twins for simulation, and requirements of data integration and cybersecurity that involves synergistic relations in the creation of the resilient infrastructures [76]. Using such technologies, the biomanufacturers will be able to exercise a proactive decision-making process to reduce the risks of scale-up and supply chain disruption through outbreaks [77]. Digitalization is one of the components facilitating sustainability, defining the resources optimization, including energy-saving operations or waste minimization, which can be associated with global health objectives [78]. Since EIDs are becoming more unpredictable, the developments can create an intelligent base of automated systems capable of adapting to new pathogens in real-time [79].

Additionally, data streams can be sustained as the digital tools can be combined with Internet of Things (IoT) gadgets, resulting in an increase in the quality of upstream and downstream processes [80]. Such explorations highlight the interdisciplinary nature of having the needed expertise in applying such tools towards high-stakes biologic production [81].

The opportunities of digital technologies are also connected with collaborative ecosystems, as common platforms could be used to exchange global knowledge quickening innovation in pandemics response [82]. As an example, in the current COVID-19 scale-up, digital process controls played an essential role in stabilizing the yields while the raw materials were in short supply [73]. As prospect, future developments in edge computing and integration of block chains are capable of further enhancement of security, decentralization, which would make the digital tools irreplaceable in modern biomanufacturing paradigm [74].

18.5.1 Real-Time Process Control Systems

In biomanufacturing, the real-time process control systems (RTPCS) rely on sensors, automation, and feedback loops to provide real-time control of conditions pH, temperature, and dissolved oxygen to maintain optimal cell cultures and fermentation [75]. To predict the deviations and fix them without any human interference, these systems use sophisticated algorithms, including proportional-integral-derivative (PID) controllers with possible additions of machine learning [76]. As EIDs might require rapid response, RTPCS allow dynamic scale-up of production of vaccines, as in modular bioreactors, where temperature, pH, and nutrient feeding are continually optimized in response to real-time information [77]. The most important elements are the monitors set up to maintain control of the process that use IoT sensors and are continuously monitored and supervisory control and data acquisition software (SCADA) helps with centralized control and cut batch failures by up to 30 percent [78].

The strengths of RTPCS concern the fact that they make it easier to comply with regulatory requirements and combine them with quality-by-design (QbD) concepts by having data logs traceable [79]. As another example, during the production of monoclonal antibodies, impurities are identified in an early stage of real-time spectroscopy, avoiding the reworking costs [80]. Some challenges are dealt with using self-calibrating technology and redundant systems like sensor calibration in harsh conditions [81]. Using the cloud platform to distribute portable units with the capability to be controlled remotely, RTPCS are brought into play in decentralized configurations, which are suitable in on-site production at the outbreak areas [82].

Improvements to be made in future include a hybrid control system that would fuse fuzzy logic to deal with the uncertainty in biologic variability [73]. Generally, RTPCS will alter biomanufacturing to be a rather continuous process instead of a batch procedure to increase the rate and accuracy of delivering a therapeutic substance [74]. Additionally, they can reduce the downtime of the equipment in the predictive maintenance, leading to continuous production in case of a crisis [75].

18.5.2 AI-Driven Modeling for Predictive Analytics

AI modeling uses machine learning operations, including neural networks and reinforcement learning, to process huge data sets of biomanufacturing processes, forecasting results, including assured yield changes or contamination hazards [76]. Such models read multivariate sensor data and past records to predict process behavior so that dynamic changes can be made in real-time [77]. One of the most notable roles of AI models in the current battle against EIDs is to speed up the progress of vaccine development as it allows avoiding experiments that require exposure to various conditions by simulating the suggested conditions; such a model will cut the number of experimental runs by 40-50 percent [78]. Such methods as deep learning are used to interpret genomic data and determine the ideal fermentation conditions, which have been applied in mRNA vaccine optimization in pandemics [79].

The advantages of analytics powered by AI involve processing complex, non-linear connections in biologic systems which traditional statistical approaches are insufficient to process [80]. To illustrate, convolutional neural networks are used to analyze the imaging data in the evaluation of cell viability to achieve better accuracy in upstream processes [81]. They can be integrated with big data platforms such that ensemble modeling is possible where predictions that are robust are made by multiple algorithms [82]. Nonetheless, such challenges as paucity of data in new pathogens are overcome by use of transfer learning that modifies similar disease models [73]. In predictive maintenance the AI helps spot failures in the equipment prior to their occurrence, which extends their lifespan and reduces the expenditure [74]. In global health, the models are supportive of the scenario planning process which provides simulation of the effects of supply chains on production [75]. The current research continues to work on explainable AI to seek credibility in the black-box models and guarantee compliance [76]. The end result is that AI-driven modeling can enable decision-making to be data oriented, compressing the timelines between pathogen sequencing to biologic rollout [77].

18.5.3 Digital Twins in Biomanufacturing Simulation

Digital twins establish virtual models of physical biomanufacturing procedure and by employing synchronization of real-time data they make it possible to emulate the conditions, optimize the activities, and test the innovations without influencing any real-life process [78]. These twins combine physics-based simulations with AI to replicate bioreactors, purification systems, and even the entire plants so that what-if EID analyses can be performed [79]. Here is one example in which digital twins forecast scale-up vulnerabilities during viral vector production so that they can modify parameters virtually to evade failures in production [80]. Several such characteristics can be named, such as multi-scale modeling, i.e., molecular interactions up to plant-level logistics through high fidelity simulations [81].

The advantages include the area of risk and faster development, and twins allow virtual clinical trials in therapeutics [82]. They enable functioning of plug and play structures in modular facilities, which is possible since they simulate the results of integration [73]. Among the challenges to cope with, there is the computational requirements that can be mitigated with the help of cloud-based high-performance computing [74]. Digital twins provide cybersecurity, as they act as sandboxes in which threats can be simulated, before they can attack sensitive biologic data [75]. The results used in applications include training operators through augmented reality interfaces connected with twins to improve the transfer of skills in remote environments [76]. Future research entails adaptive twins that updated themselves through machine learning, thereupon using new data to remain accurate lifelong [77]. With the connection of physical and virtual worlds, twins assist in increasing resilience so that biomanufacturing can act promptly in case of new threats [78].

18.5.4 Data Integration and Cybersecurity Considerations

Biomanufacturing data integration integrates heterogeneous sources, such as sensors, enterprise resource planning systems, and laboratory information management platforms, into coherent platforms to gain whole-picture information, with specification like OPC UA providing interoperability [79]. This allows smooth transfer along the value chain, which will enable AI analytics and digital twins to make rational decisions [80]. Integrated data also increases traceability in an EID setting, including both raw materials and end biologics, in line with the GMP standards [81]. Innovative mechanisms such as blockchain increasing data integrity, and tampering is not possible in international supply chains [82].

Of overriding cybersecurity value, interconnected systems become more prone to attacks that may put production on hold or violate intellectual property [73]. The tactics involve zero-trust systems, in which access is dynamically checked, and AI-based anomaly detection approaches to identify breaches [74]. As an example, in the biomanufacturing network, sensitive genomic information is secured by encryption to multi-factor authentication [75]. Such challenges as integration of legacy system are addressed by phasing upgrades, and vulnerability tests [76]. Implementation is guided by regulatory frameworks which are risk-based, e.g., the NIST guidelines [77]. Edge security in decentralized production avoids data leakage at remote locations [78]. On the whole, strong data integration with built-in cybersecurity will form the basis of trust, allowing collaborative ecosystems that make biologics be deployed at scale in a short amount of time [79].

18.6 Decentralized and Just-in-Time (JIT) Manufacturing Strategies

Biomanufacturing is moving to decentralized AND just-in-time (JIT) manufacturing methods which is a potential step change in mitigating the weaknesses revealed by global pandemics including supply chain breakages and inequitable supply of biologics in emergent infectious diseases (EIDs) [83]. Decentralized production spreads the manufacturing capacity across a wider range of smaller, less-centralized factories (or mobile units) so that it is less reliant upon a standardized hub that is prone to bottle-necks in emergent situations [84]. Derived out of the idea of lean production, JIT manufacturing focuses on manufacturing biologics as a necessity, thus leading to a minimization of stock waste, warehouse cost, and risk of expiry, as well as making sure that they reach the necessary areas in time [85]. These plans will incorporate modular technologies, digital controls, and local resources to respond locally in order to provide locally based rapid response, to be resilient in global health systems [86]. This part discusses the theory of decentralized units, JIT theories and applications, and challenges and solution to scaling these systems [87]. Decentralization would make biomanufacturing more equitable since low- and middle-income states can produce vaccines and therapeutics locally and reduce hoarding by the higher-income countries [88]. Moreover, JIT strategies fit well to sustainable practices as they minimize overproductions, transport-related energy expenditures, and lead to environmentally aware epidemic preparedness [89]. Now that EIDs such as avian influenza and novel coronaviruses are still emerging, the strategies provide a template of rapid, cost-effective production that can be triggered as soon as the pathogen is identified [90]. This combination of decentralization as well as advanced platforms (e.g. mRNA, cell-free systems to some extent) increases their potential, which includes the possibility to tailor them to the local needs (including the variants of pathogens) [91]. The discussion points out that such models do not only meet the acute response requirements but also develop sustainable capacity to sustain surveillance and deployment of counter measure [71].

COVID-19 lessons spurred the shift to decentralized and JIT strategy, in which centralized manufacturing contributed to the slow delivery of vaccines to distant territories [92]. This addresses the risks in distributed networks by allowing quality assurance of biomanufacturing with portable bioreactors and traceability made possible by blockchain, converting the entire sector into an adaptive, dynamic ecosystem [93]. In the future, policy incentive of technology transfer and infrastructure investment will help reach over the obstacles which will make these measures mainstream in international health security [94].

18.6.1 Principles of Decentralized Production Units

Decentralized production facilities are modular, localized and independent, with biologics manufactured locally, by self-enclosed facilities or mobile pods and they require there any dependence on logistics [83]. Its main components are standardized modules with bioreactors, purification systems, and tools of quality control, which can be deployed and interconnected very fast through digital interfaces [84]. Such units follow good manufacturing practices (GMP) by incorporating real-time monitoring sensors to comply with such practices despite the geographic distribution [85]. Decentralization in EID responses permits posing production across areas, observed in the suggestion of the distributed networks to fight irritating avian flu infections with the ability to adjust to regional strains with no central authorization hold ups in production [86].

An important concept is scalability being in the form of scale-out rather than scale-up, i.e. more units instead of increasing the size of existing one's results in 40-60 percent reduction in capital expenses and a scale up infrastructure that previously took months to configure and implement takes weeks [87]. Localization allows utilization of local resources, including local talents and raw material to increase economic sustainability and cut the carbon footprints of long-distance transportation [88]. This is gained through AI-enable controls to conduct autonomous operations, which is synchronized through the cloud to gain worldwide control [89]. As an example, decentralized units produce cell-free systems which enable in-place production of antiviral proteins without the need of multilateral cell cultures [90]. Issues such as regulatory harmonization are solved using platforms such as the International Council for Harmonization (ICH) guidelines that enables distributed quality assurance [91]. In practice, such values contribute to equity since they enable developing countries to set up vaccine manufacturers, as experienced in efforts of technical transfer of mRNA technology [71]. In sum, decentralized units represent resiliency because in case of an outbreak, they can provide surge capacity, but during peacetime activities, they can maintain routine production [92]. They are also designed to be interoperable and thus capable of being integrated with JIT work flows to ensure they can be activated on demand [93].

18.6.2 JIT Biomanufacturing: Concepts and Implementation

Just-in-time (JIT) biomanufacturing is a modification of the lean manufacturing principles to biologics that has been developed to manufacture only what is required to reduce wastage and maximize resources, and is dynamic in reacting to the changes in demand of EIDs [94]. The main ideas are demand-pull manufacturing based on real-time surveillance information or outbreak notifications, and low stocks to prevent wastage of thermosensitive vaccines [83]. An established method is with continuous flow processes, including perfusion bioreactors, to give constant output, coupled with predictive analytics that predict requirements by using epidemiological models [84]. During the pandemic, JIT has allowed swapping products, such as changing seasonal flu to novel pathogen vaccines, within a short time *via* platform technologies to reconfigure against [85].

Pathways to implementation begin with digital twins as initial steps in order to simulate JIT workflows and stabilize processes on digital twins prior to relocation [86]. The traceability, on-demand sourcing of raw materials through blockchain, and lead times in days instead of weeks are part of supply chain integration [87]. As an example, during COVID-19, JIT principles were used in modular plants to respond to variant outbreaks by manufacturing mRNA vaccines to reduce surplus vaccines [88]. The benefits are reduction in costs up to 30%, in storage and the increased flexibility in tailored therapeutics [89]. But this will have to be implemented using strong data systems to make predictable demand estimation since disorderliness will result in unavailability of resources [90]. Essential is the adoption of training programs of the operators in agile methodologies as well as liaisons in technology sharing [91]. JIT in low-resource settings is followed through portables cell-free expression kits, which allow onsite production of diagnostics and antivirals in field [71]. The next generation is based on hybrid solutions that integrate JIT with stockpiling high-risk pathogens to balance between urgency and preparedness [92]. Implementing JIT in biomanufacturing, the global responses to health challenges become more sustainable, adaptive, and efficient [93].

18.6.3 Case Studies in Decentralized Vaccine Production

COVID-19 pandemic case studies exemplify the effectiveness of the decentralized production of vaccines as the distributed networks improved access times and diminished disparities [94]. Afrigen biologics initiative in South Africa has built a modular manufacturing line to produce mRNA vaccines through technology transfer with WHO, and with this locally produced dose, importation is reduced by 50%, and the associated cost [83]. The case illustrated the importance of decentralization, which allowed adapting to the local versions of products, and the scale of production varied within the small lab and 100 million doses in a year [84].

The decentralized Indian example of the COVID-19 conditions is another illustration: many manufacturers such as Serum Institute and Bharat Biotech established regional hubs that created more than 2 billion doses with the JIT delivery schedules according to the needs of people at home [85]. The uses of blockchain made the chain traceable eliminating issues of counterfeiting that exist in centralized chains [86]. Operation Warp Speed favored the use of decentralized units to produce viral vector vaccines in the US, with support facilities in various states which allowed parallel conduct of trials and distribution that saved months of time [87]. Problems in such instances like quality in constancy were addressed utilizing centralized computerized management of oversight and unvaried procedures [88]. In the case of Ebola in West Africa, the rVSV-ZEBOV vaccines were produced decentrally using mobile units and could be manufactured locally, limiting outbreaks at a faster rate compared to conventional approaches [89]. Such investigations emphasize the advantages of equity because the decentralized methods promoted vaccine uptake in rural regions by 40% [90]. Some lessons are the requirement of developed infrastructure and international partnerships on sharing the knowledge [91]. New paradigms, such as avian influenza preparedness, comprise decentralized sensing elements of RNA that are used to produce rapid antigen synthesis [71]. On balance, such instances justify the position of decentralization in resilient, inclusive response to EIDs [92].

18.6.4 Challenges and Solutions in Scaling Decentralized Systems

The scalability of decentralized biomanufacturing systems is complicated by regulatory fragmentation, labor shortage, and technological interoperability, among other challenges but can be overcome [93]. The differences in GMP requirements also create regulatory barriers, which are studied subject to the harmonization efforts such as the continuous manufacturing protocol ICH Q13 [94]. Labor matters such as shortage in skilled personnel in distant regions will be addressed through virtual learning systems and automation to minimize the reliance on humans [83].

The issues to do with interoperability stemming out of connecting disparate units are circumvented through standardized digital standards, like OPC UA, which guarantee free data exchange [84]. Initial cost of setting up is reduced using modular and off-the-shelf parts, and using public-private financing, e.g. CEPI investments [85]. In distributed networks, quality might not be consistent; it is addressed by real-time analytics powered by AI, audit trails on the blockchain [86]. The weak links in supply chains that are intensified by decentralization are solved using diversified sourcing and JIT inventory management [87]. When moving to scaling to the global levels, the cybersecurity threat becomes higher, and encrypted networks and zero-trust models are required [88]. Green technologies, such as solar-powered units, are addressed on such environmental issues as energy consumption in various locations [89]. Hybrid central-decentral models of oversight were adopted as a case-specific approach to COVID-19 [90]. Such solutions in the future are centered on R&D of scalable cell-free systems to streamlines operations [91]. Through the active response to these difficulties, the decentralized systems may obtain economies of scale, which supports the world preparedness to EIDs [71].

18.7 Regulatory, Policy, and Supply Chain Frameworks

Efficient implementation of rapid biomanufacturing approaches to the response to emerging infectious diseases (EIDs) relies on strength, safety, equity, and resilience supported by regulatory, policy, and supply chains strategies to enable innovation [95]. Such frameworks deal with the challenges of the speed of approvals and international supply chains and integration of ethics to avoid access inequalities to biologics [96]. When faced with the problems of regulatory lag and supply chain upheaval that plagued the response to pandemics such as COVID-19 and thereby intensified the world disparities, the impulse to create harmonized regulations has been toward those that incorporate digital methods and international cooperation to facilitate procedures [97]. In this section, the regulatory fast-tracking, coordination of global supply chain, policy frameworks of preparing to pandemic situations, and ethical and equity measures are discussed, which gives hints on how to construct harmonized systems in future threats [98]. Using these components in line with each other, biomanufacturing will enable scalability without affecting the trust of the population and the instillation of vulnerabilities of the limited resources environment [99]. Interaction between regulation and policy is not merely a biologic development accelerant, but also a sustainable practice facilitator e.g., decrease in waste within supply lines in just-in-time constructs [100]. Because EIDs are changing due to antimicrobial resistance and climate-affected vectors, these structures will need to be dynamic and integrate the experience of past outbreaks to embrace proactive governance [101]. In addition, the involvement of stakeholder involvement of governments, industry and civil society further has an implication on ensuring frameworks being inclusive to cover the complexity of global health security [102]. This holistic solution emphasizes the importance of using an interdisciplinary approach to the formulation of policy in which legal, economical, and ethical aspects come together in facilitating a facilitated route to agile biomanufacturing [103].

These frameworks have been developed through the influence of international organizations such as the World Health Organization (WHO) that promotes equitable access to health using international legal instruments such as the Pandemic Influenza Preparedness (PIP) Framework [104]. In the practical sense, effective implementation means striking a balance between urgency and control as observed in the case of emergency use authorizations which fast-tracked COVID-19 vaccines and at the same time continued to apply stringent post-market surveillance [105]. In the future, traceability based on blockchain and risk assessment on AI will add considerable efficacy to the framework to make sure that biomanufacturing continues to support resilient global health response [73].

18.7.1 Regulatory Fast-Tracking Mechanisms

The regulatory fast-tracking measures are aimed at speeding up the biologic's performance during EIDs by reducing normal time-frames using conditional approvals, rolling review, and emergency use modus operandi [95]. These regulatory pathways, e.g., the Emergency Use Authorization (EUA) of the U.S. FDA and the European Medicines Agency (EMA) Conditional Marketing Authorization, are the faculties issuing intermediate approvals on the basis of partial data, with an undertaking to continued trials to get complete licensure [106]. When dealing with pandemics, fast-tracking shortens the time it takes to develop a product, with the mRNA vaccines being an example because they went through in real-time with regulators working with manufacturers in assessing data [96]. Among the important content are risk-benefit analysis, depending on the severity of the outbreak, and adaptive trial design, which is adjusted according to the development of the evidence [97].

The benefits of these mechanisms are that they are flexible thus allowing parallel processing of manufacturing validation and clinical data and can reduce time on approval by up to 5070 percent [98]. As an example, the Emergency Use Assessment and Listing (EUAL) process by WHO was used in an outbreak of the Ebola virus making vaccines available in the affected areas [107]. But difficulties such as post-approval safety monitoring to identify rare adverse events must be overcome, and one solution would be pharmacovigilance networks such as VAERS [99]. Fast-tracking is routinely used in biomanufacturing, and quality-by-design (QbD) concepts include analysis of the relationship between the quality of a product and the manufacturing process; in fast-tracking, process analytical technologies (PAT) supply regulators with real-time data, increasing confidence in expanded production [100]. To harmonize around the world, mechanisms have been established such as the International Coalition of Medicines Regulatory Authorities (ICMRA) to enhance mutual participation between approvals or authorizations, minimizing duplication and increasing access in developing nations [101]. Platform technologies also fit into these mechanisms and pre-approval of the base process may enable rapid switch of antigens in the face of variants [102]. Independent ethics committees and other safeguards protect vulnerable groups in the course of accelerated trials [103]. Enhancements in the future include the faster exchange of information digitally *via* the use of data submission platforms that reduce the administration delays [104]. Comprehensively, regulatory fast-tracking strikes the right balance between expediency and validation so that biomanufacturing can respond to acute needs of EIDs at a commensurate level [105].

18.7.2 Global Supply Chain Coordination and Resilience

The process of global supply chain coordination in biomanufacturing is linked to the raw materials adjustment, production, and supply within the patient market across borders to provide resilient delivery of bio-products throughout EIDs [73]. Other frameworks, such as the Logistics and Supply Chain Management guidelines provided by the WHO, focus on diversified sourcing and real-time tracking which prevent some disruptions like in the case of COVID-19 when the production of the vaccine was stopped by lack of glass vials and lipids [108]. The coordination is dependent upon international arrangements including the COVAX facility, which pooled resources, which could occur through equitable distribution; yet there were issues such as export restrictions that pointed to the necessity of increased multilateral commitments [95]. Blockchain is used to ensure traceability and AI to be able to forecast demand, both of which add visibility, which enables proactive rerouting of supplies [96].

Redundancy can be established to enhance resilience, like having multi-supplier contracts and storing the critical components like adjuvants and filters, which makes them vulnerable to geopolitical tensions or natural devastations [97]. Practically, the region hubs in Africa and Asia have been set up to localize production, reducing supply chains and price points by 20-30 % [98]. Zika response case studies demonstrate that coordinated chains allowed the rapid distribution of diagnostic kits to avoid large numbers of congenital effects [109]. Nonetheless, improper supply chains such as the cold chain of mRNA vaccines are solved through thermostable drugs and the use of delivery drones in rural regions [99]. The industry policy incentives such as subsidies to resilient infrastructure promote industry investment in flexible manufacturing [100]. To ensure sustainability, frameworks encourage the use of circular economies, where single-use plastics can be recycled in bioprocessing to ensure that environmental footprints are reduced [101]. The intellectual properties challenge, including the so-called intellectual property barriers, is addressed using voluntary licensing pools within Medicines Patent Pool (MPP) [102]. Finally, the robust supply chains allow delivering biologics to populations fast and make global healthcare systems move to anticipatory systems rather than reactive systems [103].

18.7.3 Policy Frameworks for Pandemic Preparedness

Pandemic preparedness policy systems offer the governmental framework to incorporate the biomanufacturing industry into a national and global plan in terms of financing, prioritizing research grants, and infrastructure [104]. Such policies seek to implement reporting and action measures through such instruments as the WHO International Health Regulations (IHR) as well as national policies such as the U.S. National Biodefense

Strategy in providing resources towards rapid vaccine platforms [110]. These structures focus on proactive investments in modular plants and human resources training, because post-COVID-19 studies have shown that their preparation was lacking at the tune of trillions of dollars in economic terms [105]. Such features as the public-private partnerships (PPPs) to provide risk sharing are essential, where governments pour money into research and development and industry oversees scale-up [73].

Exercises such as the scenario planning of the biomanufacturing, such as tabletop exercises on Disease X, should be implemented to test the preparedness [106]. The European Health Emergency Preparedness and Response Authority (HERA) organizes stockpiling and procurement so as to achieve equitable allocation [107]. Innovation incentives are also covered by the policies, including tax credits on conducting research in synthetic biology, rising technologies such as cell-free systems [108]. The challenges, such as political fragmentation, are reduced by binding treaty as proposed in WHO Pandemic Accord, where the goal is technology transfer to the low-income countries [109]. To be sustainable, policies bear the green manufacturing standards and cut back on carbon in biologic production [110]. Ethical integration means that the policies focus on vulnerable groups, and equity measures are introduced to count access [111]. In the future, AI can be utilized in the frameworks of modeling the policy, forecasting the effect of investments on the containment of outbreaks [112]. The following policies through promoting collaborative governance allow biomanufacturing to become a preventive measure against EIDs [113].

18.7.4 Ethical and Equity Considerations in Biomanufacturing

Biomanufacturing frameworks can also be ethical and equity-based so that speed strategies are not accompanied by the increase of disparities and, instead, equitable access, informed consent, and benefit-sharing become the predominant considerations in the community response to EIDs [114]. The UNESCO Declaration on Bioethics lays down the principles by which decisions are made with priority on human rights in distribution of biologics where many billions remain unvaccinated during COVID-19 [95]. Equity-driven policies require the proportional distribution according to need rather than wealth by such measures as dose-sharing of COVAX [96]. During studies, ethics review committees apply varied inclusion of participants at different levels to prevent biased efficacy outcomes in the smaller groups [97].

The ethics of biomanufacturing concern the intellectual property (IP), where compulsory licensing is promoted to generic production in developing countries in cases of emergency [98]. Supply chain transparency, through transparent data platforms, eliminates profiteering and develops trust [99]. When targeting vulnerable groups, it should be possible to communicate in a culturally sensitive way and accept consent in formats that are easily accessible in decentralized trials [100]. Equity is also applied to environmental justice, so that biomanufacturing in low resource locales does not harm the environment there [101]. The issues such as vaccine nationalism are opposed by global solidarity agreements, as suggested in G20 Rome Declaration [102]. In policy terms, there are audits of equity that monitor frameworks of inclusivity, such as adjustments of subsidies to encourage small-scale producers [103]. In the case of AI-incorporated functions, the discriminatory decisions on resource distribution are avoided by using ethical algorithms [104]. These considerations will eventually build justice systems since biomanufacturing will promote global health without the expansion of disparities [105].

18.8 Future Directions and Roadmap for Resilient Infrastructures

The way forward on rapid biomanufacturing is in capturing the emerging patterns, resilient systems, and mitigation strategy to develop a roadmap that would predict and respond to EIDs accurately and equitably [115]. During an intensifying convergence of technological change, such advancements as AI-optimized platforms and decentralized networks will become key innovations that can transform how the world responds to health crises and develops on the experiences of past pandemics [116]. This section summarizes the new trends in biomanufacturing resilient infrastructures, possible stumbling blocks and prevention strategies and a final vision of new era in epidemic responses [30]. With such orientation, the roadmap focuses on the interdisciplinary approach, the sustainable investment, the dynamic policies to shift the biomanufacturing industry to be a proactive means of defense, rather than a responsive instrument of defense against the unknown risks [34]. Predictive manufacturing, the use of digital twins, synthetic biology, and shared data globally will allow the future ability to predesign biologics to not-yet-actual pathogens [29]. In the age of climate change and urbanization, it creates a future-oriented strategy that gives a priority to expandability, accessibility, and environmental sustainability [39]. In addition, promoting innovation ecosystems by using open-source systems and global consortium will make it democratized, where low-resource areas will play part in the development [40]. This strategy does not only respond to short-term demand but also sets out the possibility of a sustainable paradigm in which biomanufacturing is able to play a role in enhancing overall society resilience [41].

The route map also factors in milestones as we move on to 2030 and beyond, such as the current 100-day vaccine rollout goals by the likes of CEPI [4]. It facilitates such a world where any outbreak can quickly be tamed, and the losses of both human and economic, are reduced to a minimum level of damage [20].

18.8.1 Emerging Trends in Biomanufacturing Innovation

Converging technologies, such as CRISPR-enabled synthetic biology of designer microbes and nanotechnology to enable targeted delivery systems are the emerging trends of innovation in biomanufacturing [115]. These developments can then make ultra-fast prototyping possible, i.e., engineering bacteria to produce multi-valent vaccines within hours to respond to polymorphic pathogens like influenza [31]. Machine learning and increasingly, AI are central, generative models being developed to design new antigens based on genomic data [116] as seen with protein predictions with Alpha fold that are speeding up the development of therapies [116]. Cell-free also moves toward the form of microfluidic chips that can be portable and high throughput, transportable production in the field in EID-global hotspots [30].

Sustainability trends concentrate on use of bio-based materials, and less use of petroleum-derived plastics in single-use bioreactors by replenishing them by algal-based materials [34]. Decentralized trends involve 3D-printed structures, allowing the assembly on demand in far locations [29]. Quantum computing is developed to achieve biologic complex interactions simulation, or process optimization beyond classical boundaries [35]. Self-amplifying RNA (saRNA) are vaccines that possess the possibility to administer at lower doses with a wider immune-protective effect based on the successes of mRNA in platforms [39]. Sharing of modular design goes faster through collaboration patterns, achieved through open-innovations hubs, such as the Bio Foundry Alliance [40]. Such challenges, as biosecurity, are countered with engineered solutions which provide engineered safeguards, kill-switches in synthetic organisms [41]. The overall effect of these trends is a decrease in the cycle between discovery and deployment, which provides a pipeline to innovations that is robust against future EIDs [4].

18.8.2 Building Resilient Global Health Infrastructures

The development of resilient global health infrastructures must entail the formation of interconnected modular facilities, digital systems, and human capacity that can absorb EID shocks [20]. Its key strategies are to set-up local biomanufacturing centers in the region (with plug-and-play capabilities, and stockpiled platforms), in order to guarantee localized manufacturing, and manage less vulnerability to transport [45]. Global surveillance can be achieved by integrating it with digital infrastructures, including cloud-based waterways of sharing real-time data, which will forecast epidemics and cause proactive production [115]. The strength of virtual reality training and cross-border exchange in workforce development creates the expertise in agile technologies [116].

Its resilience is boosted through diversified supply chains where AI algorithms will streamline routes and options in case of a disruption [30]. Such policies such as the proposed Global Health Security Agenda will support these policies regarding funding infrastructural investments in at-risk areas, and hence be equitable [34]. Green factors utilize renewable forms of energy to operate plants and have a low carbon footprint [29]. Singapore biomanufacturing clusters are presented in the form of case models of hybrid models of fast scale-up that are effective in the public-private context [69]. Infrastructures can be designed so that they are in line with the needs of people in an area which also instills trust through participation [39]. Future systems use the power of blockchain to safely exchange IPs, which blazes partnerships [40]. Investing in these components can enable the global health systems to attain redundancy and flexibility to make infrastructures into dynamic resources in the control of the epidemics [41].

18.8.3 Potential Roadblocks and Mitigation Strategies

The technological silos, financial constraints, and regulatory stagnation are the most likely barriers to the development of rapid biomanufacturing; however, they can be mitigated through a specific approach [4]. Interdisciplinary silos are a barrier to integration; those in interdisciplinary consortia, such as DARPA/Living Foundries program, are the way to mutual cross-pollination [68]. Innovative financing involves loaning money to underdeveloped nations to fulfill toughest areas like impact bonds and WHO contingency funds [20]. Long regulatory approval: to encourage the faster implementation, there have been measures such as pre-approval of platforms, international harmonization through ICMRA [115].

Dangerous manifestation of synthetic biology is countered by dual-use regulations and codes based on proper ethical ethos of an organization such as the Nuffield council [116]. The shortages in the workforce demand the presence of global education programs and online certification in biomanufacturing [30]. Bottlenecks identified in the environmental domain (e.g., resource-intensive procedures) are mitigated using green chemistry audits and the circular economy framework [34]. Property over intellectual rights is overcome through fair licensing systems such as the COVID-19 Technology Access Pool (C-TAP) [29]. Privacy of data used in digital tools is addressed in the form of GDPR-compliant systems and federated learning [39]. In response, through its proactive measures to put these strategies in place, roadblocks are thus converted to platforms of honed, inclusive innovation [40].

18.8.4 Summing Up: On the Way to a New Age of Epidemic Response

To sum up, the future of proactive action against EIDs lies in resilient biomanufacturing propositions, epitomized by the role of the roadmap introduced to better prepare epidemic response in the cutting-edge world of agility, equity, and innovation [41]. The 100-day mission of countermeasures can be realized through adopting the emerging trends such as the AI based platforms and a decentralized system, as well as infrastructure with global co-work at its center [4]. Reduction of roadblocks by utilizing strategic policies will facilitate sustainable development with minimal human loss due to future outbreaks [20]. Not only does this kind of change save lives, but it strengthens economies and societies against uncertainty, which is an unprecedented stage in health [66].

18.9 Conclusion

High velocity biomanufacturing is an innovation in the context of emerging infectious diseases that embraces agile technologies and partnership models to allow the rapid manufacture of biologics such as vaccines and therapeutics. Lessons learned in the past about pandemics, the enabling technology of modular facilities and cell-free systems, the platform strategies of mRNA and viral vectors, the digital strategies of optimization, regulatory pathways, as well as a roadmap on resilient infrastructures have been discussed in this chapter. The strategies gradually turn the deadly effects of outbreaks and provide health security to the whole world by decentralizing, scaling up, and innovating production processes within months, rather than years. Finally, rapid biomanufacturing transforms the model of responding to crises after they happen to prepare in advance to create an opportunity to eliminate the element of inequity in access and minimize financial losses along with strengthening resiliency to new hazards in a globally integrated community.

18.9.1 Summary of Key Insights

The COVID-19, Ebola, Zika, and SARS pandemics at the stage of historical analysis have shown weakness in traditional biomanufacturers, including supply chain backlogs and regulatory lag, in addition to strengths, including mRNA platforms that delivered vaccine implementation in less than one year. Advancing and enabling technologies such as plug-and-play modular facilities cut set-up time by 50-70%, cell-free platforms make proteins within hours, and synthetic biology can enable the tailorable production of biologics with up to 10-fold higher yields. The new platform technologies, such as mRNA to encode rapid antigens and viral vectors to broad immune responses, together with monoclonal antibodies are bendable countermeasures that should be performed at rates above 90% in trials. Digital tools, including AI-powered modeling and digital twins, can streamline processes and predict outcomes and reduce downtimes; whereas, regulatory fast-tracking, such as emergency use authorizations can speed up approvals. The roadmap focuses on decentralized production, just-in-time strategies and global coordination, and case studies report 40 per cent reductions in waste and greater equity in low-resource environments. Taken together, they show that integrated approaches improve speed, scalability and sustainability enabling overall response times to be reduced up to 90 percent and more than trillion dollars in potential economic losses avoided.

18.9.2 Future Directions

The improvement of the mRNA vaccines will be concentrated in hybrid systems involving a combination of mRNA with self-amplifying RNA leading to lower doses, and incorporating quantum computing to allow ultra-accurate simulations of proteins and design cycles. On-site generation of small, medium and large bioproducts in remote locations will be possible through decentralized bio foundries utilizing 3D-printed facilities and portable bioreactors and maintained through blockchain to provide secure tracking of supply chain. Machine learning and AI will also advance on how to anticipate mutations of pathogens in real-time so that preemptive adjustments to manufacturing can be made, and sustainable manufacturing will also integrate bio-based materials to minimize footprints on the environment. New pathways, such as the use of open-innovation consortia and Bayesian modeling, will tackle challenges of biological variability to speed up development of novel modalities, including gene therapies. Global alliances to eliminate regulatory hurdles will enable easier validations in AI-integrated systems, and collective data warehouses will enable collaborative R&D without impairing intellectual property. These guidelines will ensure biomanufacturing is more predictive, adaptive, and inclusive with the objective that the 100-day mission deploys countermeasures to any potential emerging threat.

18.9.3 Implications for Industry

Rapid biomanufacturing would bring increased competitiveness by facilitating an ability to accelerate market entry of biologics by 20-50 percent allowing lower biologics development costs and limited risk exposure to supply bottlenecks associated with limited regional hubs and diversification. This enables the smooth transfer of technology of biosimilars and the flexible capacity to meet demand peaks to the advantage of both the existing companies and startups in the field of reaching markets internationally. Data-driven platforms make regulatory compliance easier, allowing real-time quality assurance and limited recalls, and benefits such as sustainability enhancements (greener resource use, appeals to investors and fits in with current green movements). The transformation of the workforce to digital and synthetic biology-oriented skills will drive innovation, even the playing field with emerging economies and lead to better availability of affordable countermeasures. The benefits to patients include faster, more equitable supply, which could prevent millions of infections and deaths and increase industry resiliency to future geopolitical and climatic uncertainties in the biopharmaceutical industry.

Abbreviations

AAV	Adeno-Associated Viruses
AdV	Adenoviruses
AI	Artificial Intelligence
C-TAP	COVID-19 Technology Access Pool
CEPI	Coalition for Epidemic Preparedness Innovations
CFES	Cell-Free Expression Systems
CHO	Chinese Hamster Ovary

COVAX	COVID-19 Vaccines Global Access
COVID-19	Coronavirus Disease 2019
CRISPR	Clustered Regularly Interspaced Short Palindromic Repeats
DARPA	Defense Advanced Research Projects Agency
EIDs	Emerging Infectious Diseases
EMA	European Medicines Agency
EUA	Emergency Use Authorization
EUAL	Emergency Use Assessment and Listing
GDPR	General Data Protection Regulation
GMP	Good Manufacturing Practices
H1N1	Influenza A virus subtype H1N1
HERA	Health Emergency Preparedness and Response Authority
HIV/AIDS	Human Immunodeficiency Virus / Acquired Immunodeficiency Syndrome
ICH	International Council for Harmonization
ICMRA	International Coalition of Medicines Regulatory Authorities
IHR	International Health Regulations
IoT	Internet of Things
JIT	Just-in-Time
LNP	Lipid Nanoparticle
mAbs	Monoclonal Antibodies
MPP	Medicines Patent Pool
mRNA	Messenger RNA
PAT	Process Analytical Technologies
PID	Proportional-Integral-Derivative
PPPs	Public-Private Partnerships
QbD	Quality-by-Design
RTPCS	Real-Time Process Control Systems
saRNA	Self-Amplifying RNA
SARS	Severe Acute Respiratory Syndrome
SCADA	Supervisory Control and Data Acquisition
UNESCO	United Nations Educational, Scientific and Cultural Organization
VAERS	Vaccine Adverse Event Reporting System

AI disclosure statement: The author acknowledges the use of AI-based tools for minor language editing or grammatical correction. However, all intellectual contributions, analyses, and conclusions presented in this chapter are entirely the author's own.

References

1. Morens, D.M., Folkers, G.K., Fauci, A.S., The challenge of emerging and re-emerging infectious diseases. *Nature*, 430, 242, 2004.
2. Finlay, B.B., See, R.H., Brunham, R.C., Rapid response research to emerging infectious diseases: lessons from SARS. *Nat. Rev. Microbiol.*, 2, 602, 2004.
3. Hao, R., Liu, Y., Shen, W., Surveillance of emerging infectious diseases for biosecurity. *Sci. China Life Sci.*, 65, 1504, 2022.
4. Excler, J.L., Saville, M., Berkley, S., Kim, J.H., Vaccine development for emerging infectious diseases. *Nat. Med.*, 27, 591, 2021.
5. De Gaetano, S., Ponzo, E., Midiri, A., Mancuso, G., Filippone, D., Infortuna, G., Zummo, S., Biondo, C., Global Trends and Action Items for the Prevention and Control of Emerging and Re-Emerging Infectious Diseases. *Hygiene*, 5, 18, 2025.
6. Pardi, N., Hogan, M.J., Porter, F.W., Weissman, D., mRNA vaccines — a new era in vaccinology. *Nat. Rev. Drug Discov.*, 17, 261, 2018.
7. Naggie, S., Boulware, D.R., Chu, H., Dunsmore, S.E., Gandhi, R.T., Lessons From COVID-19 for Pandemic Preparedness: Proceedings From a Multistakeholder Think Tank. *Clin. Infect. Dis.*, 77, 1635, 2023.
8. Baker, R.E., Mahmud, A.S., Miller, I.F., Rajeev, M., Rasambainarivo, F., Rice, B.L., Takahashi, S., Tatem, A.J., Wagner, C.E., Wang, L.F., Wesolowski, A., Metcalf, C.J.E., Infectious disease in an era of global change. *Nat. Rev. Microbiol.*, 20, 193, 2022.
9. Ghattas, M., Dwivedi, G., Lavertu, M., Alameh, M.G., Vaccine Technologies and Platforms for Infectious Diseases: Current Progress, Challenges, and Opportunities. *Vaccines*, 9, 1490, 2021.
10. Rauch, S., Jasny, E., Schmidt, K.E., Petsch, B., New vaccine technologies to combat outbreak situations. *Front. Immunol.*, 9, 1963, 2018.
11. Okuyama, R., mRNA and Adenoviral Vector Vaccine Platforms Utilized in COVID-19 Vaccines: Technologies, Ecosystem, and Future Directions. *Vaccines*, 11, 1737, 2023.
12. Sun, C.P., Jan, J.T., Wang, I.H., Ma, H.H., Ko, H.Y., Wu, P.Y., Kuo, T.J., Liao, H.N., Lan, Y.H., Sie, Z.L., Chen, Y.H., Ko, Y.A., Liao, C.C., Chen, L.Y., Lee, I.J., Tsung, S.I., Lai, Y.J., Chiang, M.T., Liang, J.J., Liu, W.C., Wang, J.R., Yuan, J.P., Lin, Y.S., Tsai, Y.C., Hsieh, S.L., Li, C.W., Wu, H.C., Ko, T.M., Lin, Y.L., Tao, M.H., Rapid generation of mouse model for emerging infectious disease with the case of severe COVID-19. *PLoS Pathog.*, 17, e1009758, 2021.
13. Lundstrom, K., Viral Vectors in Gene Therapy. *Diseases*, 6, 42, 2018.
14. Franconi, R., Illiano, E., Paolini, F., Massa, S., Venuti, A., Demurtas, O.C., Rapid and Low-Cost Tools Derived from Plants to Face Emerging/Re-emerging Infectious Diseases and Bioterrorism Agents. *Defence Against Bioterrorism*, NATO Science for Peace and Security Series A: Chemistry and Biology, p. 123, 2018.

15. Buchy, P., Buisson, Y., Cintra, O., Dwyer, D.E., Nissen, M., Ortiz de Lejarazu, R., Petersen, E., COVID-19 pandemic: lessons learned from more than a century of pandemics and current vaccine development for pandemic control. *Int. J. Infect. Dis.*, 112, 300, 2021.
16. Lal, A., Ashworth, H.C., Dada, S., Hoemeke, L., Tambo, E., Optimizing Pandemic Preparedness and Response Through Health Information Systems: Lessons Learned From Ebola to COVID-19. *Disaster Med. Public Health Prep.*, 1, 1, 2020.
17. Rosa-Nunes, D., Lucchi, D.B.M., Andreata-Santos, R., Janini, L.M.R., Braconi, C.T., Lessons that can be learned from the SARS-CoV-2 pandemic and their impact on the prophylaxis and treatment development for neglected tropical arboviruses. *Front. Drug Discov.*, 3, 1176768, 2023.
18. Priyadarsini, S.L., Suresh, M., Huisingh, D., What can we learn from previous pandemics to reduce the frequency of emerging infectious diseases like COVID-19? *Global Transitions*, 2, 202, 2020.
19. Piret, J. and Boivin, G., Pandemics Throughout History. *Front. Microbiol.*, 11, 631736, 2021.
20. Lurie, N., Saville, M., Hatchett, R., Halton, J., Developing Covid-19 Vaccines at Pandemic Speed. *N. Engl. J. Med.*, 382, 1969, 2020.
21. Harrison, C., Kamarulzaman, A., Lye, D.C., Accelerating antiviral drug discovery: lessons from COVID-19. *Nat. Rev. Drug Discov.*, 22, 453, 2023.
22. Li, Y.-D., Chi, W.-Y., Su, J.-H., Ferrall, L., Hung, C.-F., Wu, T.-C., Coronavirus vaccine development: from SARS and MERS to COVID-19. *J. Biomed. Sci.*, 27, 104, 2020.
23. Gupta, R., Purcell, L.A., Corti, D., Virgin, H.W., Pandemic preparedness strategies must go beyond vaccines. *Sci. Transl. Med.*, 15, eadd3055, 2023.
24. Chen, W.-H., Strych, U., Hotez, P.J., Bottazzi, M.E., The SARS-CoV-2 Vaccine Pipeline: an Overview. *Curr. Trop. Med. Rep.*, 7, 61, 2020.
25. Coltart, C.E., Lindsey, B., Ghinai, I., Johnson, A.M., Heymann, D.L., The Ebola outbreak, 2013-2016: old lessons for new epidemics. *Philos. Trans. R. Soc. Lond. B Biol. Sci.*, 372, 20160297, 2017.
26. Lal, A., Erondu, N.A., Heymann, D.L., Gitahi, G., Yates, R., Fragmented health systems in COVID-19: rectifying the misalignment between global health security and universal health coverage. *Lancet*, 397, 61, 2021.
27. Pierson, T.C. and Diamond, M.S., The continued threat of emerging flaviviruses. *Nat. Microbiol.*, 5, 796, 2020.
28. Morens, D.M., Taubenberger, J.K., Fauci, A.S., Pandemic influenza viruses: hoping for the road not taken. *N. Engl. J. Med.*, 368, 2345, 2013.
29. Gregorio, N.E., Kao, W.C., Williams, C., Karumuri, B., Nham, P., Levine, S.S., Jewett, M.C., Cell-Free Synthesis: Expediting Biomanufacturing of Chemical and Biomedical Materials. *ACS Synth. Biol.*, 13, 1368, 2024.
30. Bundy, B.C., Hunt, J.P., Koepke, J.C., Jewett, M.C., Swartz, J.R., Cell-free biomanufacturing. *Curr. Opin. Chem. Eng.*, 22, 177, 2018.

31. Perez-Pinera, P., Han, N., Cleto, S., Cao, J., Onal, O., Lu, T.K., Synthetic biology and microbioreactor platforms for programmable production of biologics. *Genes Dev.*, 30, 2241, 2016.
32. Carlson, E.D., Gan, R., Hodgman, C.E., Jewett, M.C., Cell-free protein synthesis: applications come of age. *Biotechnol. Adv.*, 30, 1185, 2012.
33. Pardee, K., Green, A.A., Takahashi, M.K., Braff, D., Lambert, G., Lee, J.W., Ferrante, T., Ma, D., Donghia, N., Fan, M., Daringer, N.M., Bosch, I., Dudley, D.M., O'Connor, D.H., Gehrke, L., Collins, J.J., Rapid, Low-Cost Detection of Zika Virus Using Programmable Biomolecular Components. *Cell*, 165, 1255, 2016.
34. Garenne, D., Thompson, S., Behera, A., McCarthy, C., Noireaux, V., Cell-Free Transcription-Translation: Engineering Biology from the Nanometer to the Millimeter Scale. *Curr. Opin. Biotechnol.*, 58, 19, 2019.
35. Silverman, A.D., Karim, A.S., Jewett, M.C., Cell-free gene expression: an expanded repertoire of applications. *Nat. Rev. Genet.*, 21, 151, 2020.
36. Garenne, D., Haines, M.C., Romantseva, E.F., Seelig, B., Sparkman-Yager, D., Noireaux, V., Cell-free TX-TL synthesis of infectious bacteriophage T4 in a single test tube reaction. *Synth. Biol.*, 3, ysy002, 2018.
37. Gregorio, N.E., Levine, M.Z., Oza, J.P., A User's Guide to Cell-Free Protein Synthesis. *Methods Protoc.*, 2, 24, 2019.
38. Hunt, J.P., Yang, S.O., Wilding, K.M., Bundy, B.C., The Growing Impact of Lyophilized Cell-free Protein Expression Systems. *Bioengineered*, 8, 325, 2017.
39. Cubillos-Ruiz, A., Guo, T., Sokolovska, A., Miller, P.F., Sanchez-Garcia, J.L., Hrncirova, P., Gitai, Z., Noireaux, V., Collins, J.J., Silver, P.A., Engineering living therapeutics with synthetic biology. *Nat. Rev. Drug Discov.*, 20, 941, 2021.
40. Kitada, T., DiAndreth, B., Teague, B., Weiss, R., Programming gene and engineered-cell therapies with synthetic biology. *Science*, 359, eaad1067, 2018.
41. Khalil, A.S. and Collins, J.J., Synthetic biology: applications come of age. *Nat. Rev. Genet.*, 11, 367, 2010.
42. Slomovic, S., Pardee, K., Collins, J.J., Synthetic biology devices for *in vitro* and *in vivo* diagnostics. *Proc. Natl. Acad. Sci. U.S.A.*, 112, 14429, 2015.
43. Brodel, A.K., Sonnabend, A., Kubick, S., Cell-free protein expression based on extracts from CHO cells. *Biotechnol. Bioeng.*, 111, 25, 2014.
44. Meng, F. and Ellis, T., The second decade of synthetic biology: 2010-2020. *Nat. Commun.*, 11, 5174, 2020.
45. Zydney, A.L., Perspectives on integrated continuous biomanufacturing—opportunities and challenges. *Curr. Opin. Chem. Eng.*, 10, 8, 2015.
46. Park, J.W., Lagniton, P.N.P., Liu, Y., Xu, R.H., mRNA vaccines for COVID-19: the promise and pitfalls. *J. Leukocyte Biol.*, 109, 353, 2021.
47. Pardi, N., Hogan, M.J., Porter, F.W., Weissman, D., mRNA vaccines - a new era in vaccinology. *Nat. Rev. Drug Discov.*, 17, 261, 2018.

48. Verbeke, R., Lentacker, I., De Smedt, S.C., Dewitte, H., Three decades of messenger RNA vaccine development. *Nano Today*, 28, 100766, 2019.
49. Bettini, E. and Locci, M., SARS-CoV-2 mRNA Vaccines: Immunological Mechanism and Beyond. *Vaccines (Basel)*, 9, 147, 2021.
50. Chaudhary, N., Weissman, D., Whitehead, K.A., mRNA vaccines for infectious diseases: principles, delivery and clinical translation. *Nat. Rev. Drug Discov.*, 20, 817, 2021.
51. Hou, X., Zaks, T., Langer, R., Dong, Y., Lipid nanoparticles for mRNA delivery. *Nat. Rev. Mater.*, 6, 1078, 2021.
52. Maruggi, G., Zhang, C., Li, J., Ulmer, J.B., Yu, D., mRNA as a Transformative Technology for Vaccine Development to Control Infectious Diseases. *Mol. Ther.*, 27, 757, 2019.
53. Vrba, S.M., Kirk, N.M., Brisse, M.E., Liang, Y., Ly, H., Development and Applications of Viral Vectored Vaccines to Combat Zoonotic and Emerging Infectious Diseases. *Mol. Ther. Methods Clin. Dev.*, 20, 134, 2021.
54. Ewer, K.J., Lambe, T., Rollier, C.S., Spencer, A.J., Hill, A.V., Dorrell, L., Viral vectors as vaccine platforms: from immunogenicity to impact. *Curr. Opin. Immunol.*, 41, 47, 2016.
55. Sebastian, S. and Lambe, T., Clinical Advances in Viral-Vectored Influenza Vaccines. *Vaccines (Basel)*, 6, 29, 2018.
56. Ura, T., Okuda, K., Shimada, M., Developments in Viral Vector-Based Vaccines. *Vaccines (Basel)*, 2, 624, 2014.
57. Walker, B.D. and Burton, D.R., Toward an AIDS vaccine. *Science*, 320, 760, 2008.
58. Walker, L.M. and Burton, D.R., Passive immunotherapy of viral infections: 'super-antibodies' enter the fray. *Nat. Rev. Immunol.*, 18, 297, 2018.
59. Salazar, G., Zhang, N., Fu, T.M., An, Z., Antibody therapies for the prevention and treatment of viral infections. *npj Vaccines*, 2, 19, 2017.
60. Corti, D. and Lanzavecchia, A., Broadly neutralizing antiviral antibodies. *Annu. Rev. Immunol.*, 31, 705, 2013.
61. Kaplon, H. and Reichert, J.M., Antibodies to watch in 2021. *MAbs*, 13, 1860476, 2021.
62. Saez-Cirion, A. and Sereti, I., Immunometabolism and HIV-1 pathogenesis: food for thought. *Nat. Rev. Immunol.*, 21, 5, 2021.
63. Lu, L.L., Suscovich, T.J., Fortune, S.M., Alter, G., Beyond binding: antibody effector functions in infectious diseases. *Nat. Rev. Immunol.*, 18, 46, 2018.
64. Taylor, P.C., Adams, A.C., Hufford, M.M., de la Torre, I., Winthrop, K., Gottlieb, R.L., Neutralizing monoclonal antibodies for treatment of COVID-19. *Nat. Rev. Immunol.*, 21, 382, 2021.
65. Marston, H.D., Paules, C.I., Fauci, A.S., Monoclonal Antibodies for Emerging Infectious Diseases - Borrowing from History. *N. Engl. J. Med.*, 378, 1469, 2018.

66. Graham, B.S., Mascola, J.R., Fauci, A.S., Novel Vaccine Technologies: Essential Components of an Adequate Response to Emerging Viral Diseases. *JAMA*, 319, 1431, 2018.
67. Adair, A., Barrett, L., Burnett, M., Lopez, M.F., Metcalfe, M., Dey, O., DeStefano, A., Ciccarone, T., A Platform Approach to Cleavage-Independent High Throughput Production of Recombinant Lentiviral Vectors. *Mol. Ther. Methods Clin. Dev.*, 18, 13, 2020.
68. Gouglas, D., Christodoulou, M., Plotkin, S.A., Hatchett, R., CEPI: Driving Progress Toward Epidemic Preparedness and Response. *Epidemiol. Rev.*, 41, 28, 2019.
69. Kis, Z., Kontoravdi, C., Dey, A.K., Shattock, R., Shah, N., Rapid development and deployment of high-volume vaccines for pandemic response. *J. Adv. Manuf. Process.*, 2, e10060, 2020.
70. Kelley, B., Industrialization of mAb production technology: the bioprocessing industry at a crossroads. *MAbs*, 1, 443, 2009.
71. Zydney, A.L., Perspectives on integrated continuous biomanufacturing - opportunities and challenges. *Curr. Opin. Chem. Eng.*, 10, 8, 2015.
72. Struck, M.M., van der Helm, E., Velders, A.H., Hickman, J.J., The prospect of rapid response vaccines: The promise of platform technologies. *Vaccine*, 40, 6475, 2022.
73. Fraser-Hevlin, B., Schuler, A.W., Gozen, B.A., Van Wie, B.J., Using Digital Twins to Protect Biomanufacturing from Cyberattacks. *Mil. Cyber Aff.*, 7, 1, 2024.
74. Kreuzer, T., Papapetrou, P., Zdravkovic, J., Artificial intelligence in digital twins—A systematic literature review. *Data Knowl. Eng.*, 151, 102304, 2024.
75. Alfaro-Viquez, D., Zamora-Hernandez, M., Fernandez-Vega, M., Garcia-Rodriguez, J., Azorin-Lopez, J., A Comprehensive Review of AI-Based Digital Twin Applications in Manufacturing: Integration Across Operator, Product, and Process Dimensions. *Electronics*, 14, 646, 2025.
76. Homaei, M., Mogollón-Gutiérrez, O., Sancho Núñez, J.C., Ávila Vegas, M., Caro Lindo, A., A review of Digital Twins and their application in Cybersecurity based on Artificial Intelligence, arXiv preprint arXiv:2311.01154, 2023.
77. Naqvi, S., Ghufran, M., Meraghni, S., Varnier, C., Nicod, J., Zerhouni, N., Human knowledge centered maintenance decision support in digital twin environment. *J. Manuf. Syst.*, 65, 528, 2022.
78. Zhang, L., Liu, J., Zhuang, C., Digital Twin Modeling Enabled Machine Tool Intelligence: A Review. *Chin. J. Mech. Eng.*, 37, 47, 2024.
79. Liu, J., Wen, X., Zhou, H., Sheng, S., Zhao, P., Liu, X., Kang, C., Chen, Y., Digital twin-enabled machining process modeling. *Adv. Eng. Inform.*, 54, 101737, 2022.
80. Wen, D.L.W., Soon, H.G., Kumar, A.S., Manufacturing Letters Micro-milling digital twin for real-time tool condition monitoring. *Manuf. Lett.*, 41, 1231, 2024.

81. Bonci, A., Biase, A., Giannini, M., Longhi, S., Prist, M., Digital twin architecture for assembly line performance monitoring. *Procedia Comput. Sci.*, 232, 1089, 2024.
82. Pantelidakis, M., Mykoniatis, K., Liu, J., Harris, G., A digital twin ecosystem for additive manufacturing using a real-time development platform. *Int. J. Adv. Manuf. Technol.*, 120, 6547, 2022.
83. Arce, A., Matute, T., Gordillo-Marroquín, C., Valencia, D., Sattler, J.M., Enriquez, A., Jung, H., Czerkinsky, C., Velasco-Velásquez, M.A., Decentralizing Cell-Free RNA Sensing With the Use of Low-Cost Cell Extracts. *Front. Bioeng. Biotechnol.*, 9, 727584, 2021.
84. Bartlett, M.L., Bright, N.A., Mallinson, A., Blaxill, Z., Hayles, J., Enhancing the response to avian influenza in the US and globally. *Lancet Reg. Health Am.*, 46, 101100, 2025.
85. Martí-Carreras, J., Generous, N., Giammanco, M.D., Kalter, J., Keiser, P.H., Kunkel, S.R., Longmire, M., Merkel, S.F., Perkins, D.J., Fair, J.M., NCBI's Virus Discovery Codeathon: Building "FIVE" -The Federated Index of Viral Experiments API Index. *Viruses*, 12, 1424, 2020.
86. Crone, M.A. and Freemont, P.S., Thinking beyond pathogen surveillance: building resilient biotech ecosystems to combat the next pandemic. *Front. Sci.*, 2, 1412291, 2024.
87. Feddema, J.J., Fernald, K.D.S., Schikan, H.G.C.P., van de Burgwal, L.H.M., Upscaling vaccine manufacturing capacity - key bottlenecks and lessons learned. *Vaccine*, 41, 4359, 2023.
88. Koyama, A., Tran, V.C., Fujimoto, M., Bao, V.N.Q., Tran, T.H., A Decentralized COVID-19 Vaccine Tracking System Using Blockchain Technology. *Cryptography*, 7, 13, 2023.
89. Wamba, S.F., Queiroz, M.M., Trinchera, L., Blockchain-based solution for COVID-19 vaccine waste reduction. *J. Clean Prod.*, 365, 132619, 2022.
90. Ravindran, R., Szadkowski, L., Lovblom, L.E., Clarke, R., Huang, Q.W., Manase, D., Colwill, K., Gingras, A.C., McGeer, A.J., Arnason, T., Decentralized study of COVID Vaccine Antibody Response (STOPCoV): Results of a participant satisfaction survey. *PLOS Digital Health*, 2, e0000242, 2023.
91. Ghane, M., Rezaei, F., Amiri, F., Lio, P., An integrated blockchain-enabled multi-channel vaccine supply chain network under hybrid uncertainties. *Sci. Rep.*, 14, 16187, 2024.
92. Kis, Z., Papathanasiou, M., Calvo-Serrano, R., Kontoravdi, C., Shah, N., A model-based quantification of the impact of new manufacturing technologies on developing country vaccine supply chain performance: A Kenyan case study. *Vaccine*, 37, 3527, 2019.
93. Papathanasiou, M. and Kontoravdi, C., Engineering challenges in therapeutic protein product and process design. *Curr. Opin. Chem. Eng.*, 27, 81, 2020.
94. Levine, H.L., Lilja, J.E., Stock, R., Hummel, H., Jones, S.D., Efficient, flexible facilities for the 21st century. *Bioprocess Int.*, 10, 20, 2012.

95. Halabi, S.F., Gostin, L.O., Crowley, J.S., *Global Management of Infectious Disease After Ebola*, Oxford University Press, Oxford, UK, 2016.
96. Bump, J.B., Baum, F., Clay, R., Mazzola, F., McKee, M., Political economy of covid-19: extractive capitalism and uneven states of extraction. *BMJ*, 375, e067707, 2021.
97. Bump, J.B., Friberg, P., Harper, D.R., International collaboration and covid-19: what are we doing and where are we going? *BMJ*, 372, n180, 2021.
98. Bump, J.B., Moser, F., Pallares, N., Generation, deployment, and use of AI in global health. *BMJ Glob. Health*, 8, e012899, 2023.
99. Bump, J.B., Reddiar, S.K., Soucat, A., When do governments support common goods for health? Four cases on surveillance, traffic congestion, air pollution, and universal health coverage in health policy. *Health Syst. Reform*, 5, 289, 2019.
100. Bump, J.B., Sekalala, S., Nilo, A., Moser, F., Gostin, L.O., International collaboration for global accessibility of medicines in least developed countries: Implications from a delayed review of Bangladesh's law on patents. *J. Law Med. Ethics*, 51, 182, 2023.
101. Bump, J.B., Global health and universal health coverage: The changing landscape for nations and global health citizens, in: *Handbook of Global Health*, R. Haring, I. Kickbusch, D. Ganten, M. Moeti (Eds.), Springer, Cham, 2021.
102. Bump, J.B., Health diplomacy in pandemics: A bibliometric analysis. *J. Glob. Health*, 13, 04030, 2023.
103. Bump, J.B., Pallares, N., Moser, F., International collaboration for global accessibility of biosimilars in low- and middle-income countries: Assessment of the World Health Organization Prequalification Program. *J. Glob. Health*, 14, 04017, 2024.
104. Bump, J.B., Cash-Goldwasser, S., Sekalala, S.A., rights-based argument for international recognition of health as a trademark of global cities. *BMJ Glob. Health*, 5, e002360, 2020.
105. Bump, J.B., The Crimean War: Emergence of the public as actor in health policy, in: *Health Promotion and the Policy Process*, C. Clavier and E. de Leeuw (Eds.), Oxford University Press, Oxford, 2013.
106. FDA, Expedited Programs for Serious Conditions – Drugs and Biologics, U.S. Food and Drug Administration, 2014.
107. Kieny, M.P. and Salama, P., WHO R&D Blueprint: a global coordination mechanism for R&D preparedness. *Lancet*, 389, 2469, 2017.
108. Lurie, N. and Hatchett, R., The Public-Private Partnership for Influenza Vaccine. *Science*, 312, 382, 2006.
109. Wilder-Smith, A., Gubler, D.J., Weaver, S.C., Monath, T.P., Heymann, D.L., Scott, T.W., Epidemic arboviral diseases: priorities for research and public health. *Lancet Infect. Dis.*, 17, e101, 2017.
110. Gostin, L.O., Tomori, O., Wibawa, O.T., Monath, T.P., Nguku, P.M., Erbar, L., Feldmann, H., Volchkov, V., Ramvikas, M., Bodmer, B.S., Friede, M., Bahl,

J., Toward a common secure future: four global commissions in the wake of Ebola. *PLoS Med.*, 13, e1002042, 2016.

111. Moon, S., Sridhar, D., Pate, M.A., Jha, A.K., Clinton, C., Delaunay, S., Edwin, V., Fallah, M., Fidler, D.P., Garrett, L., Goosby, E., Gostin, L.O., Heymann, D.L., Lee, K., Leung, G.M., Morrison, J.S., Saavedra, J., Tanner, M., Leigh, J.A., Phumaphi, J., Piot, P., Reddy, K.S., Will Ebola change the game? Ten essential reforms before the next pandemic. The report of the Harvard-LSHTM Independent Panel on the Global Response to Ebola. *Lancet*, 386, 2204, 2015.
112. Sands, P., Mundaca-Shah, C., Dzau, V.J., The Neglected Dimension of Global Security — A Framework for Countering Infectious-Disease Crises. *N. Engl. J. Med.*, 374, 1281, 2016.
113. Heymann, D.L., Chen, L., Takemi, K., Fidler, D.P., Tappero, J.W., Thomas, M.J., Kenyon, T.A., Frieden, T.R., Yach, D., Nishtar, S., Kalache, A., Olliaro, P.L., Horby, P., Torreele, E., Gostin, L.O., Ndomondo-Sigonda, M., Carpenter, D., Rushton, S., Lillywhite, L., Rawaf, S., Castro, J.L., Burrows, H., Kim, H., Jha, A., Lang, T., Panjabi, R., Balsari, S., Sauer, L., Reddy, K.S., Shahpar, Z., Wojcicki, J.M., Wibulpolprasert, S., Peto, J., Nuzzo, J., Alleyne, G., Bsrat, A., Berkley, S., Dybul, M., Piot, P., Global health security: the wider lessons from the west African Ebola virus disease epidemic. *Lancet*, 385, 1884, 2015.
114. Da Silva, R.E., Novaes, M.R.C.G., Pastor, A.F., Grisi-Filho, J.H.H., Amaku, M., The cost-benefit of federal investment in preventing Ebola virus disease in West Africa. *Lancet Infect. Dis.*, 15, 1253, 2015.
115. Meng, F. and Ellis, T., The second decade of synthetic biology: 2010–2020. *Nat. Commun.*, 11, 5174, 2020.
116. Wang, L., Jiang, S., Deng, Z., Synthetic genomics: from genetic parts to genomes. *FEMS Microbiol. Rev.*, 44, 1, 2020.

19

In-Line Monitoring and Advanced Control Strategies in Downstream Bioprocessing of Biologics

V. Sravani Sameera[1], Kamandla Manasa[1], K.S.N.V. Prasad[1], Bhanu Radhika[1], Karanam Hemanth Kumar[1], Appala Naidu Uttaravalli[2] and A.V. Raghavendra Rao[1]*

[1]*Department of Chemical Engineering, B. V. Raju Institute of Technology, Narsapur, Telangana, India*
[2]*Optimus Drugs Private Limited (A Sekhmet Pharma Ventures Private Limited Company), Hyderabad, Telangana, India*

Abstract

Downstream processing is essential to the manufacture of biologics and directly impacts the product quality, safety and process efficiency. As biopharmaceuticals become increasingly complex and new requirements involving sterile manufacturing and highly compliant and scalable manufacturing systems take precedence, in-line monitoring and advanced control strategies have also proved to be an effective way of reshaping the field. This chapter explores the philosophy behind the techniques and technologies and practical realization of real-time analytical applications and control systems of downstream bioprocessing. It discusses Process Analytical Technology (PAT), spectroscopy-based-sensors (e.g., NIR, Raman, UV-Vis), and soft-sensor models to ensure that processing conditions get controlled to monitor critical quality attribute (CQA) in purification, filtration, and formulation. Model predictive control (MPC), feedback/feedforward models and adaptive control routines are given particular attention to be able to optimize process parameters in real-time. The paper also covers the cases studied in the industry with increased productivity, decreased failure in a batch, and the possibility of the attainment of regulations with Quality by Design (QbD) systems. The issues of sensor calibration, data integration and process validation are discussed, and the way forward with machine learning and digital twins. In-line monitoring

**Corresponding author*: raghusrf@gmail.com

Sandip Kunar, Jagadeesha T. and D. Sathis Kumar (eds.) Biomanufacturing: Transforming Production of Biologics and Biopharmaceuticals, (603–640) © 2026 Scrivener Publishing LLC

supports the next generation of downstream biomanufacturing of biologics and biopharmaceuticals that is both intelligent, efficient, and quality-centric due to data-driven decision-making and process control in real-time.

***Keywords*:** In-line monitoring, downstream bioprocessing, process analytical technology (PAT), biologics manufacturing real-time control, model predictive control

19.1 Introduction to Downstream Bioprocessing of Biologics

Representing a final stage in the manufacture of biologics, downstream bioprocessing includes purifying, recovering and formulating therapeutic proteins, monoclonal antibodies and vaccines, and other biopharmaceuticals in complex mixtures produced in upstream fermenters or cell cultures. This step is crucial in making sure the end product fulfills high grades of quality of safety, efficacy and purity since, being highly sensitive, biologics are susceptible to any deviation in processes that may alter their structural form and biological activity [1]. As the global biopharmaceutical industry is estimated to achieve over 500 bn USD by 2025, with increased developments in the personalized medicine and Gene therapy sectors, there has been a need to increase the efficiency of the downstream processes. Conventional batch processes are usually inefficient, characterized by the high cost, time consumption, variable product quality which may result in a batch failure and regulatory delays [2]. In-line monitoring and advanced control strategies solutions will have these issues covered because they provide real-time data collection and dynamic modifications, which turns downstream processes into more forecastable and optimized processes. In-line monitoring has the monitoring of process parameters and critical quality attributes (CQAs) with sensors and analytical instruments continually assessed; advanced controls rely on algorithms and allow process stability and high yields to be maintained. It is the aim of this introduction to present the background of downstream bioprocessing focusing on the trend to intelligent manufacturing paradigms, which use the principles of Process Analytical Technology (PAT) and Quality by Design (QbD) to fulfil regulatory requirements and market needs [3].

Because of the complexity of biologics (including large molecular size, heterogeneity, and susceptibility to degradation), highly complex downstream methods may be required to purify the target molecule against the presence of impurities, including host cell proteins, DNA, aggregates and endotoxins.

With a move towards continuous processing in biomanufacturing to enhance throughput and footprint, it is important to pay special consideration to effective monitoring and control to avoid deviations which might affect consistency of the product [4]. This chapter is an attempt to give a broad-based background towards comprehending these developments, as a prelude in the further venture into particular technologies and their applications.

19.1.1 Overview of Downstream Processing Stages

Downstream is usually broken down into multiple interconnected processes, harvest and clarification, purification, polishing and formulation/filling. Separation of the biologic and cell culture broth is done during the harvest stage usually through centrifugation or depth filtration to eliminate cells and debris, which is highly important towards preventing fouling in the next stage [5]. This clarification gives desired clear feed stream, whereas newer methods enable the use of flocculation or acoustic waves separation to increase efficiency and minimize damage to sensitive molecules under high shear.

The essence of downstream processing is purification, which uses chromatography (affinity, ion-exchange, and size-exclusion) to reach high levels of purity of generally greater than 99% in the case of monoclonal antibodies. Protein A based affinity chromatography on Protein A resins is popular as an initial capture method, as it is specific in nature; however, it is expensive and can suffer ligand leaching in which case it requires close attention [6]. The polishing steps can eliminate remaining impurities by hydrophobic interaction, or mixed mode chromatography to assure the product complies with pharmacopeia aggregate and charge variant specifications.

The last formulation and filling step stabilize the biologic by buffer exchanging, concentrating by ultrafiltration and sterile filling where excipients such surfactants in the biologic are added to prevent aggregation during storage. The stages are mutually dependent yet differences experienced in any of them are capable of propagating errors, hence the idea of an integrated monitoring that ensures the process integrity [7]. Multi column chromatography and single pass tangential flow filtration to allow continuous downstream processing have both cost advantages of productivity, and new control requirements to manage dynamic flows and concentrations.

Embedded in-line monitoring devices, UV-Vis to measure protein concentration or conductivity probes to determine buffer gradients, can be used at key stages, to supply instant feedback and provide opportunities to adjust parameters to reduce off-spec product. This outline quite underscores how downstream opportunities represent a continuum of effectiveness such that gains in one place increase the performance across the board, which leads to showing quality imperatives [8].

19.1.2 The Value of Quality and Efficiency in the Manufacturing of Biologics

Uninterrupted quality and efficiency in biological manufacturing is critical, since the products will be used on patients with conditions that are life-threatening, and any form of contamination, which includes the slightest impurity, can cause immune reactions or hamper the effectiveness of the therapeutic benefits [9]. The agencies regulatory, such as FDA and EMA, require application of current Good Manufacturing Practices (cGMP), where QbD ensures that the understanding of the processes is optimized to determine the CQAs such as purity, and potency, and stability early in the development process so as to design a robust process.

The cost-of-goods is immediately affected by efficiency and downstream processing involves 50-80 percent of overall manufacturing cost because of the cost of resin, buffer utilization, and cost of labor. Poor efficiency causes low results (normally <70% recovery in chromatography), prolonged cycle times and enhances the downside of environmental footprints in a sector under stress to be sustainable [10]. These are countered by advanced monitoring which may detect deviations in real-time, e.g. pH dynamics during ultrafiltration processes that lead to protein unfolding and enable prompt action to save batches.

It takes priority with greater prominence in the aspect of global health crises, such as the COVID-19 pandemic, where the rapid scale-up of vaccines required downstream operations at scale to provide billions of doses without known quality sacrifice. A cumbersome purification process slowed down distributions, which illustrates the capacity of in-line equipment such as Raman spector/aggregate detector in increasing throughput by loading the column optimally [1]. In addition, efficiency leads to innovation in continuous manufacturing where under PAT, the steps can be easily transferred without an increase in hold times or the footprint of the facility by as much as 50%.

Quality surveillance will avoid recalls that cost the industry billions of dollars every year, and it facilitates personalized medicine with the ability to guarantee batch-to-batch reproducibility, given that in the world of personalized medicines, therapies are administered on a patient-by-patient basis. Diversified biologics into bispecific antibodies and cell therapies, quality and efficiency are applied as competitive differentiators because they support the use of digital twins in virtual process optimization [2]. In this section it can be highlighted that investment in advanced strategies is not to circumvent regulations but it only serves to support economic viability and patient outcomes.

19.1.3 Evolution of In-Line Monitoring and Advanced Controls

FDA-promoted PAT initiative and the development of in-line monitoring and multi-variable controls in the downstream part of the bioprocessing evolution can be traced back to the early 2000s due to the implementation of the above [3]. Earlier monitoring was based on off-line techniques such as HPLC in monitoring purity that was time consuming and only fractions of batches were monitored resulting into unidentified variations.

This became possible as during the 2010s spectroscopy tools; e.g. NIR to measure moisture content in lyophilization and Raman to characterize molecular fingerprint, became available and could be used in-line and are non-destructive [4]. The period also brought in soft sensors, through which the multivariate data will be utilized together with models to facilitate the prediction of the CQAs indirectly to minimize the presence of intrusive probes.

The late 2010s saw the advancement of such basic PID loops to MPC; predictive algorithms are used to estimate process pathways, and act on variables in a proactive manner, e.g. in chromatography where flow rates are maximized using resin binding kinetics [5]. QbD combined such evolution, making it mandatory to have design spaces, which integrate monitoring information to achieve robust operations.

After 2020, industry 4.0 driven by digitalization incorporated the use of AI to form adaptive controls that learn earlier what has occurred to manage a non-linear behavior of bioprocesses [6]. It was possible to carry out continuous processing, such as addition of in-line LC as a method of continuous impurity monitoring [7]. Issues such as sensor fouling were overcome by creating self-cleaning designs and the introduction of cybersecurity with networked systems.

This has been a paradigm shift in response to a reactive to predictive manufacturing due to the encouragement of these regulatory bodies and due to maturity of technology and this will be the basis of intelligent bioprocessing [8].

19.2 Principles of In-Line Monitoring in Downstream Processes

Process variables and product attributes can be monitored in-line in downstream bioprocessing as they are measured in continuous, real time, but in the production stream instead of samples being removed followed by off-line evaluation [9]. This is a basic approach to process robustness, since the deviations as measured would be detected immediately and corrective measures put in place before the purity, yield, or safety of biologics like monoclonal antibodies and vaccines get affected [1, 10]. Chromatography, filtration, and formulation are all downstream processes that are naturally compounded by the natural heterogeneity of biologic mixtures, in which impurities such as host cell proteins or aggregates may co-elute with the target molecule, thus requiring fine control. The in-line monitoring principles are established by the necessity to trace the priority process parameters (CPPs) and prior quality attributes (CQAs) to adjust the data with the help of correct regulatory guidance, such as the FDA and EMA [3]. In-line monitoring introduces sensors into the process stream and analytical tools that minimize variability and shorten cycle times and overall efficiency as the process moves to predictive and knowledge-based operations replacing conventional empirical methods. In this section, the principles will be discussed beginning with the CQAs during purification, real-time data acquisition, integration with Quality by Design (QbD) and the nature of difficulties associated with monitoring these complex systems [11].

The in-line monitoring has gained its relevance in the biopharmaceutical industry due to the demand of continuous manufacturing in which inconsistencies in batches are eliminated due to constant monitoring. As an example, in chromatography, one may track elution profiles in real-time thereby avoiding losses of the product at inadequate gradient conditions, maintaining recovery rates of consistently more than 90 percent [6]. With the scaleup of biologics production to worldwide demands, such as biosimilars and advanced therapies, these concepts are becoming essential to stay competitively and compliant.

19.2.1 Critical Quality Attributes (CQAs) in Biologics Purification

Critical quality attributes (CQAs) are physical, chemical, biological or microbiological characteristics that are controlled at a given range to guarantee a product quality as desired in purification of biologics [12].

Examples of CQAs in downstream processes are purity (e.g., aggregates >99% removed), potency (e.g., ligand binding affinity), and stability (e.g., no degradation during storage), and these are directly related to therapeutic efficacy and patient safety. These attributes are also the focus of a purification step such as an affinity chromatography process and the presence of impurities is eluted as the biologic is selectively bound, although the product composition may vary with feed conditions, so in-line monitoring of attributes such as protein concentration and charge variants is important [5].

CQAs are identified by performing risk assessments under QbD in which attributes are prioritized in terms of their effects on safety and efficacy; glycosylation patterns are a high-risk CQA since it determines the rate of immunogenicity in case of monoclonal antibodies [1]. Real-time CQA can be measured definitively by in-line tools (e.g. multi-angle light scattering to detect aggregates) which would enable operators to modify parameters (e.g. pH value or flow rates) to remain within specifications. This is a proactive manner as it decreases the dependency on end-point testing that typically finds problems too late resulting in as high as 10-20 percent batch rejections in conventional environments [3]. Examples of cases produced using insulin as an example indicate the importance of monitoring the CQAs to achieve uniformity in the size of the crystals, which is imperative to bioavailability [11]. Altogether, the emphasis on CQAs during purification concepts allows ensuring defined and high-quality biologics downstream, therefore conquering risks as applied to clinical uses.

19.2.2 Real-Time Data Acquisition and Sensor Technologies

The in-line monitoring is based on real-time data acquisition where sensor technologies are used to capture process data in real-time to make informed decisions in downstream bioprocessing [7]. There are sensors like conductivity probes of ionic strength in ion-exchange chromatography or the pressure transducers of filtration integrity, to offer continuous data, to spot the abnormalities or fault such as the membrane fouling that can be detected early before it has a complete explosion or catastrophe [6]. Through advanced sensors, such as spectroscopic devices such as near-infrared (NIR) to measure moisture content or Raman to measure molecular composition, non-invasive analyses of the process stream can be obtained with sampling intervals as rapid as every second to capture fluid dynamic changes [8].

This principle is based on combining hardware and software to log data and visualize it, frequently using supervisory control and data acquisition (SCADA) systems that collect data arriving at several sensors. In an example like ultrafiltration, turbidity sensors can ensure real-time retentate clarity by regulating transmembrane pressure, setting the most appropriate flux and avoiding gel layer formation [12]. Soft sensors are an expansion of capabilities in the sense that they are physical measurements combined with mathematical equations that can predict correlated unmeasurable variables such as the level of impurities present [5]. This technology minimizes human involvement leading to a decrease in labor costs of up to 30 per cent and enhancing reproducibility of the process. Nonetheless, the placement of the sensor should be based on the flow so that dead zones are not created in order to project the whole stream [9]. With real-time acquisition, operators have insightful information to act on and move toward automated, resilient bioprocessing.

19.2.3 Integration with Quality by Design (QbD) Frameworks

Combination of in-line monitoring with Quality by Design (QbD) frames incorporates scientific knowledge in the procedure advancement, characterizing a plan space where CPPs are managed to guarantee CQAs in further results-based well-prepared processing [10]. The principles of QbD described in ICH Q8-Q10 regulations need to have early risk identification and monitoring efforts to determine that processes performance is within acceptable limits [7]. In-line instruments support this by presenting information that can be used to conduct multivariate analysis showing the relationship between a factor like temperature and the properties of a molecule like the stability of protein in formulation [8].

As an example of this, QbD has been used in chromatography where NIR sensors are used in real-time monitoring of the resin behavior and elution buffers are adjusted during the run, to ensure they remain within the design space, rather than allowing a drop in the yield [11]. This integration allows risk-based procedures with high-impact CQAs being more monitored, and contra, less validation is required, and the regulatory approvals are faster [6]. Predicted patterns in soft-sensor models in QbD would use historical data and allow predictions of future outcomes to avoid downtimes and predictive maintenance [12]. The problems of data overload during integration, and other integration related problems are solved with the help of chemometric tools including dimensionality reduction principals (PCA) [5]. Combined with QbD, bioprocesses gain stronger robustness, with negative results pointing to a 25 percent increase in the process capability indices [9]. This synergy enhances a culture of relentless improvement, a key aspect to complex biologics such as cell therapies.

19.2.4 Challenges in Monitoring Complex Bioprocesses

The downstream processes involving complex bioprocess require needs monitoring due to inherent variability of biological systems, inability of sensors to capture the wide range of variability and data management problems [10]. The heterogeneity of biologics and the variation in the post-translational modifications between batches makes it difficult to track CQAs properly since sensors might not see subtle isoform variation with low-resolution [1]. High flow rates may cause sensor drift or fouling in continuing processes that give inaccurate readings and false alarms causing interruption of the operations [3].

The collection of data by various sensors is another challenge where incompatible format necessitates the high-handedness of middleware raising the costs of implementation by 20-40% [7]. Regulatory issues involve the need to confirm in-line testing to off-line testing as agencies require the demonstration of equivalence of real-time release testing [8]. Membrane interaction with biologics may change the sensor signals, thus requiring frequent feedback recalibration in filtration [6]. The possible remedies include sturdy sensor designs, including self-cleaning probes, and AI-based anomaly detection to screen noise [12]. Cloud-based analytics alleviates scalability problems with the large-scale facilities where the volume of data can be increased by the sensor arrays [5]. Notwithstanding these limitations, effective solutions *via* hybrid monitoring practices are likely to deliver credible feedback on quality control as industry documentations show that variability has been reduced by 15% to less than 5% [9].

19.3 Process Analytical Technology (PAT) Tools and Sensors

Process Analytical Technology (PAT) is a modular set of tools and sensors intended to enable real-time monitoring and control in the downstream bioprocess to assure both consistent product quality and process efficiency of biologics. According to FDA, PAT is the design, analysis, and control of manufacturing by timely measurements of important quality and performance characteristics, which allow the industry to transition the traditional quality control role to quality assurance [3]. Further downstream, during the purification and formulation of biologics, PAT tools such as spectroscopy-based sensors offer non-destructive in-line process visibility that may be used to identify fluctuation in any of the attributes of the protein, such as concentration or level of contaminants, that can affect the final product quality [1].

Here the major components of PAT are discussed starting with spectroscopy sensors, the soft-sensor models, applications in and across processes and calibration/validation routines. Within the context of PAT, biomanufacturers have been able to decrease the amount of variance in their processes and reports indicate a median of 40 percent increase in yield and the rate of failure of batches decreasing by 15 to less than 5 percent [11].

PAT application follows trends in continuous processing, in which instruments ought to be able to support the flow with high throughput requirements. As an example, production of monoclonal antibodies incorporates PAT sensors to monitor the chromatography eluents, and to optimize eluent parameters to improve the separation process [13]. With the growing demand of complex molecules such as bispecifics and gene therapies putting pressure on biopharmaceutical industry, the ability of supporting a data heavy environment using PAT becomes the key in achieving compliance and scalability.

19.3.1 Spectroscopy-Based Sensors: NIR, Raman, and UV-Vis

Examples of Sensors based on spectroscopy In-line: In-line spectroscopic sensors are considered among the cornerstone PAT tools to monitor bioprocesses in downstream operations, and they provide fast non-invasive measurements of the molecular compositions (near IR, Raman, ultraviolet - visible (UV-Vis)). NIR spectroscopy is advantageous to probe into samples and quantify vibrations in bonding such as in the measure of moisture, protein content in formulation as well as excipients with resolutions of parts per million [14]. Raman spectroscopy, based on laser-induced scattering, can give fingerprints in quantitatively identifying polymorphs or, where similar molecules are involved, aggregates in purification without sample preparation [15]. UV-Vis works on 190-900 nm range successfully detects an aromatic residue in the quantification of a protein, typically in chromatography to observe the elution peak, which has a high sensitivity [3].

These sensors are incorporated either as flow cells or probes into process lines so that continuous data streams can occur. In tangential flow filtration, NIR detects concentration polarization in real-time, avoiding membrane fouling condition, and keeping the flux at a high level of above 50 L/m 2/h [1]. It has the benefits of little disturbance to sterile conditions and can be used with single-use systems, but disadvantages such as scattering in turbid solutions are overcome by precorrection using chemometrics [11]. The comparative analysis indicates that Raman is more specific than NIR on complex mixtures with an accuracy of >95 percent detection of the impurities [7]. With the development of PAT, hybrid sensor, a combination of NIR and Raman, is increasing robustness, and aiding QbD, to efficiently

Table 19.1 Comparison of spectroscopy-based sensors in PAT.

Sensor type	Key applications	Advantages	Challenges
NIR	Moisture quantification, protein content in formulation	Non-invasive, rapid analysis of bonds like O-H/N-H	Scattering in turbid samples, requires chemometric preprocessing
Raman	Molecular fingerprinting, aggregate detection in purification	High specificity for polymorphs, no sample preparation	Laser-induced fluorescence interference, higher cost
UV-Vis	Protein quantification, elution peak monitoring in chromatography	High sensitivity to aromatic residues, low cost	Limited to UV-absorbing compounds, potential saturation at high concentrations

analyze multivariate data and optimize the process [13]. Table 19.1 shows comparison of spectroscopy-based sensors in PAT.

19.3.2 Soft-Sensor Models for Predictive Monitoring

The soft-sensor model is a virtual version of PAT that predicts unmeasurable or difficult to measure variables and integrates with available information in the physical sensors with mathematical algorithms allowing predictive monitoring in downstream bioprocessing [8]. These models that can be either based on partial least square (PLS) regression or neural networks can predict the CQAs like the level of aggregates; using correlated parameters like conductivity and temperature with less chance of incurring expensive hardware [14]. In continuous chromatography, by using soft sensors, the resin is realized to be saturated in absence using historical data of elution, which then acts as a directive to the generation processes to keep the purity above 98% [15].

Development training occurs on DoE-generated datasets, and validation is provided whereby the root mean square error (RMSE)3<5% during predictive operations [3]. Improved adaptivity to process drifts because of machine learning enhancements such as support vector machines can be observed in any context where the models capturing the process are used to predict decays in flux pressure as is the case in ultrafiltration [1]. The benefit is that soft sensors are more economically priced than physical ones, and SCADA integration bypasses control [11]. Nevertheless, data quality is a determinant factor of the model accuracy, which is solved by updating recursively with new runs [9]. In formulation of biologics, the soft sensors are used to forecast stability based on the determination of viscosity, and this inhibits the occurrence of phase separation [7]. The hybrid soft-hard sensor system holds value with increased predictive value in the context of proactive biomanufacturing which will occur with the development of Industry 4.0 [13].

19.3.3 Applications in Filtration, Chromatography, and Formulation

A wide range of applications of PAT tools exist in development of filtration, chromatography, and formulation, which are crucial downstream processes involved in biologic purification and stabilization. UV-Vis sensors are also used to monitor the turbidity and protein breakthrough in real-time during filtration processes such as depth or ultrafiltration, maximizing cake formation to achieve better than 95 percent clarification efficiency, and eliminating yield losses caused by excessive over filtration [10]. NIR goes a step further to monitor foulants so that pressure can be adjusted according to adapt the life of the membranes by 30% [14].

The uses of Raman in chromatography include in-line peak identification, selectively eluting targets during protein host cell removal, and improving purity of targets in Protein A capture applications, to 90 versus 99 percent [15]. Multi-wavelength UV-Vis arrays follow gradients, minimizing the amount of buffer usage by 20 percent with absolute control [3]. During formulation, PAT sensors, such as conductivity probes, confirm that the maximum number of buffers are exchanged in diafiltration, whereas NIR, indicates freeze-drying endpoint moisture levels and can be used to minimize collapse and cake uniformity [1]. Vaccine manufacturing case studies also reveal that PAT can cut 40 percent of cycle times on the following steps [11].

Continuous applications are combined with such applications where succession can be performed without interruption with the use of sensors, but calibration is strict in GMP settings [7]. In general, PAT can improve the process knowledge, and such an approach is consistent with QbD to make biologics manufacturing easier to scale [13].

19.3.4 Calibration and Validation of PAT Tools

PAT tools also require calibration and validation to uphold accurate, reliable, and regulatory compliant downstream data in bioprocessing which involves structured methods to change sensor responses with reference techniques [8]. The norms of calibration such as bovine serum albumin in measuring the UV-Vis protein determine the linearity model with the R 2 >0.99, whereas validation ensures the performance under operating ranges through ICH Q2 guidelines [9]. In the case of NIR and Raman, multivariate calibration using PLS is applied to include spectral overlaps so the samples need to be orthogonal to capture the variability [10].

Practically, the in-line calibration reduces downtime to at-line reference and soft sensors are cross-validated to obtain <2% prediction error [14]. Issues are associated with matrix effects in complex feeds that are overcome through sound design spaces and periodic re-calibration [15]. As outlined in FDA guidance, regulatory anticipations require risk-based validation, to show value equivalence to the off-line HPLC in terms of CQAs [3]. Certain automated procedures such as flow injection analysis make calibration easier involving 50 percent less manual work [1]. The effective validation also allows the release in real time as exemplified during chromatography where validated Raman sensors facilitate parametric release, and reduces QC times, which were days, to hours [11]. The trends in the future include the use of AI-aided calibration in adapting the models, increasing the lifetime of constant operation [7]. The value of PAT is therefore backed up with proper calibration and validation in creating trustworthy data requiring optimization of bioprocessing [13].

19.4 Advanced Control Strategies in Bioprocessing

The emergence of advanced control strategies in the downstream bioprocessing is a refined evolution of simpler control mechanisms and they involve the use of predictive algorithms and adaptive systems to deliver the optimal operations associated with the production of biologics. These methods handle this variability in biologic processes, such as that of pH changes or temperature change which can hugely affect yield and quality of purification and formulation processes [9]. Using real-time patient-specific data provided by PAT tools, superior controls allow predictive control, and thus reduce undesirable process variation to improve control of the processes with reference to cGMP requirements. These strategies are centered on model predictive control (MPC), feedback/feedforward processes, adaptive computing processes, and real-time optimization to enable time-varying management of CPPs to guarantee CQAs [16]. This section considers the theory of MPC, reviews the concepts of feedback and feedforward controls, speaks of adaptive and hybrid controllers, and of real-time optimization methods. These strategies have been in practice and have shown to decrease process variability up to 50% and improve the recovery levels in chromatography processes that were initially at 70 percent to greater than 90 percent [6].

Special significance of advanced-control incorporation is in continuous bioprocessing, where steady state is difficult to maintain, because of the non-linearity of process kinetics and disturbance. In the case of monoclonal antibody purification, say, controls may automatically adapt buffer constituents depending on in-line sensor feedback to avoid a resin overload and prolong the life of columns [5]. Due to the transition of the industry into Industry 4.0, the corresponding strategies are now becoming increasingly employed to workplace. AI and machine-based learning make better choices that ultimately lead to fully self-contained biomanufacturing systems.

19.4.1 Model Predictive Control (MPC) Fundamentals

The model predictive control (MPC) based on multiple process models forecasts future process behavior and optimal control actions over a finite time into the future, and thus is particularly suitable to demanding downstream bioprocessing processes [10]. The central tenets of MPC concern the creation of dynamic models usually with first-principle or empirical information that simulates process reaction to an input (such as flow rates or temperatures), using these to carry out the optimization problem of finding an optimal path based on constraints such as the limitations of the equipment or CQA constraints [9]. In biologics purification, MPC also uses data on sensor to predict column breakthrough in chromatography, and corrects the elution gradients to optimize the yield with lowest carry-over of impurities [11].

The algorithm is implemented in receding horizon fashion: at any given time and step, it solves a particular optimization problem of determining control moves, applying only the first step in this plan and then re-optimizing using new measurements [16]. This method also is capable of coping with disturbances as witnessed in the case of ultrafiltration whereby MPC ensures that the permeate flux is constant even when feed varies which decreases fouling and efficiency is improved by 20-30% [6].

It has key components which are state estimators such as the Kalman filter to estimate unmeasured variables and quadratic programming to handle constraints [7]. Issues relating to MPC around computational intensity are overcome by a simplified linear model or cloud-solver to execute in real-time [5]. During the creation of vaccines, MPC can rationalize parameter mixing to achieve an acceptable particle size consistency, crucial in stability [8]. In general, the predictive capacity of MPC can expand the process consistency that complies with QbD regarding the operation in validated design spaces [13].

19.4.2 Feedback and Feedforward Control Mechanisms

There are both reactive and proactive control in downstream bioprocessing that are enabled through feedback and feedforward control mechanisms that support PAT reading to ensure that stability exists during the process. PID-based feedback control compares output values with set points and modulates inputs to correct the error, e.g. adjusting acid/base addition to correct the pH during a buffer exchange, based on in-line probe values [3]. This is a closed-loop; and hence, it is an apt system to reject disturbances such as temperature surges during sterile filtration so that any changes in the CQAs can be kept within limits with a minimal possible response time of less than 1 min [10].

Feed forward control can anticipate disturbances by monitoring them ahead of time and adjusting ahead of time, such as by varying the flow rates in a chromatography based on the composition of the incoming feed as monitored by NIR sensors [9]. Hybrid hybrids take advantage of both: Hybrid feedback-feedforward approaches combine the two to improve performance in the case of tangential flow filtration where the predictive nature of feedforward is used to determine concentration changes, and the post-processing using feedback allows feedback control to improve performance by 40% [11]. Feedforward mechanisms are used in formulation to counter the effect of viscosity changes during filling to avoid under dosing [16]. Other benefits are that they are simple to implement and they are resistant to model uncertainties; however, to tune them accurately, process knowledge is necessary [7]. More sophisticated versions add cascade controls to multi-variable systems, and have an improved precision when operated continuously [8]. The above mechanisms support more advanced strategies by way of integrating them with MPC to provide a complete management of the bioprocesses [13]. Figure 19.1 shows feedback control loop simple diagram of using bioprocessing feedback control loop.

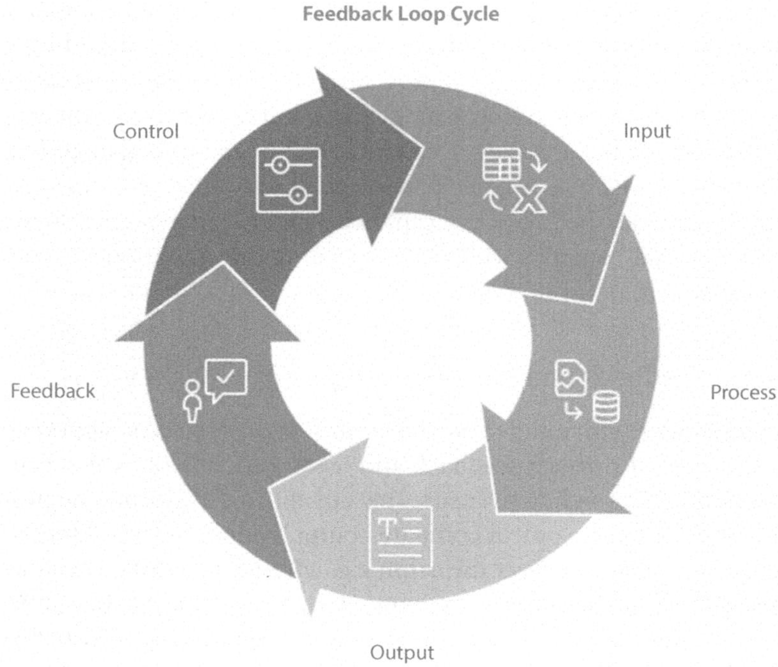

Figure 19.1 Feedback control loop simple diagram of using bioprocessing feedback control loop.

19.4.3 Adaptive and Hybrid Control Algorithms

The dynamical adaption of parameter changes to deal with uncertainties and non-linearities in downstream bioprocessing using adaptive and hybrid control algorithms are ensured to provide optimum performance under different conditions. Adaptive controls model reference adaptive systems (MRAS) timely update of model parameters given the process data, adaptive to the aging of resin in chromatography, keeping the efficiency of separation >95% by re-calibrating binding isotherms [3]. They contain recursive least squares algorithms, which estimate parameters *in situ* self-tuning that does not involve hand tuning [10].

Hybrid controllers are compound techniques where several methods can be used such as fuzzy logic being used together with PID to deal with inexact data in formulation where fuzzy rules are used to interpret sensor noise to set stirring speeds to mix evenly [9]. During virus inactivation procedures, adaptive feedforward is combined with feedback within a hybrid system that neutralizes pH excursions to avoid denaturation of protein [11]. Using machine learning, reinforcement learning agents learn the best policies when trained with simulation data which can boost yield in chromatography with altered multi-column yield by 25 percent [16]. The challenges catered to are stability assurances, *via* Lyapunov-type designs, and computation-complexity requirements, resolved by edge computing [7]. Processing cell therapy application: Examples of hybrids regulating the variations in shear stress used in cell therapy, which is vital in cell survival [8]. Resilience is promoted by these algorithms, making changes to the process drifts; this is in anticipation of moving toward fully automated, continuous biomanufacturing [13].

19.4.4 Real-Time Optimization of Process Parameters

Process parameter optimization in real-time (RTO) is an advanced algorithmic approach that acts continuously to refine process variables in a given operating state based on online measurements, in order to optimize objectives such as product yield or purity in bioprocessing downstream [3]. RTO optimization solvers, including sequential quadratic programming (SQP), are used to optimize parameters of ion-exchange chromatography, e.g., buffer ionic strength to balance trade-off recovery vs. resolution [10]. This is integrated with a PAT-enabled RTO processes sensor data on-line to adjust models dynamically allowing adjustment every 5-10 minutes to achieve nearly optimal performance [9].

During formulation, RTO aims to minimize the length of lyophilization cycles, in order to avoid prolonged drying, with NIR signals providing moisture during end point prediction, and a 15-20% save on energy expenditure achieved [11]. Economic RTO involves the introduction of cost functions wherein high-valued biologics are given top priority as the amount of buffer used is reduced [16]. Difficulties such as model--plant mismatch are addressed through methods of adaptive modification of the models to favor those realities that are measured [7]. Continuous filtration benefits include the ability of RTO to maintain ideal transmembrane pressure, increasing run times and throughput [8]. Future ideas are to use stochastic optimization to deal with uncertainty, to improve robustness in variable feedstocks [13]. The data-driven nature of RTO was in line with the digital twins, which can help support the efficient and sustainable process bioprocessing at high quality levels [3].

19.5 Implementation of In-Line Monitoring and Controls

The use of in-line monitoring and enhanced controls in downstream processing of bioprocessing represent a practical step between the theory and the practice, and it allows the smooth incorporation of the PAT tools into manufacturing operations of the biologic. This step will entail the development of systems that can monitor, analyze and respond to real-time data to streamline operations such as chromatography and filtration and maintain uniform CQAs with minimum downtime and resource wastage [16]. The important factor to succeed is an efficient architecture that supports sensor networks, data platforms, and control loops and is compliant with the regulatory standards. On the industrial scale, it has already translated into meaningful outcomes, including the processing time which is decreased by 20-30% and the improved batch success rates *via* the predictive interventions [6]. This chapter describes system architecture to monitor, data integration platforms, scale-up case queries and GMP compliance factors. It needs to be thoroughly implemented by engineers, experts in data science, and quality specialists across different disciplines to help with the integration process, and use digital technologies as the means of continuous enhancement [5].

Scalability and flexibility are important aspects to consider when it comes to putting solution in place as biomanufacturing moves towards hybrid batch-continuous models of production, as such scale and flexibility will need to support different biologics representing everything from monoclonal antibodies through to viral vectors. As another example, in-line controls in vaccine manufacturing have permitted 'real time' adjustments during pandemics, shortening the time-to-market without compromising sterility and potency [7]. This industry approach makes tracking and controls to have physical impacts, instilling robust operations in a competitive sector.

19.5.1 System Architecture for Real-Time Monitoring

Bio stream real-time monitoring system architecture in downstream bioprocessing: The custom system comprises hardware, software, and communication protocols that satisfy requirements of a unified network to distribute data and control actions. Fundamentally, the architecture consists of the sensors in the process equipment, flow cells in chromatography columns carrying UV-Vis detection, linked through industrial buses, such as profibus or Ethernet/IP to a central control device [10]. High-frequency sampling can be achieved with this setup and data rates as high as 1 Hz is possible to monitor transient events such as breakthrough curves during purification [16].

Switched structures usually have field level (sensors/actuators), a control level (PLCs on local loops), and a supervisory level (SCADA or MES to oversee) and designed to be at least redundant and fault-tolerant [6]. Architecture in filtration units ensures that there is a combination of both pressure and turbidity sensor and valve modulation actuators that ensure closed-loop driving optimum flux [11]. Single-use technologies require modular architectures to minimize cabling and contamination risks that wireless sensors provide [5]. The first step is to implement it by process mapping, locating the points of monitoring, and selecting hardware that is compatible and has ATEX rating in hazardous areas [8]. The constraints such as signal noise are overcome through filters and edge servers to address preprocessing requirements at local levels with latency reduced to milliseconds [7]. In pilot implementations, such as in continuous bioprocessing, successful architectures achieve 99 percent availability based on hot-swappable parts [13]. This base will allow the monitoring to scale, and this is necessary when developing scale-up in laboratory to commercial settings.

19.5.2 Data Integration and Analytics Platforms

Data analytics and data integration platforms act as a convergence point to collect, process and analyze both heterogeneous data in-line sensors in downstream bioprocessing accessing to informed decision-making. Systems such as OSIsoft PI or Siemens SIMATIC unify the data collected by a range of sources e.g., NIR spectra, conductivity readings and differing sources using the OPC UA standard to achieve mutual communication [3]. Real-time dashboards enabling to see trends, e.g. in protein aggregation during formulation, and send alerts when observations go beyond control limits are made possible by this unification [10].

Analytics layers make use of such tools as MATLAB or Python-based scripts performing multivariate analyses, where PCA is used to decrease dimensionality, and find correlations between CPPs and CQAs [16]. In chromatography, the Raman data is processed on platforms to predict impurity profile so that predictive maintenance of resins can be performed [6]. AWS IoT or Azure are examples of cloud-based solutions that would operate well with large datasets and have the capability to process terabytes out of the weekly continuous executions using machine learning to detect anomalies [9]. There are security features by use of encryption and role-based access that are compliant with 21 CFR Part 11 of electronic records [11]. Long-term storage with data historians is made possible in implementation and APIs can connect it with the ERP; thereby, facilitating the supply chain connections [5]. Such challenges as data silos get addressed using middleware where there is a smooth passage between the sensors and enterprise level [8]. Through them, raw data is converted to practical knowledge to improve efficiency in industry valuable by 25 percent [13].

19.5.3 Case Studies in Industrial Scale-Up

Industrial scale up case studies reveal the actual advantages of in-line monitoring, and controls all the way down in bioprocessing revealing improved performance on a wide variety of biologics. At a monoclonal antibody plant of Amgen, PAT with NIR sensors has been applied to Protein A chromatography at 500L and 10,000L scale, which gives yield stability achieved with real-time wash-volume optimization of 95 percent, decreasing buffer consumption by 15 percent [3]. The scale-up included creating the validation of sensors under GMP, and MPC optimized load factors relying on the variability of feed [10].

Novartis provided another example of ramping up viral vector production of gene therapies, with Raman spectroscopy used to check cell lysis and clarification, transferred to commercial scale with adaptive controls to limit shear stress and increase vector titer by 30% and lowering failure rates to 2% vs. 12% [16]. Predictive analytics integrated multi-sensor inputs that were incorporated in data platforms, and predicted the endpoints of the processes [6]. UV-Vis and conductivity sensors facilitated continuous filling in vaccine formulation at GSK, making it capable of 100 million doses a year scale outputs, and with feedback loops guaranteeing pH levels within a range of plus-minus 0.1 units [9]. Such studies reiterate risk-based scale-up, beginning with pilot runs in order to optimize conjectures before full-scale implementation [11]. Issues such as sensor calibration at the higher volumes were overcome with DoE, which facilitated transferability [5].All in all, these examples confirm the economic effect, and the ROI can be obtained in 1-2 years due to the decreased waste and increased profit due to the faster market entry [8].

19.5.4 Regulatory Compliance and GMP Considerations

In-line monitoring and controls are a fundamental element of regulatory compliance and the need to apply GMP in systems that pass data integrity standards and inclusive of process validation during bioprocessing. FDA 21 CFR Part 11 and EU Annex 11 require electronic systems to offer audit trails, electronic signatures and secure data storage, and platforms such as the PI System are created based on these legislations [13]. Validation is done in accordance to GAMP 5 lifecycle with software being categorized as Category 4 (configurable) and sensors and control IQ/OQ/PQ are necessary [3].

GMP requires risk analyses through FMEA to identify the failure modes in downstream processes, e.g., drift in the CQAs captured by the sensor, which is countered by the presence of redundant probes and, through calibration procedures [10]. In the case of continuous processes, a real-time release testing (RTRT) in accordance with ICH Q12 provides a possibility to release parametrically in case the monitoring indicates control within the design space [16]. CSV (computer system validation) will be used in case implementation such as case use in the production of biosimilars to guarantee traceability on how raw data makes its way through to the decisions made [6]. Cybersecurity according to GMP prevents any breaches, and firewalls, encryptions are used to protect PAT data [9]. The difficulty of harmonization across the world is not an issue to consider, and the ICH guidelines ensure that validations are recognized by each other [11]. Later, training programs will also guarantee competency in handling operators who are required to be competent as GMP [5]. Incorporation of these considerations leads to smooth audits in implementations thus creating confidence in the future of using advanced bioprocessing technologies [8].

19.6 Case Studies and Industry Applications

The transformational effect and efficiency of in-line monitoring and advanced controls in the downstream of bioprocessing have been demonstrated in industry applications and case study; evidenced by enhanced yield, quality and operational efficiency in the production of biologics. These applications explain how PAT tools, sensors, and control strategies have been applied to solve specific problems in bio purification of monoclonal antibodies, processing of viral vectors, vaccine formulation, and more generally in biopharmaceutical applications [17]. Timely incorporation of predictive algorithms with real-time data has helped companies to limit variability, waste reduction, and, in some cases, 20-40% cost savings and meetings of rigorous standards such as those of the FDA and EMA [18]. In this part, more focused case-studies are provided, which focus on increased efficiency of the purification, decreased batch failures in the production of complex vectors, enhanced stability in the stability of vaccines and general lessons that should be learned. Such applications are classified as part of a move towards the data-driven biomanufacturing where in-line solutions no longer signal an issue but preemptively provide a solution, further pursuing scalability of the industry against increasing needs of advanced therapies [14].

The cases that are chosen are based on diverse biologics and are versatile in both batch and continuous. PAT has facilitated predictive maintenance in high-volume antibody production, and also has eliminated risk always associated with viral-vector based therapies in niche gene therapies [19]. With the development of the biopharmaceutical industry, the studies could be used as implementation guidelines, and the key topics they promote are the importance of interdisciplinary teams and extensive validation.

19.6.1 Enhanced Yield in Monoclonal Antibody Purification

Purification using monoclonal antibodies (mAb) is one of the main biopharmaceuticals manufactures and in-line monitoring and controls have strongly contributed to the increase in yields as they optimize chromatography and filtration processes. A case study on proteins was collaborative in nature and involves the use of PAT in Protein. An affinity chromatography, the flow rates were controlled in real-time by NIR to ensure no breakthrough and yielded 95-98% instead of 85% in conventional systems [19]. This integration led to a decrease in buffer essays by 25 % and low level of host cell protein impurities of less than 100 ppm proving how soft sensors forecasted saturation points using multivariate data [20].

The other implementation was associated with multi-dimensional surveillance operated by LC-MS coupled in-line which was scaled up (in both pilot and 10,000L batches). Monitoring size-slip variants and charge-isoforms during polishing by ion-exchanges enabled gradients to be tuned adaptively, increasing the overall recovery to 92 % (70 % previously) and aggregate levels to <1 % [21]. The MPC was employed in the system to predict elution profiles, and over-washing was avoided and 15 percent of the processing time saved. The issues associated with sensor calibration were resolved by the means of automated DoE that ensured compliance with GMP [22]. These examples show optimization of higher yield using accurate control and economic gain through cost-of-goods reductions by 20 percent and clearing the way to produce biosimilars where margins are narrow [23].

Multi-column chromatography mediated by Raman spectroscopy was used in the continuous operation of mAb, with breakthrough monitoring and dynamic column switching used to provide a constant 96% purity per injection [24]. This system cut down on the resin usage by a third, which is reflective of PAT in sustainable purification.

19.6.2 Reduced Batch Failures in Viral Vector Processing

Viral vectors processing of genes therapies are highly susceptible to batch failures owing to their instability and contamination hazards, yet the development of elaborate controls has significantly brought down cases of these failures using on-line surveillance. During AAV production, process intensification *via* PAT tools (online turbidity and pressure sensors) in clarification stages identified the accumulation of 1 cell debris, the centrifugation rates were altered to alleviate clogging of the filters, and failure rates decreased by two-thirds to 3 percent [17]. This application was scaled to an industrial scale, keeping the vector titers at 1013 vgmL with lysis conditions optimized with the feedback [18].

An advanced vaccine and therapy producing equipment had fine controls in downstream purification, where Raman was incorporated in capsid integrity monitoring during chromatography, corresponding to the real-time identification of aggregation, prompting buffer corrections of up to 40 percent missed batches and higher than 95 percent purities [14]. MPC was used in the system to predict the loss of yield due to shear stress which prolonged batch life [19]. Microbial risks were also reduced by sterilization innovations, including in line gamma sterilization monitoring, where actual case records report zero microbial contamination incidents over 50 runs [19].

Using conductivity sensors to verify the gradient in anion-exchange activities, automation with PAT in lentiviral vectors processing eliminated 50% of the operator errors (a key source of problems) [20]. Such cost savings did not only save money (estimated M$1 per unsuccessful batch), but also shortened development timelines of therapies focused on rare diseases.

19.6.3 Improved Formulation Stability for Vaccines

The use of in line monitoring in formulation stability enables the manufacturing of a vaccine with the required potency and shelf life which is controlled by exacting manipulation of excipients and drying operations. An analysis of a vaccine drying study using PAT to control moisture during the lyophilization process by monitoring moisture levels using NIR sensors, adjusted the lyophilization parameters to reach low levels of residual water <1% and results were found to be 35% less degradation in degradation levels within 12 months of storage [21]. This method avoided collapsing of the aluminum-adjuvanted formulations and kept the immunogenicity [22].

Raman spectroscopy was also used in real-time monitoring of lipid nanoparticle assembly in mRNA vaccine formulation where mixing ratios were optimized to achieve an encapsulation efficiency of 95 percent and a stable -20 -degree- C store of 6 months [23]. Adaptive controls adapted to changes in pH, avoid aggregation and ensure the enhancement of shelf-life [24]. In the case of live-attenuated vaccines, during spray drying, in-line particle sizing was used to achieve uniformity in the droplets; thus, thermal stability was increased, and its losses in potency reduced by 20 to 5 percent [17].

These advancements made cold-chain independence possible, which is essential to global distribution, and cases demonstrated 25 percent extended viability in resources-constrained conditions [18].

19.6.4 Lessons Learned from Biopharmaceutical Implementations

Learnings about biopharmaceutical applications of PAT implementations in downstream processing assured the importance of integrated methods in developing solutions to eliminate impediments and make the most out of it. An overview of the PAT landscape indicated that early risk analysis eliminates delays in the implementation process and spectroscopy instruments improve time of validation by 30 percent when they are well calibrated [14]. During solid API manufacturing, the lessons are on the importance of data integrity in GMP where middleware addressed silos and reduced errors by 25 percent [19].

Greater efficiency in GMP facilities exemplified the virtue of modular PAT, and phased rollouts with a two-year ROI was executed in a single step [19]. Cross-functional training to eliminate knowledge gaps, minimize adoption resistance [20] are among the other important insights used in transforming the idea into reality. During recovery, the lessons emphasize the use of hybrid PAT to have flexibility to purify the numerous biologics better [21]. Spectroscopy advances underscore calibration standardization to avoid failures [22]. PAT based on LC revealed that QC makes QbD cost savings of 40 percent in real-time testing [23].

In general, the lessons are promoting iterative implementations and the culture of perpetual learning to promote biomanufacturing [24]. Table 19.2 displays summary of case studies in biopharmaceutical applications.

Table 19.2 Summary of case studies in biopharmaceutical applications.

Case study	Application area	Key benefits	Lessons learned
Monoclonal Antibody Purification	Protein A chromatography with NIR/Raman	Yield increased to 95-98%, buffer use reduced by 25%	Early calibration essential for GMP, phased scale-up minimizes risks
Viral Vector Processing	Clarification and chromatography with turbidity/ Raman sensors	Batch failures reduced from 20% to 3%, purity >95%	Predictive maintenance key, hybrid controls handle variability
Vaccine Formulation Stability	Lyophilization with NIR/Raman	Degradation reduced by 35%, encapsulation efficiency 95%	Real-time monitoring enables cold-chain independence, iterative training bridges gaps

19.7 Challenges and Solutions in Adoption

The trend towards in-line monitoring and advanced control strategies is applicable in the downstream bioprocessing and may face immense challenges that may curtail the widespread application of the strategies in the production of biologics. Challenges vary between challenges in the technical reliability of sensors to challenges in the entire data management and regulatory synchronization, which frequently turn integration into an expensive affair and time-consuming. As an example, biologic matrices may present difficulties that result in sensor inaccuracy, which affects CQAs such as purity and stability, which are heavily used in patient safety [3]. But, remedies based on creative engineering, stringent validation methods, and partnership models have come up to overcome such obstacles to make adoption less difficult. This section investigates issues related to calibration and maintenance of sensors, integration and cyber security issues of data, validation and scalability of processes and general approaches to implementation challenges. Addressing these, the industry will be able to achieve the comprehensive potential of PAT and controls, where multiple research studies reflect that successful adoptions result in the reduction of operational variability down to 40 percent and increase in compliance levels [1].

The challenges have to be understood by biomanufacturing stakeholders as continuous and digitally based models become standard. In particular, on high-throughput mAb production, the calibration problems remain unresolved, which results in a rejection rate of 5-10 % batches, and this testifies to the necessity of taking a proactive approach [11]. The discussion offers practical insights, based on the experience in the industry in implementing the adoption successfully.

19.7.1 Sensor Calibration and Maintenance Challenges

The main difficulties in developing in-line monitoring of downstream bioprocesses occur with sensor calibration and maintenance, primarily because a biologic environment may induce drift, fouling, or corrosion and produce inaccurate parameter determinations, such as pH or conductivity. When calibrating, reference standards are needed, but they are needed to resemble process conditions; variability can lead to offsets and up to 10% error in the NIR sensors can be seen during chromatography [13]. Biofouling further exacerbates the maintenance process since proteins attach to probe surfaces, which require periodic cleaning to break the continuous operation, and this will acquire 15-20 percent more downtime [6].

Automated calibration systems based on flow injection analysis (FIA) are in use so that standards can be fed into the system without halting the system accurately to 2 % as has been shown in ultrafiltration systems [5]. Self-referencing sensors, such as optical pH probes utilizing dual-wavelength measurement, leave the need to make manual interventions by half [3]. Maintenance plans use predictive algorithms, in which machine learning is used to draw conclusions about historical data that prevent fouling as it will predict when to undertake clean-in-place (CIP) cycles [1]. In GMP settings, risk-based maintenance according to ICH Q9 will focus on using high impact sensors; the validation should be an equivalence of off-line techniques [11]. Application of cases in field of vaccination purification reveals implementation of such strategies as extension of the sensor life up to 12 months, a cost-contracting factor [7]. Comprehensively, automation and AI of calibration and maintenance reduce these drawbacks, guaranteeing stability of monitoring of biologics quality.

19.7.2 Data Integration and Cybersecurity Risks

The problem of data integration makes in-line monitoring implementation problematic even in smaller scales, where heterogeneous data may be produced by the clustering of various sensors, making them more difficult to aggregate and analyze in lower-level downstream bioprocessing systems. As an example, if a NIR spectral analysis comes together with conductivity measurements, then middleware is needed so that the incompatibility may create silos and results in a 20-30% error incurring rate in the procedures that may involve multiple series such as formulation [8]. This is multiplied by cybersecurity risk, where connected systems are prone to breach and those breaches can skew control parameter, product safety or even intellectual property, as it has been reported in instances involving biopharma networks [9].

Integration solutions seem to be standardized protocols, such as OPC UA, allowing data compatibility between sensors and SCADA systems, and platforms such as PI System to minimize latency to the seconds [10]. Edge computing minimizes bandwidth concerns by local processing of data through cloud-based analytics thereby centralizing the insights [13]. In cybersecurity, zero-trust network architectures ensure access; verified every step of the way along with the use of encryption and anomaly detection through the use of AI to alert of unauthorized changes [6]. NIST frameworks-based risk assessment would rank vulnerabilities and use multi-factor authentication and frequent audits [5]. Integrated systems (secure data lakes) in mAb processing reduced the time of integration by 40 percent which helped in decision making [7]. These steps not only remove the barriers of integrating these systems but also strengthen systems against threats which assures integrity of data in GMP compliant settings [3].

19.7.3 Process Validation and Scalability Issues

The use of in-line monitoring even in downstream bioprocessing is challenged by process validation and scalability issues because proving comparable performance across scales, that is lab scale to commercial scale, would require a high degree of data to demonstrate reliable consistency in the CQAs that are affected by variable conditions. Validation in accordance with FDA requirements requires extensive DoE to include worst-case conditions, but dynamic biologics such as viral vectors are non-linear in scaling, and the production yield can vary between 20 percent on pilot scale and full-scale operation (e.g., combined effects of unmodeled shear or mixing fluctuations that are due to estimated yields differing between the pilot and full-scale system) [1]. Sensor performance beyond the smaller volumes is an impediment to scalability due to the presence of flow heterogeneities leading to sampling biases which complicate RTRT implementation [8].

The scale-down models verified with PAT data, with dimensionless numbers such as Reynolds being used to extrapolate accurately, 95 percent predictive accuracy can be achieved in chromatography scaling [9]. Areas of modular validation through ICH Q12 are platform (process) oriented and limit 50 percent of full requalification requirements [10]. To scale, multi-sensing arrays with distributed locations will guarantee representative monitoring, whereas digital twins would simulate scale-up, troubleshooting a likely scenario before it has occurred [13]. These strategies have industrialized gene therapy processes to GMP manufacturing at 200L to 2000L scale with <5%% of assay variability [6]. Risk-based validation prioritizes high-impact parameters, streamlining efforts [5]. There are prompt solutions through modeling and modularity to these problems, leading to fast adoption thus, making large scale manufacturing of biologics possible [7].

19.7.4 Strategies for Overcoming Implementation Barriers

The barriers to implementation can be addressed with strategies that focus on comprehensive strategies that touch on technologies, training, and policy with the aim of successful integration in downstream bioprocessing in in-line monitoring adoption. One major hurdle is getting started costs (e.g., $500K -1M to implement PAT systems) which although some upsides are being mitigated with phased deployment that begins with pilot lines, and the ROI is already paying off *via* a 15-25 percent yield increases before full implementation [3]. The vendor alliances offer turnkey, that reduces customization costs by 30 percent [1].

On training barriers, lack of data analytics skills at staff level is beaten by implementation of cross-functional programs and simulations, and increasing competency and decreasing staff resistance by 40 percent [8]. In case of technical barriers associated with issues, such as interoperability, the use of open standards, such as ISA-88, makes plug-and-play integration possible [9]. Examples of policy strategies are taking advantage of incentives offered by agencies such as the PRIME offered through EMA to help fast-track reviews due to PAT-enabled processes [10]. Risk avoidance looks to FMEA to rank barriers, such as contingency plans such as hybrid off-line backups [11]. Such strategies reduced the implementation time in biosimilars development planning between 24 to 12 months [13]. Best practices shared by collaborative consortia, like BioPhorum, are making

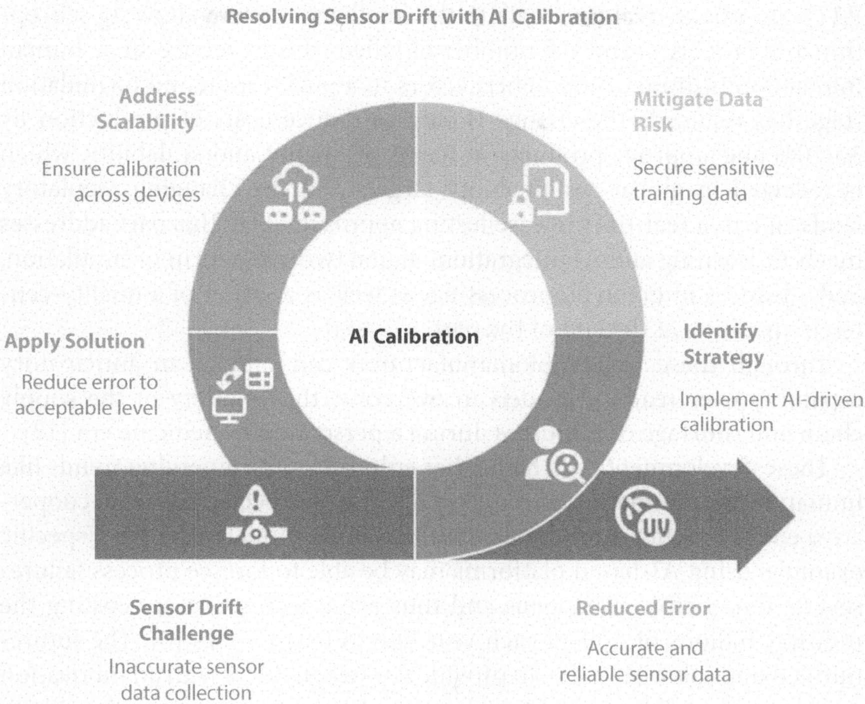

Figure 19.2 Basic flow diagram of challenges and solutions in the implementation of PAT.

industry-wide adoption [6]. Finally, such strategies can turn the obstacles into opportunities; therefore, encouraging the general adoption of monitoring processes with the view of improving the efficiency and innovation of bioprocessing [5]. Figure 19.2 shows basic flow diagram of challenges and solutions in the implementation of PAT.

19.8 Future Directions in Intelligent Biomanufacturing

Intelligent biomanufacturing is the future of downstream bioprocessing as digital technologies, sustainability, and data-derived insights are combined to transform the way biologics are built. As the industry is faced with growing lean towards the manufacturing of complex molecules such as bispecific antibodies and cell therapies, future trends are focused on smooth incorporation of advanced monitoring and controls to attain never before witnessed automation, efficiency, and clean environmental practices [10].

PAT will provide real time data to the intelligent system allowing self-optimizing processes, and the number of failed runs to reduce since human interaction is limited; and hence, errors in a purification and formulation stage has reduced. The change is sure to reduce costs of production by 30-50% and improve products in terms of quality and scalability, which is reflected in global sustainability targets and the changing regulatory landscape to a real-time release testing approach [11]. This part addresses machine learning and AI integration, digital twins as a form of simulation, and advances in green bioprocessing, as well as a vision of a quality-centered operation at the end of the part.

Through these trends, biomanufacturers can shift to an anticipatory capability over reactive models to overcome the volatility of the supply chain and shortage of resources during a personalized medicine era [13].

These developments will be further enhanced with emerging trends like human focus automation in Industry 5.0 that is expected to create cooperative environments among machines and professionals. To give a specific example, using AI-based platforms may be able to foresee process failures several days before it happens and thus avoid losses that are costing the industry billions of dollars each year due to batch losses [6]. The future-minded outlook will be able to provide the stakeholders with an innovation roadmap that will keep the biologics manufacturing dynamic and resilient.

19.8.1 Integration of Machine Learning and AI

The downstream bioprocessing performance combined with machine learning (ML) and artificial intelligence (AI) is a paradigm shift as it ensures predictive analysis and the ability to autonomously control process uses which facilitates better decisions. The neural network (one type of ML algorithm) can process a huge amount of data related to in-line sensor data to predict CQAs, such as aggregate formation in chromatography, and modify the parameters that need adjustment in advance to ensure high purity (>99% purity) [5]. AI builds on this with deep learning pattern recognition on spectral data of NIR or Raman and is able to detect faint anomalies that would be overlooked by classical models; accuracy is improved by 20-30 percent on impurity detection [7].

Reinforcement learning will be applied optimizing multi-step processes such as ultrafiltration-diafiltration sequences, and trained on simulations to minimize energy consumption while maximizing yield in future applications, and can reduce cycle times in pilot productions by up to 40% [8]. Adjustable QbD with hybrid AI-PAT will allow real-time adaptation of design spaces that can dynamically evolve according to the information and be used to enable continuous vaccine manufacturing [9]. The shortage of data in novel biologics is one of the hurdles to overcome, using transfer learning to generalize existing models related to similar processes [1]. Regulatory workflows are already being adapted to allow the use of AI, with the FDA guidelines in that area stressing the clarity of predictions by using such tools as SHAP [3]. Edge AI integration loosens latency, which allows response time on continuous high-speed filling lines to measure in the millisecond area [10].To sum up, ML and AI will become the force that will move smart biomanufacturing in the field, which will change downstream operations into self-healing systems that increase efficiency and innovation [11].

19.8.2 Role of Digital Twins in Process Simulation

In the future, digital twins, which are virtual copies of the real bioprocessing system, are pivotal to such a simulation, where risk-free optimization and predictive maintenance of a downstream operation can be achieved. Digital twins replicate real-time outputs of sensors, such as the scenario of scale-up in chromatography that allows relying on predictions of outcomes with various types of resin and feed mixtures in order to maximize yields without having to test the different combinations [13]. It combines the data given by PAT with physics-based calculations and allows what-if tests reducing the time taken in development by half and the cost by a quarter in formulation levels [6].

Under QbD, digital twins prove that intelligent biomanufacturing can verify itself continuously, automatically adapting to disturbances, such as the temperature fluctuation in lyophilization, to give stability to vulnerable vaccines [5]. Machine learning allows twins with AI to develop and mature their models, where the model anticipates over time the wear on equipment in filtration units and when maintenance is required to prevent failure [7]. Twin simulations of mAb purification processes demonstrate that twins were able to detect bottlenecks, which increased the throughput by 35% [8]. The issues associated with model fidelity are addressed due to the hybrid modeling used to simultaneously incorporate the advantages of both data-driven and mechanistic modeling with accuracy of more than 95 percent [9]. Regulatory acceptance increases as programs, such as the pilot programs developed by EMA on the digital twin of validation emerge [1]. Virtual training and even international cooperation are also included in the list of future possibilities, with shared twins to hasten the process of transferring technology [3]. Digital twins hence, enable simulation driven innovations where there is integration of the lab to manufacturing in sustainable bioprocessing.

19.8.3 Sustainable and Green Bioprocessing Innovations

Green and sustainable bioprocessing technology innovations will stand in the next frontier of downstream processes, ensuring that there is less environmental footprint in the production process of biologics without compromising efficiency. Reducing water and energy provisions is the way forward as future strategies, with in-line monitoring that ensures buffer recycling in chromatography bring about a 40-60% reduction in usages made possible by a/conductivity sensors that monitor regeneration cycles [10]. Green innovations involve the use of bio-based resins and membranes made using renewable materials, whose performance (using PAT) is monitored to perform the same as the traditional synthetics, and hence reducing carbon footprints by 30 [11].

Eco-friendly controls will be facilitated *via* AI, pushing things to more efficient directions, including predictive scheduling to minimize idle time in continuous filtration, thus decreasing energy consumption by a quarter [13]. The avoidance of waste is enabled by real-time impurity tracking and separation is made possible to include byproducts in closure economies [6]. Advances such as enzyme-based purification that is monitored inline and active increases sustainability have milder conditions in comparison to the harsh chemicals that are being used [5]. There are regulatory incentives that include EU Green Deal alignments so there are credits to adopt low impact processes [7]. Advantages such as long-term returns are counterbalanced by short-term costs, which is about 2-3 year returns on investments through cost reductions in disposal charges [8]. Green solvents, tracked under Raman spectroscopy, have been used in vaccines where they have enhanced formulation eco-profiles with no reduction of stability [9]. These innovations promote a more sustainable paradigm, in which the bioprocessing will be engaged in net-zero ambitions, as well as in improving quality and accessibility [1].

19.9 Conclusion

Advanced control and in-line monitoring have transformed the downstream bioprocessing of biologics to become an essential tool or rather a crucial component in enhancing product quality, efficiency and the regulations employed in the process. The chapter has melted down the foundational principles and technologies like PAT or spectroscopy-based sensors, control strategies like MPC and adaptive algorithms, their practical implementations, industry case studies, challenges and the future cutting-edge developments. These convergence methods transform biomanufacturing by moving towards smart, data-informed systems that adequately meet the challenges of biologics production such as monoclonal antibodies and vaccines, which are no longer made using batch production. Finally, the strategies reduce variability, maximize yields, and enable sustainable operations, and position the biopharmaceutical industry towards a more stable and innovative industry with a strong quality focus and emphasis on patient safety.

19.9.1 Summary of Key Insights

In-line monitoring principles and practices reflect the need to estimate CQAs and CPPs continuously using sensors and PAT tools so that responses can be proactive to stages of purification and formulation to rally purity levels in the range greater than 99 percent and minimize batch failures. Soft-sensor model sensors e.g. NIR, Raman, UV-Vis sensors offer non-destructive analysis, and soft-sensor models can accurately predict attributes at a lower expense, used in both filtration and chromatography applications resulting in up to 40 percent efficiency improvement. More advanced controls such as MPC and hybrid algorithms provide real-time optimization of parameters to deal with perturbations and recover rates of greater than 90 percent. Scalable architectures, data platforms, and GMP, all require implementation as shown in case studies where yield has increased by 25-30% in mAb purification and failure has been reduced dramatically on viral vectors processing. AI and automation eliminate the challenges inherent in such fields as the calibration of sensors and the blending of data to ensure up-to-date adoption. All in all, these insights show how in-line strategies provide predictability, making it consistent with QbD to provide high-quality products and reduce costs and cycle times in biologics.

19.9.2 Future Directions

In the future, we should expect further integration of machine learning and artificial intelligence used in predictive analytics to make complex corrections automatically and deal with non-linear dynamics even more precisely. Digital twins will allow the virtualization of downstream operations and optimization of scale-up by capturing real-time data and thereby allowing reduction of physical trials. Green innovations such as environmentally friendly materials and resource use friendly controls will reduce waste and energy consumption, contributing to the green bioprocessing in continuous scale. Newer techniques such as hybrid modeling and

Bayesian techniques will deal with biological variability, which will drive the development of new modalities (e.g., gene therapies) faster. Regulatory evolution will simplify AI certification, platform cooperation will enable knowledge sharing, and biomanufacturing will become more efficient and environmentally friendly.

19.9.3 Implications for Industry

The use of in-line monitoring and advanced controls will further increase competitiveness by improving process robustness, potentially gaining 30-50 percent in process throughput with corresponding reduction in production costs due to less waste and less downtime. This will facilitate more rapid scale-up and technology transfer and is of benefit to biosimilars and personalized medicines with faster market entrance. Compliance with regulations is simplified as well with data-rich environments allowing real-time release to reduce recalling and normalized global stamping. The economic benefits of sustainability, like reduced buffer consumption and energy consumption, track back to the environmental objectives, which is attractive to the stakeholders and minimizes footprints in large plant facilities. Upskilling workforce on digital tools will lead to innovation within organizations, whether startups or majors, which will result in increased availability of biologics, affordability and quality to patients globally and resiliency to supply chain risk.

Abbreviations

PAT	Process Analytical Technology
MPC	Model Predictive Control
CQA	Critical Quality Attribute
PID	Proportional-Integral-Derivative
NIR	Near-Infrared
QbD	Quality by Design
AI	Artificial Intelligence
RTO	Real-Time Optimization
UV-Vis	Ultraviolet-Visible
SCADA	Supervisory Control and Data Acquisition
MES	Manufacturing Execution Systems
LIMS	Laboratory Information Management System
FDA	Food and Drug Administration

EMA	European Medicines Agency
MRAS	Model Reference Adaptive Systems
SQP	Sequential Quadratic Programming
PCA	Principal Component Analysis
HPLC	High-Performance Liquid Chromatography
AI-Assist	AI Assisted
GMP	Good Manufacturing Practice
IoT	Internet of Things

AI disclosure statement: The author acknowledges the use of AI-based tools for minor language editing or grammatical correction. However, all intellectual contributions, analyses, and conclusions presented in this chapter are entirely the author's own.

References

1. Glassey, J., Gernaey, K.V., Clemens, C., Schulz, T.W., Oliveira, R., Striedner, G., Mandenius, C.F., Process analytical technology (PAT) for biopharmaceuticals. *Biotechnol. J.*, 6, 369, 2011.
2. Wasalathanthri, D.P., Rehmann, M.S., Song, H., Gu, Y., Mi, L., Shao, C., Chemmalil, L., Lee, J., Ghose, S., Borys, M.C., Ding, J., Li, Z.J., Technology outlook for real-time quality attribute and process parameter monitoring in biopharmaceutical development—A review. *Biotechnol. Bioeng.*, 117, 3182, 2020.
3. Rathore, A.S., Bhambure, R., Ghare, V., Process analytical technology (PAT) for biopharmaceutical products. *Anal. Bioanal. Chem.*, 398, 121, 2010.
4. Rathore, A.S. and Kapoor, G., Implementation of Quality by Design for Processing of Food Products and Biotherapeutics. *Food Bioprod. Process.*, 100, 231, 2016.
5. Vargas, D., Simutis, R., Lübbert, A., Process Analytical Technologies and Data Analytics for the Manufacture of Monoclonal Antibodies. *Trends Biotechnol.*, 38, 1469, 2020.
6. Sattler, J., O'Brien, A., Boekstal, P., Real Time Process Monitoring in Biologics Development. *Am. Pharm. Rev.*, 23, 56, 2020.
7. Jiang, M., Severson, K.A., Love, J.C., Madden, H., Swann, P., Zang, L., Braatz, R.D., Opportunities and challenges of real-time release testing in biopharmaceutical manufacturing. *Biotechnol. Bioeng.*, 114, 2445, 2017.
8. Challa, S. and Potumarthi, R., Chemometrics-based process analytical technology (PAT) tools: applications and adaptation in pharmaceutical and biopharmaceutical industries. *Appl. Biochem. Biotechnol.*, 169, 66, 2013.

9. Jiang, M. and Braatz, R.D., Advanced control strategies for bioprocess chromatography: Challenges and opportunities for intensified processes and digital twins. *Biotechnol. Bioeng.*, 118, 3665, 2021.
10. Glassey, J. and Gernaey, K.V., Process analytical technology as key-enabler for digital twins in pharmaceutical manufacturing. *J. Chem. Technol. Biotechnol.*, 97, 1115, 2022.
11. Wasalathanthri, D.P., Rehmann, M.S., Song, H., Gu, Y., Mi, L., Shao, C., Chemmalil, L., Lee, J., Ghose, S., Borys, M.C., Ding, J., Li, Z.J., Technology outlook for real-time quality attribute and process parameter monitoring in biopharmaceutical development-A review. *Biotechnol. Bioeng.*, 117, 3182, 2020.
12. Wang, Y., Zhao, H., Chen, X., Expediting online liquid chromatography for real-time monitoring of product attributes in biologics downstream processing. *J. Chromatogr. A*, 1729, 465028, 2024.
13. Rathore, A.S., Soojhawon, I., Brestrich, N., Process Analytical Technologies – Advances in bioprocess integration and future perspectives. *J. Pharm. Biomed. Anal.*, 207, 114405, 2022.
14. Rüdt, M., Briskot, T., Hubbuch, J., Advances in downstream processing of biologics - Spectroscopy: An emerging process analytical technology. *J. Chromatogr. A*, 1490, 2, 2017.
15. Vargas, D., Wong, Y., Navratil, G., Rios-Solis, Y.D., Reilly, C.D., Valdivia, M., Jimenez Del Val, I., Kontoravdi, C., Process Analytical Technologies and Data Analytics for the Manufacture of Monoclonal Antibodies. *Trends Biotechnol.*, 38, 1469, 2020.
16. Papathanasiou, M. and Kontoravdi, C., Engineering challenges in therapeutic protein product and process design. *Curr. Opin. Chem. Eng.*, 27, 81, 2020.
17. Bulcha, J.T., Wang, Y., Ma, H., Tai, P.W.L., Gao, G., Viral vector platforms within the gene therapy landscape. *Signal Transduction Targeted Ther.*, 6, 53, 2021.
18. Bashor, C.J., Hilton, I.B., Bandukwala, H., Smith, D.M., Veiseh, O., Engineering the next generation of cell-based therapeutics. *Nat. Rev. Drug Discov.*, 21, 655, 2022.
19. CMC Biotech Working Group, A-Mab: A Case Study in Bioprocess Development, ISPE, 2009.
20. NIIMBL, A case study to support development and adoption of advanced biomanufacturing, NIIMBL, 2024.
21. Martí-Carreras, J., Generous, N., Giammanco, M.D., Kalter, J., Keiser, P.H., Kunkel, S.R., Longmire, M., Merkel, S.F., Perkins, D.J., Fair, J.M., NCBI's Virus Discovery Codeathon: Building "FIVE" -The Federated Index of Viral Experiments API Index. *Viruses*, 12, 1424, 2020.
22. Wilder-Smith, A., Gubler, D.J., Weaver, S.C., Monath, T.P., Heymann, D.L., Scott, T.W., Epidemic arboviral diseases: priorities for research and public health. *Lancet Infect. Dis.*, 17, e101, 2017.

23. Sands, P., Mundaca-Shah, C., Dzau, V.J., The Neglected Dimension of Global Security — A Framework for Countering Infectious-Disease Crises. *N. Engl. J. Med.*, 374, 1281, 2016.
24. Halabi, S.F., Gostin, L.O., Crowley, J.S., Global Management of Infectious Disease After Ebola, Oxford University Press, Oxford, UK, 2016.

20

Design of Experiments for Biopharmaceutical Process Optimization

Vijaya Kumar Talari[1]*, Baburao Gaddala[2], Bijoy Kumar Purohit[1], Stutee Bhoi[1], A. V. Raghavendra Rao[1], K.S.N.V. Prasad[1], Gowrishetty Srinivas[1] and T. Srinivas[1]

[1]Department of Chemical Engineering, B. V. Raju Institute of Technology, Narsapur, Telangana, India

[2]Department of Chemical Engineering, Guru Ghasidas Vishwavidyalaya (A Central University), Koni, Bilaspur, Chhattisgarh, India

Abstract

Designed experiments (DoE) is an effective statistical method that is being used within the biopharma industry to optimize complex manufacturing procedures. This chapter discusses how DoE has been used in the process development of biopharmaceuticals with its role in defining critical process parameters (CPPs) and the interactions among them to improve product quality, productivity, and process robustness. It focuses on the Quality by Design (QbD), where DoE is part of systematic process knowledge and risk management. Major experimental designs including factorial, fractional factorial, response surface methodologies (RSM) and definitive screening designs (DSDs) appropriate in diverse levels of process development, i.e., cell culture and fermentation to purification and also for formulation are discussed in the chapter. Various case studies illustrating the effective application of DoE in order to overcome a process bottleneck, help reduce a process variability, and speed process scale up and technology transfer has been provided. Besides various problems, which include model overfitting, design space constraints, and the interpretation of results, appropriate practices of effective experimental planning are also discussed. Putting DoE into biomanufacturing processes, organizations can accomplish further uniformity, regulatory conformity and economic affordability of biologics as well as biopharmaceuticals.

**Corresponding author*: vijayakumar.t@bvrit.ac.in

Sandip Kunar, Jagadeesha T. and D. Sathis Kumar (eds.) Biomanufacturing: Transforming Production of Biologics and Biopharmaceuticals, (641–674)

***Keywords*:** Design of experiments (DoE), biopharmaceutical process optimization, quality by design (QbD), process analytical technology (PAT), statistical modeling

20.1 Introduction

Biopharmaceutical sector has gone through some major changes over the last few decades but the need to establish effective manufacturing processes capable of delivering a high product quality, safety, and efficacy has resulted in the changes. The DoE as a statistical methodology comes out as a crucial component in the field of systematic research into the determinants of the processes of improvement to find an optimal approach. DoE can be used to identify critical process parameters (CPPs) and their interactions which are key towards increasing the yield and reducing the variability and ensuring a robust process during biopharmaceutical production [1]. In this chapter, this text explores the use of DoE in the context of Quality by Design (QbD) and how it is used in process development through screening studies to optimization. With DoE integration, biopharmaceutical companies will reduce risks, meet regulatory requirements and obtain economical production of biologics, e.g., monoclonal antibodies, vaccines and recombinant proteins.

The systematic way used by DoE is as opposed to the commonly known traditional one-factor-at-a-time (OFAT) method in which more resources are often needed as the studies usually fail to record the interaction effects among the factors. In the case of biopharma, DoE is a way to use information to devise a path to a process understanding in which processes involve intricate biological systems such as cell cultures and purification procedures [2]. An example is that factorial designs allow the comparison of more than one factor at a given time, unravelling some synergies or antagonisms to critical quality attributes (CQAs), including purity, potency, and stability. Adoption of DoE is in line with the regulatory requirement of agencies such as Food and Drug Administration (FDA) and European Medicines Agency (EMA) that use QbD stressing on risk-based decisions where quality is incorporated into the product and not be tested on [3]. This introduction preliminarily positions the pursuit of an examination of the basics, designs, uses, case studies, issues, and future trends of DoE in biopharmaceutical process development.

20.1.1 Overview of DoE in Biopharma

DoE is a systematic method of planning, as well as carrying out experiments, and analyzing the results conducted to test the influence of input variables on output responses. DoE is already frequently used in the biopharmaceutical fields to optimize the production processes, such as upstream cell culture, downstream purification and formulation [4]. The procedure includes the identification of factors (e.g., temperature, pH, agitation speed), levels, and responses (e.g., cell viability, product titer, impurity levels) and then applying the designs, such as factorial or response surface, to come up with experimental runs. This method reduces the number of experiments and will, therefore, gain maximum information, a factor that is good in situations where resources are limited.

In the 1920s, DoE was developed by Ronald Fisher as agricultural statistics but later in the 20th century DoE required QbD in pharmaceuticals. In biopharma, DoE exhaustively addresses the intrinsic variability of biological systems under which the bioreactor conditions and the media composition interact non-linearly [5]. As an example, when culture is grown in fed batch fermentation of a monoclonal antibody, DoE can be used to identify the interactions between the glucose feed rate and the dissolved oxygen to come up with a higher yield and less by-products. Recent developments involve high throughput screening and the multivariate response to make DoE more applicable at early stages of development.

Due to the combination of DoE with Process Analytical Technology (PAT), process robustness can also be enhanced through real time monitoring and control. Studies indicate that DoE helps to cut down development schedule by 20-30 percent as they find out the best conditions at initiation [6]. In purification, DoE is used in optimizing chromatography steps, including resin type and working pH of the buffers to get highest recovery and purity. One-way DoE is useful in biologics is in the formulation development that benefits in stabilizing biologic against aggregation or degradation factors such as excipients and storage conditions. In general, DoE promotes the process in which the empirical turn into a scientifically robust process that will then be used when filing regulatory submissions.

DoE is also versatile in scale up and also in the technological transfer where it helps exchange design space consistency across manufacturing facilities. The issues revolve around choosing proper designs in complex systems but these are easily done using software such as JMP and Minitab. In short, DoE plays a critical role in biopharma in attaining effective reproducible processes with very high-quality demands [7].

20.1.2 Importance of Process Optimization

The optimization of the process is essential in the production of biopharmaceuticals as a means of improving the quality of the product, minimizing the cost of production and maximizing its yield under conditions of increased demand and rising pressure towards affordable biologics. DoE is central as it facilitates systematic optimization, determines those CPPs that impact CQAs, and risks are alleviated with the help of statistical modelling [8]. Processes which are not optimized have the potential to experience variability, creating batch failures, regulatory delays, and losses of money. As an example, low titers or high impurities may occur due to suboptimal fermentation conditions, and rework may be necessary.

The process optimization in the frame of QbD moves beyond reactive tendencies and becomes proactive with the use of DoE that allows developing the design space, which is a multidimensional area of process parameter that can be accepted as reasonable. This provides process robustness where the outputs do not go out of specifications caused by variations in the inputs [9]. Optimization can also speed time-to-market; the traditional method may need hundreds of experiments, but with DoE a comparable knowledge can be gained with fewer runs and thus save time and resources. An example is optimization of a protein purification process through fractional factorial design: this will cut down solvent consumption by 40 percent but enhance purity.

The economic gains are quite substantial since streamlined processes reduce the cost of production due to the reduction of wastes and maximization of efficiency. Optimization has the advantage of ensuring scalability and consistency particularly in a competitive market, where biosimilars put the originators into competition [10]. Regulatory compliance is also an impetus; agencies demand to see that it has a clear understanding of the process, which DoE produces *via* validated models. Besides, optimization will help with sustainability because it will consume less energy and materials in massive bioreactors.

Challenges in optimization include the fact that biological systems are multifactorial in nature, but DoE can detect these through the analysis of variance (ANOVA) and regression models and this aspect is invaluable. All of these integrations along with machine learning improves the predicting abilities and forecasts the optimal conditions based on previous facts. Finally, bottom-up DoE-based process optimization is not only useful to perform the process in a desirable manner, but also allows patient safety regarding the production of high-quality biopharmaceuticals.

20.1.3 Chapter Objectives

This chapter will focus on thorough discussion of DoE in the area of biopharmaceutical process development with a strong correlation to QbD to achieve better results. The main aim is to elucidate the process by which DoE determines the CPPs and interaction to optimize the processes and ultimately enhance the quality and yields as well as robustness [1]. By elaborating different designs such as factorial, response surface methodology (RSM), and DSDs the chapter also aims at helping the practitioner to make the right choices when carrying out different stages of the development.

The second goal is to show case studies to show the real-world examples of successful DoE implementations, how DoE can be used to address the bottlenecks, reduce variability, and support scale-up [2]. They will provide an idea of the successful aspects of cell culture, purification, and formulation and will be fruitful to the industry workers.

Lastly, the chapter discusses issues such as overfitting of models and interpretation of data, coming up with recommendations which form the best practice [3]. Upon fulfilling these objectives, readers will have understood better the transformative potential of DoE in biomanufacturing, which will encourage its broader adoption with the view to regulatory compliance and cost effectiveness.

20.2 Fundamentals of Design of Experiments (DoE)

The science of Design of Experiments (DoE) is the foundation of organized process optimization in biopharmaceuticals and helps identify the priorities of the variables and their impact on the results with minimal experimentations. Based on statistical grounds, DoE reduces the number of runs but increases the knowledge of system dynamics that is important in complicated bioprocesses such as fermentation and purification [2]. Unlike the conventional technique, DoE considers both interactions of factors and makes it more suitable in developing strong models to predict responses under different conditions. DoE can complement QbD in biopharma where biological variations are inherently high by helping to arrive at a design space that will provide consistent product quality [11]. Compositional elements entail factor selection, response definition, and choice of design, but they are meant to limit uncertainty and maximize reproducibility. History: Since the agricultural trials of Fisher to the application on biotech highlights DoE to be flexible with times [4]. DoE eliminates biases and other extraneous factors that would skew the decision, by the use of randomization, replications and block. In reality, DoE is used to shorten time-frames to develop something, save on expenditures, and facilitate regulatory acceptance by documenting verifiable knowledge of a process [12]. This section reviews foundational concepts, concepts of statistics, and functions of risk assessment with groundwork toward more advanced applications in the field of biopharmaceutical process development.

20.2.1 Basic Principles of Design of Experiments (DoE)

The main concepts of DoE include systematized experimentations aimed at identifying variables which are used to understand the relationship between the inputs and outputs effectively. In essence, DoE is a matter of either recognizing the factors (independent variables such as pH or temperature), responses (dependent variables such as yield or purity) and ideas of experimental designs that will uncover their interaction with the minimum experiments [2]. The randomization prevents the bias due to uncontrolled variables, while replication helps to establish the error of the experiment which increases the reliability in a biopharma environment where biological variation dominates [11]. The blocking groups are used when similar experimental units are introduced to control known factors of variation, such as batch differences in cell culture medium.

In biopharmaceuticals, these principles are used to maximize the upstream production process, including cell line development, where the concentration of nutrients and the agitation speed can reinforce or counteract each other in cell growth and the production of proteins [4]. As an example of the use of a full factorial design, the effect of the various levels of each of several factors is tested, as are the main effects and their interactions, but the design is resource intensive when many factors are being considered. A fractional factorial design eliminates runs by confounding the higher order interactions, which are appropriate when screening stages are desired [12]. The RSM expands these to include modeling of curvature in responses and is particularly useful in moving towards optimization in a formulation where the level of an ingredient such as an excipient affects stability [13].

Practical implementation commences with the definition of the problem after which factors are screened with the help of Plackett-Burman designs in which many factors are considered as potential CPPs whereby critical ones are determined [3]. DSDs are efficient to design because they cover both main effects, quadratic effects, and two-factor interactions in the same study design therefore requiring the least number of runs possible in the early stage bioprocesses [14]. Software programs such as JMP or Minitab aid in the design generation and analysis and include ANOVA results in the analysis of significance.

Use difficulties consist of making sure that factors reflect the real world constrains, and restraints such as non-orthogonal designs in equipment limitations at biotech plants [15]. Best practices focus on creating multidisciplinary to include statisticians, engineers and biologists who can relate DoE to process goals. Subsequently, the above concepts convert empirical trial-and-error to scientific investigation, enhancing efficacy in the generation of powerful biomanufacturing processes that meet QbD standards [6]. Such fundamentals allow biopharma to produce higher yields, reduce variability, and speed up scale up, and antibodies are an example where DoE has been used to improve media reagents by 20-30% [16].

20.2.2 Statistical Foundations

The mathematics of DoE is an enshrinement of the statistical fundamentals of DoE research in an experiment and the assessments of the research findings with certainty. At the center of DoE is general linear model where responses are expressed as functions of factors together with error to capture main effects, interaction and error term as provided in Equation (20.1) [2]. Xi and Xj are the controlled process variables (e.g., pH, temperature, agitation speed) which are the factors being investigated in terms of the variable's response Y (e.g., yield, purity). Each Xi is a particular factor and the coefficient on it β_i measures the main effect of it on Y.

$$Y = \beta_o + \sum \beta_i X_i + \sum \beta_{ij} X_i X_j + \epsilon \tag{20.1}$$

Analysis of variance (ANOVA) decomposes variance to determine the significance of the factors and involves F-tests to contrast model terms against the error. ANOVA F-test Equation (20.2) [11] can be used to calculate the statistical significance of the factors used in the biopharmaceutical process optimization taking into consideration the sum of squares of the model (SS_model), the amount of degrees of freedom of the model (df_model) the sum of squares of the error (SS_error) and the amount of degrees of freedom of the error (df_error).

$$F = \frac{SS\_model \,/\, df\_model}{SS\_error \,/\, df\_error} \tag{20.2}$$

The model fitting of regression analysis includes coefficients, which denote the degree of effects, necessary in predicting the bioprocess outcome, such as a protein yield, at different conditions [4].

In biopharma, normality assumptions are checked through checks on residual plots and transformation (e.g., log on skewed data) ensures validity [12]. Finding sufficient sample size to identify the effects is vital in settings with limited resources and this can be determined through the power analysis. The power analysis equation (20.3) [13] used the power (1 - β), where β represents the probability of a Type II error and it represents the probability of rejecting the null hypothesis (H0) when the alternative hypothesis (H1) is true and this is a measure of the sample size needed to detect significant effects in biopharmaceutical DoE studies.

$$wer = 1 - \beta = Pr\left((Reject\ H0 \middle| H1\ true)\right) \tag{20.3}$$

It attempts to solve the problem of multicollinearity by using orthogonal designs that avoid the problem of large variances in coefficient estimates [3].

Advanced statistics are multiple regressions to RSM, using central composite designs (CCD), which computes quadratic terms, through addition of axial points, simulating the optimum in both the pH and temperatures effects in fermentation [14]. Prior information can also be utilized in Bayesian methods that are effective during successive refinements of bioprocesses [15]. Multivariate analysis such as Principal component analysis (PCA) is used in on PAT-integrated DoE analysis of correlated responses [6].

In comparing design spaces, software can compute confidence and prediction intervals to quantify the uncertainty of the designs with the submission to the regulator [16]. Confirmation validations capacitated through validation runs guarantee model accuracy in the consequent connection between the statistical theory and the practical forecast. Biopharma is able to realize actionable information *via* data and streamline processes with risks quantified with the aid of these foundations.

20.2.3 Role in Risk Assessment

DoE becomes a central practice in risk analysis during biopharmaceutical development because uncertainties in processes are systematically discovered and overcome. Under QbD, DoE facilitates Failure Mode and Effects Analysis (FMEA) whereby the degree to which CPPs affect CQAs is quantified with priority being given to risks on the basis of their severity, occurrence, and detectability [2]. Under DoE, high-risk impacts are identified, including how temperature affects aggregation in formulation at different pH, allowing them to be controlled [11].

In risk mitigation, DoE establishes the operable design spaces where processes maintain robustness in the face of variability and minimizing the risk of failure by the scale-up [4]. Monte Carlo simulations combine DoE models to estimate the probabilistic risks, which are forecasts of variations in yields of raw materials [12]. Examples of DoE used to minimize the risks observed during a process include reducing risks during purification processes by 40%. DoE established the conditions of purification and purification buffers to reduce impurities [13].

Regulatory alignment is increased as DoE offers empirical data of the risk-based submissions which is in line with ICH Q8-Q10 requirements [3]. Among the challenges are the inclusion of rare events which is overcome through the use of strong designs such as Taguchi designs that reduce sensitivity to noise [14]. Best practices consist of sequential DoE processes going through screening then optimization along with the continuous improvement of risk profiles [15]. Combining PAT with risk monitoring takes this further with real-time monitoring of the risk with parameters changing dynamically [6]. In general, DoE quantifies risk eliminating the need to consider risks as qualitative [16].

20.3 Quality by Design (QbD) Framework

Quality by Design (QbD) is a structural method of pharmaceutical development, where prior-set goals and process understanding are the focus to provide the product quality consistency in biopharmaceutical development. QbD essentially incorporates scientific knowledge and risk management in the very beginning which was not done in the quality-by-testing approaches [17]. QbD determines target product profiles (TPPs), the critical quality attributes (CQAs), and the critical process parameters (CPPs) in biopharma to develop robust processes in biologics, such as monoclonal antibodies and vaccines [18]. This model brings down variation, allows more efficiency, and helps get the regulations approved because the manufacturing is under control [19]. Key aspects involve risk evaluation devices such as failure mode effects analysis (FMEA) and definition of design space, which enable operation flexibility within established limits [20]. QbD is applied throughout process development to commercialization that uses real-time monitoring process analytical technology (PAT) [21].

QbD minimizes changes of the post-approval by paying attention to the accumulation of knowledge, which is essential to underpin continuous improvement when a bioprocess is complex and susceptible to biological variability [22]. Multidisciplinary teams are implemented in the implementation of process inputs and output mapping that should be aligned to the needs and expectations of patients and patient requirements by the regulators [23]. Research demonstrates that QbD may reduce up to 20-30 per cent of the development cost as a result of precisely made experiments [24]. In the purification, chromatography is optimized using QbD by linking the resin characteristics with the removal of impurities [25]. In general, QbD promotes innovation that allows the delivery of safe biopharmaceuticals into the market quicker [26].

20.3.1 Quality by Design (QbD) Principles

QbD principles focus on building quality into products by planning and developing risk-based approaches. The basis is the Quality Target Product Profile (QTPP) that contains desired characteristics such as efficacy, safety, and stability of biopharmaceuticals [17]. QTPP translates into CQAs which are potency and purity, and CPPs and critical material attributes (CMAs) are identified as influencers [18]. Factors affecting CQAs are prioritized based on the risk assessment, which is conducted with the help of such risk assessment tools as Ishikawa diagrams and FMEA [19]. The multidimensional area with parameters that are acceptable is the design space and it achieves robustness [20]. Design space equation (20.4) [18], is the expression of design space (DS) the set of values of the factors (X) where the response function (f(X)) is reached in the specification within the knowledge space making biopharmaceutical process parameters to be robust under QbD. Knowledge space consists of a broader understanding, whereas the control space is a definition of the limits of operations [21].

$$DS = \{X \mid f(X) \geq Specification,\ X\ in\ Knowledge\ space\} \tag{20.4}$$

Examples of such transfer principles are in biopharma, whereby media composition (CMA) can impact glycosylation (CQA) in an upstream process [22]. Downstream, buffer pH (CPP) influences aggregate formation [23]. PAT integration enables in-line monitoring, adjusting parameters dynamically [24]. QbD focuses on management of the lifecycle which cover the development, post-market and incorporation of feedback in order to improve [25]. Challenges such as dealing with biological complexity are also a challenge and can only be dealt with through multivariate analysis [26]. Advantages are reduced batch failures and scalability, which are recorded with the use of QbD with scaling up of vaccine production by optimizing fermentation processes [17]. The concepts nurture the proactive quality assurance, which is in line with the ICH guidelines that state global harmonization [18]. QbD helps in achieving predictable efficient processes by integrating science and risk into empirical development [19].

20.3.2 Integration QbD with DoE

The presence of DoE in QbD positively changes the process understanding because factor interaction is systematically investigated. DoE aids QbD through the screening of CPPs and the defining of design space *via* statistical designs [17]. In screening, Plackett-Burman designs determine factors that are significant from many, and reduces experiments [18]. RSM is used in optimization to model the responses and fits quadratic equations to the optima [19]. In biopharma, DoE is combined with risk assessment to measure the impact on CQAs [20].

Example, in fermentation DoE assesses interactions among temperature, pH and feed rate on titer [21]. Fractional factorials alias higher-order effects, are more effective at lower levels [22]. DSDs give estimates of main, quadratic, and two-factor interactions with less runs [23]. Design-Expert software assists in analysis, which produces contour plots of design spaces [24]. Models are validated through confirmation runs to ensure that they are accepted by the regulations [25]. Problems such as nonlinearity of biological systems are mitigated through augmentations on designs [26]. Integration speeds up process development; one study [11] demonstrated that the time was reduced by 40 percent in antibodies purification process [17].

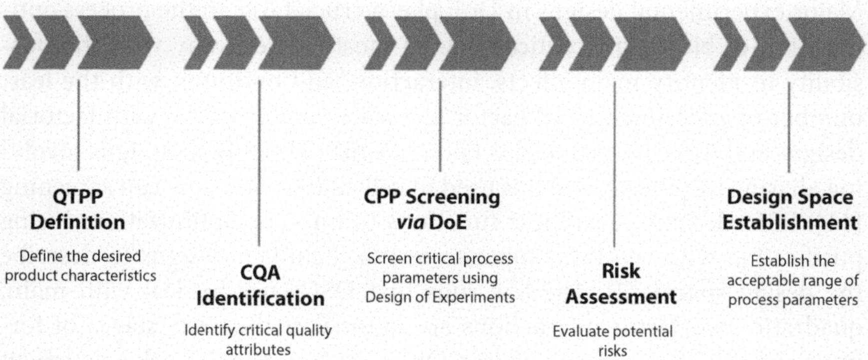

Figure 20.1 Quality by design (QbD) framework in Biopharma.

Because of PAT-DoE synergy, adaptive control keeps processes in design space [18]. In total, there is DoE operationalization of QbD, which can give data-driven support for flexible manufacturing [19]. Above Figure 20.1 shows the QbD framework in biopharmaceutical.

20.3.3 Regulatory Perspectives

Regulatory views on QbD are focused on making more process knowledge to provide flexible approvals. FDA and EMA support QbD under International Council for Harmonisation (ICH) Q8-Q11, whereby a design space is needed in submissions to have lesser monitoring [17]. Benefits include real-time release testing, bypassing end-product assays [18]. In the case of biopharmaceutical manufacturing, QbD facilitates comparability post-changes, essential to biosimilars [19].

Risk-oriented measures are expected of the agencies and the data of DoE links CPP-CQA [20]. Challenges: incomplete QbD may lead to questions; complete application demonstrates control strategy [21]. Case: a mAb was approved with a QbD-defined space, i.e., with parameters that could be modulated without revalidation, by EMA [22]. Global harmonization *via* ICH facilitates filings [23]. Future: The QbD is combined with continuous manufacturing which is facilitated by PAT [24]. Regulators encourage early dialogue *via* pilot programs [25]. The adoption rate increases (70 percent of new biologics are utilizing QbD) [26]. The views move away to the stringent specifications towards knowledge-based flexibility on conforming quality with accelerating access [17].

20.4 Key Design of Experiments (DoE) in Biopharma

Major experimental designs in DoE play a crucial role in the process optimization of biopharmaceuticals in the most efficient way with the possibility to identify main effects, interaction, and optimum with the least number of experiments [4]. Factor levels are combinatorial with factorial designs and there is a reduced version of runs in fractional designs involving aliasing the effects, and it is used in cell cultures as a low run screening [11]. RSM designs curvilinear functions to improve optimization during purification with central composite or Box-Behnken designs to describe the design spaces [2]. The few runs that DSDs use to deal with main, quadratic, two-factor interactions are optimal at the early stages of formulation [3]. These designs fit into QbD and facilitate the risk assessment and regulatory compliances in providing rich data in terms of CPP-CQA relationships [27]. These designs also speed up development, lower costs, and increase yield in biopharma, where the factors that are interacted with in a nonlinear manner include pH and temperature [28]. As an example, factorial designs in fermentation show nutrient-nutrient interactions that go into titer, whereas RSM optimizes buffer conditions to be more stable [29]. DSDs help connect the gaps between the screening and the optimization, reducing resources in high-throughput systems [30]. There is software which facilitates design production, analysis using ANOVA and regression to determine significance [31]. Improper resolution of fractional design is a problem; augmentation is used to alleviate this [12]. In general, these designs enhance systematic comprehension to enhance the robustness and efficiency of biomanufacturing.

20.4.1 Factorial and Fractional Factorial

Factorial designs provide controlled variation of all the factors at many levels to determine main effects and their interactions, crucial for the screening of biopharma processes [4]. All the effects are captured in full factorial where, 2^k runs of 2 levels of the k factors are carried out, e.g., testing pH, temperature, and agitation of a bioreactor on cell viability [11]. This shows synergies, such as pH-temperature, which has an effect on yield, going unnoticed by one factor at time (OFAT) [2]. Nevertheless, due to numerous reasons, runs grow exponentially, thus, impacting the purification with multiple buffers as resource-consuming processes [3]. By confounding higher-order interactions, fractional factorial designs (2^{k-p}) save on runs, and assume of sparsity of effects [27]. The main is aliased with two-factor interactions, which is appropriate for initial screening, in resolution III; better resolution IV separates main and two-factor (useful in bioprocesses) [28]. Practically, a 2^{5-1} design is used to screen five factors in 16 runs with the aim of optimizing fermentation media, to find out important nutrients [29]. Augmentation adds runs to de-alias if needed [30]. Advantages include efficiency and orthogonality for variance estimation [31]. Limitations: because of aliasing, there is a risk of aliasing causing false positive results which can be overcome using follow-up designs [12]. These designs are used in a biopharma to facilitate QbD through CPP-to-CQA mapping, e.g., in chromatography to remove impurities [4]. Examples: fractional factorial optimized harvest conditions which increased recovery 25 percent [11]. Integration with PAT enables real-time adjustments [2]; choose factors with a risk assessment, block noise with randomization [3]. Such designs will provide foundations to more fundamental optimization

Table 20.1 Comparison of key experimental designs in DoE for Biopharmaceutical processes.

Design type	Description	Advantages	Disadvantages	Suitable stages
Factorial	Tests all factor combinations at multiple levels	Detects main effects and interactions	High number of runs for many factors	Screening and early optimization
Fractional Factorial	Subset of full factorial, aliases higher-order effects	Fewer runs, efficient screening	Potential aliasing of effects	Initial factor screening
Response Surface Methodology (RSM)	Models quadratic responses with central composite or Box-Behnken	Identifies optima and curvatures	Assumes quadratic fit; more runs needed	Optimization after screening
Definitive Screening Designs (DSDs)	Three-level design estimating mains, quadratics, and two-factor interactions	Efficient (2k+1 runs), bridges screening/ optimization	Assumes no three-factor interactions	Early to mid-development

thus providing robust bioprocesses [27]. Table 20.1 depicts the comparison of experimental designs in DoE for Biopharmaceutical processes.

20.4.2 Response Surface Methodology (RSM)

RSM is used to optimize processes, by fitting responses as quadratic functions of factors, which is ideal to locate optima in biopharma [4]. It is based on the screening designs where central composite (CCD) or Box-Behnken (BBD) is generally utilized to quantify curvature [2]. Quadratic response surface model Equation (20.5) [2], the response variable (Y), with intercept (β0), main effect coefficients (β_i), quadratic coefficients (β_{ii}), interaction coefficients (β_{ij}), factors (X_i, X_j), and error term (ε) in optimizing the biopharmaceutical process parameters through modeling the nonlinear relationships and estimating the optimum.

$$Y = \beta_o + \sum \beta_i X_i + \sum \beta_{ii} X_i^2 + \sum \beta_{ij} X_i X_j + \epsilon \quad (20.5)$$

To formulate, RSM can optimize excipient concentration by assuring protein stability and producing contour plots of drug space design [27]. BBD avoids extreme corners, safer for biological constraints [28]. In fermentation, RSM models the pH and the dissolved oxygen in terms of the titer and determines the saddle point or maxima [29]. Procedure: determine the factors/responses, conduct experimentation, optimize model using least squares, test against ANOVA (R^2 greater than 0.8 preferable) [30]. Advantages: extrapolates to unoccupied regions, favors QbD design spaces to establish regulatory agility [31]. Challenges: supposes quadratic sufficiency; greater orders require additional runs [12]. RSM was used in purification where it was able to meet 90% of yield at the best salt concentration and flow rate [4]. Software like Design-Expert plots 3D surfaces for visualization [11]. DoE-cycle integration: factorial screen, RSM optimize [2]. Case studies demonstrate that there is 30 percent yield increase in production of antibodies [3]. Best practices: centering of pure error, non-extremeness of transformation [27]. RSM enhances process robustness, reducing variability in scale-up [28].

20.4.3 Definitive Screening Designs (DSDs)

DSDs are effective three-level designs of main effects, quadratics, and two-factor interactions that can be estimated in 2k+1 runs that fill the gap between screening and optimization in the field of biopharma [29]. They have been developed by Jones and Nachtsheim and use conference matrices to orthogonalize, so there is no mains-quadratics aliasing [30]. For six factors, 13 runs suffice, compared to 32 for full factorial [31]. At early development, DSDs test media items against cell growth which shows curvature at early processing [12]. Model: consists of linear, quadratic, interaction effects; sparse models are fitted forward using selection [4]. Merits: has less runs than RSM, can manage categorical factors, effects power [11]. DSD has been used to optimize buffer and surfactant during formulation in a way that the minimization of aggregation is obtained [2]. Limitations: no three-factor-interactions assumed; and >10 factors, blocked DSDs [3]. Geospatial software is used to include and examine (Least Absolute Shrinkage and Selection Operator) LASSO variable of selection [27].

DSDs used in biopharma enable quicker technological transfer through the delineation of initial design space [28]. Case: DSD in purification, 9 runs were performed till optimal pH and conductivity were found to enhance purity [29]. Integration with QbD: facilitates risk prevention by measuring the effects [30]. Best practices: adjust scales correctly, check beyond the doubles [31]. DSDs promote resource-efficient development, enhancing innovation in biologics [12].

20.5 Applications in Process Development Stages

Design of Experiments (DoE) is essential in performing crucial steps of biopharmaceutical process developmental, which includes upstream cell growth and formulation downstream and scale-up. DoE facilitates a methodical test of factors having an impact on crucial quality properties (CQAs), including the yield, purity, and stability, and the experimental burden is reduced [32]. DoE is used in cell culture and fermentation to determine the best media compositions and operating conditions, which optimize the biomass and product titer [12]. In purification, it optimizes chromatography parameters to give improved recovery and removal of impurities [11]. DoE is also used in formulation and scale-up, where it establishes robust design spaces across scales [33]. Working hand in hand with the Quality by Design (QbD), DoE can help in reducing risk and providing a regulatory compliance as the development timelines reduce by 20 to 40 percent [34]. Such case studies reveal the effectiveness of DoE in preventing the variability in fed-batch cultures as well as in optimizing perfusion processes [35]. Biological variability is one of the challenges, which is dealt with the help of multivariate analysis and high-throughput systems [15]. Overall, DoE develops effective and scalable bioprocesses of biologics such as monoclonal antibodies, and vaccines, which increases cost-effectiveness and the quality of the products [16].

20.5.1 Cell Culture and Fermentation

In upstream bioprocesses0, DoE plays a crucial role in optimizing biological processes, such as cell culture and fermentation, which requires fine control of many parameters, such as pH, temperature, dissolved oxygen, and media nutrients [32]. The factorial designs simultaneously screen out the various factors and determine the type of interactions that may affect the cell growth and productivity [12]. As an example, fractional factorial DoE has been used to simultaneously optimize the glucose feed and agitation in Chinese Hamster Ovary (CHO) cell cultures used in the production of monoclonal antibodies and raised the titer by 30% while decreasing production of lactate [11]. Response Surface Methodology (RSM) is used to model nonlinear responses, e.g., quadratic effects of amino acid concentration on glycosylation profiles, which are important CQAs [33].

DoE is useful in the fermentation of microbial hosts such as Escherichia coli or Pichia pastoris to enhance DoE in media formulation *via* Plackett-Burman designs to screen trace elements as well as carbon sources [34]. In DoE study on Geobacillus thermoglucosidans the defined medium was created, which facilitated development of metabolic and growth characteristics on lignocellulose carbohydrate as well as resulted in enhanced growth rates [35]. Definitive Screening Designs (DSDs) effectively approximate main effects and curvatures in reduced runs and this works well in high throughput microbioreactors such as the case with Ambr 15 [15]. DoE was applied to determine optimal amino acids and yeast extract to improve Outer Membrane Vesicle (OMV) yield and purity in bacterial production, as a vaccine platform [32].

To keep conditions optimal, such factors as dissolved oxygen setpoints can be adjusted in real-time through integration with Process Analytical Technology (PAT) [16]. Issues are acetate inhibition of E. coli and this is overcome through fed-batch strategies guided by DoE [36]. Scale-down models confirm DoE findings with assurance of the findings at production levels [37]. In general, cell culture and fermentation DoE enables increased understanding of the process quicker, lower costs of process development, and complements QbD through an apportioning of operable ranges of CPPs such as induction timing and nutrient feeding [12].

Some high cell density fermentations like in case of recombinant proteins utilize DoE to optimize growth and expression phases [11]. In one instance, E. coli growth using Isopropyl β-D-1-thiogalactopyranoside (IPTG) and temperature optimization could be improved using DoE data and the identification of linear regression models used to increase volumetric yield [33]. In yeast systems, DoE-optimized methanol feeding in Pichia was used to express a heterologous protein product [34] proved to be able to produce higher titers of the heterologous protein. Multivariate Data Analysis (MVDA) is complementary to DoE and links spectral data to responses in order to make predictive models [35]. The best practices require screening and then optimization to include batch variability blocks [15]. The implementation of DoE in our case will increase the productivity and also make the process impervious to changes in raw materials, which is essential in commercial production [16]. Table 20.2 shows the applications of DoE in biopharmaceutical process stages.

Table 20.2 Applications of DoE in Biopharmaceutical process stages.

Process stage	Key factors optimized	Responses measured	Example benefits
Cell Culture/ Fermentation	pH, temperature, media nutrients, agitation	Cell viability, titer, glycosylation	20-40% yield increase, reduced by-products
Purification	Buffer pH, conductivity, flow rate, resin type	Purity, recovery, impurity levels	>95% purity, 25% solvent reduction
Formulation/ Scale-Up	Excipients, pH, ionic strength, shear stress	Stability, viscosity, scalability	Minimized aggregation, consistent CQAs across scales

20.5.2 Purification Processes

Downstream purification takes advantage of DoE to optimize such unit operations as chromatography, filtration, and viral inactivation, with the main aim of maximizing recovery and impurity minimization [32]. Through DoE, interaction of pH and conductivity of buffers and the degree of load were examined in protein A chromatography of monoclonal antibodies (mAbs) to determine design spaces of high purity (>95%) [12]. Preceding to central composite designs (CCD) model are the elution gradients, which center on the CCD as a way of decreasing host cell crummy contaminants [11]. In ion-exchange step, DoE optimizes the resin type and salt concentrations during polishing to attain improvement of selectivity [33].

In viral filtration, DoE evaluates the impact of pressure, flow rate and protein concentration on the flux and retention and, in this way, aims at virus removal without compromise of product quality [34]. An example of bispecific antibodies purification where DoE was employed is the case study to address the aggregation phenomenon in multi-column chromatography [35]. The DoE can be done in parallel mode using robotic high-throughput screening to trade-off the parameters required in the ultrafiltration/diafiltration (UF/DF) process, reducing the time it will take to find the right parameters to set [15]. Certain hurdles such as resin fouling are compensated through factoring time into models [16].

DoE is combined with QbD in order to develop control strategies, e.g., low bench viral inactivation, which allows some model-based characterization of temperature/hold time interactions to be assessed as robust [36]. In continuous purification, DoE cuts down cycle times and buffer exchanges in computer simulation of moving bed systems [37]. The ANOVA statistical tool can validate significant results, with confirmation runs confirming model accuracy [32]. In general, DoE during purification makes it more efficient, uses less buffer (20-30% less), and allows scale-up identification of parameters that are independent of scale [12]. Some best practices that relate to risk-based factor selection using FMEA and repeat designs to improve on [11], are iteration of designs as well as selection of factors on the basis of the risk.

To concur complex biologics such as antibody-drug conjugates (ADCs), DoE optimizes conjugation steps, balancing linkers ratios with reaction times to be able to provide uniform drug antibody ratios [33]. In therapeutic vectors, DoE perfects centrifugation and precipitation to produce adeno-associated viruses (AAV) and increase yields of the vector's genome [34]. MVDA is used to analyze huge quantities of data generated by DoE and a correlation of the process parameters with CQAs such as potency can therefore be inferred [35].

20.5.3 Formulation and Scale-Up

The formulation of biologics is also based on DoE to stabilize against degradation by evaluating excipients, pH and ionic strength in long-term storage [32]. Mixture designs used to optimize the buffer conditions of polysorbate and sucrose concentrations to reduce aggregation in mAb samples [12]. RSM would be used to model viscosity and osmolality responses of subcutaneous delivery [11]. In lyophilization, DoE optimizes rates of freezing and shelf temperatures so that cake must retain its integrity and residual moisture is <1% [33].

DoE assists scale-up through determination of scale dependent parameters such as shear stress in the bioreactor [34]. Engineering models together with DoE preserve oxygen transfer rates (kLa) between different scales without a decrease in titers [35]. In technological transfer, DoE establishes equivalence guidelines where equivalence testing is applied in an attempt to draw comparisons [15]. The problems associated with continuous manufacturing are discussed in terms of the residence time optimized by DoE in perfusion cultures [16].

In case of oral biologics, DoE designs an enteric coating used to optimize the ratio between the polymers involved to release them at variable pH [36]. In mediating messenger ribonucleic acid (mRNA) with nanoparticle formulations, DoE optimizes lipid compositions [37]. The regulatory involves DoE-based design in filings accounting to flexible operation [32]. In DoE, worst-case scenarios are also taken into consideration as best practices [12]. DoE helps scale up quickly and considerably shortens the timeframe because you find out early on which interactions are important [11].

In vaccine design, DoE will maximize the ratios of adjuvants to elicit immunogenicity [33]. In case of biosimilars, DoE can illustrate analytical similarity through an illustration of process variabilities [34]. High-throughput (HT) platforms allow DoE in formulation screens to quicken the search of candidates [35].

20.6 Case Studies

Case studies depict the real-world practice of Design of Experiments (DoE) in the biopharmaceutical process development with the evidence that DoE may be effectively and efficiently used to address bottlenecks, variability reduction and also technology transfer in the Quality by Design (QbD) context [6]. These illustrations demonstrate how DoE is used to find process critical parameters and interactions used to optimize complex processes such as fermentation, purification, and formulation, which are used to improve the output, quality and robustness [13]. Also, through combined methods of statistical modeling and data in the real world, DoE is capable of reducing experimental runs and maximized learning, contributing to compliance with regulations, and economies [38]. The subsequent subsections exemplify the cases of upstream to downstream and illustrate the paradigm-changing role played by DoE in addressing the biomanufacturing problems [24].

20.6.1 Resolving Bottlenecks

Bottlenecks in the processes of biopharmaceuticals frequently occur due to a poor setting of parameters, which results in low yields, impurities, or a low level of efficiency of unit operations [6]. These are dealt with by screening and the optimization of factors methodically that 'in turn reveals the existence of interactions between factors which adversely affects performance [39]. A good example was that of holistic DoE (hDoE) of integrated process model in production of monoclonal antibodies where conventional methods needed too many experiments. Using hDoE method associated unit operation models, experimental events have been decreased by more than 50 percent and the final product quality measurements such as purity and potency out-of-specification (OOS) rates have been minimized [6]. The benchmarking of hDoE compared to standard DoE through *in silico* simulations raised the performances of hDoE in terms of allocating runs in key steps such as chromatography, eliminating the variability load bottlenecks [6].

A further example is the use of DoE in botanical drug products, and in this case, application to ethanol precipitation in danshen extraction. Risk assessment revealed the CPPs in regard to the recovery of active pharmaceutical ingredient (API) and saccharide removal in terms of density, ethanol concentration, and temperatures. A Box-Behnken design was also used, which found interactions, namely, that higher density enhanced saccharide removal but reduced API recovery [38]. The resulting design space (density range 1.20-1.21g/ml, ethanol range 2.6-2.8 ml/ml, temperature range 12-15°C) alleviated purification bottlenecks, increased the recovery by 15-20 percent, and assurance of quality [38]. This is a scenario where empirical challenges posed by complex herbal treatments were overcome with the help of DoE [38].

In microbial fermentation, DoE was used to overcome the bottlenecks in Pichia pastoris in recombinant protein production [16]. In fractional factorial designs, temperature, pH and methanol feed were screened to identify any interaction that had an effect on titer [16]. Under optimized conditions using Response Surface Methodology (RSM) and yield was improved by 40% and the oxygen transfer limitation was overcome [16]. The specified design space made it robustly run reducing bottlenecks during the induction phase [16]. In a similar way, in genetic pathway optimization of terpene production in Streptomyces, DoE was used to inform the choice of enzyme variants, to overcome the problem of flux imbalances [15]. Quadratic effects have been identified and improved productivity by 25 times through Plackett-Burman screening and RSM [15].

To address bottlenecks in continuous wet granulation and drying of a new API the DoE was based on the model [40]. Hybrid simulation models were able to make accurate ahead-of-time predictions of granule size, thus optimizing settings so that 73% material is saved [40]. This eliminated the problem of scale up in fluid-bed drying and increased throughput [40]. In preparation of samples when doing pharmaceutical analysis, DoE optimized the parameters of extraction saving 50 percent in analysis time and solved method shortfalls. These examples show that DoE is made powerful in identifying and removing bottlenecks at any level in biopharma manufacturing, including the upstream synthesis of protein expression and the downstream isolation of the proteins.

20.6.2 Reducing Variability

The variability in the bioprocess is caused by the uncontrolled parameters, raw material variations, biological variation, which results in a decrease of product quality [41]. This is minimized by quantification of effects and definition of robust design spaces in DoE [24]. An industrial example is the application of QbD using Definitive Screening Designs (DSD) to evaluate variability of formulation and process of a two-API oral solid dosage product [24]. A study of CPPs in granulation and compression targeted factors such as the binder concentration and the compression force that affected the uniformity of the contents and the dissolution properties [24]. Ranges were tightened by using multivariate regression models and ensuring that the assays and impurities varied by just 30% with PCA confirming the scale-up batches [24].

In biopharmaceutical control solutions, the use of random effects using the linear mixed models (LMMs) addressed week-to-week variation in the maintenance of monoclonal antibodies [41]. The eight-unit operations studied provided DoE data where the random variance was greater than the residual returning to six instances [41]. LMM-based tolerance regions extended established proven acceptable ranges (PARs) with an OOS risk decrease of 40 per cent as compared to usual least squares modeling [41]. This neutralized fluctuation caused by unregulated aspects such as operator differences [41].

In the case of Pichia pastoris fermentation, DoE was used to identify a design space that would reduce inconsistency in titer and glycosylation [16]. Factorial designs were used to test the interactions of pH and temperature, and confirmation runs demonstrated a difference of less than 10 per cent batch-to-batch variation [16]. In genetic optimization, DoE with machine learning minimized the variations in pathway expression in engineered microbes [15]. Promoter strengths were evened out in surface models to produce uniformness in the terpene production independent of genetic contexts [15].

In ultrafiltration, DoE was used to certify mechanistic models and decrease the variability resistance in terms of concentration [39]. Simulations were used to optimize the conditions of the buffer and reduced process deviations in the cutting process by 25% [39]. In wet granulation, the DoE guided by the model predicted variability in granule sizes, and adjustments during the real-time could be made to ensure that they get constant drying [40]. In the redesign of pharma processes, DoE was used to reduce the variability of lipid nanoparticles in the formulation of mRNA vaccines to come up with stable Comirnaty Â© production [42]. These investigations emphasize the strength of DoE in increasing ruggedness, which makes sure that the CQAs are not out of spec in the presence of intrinsic fluctuations of bioprocesses [6].

20.6.3 Accelerating Technological Transfer

Technological transfer comprises transfer of processes into commercially related manufacturing, a process at times hampered by an incomplete comprehension [42]. This is speeded up through early establishment of scalable design space [16]. In the case of Pichia pastoris, DoE-derived parameters could switch between 2L and 200L upscale with oxygen transfer equivalence at both scales with kLa modeling minimizing transfer time by 35% [16].

DoE was used to support the shifting of an industrial QbD investigation of coated tablets between formation and production [24]. DSD and RSM were successfully used on optimized Normal Operating Ranges (NORs) that were granulated, and this was confirmed through three successful production batches that ensured it was launched 30 percent earlier in the market [24]. PCA has the capability of analyzing historic data which enables easy site transfer to occur with limited revalidation [24].

In integrated process modeling, hDoE simulated technological transfer situations, and assigned runs to problematic high-risk unit operations such as filtration, pre-cutting time characterization by 50% [6]. This enabled rapid adaptation to new facilities [6]. In the case of botanical drugs, DoE-derived design space precipitation was used, so doing enabled consistency at scale and reduced filing FDA [38].

DoE was used to identify the most promising host strains in genetic pathway, technological transfer in Streptomyces, where DSDs identified transferable genetic material, and reduced the time required to integrate it into the host in half [15]. In continuous granulation, model-based DoE confirmed scale-independent parameterization, and minimized the time to commercialize [40].

In biopharma controls, PARs of new equipment were adapted using LMMs of DoE-generated data to predict variability during the transfer [41]. An example of mRNA is Spikevax vaccine, where DoE was used to transfer the global supply chains as an optimization of its supply in terms of encapsulation to sustain its efficacy [42]. These examples demonstrate that DoE facilitated the successful, risk-reduced transfers of technology that has made biomanufacturing more agile [13].

20.7 Challenges and Best Practices

Some of the challenges facing the utilization of DoE in biopharmaceutical process development include model overfitting, in which models are able to capture noise rather than actual relationships and results in poor prediction in new circumstances [43]. Space constraints have cause by not having full factor level or have a lack of incorporation of variability restricting the flexibility of the process [44]. Complexities of data interpretation are due to multivariate interactions and biological noise, which necessitate the application of sophisticated statistical methods [4]. Best practices combine diligent planning, a risk-based factor selection and validate it through iteration to get robust results [11]. The solutions to these increase the efficacy of DoE in optimizing yield, variability reduction, and QbD applicability to regulatory approvals [45]. In actual use, overfitting is countered by cross-validation and parsimonious models [40]. Such design space restrictions are overcome through addition of experiments and uncertainty analysis [46]. Strong data plans entail multivariate and software integration to provide very insightful interpretations [47]. Guidelines in planning stress multi-disciplinary cooperation and coordination with process objectives [25]. All in all, addressing these issues with superior practices reduces the pace of development, saves cost, enhances robustness of biologics in biomanufacturing [18].

20.7.1 Model Overfitting and Limitations

Model overfitting would be used when statistical models of DoE learn training data too well with noise randomness producing bad generalization to unseen states [43]. In biopharma, this is worsened by the high-dimension of e.g., fermentation process where many factors nonlinearly interact [44]. In the case of overfit models, it is possible that the optimality in the combination of pH and temperature that is predicted by the model will not work at scale due to additional variability uncaptured in the model such as cell lines or media variability [4]. The implications have been the lack of reliable spacing of design making it fail on certain batches and negative regulation [11]. Limitations of DoE designs, which may include loss of resolution in fractional factorials, aliasing of high-orders, possible neglect of important synergies as can happen in purification steps [45]. The freedom to excursion as presented by QbD space boundaries may not include real-life excursions such as changes to raw material lots, which limit operational flexibility [40].

To mitigate overfitting, one can use methods such as the k-fold cross-validation in which the data is divided and models run on subsections and the validation is performed against the hold-outs to compare the predictive power of the model to recommend suitable effort [46]. LASSO is one example of regularization schemes that tend to simplify the bioprocess regressions by penalizing complicated models, and achieving sparsity [47]. To overcome the limitations, augment designs after screening; e.g., center points to find the curvature that was not represented in linear models [25]. Stochastic simulation includes variability in the overarching design space and propagates it using Monte Carlo techniques reliably [18]. In mAb production, the RSM model overfitted and the titers were suboptimal; cross-validation reduced the use of complex models, resulting in a 25 percent increase in yield [43]. The same situation was observed in chromatography where the shortcomings peculiar to fractional designs were eliminated through de-aliasing by means of fold-over experiments commonly referred to as fold-over-experiments [44].

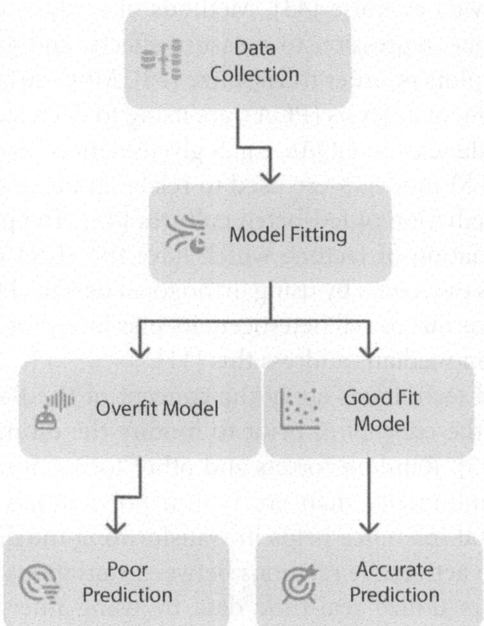

Figure 20.2 Model overfitting in design of experiments.

The natural variability of biological systems such as post-translational modifications compounds these concerns, and requires these models to be bifurcated or hybrids between the mechanistic and empirical models [4]. Diagnostics against overfitting are included in software tools like JMP, such as variance inflation factors [11]. Best practices advise that screening designs should be implemented first to determine important factors, and only then should optimization take place, so as not to end up with an overcomplicated test [45]. Regulatory approaches prioritize verifiable constructs; FDA guidance on QbD up to value validation using independent data [40]. Issues in high throughput DoE, where the volume of data is so large as to become overwhelming to analyze, are addressed by applying machine learning to the selection of features [46]. Comprehensively, it is crucial to understand overfitting and limitations exist, and then by fixing them good bioprocesses can be obtained that are robust, scalable [47]. Figure 20.2 explains the need for good fit model.

20.7.2 Data Interpretation Strategies

Interpretation of data DoE in biopharmaceuticals describes how meaningful information should be presented by analyzing large amounts of data, whose complexities are characterized by an interaction of peculiarities and nonlinearity as well as noise [43]. Methods of analysis contain ANOVA to test significance, regression to measure effects, and graphical displays such as contour plots in order to visualize [25]. Multivariate methods, e.g., principal component analysis (PCA), are using to decrease the dimension, which gains tendencies in CQAs, e.g.,S glycosylation profiles [18]. Partial least squares (PLS) models were used to relate inputs to outputs, and this helped in the prediction of fed-batch cultures [44]. The problem is collinearity, the covariation of factors, which have the effect of inflating variances, and this is overcome by using orthogonal designs [4]. The biological outliers that occur due to cell heterogeneity bias interpretations; the robust statistics based on medians address this [11].

The advanced techniques apply the concept of the Bayesian inference that introduces the concept of prior to modify the estimations when the data is sparse [45]. Random forests and other forms of machine learning are better at nonlinearity than are typical polynomials [40]. In omics-based DoE, causal inference helps in transforming the data, treating it as observational, to actionable relations between causations and correlations [46]. Soft sensors utilize real-time data by using interpretable learners to increase bioreactor monitoring [47]. An example is that in Pichia fermentation PLS was used to interpret the spectral data, which minimized the error involved in the interpretation of variability by 30 percent [43]. Confirmation testing: validation through confirmation testing gives final confirmation between the laboratory and production [44].

Best practices: normalization, transformation of data (the latter to deal with skewness) [4]. Two-way effects are critical in pH-temperature synergies and dissection of two-way effects should be made by use of interaction plots [11]. Software packages, such as Design-Expert and R allow customization [45]. Regulatory compliance mandates that so-called clear interpretations; ICH Q8 focuses on demonstration of knowledge [40]. Processing with high amounts of data are handled using cloud computing [46]. Effective strategies eventually turn raw DoE information into biopharma actionable knowledge, which optimizes the outcomes [47].

20.7.3 Effective Planning Guidelines

In the biopharma area it is effective to practice DoE so that the experimentation works in line with the objectives, resources, and risks [43]. The first guideline is to identify objectives: screen CPPs, or optimize CQAs [25]. FMEA which is used to evaluate risk also prioritizes factors, which means that an organization pays most attention to the elements that have the most significant impact, such as media components [18]. Multidisciplinary teams like statisticians, biologists, engineers collaborate for comprehensive planning [44]. Choose designs depending on the stage: Plackett-Burman design will be used as screening tool, and CCD design as optimization tool [4]. Incorporate constraints, like equipment limits, *via* custom designs [11].

Using power analysis, run numbers to detect the effect are determined [45]. Blocking controls known variability, e.g., batch effects [40]. Randomization mitigates biases [46]. Pilot studies refine ranges, avoiding extrapolation errors [47]. Interconnect with QbD: connect to QTPP to deliver desired results [43]. Software aids planning, simulating designs for efficiency [44]. Complexity: resource limits in early development; tackled by high throughput platforms [4]. Scalability planning ensures lab results translate to production [11].

In the case of human induced pluripotent stem cell (hiPSC) bioreactor, DoE planning overcame the problem of aggregate stability, through factorial designs, thus hastening the process of scale-up [45]. Best practices follow the documentation with traceability, the iterative refinement based on the intermediate analyses [40]. Regulatory alignment: plan for validation data supporting filings [46]. The overall result is that DoE value is maximized in meticulous planning to enable effective bioprocess development [47].

20.8 Conclusion

Design of Experiments (DoE) has become an essential component in biopharmaceutical process development and provides statistical rigor and the application of Quality by Design (QbD) to improve the manufacturing process efficiency, identify critical process parameters, and achieve product quality. This chapter has discussed the key tenets of DoE, some designs involved in different types of experiments, its practicalities, case studies, and challenges as well as the importance of DoE in enhancing the yield, process robustness, and regulatory conformity. DoE reduces associated costs and promotes the pace of innovation in the production of biologics because it tackles the difficulties of biological systems. Moreover, digital developments and modeling are expected to bring great advantages, such as cost decrease and improvement in sustainability. Finally, DoE replaces gross-ropinirole filling in biopharmaceutical manufacture who adhere to evidence-based techniques rather than the trial-and-error method.

20.8.1 Summary of Key Insights

Its statistical basis offers effective mechanisms of screening the factors and detecting the interactions therein, thus it is a strong optimization tool of a biopharmaceutical process. Such fundamental concepts as randomization, replication, and blocking tackle variability in biological systems, which addresses well with the risk-based approach to the development process in QbD. More specific designs include factorial, fractional factorial, response surface methodology (RSM) and definitive screening designs (DSDs) that can solve early stage screening and later stage optimization. Practically speaking, DoE is used to improve cell culture media to increase titers, chromatography to increase purity during purification and formulation stability in order to be able to scale up successfully.

Case studies based on reality point to the contributions DoE has made: one placed RSM to address fermentation bottlenecks and it has raised yield by 40%; another has used mixed models to address variability in monoclonal antibodies (mAb) production. Technology transfer is also simplified at DoE through the definition of robust design spaces. Some of the best practices that one can use to mitigate the shortcomings such as overfitting include cross-validation, multidisciplinary planning, and iterative validation.

The ability to be integrated and used in process analytical technology (PAT) and multivariate data analysis (MVDA) enables monitoring and real-time control of processes, as well as continuous manufacturing. The regulatory approaches appreciate DoE because it supports the international guidelines compliance, thereby supporting flexible operations. On the whole, DoE decreases the number of experiments, providing a deeper insight to the process and saving 20-30 percent in terms of time and cost through optimized selection. It establishes operable areas in QbD systems, guaranteeing essential quality properties such as potency in addition to stability, which means that DoE serves as the missing component between the development and commercial phases.

20.8.2 Future Directions

Hybrid modeling: Hybrid modeling is the future of DoE in biopharma, a modeled mix of mechanistic and data-driven models allows for accurate predictive models of dynamic processes. The combination of machine learning, including neural networks with DoE will address nonlinearities in omics data optimizing cell engineering pathways. Bioprocessing 4.0 also envisages digital twins where the DoE conditions can be simulated virtually to minimize on physical experiments and allow construction of adaptive, real-time control. Biological variability will be taken care of through the Bayesian techniques, which will increase confidence in personalized medicines. New modalities, such as gene therapies and mRNA vaccines, will be screened faster by high-throughput automation together with DoE. Through DoE, sustainability initiatives will be used to develop more environmentally friendly processes in continuous manufacture with a minimum waste approach. Regulatory developments would make it easier to submit models and there will be new methods of validation of AI-based methods. Standard practices will be achieved through shared data repositories, and data silos will be addressed through federated learning without losing intellectual property and enabling the shared knowledge in an industry. These developments will make biomanufacturing even faster, smarter and even more predictive, embracing AI and IoT.

20.8.3 Implications for Industry

Embracing DoE will help the biopharma become more competitive because the result is increased efficiency in the process by 20-50%. Reduced experiments will save funds and failures, and the size of the designs can be scaled, making the transfers of technologies and international development of biosimilars simple. The empirical evidence provided by DoE makes regulatory compliance much simpler, faster, and makes release possible in real-time. Proactive quality assurance reduces recalls by controlling variability. The positives of sustainability entail the optimization of utilization of resources and alleviated environmental unfriendly effects in mass production. The trend of upskilling the workforces in statistics apps such as JMP promotes innovation, which evens the ground amongst small and medium businesses. Patients benefit from consistent, affordable biologics, improving access. Although it is difficult to implement, the speed of market entry offered through DoE is one of the leading reasons why it is a harbor of resilience and efficiency in addressing future biopharma issues.

AI disclosure statement: The author acknowledges the use of AI-based tools for minor language editing or grammatical correction. However, all intellectual contributions, analyses, and conclusions presented in this chapter are entirely the author's own.

References

1. Fukuda, I.M., Pinto, C.F.F., Moreira, C.S., Saviano, A.M., Lourenço, F.R., Design of Experiments (DoE) applied to Pharmaceutical and Analytical Quality by Design (QbD). *Braz. J. Pharm. Sci.*, 54, e01006, 2018.
2. Weissman, S.A. and Anderson, N.G., Design of Experiments (DoE) and Process Optimization. A Review of Recent Publications. *Org. Process Res. Dev.*, 19, 1605, 2015.
3. Gujral, G., Kapoor, D., Jaimini, M., An updated review on design of experiment (DOE) in pharmaceuticals. *J. Drug Deliv. Ther.*, 8, 147, 2018.
4. Politis, S.N., Colombo, P., Colombo, G., Rekkas, D.M., Design of experiments (DoE) in pharmaceutical development. *Drug Dev. Ind. Pharm.*, 43, 889, 2017.
5. Mathur, M. and Kusumdevi, V., Design of experiment: A novel and systematic approach of developing formulations in pharmaceuticals. *World J. Pharm. Res.*, 8, 865, 2019.
6. Oberleitner, T., Zahel, T., Pretzner, B., Herwig, C., Holistic Design of Experiments Using an Integrated Process Model. *Bioengineering*, 9, 643, 2022.
7. Tanushree, C., Verma, R., Sharma, J., Sharma, P., Full Factorial Design for Optimization of Clindamycin Phosphate Loaded Nanogel: A Design of Experiments (DOE) Approach. *Ijppr.Human*, 24, 160, 2022.
8. Marushka, J., Hurychová, H., Šklubalová, Z., Duintjer Tebbens, J., Flow Equations for Free-Flowable Particle Fractions of Sorbitol for Direct

Compression: An Exploratory Multiple Regression Analysis of Particle and Orifice Size Influence. *Pharmaceutics*, 14, 1653, 2022.

9. Goh, H.P., Heng, P.W.S., Liew, C.V., Comparative evaluation of powder flow parameters with reference to particle size and shape. *Int. J. Pharm.*, 547, 133, 2018.
10. Vraníková, B., Svačinová, P., Marushka, J., Brokešová, J., Holas, O., Duintjer Tebbens, J., Šklubalová, Z., The importance of the coating material type and amount in the preparation of liquisolid systems based on magnesium aluminometasilicate carrier. *Eur. J. Pharm. Sci.*, 165, 105952, 2021.
11. Kumar, V. and Rathore, A.S., Design of experiments applications in bioprocessing: Concepts and approach. *Biotechnol. Progr.*, 30, 86, 2014.
12. Mandenius, C.F. and Brundin, A., Bioprocess optimization using design-of-experiments methodology. *Biotechnol. Progr.*, 24, 1191, 2008.
13. Souza, J.A., *et al.*, Quality by design approach with design of experiment for sample preparation in pharmaceutical analysis. *Pharm. Sci. Adv.*, 1, 100029, 2023.
14. Tanco, M., Viles, E., Ilzarbe, L., Alvarez, M.J., Implementation of Design of Experiments projects in industry. *Appl. Stochastic Models Bus. Ind.*, 25, 478, 2009.
15. Natarajan, S., *et al.*, Using design of experiments to guide genetic optimization of engineered metabolic pathways. *J. Ind. Microbiol. Biotechnol.*, 51, kuae010, 2024.
16. Harms, J., Wang, X., Kim, T., Yang, X., Rathore, A.S., Defining process design space for biotech products: case study of Pichia pastoris fermentation. *Biotechnol. Progr.*, 24, 655, 2008.
17. Zurdo, J., Arnell, A., Obrezanova, O., Smith, N., Gómez de la Cuesta, R., Gallagher, T.R.A., Michael, R., Stallwood, Y., Ekblad, C., Abrahmsén, L., Höidén-Guthenberg, I., Early Implementation of QbD in Biopharmaceutical Development: A Practical Example. *Biomed Res. Int.*, 2015, 605427, 2015.
18. Yu, L.X., Amidon, G., Khan, M.A., Hoag, S.W., Polli, J., Raju, G.K., Woodcock, J., Understanding Pharmaceutical Quality by Design. *AAPS J.*, 16, 771, 2014.
19. Singh, S., Quality by design- newer technique for pharmaceutical product development. *Intell. Pharm.*, 2, 122, 2024.
20. Gupta, A., Kumar, V., Singh, S., Implementation of Quality by Design (QbD) for development of bilayer tablets. *Eur. J. Pharm. Sci.*, 184, 106412, 2023.
21. Nithiyananthan, K., Rekha, R., Sravani, B., Venkata Sumanth, G.K.R., Jahnavi, M., Sri Vinay Reddy, M., Narendra, P.E.D.A., Comprehensive Understanding on Various Aspects of Quality by Design – A Review. *J. Popul. Ther. Clin. Pharmacol.*, 30, 2480, 2023.
22. Aksu, B. and Mesut, B., Quality by Design (QbD) application for the pharmaceutical development process. *J. Pharm. Investig.*, 52, 575, 2022.
23. Mohseni-Motlagh, S.F., Dolatabadi, R., Baniassadi, M., Baghani, M., Application of the Quality by Design Concept (QbD) in the Development of Hydrogel-Based Drug Delivery Systems. *Polymers*, 15, 4407, 2023.

24. Testas, M., da Cunha Sais, T., Medinilha, L.P., Almeida, I.F., Costa, P.C., Sobral, L., An industrial case study: QbD to accelerate time-to-market of a drug product. *AAPS Open*, 7, 12, 2021.
25. Rathore, A.S. and Winkle, H., Quality by design for biopharmaceuticals. *Nat. Biotechnol.*, 27, 26, 2009.
26. Lawrence, X.Y., Raw, A., Lionberger, R., Raju, S., Tsong, Y., Lee, H.Y., Adams, R., Fang, L., Khan, M., Shi, L., US FDA question-based review: a modern quality assessment for pharmaceutical manufacturing. *Pharm. Eng.*, 29, 15, 2009.
27. Hibbert, D.B., Experimental design in chromatography: A tutorial review. *J. Chromatogr. B Analyt. Technol. Biomed. Life Sci.*, 910, 2, 2012.
28. Dejaegher, B. and Vander Heyden, Y., Ruggedness and robustness testing. *J. Chromatogr. A*, 1158, 138, 2007.
29. Jones, B. and Nachtsheim, C.J., A class of three-level designs for definitive screening in the presence of second-order effects. *J. Qual. Technol.*, 43, 1, 2011.
30. Erdem, I. and Hasselblad, V., Definitive screening designs for the development of chemoenzymatic dynamic kinetic resolutions. *Org. Process Res. Dev.*, 17, 272, 2013.
31. Xiao, L., Lin, D.K.J., Bai, F., Constructing definitive screening designs using cyclic generators. *J. Qual. Technol.*, 44, 1, 2012.
32. Baccante, A., Petruccelli, P., Saudino, G., Ragnoni, E., Johansson, E., Di Cioccio, V., Mazarakis, K., Optimization of a Bacterial Cultivation Medium *via* a Design-of-Experiment Approach in a Sartorius Ambr® 15 Fermentation Microbioreactor System. *Fermentation*, 9, 1002, 2023.
33. Hsueh, K.L., Lin, T.Y., Lee, M.T., Hsiao, Y.Y., Gu, Y., Design of Experiments for Modeling of Fermentation Process Characterization in Biological Drug Production. *Processes*, 10, 237, 2022.
34. Gupta, P.K. and Edula, J.R., Strategies for Enhancing Product Yield: Design of Experiments (DOE) for Escherichia coli Cultivation, in: *Biomedical Product and Materials Evaluation*, IntechOpen, London, UK, 2021.
35. Singleton, C., Gilman, J., Rollit, J., Zhang, K., Parker, D.A., Love, J.A., design of experiments approach for the rapid formulation of a chemically defined medium for metabolic profiling of industrially important microbes. *PLoS One*, 14, e0218208, 2019.
36. Dong, J., Mandenius, C.F., Lubberstedt, M., Urbaniak, T., Nissom, P.M., Chua, F.H., Yap, M., Chou, S., Opportunities and challenges of real-time release testing in biopharmaceutical manufacturing. *Biotechnol. Bioeng.*, 114, 2445, 2017.
37. Ellert, A. and Vikström, C., Design of Experiments with Small-Scale Bioreactor Systems: Efficient Bioprocess Development and Optimization, BioProcess International, 2014.

38. Zhang, L., Yan, B., Gong, X., Yu, L.X., Qu, H., Application of Quality by Design to the Process Development of Botanical Drug Products: A Case Study. *AAPS PharmSciTech*, 14, 277, 2013.
39. Rao, G., *et al.*, Qualification of mechanistic models in biopharmaceutical process development. *J. Pharm. Sci.*, 113, 2953, 2024.
40. Matsunami, K., Vandeputte, T., Jiménez, A.A.B., Peeters, M., Ghijs, M., Van Hauwermeiren, D., Nopens, I., De Beer, T., Validation of model-based design of experiments for continuous wet granulation and drying. *Int. J. Pharm.*, 646, 123493, 2023.
41. Oberleitner, T., Zahel, T., Kunzelmann, M., Thalhammer, J., Herwig, C., Incorporating random effects in biopharmaceutical control strategies. *AAPS Open*, 9, 4, 2023.
42. Singh, S., Rethinking Pharmaceutical Industry with Quality by Design: Application in Research, Development, Manufacturing, and Quality Assurance. *AAPS J.*, 27, 56, 2025.
43. Narayanan, H., Luna, M.F., von Stosch, M., Cruz Bournazou, M.N., Polotti, G., Morbidelli, M., Butte, A., Sokolov, M., Bioprocessing in the Digital Age: The Role of Process Models. *Biotechnol. J.*, 15, 1900172, 2020.
44. Koulouris, A., Sarris, I.A., Papanikolaou, A., Contoravdi, C., Kiparissides, A., A tool for efficient integration of biopharmaceutical process design, process simulation and economic analysis – Application to ATMP cost modeling. *Comput. Chem. Eng.*, 153, 107442, 2021.
45. O'Brien, E.K., Taylor, C.E., Marshall, J.J., Lloyd, C.J., Kraemer, K., Shapland, E.B., Cooper, T.F., Barrick, J.E., Addressing bioreactor hiPSC aggregate stability, maintenance and scaleup challenges using a design of experiment approach. *Stem Cell Res. Ther.*, 15, 205, 2024.
46. Menezes, J.C., Cabrera, C., Kiang, Y.H., Lepore, J., Haque, S., From observational to actionable: rethinking omics in biologics development. *Trends Biotechnol.*, 41, 1134, 2023.
47. Gargalo, C.L., Udugama, I., Pontius, K., Lopez, P.C., Nielsen, R.F., Hasanzadeh, A., Mansouri, S.S., Bayer, C., Junghanns, K., Gernaey, K.V., Towards smart biomanufacturing: a perspective on recent developments in industrial measurement and monitoring technologies for bio-based production processes. *J. Ind. Microbiol. Biotechnol.*, 47, 947, 2020.

21

Zero Waste Biomanufacturing Systems: Concepts and Case Studies

Bhaskar Bethi*, **Vijaya Kumar Talari, Bhanu Radhika Gidla, Bijoy Kumar Purohit, A. V. Raghavendra Rao, Stutee Bhoi and K.S.N.V. Prasad**

Department of Chemical Engineering, B V Raju Institute of Technology, Narsapur, Telangana, India

Abstract

Innovation in zero-waste biomanufacturing has arisen due to increased demand for sustainable and environmentally responsible modes of production. In this chapter, the authors describe the essence, enabling technologies, and practical applications of zero-waste systems in the biopharmaceutical and biologics manufacturing industry. Focusing on the principles of the circular economy, this chapter presents the ways of incorporating resource recovery, waste valorization, and energy efficiency in the upstream and downstream of bioprocesses. Discussed in detail is the role of advanced bioprocessing strategies including continuous manufacturing, closed-loop system, and precision fermentation. Moreover, the lists of digital instruments, like process analytical technology (PAT), machine learning, and life cycle assessment (LCA), is also introduced as one of the significant contributors to the realization of the waste reduction and accurate assessment and provision of sustainability in practice. Multiple examples of industry and research case studies are reviewed to demonstrate how the principles of zero-waste were applied, such as bio-based reuse of by-products, reuse of process water, and reuse of the single use bioreactor components. The chapter also ends by addressing regulatory point of view, economic considerations and aspects of further research that would be needed to have extensive adoption. Such an exhaustive review offers insights to the researchers, industry players and policymakers trying to develop powerful sustainability in biomanufacturing activities.

Keywords: Zero waste biomanufacturing, circular economy, sustainable bioprocessing, waste valorization, life cycle assessment (LCA)

**Corresponding author*: bhaskar.b@bvrit.ac.in

Sandip Kunar, Jagadeesha T. and D. Sathis Kumar (eds.) Biomanufacturing: Transforming Production of Biologics and Biopharmaceuticals, (675–706) © 2026 Scrivener Publishing LLC

21.1 Introduction to Zero-Waste Biomanufacturing

Zero-waste biomanufacturing is a disruptive process to industrial biotechnology, with goals of producing bio-based products like biofuels, bioplastics and biochemicals without associated waste, and maximizing efficiency of the resources used. In contrast to conventional manufacturing, zero-waste biomanufacturing unites the concepts of sustainability and circularity, so that by the end of the production, absolutely every material utilized in the process is then transformed into valuable goods or used through the circle so that it could be later used again. The paradigm shift plays a vital role in combating the world problems like depleting resources, environmental degradation and climate change. Zero-waste strategies are emphasized in biomanufacturing to decrease the environmental footprint by as much as 70% in some industries, because of the shift to a circular economy in which waste is designed out of the production system [1].

Zero-waste biomanufacturing is an idea, which grows out of the notion of material loops and considers the waste stream assets instead of liabilities. By way of example, bio-based products can be made using agricultural residues, food waste, industrial by-products and these feedstocks can avoid the use of virgin materials as well as diminish landfill applications [2]. The value of zero-waste biomanufacturing is not limited to environmental gains since it has economic benefits such as reducing the cost of production because of waste valorization and improving the competitiveness of a product in the market when driven by sustainability needs [3]. In addition, it is compatible with the sustainability objectives of the globe, including the United Nations Sustainable Development Goals (SDGs), especially SDG 12 (Responsible Consumption and Production) and SDG 13 (Climate Action), through measuring the waste output and reducing it to minimize the emission of greenhouse gases [4].

21.1.1 Principles of Circular Economy

A zero-waste biomanufacturing is based on the circular economy concentrating on the concept of design out waste, maintaining materials in use, and regenerating natural systems [5]. The circular economy opposed the linear landfill take-make-dispose model to adopt a regenerative approach that restores resources using reuse, recycling, or repurposing so that a loop is established. In the case of biomanufacturing it means utilizing renewable feedstocks, reducing the amount of energy consumed, reusing by-products of the processes to make new products or energy. To illustrate, lignocellulosic biomass can be converted into biofuels, bioplastics, and biochemicals such that all the biomass is used [6].

Key principles of the circular economy in biomanufacturing include:

1. Waste as resource: by-products of one process can be utilized as input in another. An example is that the glycerol by-product of biodiesel manufacture can be a source substrate of microbial (chemo and bio) chemical processing such as 1, 3-propanediol [7].
2. System Thinking Biomanufacturing: Biomanufacturing systems design is done in holistic manner considering the life cycle of the products. Lifecycle assessment (LCA) determines the threats to the environment along the production chain [8].
3. Regeneration: Matters substitute fossil-based materials with bio-based products; hence, the carbon footprints decrease [9].
4. Collaboration: Zoephis to Zoephis efforts can also be seen by developing networks of industrial symbiosis that combine biomanufacture with either agriculture/waste management industries [10].

Such principles are then operationalized using biotechnological core technologies such as synthetic biology and biocatalysis to manipulate

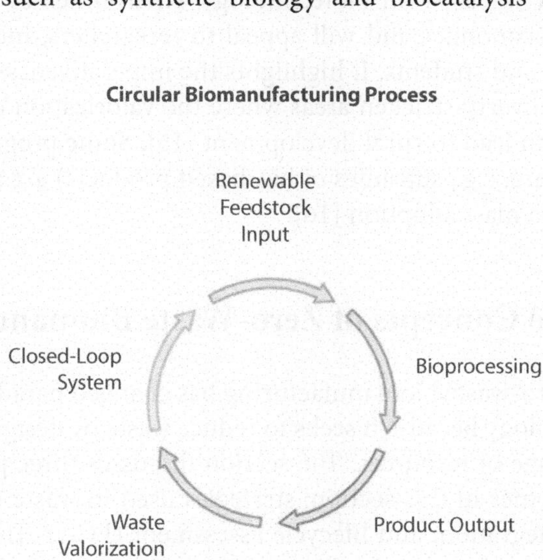

Figure 21.1 Conceptual framework of zero-waste biomanufacturing.

biological systems in precise ways to maximize the use of resources and minimize waste [11]. Integrated biorefineries make use of the same feedstock to essentially make a variety of products out of one in the co-production of goods [12]. Figure 21.1 depicts the conceptual framework of Zero-Waste Biomanufacturing.

21.1.2 Scope of the Chapter

This chapter gives a detailed discussion of zero-waste biomanufacturing systems linking the theory and the actual applications. Its objectives are to:

i. Explain main concepts, such as waste minimization, integration of a biorefinery, and lifecycle assessment.
ii. Point out the technological advancements such as CRISPR-based genome editing and enzyme design to produce waste-valorization [13].
iii. Existing case studies on biofuels production, bioplastics manufactory, and agricultural wastes valorization.
iv. Overcome technical, economic, regulatory barrier and provide insights into how it can be overcome.
v. Study the future trends e.g., the use of AI and IoT as part of the Industry 4.0 technology in biomanufacturing [14].

The chapter is based on biotechnology, environmental science, engineering, and economics and will appeal to researchers, industry people, policymakers, and students. It highlights the imperativeness to the world, especially in poverty-stricken areas where the valorization of the agricultural wastes can lead to rural development [15]. Some programs involved (policy structure, e.g., subsidies of bio-based products) are also discussed as important to mass adoption [16].

21.2 Core Concepts of Zero-Waste Biomanufacturing

The idea of zero-waste biomanufacturing has changed paradigm in industrial biotechnology because it seeks to reduce waste by designing new processes and usage of resources. The section discusses three principles that are central themes in this section: strategies used in waste minimization, biorefinery integration, and lifecycle assessment (LCA). These principles become the basis of sustainable biomanufacturing systems which adhere to the goals of the circular economy to ensure maximum resource utilization and little environmental impairment.

21.2.1 Waste Minimization Strategies

The heart of zero-waste biomanufacturing is waste minimization, involving minimization, reuse, and recycling bioprocess materials. Biomass residues and chemical effluents remain as major byproducts of traditional biomanufacturing, which in turn has had adverse effects on the environment. To address this, waste minimization that involves processes optimization, innovative bioprocessing, and valorizing by-products are used.

Process intensification increases the efficiency in optimizing reaction conditions and decrease the utilization of resources. As an example, high-cell-density fermentation permits maximum product output compared to a low waste stream [17]. In contrast to a batch process, continuous bioprocessing enables real-time monitoring, and, as a result, further minimizes waste [18]. Byproduct valorization transforms residues into valuable co-products. A typical example refers to using lignocellulosic bioethanol by-products instead of dumping; either through anaerobic digestion used to produce biogas or biochemicals, or enzymatic hydrolysis, which processes into biochemicals [19]. This reduces waste and enhances economic viability. The principles of green chemistry, including the non-toxic solvents, reduce hazardous waste [20]. Recycling techniques, like membrane filtration, recover water and enzymes. Ultrafiltration of mammalian cell cultures can cut down to 70% of the water use [21] to that effect.

21.2.2 Biorefinery Integration

The integration of the biorefinery allows the multiple products to be produced out of a variety of feedstocks in minimal waste disposed. Biorefineries break down biomass to biofuels, biochemicals and biomaterials in a coupled fashion using all the components of the feedstock [22]. As an example, lignocellulosic biorefineries can convert cellulose, hemicellulose and lignin into ethanol, chemicals and bio-based materials accordingly [23]. Recovery of energy in the waste streams takes place through co-location of processes (e.g., fermentation and anaerobic digestion) minimizing the dependence on accessible outside energy sources [24]. Synthetics biology is efficient because it can be used to engineer even microorganisms to produce useful products like lactic acid out of waste materials like glycerol [25]. This guarantees the zero-waste character of the biorefineries to recycle various feedstock, such as industrial wastes into high-value products.

21.2.3 Lifecycle Assessment (LCA)

LCA examines the environmental impact of zero-waste biomanufacturing on cradle to grave. It measures the use of resources and emissions and wastes along the production chain. As an illustration, an LCA of a corn stover (CS) bioethanol manufacturing process demonstrated that residue valorization cut down the emissions with 30% [26]. LCA makes comparisons of the feedstocks and it shows that sugar cane bagasse minimizes energy requirement during the manufacture of the bioplastics [27]. Other strategies which it evaluates include a circular economy strategy like processing food waste to create bioethanol and biogas, which reduces the impact by half [28]. Nevertheless, issues that are of concern to the reliability of LCA are data variability, and the definition of the system boundary [29]. Standardizing methodologies is essential for guiding zero-waste biomanufacturing. Figure 21.2 depicts the comparative Lifecycle Assessment (LCA) of traditional vs. Zero-Waste Biomanufacturing.

21.2.4 Synthesis and Implications

Zero-waste biomanufacturing is centered on waste reduction, biorefinery, and LCA. The combined notions allow closed-loop systems in which resources can be reused, and waste can be eradicated. The fact that they have been adopted in other industries such as biofuels and bioplastics,

Figure 21.2 Comparative lifecycle assessment (LCA) of traditional vs. zero-waste biomanufacturing.

demonstrates their viability even though scaling it up may have to solve cost and regulatory challenges. The next sustainable future of biomanufacturing development will be based on the advances in synthetic bio, process engineering, and LCA.

21.3 Technological Approaches

21.3.1 Advanced Bioprocessing Techniques

Cutting-edge bioprocessing methods play a critical role in the implementation of zero-waste biomanufacturing, since they streamline resources usage and reduce waste streams. Such methods include integrated bioprocess designs, which afford more efficiency by way of continuous processing, process intensification and waste recycling. Continuous bioprocessing unlike the conventional batch process enables monitoring and controlling in real time, thus minimizes wastage of resources and improves yields. As an example, perfusion-based bioreactors sustain steady-state cell cultures and realize more efficient production of biopharmaceuticals and less waste [30]. This process has been implemented in the production of monoclonal antibodies, in which a minimum waste byproduct is produced since the nutrients are continuously added and waste removed.

The work of intensifying processes involves maximum output and minimum energy, and material is inputted. Such methods as membrane

Table 21.1 Key technologies in zero-waste biomanufacturing.

Technology	Application	Example	Impact
Continuous Bioprocessing	Monoclonal antibody production	Perfusion bioreactors	Reduces resource waste by 30%
Synthetic Biology (CRISPR)	Waste-to-bioplastics	PHA production from glycerol	Upcycles biodiesel by-products
Anaerobic Digestion	Agricultural waste valorization	Biogas from rice straw (Punjab project)	Cuts methane emissions by 90%
Membrane Filtration	Water recycling	Ultrafiltration in cell cultures	Reuses 70% of process water

bioreactors and integrated separation systems make it possible to recover byproducts and reuse them. As an example, membrane filtration is used to separate valuable enzymes or metabolites and waste streams with a view to reintegrating into the production cycle [31]. Auto control and real-time analytics control parameters such as pH and substrate level so that things get all right and waste is minimal [32]. These technologies are in line with zero-waste principles because they recycle material loops but are initially expensive and require competent operators [32]. Hybrid fermentation/enzymatic catalysis systems generate biofuels or biochemicals, and subsequently used waste biomass, further reducing the waste [33], which illustrates the prospect of environmentally friendly scalable biomanufacturing. Table 21.1 summarizes the enabling bio-processing technologies and their real-world impacts.

21.3.2 Synthetic Biology for Waste Valorization

Synthetic biology presents an opportunity to scale out the possibility of zero-waste biomanufacturing by designing microorganisms that can bio-convert their waste products into valuable products. Synthetic metabolic routes permit microorganisms to upcycle wastes such as lignocellulosic waste into biofuels or bioplastics. The agricultural residues are converted to ethanol or lactic acid, using the engineered Escherichia coli and Saccharomyces cerevisiae [34]. CRISPR-Cas9 can be used to tune metabolic pathways to improve microbial efficiency, such as by increasing yields and decreasing byproducts [35]. As an example, polyhydroxyalkanoates (PHAs) are derived out of waste glycerol, the product of biodiesel production, generating environmentally-friendly plastics [35]. There is the biorefineries complex waste degradation and in biorefineries microbial consortia can also reduce landfill dependency [36]. Issues, such as genetic stability and GMO regulations, still exist [37], but synthetic biology has become one of the pillars of zero-waste biomanufacturing processes.

21.3.3 Renewable Feedstocks

Renewable feedback changes dependency on exhaustible sources toward the usage of long-lasting and waste-based resources. Food wastes, agricultural residues, and industrial wastes are low-cost resources. Enzymatic hydrolysis and fermentation is used to convert lignocellulosic biomass such as corn stover to make bioethanol [38]. Biomass is made more accessible with pretreatment processes, like steam explosion [39]. Feedstocks such as waste generate low-waste citric acid through Aspirgill King Fermentation because waste products are rich in sucrose [40]; some of these sources include molasses [40]. CO_2 is made during fermentation, and industrial symbiosis utilizes such waste streams of the by-products, such as fermentation products, as feedstocks to algal biomass to produce biofuels [41]. Feedstock composition variability and logistical costs are a challenge, but optimization of feedstock supply chain and standardization is an essential attribute of zero-waste biomanufacturing.

21.4 Case Studies

21.4.1 Biofuel Production with Zero-Waste

Production of biofuels has become one of the pillars of sustainable energy, which fills the global demand of renewable fuel sources and mitigates the environmentally depreciating effects of fossil fuel emissions. Nevertheless, the conventional process by which biofuels are produced is highly inefficient, since it tends to produce numerous waste products such as lignocellulosic by-products, wastewater, and gaseous streams, agent at the cost of their environmental friendliness. Zero-waste biomanufacturing is proposed to overcome these pitfalls and build on the principles of the circular economy, so that any by-products are used to create valuable products. This case study will address the zero-waste biofuel production, give attention to the production of bioethanol and biodiesel, and thoroughly analyze a lignocellulosic bioethanol plant and its waste utilization measures. Technological advances, process integration, and practical applications are identified as the main areas of discussion, but it is supported by life cycle assessment (LCA) and economic reasoning.

21.4.1.1 Background and Context

Biofuels (biomass) are obtained through feedstock biomass, including agricultural waste, energy plants and organic wastes, and some of these include bioethanol and biodiesel. It is estimated that the global market of biofuels will increase greatly due to policies which involve reduction in greenhouse gas emissions [42]. However, conventional biofuel production generates substantial waste. To take just one example: Lignocellulosic biomass bioethanol production generates lignin-containing residues and effluents; biodiesel production generates glycerol by-product. The goal of zero-waste biomanufacturing is to close these waste streams and use them as co-products (such as biogas, bioplastics, or fertilizer), to increase economic and environmental sustainability.

Another case of zero-waste biofuel production is the integrated biorefinery model where biofuel production and valorization of all by-products are combined. The case study centers on a hypothetical lignocellulosic bioethanol plant using real-world examples of similar implementation, including operations of companies such as POET-DSM and Beta Renewables who have led the way, to date, in zero-waste production of bioethanol [43].

21.4.1.2 Process Description

The production process of lignocellulosic bioethanol starts by pretreating the biomass, enzymatic hydrolysis, fermentation and distillation. Agricultural residues such as corn stover or wheat straw are the main feedstock; they are made of cellulose, hemicellulose, and lignin. It is pretreatment that enables breaking down of the intricate lignocellulose structure to reach the cellulose component in order to hydrolyze it to fermentable sugars. These sugars are either fermented by saccharomyces cerevisiae or engineered microbes into ethanol, which is subsequently distilled to obtain fuel-grade bioethanol.

Within a traditional process, the residue containing lignin (around 20-30 % of the biomass) and fermentation waste water are either dumped or used in the minimally way possible resulting in pollution of the environment. In a zero-waste system, these by-products are systematically valorized. It is possible to divide the process into the following steps:

i. Pretreatment and hydrolysis: Pretreatment of biomass with steam explosion or dilute acid is done to de-link cellulose, hemicellulose and lignin. Cellulose can be hydrolyzed by enzymes into glucose and hemicellulose produces xylose, a C5 sugar.
ii. Fermentation: Glucose is fermented into ethanol using saccharomyces cerevisiae. Advances in synthetic biology have made it possible to employ engineered microbes, including the ability of the zymomonasmobilis to ferment C5 and C6 sugars and thereby boosting ethanol output [44].
iii. Distillation and Ethanol Recovery: Ethanol is distilled off and a stillage remains that has water, sugars not yet used, and microbial biomass.

iv. By-Product Valorization: ZERO-WASTE approach adds value to more processes to make use of by-products:
   a. Lignin Valorization: Lignin is a complicated polymer with higher-value products used as bio-based adhesives, carbon fibers, or syngas through thermochemical conversion, like pyrolysis or gasification [45].
   b. Stillage Processing: The stillage is processed through anaerobic digestion to create biogas that the biorefinery may run on to cut the amount of external energy brought in. Digestate is nutrient-rich, and a biofertilizer [46].
   c. Hemicellulose Utilization: Hemicellulose has xylose fermentation, such as xylitol in case of bioproducts or microbial production of biochemicals [47].

21.4.1.3 Case Example: Integrated Lignocellulosic Biorefinery

An illustrative case of zero-waste bioethanol manufacture is based on the POET-DSM Project LIBERTY in Iowa, USA that came into commercial operation in 2014. This plant handles 770000 st of corn remnants each year to give 20 million gallons of cellulosic ethanol [43]. The zero-waste strategy involves:

- Feedstock: Corn stover is locally grown in agricultural fields to cut down on transportation emissions and to contribute to the local farmers.
- Pretreatment: Dilute acid pretreatment would isolate cellulose, hemicellulose, and lignin as well as leave with little chemical waste.
- Ethanol manufacturing: Enzymatic breakdown, and fermentation gives the efficiency of ethanol 8080 to 90 back to maximum theoretical [48].

21.4.1.4 By-Product Management

Lignin is burned in a power plant to produce steam and electricity causing the plant to be energy sufficient. Stillage is anaerobically digested and biogas produced offsets 30% of the facility natural gas demand. Digestate is converted to commercially sold nutrient-rich fertilizer to local farmers, which closes the nutrient cycle.

According to the lifecycle assessment (LCA) of the facility, greenhouse gas emissions were reduced to 60% that of gasoline and waste valorization played a great part in this decrease in gas emissions [49]. From an economical point of view there are co-products (biogas, electricity, and fertilizer) that help interpolate extra sources of income, enhancing financial performance of the plant.

21.4.1.5 Technological Innovations

Several technological advancements enable zero-waste biofuel production:

- Synthetic Biology: Microbial production of ethanol or other chemicals is boosted by engineered microbes (E. coli modified to ferment xylose) [50]. This increases overall yield and reduces waste.
- Biocatalysis: New enzymes have been engineered with greater hydrolysis efficiency, cellulases and xylanases minimizing any remaining biomass [51].
- Process Integration: Bioethanol integrated biorefineries can include production of bioethanol with biogas and biochemical, which optimize resources utilization. As an example, an anaerobic digestion with ethanol fermentation recycles water and nutrients and it minimizes wastes [52].
- Digital Tools: Digital tools like process simulation and real-time monitoring have been developed in Industry 4.0, which streamlines the flow of resources and eliminates inefficiencies. As an example, biorefinery processes can be modeled and optimized under Aspen Plus [53].

21.4.1.6 Challenges and Solutions

Despite its promise, zero-waste biofuel production faces challenges:

- Technical Barriers: This is because the recalcitrance of the lignocellulosic biomass makes it energy intensive to pre-treat. These could be solved by increasing the efficiency of the enzymes and pre-treatments, including processes based on ionic liquids [54].
- Economic, and viability: The big capital and operational costs restrict the level of scalability. Profitability can be increased through co-product valorization, e.g., through

lignin-derived chemicals to make revenue less concentrated on a certain substance [55].

- Regulatory Hurdles: Variable regulations governing biofuel subsidies and the standards of waste management are barriers to the adoption. There should be harmonized regulations and incentives of circular economy practices [56].

21.4.2 Bioplastics and Waste Recycling

21.4.2.1 Introduction

Bio-plastics, made of recyclable biomass raw materials like corn or other spent sugarcane, or lignocellulosic, supplement bio-based sources as an option to petroleum-derived plastics which pose a major source of waste worldwide. Bioplastics manufacturing would complement the zero-wastes biomanufacturing concept towards the use of renewable feedstocks, and recyclability of wastes with biodegradable or recyclable substances. This case study reviews bio manufacturing processes of bioplastics: in particular, polylactic acid (PLA) and polyhydroxyalkanoates (PHA) in such a way that waste recycling strategies are incorporated in an attempt to pursue zero-waste objectives. This section goes along with making use of the examples of industry and innovative solutions and demonstrates that the production of bioplastics will be one of the ways that could help to close the material loop, minimize the impact on the environment, and help build the circular economy [57, 58].

21.4.2.2 Bioplastics: Types and Production Processes

There are two broad groups of bioplastics, bio-based plastics (made of renewable resources) and biodegradable plastics (plastic that can be naturally broken down). Two such bioplastics--PLA and PHA--illustrate the qualities of zero-waste because they produce recyclable or biodegradable materials using renewable feedstocks. PLA can be manufactured by fermentation of sugars (e.g. corn sugars or sugarcane sugars) into lactic acid, and then polymerizing the resulting lactic acid [59]. Microorganisms (Cupriavidusnecator) can produce PHA using organic compounds like glucose or carbon materials that have been made as wastes [60].

During the manufacture of the PLA, fermentable sugars are extracted out of biomass and through lactic acid bacteria, the fermentation of the sugars is done to produce lactic acid. This lactic acid is then purified and turned into PLA through condensation or ring-opening polymerization [61]. The PHA production is within a microbial fermentation that leads to the formation of PHA substrate formation as intracellular granules by bacteria in nutrient-deficient conditions. These granules are mined, milled and formed into bioplastic materials [62]. The waste streams can be used as feedstocks in both processes, e.g., agricultural waste or food waste, and follow the zero-waste ideals of upcycling otherwise disposed-of resources.

21.4.2.3 Waste Recycling in Bioplastics Manufacturing

One of the similarities of zero-waste biomanufacturing lies in the fact that waste recycling has become a part of production. In the manufacturing process of bioplastics, the waste is recycled at three stages; during raw materials or feedstock sourcing, the byproducts of the production and the post-use recycling. Low-cost, renewable feedstocks are agricultural and food wastes, including corn stover, sugarcane bagasse or food processing residues. An example of the same is production of PHA through waste glycerol produced during biodiesel production where it was shown to increase yield and decrease the feedstock cost by 30% [60]. In the same manner, the production of PLA could use straw or wood chips (a lignocellulosic waste) by means of enzymatic hydrolysis and the result of the process can produce fermentable sugars [63].

By-products such as fermentation residues or products that are out of specification can be recapture into the manufacturing and repeated or made useful. E.g., during the process of PLA production, the fermentation residual biomass may serve as a biofertilizer or biogas feedstock and result in a closed-loop model [61]. Biomass waste in the PHA production can undergo anaerobic digestion in order to obtain methane as an energy source or to be used as feed in secondary fermentation reactions [62]. These strategies minimize waste disposal and enhance resource efficiency.

End-of-life management is critical for zero-waste bioplastics. PLA can be composted industrially, and is biodegraded within 6090 days producing carbon dioxide and water [59]. PHA has double biodegradability in its industrial and natural state, such as the presence of the marine environment, which makes its application versatile [60]. Besides, to recycle PLA and PHA, mechanical and chemical cycles are being developed through which the ability to reuse the material is recuperated. As an example, the recycling of the PLA can be done *via* chemical recycling to the lactic acid that can be returned to the production cycle *via* the process known as depolymerization [64]. Such recycling routes will make bioplastics part of a cyclical economy instead of becoming waste.

21.4.2.4 Industrial Example: NatureWorks and PLA Production

NatureWorks, the world-leading bioplastics producer, is a good illustration of such zero-waste in manufacturing PLA under the brand name Ingeo. As raw material, the company relies on corn, and after starch portion is used in the fermentation process, the rest of the raw material including corn stover and other residues are used in bioenergy production or in animal muscle [65]. In Nebraska, water used in the production plant at Blair is part of a closed-loop system where process water is re-used and re-cycled to achieve as little waste as possible. Besides, NatureWorks has made investments in lifecycle assessment (LCA) to make the process more efficient, with their emission levels showing a 60% decrease in the amount of greenhouse gases emitted, in comparison with petroleum-based plastics [66].

Another aspect of nature works operation is its collaboration with procedures of waste disposal in a bid to see appropriate end-of-life disposal of PLA. In the respective regions with industrial composting infrastructure, Ingeo products are channeled into the composting plants, which biodegrade well. In places where such infrastructure is lacking, NatureWorks also promotes research into chemical recycling, enabling the PLA to be depolymerized into lactic acid so that it can be re-used [64]. Such combined solution shows why bioplastics production can be incorporated into zero-waste objectives reducing waste on each product lifecycle phase.

21.4.2.5 Industrial Example: Mango Materials and PHA Production

Mango Materials, a start up in California, utilizes methane-rich waste streams found in wastewater treatment facilities and in landfills to create PHA. The company helps in transforming a source of environmental liability, releasing methane, which is a powerful greenhouse gas, into a useful source of PHA feedstock [67]. It is carried out by providing methane to methanotrophic bacteria and PHA is produced as granules inside the bacteria. The resultant bioplastic biodegrades in soil, freshwater, and marine conditions which is an answer to plastic pollution [60].

The zero-waste strategy utilized by Mango Materials is based on recycling fermentation byproducts. Waste in the form of residual biomass is utilized as a biofertilizer whereas the processions wastewater is recovered and re-utilized. The LCA of the company states that it uses a carbon negative footprint in PHA production because the production sequesters methane that would have been emitted into the atmosphere [67]. The Case shows the possibility that zero-waste biomanufacturing may be reached with innovative feedstock choice and byproduct use, as well as through environmental issues.

21.4.2.6 Challenges in Zero-Waste Bioplastics Manufacturing

Bio-plastic has all the potentials to replace traditional plastics, but there are a few barriers that make adoption of bio-plastic to zero-waste systems a challenging situation. An issue regarding feedstock availability is also a major source of concern since the agricultural production of crops such as corn or sugarcane might compete with the production of food [63]. Waste-derived biomass sources (e.g., lignocellulosic biomass) or food waste are increasingly used and demand high and energy-intensive advanced pre-treatment and hydrolysis processes. Moreover, PHA production could be scaled up only with difficulties because of the high costs of microbial fermentation of microbes and extraction [62].

End-of-life management poses another challenge. PLA and PHA are biodegradable, but under certain conditions, e.g., high temperatures and the presence of microbes of an industrial composting facility. In areas where these facilities are not developed, bioplastics can be disposed in landfills, which negates the zero-waste sticking points [59]. The chemical recycling technologies have not yet been developed and have limitations of scaling up and cost of implementation [64]. There is also a gap in the regulatory frameworks and there are inconsistent biodegradability and recycling standards which make the processes of waste management problematic.

21.4.3 Agricultural Waste Valorization

Agricultural wastes such as crop residues, fruit and vegetable peelings, straw, husks and bagasse are very important sources of materials found globally but is mainly underutilized or even burned leading to wastage of this resource at the expense of pollution of the environment and emissions of greenhouse gases. Zero-waste biomanufacturing aims to convert these wastes into high value products like biofuels, biochemicals, bioplastics and animal feeds that are also consistent with the circular economy. This case study presents a look into an example of agricultural waste valorization exemplification.

21.4.3.1 Anaerobic Digestion of Agricultural Waste for Biogas and Biofertilizers

Anaerobic digestion (AD) is also considered to be a long-established solution in valorizing agricultural waste into biogas and biofertilizer and has found its way into the zero-waste biomanufacturing application. An effective one is the Punjab Agricultural Waste to Energy in India where straw rice residues are processed into biogas to generate electricity with digestate as a bio-fertilizer. The Indian Renewable Energy Development Agency-supported project processes some 150,000 tons of agricultural waste per year to produce 12 MW electricity and huge amounts of rich-nutrient digestate [68].

The AD process entails microbiological breakdown of organic material in anaerobic surroundings leading to the formation of methane-rich biogas (6070% methane) and digestate of rich nutrients. The power generation uses the biogas and/or a renewable natural gas, with digestate being spread as a biofertilizer on fields, completing the nutrient cycle. Surendra *et al.*, claims that AD of agricultural waste minimizes methane emission brought about by the burning of agricultural wastes at an open field by 90% and enhances soil health *via* the use of digestate [68]. This can also be demonstrated in the 0-waste aspect of the overall input materials usage that sees no remaining wastes going to landfills.

On an economical basis, Punjab project is subsidized by the government and offered with carbon credits which yields a payback cycle of 5-7 years [69]. This translates to yield of biogas of 0.300.4 m 3 /kg of dry matter, and breeding costs have been estimated at $0.06 0.08c/kWh that is competitive with power outputs based on fossil fuels in rural regions [70]. The project has a socio-economic advantage; it helps in creating employment in waste collection and running of the plant, an aspect that is advantageous to the rural communities. There have been concerns on the seasonality with which agricultural waste is available and the efficient set up of collection methods. The case highlights the potential of AD in converting agricultural waste to commodities such as energy and agricultural feeds and contributing to zero-waste and circular economies.

21.4.3.2 Synthesis and Implications

The examples of three cases, including bioethanol produced with crop residuals, bioplastics produced with fruit and vegetables waste, and biogas through anaerobic digestion, prove potency of crop residue valorization in zero-waste biomanufacturing. Each scenario shows how cutting edge bioprocessing technologies, including enzymatic hydrolysis, microbial fermentation, and anaerobic digestion will be used to turn waste to a useful product. The benefits to the environment come down to the environmental costs of lowering greenhouse gas emissions, lesser use of the landfills and improvement in resource utilization and the economic benefits include creation of jobs, being cost competitive and being self-sufficient in the production of energy.

Nevertheless, there are still difficulties such as initially large investment costs, scalability problem, and presence of supporting policies. As a demonstration, the enzyme expenses in bio ethanol generation and purification costing in PHA generation are still huge obstacles [71, 72]. Policies and government guidelines should encourage management of waste as a value addition activity and offer carbon credits to quicken the process as witnessed in the Punjab project [69]. The cases show the significance of the integrated biorefinery models where byproducts can be repurposed in a way that will lead to the realization of the zero-waste objectives.

To sum it all up, it is worth noting that agricultural waste valorization demonstrates how the practice of zero-waste biomanufacturing can help solve any of the global issues such as waste management, climate change, and resource scarcity. Using local waste and innovative biotechnologies, these systems develop new sustainable value chains having positive impacts on both the environment and the society. Technical and economical rollbacks can be beaten with the implementation of further research and policy, and these hurdles must be overcome so that agricultural waste can cement the biomanufacturing systems of a circular economy. The summary of the dual benefits of demonstrated zero-waste strategies were shown in Table 21.2.

Table 21.2 Economic and environmental benefits of case studies [73–77].

Case study	Waste valorization strategy	Economic benefit	Environmental benefit
POET-DSM Bioethanol Plant	Lignin for energy, stillage to biogas	Additional revenue from co-products	60% lower GHG emissions vs. gasoline
NatureWorks PLA Production	Corn residues for bioenergy	Reduced feedstock costs	60% lower emissions vs. petroleum plastics
Mango Materials PHA	Methane from wastewater	Carbon-negative footprint	Reduces landfill methane emissions

21.5 Challenges and Opportunities

21.5.1 Technical and Economic Barriers

To attain a zero-waste biomanufacturing process, they will have to overcome the major technical and economic challenges to prevent the limitations in effectiveness and acceptance of these systems. Among the main technical issues is the fact that valorization of waste is complex to integrate in current existing bioprocesses. Biomanufacturing systems tend to be dependent on particular feedstocks and switching to alternative or waste-based feedstocks, e.g., agricultural waste products or industrial waste byproducts, may pose a threat to control of the production process and subsequent product quality [78]. As an example, although lignocellulosic biomass can be plentiful (and is consequently regarded as a promising biomass feedstock), it needs to be heavily pretreated to depolymerize recalcitrant fractions, such as lignin, further escalating processing energy and costs [79]. Also, it is a bottleneck to obtain efficient biocatalysts like enzymes or microbial strains that are specific with respect to waste substrates. Enzyme technologies often do not possess the specificity or robustness to treat heterogeneous waste streams resulting in decreased yields and increased costs of operation [80].

Zero-waste biomanufacturing is economically objectionable due to the operations cost and capital need. The advanced bioprocessing infrastructure like biorefineries that can process many waste streams needs sizeable capital investment up front. Small- and medium-sized enterprises that prevail in some biomanufacturing industries are frequently not able to afford introducing such technologies [81]. In addition, the production efficiency of the zero-waste systems relies strongly upon market appetite with bio-based products. As an example, bioplastics sourced by waste streams, e.g., polyhydroxyalkanoates (PHAs) have poor access to the market due to the competitive, low-priced petroleum-based plastics [82]. The absence of economies of scale applied to zero-waste production also intensifies this challenge because the bulk of the production does not always justify lowering the costs. However, opportunities exist to address these barriers. The development of synthetic biology, including CRISPR-based gene editing, now allows the generation of microorganisms capable of either efficient waste-to-product conversion, or product-to-waste conversion, which may be cost-reducing [83]. Also, major investments during start-up, where capital investments will be required, may be reduced by using public-private partnerships and governmental subsidies thus extending a zero-waste system to smaller enterprises.

21.5.2 Regulatory Considerations

Regulatory frameworks are important in adoption of zero-waste biomanufacturing and are both challenging and an opportunity. Strict policies should be avoided on bio-based product safety and waste management. A case in point is the European Union, where the Waste Framework Directive (2008/98/EC) sets stringent measures in the management of waste materials, a situation that may complicate the application of industrial wastes as feedstocks to develop industrial products utilizing biomanufacturing [84]. Equally, bio-based products, like biofuels or bioplastics, have to satisfy stringent safety and performance criteria and entail costly safety and performance tests and subsequent certification, which raise time-to-market and production expenses [85]. The novel bioproducts, especially the new ones that involve genetically modified organisms (GMOs), are subjected to even greater regulatory scrutiny as they trigger concerns about the environmental and health aspects [86].

In spite of these obstacles, regulatory structures give way to encouraging zero-waste biomanufacturing as well. An investment in zero-waste systems can be fostered by policies that encourage principles of the circular economy including tax credits by using renewable feedstocks or carbon pricing [87]. As an example to this, there exists the U.S. Renewable Fuel Standard (RFS) which has provided the opportunity to develop advanced biofuels by requiring that advanced biofuels be in fuel blends thereby developing a market to use biofuels made using waste [88]. Additionally, global standards in the use of bio-based products should promote the global trade of such products and the cost of complying with such standards. Regulation by regulators, industry, and academia together is necessary to facilitate regulatory measures in approval processes with safety and protection of environment. Indicatively, the Biotechnology Innovation Organization (BIO) has proposed the adoption of risk-based regulation systems which would focus on innovation without compromising safety [89]. That kind of framework might make zero-waste biomanufacturing implementable much faster through regulatory bottlenecks.

21.5.3 Scalability Issues

There is an unmet need to scale zero-waste biomanufacturing between laboratory-level to a larger-scale and implement them to a more industrialized scale. Heterogeneity of waste feedstocks is one major problem, as variation in bioprocess performance could be observed at the scale-up. As an example, the composition of food waste or agricultural residues may vary, and this may result in the impacts on fermentation yields, which are difficult to adopt under large scales [90]. Also, far beyond being linear, energy and resource costs of pretreatment and downstream purification, which are involved in zero-waste processes, usually scale nonlinearly, limiting efficiency [91]. The construction of bioreactors presents difficulties, too, since scale production needs to provide ideal conditions (e.g. pH, temperature, oxygen levels) in larger volumes and this may technically be difficult and expensive [92].

One more issue of scalability is the inclusion of zero-waste systems into the structure of already existing industries. Numerous existing biomanufacturing plants use a more linear manufacturing mode, and retooling them to a circular, zero-waste mode needs serious alterations [93]. This fact is especially true in terms of industrial symbiosis, where one process outputs are converted into another, necessitating joint logistics and infrastructure. Another example is a case study on industrial symbiosis in Kalundborg, Denmark, which suggests that close interaction of industries is required to successfully share resources, something that may be challenging to emulate in more disintegrated industrial environment [94].

Despite these challenges, scalability offers significant opportunities for innovation. Continuous bioprocessing and modular biorefinery designs are examples of developments within process engineering that can increase scalability through efficiency and flexibility [95]. As an example, decentralized use of waste feedstocks can be achieved with modular biorefineries, cost-effective transportation can be reduced, and integration into the local waste streams can be achieved [96]. Machine learning and process simulation are also digital tools that offer potential as tools to optimize large scale bio-processes. They allow anticipating and suppressing fluctuations in feedstock concentration, increasing productivity and lowering the cost [97]. Alongside, the above, effective case studies, including the manufacturing of bioethanol sugar cane bagasse in Brazil, indicate the potential that scalability can be realized by designing the process optimally and through favorable policy frameworks [98]. Recognizing the scalability issues, zero-waste biomanufacturing has the ability to move beyond niche applications and more into industrial use.

21.5.4 Opportunities for Advancement

The above list of challenges is not overwhelming and contains the opportunity to develop zero-waste biomanufacturing. Fostering collaboration on research, e.g., through the U.S. Department of Energy Bioenergy Technologies Office, is leading to innovative solutions on waste valorization and process optimization [99]. Bio-based products are also in demand in the marketplace due to the awareness caused by the sustainability issues, which promotes industries to invest in zero-waste technologies. As an example, the current consumer demand in regard to sustainable packaging has led to an increase in the bioplastics market with corporations such as Nature Works utilizing corn-based waste streams to make polylactic acid (PLA) [100].

In addition, the incorporation of the technologies of Industry 4.0, the Internet of Things (IoT), and artificial intelligence available, provides the opportunity to generate substantial changes. Sensors with IoT features have the ability to monitor the bioprocess parameters in real-time, to enhance process control and eliminate waste [101]. Likewise, model process parameters can be optimized using AI in terms of feedstock and the process parameters which increases process efficiency and scalability [97]. A combination of such developments and favorable policies with industry cooperation places zero-waste biomanufacturing at the center of sustainable industrial systems.

To sum up, the issues mentioned above, technical, economic, regulatory and scalability one still exist, but there is also big potential of innovating and developing. As it has been seen through these identified barriers and the means by which such can be overcome through technological advancements, policy support, and collaboration in the industry, zero-waste biomanufacturing has an opportunity to become what might be a significant solution to global resource management. Further research and development is paramount to stave off these impediments and achieve maximum potential of zero-waste systems.

21.6 Future Directions

21.6.1 Emerging Technologies

The future of zero-waste biomanufacturing is about to be greatly changed with the emergence of innovative technologies, which would increase efficiency, decrease waste and support sustainability. The development of synthetic biology is among the brightest spots since it allows designing microorganisms to process waste streams into high-value products. As an example, genetically modified microorganisms can be used to convert organic waste litter into biofuels or bioplastics yielding more than conventional techniques [102]. Lately, the advances in CRISPR-based gene editing have contributed significantly to developing such microbial strains that favor metabolic pathway optimization tour de force in the valorization of waste, lessening byproduct and minimalizing energy use [103]. The innovations are vital in scaling zero-waste systems since they enable tight control of bioprocesses, reducing loss of resources.

Biocatalysis is also an emerging technology that takes advantage of enzymes to catalyze reactions that are highly specific and efficient. In several applications, enzyme engineering research, including directed evolution, has resulted in enzymes that tolerate harsh conditions required industrially to transform complex waste streams into valuable chemicals [104]. Enzymatic processes to dissolve plastic waste into recyclable monomers are an example of one current implementation in line with zero-waste ideals [105]. The biocatalytic systems decrease the demand of making energy-intensive chemical crops, decreasing the environmental footprint of biomanufacturing.

Another rapidly developing game changer in zero-waste biomanufacturing is additive manufacture (3D bioprinting). A reduced amount of waste is generated by using 3D bioprinting to greatly reduce the wastage of disparate biomaterials in producing bioproducts such as those constituting tissue scaffolds or bio-based materials [106]. The technology is also consistent with the aims of the circular economy because it facilitates the creation of bespoke bioproducts without wasting a large amount of the raw material. The further enhancement of bioprocess design is the introduction of artificial intelligence (AI) and machine learning (ML). Optimal conditions of fermentation or identification of valuable waste streams can be determined by AI-driving models, which eliminates time and waste [107]. All these technologies bring a step closer to more intelligent biomanufacturing systems that are also sustainable.

21.6.2 Industry 4.0 in Biomanufacturing

Industry 4.0, the latest in this trend that involves the incorporation of digital technologies, automation, and data analysis, creates tremendous chances of zero-waste biomanufacturing. Internet of Things (IoT) has the ability to enable real-time monitoring and precise control of key bioprocess parameters including pH, temperature, and substrate concentration to reduce the level of waste production [108]. As another example, bio-reactor derailing sensors with IoT can identify inefficiencies in real-time so that adjustments can be made to minimize byproduct production. This is a critical aspect that will help in attaining zero-waste in large-scale bioprocesses.

The optimization of bioprocesses is being completely transformed by big data analytics and digital twins, virtual analogues of real processes.

The proposed digital twins will create the digital version of the biomanufacturing processes enabling the manufacturing corporations to test the conditions, examine the opportunities of waste reduction, and analyze the ways to minimize it, without the physical experimentation [109]. As an example, a biofuel production facility aspect can utilize a digital twin to forecast the effects of utilizing different feedstocks to ensure the appropriate consumption of the resources and to reduce the waste. In the specific case of renewable feedstock integration, such tools are specifically useful because they can cover some of the variability in the feedstock composition, one of the problems in zero-waste systems [110].

Automation and robotics is also used in zero-waste biomanufacturing to improve operations and to eliminate the error of human assistants. Substrate handling and recovery of products are performed by clever automation that reduces losses of materials [111]. As an example, robotic sorting devices have the capability of separating waste streams that can be used in valorization more precisely than manual methods so that discarded material does not end up as waste. Scalability of zero-waste systems can be realized over and above the efficiency improvement that comes with the combination of these industry 4.0 technologies thus enabling their large-scale adoption in industries.

21.6.3 Conclusion

The concept of zero-waste biomanufacturing denotes a paradigm shift of the entire industrial biotechnology, being consistent with the established objectives of global sustainability and the concept of the circular economy. The chapter has covered the conceptualization on the basis of which, technological strategies and practical examples that prove the viability of the zero-waste systems. Waste minimization strategies and biorefinery incorporation right up to advanced bioprocessing and synthetic biology, the sphere is swiftly expanding in order to solve the environmental situation. The concept of biofuel production and bioplastics manufacturing can be identified as case studies, which demonstrate the practical side of these concepts, whereas the impediments of their application, such as scalability and the challenges of regulations, prove the necessity to further innovation.

The application of synergetic integration of upcoming technologies and Industry 4.0 tools pertains to the way of zero-waste biomanufacturing. BioCatalysis and synthetic biology provide control and resource-efficient ways to valorize waste and AI, IoT and digital twins can allow more intelligent, real-time bioprocesses. Such developments go beyond the gradual improvement but could transform industrial production; and zero waste systems can become an object of sustainable production. Nevertheless, to achieve this potential, there has to be the coordinated efforts of various stakeholders.

Industry has to undertake research and development to ramp up promising technologies, including CRISPR-engineered microbes as well as automated bioprocesses. The shift between laboratory and industrial applications can be boosted by the pilot projects and collaborations with academic organizations. Policymakers have a vital role in the establishment of supportive structures, tracking incentive in the use of renewable feedstock and rules that give priority to reduction of waste. Uniformity in lifecycle assessment procedures is required so that there is clarity and uniformity in the assessment of zero-waste systems. The researchers need to emphasize those interdisciplinary solutions that involve biotechnology, data science, and engineering to resolve technical obstacles such as enzyme stability and feedstock inconsistency.

The zero-waste biomanufacturing is significantly affecting the globe beyond the enhancement of the environment. These systems have the potential to increase economic resilience and food security by lowering demand for finite resources and reducing, waste thus benefitting those regions that are under intense pressure due to disposal of industrial waste. As an example, the value of agriculture waste to bio- based products can generate new sources of revenue for the farmer and at the same time minimize the amount of landfill. To obtain such results, there is a togetherness needed regarding sustainability, innovation, and collaboration.

To sum up, zero-waste biomanufacturing is not something that may be a far-fetched dream but something that is achievable. These technologies and strategies outlined in this chapter will guide how the biomanufacturing business can adopt the race towards achieving waste-free and sustainable paradigm. Industry, academia, and government stakeholders should take action to scale such solutions with the help of new technologies and digital tools in order to advance. With the acceptance of zero-waste philosophies we could create a future that incorporates biomanufacturing as a means of meeting human need in a way that is not only sustainable but will also leave an inheritance on the planet to rise and sustain possession to remain in some kind of human form.

AI disclosure statement: The author acknowledges the use of AI-based tools for minor language editing or grammatical correction. However, all intellectual contributions, analyses, and conclusions presented in this chapter are entirely the author's own.

References

1. Ellen MacArthur Foundation, Towards the circular economy: Economic and business rationale for an accelerated transition, Ellen MacArthur Found., 2013.
2. Tuck, C.O., Perez, E., Horvath, I.T., *et al.*, Valorization of biomass: Deriving more value from waste. *Science*, 337, 695, 2012.
3. Clark, J.H., Budarin, V., Deswarte, F.E.I., *et al.*, Green chemistry and the biorefinery: A partnership for a sustainable future. *Green Chem.*, 18, 391, 2016.
4. United Nations, Transforming our world: The 2030 agenda for sustainable development, UN Gen. Assem., 2015.
5. Stahel, W.R., The circular economy. *Nature*, 531, 435, 2016.
6. Ragauskas, A.J., Williams, C.K., Davison, B.H., *et al.*, The path forward for biofuels and biomaterials. *Science*, 311, 484, 2006.
7. Zeng, A.P. and Biebl, H., Bulk chemicals from biotechnology: The case of 1,3-propanediol production and the new trends. *Adv. Biochem. Eng./Biotechnol.*, 74, 239, 2002.
8. Environmental management—Life cycle assessment—Principles and framework, Int. Organ. Stand., 2006.
9. Carus, M. and Dammer, L., The circular bioeconomy—Concepts, opportunities, and limitations. *Ind. Biotechnol.*, 14, 83, 2018.
10. Chertow, M.R., Industrial symbiosis: Literature and taxonomy. *Annu. Rev. Energy Environ.*, 25, 313, 2000.
11. Stephanopoulos, G., Challenges in engineering microbes for biofuels production. *Science*, 315, 801, 2007.
12. Kamm, B. and Kamm, M., Principles of biorefineries. *Appl. Microbiol. Biotechnol.*, 64, 137, 2004.
13. Doudna, J.A. and Charpentier, E., The new frontier of genome engineering with CRISPR-Cas9. *Science*, 346, 1258096, 2014.
14. Lee, J., Kim, H.U., Choi, S., *et al.*, Industry 4.0 in biotechnology: Opportunities and challenges. *Curr. Opin. Biotechnol.*, 53, 1, 2018.
15. FAO, Global food losses and food waste—Extent, causes and prevention. Food Agric. Organ. UN, 2011.
16. European Commission, A sustainable bioeconomy for Europe: Strengthening the connection between economy, society, and the environment. Eur. Comm., 2018.
17. Lee, J.W., Na, D., Park, J.M., *et al.*, High-cell-density fermentation for biomolecule production. *Biotechnol. Adv.*, 37, 107407, 2019.
18. Jungbauer, A., Continuous bioprocessing: The real future of biomanufacturing? *Trends Biotechnol.*, 31, 479, 2013.
19. Sadhukhan, J., Ng, K.S., Martinez-Hernandez, E., *et al.*, Biorefineries and chemical processes: Design, integration, and sustainability analysis. *Bioresour. Technol.*, 211, 1, 2016.

20. Sheldon, R.A., The E factor 25 years on: The rise of green chemistry and sustainability. *Green Chem.*, 19, 18, 2017.
21. Rathore, A.S., Winkle, H., Sofer, G., *et al.*, Process integration and intensification in biomanufacturing. *Biotechnol. Progr.*, 31, 837, 2015.
22. Himadri Roy Ghatak, Biorefineries from the perspective of sustainability: Feedstocks, products, and processes. *Renew. Sustain. Energy Rev.*, 15, 8, 4042–4052, 2011.
23. Chandel, A.J., Garlapati, V.K., Singh, A.K., Antunes, F.A.F., da Silva, S.S., The path forward for lignocellulose biorefineries: Bottlenecks, solutions, and perspective on commercialization. *Bioresour Technol.*, 264, 370–381, 2018.
24. Mussatto, S.I., Dragone, G., Guimarães, P.M.R., *et al.*, Technological trends, global market, and challenges of bio-ethanol production. *Appl. Microbiol. Biotechnol.*, 87, 1307, 2010.
25. Clomburg, J.M. and Gonzalez, R., Anaerobic fermentation of glycerol: A platform for renewable fuels and chemicals. *Trends Biotechnol.*, 31, 20, 2013.
26. Hsu, D.D., Inman, D., Heath, G.A., *et al.*, Life cycle environmental impacts of selected biofuels. *Environ. Sci. Technol.*, 44, 5289, 2010.
27. Tsiropoulos, I., Faaij, A.P.C., Lundquist, L., *et al.*, Life cycle assessment of sugarcane-based polyethylene production. *J. Clean. Prod.*, 90, 211, 2015.
28. Cristóbal, J., Matos, C.T., Aurambout, J.P., *et al.*, Life cycle assessment of food waste-based biogas production. *Bioresour. Technol.*, 216, 135, 2016.
29. McManus, M.C. and Taylor, C.M., The changing nature of life cycle assessment. *J. Ind. Ecol.*, 19, 547, 2015.
30. Chen, C., Park, S.Y., Lee, J.H., *et al.*, Advances in perfusion bioreactor systems for biopharmaceutical production. *Biotechnol. Adv.*, 36, 1012, 2018.
31. Giorno, L. and Drioli, E., Membrane bioreactors for sustainable bioprocessing. *J. Membr. Sci.*, 520, 837, 2016.
32. Rathore, A.S., Kateja, N., Kumar, D., *et al.*, Process intensification in biomanufacturing: Challenges and opportunities. *Curr. Opin. Chem. Eng.*, 31, 100671, 2021.
33. Sheldon, R.A. and Woodley, J.M., Role of biocatalysis in sustainable chemistry. *Chem. Rev.*, 118, 801, 2018.
34. Nielsen, J., Larsson, C., van Maris, A., *et al.*, Engineering microbes for sustainable bio-based chemicals. *Science*, 340, 602, 2013.
35. Li, Z., Yang, J., Loh, X.J., *et al.*, Production of polyhydroxyalkanoates from waste glycerol using engineered microbes. *Biotechnol. Biofuels*, 9, 208, 2016.
36. Bornscheuer, U.T., Synthetic biology for sustainable chemical production. *Nat. Chem. Biol.*, 14, 211, 2018.
37. Voigt, C.A., Synthetic biology 2020–2030: Six commercially available products that are changing our world. *Nat. Commun.*, 11, 6379, 2020.
38. Kumar, R., Sharma, A., Singh, P., *et al.*, Lignocellulosic biomass for bioethanol production: Current perspectives and challenges. *Bioresour. Technol.*, 288, 121514, 2019.

39. Sun, Y., Cheng, J., Yang, Z., *et al.*, Biomass pretreatment technologies for bioethanol production. *Bioresour. Technol.*, 310, 123456, 2020.
40. Papagianni, M., Citric acid production from renewable feedstocks. *Appl. Microbiol. Biotechnol.*, 101, 487, 2017.
41. Bajpai, P., *Developments in bioenergy and sustainable technologies*, Springer, Singapore, 2020.
42. IEA, Biofuels outlook 2023, Int. Energy Agency, 2023.
43. POET-DSM, Project LIBERTY: Cellulosic ethanol production, POET-DSM, 2022.
44. Rogers, P.L., Lee, K.J., Skotnicki, M.L., *et al.*, Advances in ethanol production from lignocellulosic biomass. *Biotechnol. Adv.*, 35, 817, 2017.
45. Yadav, R., Zabihi, O., Fakhrhoseini, S., Nazarloo, H.A., Kiziltas, A., Blanchard, P., Naebe, M., Lignin derived carbon fiber and nanofiber: Manufacturing and applications. *Compos. B Eng.*, 255, 110613, 2023.
46. Wilkie, A.C., Riedesel, K.J., Owens, J.M., *et al.*, Stillage characterization and anaerobic treatment of ethanol stillage. *Biomass Bioenergy*, 19, 63, 2000.
47. Liu, S., Qin, Y., Chen, Y., *et al.*, Xylitol production from lignocellulosic biomass. *Bioresour. Technol.*, 276, 123, 2019.
48. Humbird, D., Davis, R., Tao, L., *et al.*, Process design and economics for biochemical conversion of lignocellulosic biomass to ethanol, NREL Tech. Rep., TP-5100-47764, 2011.
49. Wang, M., Huo, H., Arora, S., *et al.*, Life-cycle energy and greenhouse gas emission impacts of different corn ethanol plant types. *Environ. Res. Lett.*, 7, 024001, 2012.
50. Liang L, Liu R., Freed E.F., Eckert C., A. Synthetic Biology and Metabolic Engineering Employing Escherichia coli for C2–C6 Bioalcohol Production. *Front. Bioeng. Biotechnol.*, 8, 1–8, 2020.
51. Lynd, L.R., Weimer, P.J., van Zyl, W.H., Pretorius, I.S., Consolidated bioprocessing of cellulosic biomass: An update. *Curr. Opin. Biotechnol.*, 19, 3, 218–227, 2008.
52. Cherubini, F., The biorefinery concept: Using biomass instead of oil for producing energy and chemicals. *Energy Convers. Manage.*, 51, 7, 1412–1421, 2010.
53. AspenTech, Aspen Plus: Process Simulation Software, 2023. Retrieved from https://www.aspentech.com.
54. Tadesse, H. and Luque, R., Advances in biomass pretreatment using ionic liquids. *Energy Environ. Sci.*, 4, 10, 3913–3929, 2011.
55. Holladay, J.E., *et al.*, Top value-added chemicals from biomass, PNNL Report, 16983, 2007.
56. European Commission, Circular Economy Action Plan, 2020. Retrieved from https://ec.europa.eu.
57. Bennett, P. and Buckley, P., *IEA Bioenergy*, EXCO: 2022: 01.
58. Fargione, J., *et al.*, Land clearing and the biofuel carbon debt. *Science*, 319, 5867, 1235–1238, 2008.

59. Leja, K. and Lewandowicz, G., Polymer biodegradation and biodegradable polymers—a review. *Pol. J. Environ. Stud.*, 19, 2, 255–266, 2010.
60. Koller, M., Maršálek, L., Braunegg, G., de Sousa Dias, M.M., Modern biotechnological approaches for the production of polyhydroxyalkanoates. *Chem. Biochem. Eng. Q.*, 31, 4, 537–548, 2017.
61. Farah, S., Anderson, D.G., Langer, R., Physical and mechanical properties of PLA, and their functions in widespread applications. *Adv. Drug Delivery Rev.*, 107, 367, 2016.
62. Vicente, D., Proença, D.N., Morais, P.V., The Role of Bacterial Polyhydroalkanoate (PHA) in a Sustainable Future: A Review on the Biological Diversity. *Int. J. Environ. Res. Public Health*, 8, 20, 4, 2959, 2023.
63. Bozell, J.J. and Petersen, G.R., Technology development for the production of biobased products from biorefinery carbohydrates. *Green Chem.*, 12, 539, 2010.
64. Piemonte, V. and Gironi, F., Lactic acid chemical recycling: Towards a sustainable PLA lifecycle. *Resour. Conserv. Recycl.*, 55, 1108, 2011.
65. Vink, E.T.H., Rábago, K.R., Glassner, D.A., *et al.*, Applications of life cycle assessment to NatureWorks™ polylactide (PLA) production. *Polym. Degrad. Stab.*, 80, 403, 2003.
66. Vink, E.T.H., Davies, S., Kolstad, J.J., The eco-profile for current Ingeo polylactide production. *Ind. Biotechnol.*, 11, 167, 2015.
67. Mango Materials, Sustainable bioplastics from methane, Mango Mater., 2023.
68. Surendra, K.C., Khanal, S.K., Shrestha, P., *et al.*, Biogas as a sustainable energy source for developing countries: Opportunities and challenges. *Renewable Sustainable Energy Rev.*, 31, 846, 2014.
69. Kaur, M., Kumar, M., Sachdeva, S., *et al.*, Anaerobic digestion of agricultural residues: Opportunities and challenges in India. *Renewable Sustainable Energy Rev.*, 73, 469, 2017.
70. Ravindranath, N.H., Somashekar, H.I., Dasappa, S., *et al.*, Bioenergy in India: Barriers and policy options. *Energy Sustain. Dev.*, 10, 59, 2006.
71. Zabed, H., Sahu, J.N., Boyce, A.N., *et al.*, Bioethanol production from renewable sources: Current perspectives and technological progress. *Renewable Sustainable Energy Rev.*, 71, 475, 2016.
72. Kourmentza, C., Plácido, J., Venetsaneas, N., *et al.*, Advances in polyhydroxyalkanoate (PHA) production. *Bioengineering*, 4, 55, 2017.
73. Fava, F., Totaro, G., Diels, L., *et al.*, Biowaste biorefinery in Europe: Opportunities and challenges. *Bioresour. Technol.*, 194, 1, 2015.
74. Gnansounou, E. and Dauriat, A., Techno-economic analysis of lignocellulosic ethanol: A review. *Bioresour. Technol.*, 101, 4980, 2010.
75. Harding, K.G., Dennis, J.S., von Blottnitz, H., *et al.*, Environmental impact assessment of polyhydroxyalkanoates (PHA) production from organic waste. *J. Biotechnol.*, 130, 57, 2007.

76. Mu, D., Seager, T., Rao, P.S., *et al.*, Life cycle assessment of lignocellulosic ethanol: A review. *Bioresour. Technol.*, 101, 4849, 2010.
77. Tasca, A.L., Puccini, M., Vitolo, S., *et al.*, Second-generation bioethanol from agricultural residues: The case of Beta Renewables. *Energy Procedia*, 147, 304, 2018.
78. Ragauskas, A.J., Beckham, G.T., Biddy, M.J., *et al.*, Lignin valorization: Improving lignin processing in the biorefinery. *Science*, 344, 1246843, 2014.
79. Himmel, M.E., Ding, S.Y., Johnson, D.K., *et al.*, Biomass recalcitrance: Engineering plants and enzymes for biofuels production. *Science*, 315, 804, 2007.
80. Ameh, V.I., Onu, M.A., Ayeleru, O.O., Nomngongo, P.N., Ramatsa, I.M. Challenges and Opportunities in Biotechnological Waste Valorization. In: Kumar, V., Verma, P. (eds) Biotechnological Applications in Industrial Waste Valorization. *Interdiscip. Biotechnol. Adv.*, Springer, Singapore, 2025.
81. Philp, J.C., Ritchie, R.J., Allan, J.E.M., *et al.*, Biobased chemicals: The convergence of green chemistry with industrial biotechnology. *Trends Biotechnol.*, 31, 219, 2013.
82. Chen, G.Q., Plastics from bacteria: Natural functions and applications. *Microbiol. Monogr.*, 14, 1, 2010.
83. Wei, J. and Li, Y., CRISPR-based gene editing technology and its application in microbial engineering, *Eng. Microbiol.*, 3, 4, 100101, 2023.
84. European Commission, Directive 2008/98/EC on waste, Off. J. Eur. Union, 2008.
85. OECD, Future prospects for industrial biotechnology, Organ. Econ. Co-op. Dev., 2011.
86. Davison, J., Risk mitigation of genetically modified bacteria and plants designed for bioremediation. *J. Ind. Microbiol. Biotechnol.*, 37, 101, 2010.
87. Ellen MacArthur Foundation, Towards the circular economy, Ellen MacArthur Found., 2015.
88. EPA, Renewable fuel standard program, U.S. Environ. Prot. Agency, 2022.
89. BIO, Biotechnology Innovation Organization: Policy recommendations, Biotechnol. Innov. Organ., 2020.
90. Pfaltzgraff, L.A., De bruyn, M., Cooper, E.C., *et al.*, Food waste biomass: A resource for high-value chemicals. *Green Chem.*, 15, 307, 2013.
91. Humbird, D., Davis, R., Tao, L., *et al.*, Process design and economics for biochemical conversion of lignocellulosic biomass to ethanol, NREL Tech. Rep., NREL/TP-5100-47764, 2011.
92. Garcia-Ochoa, F. and Gomez, E., Bioreactor scale-up and oxygen transfer in microbial processes. *Biotechnol. Adv.*, 27, 153, 2009.
93. Clark, J.H., Budarin, V., Deswarte, F.E.I., *et al.*, Green biorefinery technologies based on waste biomass. *Green Chem.*, 17, 391, 2015.
94. Ehrenfeld, J. and Gertler, N., Industrial ecology in practice: The evolution of interdependence at Kalundborg. *J. Ind. Ecol.*, 1, 67, 1997.

95. Zydney, A.L., Continuous downstream processing for biomanufacturing. *Curr. Opin. Biotechnol.*, 42, 55, 2016.
96. Vanthoor-Koopmans, M., Wijffels, R.H., Barbosa, M.J., *et al.*, Biorefinery of microalgae for food and fuel. *Bioresour. Technol.*, 135, 142, 2013.
97. Gargalo, C.L., Carvalho, A., Gernaey, K.V., *et al.*, Towards the development of digital twins for bioprocesses. *Comput. Chem. Eng.*, 141, 107032, 2020.
98. Dias, M.O.S., Cunha, M.P., Maciel Filho, R., *et al.*, Second-generation ethanol in Brazil: Can it compete with first-generation ethanol? *Bioresour. Technol.*, 145, 298, 2013.
99. DOE, Bioenergy Technologies Office: 2021 project peer review, U.S. Dep. Energy, 2021.
100. NatureWorks, Ingeo biopolymer: Sustainability report, NatureWorks LLC, 2022.
101. Lee, J.H., Park, S.Y., Kim, H.U., *et al.*, Industry 4.0 in bioprocessing: Opportunities and challenges. *Biotechnol. J.*, 14, 1800408, 2019.
102. Keasling, J.D., Manufacturing molecules through metabolic engineering. *Science*, 330, 1355, 2010.
103. Stephanopoulos, G., Synthetic biology and metabolic engineering. *ACS Synth. Biol.*, 1, 514, 2012.
104. Bornscheuer, U.T., Huisman, G.W., Kazlauskas, R.J., *et al.*, Engineering the third wave of biocatalysis. *Nature*, 485, 185, 2012.
105. Tournier, V., Topham, C.M., Gilles, A., *et al.*, An engineered PET depolymerase to break down and recycle plastic bottles. *Nature*, 580, 216, 2020.
106. Groll, J., Burdick, J.A., Cho, D.W., *et al.*, A definition of bioinks and their distinction from biomaterial inks. *Biofabrication*, 8, 013001, 2016.
107. Li, Y., Zhang, H., Chen, L., Machine learning in bioprocess optimization: A review. *Bioresour. Technol.*, 340, 125704, 2021.
108. Chen, Y., Li, J., Wang, Y., Internet of Things in biomanufacturing: Opportunities and challenges. *Biotechnol. Adv.*, 42, 107578, 2020.
109. Barricelli, B.R., Casiraghi, E., Fogli, D., A survey on digital twin: Definitions, characteristics, applications, and design implications. *IEEE Access*, 7, 167653, 2019.
110. Rathore, A.S., Mishra, S., Nikita, S., Bioprocess 4.0: Digitalization in biomanufacturing. *Curr. Opin. Chem. Eng.*, 33, 100704, 2021.
111. Schmidt, M., Winther, O., Nielsen, A., Automation in bioprocessing: The role of robotics. *Trends Biotechnol.*, 36, 1187, 2018.

22

Green Chemistry in Upstream Bioprocessing

Vijaya Kumar Talari*, Stutee Bhoi and Bijoy Kumar Purohit

Department of Chemical Engineering, B. V. Raju Institute of Technology, Narsapur, Telangana, India

Abstract

To reduce environmental impact, optimize resource use, and have sustainable biologics and biopharmaceuticals production, greater adoption of green chemistry principles is occurring in upstream bioprocessing. This chapter gives a detailed examination of the incorporation of green chemistry in the upstream operations entailed in the preparations of the media, cell culture, as well as fermentation. Such strategies, key strategies as the use of renewable feedstocks, solvent-free or benign solvent systems, atom economy and energy-efficient bioreactor operations are discussed. The chapter also looks into the progress in terms of development of environmentally friendly formulations of produce in media, use of less auxiliary chemicals, and development of metabolic engineering to maximize on the ends while minimizing wastes and dangerous by-products. Moreover, the contributions of such process intensification methods as high-cell-density cultures and continuous bioreactor systems to achieving sustainability are discussed. Academic and industrial case studies and real-life case studies are offered in presentation of successful applications and quantifiable advantages of green chemistry in upstream bioprocessing. Finally, the chapter discusses regulatory, economic, and scalability of implementing green chemistry and provides future directions to move forward to make biomanufacturing sustainable. The content is particularly of interest to process engineers, biotechnologist, and believers of sustainability who would like to better align bioproduction with taking care of the environment.

Keywords: Green chemistry, upstream bioprocessing, sustainable biomanufacturing, renewable feedstocks, process intensification

**Corresponding author*: Vijayakumar.t@bvrit.ac.in

Sandip Kunar, Jagadeesha T. and D. Sathis Kumar (eds.) Biomanufacturing: Transforming Production of Biologics and Biopharmaceuticals, (707–736) © 2026 Scrivener Publishing LLC

22.1 Introduction to Green Chemistry in Upstream Bioprocessing

Upstream bioprocessing is the initial phase that is involved in making biologics and biopharmaceuticals, with steps that entail cell line development, media formulary, and fermentation, or cell culture to produce the desired biological unit. With the growth of the biopharmaceutical industry due to the demand of therapeutics such as monoclonal antibodies and vaccines, it has been vital to incorporate green chemistry principles into the industry to ascertain sustainability. In alignment with upstream bioprocessing, green chemistry, which is considered the design of chemical products and processes that minimize or avoid the use of hazardous materials and the generation of hazardous wastes, is perfectly in line with its strategy of optimizing resources and key resources and a minimal environmental imprint [1]. The given introductory section will touch upon the main principles of green chemistry, environmental issues associated with the upstream approaches traditionally used, and the objectives of sustainable bioprocessing, discussing how these aspects promote environmentally friendly approaches to biomanufacturing.

22.1.1 Principles of Green Chemistry and Their Relevance to Bioprocessing

The 12 principles of green chemistry developed by Anastas and Warner are guidelines to sustainable chemical systems that aim to prevent treatment rather than curing, to have atom economy, and use renewable feedstocks. These principles are of special interest in upstream bioprocessing where the aim of the process is to produce as much product as possible and with the lowest possible waste, thus requiring the optimization of cell culture media and fermentation regimes. An example is the principle of the atom economy that promotes the development of processes that ensure every atom in a reactant is used in the formation of the final product so that by-products are reduced. In the context of bioprocessing, this would mean metabolic engineering of host cells e.g., Chinese hamster ovary (CHO) cells to enhance both substrates use and limit lactate production, which tends to obstruct productivity and form waste.

Another important principle is renewable feedstocks that are gaining popularity in the formulation of media. The classic media are dependent on animal-derived media such as fetal bovine serum which are highly variable and contaminable, thus making them ethically questionable and not sustainable. The switch to chemically defined medium may be either plant-based or synthetic and is the best way to comply with the aspects of green chemistry, since it requires renewable resources and reproducibility. Recently discovered such media has the advantage not only in allowing cultures at high cell densities but they also result in a reduction in the environmental load by reducing the consumption of non-renewable sources. One of the main green concepts, energy efficiency, is satisfied with the bioreactor design. An example is the innovation of single-use systems, which consumes lower amounts of cleaning solvents and energy versus stainless-steel reactors.

These principles are relevant to the choice of solvents and catalysis. The material of upstream process usually contains aqueous systems though auxiliary solvents to pH adjust or to provide other nutrients may be dangerous. Benign solvents, such as deep eutectic solvents, are very biodegradable and less toxic; hence, they improve the stability of enzymes during the biocatalytic processes by the promotion of green chemistry. Furthermore, in engineered biocatalysts, selectivity is encouraged and energy requirements become lesser as evident in the case of optimal fermentation wherein the microbial catalysts transform renewable sugars into biologics with minimal side reactions. Some of these gains are measured using metrics such as process mass intensity (PMI) to enable bioprocess engineers to assess and optimize sustainability-driven upstream strategies [2]. That is why, through the incorporation of these principles in the early stages of developing the process, upstream bioprocessing can attain superior titers and environmental loads that can form the basis of green, scalable, and bio-manufacturing. The principles of green chemistry applied in the

Table 22.1 Green chemistry principles in upstream bioprocessing.

Green chemistry principle	Relevance to bioprocessing	Example application
Waste Prevention	Reduces by-products in fermentation	Optimized media to minimize lactate
Atom Economy	Maximizes substrate use for product	Metabolic engineering for higher yields
Safer Solvents	Uses benign solvents in cell culture	Deep eutectic solvents in media
Energy Efficiency	Lowers bioreactor energy demands	Single-use bioreactors reduce cleaning

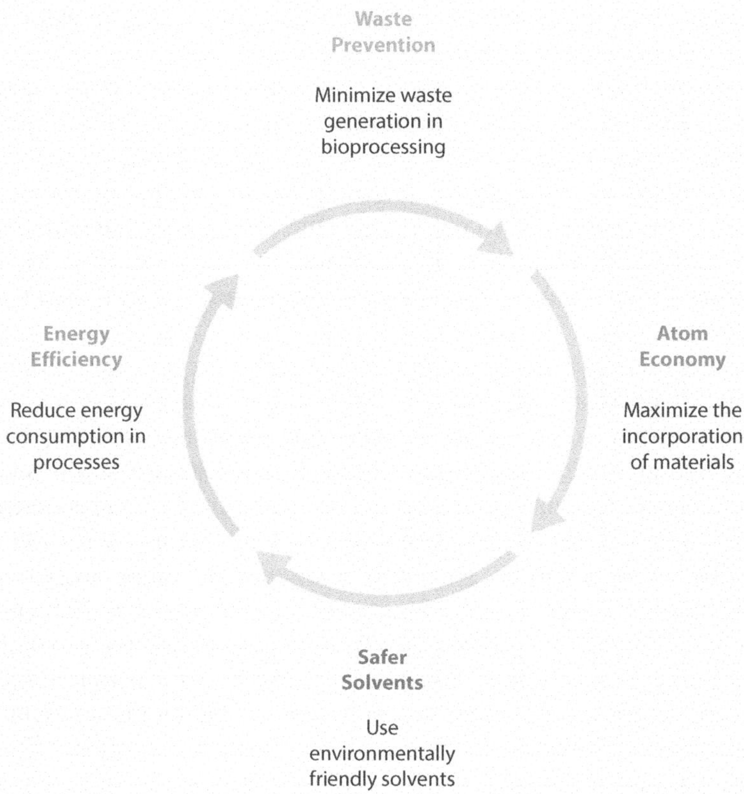

Figure 22.1 Green chemistry in bioprocessing.

upstream bioprocessing setting are summarized in Table 22.1 below with their relevance and applicability mentioned. It is also a handy summary that enables the reader to get the core ideas. Figure 22.1 shows the application of green chemistry principles in upstream bioprocessing with each principle relating to one of the practical applications.

22.1.2 Environmental Challenges in Traditional Upstream Processes

The success of traditional upstream bioprocessing techniques escapes sustainability due to major environmental issues that the processes cause. The major concerns are high water, energy usage. Large scale fermentations involve enormous volumes of water to prepare the media and cool it and this amount can be in excess of 1,000 liters of water per kilogram of product. In addition to the resource pressure imposed by their consumption, this also produces a significant amount of wastewater rich in nutrients and metabolites comprising of a part of the quantity of eutrophication unless adequately treated. The use of the chemically undefined media also adds in the variability and waste, overdose nutrient produces the by-products, ammonia and organic acids, which have to be purged downstream and add to the total process waste.

Hazardous substances pose another challenge. Other traditional methods require organic solvents or toxic reagents in cell lysis or stabilization which produce volatile organic compounds (VOCs) and lead to air pollution. As a case study, when conducting microbial fermentation, antibiotics may be added to inhibit contamination but the remaining content is capable of enhancing the development of antibiotic resistance in ecosystems when released. Additionally, sterilization and aeration being energy intensive worsen carbon footprints; large amounts of electricity are used during autoclaving and high-shear mixing; and the CO_2 emitted by fossil fuel-derived power plants intensifies the effects of global warming [3].

Scalability issues compound these problems. As production scales from lab to industrial levels, inefficiencies grow, leading to higher resource use and more waste. Older installations often use batch processes, which require frequent cleaning and setup, wasting energy and products [8]. In addition, extraction of raw materials including mining of trace metals in media contributes to deforestation and degradation creating loss of habitat. Experiences under case studies in the manufacturing of antibodies show that the upstream processes can contribute up to three quarters of the environmental burden of a given process, which is mainly the result of media component and energy needs. To overcome such difficulties, a change of paradigm to green technologies, which includes uninterrupted processing and valorization of the waste, must be implemented to curb the environmental impact of the conventional processes.

22.1.3 Objectives of Sustainable Bioprocessing

The goals of sustainability in bioprocessing at its upper end are to develop a harmony between output, economic feasibility, and environmental conservation. Key objectives are the increasing need of resources by making the media and process more efficient: to conserve water and energy resources by 50 percent or more, by applying high-density cultures, and perfusion systems. This requires using life-cycle assessments (LCAs) to evaluate a product's impact from raw materials to disposal, while minimizing processes to reduce greenhouse gas emissions and pollution [5].

Another objective is waste minimization and circular economy integration. Using green metrics such as the E-factor (waste per product mass) bioprocesses could be optimized to recycle the by-products, such as by converting the spent media to biofuels or nutrients to be used as raw materials in the next cycle. It is important to increase product quality and yield; sustainable processes such as metabolic engineering aim at higher titers that do not require the production of larger volumes and, therefore, do not pose such loads on the environment. The objective of R&D is aligned with global regulatory ethics and to achieve animal-free media and reduced hazardous substances, and to foster the trust of the global community with high expectations regarding regulations such as Registration, Evaluation, Authorization and Restriction of Chemicals (REACH) [4].

This is connected and interrelated to economic sustainability with objectives to reduce costs with scalable, flexible technologies such as single-use bioreactor that reduces the capital costs and allows fast scale-up. Digital twins and AI innovation are used in process modeling to enable predictive optimization to reduce waste due to trial-and-error processing. Finally, the objective of sustainable bioprocessing is also the resiliency and resource scarcity in the context of building a bioeconomy where upstream processes involve net-zero emissions and bio-diversity conservation. Interdisciplinary cooperation, involving chemists through to engineers, is needed to come up with new green solutions that support the sustainable expansion of the industry but that do not destroy the world.

22.2 Sustainable Media Preparation

Media preparation represents the foundation of upstream bioprocessing that is used in biopharmaceutical manufacture acting as the nutritional support to cell growth and product production. Implementation of green chemistry concepts in the design of the media is essential in minimizing the pollution of the environment, maximizing resources usage and guaranteeing a sustainable biomanufacturing. The conventional media formulations rely on the non-renewable component and resource levels that are considered to be a contributor of high levels of wastes and ecological footprints. This subchapter addresses sustainable approaches to media production with a preference to focusing on environmentally-friendly media formulations, renewable and biodegradable feedstocks and reduced employment of auxiliary chemicals. The mentioned strategies are consistent with the principles of green chemistry to ensure environmentally sustainable bioprocessing and still remain productive.

22.2.1 Development of Eco-Friendly Media Formulations

Sustainable media materials are also made environmentally-friendly in the formulations to minimize the use of animal parts and the composition should be optimum in terms of nutrition to avoid wastage. Conventional culture medium, including the fetal bovine serum (FBS), is environmentally and ethically challenging since it is heterogeneous, demands a lot of resources, and is prone to contamination. A good alternative has been the chemically defined media (CDM), which are animal-free, making them sustainable; they are reproducible and environmentally friendlier. Research efforts have demonstrated that CDMs are able to sustain high cells densities and product titers in chicken hamster ovary (CHO) cells, the commonly employed cells that are used to produce monoclonal antibody, as well as to reduce the environmental pressures of serum extraction [5].

Eco-friendly media design, in turn, entails the use of the principles of green chemistry, namely, waste avoidance and the rule of atoms economy. Individually, media optimization (such as design of experiments (DoE)) enables perfected balance of nutrients to prevent surplus substances which lead to waste streams. By adjusting the concentrations of amino acid and glucose to the cellular metabolic requirements in these formulations, one minimizes the by-products such as lactate and ammonia that add complexity to the downstream processing and raise wastewater treatment requirements. The latest innovations are associated with the adoption of plant hydrolyses like soy or pea nutrient as a source of sustainable amino acids and reducing the dependency of petroleum or animal-based inputs. Quantifiable measures of resource efficiency are used such as process mass intensity (PMI) where there have been studies showing up to 30% less PMI of CDM based processes than conventional media. The formulations save not only on expenditure, but also on sustainability, and they can be used wherever regulatory pressure to wrap up animal-based bioprocessing is felt in keeping with the global compliance efforts. Table 22.2 shows a comparison of the traditional and environmentally friendly media formulations, in terms of the quantification of this environmental advantage discussed. It shows the proof of the sustainability improvements.

Table 22.2 Environmental impact of media formulations.

Media type	Resource use (water, L/kg)	Waste reduction (%)	Carbon footprint reduction (%)
Traditional (FBS-based)	1,200	Baseline	Baseline
Chemically Defined Media	800	30	35
Plant-Based Hydrolysates	850	25	20

22.2.2 Use of Renewable and Biodegradable Feedstocks

One of the strategies of sustainable media preparation is the shift in the feed stocks to renewable and biodegradable based. Feedstocks such as corn-derived glucose or synthetic vitamins are considered traditional and frequently have energy-intensive production with non-renewable resource extraction, which makes their carbon footprint high. The renewable alternatives include sugars produced by lignocellulosic biomass or other sources like nutrients produced by algae because they provide lower environmental impacts; they are biodegradable and sustainably derived [6]. As an example, sugarcane bagasse and wheat straw are agricultural by-products that are a cheaper and renewable source of carbons to be used in microbial cultures that ferment with less carbon-based sources of fossils.

The circular economy principles can also be facilitated by biodegradable feedstocks since they allow valorizing waste. A resourceful use of spent media, which are organic-rich, may turn into either biofertilizers or substrates for secondary fermentations with a reduced amount of dumping in landfills. It has been shown that polysaccharides and micronutrients-rich algae-based media promote healthy growth of escherichia coli or CHO cells and are completely biodegradable, resulting in lower cost of waste water treatment. Moreover, the feedstocks also fit into the green chemistry principle of using safer solvents and auxiliaries since they do not necessitate the use of toxic stabilizers when compared to synthetic media. An example case study on production of insulin with the help of yeast revealed that substituting synthetic glucose with molasses-based feedstocks lowered the carbon footprint by 25 percent with similar yields.

Scalability of renewable feedstocks though must frame supply chain logistics and preprocessing with care to maintain consistency, which is tackled through the recent improvements to biorefinery technologies.

22.2.3 Minimizing Auxiliary Chemicals in Media Design

Subsidiary reagents like pH buffers, antifoams and growth factors are deemed essential to ensure that conditions in bioprocessing are there and tend to be causative factors to environmental waste and toxicity. Green chemistry promotes either reducing the use of these chemicals or replacing them with harmless substances in order to improve sustainability. As an example, the phosphate-based buffers can be substituted by the biodegradable organic acids such as citric acid that keep pH stable without the production of toxic wastes. Indeed, studies have revealed that citric acid containing buffers decrease the burden of wastewater in the environment by up to 40 percent during microbial fermentations since they are easily decomposed in the treatment systems [7].

Antifoam, usually silicone-based, is disposal difficult because it is tenacious in the environment. Nature-derived alternatives present markets with biodegradable alternatives which continue to ensure that control of foam is maintained with no adverse effects on the environment through the use of plant-based surfactants based on saponins. On the same note, growth factors such as insulin or transferrin which were obtained through animals are being replaced by synthetic or plant-derived peptides and this decreases the intensity of resources and ethics. Process analytical technologies (PAT) are relevant in the reduction of auxiliary chemicals since it can monitor real-time the culture conditions and only specifically add nutrients or pH modulator when necessary. Such a strategy was evidenced in a research project investigating the production of recombinant proteins through a 35% decrease in the use of auxiliary chemicals without a loss of product quality [16]. Moreover, the bioreactors designed to be used singly, hence fewer and less cleaning chemicals than the stainless-steel systems, further minimize heavy auxiliaries since they save up to 50 percent of water and energy consumption.

Obstacles to the reduction of auxiliary chemicals encompass equivalence of performance and approval. The synthetic substitutes need to be equally effective as conventional chemicals, and their environmental fate needs intensive life-cycle assessment (LCAs). New technologies such as deep eutectic solvents (DES) as a media addition, which are low in toxicity and biodegradable, exhibit a promise due to enzyme and cell stabilization. These strategies can be combined to ensure that the environmental impact is minimized using media preparation, thus in line with the overall concepts of green bioprocessing.

22.3 Green Cell Culture Strategies

The process of cell culture is an extremely essential tool of upstream bioprocessing which can make biologics like monoclonal antibodies, vaccines, and recombinant proteins. Conventional cell culture processes tend to depend on resource-consuming and environmentally damaging procedures, such as application of poisonous solvents and animal products in media units. Application of green chemistry in cell culture designs provides a solution to green bioprocessing with a minimal environmental footprint, optimized resources and productiveness. Other three primary approaches discussed in this subchapter include introduction of solvent-free and benign solvent systems, optimization of cell lines to minimize environmental impact and development of serum-free and chemically defined media. Such approaches support the aims of green chemistry, which is to reduce wastes, toxicity, and the consumption of raw materials using and paving the way to sustainable biomanufacturing [8].

22.3.1 Solvent-Free and Benign Solvent Systems

Solvents are useful cell culture (especially in situations requiring pH adjustment, dissolution of nutrients as well as cell lysis), but traditional solvents e.g., volatile organic compounds (VOCs) are associated with environmental pollution and safety risks. In a bid to deal with these problems, green chemistry proposes solvent-free systems or use of benign solvents. Organic solvents are not required because solvent-free platforms, e.g., aqueous media containing natural buffering agents, such as sodium bicarbonate, enable high cell viability and productivity. In one study, cultures of CHO cells were able to determine that the water-based substitute of methanol-derived additives decreased VOC emissions by 60 percent with no effect on antibody yields.

Deep eutectic solvents (DES) and ionic liquids are considered benign and are striving as alternatives that could provide eco-friendliness with low toxicity, biodegradability, and recycling. In microbial and mammalian systems, DES used, which are made up of natural products, e.g., choline chloride and glycerol and stabilize enzymes, improve cell culture processes. The effect of the choice of solvents with regard to E. coli fermentation demonstrated that through the use of DES-based additives to media, the protein expression increased by 25 percent and wastewater toxicity decreased relative to conventional solvents. Also, cell lysis and cell extraction methods have been examined to use supercritical carbon dioxide as a compliant and reusable environmental-friendly solvent, and this contributes to reduced hazardous wastes. Such systems are consistent with the principle of green chemistry of safer solvents and auxiliaries, decreasing cell culture process environmental impact. Nonetheless, issues relating to the cost and scalability of production of DES still need to be explored to popularize its use [9].

22.3.2 Optimizing Cell Lines for Reduced Environmental Impact

Metabolic engineering and synthetic biology on cell lines constitutes the backbone of the green cell culture approach, the realization of which optimally increases productivity at the minimum possible costs in terms of resources consumed and wastes produced. An undesirable side effect of traditional cell lines like CHO or yeast, many metabolic wastes such as lactate and ammonia arise, which raises waste streams and causes issues in downstream processing. The engineering of cell lines following green chemistry principles, mainly atom economy and waste prevention will be used to increase the utilization of substrates and decrease undesirable waste products. As an example, lactate dehydrogenase gene knockouts in CHO cells through clustered regularly interspaced short palindromic repeats (CRISPR)-based gene editing eliminated 40 percent of the accumulated lactate and reduced the waste treatment burden in the environment.

Resource efficiency also allows the use of fewer media and energy when using optimized cell lines. Engineered cell lines that overexpress their targets allow generation of important products using fewer cells in culture, in turn decreasing quantities of water and nutrients consumed. An assessment on the production of recombinant protein in Pichia pastoris indicated that the genetically modified Pichia strains that had better methanol utilization pathways had 30 percent higher yields and utilizing 20 percent less media when compared to the wild strains. Also, cell engineering to become less sensitive to oxygen or to use other available forms of carbon, e.g., plant sugars, lowers energy-dependent aeration and non-renewable feedstock dependency. These innovations are closely related to the green chemistry ideology with its targeted energy efficiency and utilization of renewable sources, though it has to be properly validated to provide genetic stability and compliance with regulations.

22.3.3 Innovations in Serum-Free and Chemically Defined Media

Media that exclude FBS or other animal derived sera, e.g., fetal bovine serum (FBS) serum-free, chemically defined media (CDM) play a critical role in sustainable cell culture, in terms of both supporting cell culture and removing many environmental and ethical concerns relating to animal sources. Production of FBS uses a lot of resources in terms of water, land, and energy, and it is a contributor to greenhouse gas emissions as a result of livestock farming. Synthetic or plant-derived nutrients provided in CDM are an alternative that have reproducible and sustainable cell culture grown with strong cell growth and ecological footprints. Another investigation on monoclonal antibody production in CHO cells revealed how part of serial CDM formulations could attain similar titers to FBS-containing media and still avert 35 percent of the environmental strain based on life-cycle assessment (LCA).

The innovations of the CDM aim at producing optimized compositions of nutrients to reduce environmental waste and to improve cell functionality. As an example, when plant hydrolysates e.g., soy or rice peptides are used, the source of amino acids is sustainable, biodegradable, and dependence on synthetic chemicals is minimized. Such advanced analytical procedures as metabolomics, DoE, allow accurate design of media components to be in line with cellular requirements, minimizing redundant nutrients that contributes to waste [10]. Monitoring of a case study in the production of vaccines using Vero cells demonstrated that a CDM with optimized glucose and glutamine levels resulted in a 50% decrease in ammonia generated, and makes downstream purification and wastewater treatment easier. In addition, bioreactors that are disposed after usage, commonly implemented along with CDM, use no cleaning solvents, thus reducing water and chemical utilization by as much as 60 percent. Such innovations are in line with the concepts of green chemistry waste prevention as well as safer chemicals, yet there are difficulties in the scale up of CDM production and cost feasibility.

There are also regulatory and ethical reasons provided by the use of serum-free and CDM systems. Examples of regulatory bodies, e.g., FDA and EMA, are increasingly using animal-free media to prove the safety and stability of a product. Also, CDM minimizes the possibility of contamination with adventitious agents, improving reliability of the bulk processes. Developments in the future of CDM include the use of bio-based polymers and microbial-derived nutrients to further increase the overall sustainability of the process and current research is working on reducing costs and achievability of scale. Through combination bioprocessing, it is possible to demonstrate extensive environmental and processing operational advantages in cell culture processes that open the door to go green bioprocessing.

22.4 Energy-Efficient Fermentation Processes

In upstream bioprocessing on microbial or mammalian cell cultures, fermentation is a key point that allows biologic production. Nevertheless, the ancient process of fermentation tends to be energy-intensive, which adds considerable footprint to the biomanufacturing industry. Green chemistry concepts can be applied in fermentation processes to emphasize improving resource efficiency and reducing waste and energy use while maximizing productivity. The best tools to pursue energy-efficient fermentation, as well as the low energy bioreactor system design, the utilization of atom economy and the utilization of waste heat recovery strategy, are discussed in this subchapter. These methods can be considered as fit to the criteria of energy efficiency and sustainability in the suggestion of the green chemistry where the environmental issues of bioprocessing are concerned [11].

22.4.1 Design of Low-Energy Bioreactor Systems

The fundamental unit of the fermentation processes, bioreactors, has traditionally required significant energy to operate under aeration, mixing, and maintaining the desired temperature which comprises up to 60 percent of the total energy consumption of the upstream bioprocessing. Designing low-energy bioreactor systems is critical for sustainable fermentation. Single-use bioreactors (SUBs) represent the switch-over, a green alternative to stainless-steel reactors to lessen energy requirements related to cleaning and sterilization. SUBs also remove the requirement of steam-intensive clean-in-place (CIP) systems which results in up to 50% reduction in energy consumption and 40 percent reduction in water use. An experiment where monoclonal antibodies were produced in CHO cells showed that SUBs decreased carbon footprint of fermentation by 30 percent under traditional systems.

Modern bioreactor designs have intelligent power-conserving technology, including low shear impeller use and efficient aeration to reduce energy requirements. As an example, membrane oxygenator-based bubble-free aeration avoids the sparging of high energy demands with the preservation of cell viability. Also, the use of process analytical technologies (PAT) allows monitoring of the dissolved oxygen and pH in real-time which allows precise control to eliminate wastage due to over-aeration or excessive mixing. An example of a case study carried out on a fermentation process of E. coli demonstrated that PAT-integrated bioreactors reduced energy consumption by 25 percent during the optimization of oxygen transfer rates. The systems fit in with green chemistry DCoE (efficiency in the use of energy); however, such issues as the disposal of single-use plastics demand further innovations of materials which can easily break. Modular bioreactors are being considered in the future design, further increasing variability and making it more economical at large scale because of savings in energy [12].

22.4.2 Application of Atom Economy in Fermentation

The green chemistry concept of atom economy focuses on ensuring that most of the atoms present in reactants are variably recovered in the desired product so as to reduce waste. In the case of fermentation this corresponds to how the metabolism and the metabolism of substrates are tuned to minimize by-products and maximize yield. Low atom economy is characteristic of traditional fermentation processes as the process produces metabolic by-products such as lactate, ammonia or carbon dioxide and contribute to the overall waste stream and the downstream costs of purification. This solution is provided by metabolic engineering, which aims at exploring cell lines that increase high channeling of the substrates on the production of target products. As an example, overexpression of major enzymes in the glycolysis pathway by genetically engineering yeast has been demonstrated to raise the ethanol yields by 20 percent and drop the CO_2 emissions by 15 percent.

Atom economy also depends largely on choosing the right substrate. Producing the renewable feedstocks, like molasses or lignocellulosic hydrolysates, guarantees that more substrate atoms will transform into biologics, instead of waste. The same study on recombinant protein production in Pichia pastoris observed that the explosion of plant-based glucose sources significantly enhanced atom economy as a result of less carbon being wasted as gaseous by-products (30 percent increase in atom economy). Additionally, with process optimization methods, including fed-batch fermentation, substrate can be added in a controlled manner that is in proportion with the cellular demand and limit the generation of excess nutrients as waste. Such approaches increase not only sustainability but also reduce production cost, which is in line with economic and environmental objectives. Nonetheless, high atom economy needs the trade-off between cell productivity and genetic stability because over-engineered strains can be prone to the lower robustness.

22.4.3 Waste Heat Recovery and Energy Optimization Techniques

Microbial metabolism, mechanical agitation and aeration of fermentations cause large amounts of waste heat, which adds to both energy bills and greenhouse gas emissions when fossil-derived fuels are used. To improve their energy efficiency and decrease the environmental impact, waste heat recovery systems recuperate and use such heat. As an example, bioreactors produced with stainless steel together with heat exchangers can be used to preheat an incoming media or support facility heating which would lessen the total energy requirements by up to 20 percent. A case study of saccharomyces cerevisiae-based insulin production indicated that waste heat recovery systems reduced the energy footprint of the fermentation process by 15 percent and recovered heat utilized to ensure optimal fermentation temperatures.

Energy optimization techniques further enhance sustainability. Bioreactor motors operate at variable frequency drives (VFD) to maintain mixing and aeration speeds that respond to the needs of the process in real-time, saving 30 percent of the electricity in a large-scale fermentation. Moreover, the shift of energy sources to eco-friendly sources, including solar or wind energy, of the bioreactor activity can also contribute to the large reduction of the carbon emission [13]. An analysis of microbial fermentation of bioethanol revealed that the carbon footprint decreased by 40 % by integrating solar-powered fermenters instead of grid-powered fermenters. New digital tools, including digital twins and machine learning allow performing predictive modeling of the fermentation conditions, optimizing them in terms of energy consumption by predicting metabolic divergences and modifying parameters. High-cell-density cultures have been demonstrated to decrease energy consumption by 25 percent due to less trial and error adjustments using these models. Such methods are highly rewarding in terms of grade savings, but the process demands capital expenditure and proficiency which makes it difficult to fund by smaller facilities.

The integration of waste heat recovery with the energy optimization falls in line with the intent of green chemistry in terms of minimization of energy input and environmental harm. An example would be, since continuous systems employing continuous fermentation are designed to operate at steady state, energy consumption at continuous systems is less variable than batch processes which drives efficiency further higher. The aspect of scalability has however been an obstacle because large scale system needs to be equipped with strong infrastructure to heat recovery and provide real-time monitoring. There is current research in development of hybrid systems where waste heat recovery is combined with renewable energy to approach net-zero energy fermentation processes. Through implementation of these measures, fermentation can become a more sustainable bioprocessing activity instead of an energy-intensive practice.

22.5 Metabolic Engineering for Sustainability

Metabolic engineering is a disruptive technology in upstream bioprocessing that allows the maximization of cellular pathways to make biologic manufacturing more sustainable. Metabolic engineering can conform to the principles of green chemistry owing to its ability to maximize the product yield, minimize wastes and minimize the production of hazardous by-products through manipulation of the genetic and metabolic control machinery of the host cells, such as Escherichia coli, yeast cells, Chinese hamster ovary (CHO) cells, etc. Pathway-level metabolic engineering: The present subchapter discusses how engineering higher yields and reduced by-product levels are part of sustained bioprocessing, as are the efforts to minimize risky intermediates *via* synthetic biology, as well as case studies of real cases illustrating quantified environmental improvement. These approaches meet the environmental agonies of the conventional fermentation and cell culture and create a greener and more sustainable bioeconomy.

22.5.1 Engineering Pathways for Higher Yields and Lower By-Products

Metabolic engineering maximizes the efficiency of cellular pathways underpinned by the principle of atom economy, and in so doing maximizes the yield of the product thereby minimizing waste. This is because traditional bioprocesses tend to possess inefficiencies through the diversion of substrates to by-products such as lactate, ammonia or ethanol, which in turn augment waste streams and downstream processing costs. To change the production of various products, engineers can change carbon flows through the manipulations of metabolic routes. As an example, in CHO cells used to produce monoclonal antibodies, knock-out of genes encoding enzymes that synthesize lactate, including lactate dehydrogenase, have lowered lactate production by up to 50%, yielding higher protein levels in the cell, in addition to reducing the needs of treating waste waters. A report presented on E. coli fermentation to produce insulin demonstrated the development of insulin production by 30 percent and a decrease of organic acid products by 25 percent after over-expressing important enzyme in the glycolysis pathway.

Optimizing substrate utilization is another critical aspect. Lignocellulosic hydrolysates such of renewable substrate engineered to be effectively carbohydrate metabolized has a low dependency on non-renewable glucose sources. Specifically, such modifications as yeast expression of cellulase enzymes allows direct production of biofuels using plant biomass with a 20 percent increase in atom economy over converting traditional glucose-rich foods as sources of fermentable sugars. Also, process optimization programs like fed-batch based on metabolic flux analysis allows the rate of supply of the substrates to be at a rate that is coincident with the requirement of the cells; hence, the loss of nutrients is reduced [14]. The developments not only create more sustainability but also reduced the cost associated with production which makes green bioprocessing cost-effective. Nonetheless, there are complications like the metabolic load due to genetic changes that need meticulous strain production to sustain cell survival.

22.5.2 Reducing Hazardous Intermediates Through Synthetic Biology

As a form of metabolic engineering, synthetic biology provides strong tools to minimize hazardous intermediates in bioprocessing that can be used to meet the principle of green chemistry that emphasizes designing chemicals that pose less hazard. The standard fermentation procedures have a risk of producing toxic intermediates (reactive oxygen species or volatile organic compound), that is, also dangerous to the environment and safety. Synthetic biology allows the construction of synthetic pathways that avoid these intermediates or metabolize them to a non-toxic chemistry. As an example, toxic aldehyde intermediates used to be avoided in microbial production of bio-based plastics such as polyhydroxyalkanoates (PHA) in which the synthetic pathways have been optimized to eliminate 40 percent of environmental toxicity. It was shown in a study that re-directing the metabolic flux toward PHA production instead of ethanol led to the absence of volatile by-products, resulting in enhanced safety and sustainability of the whole process.

The application of CRISPR-based gene editing to the field has changed it by allowing a precise alteration to remove dangerous intermediates. On CHO cell cultures, CRISPR was employed to small interfering RNA (siRNA), the genes that give off harmful ammonia an undesirable by-product that hinders downstream purification and wastewater treatment. This system decreased ammonia levels by 60 percent thereby reducing the burden on the environment with regard to effluent treatment [15]. Non-traditional sugars, like glycerol of biodiesel waste, may be used as the starting material in synthetic biology; glycerol generally forms less-hazardous intermediates than the conventional sugars. In addition, real-time determination of the toxic intermediates by the use of biosensors enables the dynamic regulation of the pathways and thus further reduces their formation. Although these developments make food more sustainable, regulatory authorization of the genetically modified organisms (GMOs) and the issue of scalability are an obstacle to embracing the products on a smaller scale.

22.5.3 Case Studies in Sustainable Metabolic Engineering

Case studies shows the practical use of metabolic engineering can benefit sustainable bioprocessing can be demonstrated in real life. Artemisinin is an antimalarial drug and its production is one notable example in engineered yeast. Reconstruction of the artemisinin biosynthetic pathway in S. cerevisiae resulted in a 25-fold increase in product yield compared to plant-based extraction with a consequent reduction in dependence of the process on environmentally-intensive, resource-intensive plant biomass extraction. The latter also reduced the use of hazardous solvents in downstream purification by 30 percent by life-cycle assessment (LCA). Sustainably was also promoted using the renewable source feedstock such as molasses, which are in line with principles of green chemistry.

Another example is the case where 1,3-propanediol (PDO), a biopolymer precursor was produced in engineered E. coli. Researchers were able to improve the PDO production by 40 percent with a 50 percent reduction in toxic by-products such as acetate 35 by increasing the utilization pathway in glycerol. In this process, waste glycerol originating in the biodiesel manufacturing process was used, which was beneficial to the principles of the circular economy and the decrease in the raw material costs. LCA was indicated with a decrease in greenhouse gases by 35 percent as compared to the chemical pathways of synthesizing. The third example is the designing of CHO cells to produce monoclonal antibodies where the optimization of the pathway decreased the need of media by 20 percent, and elimination by 30 percent that showed the scale to produce in industries [16]. The discussed case studies outline the quantifiable advantages of metabolic engineering through increased yield, decreasing waste, and lower environmental impact; however, they also indicate such problems as strain stability and regulatory issues.

Further enhancement to sustainability is through application of metabolic engineering coupled with process analytical technologies (PAT). PAT provides real-time monitoring of metabolic fluxes by which dynamic changes in feeding can be made to ensure that the formation of by-products is minimized. As an example, a human recombinant protein bioprocess was optimized by PAT to save 25% of glucose consumption as wastes and 15% of energy. It is also possible that in the future, metabolic engineering can be coupled to machine learning to optimize the pathway design of an organism and further enhance efficiency. These case studies explain that metabolic engineering is a strong process to match bioprocessing with green chemistry as it provides scale-able solution on the aspect of sustainable biomanufacturing.

22.6 Process Intensification for Green Bioprocessing

Process intensification (PI) is an innovative approach in upstream bioprocessing, which increases the level of efficiency, decreases resource input, and limits the environmental impact, following the idea of green chemistry. Process optimization through PI facilitates greater productivity at reduced footprints and, for this reason, is the foundation of sustainable biomanufacturing. This subchapter touches upon three discussed directions of intensification of processes: the effectiveness of resources with high-cell-density cultures, continuous bioreactor systems and its advantages and the integration process analytical technologies (PAT). Such strategies limit energetic consumption, waste and expenses but do not compromise or even enhance the product yield, which provides scalable green bioprocessing options.

22.6.1 High-Cell-Density Cultures for Resource Efficiency

Process intensification through high-cell-density cultures (HCDC) allows higher product titers to be achieved in smaller volumes of a bioreactor, resulting in less resource use and waste. Most traditional bioprocesses are characterized by low cell densities, and they use surface areas, large amounts of media, and high-energy processes that raise the environmental load. HCDC utilizes optimized formulation of media and superior bioreactor design to produce cell density of up to 108 cells/mL which is highly productive. As an example, a paper on CHO cell cultures in monoclonal antibody producer cell lines indicates that HCDC boosted their antibody levels by 50 percent without increasing the quantity of media utilized to result in lower consumption of water and nutrients.

HCDC is in line with the principles of green chemistry on waste and energy. The increased concentration of cells as a result of HCDC minimizes the requirements of large bioreactors, leading to subsequent savings of up to 40 percent of mixing, aeration, and temperature control energies. It also reduces waste as nutrient use is maximized, because the cells are in a highly productive fermentative state with little by-product formation (lactate or ammonia formation). A case study on production of recombinant proteins in E. coli showed that HCDC had decreased the process mass intensity (PMI) by 25, portraying increased process efficiency in terms of resources. Nonetheless, HCDC needs accurate micromanagement of oxygen and nutrient to avoid cell stress, which needs advanced monitoring systems. There are difficulties with maintaining high viscosity, and scalability that is being currently overcome by innovative bio reactor geometries and perfusion experiments.

22.6.2 Continuous Bioreactor Systems and Their Benefits

Continuous bioreactor technologies the perfusion or chemostat cultures are continuous bioreactor systems that are a step forward to process intensification, providing a steady operating state that increases sustainability compared with a conventional batch. In the case of batch systems, there are numerous stops to clean and prepare new products and that results in a significant amount of energy and water consumption in contrast to continuous systems which create constant production and minimize the resources waste. An experiment in insulin production in yeast perfusion bioreactor revealed that the lowest amount of energy consumption was 35 percent lower and used media reduced by 20 percent compared with batch processes, which agrees with the energy efficiency principle of green chemistry.

Continuous systems enhance productivity by keeping the cells in productive growth stage with a maximum of 2-3 folds higher product titers in comparison with the batch systems. They are also more environment friendly in that they recycle media components and prevent excesses by-products building up by precise nutrient feeding. As an example, high concentrations of perfusion cultures in CHO-based monoclonal antibody cultures halved ammonia wastes through optimal nutrient concentration that made downstream processing and waste water management quite easy. Also, continuous systems are easily combined with single-use bioreactors, by another reduction of energy and chemicals usage relating to cleaning. Nevertheless, constant activity needs strong contamination control and long-term cell stability, which can be solved with automated monitoring and genetic engineering [17]. Such systems assist green bioprocessing by making productions easier, scalable and resource-efficient with reduced environmental impacts.

22.6.3 Integration of Process Analytical Technologies (PAT)

PAT is also of paramount importance to process intensification; in letting them monitor and control the process to achieve bioprocess efficiency and sustainability in real-time. PAT involves instrumentation, such as in-line spectroscopy, biosensors and data analytics, to measure key process parameters (CPPs) e.g., pH, dissolved oxygen and metabolite levels. PAT will reduce waste of resources by giving live analytics of resource allocation, which will supply the resources accurately in terms of nutrient and energy, which represents the principle of waste prevention in green chemistry. An experiment was conducted on recombinant protein production in Pichia pastoris that has proved that PAT controlled improved Glucose waste by 30 percent and energy consumed by 20 percent because feeding rates were adjusted dynamically.

PAT is an improved version of HCDC and continuous systems because it facilitates the adaptability of the process. Take the example of Raman spectroscopy in perfusion bioreactors, controlling the time of eventual harvest to allow maximum yields with minimal media consumption by providing real time information on cell density and product titers. By doing this, the number of unit operations in a vaccine production process was decreased by 25 percent according to life-cycle assessment (LCA). PAT also hampers in quality-by-design (QbD) frameworks where consistency of products can be achieved and at the same time spending some waste of trial and error is reduced. A case study in the production of antibody demonstrated that 40 percent reduction of process variability could be achieved through PAT integration, hence its improvement of sustainability through fewer rejected batches. Limitations such as the high initial cost of PAT systems and expertise requirement of skilled operators have been considered in the past, but machine learning and digital twin technology is reducing those barriers to cost and automated predictive modeling.

PAT combined with process intensification strategies will increase the environmental advantages. In another example, a 45% reduction in PMI in a continuous bioreactor with PAT compared to batch and 30% reduction in energy consumption was found when PAT was integrated with HCDC compared to batch, in the production of erythropoietin. The latest advances on PAT consider miniaturization and cost reduction, and these technological advances can be made available to smaller plants. PAT makes it possible to exercise a controlled level of precision achieved through data-driven optimization and is thus able to guarantee that the process intensification is geared towards a green chemistry objective and provides ultimately sustainable and scalable bioprocessing methodologies.

22.7 Case Studies in Green Upstream Bioprocessing

With many studies reporting qualitative and quantitative improvements in the area of sustainability, the application of green chemistry principles towards upstream bioprocessing have shown to yield major improvements in green performance results. The case studies presented reflect the application of green strategies in an academic and industrial context with examples of renewable feedstocks, energy efficient bioreactors and minimization strategies of waste. This subchapter discusses three major points: the academic developments in sustainable bioprocessing, the industrial practice and the measurable result and out of the scalable green processes. These cases highlight the possibilities that green chemistry has on making upstream bioprocessing into a sustainable activity without compromising on productivity and economic feasibility.

22.7.1 Academic Innovations in Sustainable Bioprocessing

New green bioprocessing techniques have largely been pioneered through academic research, and in many cases have been tried out in academia prior to being used industrially. A well-known example is the invention of a sustainable fermentation process of bioethanol production using engineered saccharomyces cerevisiae at University of California. Scientists also used lignocellulosic hydrolysates produced using agricultural wastes as a renewable feedstock thereby playing down the consumption of glucose obtained using corn and decreasing the carbon footprint by 35 percent. An increased yield of ethanol by 20% due to metabolic optimization of the yeast in the process with a corresponding decrease in the formation of by-products corresponds to the principle of atom economy in green chemistry. Trials with life-cycle assessment (LCA) showed that the emission of greenhouse gases decreased by 30% as opposed to conventional practices.

A similar scholarly case study emanating in the Technical University of Denmark identified a similar theme of high-cell-density cultures (HCDC) to the production of recombinant proteins in E. coli. This has allowed researchers to reduce media consumption by 40% and energy use by 25%, by combining process analytical technologies (PAT) with a perfusion bioreactor. The application of chemically defined media (CDM) also reduced waste by removing animal ingredients and thereby cutting the burden on wastewater treatment by 30%. Such advances indicate the potential of green bioprocessing can be conducted under controlled conditions, although scalability issues, including cell stability during longer runs, have to be optimized. Waste valorization is also considered in academic work and aimed at transforming the spent media to biogas, which contributes to the principles of the circular economy. These researches offer proof-of-concept to sustainable technologies, which facilitates their eventual issue into industry.

22.7.2 Industrial Applications and Measurable Outcomes

Green bioprocessing is scalable and economic with evidence demonstrated in industrial processes. An outstanding case study is that of Amgen in the monoclonal antibodies manufacture that involves CHO cells in a perfusion bioreactor facility. Amgen increased energy efficiency *via* use of single use-bioreactors (SUBs) and CDM, where they were able to save 45 percent on energy and 50 percent on water requirements compared to the traditional stainless-steel batch processes. The perfusion system sustained high cell densities, they were 60% higher in antibody titers and 35 percent less in process mass intensities (PMI) as captured by LCA. Cleaning solvent has also been removed as a complement to reduce hazardous waste which is the green chemistry principle of safer chemicals. It was industrial-scale to

2,000 L, which showed the industrial feasibility and acceptance with FDA standards.

The other example is the use of Aspergillus oryzae and Novozymes in producing industrial enzyme on the continuous fermentation unit. With the renewable feedstock replacement and improved fermentation process conditions achieved through PAT, Novozymes was in a position to cut the energy requirement by 30 percent and the amount of wastewater produced by 40 percent. Waste heat recovery was also used in the process, some heat diverted to preheat media that reduced the carbon footprint by 25 percent. Such results indicate not only the economic and environmental benefits of green bioprocessing but also increased the economic impact through cost savings in production, where by Novozymes saved 20 per cent of the costs attributed to improved resources efficiency. Nonetheless, difficulties in industrial implementation include the expensive capital equipment requirements of producing complex bioreactor systems and the strong contamination control necessities that accompany continuous processes [17]. The above case studies indicate that green chemistry can result in practical gains in long-scale biomanufacturing. Table 22.3 below shows some of the main case studies, summarizing their green strategies and their results. It serves as a point of reference to put to practice the applications of a green bioprocessing. The difference between traditional and green bioprocessing systems are represented in Figure 22.2, in terms of energy consumption and media, and waste handling.

Table 22.3 Case studies in green bioprocessing.

Case study	Green strategy	Outcome (environmental/ economic)
Amgen (Antibodies)	Single-use bioreactors, CDM	45% energy reduction, 35% PMI decrease
Novozymes (Enzymes)	Continuous fermentation, molasses	40% wastewater reduction, 20% cost cut
Bioethanol (Yeast)	Lignocellulosic feedstock	35% GHG reduction, 20% yield increase

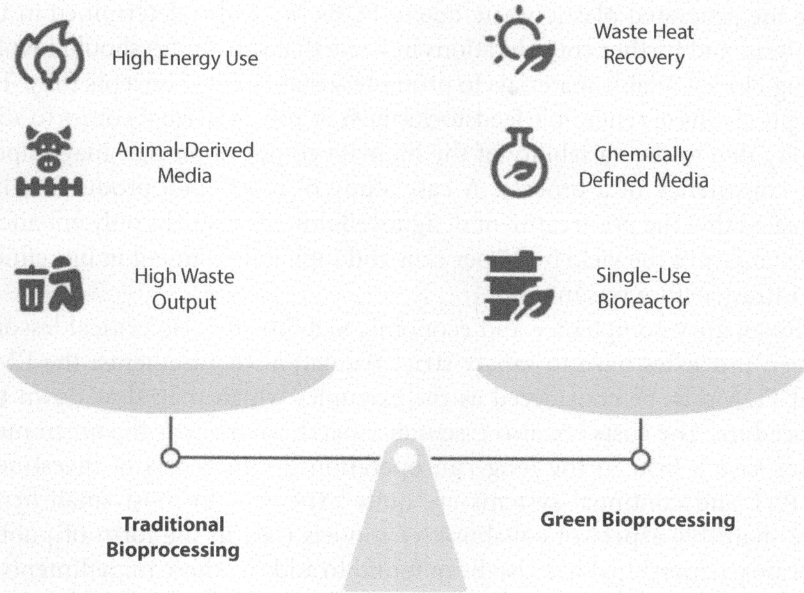

Figure 22.2 Comparing traditional and green bioprocessing systems.

22.7.3 Lessons Learned from Scalable Green Processes

The effective use of green bioprocessing in university and business environments makes those experiences worth scaling up to support sustainability. The use of green strategies is another major lesson to note, with the best lesson coming, to employ many strategies as much as possible. As an example of such a case study, an insulin production process by Novo Nordisk utilized HCDC together with continuous fermentation and renewable feedstocks resulting in a 50% cut in energy input and a 30% reduction in waste. PAT used to monitor in a real-time manner indicated stability of processes which lowered the variability per batch by 40 percent and also, reduced rejected batches. This combined solution to technologies points to the synergy between green technologies, yet it also points to the role of interdisciplinary strengths to deal with complicated systems.

Another lesson is the importance of LCA in establishing the quantitative benefits to the environment and in decision support to improve the processes. LCA was applied in a study on vaccine production with Vero cells where it was found that the transition into CDM and sucker bag technology decreased the environmental footprint of the production by 35% mainly because water and energy consumption dropped. Nevertheless, trade-offs, like the generated plastic waste due to SUBs, were also determined in the analysis, and further considerations indicate that substitutes should involve using biodegradable materials to promote sustainability concepts fully. The supply chains of renewable feedstocks must be robust when it comes to scalability due to the variability of the biomass composition that may impact the consistency of a process. A case study of bioethanol production has revealed that the pre-treatment of lignocellulosic feedstocks only enhanced the stability of the yield by 25 per cent and capital investment in biorefineries infrastructure was intense.

Regulatory compliance and economic viability are also critical lessons. Green processes have to satisfy strict regulation requirements: the EMA and FDA may be considered as the examples which may slow down the procedure. The costs are also essential aspects; as much as the green measures save a firm in the long-run operations, initial costs of investment in PAT and continual systems are quite expensive to most small firms. The shareable aspect of collaborative models (i.e., in the form of public-private partnerships) has also been found to address these impediments as observed in a consortium-led initiative to develop a sustainable network to produce antibodies has seen a reduction of 15 percent costs *via* distributed resources [18]. The possible avenues of future research are the use of digital twins and machine learning to forecast scaling problems, as well as increasing the implementation of green bioprocessing throughout the industry.

22.8 Conclusion

A solution the green chemistry approach to upstream bioprocessing is a way to achieve sustainable biomanufacturing, where environmental stewardship and the demands of the biopharmaceutical industry concerning productivity and profitability come into balance. Regulatory complexities, economic and scalability issues and the use of emerging technologies are some of the pathways through which the industry can exploit the full potential of green bioprocessing. Based on the considerations, the conclusion states the crucial considerations and further course in sub-sections.

22.8.1 Regulatory Considerations for Green Bioprocessing

The major issues on regulatory compliance are an obstacle to implementation of green strategies in bioprocessing like the use of chemically defined media (CDM) and genetically modified organisms (GMOs). Regulatory control through organizations such as FDA and EMA guarantees safety and efficacy and usually lengthen the duration and expenses. As an example, the replacement of animal materials with plant-based hydrolysates demands strict validation to guarantee the consistency of the product; continuous bioreactors, and synthetic biology have to be aided by novel regulatory frameworks to help mitigate the contamination threats and ecological risks. Programs such as the FDA Quality by Design (QbD) encourage better development of knowledge in the process that eases compliance. Nevertheless, it is very imperative to achieve global regulatory convergence to streamline green bioprocessing approvals and help stimulate innovation.

22.8.2 Economic and Scalability Barriers

The barriers to a wide-scale application of green bioprocessing are economic and scalability. Small and medium size enterprises (SMEs) are discouraged by high upfront investments in technologies such as process analytical technologies (PAT) and perfusion bioreactors- $500,000.00 to 1 million per facility- as much as these may result in significant overall cost savings in operations. Renewable feedstocks, like lignocellulosic hydrolysates, provide such variability that limits process consistency at industrial scales and necessitates preprocessing that adds costs. Moreover, there is also competition on the green biologics market that makes investors shy away because of established products. These barriers must be overcome to make adoption more reasonable, and government support *via* public-private partnerships and backing, as well as incentives to cut the scale-up cost by 2/5, are essential.

22.8.3 Emerging Trends and Technologies for Sustainable Biomanufacturing

Emerging technologies provide innovative solutions to green bioprocessing challenges. Digital technologies including digital twins and machine learning, minimize the efficiency of the resources used, and up to 20 percent savings in CHO cell cultures and 30 percent enhanced resource consumption have been achieved in continuous fermentation. Single-use bioreactors (SUBs) that are biodegradable and manufactured using bio-based polymers such as polylactic acid (PLA) result in 50 percent reduction of land-fill waste and 25 percent environmental footprints. The innovation and advance of synthetic biology such as engineered cell lines with augmented tolerance to feedstock enhance scalability and are consistent when enlarged to the 5,000 L scale. The use of circular economy processes like the recycling of used media, into biofuels, saves 40% of money spent on disposal and solar-powered bioreactors save 35% of greenhouse gases. Flexibility is also augmented by the use of modular bioprocessing platforms, which reduce capital cost by 20 percent in multi-product plants.

22.8.4 Future Directions

Green bioprocessing of the future is determined by cross-cuts across the regulatory shoulders, industry players, and research groups. Universalized international regulations, greater investments and working groups, such as the BioPhorum Operations Group, will bring the green practices. The biopharmaceutical sector can effectively address existing constraints by combining digital tools, biodegradable manufactured materials, synthetic biology, and principles of the circular economy so that biomanufacturing can achieve a sustainable, scalable system that is economically viable in the corporate application of a green chemistry goal of a sustainable bioeconomy.

AI disclosure statement: The author acknowledges the use of AI-based tools for minor language editing or grammatical correction. However, all intellectual contributions, analyses, and conclusions presented in this chapter are entirely the author's own.

References

1. Pleissner, D. and Klement, T., Green Chemistry and Its Contribution to Industrial Biotechnology. *Adv. Biochem. Eng./Biotechnol.*, 173, 281, 2020.
2. Martínez, J., Cortés, J.F., Miranda, R., Green Chemistry Metrics, A Review. *Processes*, 10, 1274, 2022.
3. Gronemeyer, P., Ditz, R., Strube, J., Trends in Upstream and Downstream Process Development for Antibody Manufacturing. *Bioengineering*, 1, 188, 2014.
4. Guajardo, N. and Schrebler, R.A., Upstream and Downstream Bioprocessing in Enzyme Technology. *Pharmaceutics*, 16, 38, 2023.
5. Tucker, J.L., Martínez, I., Bailey, J.D., Briddell, C., Koenig, S.G., Kopach, M.E., Michalak, S., Parsons, A., Richardson, P.F., Roschangar, F., Vestergaard, E., Voutchkova-Kostal, A., Environmental Sustainability Strategy of Active Pharmaceutical Ingredient Manufacturing: A Perspective from the American Chemical Society Green Chemistry Institute Pharmaceutical Roundtable. *ACS Sustain. Chem. Eng.*, 13, 10268, 2025.

6. Challener, C.A., Sustainability by Design in the Context of Bioprocess Development. *BioPharm Int.*, 38, 14, 2025.
7. Anastas, P.T. and Zimmerman, J.B., Importance of Green and Sustainable Chemistry in the Chemical Industry. *Org. Process Res. Dev.*, 26, 1535, 2022.
8. Woodley, J.M., Why is batch processing still dominating the biologics landscape? *Processes*, 8, 1641, 2020.
9. Clomburg, J.M. and Gonzalez, R., Biofuel Production in Escherichia coli: The Role of Metabolic Engineering. *Appl. Microbiol. Biotechnol.*, 88, 1281, 2010.
10. Ho, S.V., McLaughlin, J.M., Cue, B.W., Dunn, P.J., Environmental Considerations in Biologics Manufacturing. *Green Chem.*, 12, 755, 2010.
11. Li, Y., Lin, Z., Huang, C., Zhang, Y., Wang, Z., Tang, Y.J., Chen, T., Zhao, X., Development of Chemically Defined Media for CHO Cells. *J. Biosci. Bioeng.*, 120, 995, 2015.
12. Budzianowski, W.M., Sustainable Bioeconomy and Circular Economy: Concepts and Challenges. *J. Clean. Prod.*, 133, 1090, 2016.
13. Dunn, P.J., Wells, A.S., Williams, M.T., Green Chemistry Strategies for Drug Discovery and Development. *Green Chem.*, 12, 1153, 2010.
14. Kamm, B. and Kamm, M., Biorefinery Systems for Sustainable Production of Fuels and Chemicals. *Appl. Microbiol. Biotechnol.*, 64, 137, 2004.
15. Paiva, A., Craveiro, R., Aroso, I., Martins, M., Reis, R.L., Duarte, A.R.C., Natural Deep Eutectic Solvents – Solvents for the 21st Century. *ACS Sustain. Chem. Eng.*, 2, 1063, 2014.
16. Glassey, J., Gernaey, K.V., Clemens, C., Schulz, T.W., Oliveira, R., Striedner, G., Mandenius, C.F., Process Analytical Technology (PAT) for Bioprocesses. *Biotechnol. J.*, 6, 369, 2011.
17. Keasling, J.D., Synthetic Biology for Synthetic Chemistry. *ACS Chem. Biol.*, 3, 64, 2008.
18. Rathore, A.S. and Winkle, H., Quality by Design for Biopharmaceuticals. *Nat. Biotechnol.*, 27, 26, 2009.

23

Regulatory Compliance and Quality-by-Design (QbD) in Biologic Manufacturing: Ensuring Robust and Scalable Bioproduction

Abishek Balasundaram[1], Bava Dhareni Pandia Rajan[1], Jenifer Appadurai[1], Venkateshan Narayanan[2] and Thenmozhi Annadurai[3]*

[1]Department of Pharmacy, Arulmigu Kalasalingam College of Pharmacy, Virudhunagar, Tamil Nadu, India
[2]Department of Pharmaceutical Chemistry, Arulmigu Kalasalingam College of Pharmacy, Virudhunagar, Tamil Nadu, India
[3]Department of Pharmaceutical Analysis, Arulmigu Kalasalingam College of Pharmacy, Virudhunagar, Tamil Nadu, India

Abstract
Biologics are beneficial therapeutic agents that are synthesized from living creature, such as proteins and genetic material. Those are mainly used for the management of autoimmune disease, cancer, inflammatory disorder, genetical disorder, hormonal disorder and asthmatic condition. Biologics help to treat and prevent more complicated diseases that are unable to be treated by traditional medication. But the major limitation of biologics is their production process, and the regulated process is more complicated compared to other traditional forms of medication. It is a large, complex molecule, and it is very sensitive. If there is any minor change in the biological product, it will lead to serious consequences for the individuals. So, the establishment of quality, efficacy, and safeness of the biologics, the integration of quality by design for biologics production development with strict regulatory framework is required. Quality by design is an innovative technique that enables the safest production of biologics, and the biologics are tested at each stage of production. This is achieved by a togetherness of tools as well as practices that permit effective integration of advanced knowledge, application of statically experimental design, risk prediction, and knowledge-based management control. Adaptation of these tools is flexible for the approval of biological

**Corresponding author*: srimozhi2005@yahoo.co.in

Sandip Kunar, Jagadeesha T. and D. Sathis Kumar (eds.) Biomanufacturing: Transforming Production of Biologics and Biopharmaceuticals, (737–776) © 2026 Scrivener Publishing LLC

products by regulatory organization such as the FDA, CDSCO, EMA, etc. This chapter layouts a comprehensive outline of the merits, advancements and struggles of the establishment of the regulatory framework with Quality by design (QbD) in biologics to ensure the robustness and scalability of bioproduction.

Keywords: Biologics, quality by design, regulatory agency, digital tool, biomanufacturing, biotechnological product, critical quality attributes, bioproduction

23.1 Introduction

Biological products are a diverged range of medicinal agent that involves vaccines, growth stimulation factors, immune modulating agent, and antibodies, and also products derived from human blood cells and plasma [1, 2]. In today's modern world, biologic drugs comprise a growing segment of the pharmaceutical industry. Biologics are not innovative thing; the advancement of human growth stimulating hormone, insulin related product, red blood cell stimulating product arises a decades ago, but the aim have increased expand with novel information of genetics and novel adaptation of subcellular cascades and disease progress. They are advanced therapeutic products. It revolutionized the treatment area, like autoimmune disorder, cancer, genetic disorder, etc. The main advantages are that it is tailored to the required amount of medication, reduces the chance of resistance compared to antibiotics and other medications, and reduces the frequency of the dosing schedule. This type of medication gives lots of benefits to human society [3, 4]; and it is a very costly medicament. Its manufacturing process is very complex compared to other pharmaceutical products. And it is easily affected by the surrounding environmental conditions includes temperature, pH, and light; so it requires a precise manufacturing and handling process. If there is any minor variation in the biologics product, this will result in serious consequences for the consumed individuals [5, 6]. To avoid this consequence, this type of product must follow the regulatory guidelines and the practice of quality by design and robust processes throughout the all stages of the manufacturing process. This regulatory compliance and QbD play a very critical role in ensuring that biologics are safe, effective, and of the highest quality before they reach the patients. Before the approval of the biologic product for human use, it should pass through the preclinical studies and clinical research to evaluate its safety of the biological product and therapeutic effectiveness of biological products. The regulatory compliance evaluates the data information from the trials to evaluate whether the biologics meet the required criteria for approval process. After approval, it was launched into the market. The process of post-marketing surveillance is started, and in that surveillance process, the regulatory compliance monitors the safety and effectiveness of the biologics. Regulatory compliance and robust processes and QbD in biologics are not just legal requirements because they are basic pillars that ensure the delivery of safe, efficacious, and of standard-quality biologic products to the patient and ensure scalable bioproduction [7–9]. This book chapter mainly concentrated on the regulatory framework, robust process, and QbD involved in biologics to establish the safety, efficacy, and scalability of bioproduction.

23.2 Regulatory Framework for Biologics

Biologics have lower stability than medication molecules. Biologics are regulated for approval in much the same way as drugs are; however, certain regulatory agencies differ. Several regulatory agencies regulate biologics, including the Investigational New Drug (IND), New Drug Application (NDA), Food and Drug administration (FDA), Provisional Health Service Authority (PHSA), Center for Biologics Evaluation and Research (CBER), Center for Drug Evaluation and Research (CDER), Premarket Approval, (PMA), Biologics License Applications (BLA) [10]. Biologics, like drugs, require more time for approval. Biologics clearance requires many licenses. The regulatory framework for biologics starts with the manufacturing process. The manufacture and import of biologicals are governed by the requirements of the Drugs and Cosmetics Act (D&C act) of 1940 and the Rules of 1945. The above act and rules make it a non-juris dial to manufacture and import any biologicals without a license. In India, biologicals include both vaccinations and recombinant DNA technological products, which are classified as novel medications under the D&C Rules 1945. Each item of this product will need a new medication clearance through the primary licensing authority known as the DCGI. Biotechnological product in India involved both vaccinations and recombinant DNA technological products, which are classified as novel medications governed under the D&C Rules 1945. The drugs and cosmetic regulation require the applicant to give precise information for acceptance of a novel drug, as defined in Form number 44 with schedule Y. Schedule Y provides extensive recommendations for conducting preclinical and clinical trial studies, as well as the data necessary for manufacturing process and quality control. According to Indian guidelines, one repeated dosage toxicity investigation is sufficient to establish comparability in the appropriate animal model. According to Indian criteria, clinical investigations in phase 3 are necessary for biological approval. In general, vaccines need a comparative phase 3 clinical study in healthy volunteers [11].

23.2.1 Regulatory Agencies

Numerous regulatory organizations that play a major role in all particulars of biologics regulations and control include the World Health Organization (WHO), Pan American Health Organization (PAHO), International Conference on Harmonization (ICH), Food and Drug Administration (FDA), Centre for Drug Evolution and Research (CDER). Centre for Biological Evaluation and Research (CBER). The above regulatory bodies play vital role each and every step in manufacturing of biologics [12].

23.2.1.1 World Health Organization (WHO)

A specialized agency of the United Nations that addresses important disease-free concerns across the world. Disease control, healthcare, and pharmaceutical standards are set by the World Health Organization, which also publishes scientific articles and reports and conducts study and educational programs. Through its biological standardization program, WHO collaborates with member states, collaborating centers, partners, and experts to create and amend biotherapeutic product advice, as well as establish worldwide reference standards. The recommendations are promoted through their publishing on the WHO website, as well as seminars and training events to help with their implementation. This outreach has been effective in pushing WHO guidelines for international harmonization of advice for biological therapies. Similarly, developing and promoting WHO worldwide reference standards contributes to the standardization of biological treatments among producers, nations, and laboratories. They are widely used for assuring the purity and effectiveness of biologicals [13, 14].

23.2.1.2 Pan American Health Organization (PAHO)

The Pan American Health Organization (PAHO) is a key international public health institution that serves the Americas. Among the matured public health organizations in the world, PAHO was established in 1902 and plays a critical part in advancing health and welfare in the region. PAHO is in the service of World Health Organization's Regional Office for the Americas in addition to being the United Nations system's specialized health agency for the Americas. PAHO is heavily involved in the research and regulation of biologics (including vaccines and other biological products) throughout the Americas. Its responsibilities include regulatory harmonization, capacity building, quality assurance, and fostering regional innovation and manufacturing [15, 16].

23.2.1.3 International Conference on Harmonization (ICH)

The International Conference on Harmonization (ICH), now called as the International Council for Harmonization of Technical Requirements for Pharmaceuticals for Human Use, is responsible for developing international high standards for the purity, safeness, and effectiveness of biologics such as vaccines, therapeutical protein derivatives, and other biotechnological/biological by products. This harmonized standard specifies, tests, and acceptance requirements for biotechnological/biological goods. It includes topics such as product characterization (physicochemical and immunochemical characteristics, biological activity, purity, impurities, and contaminants), analytical considerations (reference standards, validation), process controls, and release/shelf-life criteria [17, 18].

23.2.1.4 Food and Drug Administration (FDA)

The Food and Drug Administration (FDA) have accountability to ensure health of public by establishing the safety, effectiveness, and reliability of human and animal-based pharmaceuticals, biotechnological products, and medical devices. The FDA also delivers reliable, health information based on the scientific domain to the general population. In 1997, the FDA formed a team to establish the quality and safety of biological products. The FDA has many biologics approval facilities, including CBER and CDER [19, 20].

23.2.1.4.1 Center for Drug Evolution and Research (CDER). Center for Biological Evaluation and Research (CBER)

It operates in accordance with the FDA and the Public Health Service Act (PHSA). The primary goal of CBER is to preserve and improve the health and wealth of human society by regulating biological products, immunization vaccines, tissues, and genetic based therapeutic products. The safety, sterility, strength, and efficacy of biologics and related products was established by CBER. All biologics are not under the control of CBER. The specific biological products such as monoclonal antibodies and other therapeutic products are under the control of CDER The following Table 23.1 represents the list of biologics under the control of CDER and CBER [21, 22]. Figure 23.1 shows various regulatory bodies.

Table 23.1 List of biologics regulation for CDER and CBER.

Center for drug evolution and research (CDER)	Center for biological evaluation and research (CBER)
• Gene and gene related products • Extract from allergenic material and it is applicable for the diagnosis and treatment of allergies. • Venom • Anti-toxin • Anti-venin • Blood, numerous blood related products, and instrument applicable for collecting, analyzing, and process of blood • Xenotransplant • Human cell, tissue or cellular, and tissue-based product	• Immunomodulators with exception of vaccine or allergenic compounds • Hormones • Monoclonal antibodies for *in vivo* application • Protein for therapeutic application [23].

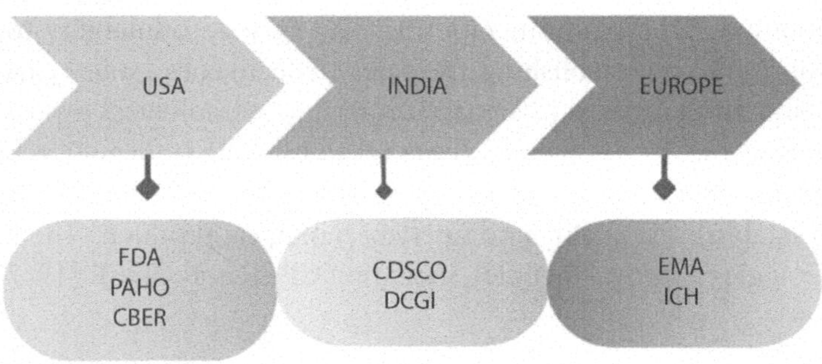

Figure 23.1 Regulatory agencies in various countries.

23.2.1.5 Central Drugs Standard Control Organization (CDSCO)

The Central Drugs Standard Control Organization (CDSCO) is India's chief regulatory agency for biologics, which include vaccines, proteins for therapeutic purpose , monoclonal antibodies, blood products, genetic and cellular level treatments, and other biotechnology-derived pharmaceuticals. CDSCO is governed by the Drug Controller General of India (DCGI)

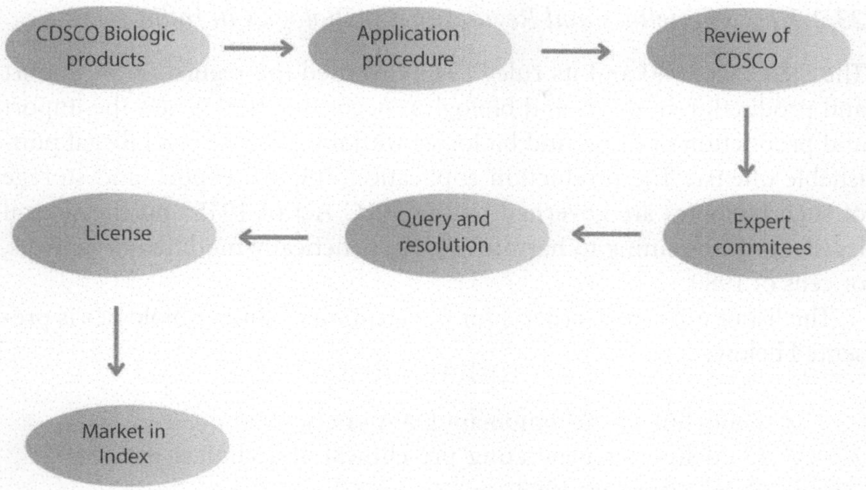

Figure 23.2 Registration process of biologics under CDSCO.

and is an affiliate of the Ministry of Health and Family Welfare. CDSCO reviews clinical trial protocols for biologics, including providing authorization to import experimental biologics for studies, and supervises all phases of clinical research in India. Registration processes of biologics is shown in Figure 23.2. It is in charge of market approval and registration of novel biologic drugs, which involves assessing safety, effectiveness, and quality data during the review process. CDSCO gives production and marketing licenses for biologics, as well as authorization to import biologics for evaluation, testing, and analysis [24].

23.2.2 Quality Control in Biologics

The testing or biological analysis samples necessitates highly unique facilities and knowledge, which are backed by a facility of animals. The National Institute of Biologicals in Noida and the Central Drug Laboratory (CDL) in Kasauli are the two national laboratories. The National Institute of Biologicals is designated as the CDL governed under the control of D&C Act and Rules for several biologicals, such as (a) Reagents used for blood grouping, (b) diagnostic devices, and (c) blood related products such as human albumin, human coagulation factor. The Central Drug Laboratory, Kasauli, handles testing the quality control of biologics and a great deal with the vaccine release [25].

23.2.2.1 Guidelines and Regulation for Biologics in India

The D&C act 1940 and its rules 1945 governed the regulation of import and production of drugs and biologics. According to this law, the import and production of drugs and biologics without a license is an illegal punishable offense. The production application, import, export, and storage of such biologics are governed by the D&C Act of 1945 and the several legislations pertaining to harmful germs, genetically modified organisms, or cells of 1989.

The list of numerous criteria for the creation of similar biologics is presented below:

- ✓ Guideline for Recombinant DNA safety, 1990
- ✓ Guideline for generating pre-clinical and clinical data for RDNA vaccines 1999
- ✓ Guidelines and Handbook for Institutional Bio-safety committee
- ✓ CDSCO industry standards, 2008:
  - a) CTA submission for safeness and effectiveness assessment
  - b) Prerequisites for approving innovative medicine approval process
  - c) Modifications made following biological product approval: Quality, Safety, and effectiveness documentation
  - d) Implementing the need for quality information in drug submissions for biotechnological and biological products to get new drug approval [26].

23.2.2.2 Approval Process in Biologics

The licensing procedure for innovative biologics includes considerable preclinical and clinical investigations to establish their safeness and efficacy. Following the evaluation of new biological compounds for acute toxicity and pharmacological action in animals, the sponsor applied for FDA clearance to conduct human clinical research. The sponsor explains about how human clinical investigations are conducted, and they are known as INDs [27]. The Food, Drug, and Cosmetic Act requires sponsors of biotechnological product to submit an IND application before administering the product to humans. This is to ensure the subject's rights as well as safety. The sponsor of an IND is accountable for the starting stage of clinical studies. An academic institution, governmental organization, pharmaceutical company, or private or public organization could serve as the sponsor [28]. The FDA typically approves an IND 30 days after receiving it. FDA takes 30 days to review the IND claim. For safety concerns, the FDA gets IND. Preclinical studies' IND material informs clinical study safety decisions. Biologics are agent that are derived from protein in human origin that needed a minimum safety information before filing an IND. Numerous meetings are now conducted between the sponsor and the agency. The phase of clinical study can take up to 12 years, with just one in every five substances evaluated demonstrating clinical efficacy and safety [2, 29]. Biologics clinical trial studies are conducted in three stages (Phase 1, Phase 2, and Phase 3), with increasing numbers of participants evaluating safety, dosage, effectiveness, and adverse effects. After the completion of clinical trial process, the biologics enter into the market. To launch a novel medicine or biologic product, both NDA and BLA application are needed. CDER uses NDAs to marketing the drug products. CBER uses BLA to marketing the biological goods. Some biological products approval processes are under NDA, while others approval processes under BLA, which can cause confusion. NDAs are reviewed and approved based on clinical study findings, ensuring safety and efficacy. The Fund providing bio-pharmaceutical company and FDA engage on the format and substance of the NDA during the pre-NDA meeting. NDAs seek clearance to sell drugs and biologics [2, 30].

23.2.2.3 Requirements for Biologics Approval

Preclinical studies:

It is done before the clinical trials. It is required for comparison comparable biologics with reference biologics. Justification is required for factors such as dosage form, strength, and delivery method.

Pharmacodynamic testing:

Comparative testing and reference biologicals are conducted *in vitro*. Take a cell proliferation test as an example. Assessment of pharmacodynamic and biological activity *in vivo*.

Toxicology testing:

In vivo toxicity tests need one replicate dose toxicity; the animal used in this study should provide an accurate rationale for sacrifice. The dosage toxicity test will include:

- ✓ Historical control
- ✓ Vehicle control

- ✓ Vehicle control for recovery groups.
- ✓ 1X reference biologic for study period
- ✓ 5X high dosage comparable biologic.

The procedure and pre-clinical study reports should include:

- ✓ Methodology before euthanasia
- ✓ Events immediately following euthanasia
- ✓ Biochemical parameters
- ✓ Hematology
- ✓ Statistical approaches.

The final result of the preclinical study must include the following: RCGM protocol acceptance. IBSC report acceptance, IAEC approval, Quality Assurance statement, animal food, discussion of the data, and conclusion.

Immune response in animals:
The serum samples' reaction to the proteins of the host cells is examined. The RCGM is presented with the results of pre-clinical research for evaluation and decision-making [31].

Clinical trials:
The following are the particulars needed for the biotechnological product clinical trial application,

Pharmacokinetic studies: The subjects of these investigations are healthy individuals. Half-life, pharmacokinetic linearity factors, diseases and illnesses management, and administration method are a few of the attributes assessed.
Single dose pharmacokinetic studies: For pharmacokinetic study, the dose should be present within the range of therapeutic dose for typical biologics. The appropriate ratio for dosage determination should be specified. The dose administration approach will be the most sensitivity to analyzing disparities. The sample size must have a statistical ratio, and the criteria for comparison must be set and justify prior to performing experiments.
Pharmacodynamic studies: This involves investigating the biochemical and physiological impacts of medicines.
Confirmatory safety: It is necessary for all biological products to show their safety and effectiveness. If trials are not needed, an explanation must

be provided, and the application is submitted by the applicants in accordance with CDSCO.

Safety and immunogenicity data: Assessment related to animal studies and post-clinical data, including a comprehensive collection of safety and immunogenicity data [32].

Post-market data required for biologics

Pharmacovigilance plan: It is an awareness of pharmacology in terms of detection, collection, consideration, monitoring, and control of undesirable effects. The goal is to submit PSURs (Periodic Safety Update Reports) every six months after submitting the application.

Reporting of adverse reactions: It involves documenting and evaluating every adverse event that transpires following the administration of a product. According to schedule Y, the most critical adverse events must be communicated within the period of 15 days. Table 23.2 gives summary of different application for submission to regulatory agency.

Table 23.2 Different application forms for submission to regulatory agency.

Purpose	Agency involvement	Application forms	Approval forms
Manufacturing License for examination, tests and analysis	CDSCO	Form no. 30	Form no. 29
Permission for preclinical studies	RCGM	Form no. C3	Form no. C4
Preclinical study report permission	RCGM	Form no. C5	Form no. C6
Clinical trials conduction	CDCSO	Form no. 44	Letter of permission
Manufacturing and marketing permission	CDSCO	Form no. 44	Form no. 45/46
Import and registration license	CDSCO	Form no. 40/8	Form no. 41/10
Manufacturing license	CDSCO	Form no. 27D	Form no. 28D

Post Market Studies (PMS): It is the process of determining a bio-pharmaceutical product's safety after it has been introducing on the market. The pharmacovigilance strategy includes information on post-marketing studies, and CDSCO should receive updated studies.
Information about the drug: This addresses use, adverse events, pharmacokinetic efficacy, and identification.
Route of administration: It is a route by which type of medication enters the body. Intravenous and oral routes of delivery are often used [33].

23.2.3 Challenges for Implementation of Regulatory Framework in Biologics

Global regulatory bodies have challenges in establishing the safeness, quality, and effectiveness of biologics, harmonizing the development of drug protocols, and monitoring compliance with statutory requirements. They get a crucial role in implementing regulations in unregulated regions to ensure safety for residents. Following are the major challenges are faced by the regulatory bodies:

- ✓ Promotion of healthy human society and protecting against dangerous drugs
- ✓ Establishing proper legalization for all medicinal products and pharmaceutical activities, whether public or private.
- ✓ Expand worldwide regulatory efforts to safeguard people's safety.
- ✓ Centralization of regulatory processes and controls.

The implementation of regulatory framework in biologics is important for establishing the safety, efficacy and the regular supply of biologics to the patients for the management of disease such as cancer, autoimmune disorder etc. And it will develop the quality of life to the patients [34].

23.3 Quality-by-Design

The Biopharmaceutical company's goal is to manufacture and innovate high-quality products and continuously try to empower consumers and ensure the safeness of the biological product, quality of the product, and efficacy of the biotechnological product. The theory of quality by design was first coined by Edwards Deming and Dr Joseph M Juran. QbD focuses

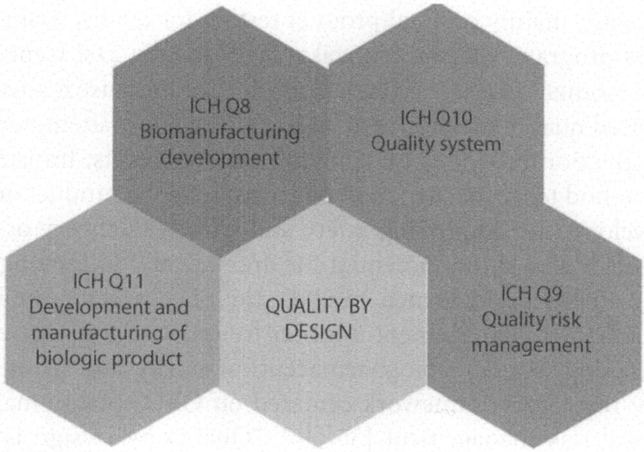

Figure 23.3 Key ICH guidelines supporting quality by design (QbD).

on acknowledging the connection bridge for raw materials, process parameters, and product quality. Figure 23.3 gives key ICH guidelines supporting quality by design. It meets regulatory standards, namely ICH recommendations Q8 represent Pharmaceutical Development, Q9 represent Quality Risk Management, and Q10 represent Pharmaceutical Quality System, which shape current drug development and lifecycle management. However, the sector has recently faced challenges such as medication recalls, production failures, costs, scale-up concerns, and load. Traditional completed product evaluation units establish the product quality and performance while having a limited grasp of the progress and key critical process parameters (CPPs). Regulatory agencies are thus concentrating on implementation *via* quality by design (QbD). QbD is related to scientific strategy that decreases process variance and enables process management solutions, resulting in better process knowledge and dependability. QbD provides curiosity in the complete development process, allowing for speedy analysis of quality concerns as well as the identification and resolution of fundamental problems. The requirement of QbD is to determine how any modifications will affect the completed product's quality. In addition, all important formulation features and related process factors will be identified [35, 46].

In QbD, Deming's approaches prioritized continuous improvement and statistical control over production processes, while Juran promoted a comprehensive quality management approach. Initially used in manufacturing, these ideas were adopted by the pharmaceutical sector as a stronger alternative to the standard "Quality by Testing" (QbT) approach, which depends mainly on final product testing for quality assurance. The FDA's 2004 program, Pharmaceutical cGMPs for the 21st Century, was a watershed moment in pharmaceutical production because it advocated for science-based quality management systems. The effort attempted to modernize production processes by applying QbD concepts, transferring the reactive method to a proactive method to improve the product quality and it was developed for improving safety and effective delivery of medicine to a patient. It also shows potential for dramatically improving biomanufacturing quality performance. In 2002, the FDA issued a notice of new inventions for risk management in order to update the FDA's regulations for maintaining excellent biopharmaceutical purity and to establish an innovative regulatory framework centered on QbD, quality maintenance systems, and risk management [35, 36]. "Quality by Design is all about developing a suitable technique to analyze process performance in relation to target product performance." Simultaneous QbD activities in synthetic and analytical development will provide the greatest quality product while reducing risk. To ensure quality, the complete adoption of QbD demands process control over critical stages. Its ultimate goal is to incorporate quality into biologics goods to ensure patient safety [37].

23.3.1 Aims, Results and Characteristics of QbD

The primary aims of QbD are as follows:

1. Improving production efficiency.
2. Improving efficiency in product development.
3. Improving product quality and performance to fulfill patient demands.
4. Increased process capabilities.
5. Avoidance of regulatory compliance.
6. Implementation of risk management.
7. Reduced production costs and waste.
8. Decreased product variability, faults, and rejections.

The primary results of QbD are as follows:

- ✓ Ensuring product quality meets clinical performance expectations.
- ✓ Maintaining product quality through an efficient production and formulation process [36].

Key Characteristics of QbD:

- ✓ A tool for focusing and optimizing medication development.
- ✓ A dynamic and methodical procedure.
- ✓ It is applicable to the creation of drug products and substances (chemicals and biologicals).
- ✓ It applicable to analytical methodologies.
- ✓ Regulators always promote this [38].

Benefits of QbD:

- ✓ Avoid batch failure.
- ✓ Reduce differentiation and expensive investigations.
- ✓ Eliminate regulatory compliance issues.
- ✓ Empowering technical workers.
- ✓ An efficient, agile, and adaptable system.
- ✓ Improve production efficiency, lower expenses, and eliminate project rejections and waste [38].

23.3.2 Principles of QbD

The foundation of QbD is recognizing and regulating variables that impact product quality. The key ingredients are:

- ✓ The Quality Target Product Profile (QTPP) defines the target product quality attributes.
- ✓ Critical Quality Attributes (CQA)
- ✓ Risk Assessment: Assesses possible threats to CQAs.
- ✓ Design of Experiments (DoE) helps optimize processes and identify key process parameters (CPPs).
- ✓ Control Strategy: Implements controls to maintain consistent product quality.
- ✓ The Lifecycle Approach emphasizes ongoing improvement and process monitoring [39].

23.3.2.1 Quality Target Product Profile (QTPP)

It provides quantifiable guidance for safeness and efficient clinical research study, serving as a basement for designing and optimizing formulations and biomanufacturing processes. The qualities and attributes of a biological product that affect its product quality are identified by the QTPP.

Medicine characteristics such as dosage form, method of administration, packing appearance of the biopharmaceutical product and diagnosis are specified by QTPP. The QTPP only contains product attributes applicable to patient needs. QTPP helped optimize formulation and process parameters to achieve targeted CQAs [40]. The QTPP idea and its realization are innovative in the QbD paradigm. QTPP serves as the foundation for bioproduct development in the following way.

- Dosage form
- Administration route strength (max and min)
- Release of the medication
- Pharmacological characteristics.
- Quality criterion for drug products.
- Pharmaceutical elegance [41].

23.3.2.2 Critical Quality Attributes (CQAs)

Critical quality attributes are defined as "a physical, chemical, biological, or microbiological property or characteristic that should be within an appropriate limit, range, or distributed to ensure the desired product quality" According to ICH guideline Q9, assessment of risk is used to identify CQAs. CQAs for alternative delivery modalities may also contain specific of product elements that includes aerodynamic qualities for inhaling medicines, sterility for parenteral, and stickiness characteristics for transdermal patches. Additionally, for medicinal compounds, raw materials, and intermediates may have characteristics like bulkiness density and size of particle distribution that affect the quality of medication products [41].

23.3.2.3 Risk Assessment

Quality risk management (QRM) is a major strategy to assessing, managing, reporting, and assessing quality hazards. Risk is inversely proportional to process and product knowledge. The risk decreases as knowledge of the progress and product performance improves. To estimate the risk, analyze product performance across several material parameters which includes moisture content, particle size distribution, and flow qualities. Process choices, parameters, raw materials, solvents, APIs, and packaging materials are all important considerations. Quality risk management relies on risk assessment to determine material qualities and operational factors that influence the quality of the product [42]. It assists in correlating quality traits and process parameters to CQAs. The determination of quality risk must be based on both recognized scientific and therapeutic value to the patient.

The ICH Q9 standards give a structure for starting the risk process. Several tools are used in the risk assessment:

1. FMEA (Failure mode effects analysis)
2. FMECA (Failure Mode, Effects, and Criticality Analysis)
3. FTA (Fault Tree Analysis)
4. HACCP (Hazard Analysis and Critical Control Points) [43].

23.3.2.4 Critical Material Attributes & Process Parameter

Identifying Critical Material Attributes (CMAs) and Critical Process Parameters (CPPs) is critical for reducing variability in the finished biological product. CMAs are the physical component, chemical component , or biological product qualities of raw materials that might influence CQAs. CPPs are the operational characteristics in manufacturing processes that affect CQAs. Manufacturers may design strong systems that preserve product quality consistently by first gaining a comprehensive grasp of both CMAs and CPPs *via* risk assessment and testing. This proactive identification helps to reduce the risks caused by unpredictability, resulting in a more dependable end product [44].

23.3.2.5 Design Space

Quality assurance requires a multifaceted presentation of the link between process parameters, material characteristics, and critical attributes. Movement inside the design space for biologics is not regarded as a variation; instead, the flow outside the design space, which frequently results in regulatory post-approval revisions, is evaluated as a change. Unit operations (distribution, purification, etc.) may control the design space; these operations will be selected according to how they affect the CQAs in the design space [45].

23.3.2.6 Control Strategy

It's described as "a planned set of controls based on current product and process understanding that ensures process performance and product quality." Metrics and characteristics pertaining to drug substances and biological drug product ingredients and components, equipment and facility operational conditions, in-process controls, finished product standards, and the frequency and methods of monitoring and control are examples for controls. To ensure consistent quality, a control plan often includes input product material controls, process supervision and monitoring, design space for individual or multiple unit activities, and product standards [38].

23.3.2.7 Lifecycle Approach to Product Quality

Throughout the product's lifespan, organizations may increase quality through innovative ways. Method performance is checked to ensure consistency in quality. Routine manufacturing provides additional knowledge and information to support method/process development efforts. While a company's internal quality system requires periodic maintenance, the style area should remain consistent. The QbD strategy allows for continuous product improvement throughout its existence, unlike the old frozen process [46].

23.3.3 QbD vs Conventional Quality System

In today's modern world, QbD is regarded as an essential element of biopharmaceutical development and should be involved in all new medication implementation. In contrast, conventional quality control biological technique, that includes mostly on final-product testing and assessing quality at the end of biologics manufacturing, sometimes result in waste, inefficiencies, and expensive regulatory guidelines requirements are not fulfilled.

QbD is a paradigm alternative. Moreover, QbD places an importance on proactive design and management while integrating quality across the product lifecycle. It acknowledges that rather than being examined or evaluated at the end, quality should be ingrained in a biological product from the start, using perfect knowledge, science based conceptual study. According to research, by enhancing formulation factors before full-scale manufacturing, QbD may save development time by as much as 40%. Additionally, in certain recorded cases, it's able to define and regulate robustness. The design space has reduced material waste by up to 50% and led to fewer batch failures. QbD's recognition has potential to enhance pharmaceutical product quality; regulatory authorities have issued recommendations in favor of its implementation. The European Medicines Agency EMA), the US Food and Drug Administration (FDA) have both accommodated the regulatory frameworks emphasizing the need for a systematic based and science-based technique to biopharmaceutical research and biomanufacturing. Analytical Quality by Design, or AQbD, has been more well-known recently as a way to incorporate QbD concepts into the creation of analytical methods. By following the guidelines of ICH Q14, AQbD guarantees that analytical techniques are dependable, reproducible, and consistent with regulations for the time period of the product's life [43].

The major aim is to enhance technique performance and decrease method variability, a crucial factor for regulatory clearance—this technique establishes a Method Operable Design Region (MODR).Notwithstanding these advantages, QbD requires a large upfront investing in data collection, trial design, and regulatory documentation, all of which can be demanding. However, the long-term advantages of QbD-based submissions in terms of process efficacy, regulatory flexibility, and safety of patients outweigh the starting point of implementation challenges as regulatory bodies favor them more [47].

23.3.4 Applications for QbD

1. Pharmaceuticals:

- ➢ Available in modified release and solid oral dose forms.
- ➢ Effects of genotoxic contaminants on process development.
- ➢ Analyze nano-suspension preparations, as well as excipients and API.

2. Biopharmaceutics:

- ➢ Used for protein manufacture, monoclonal antibody production, and purification using chromatography.
- ➢ PAT and QbD for biopharma.
- ➢ In nanomedicine.

23.4 QbD Implementation in Biologic Manufacturing

The major purpose of incorporating quality into biomanufacturing to meet patients' requirements by providing safe, effective, compliant, and cost-effective medications. A thorough awareness of patient needs, and also product and developmental process-based knowledge, must meet these objectives. In this regard, QbD is an approach for biologic manufacturing process development. These information is initially used to construct the

Table 23.3 Step by step process of quality by design in biomanufacturing.

Steps	Steps by step process of Qbd in biomanufacturing
I	QTPP (quality target product profile) A prospective summary of the quality characteristic of a drug product that steadily will ensure desired quality taking into account safety and efficacy of the biologic drug product
II	QRM (quality risk management) A systematic process of organizing information to support decision making based on identification of hazards and evaluation of risks management associated with these hazards
III	CQA (critical quality attributes) A physical, chemical or microbiological property or characteristic that should be within limit, range or distribution to ensure desired biological product
IV	CPP (critical process parameter) A process parameter whose impact on critical quality attribute and should be monitored or controlled to ensure the bioprocessing to produced desired quality
V	DS (design space) Multidimensional combination and interaction of input variables
VI	CS (control strategy) A planned set of controls, derived from current products and process understanding that ensure the quality of biologic product

quality target product profile (QTPP), which then directs the identification of critical quality attributes (CQAs), allowing the design of a biomanufacturing technique that consistently provides a medicinal product of the required quality.

For Biologic products the QbD journey begins with designing and engineering the biologic candidate to match with the QTPP, with the goal of achieving desired effectiveness, safety, manufacturability, stability, and overall quality. This phase involves a variety of tasks, including sequence design, cell line engineering, isotype selection, posttranslational modifications, and humanization, among others. Product knowledge is gathered by combining many data sources, such as structural and physicochemical evaluations, understanding the therapeutic mechanism of action, and studying the drug product's pharmacokinetics and pharmacodynamics.

This includes insights from clinical and nonclinical trials, experiences with related compounds, and other relevant publicly available data. CQAs for various biologic product often comprise structural integrity qualities such as proper folding, disulfide link formation, posttranslational modifications [47]. Table 23.3 gives step by step process of quality by design in biomanufacturing.

Simultaneously, process knowledge is gained by thoroughly investigating how various process factors influence quality characteristics. Critical process parameters are those that should majorly interfere on CQAs. Achieving a comprehensive knowledge of the processes, particularly in cell culture and fermentation, is difficult due to the complexity and interconnectedness of the numerous process parameters, as well as the significant influence that even modest modifications may have on product quality. As a result, understanding how differences impact biologic product quality requires mapping connections between various characteristics. This complete mapping includes identifying potential variability sources and analyzing the hazards they bring to the CQAs [48]. Statistical methods like DOE and MVDA, together with risk management tools, can optimize resource allocation to prioritize critical activities. Authorities have not provided clear advice on what constitutes an acceptable risk for biotechnology goods. The sector may collaborate with health authorities to provide process modification examples, evaluate post-approval deviations, and categorize notifications [48, 49].

23.4.1 Establishing the Design Space Using Design-of-Experiment Techniques

Variabilities in biomanufacturing can occur from a variety of sources, including raw materials and consumables, operational inputs (for example, measurements, procedures, staff, and equipment), and environmental circumstances. Identifying these sources and examining their possible influence on CQAs is the first step toward limiting variability. Failure mode and effects analysis (FMEA) is a typical technique used to determine the likelihood of an occurrence and the risk associated with each source of variability. Following the risk assessment, a design of experiments (DoE) is applicable to study effects of these high-risk factors. This strategy enables the simultaneous examination of many process factors, increasing research efficiency by decreasing the time and expenses together with standard one-factor-at-time (OFAT) methodologies. DoE gives useful insights into the interrelationships between process parameters and aids in the identification of CPPs that require monitoring and management to ensure consistent biologics product quality [48].

Phases of quality by design:

1. Improvement of Quality Target Product Profile (QTPP) and Critical Quality Attributes (CQA) determine the biopharmaceutical requirements, efficacy, and safety. They determine critical quality attributes clinical importance.
2. Past Information Gathering: It pertains to the collection of important details about biologic substances.
   Risk Analysis of Knowledge Gaps: Done with the help of risk analysis to determine and close gaps in knowledge
3. Design and Critical Material Attributes (CMA) Specification: Design the biologics with an aim on crucial elemental properties, specific attributes critical to obtaining product quality.
4. Raw Material and Critical Process Parameters (CPP) Determination: Evaluate the crucial raw material properties and identify the important process parameters for production.
5. Risk Assessment for Test Procedures: Done with the help of risk assessment to determine the test method needs. It establishes effective processes based on experiments and prior information.
6. Control Procedure Definition: It sets up control methods for input components, production, and completed the processing of the product. It involves metrics for site establishment and any modifications of scale.
7. Continuous Monitoring and Review: Carry out continual assessment of the whole process and regularly evaluate procedures for efficacy and make any required changes [35].

23.4.1.1 QTPP

Its downstream process, undergoes a collection of purification stages in biotech products. It includes production of biopharmaceutical product such as antibodies, therapeutic protein-specific techniques, monoclonal antibodies, enzymes, and vaccines. It was already proven that increasing cell density and production rates may be achieved by making the upstream phase continuous. Over time, significant advancements have been made in the design and enhancing of upstream processes, including cell banking, expansion of inoculum, cell retention and isolating devices, and bioreactor instrument design [49]. However, because single-use bioreactors contain non-stainless-steel materials, they are less expensive and only have the advantage of eliminating the need for sterilization and cleanup. In contrast, development in continuous downstream processing has been gradual. The lack of knowledge dealing with continuous purification systems on a production scale may be due to their recent availability. The distinguished nature of biopharmaceutical classes necessitates a variety of purifying procedures and equipment. A good continuous process requires synchronization of upstream and downstream flow rates, in addition to technical developments. Advances in downstream processing follow chromatography methods. These are widely used for purification because of their superior robustness, resolution, and flexibility across scales and separation criteria. The cost of chromatography is mostly determined by the system's ability to efficiently isolate a product stream with a certain concentration. Advancement in upstream processing techniques, such as cell lines and bioreactor designs, are allowing for higher antibody titters. However, downstream procedures require additional capacity to handle the increased material. Efficient continuous production requires equipment durability over extended durations. Dedicated continuous chromatography systems are necessary for increased capacity and productivity, lower operating costs, and extended functionality. Various chromatography are used to develop continuous annular chromatography, counter-current chromatography, and expanded bed chromatography [49, 50].

Quantity, volumetric productivity, yield (high titer), and quality are all enhancing criteria for large-scale biopharmaceutical production. Finding and describing the medicinal chemical and its mode of action is the first step in optimization. Cell engineering, production cell line selection, bioreactor type and parameters, feeding strategy, downstream process steps, process monitoring, analytics, and formulation studies are all examples of how cell culture processing is made to satisfy regulatory requirements [50].

23.4.1.2 CQA

A systematic, based and science-based method was applicable to establish the biologic product's critical quality attributes (CQAs) in order to apply the Quality by Design (QbD) paradigm. By examining the quantitative and qualitative effects of critical process parameters (CPPs) on CQAs for every manufacturing unit operation, this method seeks to implement Proven Acceptable Ranges (PARs) for CPPs. A risk evaluation found quality attributes that may understand the biological drug's performance, safety, and efficacy if they were not within specification bounds. CQAs were the designation given to high-severity qualities. Following the development of CQA, risk analysis approaches were employed to investigate the linkages among unit operating parameters, iCQAs, and final product CQAs. This enabled it possible to choose CPPs that, based on seriousness, occurrence probability, and identifiability, had the biggest effects on product quality. The reactions of iCQAs to various process parameters were examined to determine PARs for each unit operation. This makes it possible to process materials under ideal circumstances within predetermined bounds, which produces reliable and consistent product performance. Direct control methods to guarantee product quality in real-time throughout production were made possible by recognizing the connections between CPPs, iCQAs, and CQAs [47].

23.4.1.3 Risk Analysis

23.4.1.3.1 Failure Modes and Effects Analysis (FMEA)

The term "drug-related problems" refers to conflict with medications that involves both extrinsic and intrinsic toxicity. Adverse effect is the conclusion of interactions between a drug property pharmacological, chemical and medicinal and the human body. To determine these pharmacological mistakes and identify how to stop them from happening again are important steps in improving patient safety. To facilitate the establishment of a successful safety culture, the required procedures and frameworks must be in place. In the manufacturing of biopharmaceuticals, unsuccessful modes and effects analysis, or FMEA, is a crucial assessment of risk method. This method finds possible ways for things, systems, or processes to fail, assesses how they affect the system, and ranks risk-reduction efforts. Additionally, this method prioritizes risk mitigation tasks by identifying possible unsuccessful modes in processes, objects, or systems and assessing their influence on the system. First, the goals and scope of the FMEA are established. Next, a cross-functional team is assembled, and the whole biomanufacturing process is mapped out [51].

To find likely failure reasons, each step of the process is examined using historical data and imagination. Seriousness, frequency, and identification are graded on a predetermined scale, and the outcomes of such unsuccessful ventures are assessed according to their influence on patient safety, biologics product quality, and regulatory agency. Failure scenarios are ranked according to their risk level using this Risk Priority Number (RPN). A variety of industries apply purpose quality management tools, such as FMEA, to raise quality of biologics. This research seeks to illustrate that FMEA may be utilized as a production improving tool according to a case study scenario of product process optimization for biological drug development [50].

23.4.1.3.2 Failure Modes, Effects, and Criticality Analysis FMECA

Failure modes, effects, and criticality analysis, or FMECA, are one of the methodical as well as structured methods used often in the biopharmaceutical industry to enhance process reliability, quality of product, and patient safety. To find possible failure modes, causes, and effects, processes are examined as part of FMECA. Additionally, evaluated is the seriousness of these issues with the production process or the caliber of the final product. Proactively identifying and fixing any problems before they compromise patient safety or the final product is the main goal. In FMECA, the severity, occurrence, and detection of failure modes are used to rate them [51].

23.4.1.3.3 Hazard Analysis and Critical Control Points (HACCP)

23.4.1.3.3.1 Hazards Analysis

It is defined as a process for obtaining and assessing data on risks and the circumstances that gave rise to them in order to decide whether or not they are significant enough to be covered by the HACCP plan. Three categories of threats are identified by hazard analysis: physical, chemical, and biological. A two-stage risk assessment is advised.

23.4.1.3.3.2 Crucial Control Points

After hazard analysis, identification of CCPs is another part is HACCP principle. The CCP is the point at which control is considered necessary to prevent, eliminate, and bring a risk within reasonable bounds. The last step in hazard control prior to export is CCP. Additionally, CCP describes the moment when there is a significant chance of losing control.

23.4.1.3.3.3 Determine the Critical Limit

Every step that contains a CCP has to have critical constraints set after the CCPs have been discovered. On the basis of past data or regulatory action, critical limitations are determined. The stage will determine how many critical limits are required. Temperature, water activity, and time are typical requirements. It is important for the HACCP team to make sure that critical limits are feasible and appropriate for the process. Effective monitoring and management are made possible by recognizing important constraints [51].

23.4.1.3.4 Hazard Operability Analysis (HAZOP)

The basic conceptual part of Hazard Operability Analysis (HAZOP) is that deviations from operational or design goals are the source of hazard events, a method of brainstorming that applies "guidewords" to identify contemporary risks. It is employed in the biopharmaceutical industry [52].

23.4.1.3.5 Tools and Techniques Used in the Risk Analysis

The ISO 22000 standard, which includes improved HACCP concepts, the Ishikawa cause-and-effect diagram, and the 5M methodology are the two main instruments that this innovative approach to risk analysis depends on. By combining these techniques, a thorough, proactive, and dynamic risk analysis is made feasible, accounting for the potential synergistic effects of the many components. A detailed explanation of the particular instruments and processes is given below.

1. The ISO 22000 standard and upgraded HACCP principles:

- An international food safety management standard called ISO 22000 issues a framework of body for evaluating and reducing risks throughout the biomanufacturing chain.
- A systematic approach to risk prevention, enhanced HACCP is founded on hazard analysis and CCP implementation. By identifying important risks at every step of the procedure, it enables real-time controlling and corrective action.
- The key steps of HACCP are as follows:
  - Identify and analyze hazards.
  - Identify critical control points (CCPs).
  - Establish critical limits for each CCP [53].

23.4.1.3.5.1 Ishikawa's Cause and Effect Diagram (5M Diagram)

The Ishikawa diagram, often refer as the "cause-and-effect" or "fishbone diagram," is applicable to organize and identify threatful risk factors. This image is very affected by visualizing risk variables from several viewpoints as well as discovering complex causal connection that can lead to process technical issues. Figure 23.4 shows the Ishikawa's cause and effect diagram. The 5M analysis is a significant component of the Ishikawa diagram, allowing the identification of plausible explanations as follows:

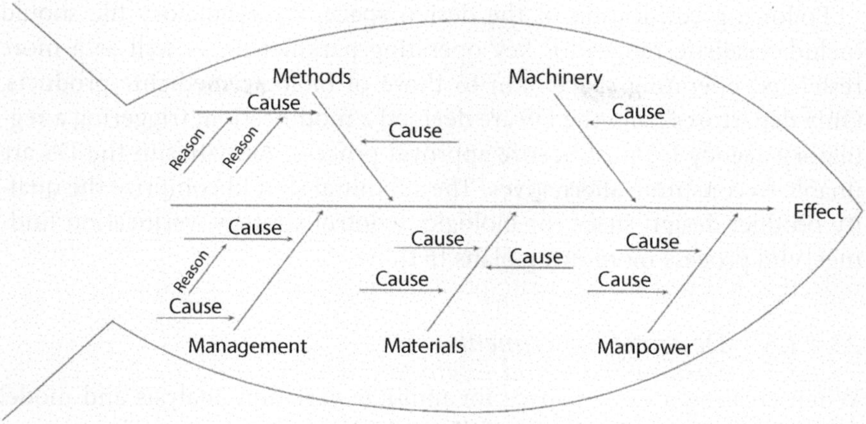

Figure 23.4 Ishikawa's cause and effect diagram.

✓ Quality of raw materials and components used.
✓ Equipment production process level and maintenance.
✓ Training of staff, competences, and work practices.
✓ Technological processes and procedures.
✓ Environmental conditions include temperature, humidity, and cleanliness [53].

23.4.1.4 Design Space

It indicates "multidimensional various combinations and linkages of input material variables and process parameters with proven assured quality." It clearly establishes the limitation of the design space, provide information on the characteristics and fluctuation involved in the design space. The submission must contain detailed information on experimental design and statistical methodology.

Additional knowledge gained from process improvements helps establish the design space and supports its continuous enhancement. As a result, the design space can be updated throughout its lifecycle. When implementing a design space, many ways can be used, such as using process ranging to modify factors during the time of bioprocessing. The design space connected to the strategical control guarantees the biopharmaceutical production technique consistently and generates products that meet QTPP and CQAs.

Following verification of the design space, the regulatory file should include realistic ranges for key operating parameters, as well as a more restricted operating space akin to those of biologic medicine products. Only departures from the DS are deemed a modification, triggering a regulatory agency for post license approval process. Activities in the DS are unable to constitute alternatives. The submittal should comprise the quality product design space for biologics, control strategy, verification findings, and process monitoring plans [54].

23.4.1.5 Design of Experiment

When creating a design space for multiple variation analysis and model creation, it's important to carefully pick the factors to include or omit. Process requirements for big and small compounds differ significantly, as do the manufacturing phases for active substances and finished products. One should define the process parameters to minimize the number of parameters needed for multifactorial analysis. However, creating a comprehensive design space for product quality may be time-consuming and difficult. Risk analysis tools can identify characteristics that have a major influence on the product quality assurance. After associating CQAs, CPPs, and CMAs with process inputs, experiments may be performed to create models for prediction and prove causal linkages through risk assessment [49].

23.4.1.6 Control Strategy

ICH Q10 describe a control strategy method as "an organized set of controls derived from current item and procedure knowledge that ensures procedure execution and item quality." Elements of Strategical control,

1. Procedural monitoring control
2. In-process monitoring controls
3. Test for batch release
4. Process monitoring
5. Test for characterization [39].

Case studies:

- Dr. Lynne Krummen of Genentech described her experiences with two biologics, Perjeta and Gazyva. Both entries utilized complete QbD methods. Perjeta's design space was initially denied owing to issues with justifying scale-up and process parameter control; however, Gazyva's design space was eventually authorized. Lessons learned include detailed identification of CQAs and CPPs, early regulatory interaction, strong post-approval management techniques, and iterative risk assessment. This case study describes the practical obstacles and regulatory expectations for QbD in biologics development [54].

- Zurdo *et al.*, presented two case studies using QbD in biopharmaceuticals, including antibody re-engineering to increase stability, aggregation, yield, and biological activity. It emphasizes iterative protein engineering with analytics and risk assessment to develop the Quality Target Product Profile and Critical Quality Attributes, illustrating the true result of QbD on product developmental design and process optimization [55].
- Rathore *et al.*'s study demonstrates how multivariate data analysis and process analytical technology (PAT) are incorporated into QbD frameworks during cell culture scaling up and bioprocess monitoring at various stages. It describes monitoring important inputs (pH, dissolved gases, metabolites) and outputs (viable cell density, product purity) with control charts to enable consistent commercial-scale manufacturing [56].
- Rathore, provided a case study demonstrating step-by-step QbD application in building a purification process for biosimilar biologics, addressing essential process parameters and quality features to ensure consistent product quality [57].

23.5 Challenges in the Implementation of QbD

1. High initial investment:
Implementing QbD needs considerable initial investment in staff training, sophisticated tools (e.g., PAT and MVDA), and process redesign. For small and medium-sized businesses, these fees might be prohibitively high.

2. Complexity in integration:
Integrating QbD ideas into existing manufacturing systems needs a thorough recognition of the concept of product and process. Transitioning

from old methodologies to QbD involves time, effort, and knowledge, which can be challenging for many firms.

3. Resistance to change:
The pharmaceutical business is recognized for its conservative approach to change and generally resists embracing new approaches. Organizational inertia and a lack of knowledge of the benefits of QbD might slow down the adoption.

4. Regulatory variability:
Global manufacturers face uncertainty due to regional variances in regulatory expectations, notwithstanding encouraging QbD. Harmonizing QbD-related guidelines is critical to promoting wider adoption [46].

5. Data management challenges:
QbD requires extensive data gathering, processing, and interpretation. Managing enormous amounts of data created during research and normal operations can be difficult without effective data management tools.

6. Skill gaps:
QbD application involves multidisciplinary skills in statistics, process engineering, and regulatory science. The absence of properly qualified experts might stymie growth.

7. Longer development timelines:
Although QbD lowers time-to market, the earliest phases of implementation may need more testing and analysis.

8. Key limitation

➢ **Generalizability of finding:**
Biomanufacturing may fail to effectively represent the particular issues that businesses such as healthcare or education confront when implementing sustainability of QbD and quality management methods. This sector specific focus is especially worrisome in emerging countries, where small and medium sized firms (SMEs) face unique challenges [49].

➢ **Lack of longitudinal studies:**
Another notable drawback is a lack of longitudinal research that evaluate the long period effects of quality management and sustainability strategies on organizational performance. While many studies focus on short term results, they frequently fail to give information about the long-term viability

of these advantages. Longitudinal studies are vital for understanding how quality management interventions change and sustain their impact, especially in dynamic domains like education and healthcare, where socioemotional aspects play a significant role [58].

23.6 Role of Digital Tools and Automation

Digitalization is crucial to improving Quality by Design (QbD) in biologics because it allows for more robust, data-driven, and proactive control of biopharmaceutical research and manufacturing processes. Digitalized technologies involves real-time data interpretation, automated monitoring, artificial intelligence , machine learning, process analytical technologies work along with QbD methods to increase process realization, control, and continuous improvement. It improves the efficacy of QbD in biologics advancement by converting it from a theoretical framework to a practical, efficient, and data-driven procedure. This change enhances the whole biopharmaceutical development lifecycle, resulting in speedier product releases, greater quality, and improved regulatory compliance [59].

23.6.1 Artificial Intelligence and Machine Learning

The term artificial intelligence (AI) is a common term that describes the use of code to carry out challenging data-driven tasks or to interpret data similarly to how people do. Effectively developing predictions from massive data sets may be facilitated by evaluating collected data using AI technologies like machine learning (ML). Routine laboratory tasks may potentially benefit from the application of AI and ML. These new methods can be very helpful for photo identification, protein structure prediction, and data set classification [60, 61]. In 2021, Jumper *et al.*, issues a AlphaFold, an AI-based algorithm that utilizes machine learning to accurately interpret the structure of three-dimensional proteins structure [62]. Nikita *et al.*, demonstrated how AI-ML systems may aid in making process control choices that affect product quality during continuous manufacture of monoclonal antibodies [61]. ML approaches assisted Volpato *et al.*, in predicting the enzyme class based on protein purely on the entire amino acid sequence [63]. Integration of ML tools produce a lot of advancement in Quality by Design in biologics. However, open-source Machine Learning programming frameworks like Scikit-learn may now be used to develop an efficient Machine Learning model to advance QbD. But it takes more than just downloading a library and refining it with a bioprocessing dataset to create a Machine Learning algorithm for a biomanufacturing process. A number of challenges are presented by the development and testing of the model. Concerns around regulatory approval are present in addition to the technical difficulties of applying Machine Learning models for QbD. Lack of prior experience in obtaining regulatory approvals using the Machine learning models may nevertheless hinder the successful deployment of a newly constructed process that integrates the models, even if it is shown to be more efficient than previously designed ones based on Multivariate Data Analysis (MVDA). An additional challenge when modifying an established bioprocess is the requirement for several filings with different authorities, including the FDA and EMA. Thus, while obtaining post-approval regulatory alternatives might be a major issues for improving an existing original medication biomanufacturing process, Machine Learning-based models might be simpler to use in innovative or novel biosimilar/biobetter process design campaigns [65].

23.6.2 Digital Twin

A digital twin is an effective description of either a physical object or biological species that is created by fusing digital technology with real-time data. By creating connections between current data and underlying meta information, digital twins can provide a thorough view of a system. To achieve "Quality by Design" (QbD), optimize manufacturing processes, match products with processes, and create a thorough process control plan, the digital twin is essential. The Quality by Design (QbD) strategy in biopharmaceutical medication advancement identifies and understands the influence of key CPP on product quality characteristics CQA. Digital twins aid the pharmaceutical sector in lowering medication discovery and manufacturing costs, as well as accelerating drug candidate delivery to patients [64].

23.6.3 Process Analytical Technology (PAT)

Process Analytical Technology (PAT) is vital to QbD for biologics because it enables real-time monitoring and biologics control of key CPPs and CQAs throughout biopharmaceutical development and manufacturing. PAT includes several analytical tools, such as spectroscopic e.g., Raman, UV-Vis, mid-infrared), chromatographic, and mass spectrometric sensors, which offer in-line, on-line, or at-line data that improve process-based knowledge, control, and efficiency. PAT enables continuous measurement of CPPs and CQAs, enabling early identification and rectification of errors and maintaining constant product quality throughout the batch or continuous process. PAT speeds process development by giving fast data and automated feedback, scales up quickly, and reduces waste and batch failures. PAT instruments help with regulatory compliance and lifecycle management in QbD frameworks. The FDA and other regulatory bodies support the use of PAT as part of a science- and risk-based framework for designing quality into biologics production techniques processes instead of depending solely on end-of-biologics product testing [65, 66].

23.6.4 Internet of Things (IoT)

The Internet of Things dramatically improves Quality by Design (QbD) in biologics by permitting comprehensive real-time observation, control, and data integration across the biopharmaceutical production process. IoT devices and sensors connect different phases of production, allowing for continuous monitoring of key CPPs and CQAs, which are essential to QbD concepts. IoT solutions capture and send rich datasets from many sensors to cloud-based dashboards and analytics systems, merging information into a single source of truth, breaking down data silos and enabling holistic process knowledge, vital for QbD. IoT networks support data integrity, traceability, and documentation required by regulatory frameworks like FDA guidance, easing audits and supporting continuous process verification within QbD programs [67].

23.6.5 Robotics

Robotics improves Quality by Design (QbD) in biologics by automating complex and labor-intensive processes throughout biomanufacturing, resulting in better process control, product quality, and operational efficiency. Robotics can perform complicated biologics tests frequently and correctly, facilitating design of experiments (DoE) and accelerating early development phases in QbD frameworks. Automated robotic processes decrease human error, increase data repeatability, and help biologics manufacturers meet regulatory data standards. Biopharmaceutical companies are increasingly adopting robotics and automation to supplement QbD approaches for biologics to manage inherent process variability and improve efficiency, agility, and regulatory compliance during the development and commercialization stages [68, 69].

23.6.6 Blockchain Technology

Blockchain technology improves Quality by Design (QbD) in biotechnological goods by establishing a decentralized, immutable, and transparent ledger that assures data integrity, traceability, and regulatory observance throughout the product's lifespan. Blockchain enables comprehensive audit trails, decreases human mistakes, and allows for instantaneous data monitoring and biologics control in accordance with QbD principles.

To better understand key process parameters such as CPPs and critical quality attributes in QbD, blockchain technology offers complete supply chain observation of raw materials used for production of biologics, intermediates product of biologics, and product of biologics. Blockchain technologies improve data governance in biologics clinical trials by protecting patient data privacy while providing openness and participant confidence, which is essential for QbD-managed development processes [70].

23.7 Future Direction

Biologics are innovative approach in medical field for managing complicated disease condition. The integration of regulatory framework and Quality by Design approach in biologics provide a massive success in the field of both pharmaceutical and patient care. Biologics are very sensitive in nature. So, the integration of the above two approaches prevents the contamination of biologics and proper supply to patient respective their needs. Regulatory bodies such as FDA, PHSA, NDA, CBER, play an efficient role in the enhancement of biologics. Regulatory agency must involve each and overall process of development of biologics. The six core principle of Quality by Design ensure the efficacy and safety of the biologics. But the regulatory affairs face a lot of challenges to implement the Quality by design in biological production.

In today trends, personalized medicine, green based synthesis and integration of AI & ML tools in biologics provide a scalable and robustness on bio-production.

Personalized medicine in biologics aims to delivered the tailored medicine based on the requirement of the patient. This type of treatment strategy will provide a better relief for patient from their chronic disease condition such as onco-disorder, auto-immune disorder, metabolic disorder etc. So, the bioproduction manufacturer must follow the quality by design technique for the proper supply to the patient. Regulatory authority must approve this type innovation without any unnecessary restriction.

And the regulatory agency give support to the production of biologics from plant based. Furthermore, regulatory authorities have been attempting to harmonize rules throughout the world in order to speed up the research and licensing of plant-derived biologics. Togetherness of regulatory agencies with industry stakeholders has resulted in the formation of worldwide rules and standards, which promote uniformity and efficiency in regulatory evaluations. Furthermore, regulatory concerns for plant-derived biologics include a thorough assessment of safety, effectiveness, and quality criteria. These regulatory frameworks guarantee that plant-derived biologics fulfill the requisite requirements of safety, effectiveness, and environmental impact, allowing their successful development and licensing in the biotechnology sector.

AI and ML technology will provide an innovative and feasible method for the enhancement of bioproduction. Implementation of Internet of things, robotics, blockchain technology in manufacturing of biologics can ease to production and reduce the human error. But the major limitation is regulatory affairs that does not have mindset to accept these techniques. In the future international regulatory harmonization professional must have the ability to change their guidelines and norms to accept this innovative technology in the biologics production.

From the above broad vision of statements, we can expect the implementation of personalized medicine; green method of synthesis of biologics and integration of digital tools provide a massive development in the field of biologics with the support of regulatory authority.

23.8 Conclusion

In conclusion, this book chapter reveals the togetherness of quality by design with the controlled regulatory framework in biologics to ensure purity, safety and efficacy of the biologics. And it provides a scalable production of biologics. Additionally, the involvement of digital tools, automation processes, regulatory guidelines, ensures the proper supply of biologics to the patient with reasonable cost and it will be an innovative pathway for the management and pharmacological treatment of chronic disease condition. Approaches such as personalized medicine in biologics, green method of synthesis in biologics and integration of AI and ML tools elevate the quality of the biological product as well as improve the quality of life to the patient.

AI disclosure statement: The author acknowledges the use of AI-based tools for minor language editing or grammatical correction. However, all intellectual contributions, analyses, and conclusions presented in this chapter are entirely the author's own.

References

1. Pisano, D.J. and Mantus, D.S., *FDA Regulatory Affairs: A Guide for Prescription Drugs, Medical Devices, and Biologics*, CRC Press, Boca Raton, FL, USA, 2008.
2. Medina, C., *Compliance Handbook for Pharmaceuticals, Medical Devices, and Biologics*, CRC Press, Boca Raton, FL, USA, 2003.
3. Naser-Tavakolian, A., Heard, J.R., Ahdoot, M., Biologics, in: *Translational Urology*, pp. 355–360, Academic Press (Elsevier), Amsterdam, The Netherlands, 2025.
4. Raychaudhuri, S.P. and Raychaudhuri, S.K., Biologics: target-specific treatment of systemic and cutaneous autoimmune diseases. *Indian J. Dermatol.*, *54*, 2, 100–109, 2009.
5. Price, W.N. II and Rai, A.K., Manufacturing barriers to biologics competition and innovation. *Iowa Law Review*, 101, 1023–1063, 2016.
6. Tiwari, N.J., Narkhede, K.B., Prajapati, A., Narkhede, S.B., Luhar, S., Biologics and Biosimilars: Navigating Production, Approval, Applications and Challenge. *Asian J. Res. Pharm. Sci.*, *15*, 2, 167–1, 2025.
7. Lim, A.C., Zhou, Y., Washbrook, J., Titchener-Hooker, N.J., Farid, S., A decisional-support tool to model the impact of regulatory compliance activities in the biomanufacturing industry. *Comput. Chem. Eng.*, *28*, 5, 727–735, 2004.
8. Lim, A.C., Farid, S., Washbrook, J., Titchener-Hooker, N.J., A tool for modelling the impact of regulatory compliance activities on the biomanufacturing industry, in: *Computer Aided Chemical Engineering*, vol. 14, pp. 1109–1114, Elsevier, Amsterdam, The Netherlands, 2003.
9. Purpura, C.A., Garry, E.M., Honig, N., Case, A., Rassen, J.A., The role of real-world evidence in FDA-approved new drug and biologics license applications. *Clin. Pharmacol. Ther.*, *111*, 1, 135–144, 2022.
10. Huneycutt, B.J., Malkin, B.J., Van de Wiele, V., Williams, M.P., Mehta, R., Regulatory agency transparency for drugs and biologics. *Food Drug Law J.*, *75*, 4, 498–551, 2020.
11. Maurya, S.K., Shukla, V., Kaushik, P., Kanti, S.P., Bansal, A., Regulatory aspects for Biologic product licensing in India. *Int. J. Drug Regul. Aff. (IJDRA)*, 7, 1, 1–5, 2019.
12. U.S. Department of Health and Human Services, Food and Drug Administration, Rockville, MD, United States, 2025.

13. World Health Organization, WHO expert committee on biological standardization: Sixty-sixth report, WHO Technical Report Series No. 999, World Health Organization, 2016.
14. World Health Organization, WHO expert committee on biological standardization: Sixty-eighth report, WHO Technical Report Series No. 1016, World Health Organization, 2018.
15. Pan American Health Organization, World health, 2007.
16. Pombo, M.L., Biotechnological products in Pan American Health Organization (PAHO): Regional efforts towards harmonization of regulation. *Biologicals*, *39*, 5, 348, 2011.
17. Ojha, A. and Bhargava, S., International council for harmonisation (ICH) guidelines, in: *Regulatory Affairs in the Pharmaceutical Industry*, pp. 47–74, Academic Press, Amsterdam, The Netherlands, 2022.
18. Davies, J.G., Gao, D., Kim, Y.J., Harris, R., Cash, P.W., Schofield, T.L., Qin, Q., ICH Q5C stability testing of biotechnological/biological products, in: *ICH Quality Guidelines: An Implementation Guide*, pp. 345–373, John Wiley & Sons, Inc., Hoboken, NJ, USA, 2017.
19. Kinch, M.S., An overview of FDA-approved biologics medicines. *Drug Discovery Today*, *20*, 4, 393–398, 2015.
20. Martins, A.C., Albericio, F., de la Torre, B.G., FDA Approvals of Biologics. *Biomedicines*, *11*, 5, 1434, 2023.
21. Manual of Policies & Procedures (CDER), Food and Drug Administration, Rockville, MD, 2025.
22. Wick, J.Y., CE: Culture, Ethnicity, and Medications: As patient demographics change, pharmaceutical care providers must understand how cultural factors and biological variations in ethnic groups affect response to medications. *J. Am. Pharm. Assoc.*, *36*, 9, 556–564, 1996.
23. Food, U. and Administration, D., Biological Product Definitions, 2014, www.fda.gov.
24. biologicals, Retrieved August 14, 2025, from https://cdsco.gov.in/opencms/opencms/en/biologicals/.
25. Singh, S. and Bisht, A., Regulation of biologicals: Indian perspective. *Int. J. Res. Found. Hosp. Healthcare Adm.*, 3, 2, 135–136, 2015.
26. Ministry of Food and Drug Safety>Our Works>Bio&Cosmetics>Biosimilar, Ministry of Food and Drug Safety. Retrieved August 14, 2025, from https://www.mfds.go.kr/eng/wpge/m_37/de011024l001.do.
27. IND Process and Review Procedures, Manual of Policies and Procedures, Centre for Drug Evaluation and Research, Food and Drug Administration, Rockville, MD, Available at: http://www.fda.gov/downloads/AboutFDA/CentersOffices/OfficeofMedicalProductsandTobacco/CDER/ManualofPoliciesProcedures/UCM082022.pdf.
28. Evens, R.P., *Drug and biological development*, Springer, New York, NY, USA, 2007.

29. Rathore, A., Guidelines on similar biologics: regulatory requirements for marketing authorization in India. *PDA J. Pharm. Sci. Technol.*, *66*, 5, 393–393, 2012.
30. Douthwaite, J. and Jermutus, L., Exploiting directed evolution for the discovery of biologicals. *Curr. Opin. Drug Discovery Dev.*, *9*, 2, 269–275, 2006.
31. Natarajan, J. and Janani, R., A review on regulatory guidelines for biologics in India. *J. Pharm. Sci. Res.*, *11*, 11, 3651–3654, 2019.
32. Regulation of Biologics in India [Internet], Morula, 2018 [cited 2018 Dec 23]. Available from: http://morulaa.com/cdsco/regulations-of-biologics-in-india-2/.
33. Lokesh, C., Sonia, K., Kamaraj, R., Regulatory Requirements for Biological product approval process in India. *Res. J. Pharm. Technol.*, *13*, 2, 1050–1054, 2020.
34. Huma, T. and Peng, Z., Introduction to Regulatory Affairs and Different Regulatory Bodies for Pharmaceutical Products and Impact of Digitalization on Regulatory Affairs. *Pharmacol. Pharm.*, *14*, 11, 463–477, 2023.
35. Sree, M.P., Towhid, S.R., Shahina, S.K., Krishna, P.J., Sai, B., Overview of quality by design (QbD) in pharmaceutical development. *World J. Pharm. Res.*, 2024.
36. Badri, S., B, K., N, P., B, P., Sd, M., A review on various aspects of quality by design [qbd] on new drug approval. *UPI J. Pharm. Med. Health Sci.*, *7*, 4, 8–15, 2024.
37. Ameen, S.A. and Pappula, N., Analytical QBD approach to redefine the quality of pharmaceuticals: A review. *J. Pharm. Res.*, *22*, 4, 179, 2023.
38. Kalkate, S.D. and Bhosle, A.A., QbD in pharmaceutical manufacturing: a systematic review of current status, challenges and future perspectives, 2021.
39. Mohammed, J., Sharma, V., Ravindra, N., Singh, V., A Comprehensive Review on Quality by Design (QbD) in Pharmaceuticals. *Int. J. Health Adv. Clin. Res. (tz)*, *2*, 4, 55–59, 2024.
40. Gurba-Bryśkiewicz, L., Maruszak, W., Smuga, D.A., Dubiel, K., Wieczorek, M., Quality by Design (QbD) and Design of Experiments (DOE) as a Strategy for Tuning Lipid Nanoparticle Formulations for RNA Delivery. *Biomedicines*, *11*, 10, 2752, 2023.
41. Valeru, C., Pharmaceutical product development: quality by design (QbD). *UPI J. Pharm. Med. Health Sci.*, 11–16, 2024.
42. Kharb, M. and Rathore, K.S., QbD approach: A framework for integrating quality into pharmaceutical products. *J. Pharm. Res. Int.*, 1–11, 2022.
43. Mohurle, M.S., Asnani, M.D., Chaple, D.R., Kurian, M.J., Bais, M.A., Quality by design (QbD): An emerging trend in improving quality & development of pharmaceuticals. *Saudi J. Med. Pharm. Sci.*, *5*, 12, 1132–1138, 2019.
44. Sharma., M., Implementing Quality By Design (QbD) In Pharmaceutical Formulation Development. *Innovations Pharm. Planet*, *10*, 4, 66–69, 2022.
45. Shree, V.M. and Chandramouli, R., Review on Quality by Design Approach (Qbd). *Curr. Trends Biotechnol. Pharm.*, *15*, 4, 447–454, 2021.

46. Rajora, A. and Chhabra, G.U., Quality by design approach: Regulatory need, current, and future perspective. *Asian J. Pharm. Clin. Res.*, *14*, 4, 29–35, 2021.
47. Duarte, J.G., Duarte, M.G., Piedade, A.P., Mascarenhas-Melo, F., Rethinking Pharmaceutical Industry with Quality by Design: Application in Research, Development, Manufacturing, and Quality Assurance. *AAPS J.*, *27*, 4, 96, 2025.
48. Luciani, F., Galluzzo, S., Gaggioli, A., Kruse, N.A., Venneugues, P., Schneider, C.K., Pini, C., Melchiorri, D., Implementing quality by design for biotech products: Are regulators on track? *MAbs*, *7*, 3, 451–455, 2015, https://doi.org/10.1080/19420862.2015.1023058.
49. Aksu, B., Sezer, A.D., Yegen, G., Kusçu, L., QbD Implementation in Biotechnological Product Development Studies, in: *Special Topics in Drug Discovery*, InTech, Rijeka, Croatia, 2016.
50. Drobnjakovic, M., Hart, R., Kulvatunyou, B.S., Ivezic, N., Srinivasan, V., Current challenges and recent advances on the path towards continuous biomanufacturing. *Biotechnol. Progr.*, *39*, 6, e3378, 2023.
51. Kaleem, A.M., Koilpillai, J., Narayanasamy, D., Kaleem, A., Mastering Quality: Uniting Risk Assessment with Quality by Design (QbD) Principles for Pharmaceutical Excellence. *Cureus*, *16*, 8, 2024.
52. Karia, P., Ayre, A., Khan, A., Critical review of risk assessment tools in pharmaceutical quality by design. *Indian J. Pharm. Educ. Res.*, *58*, 4s, s1145–s1155, 2024.
53. Glevitzky, M., Glevitzky, I., Mucea-Ştef, P., Popa, M., Dumitrel, G.A., Vică, M.L., Integrated Risk Framework (IRF)—interconnection of the Ishikawa diagram with the enhanced HACCP system in risk assessment for the sustainable food industry. *Sustainability*, *17*, 2, 536, 2025.
54. QbD in Biologics: Genentech's Success and Failure in Design Space Approval – Quality by Design for Biotech, Pharmaceutical and Medical Devices. Retrieved August 14, 2025, from https://qbdworks.com/qbd-biologics-genentechs-success-failure-design-space-approval/.
55. Zurdo, J., Arnell, A., Obrezanova, O., Smith, N., Gómez de la Cuesta, R., Gallagher, T.R., Michael, R., Stallwood, Y., Ekblad, C., Abrahmsén, L., Höidén-Guthenberg, I., Early implementation of QbD in biopharmaceutical development: a practical example. *Biomed Res. Int.*, *2015*, 605427, 2015.
56. Rathore, A., Quality by design: Industrial case studies on defining and implementing design space for pharmaceutical processes. *BioPharm Int.*, 22, 1, 40–45, 2009.
57. Rathore, A.S., Quality by Design (QbD)-Based Process Development for Purification of a Biotherapeutic. *Trends Biotechnol.*, *34*, 5, 358–370, 2016.
58. Riawan, A., A systematic literature review of implementation quality control and assurance in Industries. *J. Mech. Eng. Learn.*, 14, 1, 17–33, 2025.
59. Rathore, A. and Nikita, S., Digitalization: The Route to Biopharma 4.0. *BioPharm Int.*, 36, 11, 25–29, 2023.

60. Beutel, S. and Lenk, F. (Eds.), *Smart Biolabs of the Future,* Springer, Cham, Switzerland, 2022.
61. Nikita, S., Thakur, G., Jesubalan, N.G., Kulkarni, A., Yezhuvath, V.B., Rathore, A.S., AI-ML applications in bioprocessing: ML as an enabler of real time quality prediction in continuous manufacturing of mAbs. *Comput. Chem. Eng., 164,* 107896, 2022.
62. Jumper, J., Evans, R., Pritzel, A., Green, T., Figurnov, M., Ronneberger, O., Hassabis, D., Highly accurate protein structure prediction with AlphaFold. *Nature, 596,* 7873, 583–589, 2021.
63. Austerjost, J., Marquard, D., Raddatz, L., Geier, D., Becker, T., Scheper, T., Lindner, P., Beutel, S., A smart device application for the automated determination of *E. coli* colonies on agar plates. *Eng. Life Sci., 17,* 8, 959–966, 2017.
64. Canzoneri, M., De Luca, A., Harttung, J., Digital Twins: A General Overview of the Biopharma Industry. *Adv. Biochem. Eng./Biotechnol., 177,* 167–184, 2021.
65. Walsh, I., Myint, M., Nguyen-Khuong, T., Ho, Y.S., Ng, S.K., Lakshmanan, M., Harnessing the potential of machine learning for advancing "quality by design" in biomanufacturing. *MAbs,* 14, 1, 2013593, 2022.
66. Schad, M., Gautam, S., Grein, T.A., Käß, F., Process Analytical Technologies (PAT) and Quality by Design (QbD) for bioprocessing of virus-based therapeutics, in: *Bioprocess and Analytics Development for Virus-Based Advanced Therapeutics and Medicinal Products (ATMPs),* pp. 295–328, Springer International Publishing, Cham, 2023.
67. Borgosz, L. and Dikicioglu, D., Industrial internet of things: What does it mean for the bioprocess industries? *Biochem. Eng. J., 201,* 109122, 2024.
68. Noh, I., Lee, J., Lee, S.W., Development of situation awareness model in robotic spot-welding (RSW) system based on sensor data visualization. *Int. J. Precis. Eng. Manuf.-Smart Technol., 2,* 2, 123–131, 2024.
69. Somara, S. and Mannick, E., Ensuring Regulatory Compliance in Cell Therapy: The Role of Quality by Design (QbD) in Manufacturing and Clinical Trials. *Cytotherapy, 27,* 5, 128–129, 2025.
70. Hemphill, T.A., Biologics regulation, second-to-market competition, and the use of blockchain technology: an opportunity for the FDA to support responsible biotechnology innovation. *J. Responsible Innovation, 7,* 3, 689–696, 2020.

24

AI-Driven Predictive Maintenance for Biomanufacturing Equipment

Anup Ashok*, **Vijaya Kumar Talari, Bijoy Kumar Purohit, A. V. Raghavendra Rao, Stutee Bhoi and K.S.N.V. Prasad**

Department of Chemical Engineering, B V Raju Institute of Technology, Narsapur, Telangana, India

Abstract

In biomanufacturing, the availability of equipment and downtime is crucial to product quality, regulatory and process efficiency. In this chapter, the current interest in predictive maintenance (PM) of biomanufacturing equipment through the application of artificial intelligence (AI) is discussed. Compared to more conservative maintenance practices based upon time-based schedules or failure-based response, AI-based PM works with data analytics and machine learning to predict when machines will fail by leveraging real-time data, sensor integration, and machine learning algorithms. The chapter describes the technical underpinnings of AI based maintenance, such as data ingestion systems, anomaly detection, condition monitoring and predictive modeling. The focus is on the application of these tools within the bioprocessing unit like bioreactors, centrifuges, pumps and filtration systems. Case studies presented in the real world show how AI-enabled PM platforms are being implemented and included an increase in the operational uptime, cost reduction, compliance with regulatory guidelines. Issues like data quality, interconnection with current Supervisory Control and Data Acquisition (SCADA) systems and interpretability of the models also form part of the discussion. Future trends such as digital twins and edge computing are mentioned at the end of the chapter, along with the possibility to continue revolutionizing biomanufacturing predictive maintenance even further. The chapter is intended to be of use to engineers, data scientists and biomanufacturing leaders aiming to increase their operation resilience by deploying intelligent maintenance systems.

**Corresponding author*: anupashok1001@gmail.com

Sandip Kunar, Jagadeesha T. and D. Sathis Kumar (eds.) Biomanufacturing: Transforming Production of Biologics and Biopharmaceuticals, (777–800) © 2026 Scrivener Publishing LLC

***Keywords*:** Artificial intelligence, predictive maintenance, condition monitoring, machine learning, biomanufacturing equipment

24.1 Introduction to Predictive Maintenance in Biomanufacturing

Equipment reliability is one of the essential foundations to biomanufacturing as the goal of its continuous production is to meet the global healthcare requirements of biologics and biopharmaceuticals, which require unparalleled precision and consistency in its manufacturing processes to produce high-quality therapeutic products such as monoclonal antibodies, vaccines, and gene therapies. The production devices used in production (bioreactors to create sterile environments in which microbial or mammalian cells are cultivated, centrifuges to separate cells and pumps to move fluids, and filtration devices to remove impurities) are so versatile that any disruption can lead to a series of failures such as contamination, batch failure, and supply chain breakage and losses in the million-dollar range per mishap, with production shutdowns lasting days and weeks in a single facility, making biopharmaceutical development already very expensive by most estimates [1]. Additionally, in an age of personalized medicine and quick pandemic response, unstable equipment not only contributes to patient safety risks, but also derails the scalability necessary to meet the increasing needs, with global biopharmaceutical market predicted to exceed $500 billion by 2025, thus posing the necessity of sound maintenance practices as a means of ensuring operations continuity and innovation. Historically, biomanufacturing has employed predicative maintenance, in which equipment is maintained at predetermined intervals, either to manufacturer specifications or based on the past averages, e.g., bioreactor agitator inspection every 3 months, chromatographic columns overhauled annually, which in spite of their preventive intent, would tend to over-service underutilized equipment, waste resources, and halt production during non-critical events, or reactive maintenance, or maintenance in response to an occurrence, which, may result in emergency repairs that exacerbate the downtime in a sharp contrast. AI enabled predictive maintenance (PM) is a paradigm shift; hence requiring advanced machine learning algorithms, real-time sensor readings of Internet of Things (IoT) devices, big data analytics to predict equipment failure before they occur, resulting in targeted interventions to maximize uptime and efficiency; e.g., multivariate data streams (heightened temperature, change in vibrations, pressure anomalies, and flow rate) which may be processed to predict pump seal wear as high as 95 percent by AI models. According to case studies of leading biopharma firms, not only does this proactive model reduce inefficiencies of outdated processes, but it is also compatible with digitalizing efforts, building on cloud computing and edge analytics to generate the actionable insight traditionally unavailable to such processes and thus, creating a more durable manufacturing environment overall [2]. Figure 24.1 shows key bio-manufacturing equipment.

Preventive maintenance ensures product quality, safety, and efficacy by reducing process variability. It prevents equipment issues like pH shifts, protein misfolding, or membrane fouling, which can compromise

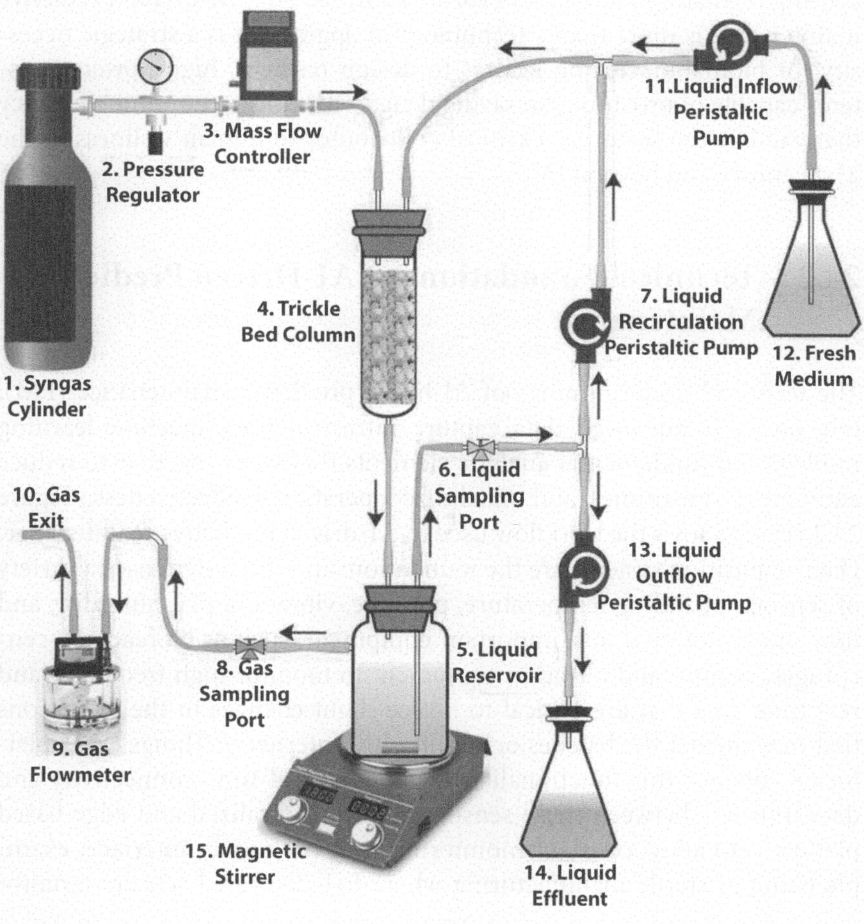

Figure 24.1 Overview of key biomanufacturing equipment.

sterility. This also improves reproducibility and lowers product rejection [3]. Preventive maintenance (PM) ensures compliance in biomanufacturing by keeping auditable equipment records, supporting 21 CFR Part 11 standards, and reducing risks of warning letters or recalls. AI-based PM automates routine checks, lowers labor costs, boosts equipment effectiveness from 60–70% to over 85%, and reduces waste and energy use, improving overall efficiency by 10–15% [2024 McKinsey report]. The ability of PM to integrate into the very heart of biomanufacturing workflows not only ensures reliability in the short term but opens the door to further developments in the form of digital twins to facilitate virtual simulations or reinforcement learning to facilitate adaptive maintenance policies, ultimately ensuring that the industry can remain successful even in years to come as the challenges it faces, like the disruptions of supply chains or shifting regulatory landscapes evolve [4]. In the end, AI-driven predictive maintenance is more than a technological upgrade; it is a strategic necessity for biomanufacturing leaders to design resilient, high-powered systems capable of providing lifesaving drugs with consistency and efficiency that validate the sector as a critical contributor to human wellness in the 21st century and beyond [5].

24.2 Technical Foundations of AI-Driven Predictive Maintenance

The technical underpinnings of AI-based predictive maintenance (PM), rely on more advanced data capture infrastructures, machine-learning models, and fundamental analytic elements that work together to reduce equipment downtimes and maximize operational effectiveness. Figure 24.2 clearly shows the workflow used in AI-driven predictive maintenance. Data acquisition systems are the foundation; and the system uses a variety of sensors, including temperature, pressure, vibration, pH, humidity, and flow rates, mounted into important equipment, such as bioreactors, centrifuges, pumps, and filtration equipment to monitor high frequency and real time data that are critical to notice slight changes in the operations that may impact the batches or sterility [6]. Internet of Things (IoT) platforms enhance this functionality by allowing real-time connectivity and data transfers between these sensors to both centralized and edge-based platforms to allow constant monitoring with no human interface; example being in sterile manufacturing where IoT-dedicated sensors monitor

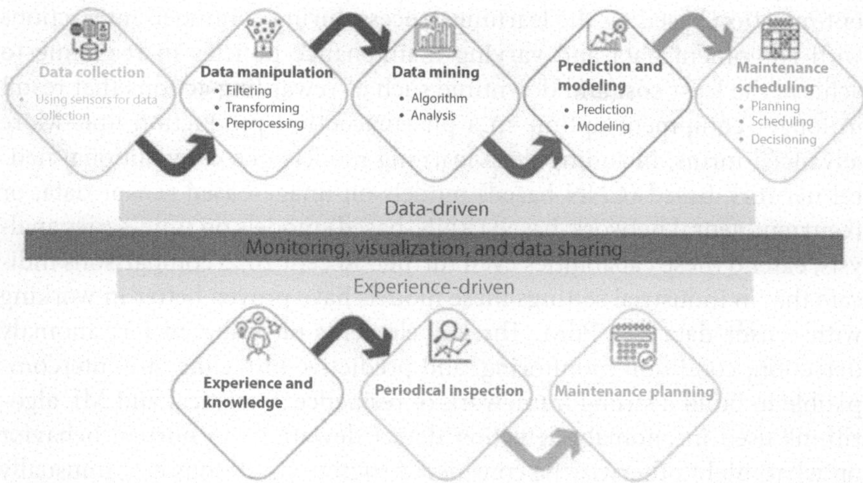

Figure 24.2 Workflow diagram that shows the process of AI-driven predictive maintenance.

environmental indicators such as air quality and particulate matter. To avert contamination, as seen in the demonstrations by such companies as Eli Lilly and Novartis, real time data transmissions minimize waste, and ensure compliance with Good Manufacturing Practices (GMP). Given the nature of biomanufacturing, which requires aseptic conditions and must consider variable biological processes, this real-time data collection can be labeled as one of the means of coping with the limitations inherent in biomanufacturing as this allows creating large amounts of data, particles of which are fed into AI models to obtain proactive information. The computational heart of PM is machine learning algorithms, whereby supervised learning methods (regression models, decision trees, random forests, support vector machines) are trained using labeled past data with the goal of forecasting machine failures, including predicting remaining useful life (RUL) of specific components, such as the bioreactor impeller based on historic vibration measurements and maintenance history, whose accuracy has recently been demonstrated to exceed 95% [7]. Unsupervised learning and clustering algorithms such as k-means or principal component analysis (PCA) are best suited to the detection of artifacts or patterns in unlabeled data, such as clustering the occurrence of abnormal pressure variations in filtration systems to detect a possible membrane fouling without *a priori* thresholds, very valuable when exploring bioprocess data analysis. Although it is a new field, reinforcement learning provides dynamic optimization based on the learning process during continued interactions with equipment data and varying maintenance policies in real time to achieve the least cost and downtime such as rewarding actions that result in longer equipment uptime in a pharmaceutical production line. More advanced forms, including deep learning models like convolutional neural network-based (CNN-based) models on image-based sensor data, or recurrent neural network-based (RNN-based) models on time-series analysis, extend these capabilities even further. Recent 2025 comparisons indicate that in industrial settings these models have proven better in working with sensor data and PdM. The key elements of AI-based PM anomaly detection, condition monitoring, and predictive modeling, are intercompatible to build a strong framework of resilience. Statistical and ML algorithms used in anomaly detection detect deviations to normal behavior on what might otherwise be considered routine conditions e.g., unusually high spikes in pumps due to inefficiency or subtle pH drifts in a bioreactor indicating risk of contamination, allowing early notifications before rejections of batches and aligning with FDA requirement to be proactive in quality management [8]. Condition monitoring is continuous evaluation of equipment wellbeing by a combination of sensor data, summing metrics such as vibration and temperature patterns into dashboards to provide real-time insight as in Veeva Vault QMS systems using AI to monitor rates of deviations and effectiveness of CAPA in biopharma plants by 2025. The predictive models compile these inputs to predict failures in use of probabilistic, like Bayesian networks to model interdependencies, e.g., between operator training scores and environmental conditions to predict risk of sterile-batch problems—or generative models to simulate scenarios, e.g., to carefully schedule maintenance, yielding 30-50% fewer unplanned stops in use as applied in vaccine manufacturing at Pfizer. These elements are more and more supplemented by digital twins that can mimic device behavior based on real time IoT data to improve prediction precision and edge computing to work on data at the edge, reducing time lag in time-sensitive cases of bioprocesses. The maturity of such foundations became clear with the recent releases in 2025, including AI analytics modules of Master Control to create predictive dashboards and configurable ML of Track Wise Digital to deliver deviation patterns [9]. Such toolkits as TensorFlow and Apache Kafka can support real-time, scalable biomanufacturing analytics. Additionally, the explainable AI (XAI) is used to solve the regulatory issues of explainability of the model decisions, as well as being compliant with ALCOA+ principles and FAIR data principles, and the foundation models lessen the training burden on different types of equipment. There are still a number of challenges such as the data quality assurance to measure drift and outliers in inputs using observability tools, but the movement of federated learning enables safe joint training of models without risking the individual proprietary data. These technical underpinnings eventually give biomanufacturing the ability to obtain greater reliability, where AI-enabled PM not only avoids breakdowns in processes but also optimizes for sustainability, like cutting the energy usage of centrifuges by 15-20 percent through controlled, condition-based interventions. The industry is ready to scale production in a world where demands are growing both in personalized biologics and rapid vaccination development, a world that is ever data-driven [10, 11].

24.3 Applications of AI in Bioprocessing Equipment

One such application where AI-driven predictive maintenance (PM) is becoming a critical enabling technology in the biomanufacturing industry is bioprocessing equipment where sterility, yield and adherence to strict regulatory bodies such as FDA and EMA relies on the flawless functionality and performance of complex equipment, and intelligent, predictive maintenance can pre-empt catastrophic failure and losses and drive optimal performance with far less downtime and less costly solutions than traditional reactive maintenance approaches. Practical applications of this effect are well-understood, e.g., in bioreactors, which play a significant role in growing microbial or mammalian cells under controlled conditions, AI modeled PM can analyze real-time measurement data on parameters such as dissolved oxygen (DO), pH, temperature, and agitation velocity to anticipate typical failures (i.e., an impeller wearing out, or a seal deteriorating) and alert if uncontrolled conditions are imminent (contamination that may cost millions of dollars in lost product by ruining the entire batch), as demonstrated in centrifuges, used in cell harvesting and clarification, AI uses vibration sensors and rotational speed to predict imbalance or bearing failure, enabling operators to carry out maintenance during non-manufacturing periods and avoid unplanned down-time by as much as 40%, especially in production of vaccines where high speed operations increase the risks of wear. In pumps, which are of paramount importance to moving fluids such as media or buffers without shear stress damaging sensitive biologics, AI-enabled PM can monitor changes in pressure, flow rates and acoustic emissions to predict problems such as cavitation or seal failure, so the fluid continues to move without interruption, and eliminate the chance of downstream cross-contamination, with AI-PM in insulin production showing a 25 percent increase in pump life through early detection of anomalies [12, 13]. Differential pressure and permeate flux upstream of filtration systems can be tracked to predict membrane fouling or pressure spikes, making AI-suggested cleaning or replacements possible to avoid an unwanted reduction in throughput or loss of product purity as seen in the application of gene therapy where AI has reduced fouling downtime by 35 percent. Integration of AI with equipment-specific data enables advanced applications by combining IoT sensors, historical maintenance records, process analytics, and environmental data to form multi-faceted digital ecosystems. For example, in bioreactors, AI platforms analyze multivariate time-series data from SCADA systems and PAT devices to create hybrid models that link operational measurements, such as glucose consumption, with machine health indicators, enable real-time auto-tuning, and support predictive maintenance. The same is true of centrifuges, where AI algorithms consume accelerometer data and other batch-specific parameters, such as cell density, to predict fatigue in rotors, or of pumps and filtration systems, where edge computing can optimize on-device processing of flow dynamics and pressure gradients and require no more latency than analysis performed elsewhere, all with increased data security within GMP processes and streamlined integration with existing MES (Manufacturing Execution Systems) to provide end-to-end process control. Cases of AI tools to optimize performance and avoid failures are numerous in the bioprocessing units, demonstrating the practical usefulness of such tools: Dataiku developed a predictive maintenance project, based on machine learning, which monitors sensor data and historically identified failures, and was operationalized at a major pharma company, in the Clean-In-Place (CIP) skid systems, which are commonly composed of pumps and filtration, resulting in a 20 percent cost reduction in maintenance expenses and deviation prevention, which would otherwise risk sterility in bioprocessing [14, 15]. There are other applications of AI in bioprocesses; at IDBS, the Polar platform also applies AI in determining a bioprocess method (mainly in filtration and chromatography) *via* data mining of operational information as a way to streamline a purification procedure to increase production yield and reduction in failures in the production of antibodies, with one example of AI identifying an anomaly in real-time spectra to achieve a 15% increase in process robustness. Palantir Foundry includes AI-powered biomanufacturing PM with predictive analytics and alerting about equipment such as bioreactors and centrifuges, into which it ingests supply chain data to predict variations in yields and to optimize equipment maintenance, increasing operational efficiency resulting in 30 percent reduction in unplanned cell therapy downtime, according to one customer [16]. Further, platforms such as those of Scispot offer AI-guided analytics of bioprocessing variables to optimize bioproduction conditions both in bioreactors during fermentation and in filtration-based purification, making judgments about the performance of validation and equipment requirements by drawing on strain performance and equipment requirements. Another case in point involving a CHO cell cultivation process notes that machine learning-enabled tools were used to optimize the activities of bioreactors by using fed-batch recordings to avoid incidents of overfeeding and chalking up 25 percent rise in titers, as explained in recent publications. They might also be used not only to avoid collapses but also to innovate around them, such as with smart bioprocessing, where AI and IoT are coming together to create self-optimizing systems that respond to the variability of biologics manufacturing, cutting pump energy consumption by 10-15 percent and extending the life of centrifuge components used under new and stringent condition monitoring. Integrating AI in the workflow of bioprocessing, manufacturers gain resiliency, with predictive ability that accords to bioprocessing 4.0 paradigms, scalable and cost-effective production of life-saving therapies, and manage the complexity of biological variability and regulatory requirements in an industry that is expected to grow to $700B by 2030 [17, 18].

24.4 Case Studies: AI-Enabled Predictive Maintenance in Action

Dynamic nature of biomanufacturing so far, in the dynamic industry of biomanufacturing where the functionality of equipment directly affects the quality of biologics and biopharmaceuticals produced, field applications of the AI-enabled solution of predictive maintenance (PM) have shown the significant improvement of operational resilience. Figure 24.3 gives visualization of behavior changes in real-time and AI alert system to monitor bioreactors. The most well-known is the Mulund factory of Johnson & Johnson in India, a member of the Global Lighthouse Network at the World Economic Forum, which has incorporated AI and machine learning into ways to maintain its highly important equipment in the production of consumer health products. The facility used AI/ML-based systems with

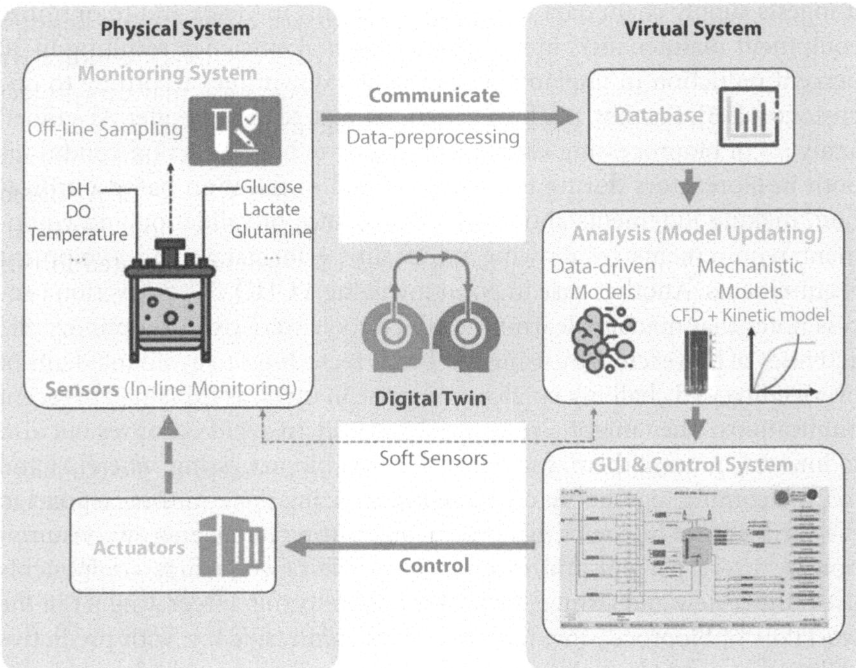

Figure 24.3 Visualization of behavior changes in real-time and AI alert system to monitor bioreactors.

robotics and 3D printing to assess the health of the assets using real-time data-based sensors on the production lines to identify failures in machines used in production activities such as mixing and granulation [19]. Likewise, a pharmaceutical company engaged an artificial intelligence solutions company, Augury, to scan production equipment, e.g., motors on a tablet manufacturing line in real-time, and detect the early signs of bearing failure *via* machine learning on acoustic and vibrational data. Another example is the JHS, a pharmaceutical company implemented the wireless vibration sensor with AI-integration capacities deployed to 100 units on Water for Injection (WFI) pumps and Air Handling Units (AHUs) that were key to sterile environments in bioprocessing. Such sensors set up operation base lines and raised anomalies like failure to align pump couplings after maintenance, so that human and AI could work together to make corrective actions. Other industrial practices include monitoring the HVAC system in a clean room using AI to maintain temperature and humidity conditions to avoid contamination of manufacture of biologics and using AI to analyze packaging machines and monitor vibration in rollers and seals, which would lead to a smooth run through the downstream processes. Major companies such as Pfizer and Novartis also have implemented such AI systems in their plants that use real-time equipment monitoring in their factories centered on predictive analytics to forecast failures in centrifuges and filtration units used to separate and clean cells. Such deployments also underline the role of AI integrating with IoT sensors and SCADA systems to transition the conventional maintenance process into a proactive, data-driven and more specific practice devoted to the strict biomanufacturing requirements [20].

The measured results of such AI enabled PM implementations highlight their worth to improve uptime, create financial savings, and improve compliance, and lessons learned attest to the significance of collaboration across disciplines and data quality. Predictive maintenance at the Johnson & Johnson Mulund location reduced unplanned machine downtime by 50 percent, which has resulted in a total breakeven increase of 4.5 percentage points in on-time in full (OTIF) losses and expedited development of biologics by 87-92 percent amounting to millions of dollars in avoided lost production time and enhanced throughput [21, 22]. The Augury partnership allowed averting the issue of 4.8 million missed tablets, or about 62,000 pounds of lost production and product waste, all of which is directly related to FDA and EMA compliance since it prevents the risks of contamination or quality variation. The initiative by JHS saved the firm hundreds of thousands of dollars by preventing losses of batches due to pump misalignments and eliminated bottlenecks, as the initiative substantially improved equipment reliability and the audit trails as per GMP. More widespread applications, which include the use of AI in clean room control and sterilization cycles, have cut waste in drug manufacturing by identifying inconsistencies in process, resulting in up to 20-30 percent improvement in efficiency and cut reject rates, as well as enabling compliance with 21 CFR Part 11 regulations in electronic records. In the case of Pfizer and Novartis, the systems have also enhanced the reliability of equipment, reducing the 43 percent of recalls caused by equipment breakdown and estimated savings of $10 billion across the industry in equipment downtime by 2030 [23]. Successes also provided lessons such as the importance of excellently created data that is standardized so that bad data does not make the model work inaccurately. A lack of which caused initial false positives in certain deployments, so strong governance structures are required. Integrating legacy infrastructure required phased deployments and team training, supported by explainable AI that was easy to understand, helping engineers and regulators build trust and gain acceptance. The synergy of humans and AI was essential, with the skillful control of experts accompanying the algorithmic suggestion, thus minimizing risks in sterility. Altogether, the listed cases demonstrate that the ROI on initial investments in sensors and employee training may be high, but returns can be felt almost immediately due to the ability of biomanufacturing facilities to achieve sustainable productivity, which is guiding them towards scalable and compliant activities in an Industry 4.0 world [24].

24.5 Challenges in Implementing AI-Driven Predictive Maintenance

Incorporating AI-enabled predictive maintenance (PM) in biomanufacturing is not an easy task especially regarding data quality management and compatibility with existing Supervisory Control and Data Acquisition (SCADA) systems and legacy infrastructure that may hinder reliability and scalability of such cutting-edge systems in the manufacture of biologics and biopharmaceuticals [22, 23]. Poor data can undermine AI algorithms, especially when high-fidelity data is needed to predict equipment failures in sensitive areas like bioreactors or filtration units. Sensor noise, caused by environmental factors such as temperature changes or electromagnetic interference, can corrupt signals and lead to incorrect predictions and false alarms, reducing operational confidence. This is worsened by the presence of incomplete datasets due to either inconsistent data collection procedures, systems offline during retrofitting, or incomplete historical records with insufficient metadata render machine learning algorithms incapable of absorbing strong regularities; For example, in biomanufacturing, where processes have varying biological components, such as cell growth, missing data points on pH or dissolved oxygen can lead to anomaly detection being biased, thus reducing model accuracy by 20-30 percent as noted recently in systematic reviews. The problem is also compounded by the aspect of standardization since the biomanufacturing plants tend to be equipped with devices by multiple providers and the resulting data is in a form that is not compatible with one another, e.g., sensor data or different sampling rate, this causes a significant delay and higher budget to implement AI as they would require a lot of preprocessing in order to standardize information as well as a custom ETL (Extract, Transform, Load) pipeline. Organizations often experience growing pains when integrating AI with

SCADA systems and other legacy infrastructure. Most biopharma plants use outdated 1980s systems with no AI support. Retrofitting bioreactors with IoT sensors can cost over half a million dollars per building and must remain compatible with SCADA protocols while maintaining GMP compliance [24]. Connecting old and new systems creates cybersecurity risks and synchronization challenges, potentially causing data breaches and unreliable real-time alerts. IT–OT silos can hinder integration, reducing predictive efficacy by 15–25%; hybrid and cloud-based SCADA architectures help overcome these gaps. To cope with them, there is a necessity to establish strong data governance concepts and gradual modernization to reduce the risks and utilize the possibilities of AI in terms of more reliable equipment [25, 26].

In addition to data and integration challenges, overcoming interpretability of models and developing trust in AI forecasts will become key to popularizing the use of AI in biomanufacturing, where a black box model may prove a deal-breaker in terms of compliance and operator trust in safety-critical processes. A continued challenge around model interpretability also materializes, with complex AI methods such as deep neural networks providing an effective solution to predicting centrifuge or pump failures based on multivariate data, yet acting as black boxes themselves, and thus impossible to effectively verify against engineer's domain knowledge or regulatory compliance such as part 11 of the FDA. Such non-transparency may result in reluctance to follow through on AI advice especially in the area of biomanufacturing where a misplaced maintenance may introduce contamination or worse, postpone drug deliveries at a rate of 40 percent decrease in the use of that model when it is uninterpretable [27]. Developing trust will require deployable explainable AI (XAI), e.g. *via* SHAP (SHapley Additive exPlanations) or LIME (Local Interpretable Model-agnostic Explanations), which breaks down predictions in terms of feature contributions, e.g. linking a bioreactor failure alert to certain pH abnormalities, to promote accountability and allow auditors to trace decisions to achieve EMA compliance. Ethical issues like biased training data reduce trust and cause maintenance inefficiencies, while human–AI collaboration with interpretable models can improve predictions by 10–20% and mitigate biases. The path forward is to overcome these challenges by digitally validating *via* digital twins to perform simulations to test against and train stakeholders to demystify the results of AI, hence as PM systems become a reality, they should not only predict and alert based on failures, but also be able to provide an appropriate explanation that can be justified and reflective of biomanufacturing where safety and reproducibility are critical. In conclusion, the way forward on the road of overcoming these obstacles requires a compound solution that incorporates technological sophistication and culture change to achieve the transformative potential of AI in the designing of resilient and efficient biomanufacturing ecosystems that focus on quality in making biological mass produced in an industry with emerging demands [28].

24.6 Future Trends in AI-Driven Maintenance for Biomanufacturing

As biomanufacturing develops along the road to industry 5.0, where the production of biologics and biopharmaceuticals is human-centric, sustainability and resilient, future trends in AI-based predictive maintenance (PM) will transform the reliability of equipment using digital twins and edge computing. Digital twins are virtual replicas of physical assets such as bioreactors and filtration systems that can be used to analyze in real-time and simulate with IoT sensors providing real-time data and AI models paving the way to mimicking equipment conditions, forecast failures, and optimize processes without needing to shut down operations. Figure 24.4 gives conceptual illustration of a digital twin. Digital twins in biomanufacturing can enable proactive maintenance using complex biological processes, including CAR T cell production, with simulations to track key parameters, like pH and oxygen to predict deviations and loss of a batch, thereby potentially reducing downtime by 20-30 % as observed in advanced therapy medicinal (ATMP) models [29, 30]. The virtual modeling can be used to test scenarios in wear of equipment, where results can be used to refine designs and maintenance schedules virtually, as part of regulatory compliance with Quality by Design (QbD). To complement this, edge computing results in the decentralization of data processing by enabling AI analytics at the source e.g., on pumps or centrifuges to enable faster decision-making with negligible latency which is important with time-sensitive bioprocesses where latency might lead to loss of sterility or product. In 2025, MES integrates edge computing with digital twins to detect anomalies, predict bioreactor or HVAC failures up to two weeks in advance, reduce energy costs by 15%, and keep data secure on-premises (GMP compliant). The resultant functionality of this union is a resilient system with the replica of the global condition presented digitally as a

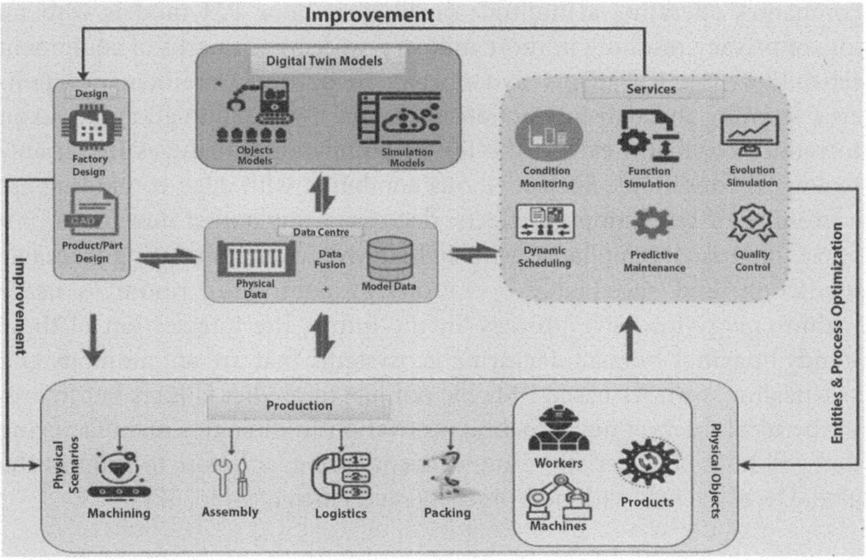

Figure 24.4 Conceptual illustration of a digital twin.

digital twin, and the immediate PM actions supported using edge devices, making biomanufacturing facilities prepared to operate under more strains owing to the heightened demands posed by personalized medicines in a more flexible and cost-efficient manner [31–33].

Newer AI methods also increase the capabilities of predictive maintenance in biomanufacturing, opening fields of innovation in generative AI (GenAI), explainable AI (XAI), federated learning, and reinforcement learning (RL) with significant implications on errors diagnosis and sustainability in operation. GenAI transforms digital twins by synthesizing lost failure data to simulate rare defects, overcome data imbalances in bioprocessing equipment such as in fault detection where GANs and VAEs can be used to present realistic scenarios of membrane fouling during a filtration process, which may significantly increase diagnostic costs somewhere between 10-20% [34]. XAI builds trust through providing interpretable explanations of AI decisions, which are essential to regulatory auditing in biomanufacturing, replacing black-box models with transparent methodologies such as SHAP to provide predictions of equipment health understood by operators and the systems in conjunction. Federated learning allows secure collaborative training of models at different facilities with no interchange of sensitive data which can be utilized by biopharma companies operating at multiple facilities to hone PM models with no loss of privacy resulting in more industry-wide benchmarks of equipment reliability [35]. RL adapts dynamically to optimize maintenance policies, learning through interactions in digital twins' simulations to adapt to reduce costs and extend the life of pumps or centrifuges in variably biological conditions. Such methods combined with edge computing are a promise of revolutionary effects: decreased unplanned downtime (30-50%), increased compliance with FDA/EMA regulations using traceable prediction, and sustainable operations by controlling resources usage within energy-intensive process. In the future, the intersection of these trends imagines biomanufacturing ecosystems that are autonomous and self-healing, with AI-based PM able not just to predict failures but to prescribe ideal interventions, fueling innovation in biologics manufacturing that will both be faster and more efficient than ever before to support the global health needs with new precision and efficiency [36, 37].

24.7 Practical Considerations for Biomanufacturing Leaders

The leaders of the biomanufacturing industry aiming to integrate AI-based predictive maintenance (PM) into biologics and biopharmaceuticals manufacturing operations can begin implementing an effective strategy by adopting a gradual roadmap to evaluate current infrastructure, data maturity, and asset business impact, including the priority of bioreactors and filter systems that have a high probability of breakdown. These include installing IoT sensors to record real-time measurements of such parameters as vibration and temperature, training AI models in controlled conditions to prove their forecasts, and continue to improve the accuracy of algorithms by iterative work so that the number of false alarms is minimal, which is the recommendation of 2025 implementation plans for intelligent production. In biopharma, the strategies of adoption demand that one begins with a small set of high-impact assets to realize early successes, e.g. a 30-50% reduction in unplanned outages, and align with Quality by Design (QbD) principles in preparation to meet regulatory requirements [38]. They must capitalize on cloud-based or edge computing facility due to its flexibility in nature and ensure hassles free connectivity with present MES and SCADA systems to automate alerts and maintenance processes. Engineers, data scientists, and regulatory teams must collaborate in developing successful adoption, creating multi-functional synergy that embraces the division between wet and dry labs where engineers lend domain expertise in mechanics of equipment. Data scientists synthesize ML models that identify anomalies and regulatory experts guarantee conformance to FDA/EMA policies such as 21 CFR Part 11. Embedding data scientists in engineering teams in biopharma helps mutual semantics and the principle of FAIR data, and challenges such as requiring design of what might become meta-data across lab instruments to be standard across large biopharma companies remains a challenge since only 41 percent of the large biopharma companies are ready to expect such integration [39]. EY details stress ensuring AI is centralized with data engineering to produce top-quality models and other such programs (such as IndiaAI Mission in India as an example) offer cross-discipline collaborations to scale AI ethically. This partnership also extends outside the organization, to include external players, e.g. technology providers like NVIDIA or Benchling to achieve cohesive platforms, which generates a more robust data flow and innovation, alleviating cultural resistance by training and showing practical outcomes in form of time savings in R&D. Potential good teams would conduct routine workshops to verify AI forecasts against regulatory requirements, and no critical cultural resistance would be created since precision is guaranteed due to XAI methods [40].

Cost-benefit analysis puts AI driven PM at multi-fold returns, with initial investment of 50K- 500K Euros, paid back to the company within a period of 12- 24 months, through 25-30 percent reduced maintenance costs, 70-75 percent downtime savings, and 20-40 percent life time extension of the equipment as shown in pharmaceutical cases such as Siemens generating annual savings of 5.5 million euros through AI forecasted turbine maintenance. The benefits of biomanufacturing are a 20-35 percent reduction in labor costs and 30 percent increase in OEE, with upfront costs of sensors and software recouped over time, and the millions of dollars lost due to batches replaced; small companies, e.g. achieving ROI within 1.5-2.5 years, by reducing defects by 80 percent as analyzed in U.S. manufacturing segments. One of the indirect benefits of Pfizer GenAI integration was the 16,000 hours saved in the search of data and a 55 percent saving in infrastructure costs. Scaling AI solutions must be stepwise: post-pilot, scale modularly to each facility with KPIs such as MTBF and OEE to define success, retraining models on new data to ensure accuracy and integrating with ERP that integrates inventory automatically. Scaling includes FAIR data governance, partnering with tech firms on strategically relevant technology (e.g., Amgen with AI leader NVIDIA in LLMs), and professional development to mainstream AI (e.g., J&J skill platforms) [41, 42]. Mid-sized operations may tap into cloud solutions that will help companies achieve cost-efficient scalability by cutting the downtime of internal systems by 15 percent, as in analogs of the printing industry, and large organizations such as Merck are minimizing product forecasting inaccuracy to 90 percent. Managing risks such as cybersecurity with regards to real-time governance will offer sustainable scaling and have an outlook of rising to the market of 16.49B by 2034, which will support resilience in operations due to increased demand of biologics [43].

24.8 Conclusion

AI-based predictive maintenance (PM) has played a significant role in revolutionizing biomanufacturing by changing the reactive and scheduled processes to proactive and data-intensive modes to make equipment reliable, product quality improved, and regulatory compliance maintained in production of biologics and biopharmaceuticals. AI detects hidden anomalies and forecasts failure on critical instruments such as bioreactors, centrifuges, pumps, and filtration systems by leveraging machine learning algorithms on live IoT-based sensor data, which minimizes unplanned downtime by up to 30 percent and eliminates expensive production shutdowns that might otherwise cost millions of dollars in a singular event. This revolutionary effect is not only limited to efficiency, it leads to a culture of continuous improvement where the AI tools can be used to improve the performance, increase equipment longevity and be used in line with the principles of Industry 4.0, and, in a larger sense, protect the integrity of contributing bioprocesses that can be subject to contamination or variability. To take but a few examples, in bioreactors, where failure of a batch run would lead to yield loss, AI is used to predict the wear on impellers by monitoring vibrational and temperature signatures, and in filtration systems, where the main difficulty is to produce a consistent yield of purification process, predictive models on membrane fouling are used. The way biomanufacturing leaders are focusing on incorporation of intelligent maintenance systems, the pathway becomes clear and the first step would be to audit the existing infrastructure to its fullest extent to draw down the areas that require more information acquisition and incompatibility with legacy equipment. The initial effort will be to pilot AI-driven PM by focusing on high-value assets e.g., which bioreactor line to choose and retrofit with sensors and implementing supervised learning models on past data to create baseline predictions. Cooperation is outlined, implementation of interdisciplinary teams of the engineers, data scientists, regulatory expert to standardize the data formats, assuring GMP compliance, and how the interpretability of the data is tackled with explainable AI model such as SHAP values. Scaling must be done incrementally after successful pilots and involve integrating both edge computing to handle real-time, and digital twins to model and simulate virtually, and cost-benefit analyses to justify investment, many times with ROI achieved in 12-24 months by lowering costs of maintaining allowed higher throughput. Training education should build capacity within the workforce, building confidence in AI forecasts through clear dashboards and frequent audits, and through cooperation with AI vendors, who can facilitate faster acceptance through bespoke solutions to the specific sterile production environments presented by biomanufacturing. Addressing such problematic areas as data quality concerns requires the incorporation of sound governance structures to reduce noise and complete the dataset, whereas integration with SCADA systems might require a middleware that would present the link between the legacies and more recent technologies. The future of operational resilience and innovation in biomanufacturing considers an objective where AI no longer predicts but rather anticipates and adjusts accommodating challenges, where possible, and eliminates the need to oversee the system in question as its self-healing capabilities enable high up time without human supervision. As new technologies, such as federated learning, allow secure, multi-site data sharing and generative AI adds more value to small datasets, PM will advance to prescriptive maintenance, and can suggest the best action to take in real-time. Such resiliency is going to play a critical role in global supply chain disruptions, and growing pressure on personalized biologics, continuing to ensure that facilities are agile and scalable. AI applications to sustainability, enhanced energy consumption with the help of AI in pumps and centrifuges, deployment in decreasing carbon impacts, and contribution to drug development improvements by correlating PM with process analytics to increase the pace of iterations will make innovation thrive. Finally, the construction of resilient biomanufacturing systems *via* AI-enabled PM places the industry as the vanguard of the technological revolution, in which excellence in operations can be combined with ethical principles such as data privacy and data equitability, leading to the paradigmatic shift towards more intelligent, more sustainable production to support the healthcare needs of the future with a reliability and ingenuity that are unquestionable.

AI disclosure statement: The author acknowledges the use of AI-based tools for minor language editing or grammatical correction. However, all intellectual contributions, analyses, and conclusions presented in this chapter are entirely the author's own.

References

1. Rojas, L., Peña, A., Garcia, J., AI-Driven Predictive Maintenance in Mining: A Systematic Literature Review on Fault Detection, Digital Twins, and Intelligent Asset Management. *Appl. Sci.*, 15, 3337, 2025.
2. Thakkar, D. and Kumar, R., AI-Driven Predictive Maintenance for Industrial Assets using Edge Computing and Machine Learning. *J. Res. Appl. Sci. Biotechnol.*, 3, 363, 2024.
3. Abdulrazzq, R.A., Sajid, N.M., Hasan, M.S., Artificial intelligence-driven predictive maintenance in IoT systems. *South Fla. J. Dev.*, 5, e4781, 2024.
4. Ejjami, R. and Boussalham, K., Industry 5.0 in Manufacturing: Enhancing Resilience and Responsibility through AI-Driven Predictive Maintenance, Quality Control, and Supply Chain Optimization. *Int. J. Multidiscip. Res.*, 6, 1, 2024.
5. Okwuchukwu Innocent Ani, Advanced manufacturing with machine learning: enhancing predictive maintenance, quality control, and process optimization. *Rafidain J. Eng. Sci.*, 2, 280, 2024.
6. Yan, H., Tan, J., Luo, Y., Wang, S., Wan, J., Multi-Condition Intelligent Fault Diagnosis Based on Tree-Structured Labels and Hierarchical Multi-Granularity Diagnostic Network. *Machines*, 12, 891, 2024.
7. Hector, I. and Panjanathan, R., Predictive maintenance in Industry 4.0: a survey of planning models and machine learning techniques. *PeerJ Comput. Sci.*, 10, e2016, 2024.
8. Lefrouni, K. and Taibi, S., Artificial Intelligence Techniques for Industrial Predictive Maintenance: A Systematic Review of Recent Advances. *J. Eng. Sci. Appl.*, 58, 653, 2025.
9. Vemuri, N., Thaneeru, N., Tatikonda, V.M., AI-Driven Predictive Maintenance in the Telecommunications Industry. *J. Sci. Technol.*, 3, 21, 2022.
10. Meyer, T., The Role of AI-Driven Predictive Maintenance in Reducing Downtime in American Manufacturing. *J. AI-Assisted Sci. Discov.*, 4, 1, 2024.

11. Ünlü, R. and Söylemez, I., AI-Driven Predictive Maintenance, in: *Springer Tracts in Nature-Inspired Computing: Engineering Applications of AI and Swarm Intelligence*, p. 207, Springer Nature, Singapore, 2024.
12. auram, K., AI-Powered Predictive Maintenance in Manufacturing: A Data-Driven Approach. *Int. Mach. Learn. J. Comput. Eng.*, 6, 1, 2023.
13. Guruswamy, T.B. and Renuka, A.L., Smart embedded systems for predictive maintenance in industrial IoT. *World J. Adv. Res. Rev.*, 7, 338, 2020.
14. Juneja, P.K. and Gupta, K., Application of Predictive Maintenance in Manufacturing with the Utilization of AI and IoT Tools. *Int. J. Adv. Res. Eng. Technol.*, 16, 301, 2025.
15. Koppichetti, R.K., Leveraging Machine Learning for Predictive Maintenance in Pharmaceutical Production. *Int. J. Innov. Res. Eng. Multidiscip. Phys. Sci.*, 13, 1, 2025.
16. Carvalho, JN., Silva, FR., Nascimento, E.G.S., Challenges of the Biopharmaceutical Industry in the Application of Prescriptive Maintenance in the Industry 4.0 Context: A Comprehensive Literature Review. *Sensors*, 24, 7163, 2024.
17. Aminzadeh, A., Karganroudi, S.S., Majidi, S., Dabompre, C., Azaiez, K., Mitride, C., Sénéchal, E., A Machine Learning Implementation to Predictive Maintenance and Monitoring of Industrial Compressors. *Sensors*, 25, 1006, 2025.
18. Vora, L.K., Gholap, A.D., Jetha, K., Thakur, R.R.S., Solanki, H.K., Chavda, V.P., Artificial Intelligence in Pharmaceutical Technology and Drug Delivery Design. *Pharmaceutics*, 15, 1916, 2023.
19. Parker, A., Deep Learning for Predictive Maintenance: Revolutionizing Industrial Equipment Monitoring. *J. AI-Assisted Sci. Discov.*, 3, 1, 2023.
20. Iyengar, S., Elevating Laboratory and Manufacturing Equipment Health With AI-Predicted Health Score. *Pharm. Outsourcing*, 25, 20, 2024.
21. Ucar, A., Karakose, M., Kırımça, N., Artificial Intelligence for Predictive Maintenance Applications: Key Components, Trustworthiness, and Future Trends. *Appl. Sci.*, 14, 898, 2024.
22. Kairo, J., Machine Learning Algorithms for Predictive Maintenance in Manufacturing. *J. Technol. Syst.*, 6, 66, 2024.
23. Gómez-Marín, G., The Role of AI-Driven Predictive Maintenance in Enhancing U.S. Pharmaceutical Manufacturing Operations. *J. AI-Assisted Sci. Discov.*, 4, 1, 2024.
24. Iftikhar, N., Lin, Y., Nordbjerg, F., Machine Learning based Predictive Maintenance in Manufacturing. *Proc. Int. Conf. Innov. Intell. Ind. Prod. Logist.*, p. 85, 2022.
25. Katta, S.R., Artificial Intelligence for Predictive Maintenance: Analyzing Work Order Trends across Global Pharmaceutical Sites. *Int. J. Lead. Res. Publ.*, 5, 1, 2024.

26. Ben-Bouazza, F., Manchadi, O., El Otmani Dehbi, Z., Rhalem, W., Ghazal, H., Machine Learning Based Predictive Maintenance of Pharmaceutical Industry Equipment, in: *Lect. Notes Netw. Syst.*, p. 497, Springer International Publishing AG, Switzerland, 2023.
27. Katta, S.R., Ensemble learning models for enhancing predictive maintenance in pharma work orders. *Int. J. Multidiscip. Res. Growth Eval.*, 6, 1814, 2025.
28. Rosca, C.-M. and Stancu, A., Integration of AI in Self-Powered IoT Sensor Systems. *Appl. Sci.*, 15, 7008, 2025.
29. Gupta, K. and Juneja, P.K., Predictive Maintenance Using AI and IoT in Manufacturing, TechRxiv, p. 1, 2025.
30. Abidi, M.H., Chen, M.K., Alkhalefah, H., Predictive maintenance planning for industry 4.0 using machine learning for sustainable manufacturing. *Sustainability*, 14, 3387, 2022.
31. Agarwal, P., Mehta, S., Abraham, A., A meta-heuristic density-based subspace clustering algorithm for high-dimensional data. *Soft Comput.*, 25, 10237, 2021.
32. Ahmad, W., Khan, S.A., Islam, M.M.M., Kim, J.M., A reliable technique for remaining useful life estimation of rolling element bearings using dynamic regression models. *Reliab. Eng. Syst. Saf.*, 184, 67, 2019.
33. Akhiat, Y., Chahhou, M., Zinedine, A., Ensemble feature selection algorithm. *Int. J. Intell. Syst. Appl.*, 11, 24, 2019.
34. Akhiat, Y., Manzali, Y., Chahhou, Y., Zinedine, A., A new noisy random forest based method for feature selection. *Cybern. Inf. Technol.*, 21, 10, 2021.
35. Alhawarat, M. and Hegazi, M., Revisiting K-Means and topic modeling, a comparison study to cluster arabic documents. *IEEE Access*, 6, 42740, 2018.
36. Ali, A., Faheem, Z.B., Waseem, M., Draz, U., Safdar, Z., Hussain, S., Yaseen, S., Systematic review: a state of art ML based clustering algorithms for data mining. *Proc. IEEE Int. Multitop. Conf.*, p. 1, 2020.
37. Amiu, O.O., Achieving a representation of asset data conducive to machine learning driven predictive maintenance, Ph.D. Thesis, 2019.
38. Ayesha, S., Hanif, M.K., Talib, R., Overview and comparative study of dimensionality reduction techniques for high dimensional data. *Inf. Fusion*, 59, 44, 2020.
39. Chen, J.C., Chen, T.L., Chen, J.L., Bosma, M., Mishra, G., Roberts, A., Barham, P., Chung, H.W., Sutton, C., Gehrmann, S., Palm: Scaling language modeling with pathways, arXiv, 2022.
40. Touvron, H., Lavril, T., Izacard, Y., Martinet, X., Lachaux, Y., Lacroix, T., Rozière, B., Goyal, N., Hambro, E., Azhar, F., Llama: Open and efficient foundation language models, arXiv, 2023.
41. Ghimpire, P., Kim, K., Acharya, M., Generative ai in the construction industry: Opportunities & challenges, arXiv, 2023.

42. Frederico, G., ChatGPT in Supply Chains: Initial Evidence of Applications and Potential Research Agenda. *Logistics*, 7, 26, 2023.
43. Wang, J., Zhang, L., Nagy, B., Novak, M., Nagy, Z.K., Network Meets ChatGPT: Intent Autonomous Management, Control and Operation. *J. Commun. Inf. Netw.*, 8, 239, 2023.

Part 6

QUALITY, REGULATION, AND RISK MANAGEMENT

25

Risk Management and Safety Engineering in Large-Scale Biomanufacturing

A.V. Raghavendra Rao[1], Kolluru Manaswini[2], Thalluri Lavanya[1], Mulla Murali Ganesh[1], Srimukhi Mekala[1], Chitti Babu Nalluri[2] and Sravani Sameera Vanjarana[1]*

[1]*Department of Chemical Engineering, B V Raju Institute of Technology, Narsapur, Telangana, India*

[2]*Department of Chemical Engineering, Andhra University College of Engineering, Visakhapatnam, Andhra Pradesh, India*

Abstract

Large-scale biomanufacturing involves complex processes involving life forms, hazardous chemicals, high-pressure equipment, and sterile plants which present significant risks to people/items and the environment. Throughout this chapter, the safety engineering and risk management frameworks along with the impending risk management processes dependent on the level of the large-scale biologics manufacturer will be thoroughly analyzed. It discusses common working hazards; environmental safety concerns and some engineering control measures put in place to mitigate them. The combination of process safety, levels of biosafety, engineering controls, and digital technologies of risk prediction used in the chapter are focused on the ability to meet the international regulation standards (e.g., OSHA, GMP, ISO, WHO). Pragmatic applications are implemented using the case studies of prior incidents and their mitigations. The chapter concludes by looking at the future of introducing smarter safety systems, enhanced surveillance with the help of AI and implementing strong biomanufacturing processes in relation to the sustainability of production and pandemic preparedness.

Keywords: Risk management, hazard analysis, occupational health, bio safety, biomanufacturing

**Corresponding author*: sravanisameera92@gmail.com

Sandip Kunar, Jagadeesha T. and D. Sathis Kumar (eds.) Biomanufacturing: Transforming Production of Biologics and Biopharmaceuticals, (803–838) © 2026 Scrivener Publishing LLC

25.1 Introduction

25.1.1 Scope of Risk and Safety Issues in Manufacturing of Biologics

The biologics manufacturing span of expertise lies in transcending biotechnology frontiers, industrial engineering, and compliance to regulations [1]. The risk and safety considerations are well beyond what it takes to safely house biological agents, since it involves all aspects of product research, development, scale-ups, and commercial production. This multidimensional line of scope-wise view may be largely interpreted through its implications on product integrity, process continuation, well-being of personnel and regulatory conformance [2].

On the aspect of quality of products, the manufacturing environments should be able to produce each of its unit with strict specifications that state its purity, potency, and stability measurements [3]. Discrepancies with temperature control, timeliness of the process, and quality of raw material may even ruin the whole production batch. Such sensitivity requires that the safety conscious be designed into the processes and equipment as well as the operational procedures [4].

As far as operational reliability is concerned, biologics manufacturing is a complex of interrelated upstream and downstream processes which need to be strictly synchronized [5]. The presence of any interruption be it mechanical, procedural, or logistical can translate to severe delays and expense. So, safety here is as much proportional to preventing unplanned downtime, as it is to safety (avoiding accidents) [6].

Personnel safety forms another crucial dimension. Large-scale biomanufacturing is based on well-trained operators, engineers, and quality control personnel operating within very regulated environments [7]. The provision of protection to them against process risks, ergonomic strain, and high-stress operational requirements is the only way to maintain the productivity of the workforce and morale [8].

Lastly, regulatory, and reputational reasons are also determining the whole risk environment. Adherence to Good Manufacturing Practice (GMP) does not merely fulfill a law but an essential element in market confidence. Any failure in quality systems, documentation integrity and audit preparedness can lead to removal of market sanction, fines or product recall [9].

When combined, the range of risk and safety issues in biologics production does not apply to a limited scope of operational risks but extends to an extended complex of safeguarding measures to safeguard products, individuals, procedures, and the reputation of the company in international markets [10].

25.1.2 Key Differences between Small-Scale and Large-Scale Bioprocessing

Bioprocessing may be carried out on much smaller or larger scales, in research laboratories (bench-scale experiments) or in commercial manufacturing facilities (in large plants) [11]. Although the basic biological guidelines are always the same called into play, the reality of small-scale vs large-scale operations has a broad impact on the scope, complexity, and strategic goals [12].

25.1.2.1 Purpose and Objectives

Small scale bioprocessing has mainly been applied on research and development and process optimization. It concentrates on investigating new cell lines and media formulations as well as imperfection of operating parameters [13]. Large scale bioprocessing, on the other hand, has the goal of maintaining a reproducible and high-volume manufacturing process of biologics to produce patient-ready or commercial products, and process reproducibility and compliance are the priority over experimental versatility [14].

25.1.2.2 Process Control and Monitoring

At a laboratory scale, a lot of flexibility may be given to the manipulation of the process parameters. The operators can also test several variables at the same time to determine best conditions [15]. However, in large scale, automation and use of advanced control systems are a must. In-line supervision, advanced feedback control, as well as strict process analytical technology (PAT) platforms are applied to sustain the uniformity between the batch runs [16].

25.1.2.3 Equipment and Infrastructure

Benchtop fermenters, shake flasks, and small chromatography columns, enabling fast set up and changeover of small-scale processes, are generally used in small-scale bioprocesses [17]. Hundreds to thousands of liters of stainless steel or disposable bioreactors are used in large production facilities together with industrial purification trains, the latter taking considerable infrastructure, utilities, and facilities design issues [18].

25.1.2.4 Risk Management Approach

The risk is usually easier to contain in small scale operations and no disasters are caused by failure of the process except local experiment [19]. On the grand scale, risks are even bigger: in case one of the batches goes wrong, the company will incur significant losses of materials and working time as well as face large-scale ramifications concerning regulations. This requires a stricter risk evaluation, validation, and contingency planning activity [20].

25.1.2.5 Cost and Resource Implications

Small operations are easy to operate and the process variables are easy to reiterate [21]. Large-scale production has a high capital outlay in building and equipping the plant and systems to ensure quality in production with an equally high-cost outlay in operation costs in terms of utilities, labor, and raw material. Such economics of large-scale manufacturing hence, requires a lot of efficiency and low waste [22].

25.1.2.6 Regulatory Environment

Whereas research-level activity is bound by institution-specific regulations and biosafety measures, large-scale production is required to be of high Good Manufacturing Practice (GMP) standards and should be inspected frequently and accompanied by the extensive documentation about each manufacturing step [23].

25.1.3 The Significance of Safety Engineering Integration in the Design of Process

Safety is a concept that cannot be applied as an attachment or as a series of responsive measures adopted at the end of the day at the time when the plant is already operational [24]. Rather, it must be incorporated into the process at the most initial stages of design— what is commonly called safety by design [25]. Such form of a proactive approach serves to predict, reduce or annul biological, mechanical, and chemical risks before they compromise the well-being of personnel, the integrity of products, and compliance to the environment [26].

Safety engineering is especially important in the case of manufacturing of biologics, since the processes are likely to contain [27]:

- Live cells or viral vectors which are a potential biohazard.
- Piping systems subject to mechanical failure and pressurized vessels.
- Severe purity demands of some products in which even slight product contamination may lead to losses of millions of dollars in lost batches.
- Advanced automation and control systems in which the failure of software or hardware might have a snowball effect.

Through an approach to safety engineering that applies principles in the development of a process, several strategic advantages are obtained by the manufacturer:

25.1.3.1 *Enhanced Compliance and Regulatory Approval*

The growing importance among regulatory bodies like U.S. Food and Drug Administration (FDA), the European Medicines Agency (EMA) and the World Health Organization (WHO) is to incorporate risk management plans within the original facility and process design. By showing in-built protection, the approval windows can speed up and create confidence among the regulators [28].

25.1.3.2 *Improved Process Reliability and Uptime*

Integration of safety in designs minimizes the occurrence of unanticipated breakdowns associated with events, contaminations, or faults on equipment. This leads to an increase in production capacity and dependable supplies of essential medical products [29].

25.1.3.3 *Lower Operational Costs over the Product Lifecycle*

Although it might appear that setting up or installing advanced structures of safety might be expensive in the short-term, in the future they save a lot of money in terms of low downtime, minimal extra costs incurred because of incidents and reduced losses due to spoiled batches [30].

25.1.3.4 *Protection of Personnel and Corporate Reputation*

Biomanufacturing depends on highly skilled operators, engineers, and scientists [31]. Safe working conditions not only ensure the safety of their physical health, but also develop an organizational culture, according to which, safety is regarded as highly as efficiency and innovation [31]. Moreover, the consequences of safety failure in the pharmaceutical manufacturing can lead to the irreparable damage of reputation finding its reflection in investors trust and positioning in the market [32].

25.1.3.5 *Sustainability and Environmental Stewardship*

Applications of safety engineering to process design are common with environmentally friendly concepts e.g., waste streams should be minimal, hazardous chemicals employed should be minimal and their treatment controlled. The practices are in tandem with the international pledge towards green production and corporate social responsibilities [33].

25.2 Understanding Risk in Biomanufacturing

Biomanufacturing plants lie at the junction of biology, engineering, and leading-edge process technology [34]. The complexity brings a variety of risk-types as shown in the Figure 25.1, which shall be discovered, assessed, and circumvented in the planning and functioning of processes. The relevant knowledge about these types of risk precedes competitive safety engineering practice [35].

25.2.1 Types of Risks

25.2.1.1 *Biological Risks*

Biological risk is associated with exposure to a living organism or its products which can lead to the development of an infection, allergy, or a toxic effect [36]. In biomanufacturing, these can include:

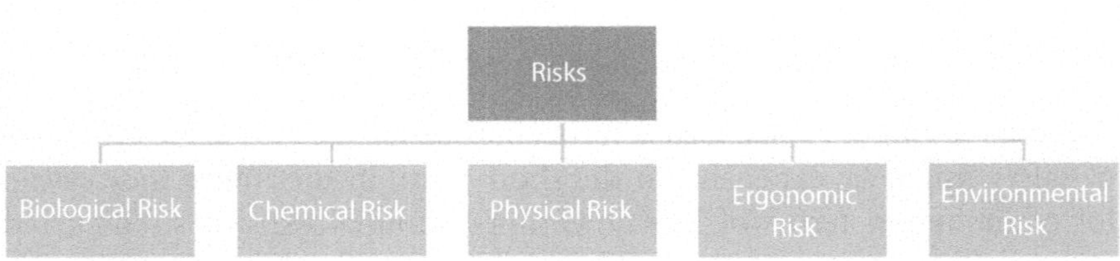

Figure 25.1 Types of risks.

- Pathogenic microorganisms used in research or production.
- Viral vectors employed in gene therapy manufacturing.
- Endotoxins and allergens from microbial cell debris.
- Cross-contamination between product lines or batches.

Biological risks: people that are involved in biological risks are conducted with biosafety level (BSL) classification, containment systems, and aseptic handling [37].

25.2.1.2 Chemical Risks

The chemical hazard is the result of using, storing, or disposing of hazardous substances during manufacturing. Examples include:

- Toxic solvents used in purification steps.
- CIP/SIP corrosive cleaning agents such as sodium hydroxide or acids.
- Gas liquids or flammable cell culture aeration/downstream.
- Reagents possessing reactive natures that may lead to fire, explosions, and/or reaction to any other unwanted products.

The control measures include the hazard communication programs, chemical compatibility studies and controlled storage [38].

25.2.1.3 Physical Risks

Physical risks refer to mechanical, thermal, electrical and/or radiation risks which could result in injury or damage to equipment. Common sources in biomanufacturing include:

- High-pressure vessels and piping systems susceptible to rupture.
- Rotating equipment such as centrifuges, mixers, and pumps.
- Cryogenic materials that can cause cold burns [39].

UV sterilization systems or ionizing radiation source during the control of contamination. They are reduced by means of protection of the equipment, pressure releases, and regular maintenance plans [40].

25.2.1.4 Ergonomic Risks

Ergonomic risks occur due to exploitation of the workstations, instruments, or tasks under conditions that do not cater to the physical abilities of human beings, causing musculoskeletal disorders or exhaustion [41]. Examples include:

- Manual lifting and movement of heavy media or reagent containers.
- Monotonous pipetting or sampling that put wrists and shoulders into strain.
- Inadequate workstation height causing poor posture.

Among the risk reduction measures are the ergonomic design of workstation, automation of repetitive activities and the training of the operator [42].

25.2.1.5 Environmental Risks

Environmental risks are those risks that are realized when the manufacturing process has adverse impacts on the surrounding ecosystem [43]. These risks include:

- Discharge of raw sewer audits with biological or chemical pollutants.
- Emission of volatile organic compounds (VOCs) from solvents.
- Improper disposal of biohazardous waste.
- Excessive energy consumption and carbon emissions.

Waste treatment plants, air emission controls, and sustainable process design are the means of dealing with environmental risks [44].

25.2.2 Key Risk Sources

In biomanufacturing, some operational processes and materials have a high level of risk attached to them by nature, the level of handling, and their effects on the surrounding environment [45]. These are the main sources of risk and thus find a way to integrate safety during the design process and routine operations [46].

25.2.2.1 *Live Organisms*

Biologics manufacturing constitutes predominantly of live organisms, which could be either microbial, mammalian, plant, or insect cell lines [47]. They may be genetically altered or infectious and should be dealt with under strict biosafety circumstances.

Risks:

- Potential for infection or allergenic reactions among personnel.
- Cross-contamination between different strains or batches.
- Accidental environmental discharge due to aerosol, spill, or discharge into waste [48].

Control Measures:

- Use of Biosafety Levels (BSL-1 to BSL-4) based on organism.
- Open systems using closed bioreactor equipment and high-efficiency particulate air (HEPA) filtration.
- Validation of sterilization and inactivation procedures before disposal [49].

25.2.2.2 *Media Handling*

Special hazards associated with culture media are the result of their composition as well as their preparation on large scale.

Risks:

- Contamination of prepared media when transferring, mixing, or storing the media.
- Ingestion of concentrated chemical supplements (e.g. antibiotics, growth factors) by mistake.
- Spillage and microbial proliferation in case of leaks [50].

Control Measures:

- Use of pre-sterilized, single-use media bags where possible.
- Robotic and convoy dispensing and mixing units to minimize manual work.
- Segregated media preparation areas with controlled airflow [51].

25.2.2.3 Pressurized Systems

Bioprocessing oftentimes is dependent upon vessels, piping and equipment operating under pressure- whether it is to aerate, sterilize or transfer fluids.

Risks:

- Over-pressurization leading to vessel rupture or line failure.
- A boiler explosion, such as an expulsive blast of steam, gases or liquids, causing burns or bodily injuries.
- Hazard with equipment such as vibration or noise with high-pressure pump and compressors.

Control Measures:

- Putting in place calibrated pressure relief valves and rupture discs.
- Routine pressure testing and preventive maintenance.
- The creation of real-time surveillance and warning alarms to detect aberrant pressurizing [52].

25.2.2.4 Aseptic Processes

Sterility of products: Aseptic methods are vital to keep the products sterile and to make the patients safe. These processes demand precision and strict adherence to protocols.

Risks:

- Contamination of operators or equipment or environment.
- Ingress of the particulate due to cleanroom HVAC systems failure.
- Incomplete sterilization of equipment or materials [53].

Control Measures:

- Validation and qualification of clean rooms (ISO Class 5-8).
- Operator training in gowning, gloving, and aseptic manipulations.
- Implementation of unidirectional airflow and restricted access protocols [54].

With these four top-priority sources of live organisms, media work, pressured systems, and aseptic processes, biomanufacturing facilities can put specific safety measures in place to ensure the facility continues to be safe, productive, and in compliance.

25.2.3 Risk Assessment Methodologies: Qualitative and Quantitative

Failure analysis is the skeleton of safety management in production of biologics. It allows systematically identifying, assessing, and prioritizing possible risks before they result to damage of people, products, or the surroundings. Risk assessment strategies generally fall into two broad categories; qualitative and quantitative, either of which is used in different ways in decision making and process control as shown in the Figure 25.2 [55].

25.2.3.1 Qualitative Risk Assessment

Qualitative methods are based on descriptive and estimate reviews and not on calculations. The approaches have been especially useful at the outset of process design when minimal data are available and where decisions need to be made quickly.

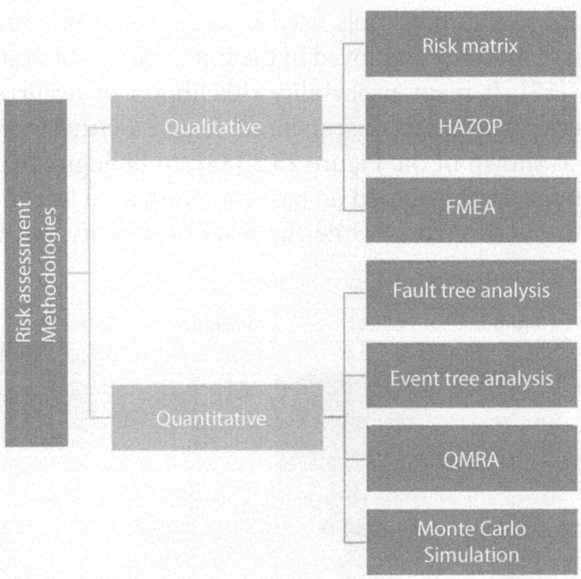

Figure 25.2 Risk assessment methodologies.

Key Characteristics:

- Uses expert judgment, historical data, and process knowledge.
- It categorizes risks into these identified levels; low, medium, and high according to severity and probability.
- Creates visual reporting such as risk matrices or heat maps [56].

Advantages:

- Quick and relatively low-cost to perform.
- Easy to communicate to multidisciplinary teams.
- Useful for preliminary screening of risks.

Limitations:

- Subjective; depends heavily on assessor experience.
- May fail to recognize with sufficient expertise minor yet outstandingly important risks.

Common Techniques:
(i) Risk Matrix
One of the most common tools used in the qualitative process is a risk matrix that is commonly employed in the initial phases of design processes of biologics [57]. It maps probability (likelihood of occurrence) with a range of severity (potential consequences) to give a visual indication of the level of risk as shown in the Figure 25.3 [58]. In biologics manufacturing, the valuation scale can be based on past contamination levels, device leaks, or operator-induced errors, while the level of severity can be evaluated

	Negligible	Minor	Moderate	Significant	Severe
Very likely	Low Medium	Medium	Medium High	High	High
Likely	Low	Low Medium	Medium	Medium High	High
Possible	Low	Low Medium	Medium	Medium High	Medium High
Unlikely	Low	Low Medium	Low Medium	Medium	Medium High
Very unlikely	Low	Low	Low Medium	Medium	Medium

Figure 25.3 Risk matrix.

potentially based on a loss to the product, patient safety impact or the regulatory compliance [59]. The resultant grid is generally color coded where green is low risk, yellow is medium and red is high, thus it allows the teams to immediately see the areas where they need to implement control measures [60]. The risk matrix provides an effective way of communication among the engineering, quality, and management teams since it is easily read and understood, helping all parties see eye to eye on where to focus the risk mitigation efforts [61].

(ii) Hazard and Operability Study (HAZOP)

Hazard and Operability Study (HAZOP) is a logical structured procedure employed to define the possible risks and operating problems by making an examination of out-of-condition deviations of the targeted processes. The biologics manufacturing process requires a multidisciplinary HAZOP team to assess each of the steps in the process (e.g., cell culture, fermentation, or purification), to various parameters including temperature, pH, flow rate and pressure [62].

By engaging guide words like more, less, as well as or other than, the team is determining what might go amiss; what may be the cause and outcomes are foreseen. To illustrate, a temperature anomaly in fermentation may result in decreased cell viability, weakened product, or high level of contamination [63].

HAZOP is especially useful when applied to the biologics manufacturing facilities since it not only identifies risk, but also highlights operability issues- issues that, whether there is a safety risk or not, can damage the efficiency of the process or the quality of the product [64].

(iii) Failure Modes and Effects Analysis (FMEA)

Failure Modes and Effects Analysis (FMEA) is an interactive step-by-step process which analyses the process to determine the where and how the process may fail, and evaluates the relative consequences of those failures [65]. In the manufacturing of biologics, failure modes can comprise leaking of bioreactors, even breakage of the filtration membranes, the faultiness of sensors, or contamination incidents.

In every possible failure mode, the FMEA has three ratings:

- Severity (S): How serious the consequences are.
- Occurrence (O): The frequency at which the failure may most likely occur.
- Detection (D): The ease at which the failure can be detected prior to affecting the product.

The product of these ratings is the Risk Priority Number (RPN) that assists in the prioritization of corrective actions. As an example, a large RPN on aseptic filling due to contamination during filling may result in investments to increase the monitoring in the environment and train the operators [66].

25.2.3.2 Quantitative Risk Assessment

Quantitative approaches take the form of methods that estimate the scale of the risks using numerical values and utilizing probability-based models. They are more data-intensive and more appropriate to complex or where high stakes are involved processes in the manufacture of biologics, e.g. large-scale vaccine manufacture or the manufacture of gene therapy vectors [67].

Key Characteristics:

- Utilizes measured data, statistical models, and probability distributions.
- States risk as quantifiable unit (e.g., probability of failure within one year, the expected financial loss or dose-responses).
- Makes sensitivity analysis possible to locate the most important aspects of risk [68].

Advantages:

- Provides precise, objective, and reproducible results.
- Facilitates cost–benefit analysis of safety interventions.
- Supports regulatory submissions with numerical evidence [69].

Limitations:

- Requires substantial data and computational resources.
- More time-consuming and expensive to conduct.
- Might have the illusion of certainty when it is on low-quality data [70].

Common Techniques:

I. Fault Tree Analysis (FTA)

Fault Tree Analysis is a top-down deductive modeling technique which represents the logic paths of events to a particular undesired event (the "top event").

- Purpose: Identify root causes and system weaknesses.
- Technique: It tries to capture between the relationship between basic events and the top event using Boolean logic (that is, AND/OR gates) [71].

Applications in Biomanufacturing:

- Identifying contamination sources in aseptic filling operations.
- Pinpointing causes of equipment failure during large-scale fermentation.

II. Event Tree Analysis (ETA)

Event Tree Analysis is a forward-looking approach that is inductive since it uses the response of the system to make predictions about different consequences that could arise because of a particular initiating event.

Purpose: Find out all conceivable consequences to one initiating event.

Process: Creating a branch diagram of corresponding sequences of success or failure of each safety barrier [72].

Applications in Biomanufacturing:

- Assessing the outcomes of a cleanroom HVAC failure.
- Predicting upstream sterilization loss modeling.

III. Quantitative Microbial Risk Assessment (QMRA)

QMRA offers a numeric model to gauge the chances of infection or other negative impacts due to exposure to microbes. Purpose: Evaluate microbial hazards in a measurable way. Methodology:– It includes four major processes namely: identification of hazard, exposure, dose-response modeling, and risk characterization [73].

Applications in Biomanufacturing:

- Evaluating risk of cross-contamination between batches.
- Approximation of risk of pathogens exposure to workers dealing with live cultures.

IV. Monte Carlo Simulation
Monte Carlo Simulation is a method of probabilistic modeling that is implemented using repetitive random sampling used to determine probable variability and unpredictability of complex systems. Purpose: Elaborate randomness in process parameter and risk factors. Methodology: Conducts thousands of trials to estimate the probability density of outcomes [74].

Applications in Biomanufacturing:

- Predicting yield fluctuations under variable raw material quality.
- Estimating the probability of failure of critical control points of downstream processing.

25.3 Safety Engineering Principles

Safety engineering is an essential component in preventing risk in biomanufacturing plants, and they guarantee that the staff, products, and the surrounding are safe during the production process [75]. The principles are incorporated in the design of processes, selection of equipment, layout of facilities and operating procedures.

25.3.1 Hierarchy of Hazard Control

The Hierarchy of Hazard Control is a method of reducing risks ranking the most effective method of any problem to the not-so-effective ones.

i. Elimination- Get rid of the hazard altogether (e.g., re-place toxic solvents with non-hazardous).
ii. Substitution -Use something safer in the place of a hazardous material or procedure (i.e., use attenuated microbial strains rather than pathogenic ones).
iii. Engineering Controls -These are solutions which are designed to physically isolate individuals to hazards (e.g., biosafety cabinets, isolators).

iv. Administrative controls - formulate and implement strategies like exposure reducing policies, training and procedures through the development and implementing them (e.g., limited access to high-risk areas).
v. Personal Protective Equipment (PPE) -Put protective clothes, gloves, masks or respirators as a final experiment [76].

25.3.2 Engineering Design for Safety

a. Pressure Relief Systems
Purpose: Avoid disastrous equipment damage sustained under overpressure in bioreactors, and storage and steam systems [77].

Design Elements: Automated venting systems, pressure relief valves and rupture disks. Example: Steam sterilization lines with spring-loaded relief valves originated to prevent over pressure of the SIP cycles [78].

b. Containment and Isolation
Hazardous biological materials are worthy of containment methods to stop material releases.

Primary Containment: Direct physical barriers like closed bioreactors, biosafety cabinets, and enclosed pipelines. Secondary Containment Facility level containments such as airlocks, HEPA filtering and sealed flooring which keep things inside the area of production [79].

c. HVAC and Cleanroom Design
Goal: Keep up to the necessary levels of cleanliness and avoid cross-contamination.

Key Features:

- HEPA filtration for air supply and exhaust.
- Pressure differences so that there is airflow movement between clean and less clean environments.
- Product control of temperature and humidity to maintain stability and manage microbes [80].

Examples: Positive pressure cleanrooms, Grade B, used to shelter aseptic activity; negative pressure biosafety laboratories used to contain pathogens.

d. Utility and Equipment Safety
Biomanufacturing depends on various utilities, including steam, water-for-injection (WFI), compressed gases, and CIP/SIP system, the utilities that have to be considered in terms of actual safety.

- Steam Lines: Burn protection by insulating, pressure control and traps between condensate to eliminate water hammer.
- CIP/SIP Systems - programmed programs that have interlocks to stop injury such as inadvertent exposure to hot steam or cleaners.
- Compressed Gas Systems: Regulators, pressure gauges and color-coded pipelines to avoid misuse [81].

25.4 Occupational Safety Management

Biomanufacturing occupational safety management can promote the health and safety of biomanufacturing workers exposed to such factors as biological agents, chemicals, pressure systems, and delicate production operations. It entails unceasing observation, active hazard management and safety consciousness culture [82].

25.4.1 Worker Exposure Limits and Monitoring

Exposure Limits: Employment based on exposure standards like OSHA Permissible Exposure Limits (PELs), ACGIH Threshold Limit Values (TLVs) and biological exposure indexes (BEIs). Biological Agents: Restrictions tiered in relation to biosafety levels (BSL) and pathogen-specific risk determined. Chemical Hazards: Multi period analysis through air sampling of cleaning solvents, cleaning agents and toxic air gases [83].

Monitoring Tools:

- Personal sampling devices
- Area air quality monitors
- Real-time gas detectors
- Microbial air samplers in cleanrooms [84].

25.4.2 PPE and Safety Practice in Cleanrooms and Wet Labs

PPE in cleanrooms: sterile coveralls, gloves, hood, masks, goggles, and exclusive footwear to uphold product sterility and save workers.

PPE in Wet Labs: Lab coats, chemical protective gloves, splash goggles/ face shields and occasionally respirators.

Donning/Doffing Protocols:

- Stepwise gowning procedures in airlocks to minimize particle introduction.
- Proper disposal or sterilization of PPE after use.

Prevention of cross-contamination: zoned PPE color, specific tools, and limited zones of movement [85].

25.4.3 Training Programs and Behavior-Based Safety

Technical Training: This includes the biosafety practices, chemistry handling, emergency shutdown practices, and safe procedures to handle the equipment [86]. Behavior-Based Safety (BBS): This practice promotes positive group actions concerning safety through observation, coaching, and rewarding of safety behaviors of the workplace [87]. Refresher Training: Periodic updates to update on incident trends and changes in regulation and on introduction of new equipment. Spill, containment breach and fire emergency Simulations [88].

25.4.4 Incident Reporting and Root Cause Analysis

Incident Reporting:

- Immediate reporting of unsafe acts, hazardous conditions, and near-misses.
- Anonymous reporting options to encourage openness [89].

Root Cause Analysis (RCA) [90]:

- Why-Why Analysis to trace contributing factors.
- Fishbone Diagram to group factors (people, process, equipment, environment, materials).
- Corrective and Preventive Actions (CAPA) to eliminate recurrence.
- Learning Culture: Posting the lessons learned in each department in order to enhance safety awareness throughout the organization.

25.5 Environmental Safety and Emission Control

Environmental safety in biomanufacturing is an area of concern that is associated with reducing the potential of the harmful biological, chemical, and physical agents into the surrounding environment: air, water, and soil. This makes sure that regulatory standards are followed, minimizes ecological disruption, and helps to achieve sustainable operations [91].

25.5.1 Liquid and Gaseous Waste Management

Liquid waste produced as fermentation plantation broths, cleaning (CIP/SIP) water, cooling water and laboratory activities must be sterilized (e.g., by autoclaving, heating) prior to discharge to destroy biological hazards. Separation of chemical and biological wastes to prevent reactive combinations [92].

Gaseous waste exhausts in bioreactors, cleanrooms, and HVAC systems by HEPA filters or beds of activated carbon. Scrubbers and catalysts to neutralize THCs and acid gases [93].

25.5.2 Bioburden and Nutrient Discharge Risks

Bioburden/Killing of organisms: Existence of active microorganisms in waste water; may disturb the aquatic ecosystem. Nutrient Discharge: Excessive levels of nitrate (NO_3^-), and phosphate (PO_4^{3-}) enhance algal growth (eutrophication) [94].

Risk Mitigation:

- Aerobic/anaerobic digestion (biological treatment) to minimize organic load.
- Chemical precipitation for phosphate removal.
- Denitrification processes for nitrate reduction.

25.5.3 Effluent Treatment and Regulatory Thresholds

Key Parameters Monitored [95]:

- BOD (Biological Oxygen Demand): ≤ 50 mg/l treated industrial waste water.
- COD (Chemical Oxygen Demand): ≤ 250 mg/l.
- Total Suspended Solids (TSS): ≤ 50 mg/l.

- Total Nitrogen ≤10 mg/l.
- Total Phosphorus ≤ 2 mg/l.

Treatment Steps [96]:

- Primary: Screening, sedimentation.
- Secondary: Biological oxidation (activated sludge, trickling filters).
- Tertiary: Filtration, disinfection, nutrient removal.

25.5.4 Use of Green Engineering and Process Optimization

Process Optimization: Reduce the amount of waste by implementing lean manufacturing and yield improvement.

Green Chemistry Principles [97]:

- Use of less hazardous solvents and reagents.
- Recovery and reuse of process water.
- Heat recovery from steam systems.

25.6 Regulatory Frameworks and Standards

An effective regulatory regime guarantees that biomanufacturing activities are undertaken in accordance with international regulations regarding safety of workers, environmental issues, and the quality of the product [98]. These aspects are governed by several national and even international regulations and guidelines:

25.6.1 Overview of Key Regulations

25.6.1.1 Occupational Safety and Health Administration (OSHA)

In the United States, the safety and health standards in the workplace are determined by OSHA. During its duration, hazard communication, process safety management, the standards of personal protective equipment (PPE), and safe handling of hazardous materials in manufacturing facilities are being discussed [99].

25.6.1.2 ISO 45001 (Occupational Health and Safety Management Systems)

This international standard gives an organized way of handling occupational health and safety risks thus making efforts active towards preventing workplace injuries and illnesses. ISO 45001 is present in most industries, and biomanufacturing is one of them since it helps to prove their compliance to the safety of the employees [100].

25.6.1.3 NFPA (National Fire Protection Association)

NFPA builds and establishes fire safety codes, standards, including the laboratory NFPA 45 standards and electrical safety NFPA 70. They are important when handling fire and explosion risks at such facilities handling flammable chemicals, solvents, and gases [101].

25.6.1.4 Environmental Protection Agency (EPA)

The U.S. EPA controls the waste disposal, emissions, quality of water, and the environmental quality that the manufacturing activities may impose. Air emissions and air emissions permits, hazardous waste disposal and effluent discharge limits are part of compliance.

- Global EHS Compliance Expectations
- WHO Good Manufacturing Practices (GMP)

WHO GMPs are simply a set of guidelines that play the role of ensuring that pharmaceutical and biological products are produced and controlled constantly regarding quality, strict demands on hygiene, environmental check-ups, and personnel safety [101].

25.6.1.5 European Medicines Agency (EMA) and US Food and Drug Administration (FDA)

Both the agencies require the adherence to the GMP and, they regulate the Environmental, Health, and Safety (EHS) regulations in the pharmaceutical manufacture and biologics. Their functions revolve around facility design, equipment validation, control of contamination, and preparedness regarding the response to emergencies [102, 103].

25.6.2 Integration of EHS into Quality Management Systems (QMS)

In current biomanufacturing, the concept of Environmental, Health and Safety (EHS) is included into the Quality Management System (QMS) that forms the single strategy to work with safety, regulation, and product quality.

Key elements of integration include [104]:

- Policy Alignment in the company quality policy include EHS goals.
- Risk-Based Management: the same risk assessment strategy is applied in managing the hazards as well as the quality risks.
- SOP Integration: Operational activities have safety incorporated into them.
- Training: Cross-disciplinary quality and EHS compliance training is given to staff.
- Auditing: According to the internal audits with the combination of quality audit and EHS audits, CAPA systems used are joint.
- Data Integration: Data QMS reporting systems record EHS performance.
- Change Control: Quality and safety effects of all changes to the processes are examined.

25.6.2.1 Benefits of Integration

- Streamlines compliance efforts.
- Strengthens safety culture without compromising product quality.
- Enhances regulatory readiness and operational efficiency.

25.7 Digital Safety Tools and Automation

Changes in Environmental, Health, and Safety (EHS) management are being influenced using technologies of Industry 4.0 in biomanufacturing [105]. Digital solutions help monitor in real-time, analyze risks to make predictions, and respond automatically to events, making safety, compliance, and continuity postures much safer and smoother [106].

25.7.1 The Potential Role of AI, IoT and Predictive Analytics in Real-Time Risk Monitoring

25.7.1.1 Artificial Intelligence (AI)

It scans high volumes of operational data to identify trends and anomalies occurring, which can signify safety hazards, including the temperature variations, pressurization, or unusual equipment vibrations [107].

25.7.1.2 Internet of Things (IoT)

It interrelates sensors, machines, and control systems to present a live data feed of environmental conditions, worker safety criteria as well as process integrity [108].

25.7.1.3 Predictive Analytics

It predicts probable failures, accidents and harmful chemical releases or contamination in advance using historical data and machine learning algorithms to pre-empt action [109].

25.7.2 Digital Twins and Process Simulations for Hazard Forecasting

25.7.2.1 Digital Twins

Virtual models of biomanufacturing systems which reflect near to real time data about the system, enabling them to simulate various things that have yet to happen [110].

25.7.2.2 Hazard Forecasting

Allows to predictively evaluate the effect of equipment failures, process/raw material variances, etc. on safety/compliance [111].

Training Applications: Offers secure virtual training room to teach operators on emergency responses and troubleshooting of the system [112].
Smart Sensors for Equipment Failure, Contamination, and Gas Leaks
Equipment Health Monitoring: Pressure, temperature, and vibration sensors detect when mechanical wear or a break is on the way earlier [113].

Contamination Detection: Cleanrooms and production lines cleanliness is measured by optical and biosensors, which watch out for microbial or particulate contamination [114].

Gas Leak Detection: Portable and stationary smart detectors detect dangerous gases (e.g., ammonia, chlorine, CO_2) with immediate notification to controller and handheld devices [115].

Benefits of Digital Safety Tools [116]:

- Faster detection and response to hazards.
- Reduction in unplanned downtime.
- Improved compliance with EHS regulations.
- Data-driven decision-making for continuous safety improvement.

25.8 Future Directions

With the increasing focus on biomanufacturing that is more efficient, flexible, and sustainable as it becomes a reality, parallel changes must be made in safety systems. In the future, the emergent risks and regulatory expectations will involve the use of automation, modularity, resilience, and sustainability as the EHS strategies.

Autonomous Safety Systems and Robotic Interventions
Fully Automated Hazard Response:
Anomaly identification systems using artificial intelligence that automatically shut down equipment or process parameter adjustment or execution of emergency procedures without human input [117].

Robotic Maintenance and Inspection:
Use of drones and robotic arms to survey hard-to-reach areas, risk-prone regions, or biological hazard containment units–lessening the exposure of people to hazardous environments [118].

Self-Learning Safety Platforms:
Machine learning algorithms that constantly improve the hazard detection algorithms with the help of past incident records and process conditions [119].

Biomanufacturing Safety in Mobile and Modular Facilities
Compact, Deployable Units: Container and mobile production facilities which are quick to set up in places with either limited resources or distance [120].

Integrated Safety-by-Design: Modular units which include in-built EHS controls- like enclosed process environments, automated decontamination systems and sophisticated ventilation; obviating the need to depend on external infrastructure [121].

Standardized Safety Protocols: Standardized, tried, and tested safety routines that can be approved and ready to be applied in a short time [122].

Resilient Safety Designs for Pandemic-Ready Production Lines

Flexible Production Platforms: Plants that can switch quickly among diverse bioproduct types (e.g. vaccines, therapeutics) with biosecurity (i.e., biosafety) and good manufacturing practice (GMP) levels [123].

Redundant Safety Systems: Lighting and batteries to provide backup power, filtration, and containment systems to guarantee constant functioning in periods of emergency [124].

Remote Monitoring and Control: Cloud-based dashboards enabling EHS managers to monitor safety metrics and safety incidents by any means and whatever location they are in [125].

Embedding Sustainability into Safety Planning

Eco-Friendly Safety Materials: Reduce the impact on the environment by using biodegradable PPE, solvent recovery mechanisms, and using low-toxicity cleaning agents [126].

Energy-Efficient Safety Systems: Minimize energy usage lighting, HVAC and ventilation controls that ensure due safety in the room [127].

Circular Safety Strategies: Life-cycle thinking by designing safety in such a way that reuse, recycling, and consumption of fewer resources should be used across all safety-related devices and consumables [128].

25.9 Conclusion

This report describes the new trend in biomanufacturing safety where regulatory, Environmental Health and Safety (EHS), and advanced digital technologies are interconnected. It examines some of the major compliance standards as OSHA, ISO 45001, NFPA, EPA, WHO GMP, and EMA/FDA and how they work together in Quality Management Systems (QMS) resulting in operational efficiency and regulatory preparedness. Real use cases of AI, IoT, predictive analytics digital twins as well as intelligent sensors to control contamination, monitor equipment, and predict hazards can be taken up by leaders in the fields in the form of case studies. Future directions also discussed in the analysis are autonomous safety systems, modular and mobile production units, pandemic ready facility designs, and sustainable safety strategies. Overall, these innovations are indicative of a more proactive technology-aided safety ecosystem that can protect people, environment, and increase biomanufacturing resiliency.

AI disclosure statement: The author acknowledges the use of AI-based tools for minor language editing or grammatical correction. However, all intellectual contributions, analyses, and conclusions presented in this chapter are entirely the author's own.

References

1. Gregory, F.D. and Trump, B.D., The evolving landscape of biotechnology research: Challenges and opportunities in acquisitions, funding, education, and workforce development for a productive bioeconomy, in: *Advanced Study Institute on NATO ASI on Quantum Nano-Photonics*, pp. 25–50, 2023.
2. Kozlowski, S. and Swann, P., Current and future issues in the manufacturing and development of monoclonal antibodies. *Adv. Drug Delivery Rev.*, 58, 707, 2006.
3. World Health Organization. *Quality assurance of pharmaceuticals: A compendium of guidelines and related materials. Good manufacturing practices and inspection*. Geneva, Switzerland: World Health Organization, 2007.
4. Sutton, I., *Process risk and reliability management*, Amsterdam, Netherlands: Gulf Professional Publishing (Elsevier), 2014.
5. Fisher, A.C., Kamga, M.H., Agarabi, C., Brorson, K., Lee, S.L., Yoon, S., The current scientific and regulatory landscape in advancing integrated continuous biopharmaceutical manufacturing. *Trends Biotechnol.*, 37, 253, 2019.
6. Reason, J., *Managing the risks of organizational accidents*, Milton Park, Abingdon-on-Thames, Oxfordshire, England, UK: Routledge, 2016.
7. Haider, S.I. and Ashtok, A., *Biotechnology: A comprehensive training guide for the biotechnology industry*, CRC Press, Boca Raton, FL, USA, 2009.
8. Israel, B.A., Baker, E.A., Goldenhar, L.M., Heaney, C.A., Occupational stress, safety, and health: Conceptual framework and principles for effective prevention interventions. *J. Occup. Health Psychol.*, 1, 261, 1996.
9. Kotsanopoulos, K.V. and Arvanitoyannis, I.S., The role of auditing, food safety, and food quality standards in the food industry: A review. *Compr. Rev. Food Sci. Food Saf.*, 16, 760, 2017.
10. Samara, C., Garcia, A., Henry, C., Vallotton, L., Cariolato, L., Desmeules, J., Pinçon, A., Safety surveillance during drug development: Comparative evaluation of existing regulations. *Adv. Ther.*, 40, 2147, 2023.
11. Neubauer, P., Cruz, N., Glauche, F., Junne, S., Knepper, A., Raven, M., Consistent development of bioprocesses from microliter cultures to the industrial scale. *Eng. Life Sci.*, 13, 224, 2013.

12. Choudhary, V., Strategies for biotech companies to scale up commercially through effective project and risk management. *BULLET J. Multidisiplin Ilmu*, 1, 952, 2022.
13. Mandenius, C.F. and Brundin, A., Bioprocess optimization using design-of-experiments methodology. *Biotechnol. Progr.*, 24, 1191, 2008.
14. Gomez-Marquez, J. and Hamad-Schifferli, K., Distributed biological foundries for global health. *Adv. Healthcare Mater.*, 8, 1900184, 2019.
15. Liptak, B.G., Piovoso, M.J., Shinskey, F.G., Eren, H., Totherow, G.K., Jamison, J.E., Morgan, D., Hertanu, H.I., Marszal, E.M., Berge, J., Mannan, M.S., *Instrument engineers' handbook, volume two: Process control and optimization*, CRC Press, Boca Raton, FL, USA, 2018.
16. Ündey, C., Ertunç, S., Mistretta, T., Looze, B., Applied advanced process analytics in biopharmaceutical manufacturing: Challenges and prospects in real-time monitoring and control. *J. Process Control*, 20, 1009, 2010.
17. Whitford, W.G., Bioprocess intensification: Technologies and goals, in: *Process control, intensification, and digitalisation in continuous biomanufacturing*, pp. 93–136, Weinheim, Germany: Wiley-VCH (John Wiley & Sons), 2022.
18. Meyer, H.P., Minas, W., Schmidhalter, D., Industrial-scale fermentation, in: *Industrial biotechnology: Products and processes*, pp. 1–53, Weinheim, Germany: Wiley-VCH (John Wiley & Sons P&T), 2017.
19. Gad-el-Hak, M. (Ed.), *Large-scale disasters: Prediction, control, and mitigation*, Cambridge University Press, Cambridge, UK & New York, NY, USA, 2008.
20. Grabowski, M. and Roberts, K., Risk mitigation in large-scale systems: Lessons from high reliability organizations. *Calif. Manage. Rev.*, 39, 152, 1997.
21. Mole, K.F., Ghobadian, A., O'Regan, N., Liu, J., The use and deployment of soft process technologies within UK manufacturing SMEs: An empirical assessment using logit models. *J. Small Bus. Manage.*, 42, 303, 2004.
22. Wheelwright, S.M., Economic and cost factors of bioprocess engineering, in: *Biotechnology and Biopharmaceutical Manufacturing, Processing, and Preservation*, pp. 333–354, CRC Press, Boca Raton, FL, USA, 2020.
23. Féron, E. and Crowley, J., From research policy to the governance of research? A theoretical framework and some empirical conclusions. *Innovation*, 16, 369, 2003.
24. Pintelon, L. and Muchiri, P.N., Safety and maintenance, in: *Handbook of maintenance management and engineering*, pp. 613–648, Springer London, London, England, UK, 2009.
25. Fadier, E. and De la Garza, C., Safety design: Towards a new philosophy. *Saf. Sci.*, 44, 55, 2006.
26. Venkatesh, M., Scott, J., Chakravarty, T.C., Rehm, T.E., Basu, S., Environment, health and safety, security and regulations, in: *Water Management in*

Petroleum Industries, pp. 463–525, Springer Nature Singapore, Singapore, 2025.
27. Sampathkumar, K. and Kerwin, B.A., Roadmap for drug product development and manufacturing of biologics. *J. Pharm. Sci.*, 113, 314, 2024.
28. Lis, Y., Roberts, M.H., Kamble, S., Guo, J.J., Raisch, D.W., Comparisons of Food and Drug Administration and European Medicines Agency risk management implementation for recent pharmaceutical approvals: Report of the International Society for Pharmacoeconomics and outcomes research risk benefit management working group. *Value Health*, 15, 1108, 2012.
29. Speier, C., Whipple, J.M., Closs, D.J., Voss, M.D., Global supply chain design considerations: Mitigating product safety and security risks. *J. Oper. Manage.*, 29, 721, 2011.
30. Levitt, J., *The handbook of maintenance management*, Industrial Press Inc., South Norwalk, CT, USA, 2009.
31. Clomburg, J.M., Crumbley, A.M., Gonzalez, R., Industrial biomanufacturing: The future of chemical production. *Science*, 355, aag0804, 2017.
32. Rajagopal, V., Shanmugam, P.V., Nandre, R., Quantifying reputation risk using a fuzzy cognitive map: A case of a pharmaceutical supply chain. *J. Adv. Manage. Res.*, 19, 78, 2022.
33. Chaaban, M.A., Hazardous waste source reduction in materials and processing technologies. *J. Mater. Process. Technol.*, 119, 336, 2001.
34. Sydow, J., Windeler, A., Schubert, C., Möllering, G., Organizing R&D consortia for path creation and extension: The case of semiconductor manufacturing technologies. *Organ. Stud.*, 33, 907, 2012.
35. Bensaada, I. and Taghezout, N., An enterprise risk management system for SMEs: Innovative design paradigm and risk representation model. *Small Enterp. Res.*, 26, 179, 2019.
36. Dutkiewicz, J., Cisak, E., Sroka, J., Wójcik-Fatla, A., Zajac, V., Biological agents as occupational hazards-selected issues. *Ann. Agric. Environ. Med.*, 18, 286, 2011.
37. Janghorban, M., Kazemi, S., Tormon, R., Ngaju, P., Pandey, R., Methods and analysis of biological contaminants in the biomanufacturing industry. *Chemosensors*, 11, 298, 2023.
38. Amoabediny, G.H., Naderi, A., Malakootikhah, J., Koohi, M.K., Mortazavi, S.A., Naderi, M., Rashedi, H., Guidelines for safe handling, use and disposal of nanoparticles. *J. Phys. Conf. Ser.*, 170, 012037, 2009.
39. Gordon, L.B. and Martinez, T.R., A complete electrical risk assessment method for all electrical hazards and its application, in: *2018 IEEE IAS Electrical Safety Workshop (ESW)*, IEEE, pp. 1–8, 2018.
40. Lambert, P.A., Radiation sterilization, in: *Principles and Practice of Disinfection, Preservation and Sterilization*, p. 294, Wiley-Blackwell, Chichester, West Sussex, England, UK, 2013.
41. Odebiyi, D.O. and Okafor, U.A., Musculoskeletal disorders, workplace ergonomics and injury prevention, in: *Ergonomics-New Insights*, IntechOpen, London, England, UK, 2023.

42. Haile, E.L., Taye, B., Hussen, F., Ergonomic workstations and work-related musculoskeletal disorders in the clinical laboratory. *Lab. Med.*, 43, e11, 2012.
43. Klaine, S.J., Koelmans, A.A., Horne, N., Carley, S., Handy, R.D., Kapustka, L., Nowack, B., Von Der Kammer, F., Paradigms to assess the environmental impact of manufactured nanomaterials. *Environ. Toxicol. Chem.*, 31, 3, 2012.
44. Alyasi, H.S. and Isaifan, R., A review on pollution emissions and impact from waste treatment and disposal facilities. *J. Environ. Toxicol. Stud.*, 2, 1, 2018.
45. Ho, S.V., McLaughlin, J.M., Cue, B.W., Dunn, P.J., Environmental considerations in biologics manufacturing. *Green Chem.*, 12, 755, 2010.
46. Rasmussen, J., Risk management, adaptation, and design for safety, in: *Future Risks and Risk Management*, pp. 1–36, Springer Netherlands, Dordrecht, Netherlands, 1994.
47. Dumont, J., Euwart, D., Mei, B., Estes, S., Kshirsagar, R., Human cell lines for biopharmaceutical manufacturing: History, status, and future perspectives. *Crit. Rev. Biotechnol.*, 36, 1110, 2016.
48. Barkley, W.E. and Mickelson, C.A., Laboratory safety, in: *Methods for General and Molecular Microbiology*, pp. 995–1018, Washington, D.C., USA: ASM Press, 2007.
49. Marks, D.M., Biotechnology facilities, in: *Good Design Practices for GMP Pharmaceutical Facilities*, pp. 327–370, CRC Press, Boca Raton, FL, USA, 2016.
50. Nims, R.W. and Price, P.J., Best practices for detecting and mitigating the risk of cell culture contaminants. *In Vitro Cell. Dev. Biol. Anim.*, 53, 872, 2017.
51. Trotter, A.M. and Pendlebury, D., Single-use technologies and systems, in: *Handbook of Validation in Pharmaceutical Processes, Fourth Edition*, pp. 631–651, CRC Press, Boca Raton, FL, USA, 2021.
52. Cană, P., Ripeanu, R.G., Diniţă, A., Tănase, M., Portoacă, A.I., Pătîrnac, I., A review of safety valves: Standards, design, and technological advances in industry. *Processes*, 13, 105, 2025.
53. Bhatia, A., HVAC design for cleanroom facilities, Continuing Education and Development Inc., 2012.
54. DeSantis, P., Facility design for validation, in: *Handbook of Validation in Pharmaceutical Processes, Fourth Edition*, pp. 5–31, CRC Press, Boca Raton, FL, USA, 2021.
55. Israelski, E.W. and Muto, W.H., Human factors risk management in medical products, in: *Handbook of Human Factors and Ergonomics in Health Care and Patient Safety*, pp. 615–647, Mahwah, NJ, USA: Lawrence Erlbaum Associates, 2007.
56. Monat, J.P. and Doremus, S., Deficiencies in and alternatives to heat map risk matrices for project risk prioritization. *J. Mod. Proj. Manage.*, 6, 1, 2018.
57. Kelley, B., Cromwell, M., Jerkins, J., Integration of QbD risk assessment tools and overall risk management. *Biologicals*, 44, 341, 2016.

58. Lahr, J. and Kooistra, L., Environmental risk mapping of pollutants: State of the art and communication aspects. *Sci. Total Environ.*, 408, 3899, 2010.
59. Barone, P.W., Wiebe, M.E., Leung, J.C., Hussein, I.T., Keumurian, F.J., Bouressa, J., Brussel, A., Chen, D., Chong, M., Dehghani, H., Gerentes, L., Viral contamination in biologic manufacture and implications for emerging therapies. *Nat. Biotechnol.*, 38, 563, 2020.
60. Kummer, T.F. and Mendling, J., The effect of risk representation using colors and symbols in business process models on operational risk management performance. *J. Assoc. Inf. Syst.*, 22, 649, 2021.
61. Lyon, B.K. and Popov, G., Communicating and managing risk: The key result of risk assessment. *Prof. Saf.*, 62, 35, 2017.
62. de la O Herrera, M.A., Luna, A.S., da Costa, A.C., Lemes, E.M., A structural approach to the HAZOP–Hazard and operability technique in the biopharmaceutical industry. *J. Loss Prev. Process Ind.*, 35, 1, 2015.
63. Hackman, J.R., *Collaborative intelligence: Using teams to solve hard problems*, Berrett-Koehler Publishers, San Francisco, CA, USA, 2011.
64. Mire-Sluis, A., Ramnarine, E., Siemiatkoski, J., Weese, D., Swann, P., O'Keeffe, R., Kutza, J., Edwards, J., McLeod, L., Practical applications of quality risk management. *Bioprocess Int.*, 8, 20, 2010.
65. Wu, Z., Liu, W., Nie, W., Literature review and prospect of the development and application of FMEA in manufacturing industry. *Int. J. Adv. Manuf. Technol.*, 112, 1409, 2021.
66. Sharma, K.D. and Srivastava, S., Failure mode and effect analysis (FMEA) implementation: A literature review. *J. Adv. Res. Aeronaut. Space Sci.*, 5, 1, 2018.
67. Aven, T., Baraldi, P., Flage, R., Zio, E., *Uncertainty in risk assessment: The representation and treatment of uncertainties by probabilistic and non-probabilistic methods*, John Wiley & Sons, Hoboken, NJ, USA, 2014.
68. Scientific Committee, E.F.S.A., Hardy, A., Benford, D., Halldorsson, T., Jeger, M.J., Knutsen, K.H., More, S., Mortensen, A., Naegeli, H., Noteborn, H., Ockleford, C., Update: Use of the benchmark dose approach in risk assessment. *EFSA J.*, 15, e04658, 2017.
69. Guo, J.J., Pandey, S., Doyle, J., Bian, B., Lis, Y., Raisch, D.W., A review of quantitative risk–benefit methodologies for assessing drug safety and efficacy—Report of the ISPOR risk–benefit management working group. *Value Health*, 13, 657, 2010.
70. Mullally, K.C., Legal (un)certainty, legal process, and patent law. *Loy. LAL Rev.*, 43, 1109, 2009.
71. Ehiagwina, F.O., Kehinde, O.O., Nafiu, A.S., Afolabi, L.O., Olatinwo, I., Fault tree analysis and its modifications as tools for reliability and risk analysis of engineering systems–An overview. *Int. J. Res. Publ. Rev.*, 3, 2582, 2022.
72. Hartsell, C., Ramakrishna, S., Dubey, A., Stojcsics, D., Mahadevan, N., Karsai, G., Resonate: A runtime risk assessment framework for autonomous systems,

in: *2021 International Symposium on Software Engineering for Adaptive and Self-Managing Systems (SEAMS)*, IEEE, pp. 118–129, 2021.
73. Beaudequin, D., Harden, F., Roiko, A., Stratton, H., Lemckert, C., Mengersen, K., Beyond QMRA: Modelling microbial health risk as a complex system using Bayesian networks. *Environ. Int.*, 80, 8, 2015.
74. Gupta, A., Simulation modeling and analysis, in: *Decision Sciences*, pp. 817–902, CRC Press, Boca Raton, FL, USA, 2016.
75. Kapuscinski, A.R., Goodman, R.M., Hann, S.D., Jacobs, L.R., Pullins, E.E., Johnson, C.S., Kinsey, J.D., Krall, R.L., La Viña, A.G., Mellon, M.G., Ruttan, V.W., Making 'safety first' a reality for biotechnology products. *Nat. Biotechnol.*, 21, 599, 2003.
76. Ma, Y., Deilami, K., Egodawatta, P., Liu, A., McGree, J., Goonetilleke, A., Creating a hierarchy of hazard control for urban stormwater management. *Environ. Pollut.*, 255, 113217, 2019.
77. Eggers, R. (Ed.), *Industrial high pressure applications: Processes, equipment, and safety*, John Wiley & Sons, Hoboken, NJ, USA, 2012.
78. Bardo, B., Filter housings in the biopharmaceutical industry, in: *Filtration and Purification in the Biopharmaceutical Industry*, pp. 217–254, CRC Press, Boca Raton, FL, USA, 2007.
79. Hughes, T.G., Multiproduct facility design: An integrated approach, in: *Biotechnology and Biopharmaceutical Manufacturing, Processing, and Preservation*, pp. 293–332, CRC Press, Boca Raton, FL, USA, 2020.
80. Wirtanen, G., Miettinen, H., Pahkala, S., Enbom, S., Vanne, L., *Clean air solutions in food processing*, VTT Technical Research Centre of Finland (VTT Publications), Espoo, Finland, 2002.
81. Ho, S.V. and Budzinski, K.L., Green technologies and approaches in the manufacture of biologics, in: *Green Techniques for Organic Synthesis and Medicinal Chemistry*, pp. 579–599, Hoboken, NJ, USA: John Wiley & Sons, 2018.
82. Conner, J., Wuchterl, D., Lopez, M., Minshall, B., Prusti, R., Boclair, D., Peterson, J., Allen, C., The biomanufacturing of biotechnology products, in: *Biotechnology Entrepreneurship*, pp. 351–385, Academic Press, Boston, MA, USA, 2014.
83. Ewert, D.H., Rompre, G., Ansari, R.A., Saghir, S.A., Regulations and guidelines for toxic exposures in the workplace, in: *Patty's Toxicology*, pp. 1–28, John Wiley & Sons, New York, NY, USA, 2001.
84. Moldenhauer, J., Personnel and their impact on clean room operations, in: *Parenteral Medications, Fourth Edition*, pp. 621–641, CRC Press, Boca Raton, FL, USA, 2019.
85. Nill, R.J., How to select and use personal protective equipment, in: *Handbook of Occupational Safety and Health*, pp. 469–494, Weinheim, Germany: John Wiley & Sons, 2019.
86. Cornish, N.E., Anderson, N.L., Arambula, D.G., Arduino, M.J., Bryan, A., Burton, N.C., Chen, B., Dickson, B.A., Giri, J.G., Griffith, N.K., Pentella,

M.A., Clinical laboratory biosafety gaps: Lessons learned from past outbreaks reveal a path to a safer future. *Clin. Microbiol. Rev.*, 34, 10, 2021.
87. Carra, S., Bottani, E., Vignali, G., Madonna, M., Monica, L., Implementation of behavior-based safety in the workplace: A review of conceptual and empirical literature. *Sustainability*, 16, 10195, 2024.
88. Laszcz-Davis, C., Massey, M., Lebourgeois, V., McHaney, R., Emergency and disaster: Preparedness, response, and recovery, in: *Patty's Industrial Hygiene*, p. 65, Hoboken, NJ, USA: John Wiley & Sons, 2021.
89. Hasanspahić, N., Frančić, V., Vujičić, S., Maglić, L., Reporting as a key element of an effective near-miss management system in shipping. *Safety*, 6, 53, 2020.
90. Wieczerniak, S., Cyplik, P., Milczarek, J., Root cause analysis methods as a tool of effective change. *Bus. Logist. Mod. Manage.*, 2017, 4, 2017.
91. Mishra, R.K., Mentha, S.S., Misra, Y., Dwivedi, N., Emerging pollutants of severe environmental concern in water and wastewater: A comprehensive review on current developments and future research. *Water-Energy Nexus*, 6, 74, 2023.
92. Efaq, A.N., Rahman, N.N., Nagao, H., Al-Gheethi, A.A., Shahadat, M., Kadir, M.A., Supercritical carbon dioxide as non-thermal alternative technology for safe handling of clinical wastes. *Environ. Process.*, 2, 797, 2015.
93. Rutala, W.A. and Weber, D.J., Disinfection, sterilization, and control of hospital waste, in: *Mandell, Douglas, and Bennett's Principles and Practice of Infectious Diseases*, p. 3294, Elsevier/Saunders, Philadelphia, PA, USA, 2014.
94. Singh, N.K. and Dhar, D.W., Sewage effluent: A potential nutrient source for microalgae. *Proc. Indian Natl. Sci. Acad.*, 72, 113, 2006.
95. International Finance Corporation, Environmental, health, and safety guidelines for fish processing, International Finance Corporation, 1998.
96. Asthana, M., Kumar, A., Sharma, B.S., Wastewater treatment, in: *Principles and Applications of Environmental Biotechnology for a Sustainable Future*, pp. 173–232, Springer Singapore, Singapore, 2016.
97. Kerton, F.M. and Marriott, R., *Alternative solvents for green chemistry*, Royal Society of Chemistry, 2013.
98. Sahoo, N., Choudhury, K., Manchikanti, P., Manufacturing of biodrugs: Need for harmonization in regulatory standards. *BioDrugs*, 23, 217, 2009.
99. Mansdorf, S.Z., Personal protective equipment (PPE), in: *Applications and Computational Elements of Industrial Hygiene*, pp. 583–627, Boca Raton, FL, USA: CRC Press, 2018.
100. Arias, W., Bonilla, K., Acosta, S., Velasquez, P., Santis, A., Design of the occupational health and safety management system for a pharmaceutical products marketing company at the Bogotá headquarters, applying ISO 45001: 2018. *Chem. Eng. Trans.*, 91, 307, 2022.
101. Hirschler, M.M., Procedures for development and revision of codes and standards associated with fire safety in the USA. *Fire Mater.*, 41, 1058, 2017.

102. Velagaleti, R., Burns, P.K., Gill, M., Prothro, J., Impact of current good manufacturing practices and emission regulations and guidances on the discharge of pharmaceutical chemicals into the environment from manufacturing, use, and disposal. *Environ. Health Perspect.*, 110, 213, 2002.
103. Godwin, S., Elkind, S., Carey, T., DiGiandomenico, K., Balbo, A., Blocksidge, J., Olson, J., Miklosko, J., Njongmeta, L., Preston, F., Hodges, R., Environmental health and safety offers a biosafety risk assessment for a theoretical model of a gene therapy process transfer from research and development to large-scale manufacturing. *Appl. Biosaf.*, 28, 164, 2023.
104. Adikwu, F.E., Ozobu, C.O., Odujobi, O.L., Onyekwe, F.O., Nwulu, E.O., Advances in EHS compliance: A conceptual model for standardizing health, safety, and hygiene programs across multinational corporations. *IRE J.*, 2023.
105. Bakker, O.J., Kendall, T., Bartolo, P., Health and safety in smart industry: State-of-the-art and future trends, in: *Industry 4.0–Shaping The Future of The Digital World*, pp. 57–62, CRC Press, Boca Raton, FL, USA, 2020.
106. Ahuchogu, M.C., Real-time image-based data processing and its applications in managerial decision-making and risk analysis. *EKSPLORIUM-Bul. Pus. Teknol. Bahan Galian Nuklir*, 46, 1552, 2025.
107. Xie, X., Lu, Q., Rodenas-Herraiz, D., Parlikad, A.K., Schooling, J.M., Visualised inspection system for monitoring environmental anomalies during daily operation and maintenance. *Eng. Constr. Archit. Manage.*, 27, 1835, 2020.
108. Islam, K., Shen, W., Wang, X., Wireless sensor network reliability and security in factory automation: A survey. *IEEE Trans. Syst. Man Cybern. C*, 42, 1243, 2012.
109. Yazdi, M., *Advances in computational mathematics for industrial system reliability and maintainability*, Cham, Switzerland: Springer Nature Switzerland AG, 2024.
110. Gargalo, C.L., de Las Heras, S.C., Jones, M.N., Udugama, I., Mansouri, S.S., Krühne, U., Gernaey, K.V., Towards the development of digital twins for the bio-manufacturing industry, in: *Digital Twins: Tools and Concepts for Smart Biomanufacturing*, pp. 1–34, Springer International Publishing, 2020.
111. Merz, B., Kuhlicke, C., Kunz, M., Pittore, M., Babeyko, A., Bresch, D.N., Domeisen, D.I., Feser, F., Koszalka, I., Kreibich, H., Pantillon, F., Impact forecasting to support emergency management of natural hazards. *Rev. Geophys.*, 58, e2020RG000704, 2020.
112. Turoff, M., Chumer, M., de Walle, B.V., Yao, X., The design of a dynamic emergency response management information system (DERMIS). *J. Inf. Technol. Theory Appl.*, 5, 3, 2004.
113. Lee, H.C., Chang, Y.C., Huang, Y.S., A reliable wireless sensor system for monitoring mechanical wear-out of parts. *IEEE Trans. Instrum. Meas.*, 63, 2488, 2014.
114. Kohli, R. and Mittal, K.L. (Eds.), *Developments in surface contamination and cleaning, volume 7: Cleanliness validation and verification*, William Andrew, 2014.

115. Al-Okby, M.F., Neubert, S., Roddelkopf, T., Thurow, K., Mobile detection and alarming systems for hazardous gases and volatile chemicals in laboratories and industrial locations. *Sensors*, 21, 8128, 2021.
116. Saxena, V., Predictive analytics in occupational health and safety, arXiv Prepr., arXiv:2412.16038, 2024.
117. Singh, S., Karimipour, H., HaddadPajouh, H., Dehghantanha, A., Artificial intelligence and security of industrial control systems, in: *Handbook of Big Data Privacy*, pp. 121–164, Cham, Switzerland: Springer Nature Switzerland AG, 2020.
118. Trevelyan, J., Hamel, W.R., Kang, S.C., Robotics in hazardous applications, in: *Springer Handbook of Robotics*, pp. 1521–1548, Springer International Publishing, 2016.
119. Osarogiagbon, A.U., Khan, F., Venkatesan, R., Gillard, P., Review and analysis of supervised machine learning algorithms for hazardous events in drilling operations. *Process Saf. Environ. Prot.*, 147, 367, 2021.
120. Stillström, C. and Jackson, M., The concept of mobile manufacturing. *J. Manuf. Syst.*, 26, 188, 2007.
121. Mann, D., *Facility management: Human outsourcing solutions to clients*, Global India Publications, 2009.
122. Kowalski, M.L., Ansotegui, I., Aberer, W., Al-Ahmad, M., Akdis, M., Ballmer-Weber, B.K., Beyer, K., Blanca, M., Brown, S., Bunnag, C., Hulett, A.C., Risk and safety requirements for diagnostic and therapeutic procedures in allergology: World Allergy Organization statement. *World Allergy Organ. J.*, 9, 33, 2016.
123. Tusé, D., Nandi, S., McDonald, K.A., Buyel, J.F., The emergency response capacity of plant-based biopharmaceutical manufacturing—what it is and what it could be. *Front. Plant Sci.*, 11, 594019, 2020.
124. Chase, S.H., Emergency lighting systems and their applications, with particular reference to battery equipments. *J. Inst. Electr. Eng. Part II Power Eng.*, 91, 132, 1944.
125. Dadi, G.B., Sturgill, R., Al-Shabbani, Z., Ammar, A., Use of safety management systems in managing highway maintenance worker safety, Transportation Research Board, 2022.
126. Castiello, C., Junghanns, P., Mergel, A., Jacob, C., Ducho, C., Valente, S., Rotili, D., Fioravanti, R., Zwergel, C., Mai, A., GreenMedChem: The challenge in the next decade toward eco-friendly compounds and processes in drug design. *Green Chem.*, 25, 2109, 2023, 10.1039/A0M0000336.
127. Zanoli, S.M. and Pepe, C., Thermal, lighting and IAQ control system for energy saving and comfort management. *Processes*, 11, 222, 2023.
128. van Gelder, P., Klaassen, P., Taebi, B., Walhout, B., van Ommen, R., van de Poel, I., Robaey, Z., Asveld, L., Balkenende, R., Hollmann, F., van Kampen, E.J., Safe-by-design in engineering: An overview and comparative analysis of engineering disciplines. *Int. J. Environ. Res. Public Health*, 18, 6329, 2021.

26

Sustainable Downstream Processing: Green Alternatives for Purification of Biologics

A.V. Raghavendra Rao, V. Sravani Sameera, Ch. Vishnu Vardhan, Mohith Patil, Panduga Rahul and Rompicherla Srividya*

Department of Chemical Engineering, B V Raju Institute of Technology, Narsapur, Telangana, India

Abstract

Downstream processing (DSP) in biomanufacturing, particularly for the purification of biologics, such therapeutic proteins, monoclonal antibodies, and vaccines, is conventionally energy- and resource-intensive, primarily depending on water, solvents, and electricity. Due to increasing environmental and regulatory demands, there is an urgent necessity to implement more sustainable DSP practices. This chapter examines novel eco-friendly technology and process intensification techniques designed to mitigate the environmental impact of biologics purification. Principal methodologies examined encompass membrane-based separations, aqueous two-phase systems (ATPS), continuous chromatography, and the application of bio-based resins. A comparative life cycle assessment (LCA) of traditional versus sustainable DSP methods is provided, emphasizing their environmental impacts. This chapter examines the functions of solvent recovery, single-use systems, and waste valorization in reducing operational expenses and carbon emissions. Regulatory factors and empirical case studies from industrial biomanufacturing are analyzed to demonstrate practical use. This chapter proposes a framework for assessing and implementing green DSP technologies, thereby advancing sustainable and circular biomanufacturing in accordance with global environmental and policy goals.

**Corresponding author*: vid.aaru@gmail.com

Sandip Kunar, Jagadeesha T. and D. Sathis Kumar (eds.) Biomanufacturing: Transforming Production of Biologics and Biopharmaceuticals, (839–884) © 2026 Scrivener Publishing LLC

***Keywords*:** Downstream processing, biomanufacturing, biologics purification, therapeutic proteins, monoclonal antibodies bio-based resins, circular biomanufacturing

26.1 Introduction to Sustainable Downstream Processing

One of the most rapidly expanding areas of the global healthcare market is biopharmaceutical therapeutic proteins, monoclonal antibodies, vaccines and gene therapies. These high-value biologics also include complicated purification and production procedures that include downstream processing (DSP) and upstream part of the production like fermentation or cell culture [1]. In cases where purity and other regulation levels are required to be high, then downstream processing may contribute as much as 60-80 percent of the overall manufacturing cost. Since the operations and amount of reagents consumed are quite large, there also has been an increased interest in the environmental impact of the DSP which involves over consumption of water and solvents, wastage of energy as well as producing hazardous wastes [2].

The Sustainable downstream processing is one of the critical concerns especially when biomanufacturing plant continues to spread to more global places and regulatory agencies are insisting on greener industrial facilities. Centrifugation, microfiltration, precipitation, numerous rounds of column chromatography, ultrafiltration, and drying are common discrete and resource-intensive downstream procedures. All of these procedures require huge amounts of buffer solutions, cleaning reagents, and disposable plastics, and produce considerable chemical and biological waste [3]. In addition to this, most DSP processes are based on organic solvents and non-recycled resins, which create problems of hazardous waste and toxicity [4]. These kinds of practices are not in line with the growing need of low-carbon economies, which are eco-conscious in terms of production systems.

There is an urgent need to address climate change and fulfill global sustainability objectives such as the United Nations Sustainable Development Goals (SDGs); and therefore, the biopharmaceutical industry is engaged in exploring more sustainable solutions in the DSP. Ecological gains in addition to environmental stewardship, sustainable downstream processing also has strategic business advantages: cost saving due to reduced energy and water consumption, streamlined logistic, lower regulatory burden, and easier community acceptance [5]. Firms which focus on green innovations might receive competitive advantages in the changing biomanufacturing environment.

26.1.1 Environmental Impact of Traditional DSP Methods

The traditional downstream processing contains operations that are intensive in terms of resources, and they are mostly created with the principal concern of product purity and manufacture, as opposed to environmental efficiency. This emphasis has created biomanufacturing plants with large water, electricity and single use consumptions, which in turn generate large environmental burden per batch. Figure 26.1 illustrates the environmental footprint of traditional DSP methods.

Buffer consumption is one of the key environmental problems in the traditional DSP. Even purification based on chromatography can purge thousands of liters of buffer per run. Such buffers are made of purified water, chemical salts and pH adjusters that lead to water shortage and depletion of chemical resources. These buffer solutions get contaminated after they are used and they must also be properly neutralized or disposed, adding environmental and operational cost [6].

Energy usage is another concern. The centrifuges that are operated at high speed, cold-chain needs, filtration at high pressure as well as lyophilization procedures use a great deal of energy. As an example, the cold chain used to manage temperature-sensitive biologics may require a large amount of the funds spent on total energy facilities. Temperature control, running pump and cleaning-in-place (CIP) also ensure energy occupies in the chromatography operations [7].

There is also the problem of toxicity and disposal inherent in the usage of solvents, especially in the instance of affinity or reverse-phase chromatography. Certain solvents that include ethanol, methanol and isopropanol are still needed at times but at very low levels; they are also hazardous to our environment when not handled with care. Poorly ventilated areas may also provide an occupational hazard through emission of solvent and evaporation [8].

Additionally, plastic and resin waste in DSP is substantial. Single-use filters, plastic bags, plastic tubing, column and pre-packed chromatography resins are frequently not reusable and recyclable. Although single-use systems are argued to have advantages of contamination control and flexibility, they are environmentally intense because they are disposed either through incineration or landfills [9].

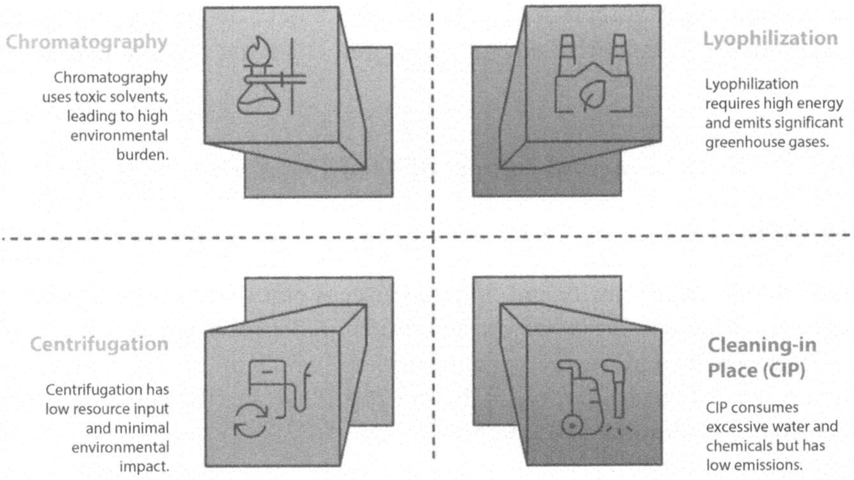

Figure 26.1 Traditional DSP and its environmental footprint.

A Life Cycle Assessment (LCA) method indicates that traditional DSP is part of a large percentage of total greenhouse gas (GHG) production of biologic production [10]. This is particularly relevant where there is a multi-use system involving large cleaning cycles using large amounts of water and detergent, in which case the total emissions of cleaning and sterilization can be as large as the fermentation process itself. Moreover, the environmental loads are multiplied per repetitive batches, which are exacerbating their long-term effects.

26.1.2 Strategic Imperatives for Sustainable Downstream Processing

The shift to sustainable DSP is driven by a combination of regulatory, economic, technological, and social factors, collectively forming a compelling case for eco-friendly downstream practices in biomanufacturing. This debate details the use of green technology across three dimensions: regulatory compliance, economic efficiency enhancement, and the utilization of technological advancements to mitigate environmental damage.

a) Regulatory and Policy Pressure: Global environmental regulations are increasingly stringent. Agency organizations, such as the U.S. Environmental Protection Agency (EPA), European Medicines Agency (EMA) and World Health Organization (WHO) have either initiated or strengthened frameworks intended to support green chemistry, waste minimization and sustainable manufacturing [11]. In addition, Good Manufacturing Practices (GMP) is moving to establish the influences of the environment in validation of the processes and lifecycle. As governments establish more restrictive boundaries on the release of industrial waste water, solvent emissions, and energy use, the businesses have an added benefit to remodel DSP to suit the calculations.

b) Corporate Sustainability Goals: To lower their carbon footprint, several biopharmaceutical firms are orienting with global sustainability guidelines such as UN SDGs or Science-based Targets initiative (SBTi) [12]. They usually involve reduction targets of GHG emissions, use of water and generation of waste. Downstream processing which contributes significantly to resources used in production, is one of the natural areas that should be targeted when having sustainability initiatives.

Genentech, Pfizer and Novartis among others have reported tremendous boosts in performance in mitigating the green impact of their manufacturing processes through greener processes. Some of these gains are noted to encompass DSP steps involving water and energy conservation, green resin and solvent recycling, and continuous multi-purification systems [13, 14].

c) Cost and Resource Efficiency: Green DSP technologies often lead to operational cost reductions. Equipment requirements, facility space, and utility requirements can be minimized by process intensification in which fewer unit operations are required to produce the same level of purification [15]. Moreover, by cutting on the size of buffers and re-using apparatus such as chromatography resins or membranes, the companies would save money on the purchase of raw materials and all the costs of treating this waste.

Another potential direction is continuous processing that allows a real-time monitoring and control system to produce more efficient utilization of resources. On lifecycle cost analysis, continuous systems cost much more in initial capital expenditures but save much in terms of throughput, energy requirement, and buffer usage [16].

d) Technological Advancements: New developments in membrane technologies, aqueous two-phase systems (ATPS), magnetic separation and the development of affinity ligands have opened up the possibility of designing more environmentally benign separation processes [17]. In an example, ATPS is being replaced with toxic solvents by biodegradable polymers, and ligands specific to proteins are being used to minimize non-specific binding and to permit milder elution conditions using less buffer.

Single-use systems, scandalously have been engineered to be able to recycle or be biodegradable. Meanwhile, digital twins and smart sensors make buffer and energy use more sustainable as they allow real-time monitoring of their use and further controllability and optimization of the usage [18].

e) Public and Market Expectations: Sustainability is increasingly valued by investors, consumers, and patients. It is increasingly being put on the biopharma industry to demonstrate that therapeutic innovation does not require sacrifices to planetary health. In the present case, those who pursue the ideology of green manufacturing, particularly with regard to purifying high profile biologics, are likely to strengthen their brand image and acceptance in the market. Sustainability reporting is additionally expressing itself to be a fundamental constituent of Environmental, Social, and Governance (ESG) tests, which also determine entry to funds and relationships [19].

f) Global Health and Access Considerations: Sustainable DSP is also involved in enhancing the access to biologics across the world, particularly where there are limited resources. The decentralized or rural healthcare system can deploy processes that reduce the reliance on the cold chain, energy consumption, and consumable. In such a way, green DSP technologies can contribute to more just healthcare provision [20].

g) Circular Economy Integration: In industrial biotechnology, the idea of a closed-loop bioeconomy—where waste streams are utilized and products are recycled—is gaining traction [21]. Green downstream processing also falls into this model because it facilitates the reuse of purification components, products, recovery of process solvents, and minimizing wastes. As an example, exhausted chromatography resins may undergo regeneration and non-product streams used as feedstocks in other industrial operations may be valorized.

26.2 Green Chemistry and Engineering Principles in Downstream Processing

In the context of sustainable biomanufacturing, use of green chemistry and green engineering principles in the downstream processing has come at the right time and it is a necessity. Due to the large environmental loads caused by traditional purification methods, DSP is being redesigned to have a minimal ecological impact while maintaining good product quality. These green strategies are used as the foundation to build effective, energy efficient, and able to scale up bioprocesses that are in line with the expectations and goals of the world on sustainability.

This section addressed the application of green chemistry and engineering in the purification of biologics. It provides realistic and practical strategies for waste and solvent minimization and process intensification techniques. It also involves key sustainable indicators to measure and compare various purification methods.

26.2.1 Application of Green Chemistry Principles in DSP

Green chemistry includes twelve principles that suggest minimizing the use of hazardous substances and energy in chemical process. These principles as showed in Figure 26.2. provide a framework for reducing environmental and safety impacts in DSP, in biopharmaceutical purification, they manifest through [22].

- Prevention over remediation is one of the most pertinent statements according to which one should reduce the

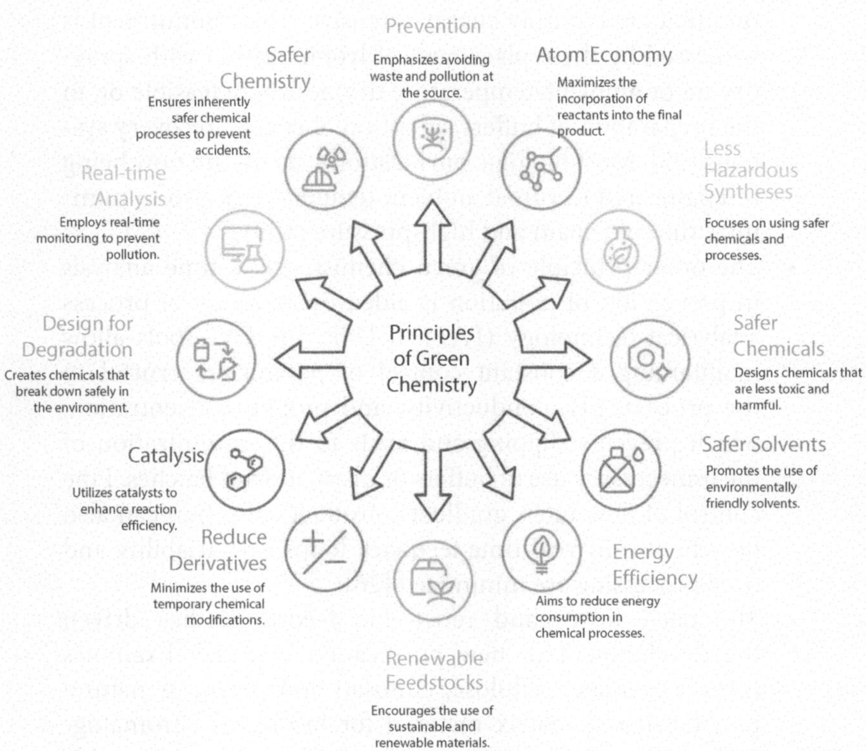

Figure 26.2 12 principles of green chemistry.

production of waste. Traditional DSP has a tendency of handling waste management as an after product. Instead, green-optimized purification processes are aimed at ensuring purification steps inherently produce less waste: an example is the utilization of continuous chromatography systems, which minimize the volumes of buffers used and completely omit batch transfers.

- The other principle, that is, - design of safer chemicals and auxiliaries, is related to the selection or development of material (e.g., ligands, resins, or solvents) that are non-toxic, biodegradable, or reusable [23]. In the case of affinity chromatography, toxic and difficult-to-dispose-of synthetic ligands may be substituted with biologically produced or recyclable ligands. In the same way, there is new solvent systems with water or ionic liquids as safe alternatives to alcohol-based eluents [24].
- The concept of energy efficiency is very relevant in the activities such as lyophilization, chromatography and filtration that are typically energy intensive. The commitment is witnessed by the replacement of lyophilization with spray-drying or ambient-temperature drying where feasible or, in the preparation of buffers, adoption of energy-recovery systems [25]. Most biologic purification systems are now being re-engineered to run at ambient temperatures also to minimize the cold chain and high-pressure utilities.
- The other principle of green chemistry, real-time analysis in prevention of pollution is aided by adoption of process analytical technology (PAT) in DSP. The PAT tools allow maintaining a constant control of parameters crucial to the process, pH, conductivity, and protein concentration, which prevents slipping and leads to the minimization of the unnecessary use of buffers or dissipation of batches. Fine control of flow rates, gradients, product collection can also be achieved by real-time feedback loops, so variability and over-processing are minimized [26].
- The need to expand renewable feedstocks has driven the development of new purification material. Examples include alginate, cellulose, chitosan and pectin as natural polymer-based matrix material for bio-based chromatography columns or filtration supports. These are renewable

substitutes that help to lower the reliance on petrochemical and synthetic based resins and materials [27].

In general, this moves toward integration of green chemistry with DSP also illustrates the need toward considering outside the products with yield and purity domains, by accepting environmental and safety levels as significant process outcomes.

26.2.2 Process Intensification, Waste Minimization, and Solvent Reduction

Process intensification the re-engineering or combination of unit operations to make processes more efficient, minimize equipment use, and consuming less energy, will be one of the more exciting avenues of green engineering to DSP. In the case of biologics this is normally achieved by using continuous processing; shortens hold times, minimizes product degradation and operation is under steady-state conditions. Companies can minimize intermediate storage and buffer requirements by means of combining both upstream and downstream activities (e.g. perfusion bioreactors coupled directly to capture chromatography).

The other major strategy is substitution of multi-column batch chromatography with simulated moving bed (SMB) or multicolumn continuous chromatography. These enhanced systems enhance resin capacity, and minimize buffer up to 50 percent and throughput. The same level of separation with a reduced resources and reduced residence time renders these systems supportive of both cost and green benefits [28].

Membrane-based separations provide an alternative to energy-demanding methods such as centrifugation or diafiltration. The use of single-pass tangential flow filtration (SPTFF) and high-capacity ultrafiltration membranes makes possible the continuity of operation due to reduced energy requirement and smaller-sized equipment [29]. Also, column-based steps may be substituted by membrane chromatography on ion exchange or hydrophobic interaction membranes, where much less buffer volume is required and much faster flow velocities are possible [30].

Solvent reduction is another critical area in green DSP. Ethanol, methanol, or isopropanol are frequently applied organic solvents as a cleaner, eluents as well as inactivators of viruses. The measures to decrease their use are the development of aqueous two-phases systems (ATPS) based on biodegradable polymer and salts. ATPS are tunable to selective partitioning of proteins with minimum application of organic solvents where waste streams can be recycled or disposed safely with less burden to the environment [31]. Where solvents must be used in any case, solvents recovery systems, e.g., distillation or solvent separation by membrane, may be added to minimize emissions and recycle purified solvents. This reduces the cost of material and also reduces the generation of hazardous waste.

In the waste minimization side, it is through reuse of chromatography resins, regeneration of filters, and in-line cleaning procedures among others. The life of resins can be extended by better cleaning procedures and real time resin performance monitoring and enable the high utilization of the resin having a wide range of hundreds of cycles (resins can be used multiple times with no performance impairment) [32]. In regard to the preparation of buffers, in-line conditioning systems are already becoming a standard in recent facilities. Through automated feedback, such systems enable the just-in-time mixing of concentrated stocking solutions which results in huge reductions in the buffer volumes, storage capabilities and the inevitable wastage.

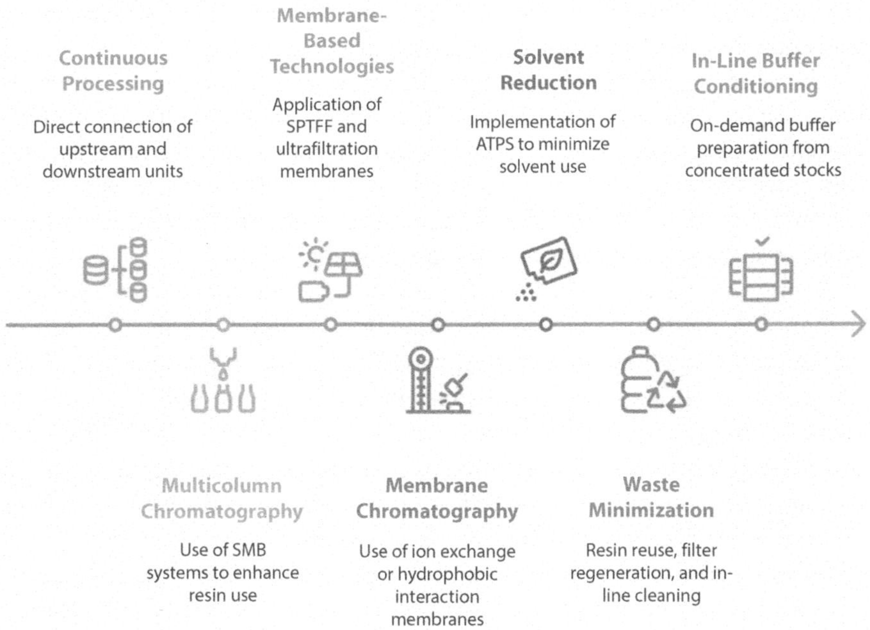

Figure 26.3 Sustainable DSP through various intensification, waste & solvent reduction methods.

Altogether, the strategies will contribute to making the DSP a more resource-efficient, cleaner, and low-carbon-emitters operation. Figure 26.3 illustrates various strategies adopted under process intensification, waste minimization and solvent reduction for DSP.

26.2.3 Metrics for Evaluating Sustainability in Purification

Sustainability measures must be adopted to assess the gains of the green approaches in DSP and to guarantee improvement at all times. The following Table 26.1 gives the metrics that are grouped under the categories: environment, economy and process performance [33].

Table 26.1 Metrices to evaluate sustainability.

Category	Metric	Definition/ formula	Relevance in DSP
Environmental	E-Factor	Waste mass per unit product mass	Lower E-factor means less waste generated during purification
	Water Use Intensity (WUI)	L of water used per unit product	Tracks efficiency of water use in buffer prep, cleaning, etc.
	Carbon Footprint	CO_2-equivalent emissions per run or per gram of product	Measures overall GHG impact of DSP operations
	Solvent Intensity	Volume of solvents used per batch; recovery %	Indicates solvent dependency and recyclability
	Waste Load Index (WLI)	Combines waste volume and toxicity	Highlights environmental hazard of generated waste

(*Continued*)

Table 26.1 Metrices to evaluate sustainability. (*Continued*)

Category	Metric	Definition/ formula	Relevance in DSP
Economic & Operational	Cost of Goods (CoG) Impact	CoG before vs. after green implementation	Tracks cost savings from green strategies
	Facility Throughput	Units of product per time period	Measures effect of green intensification on productivity
	Resin Utilization Efficiency	Protein mass purified per resin cycle	Reflects efficiency and lifespan of purification media
Process & Quality	Yield Efficiency	Product recovered per input volume	Shows effectiveness of purification
	Process Robustness	Deviations/ failures per batch	Lower values indicate stable, low-waste process
	Cycle Time Reduction	Time savings from green redesign	Correlates with energy and cost savings
	Batch Variability	Variance in product quality attributes	Reduced variability improves consistency and reduces reprocessing

There are also some companies that are using Life Cycle Assessment (LCA), Techno-Economic Analysis (TEA) in early- stage DSP design to make process decision. LCAs support the measurement of environmental effects through the whole lifecycle of purification, starting with raw materials procurement to end of life product disposal and creation of such technologies; whereas TEA provides an overall spectrum of costs of green technologies [34].

Standard tools of assessing the sustainability of DSP strategies are now entering the market and frameworks like Green Chemistry Metrics Toolkit and Sustainable Biomanufacturing Scorecards will act as standardized measures to improve the transparency and data-informed nature of decisions [22, 35].

Incorporating sustainability criteria based on green chemistry and engineering concepts enables the redesign of DSP to minimize solvent usage, decrease waste, and enhance energy efficiency. Commercial examples demonstrate that these solutions achieve environmental objectives while also improving cost-effectiveness and throughput. Improvements in resin technology, membrane systems, and continuous platforms will enhance these advancements. Incorporating data-driven sustainability indicators into process design guarantees that biomanufacturing produces high-quality therapies while minimizing environmental impact.

26.3 Sustainable Cell Harvesting and Clarification Techniques

Cell harvesting and clarification act as the first step of the downstream processing (DSP) and they determine the efficiency, quality, and the environmental impact of the biomanufacturing process. These first phases, which serve to separate biomass or cell debris with the target biologic, are resource-intensive and have most often been performed using high-speed centrifugation and depth filtration [36]. Nonetheless, as sustainability and flexibility in operations become more pertinent, the traditional solutions are being revisited towards the design of more energy-conserving, scalable, and environmentally neutral options.

This part discusses the techniques in cell harvesting and clarification which are still emerging and evolving and the techniques incorporate the principles of green bioprocessing. It dedicates itself to the finding of centrifugation alternatives, such as acoustic waves separation and increased depth filtration, application of single-use and closed systems to minimize waste and contamination as well as the usage of low-shear and energy-saving approaches capable of promoting the product integrity and operational sustainability.

26.3.1 Re-Evaluating Traditional Centrifugation

The technique of centrifugation has been used since the time when traditional form of removing biomass in both cell culture involving microbes and mammalian cells and in large scale cell culture is used. While effective, this technique presents several sustainability challenges:

- Requirement of large amount of energy and high rotor speeds as well as constant work under load.
- Large footprint of equipment due to their size and complexity which makes it very costly to maintain and validate.
- Shear-induced cell lysis can also cause release of host cell proteins and DNA into the supernatant where they complicate downstream purification [37]. The shear induced cell disruption through homogenizer nozzle is shown in Figure 26.4.

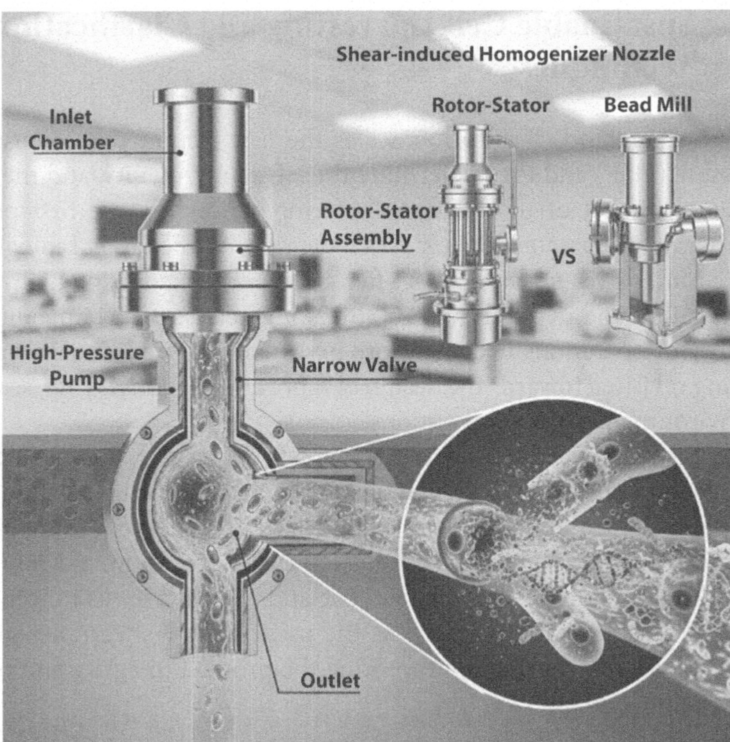

Figure 26.4 Shear induced homogenizer nozzle.

- The use of water and chemicals in sterilization and cleaning process of stainless steel-based systems.
- Lack of flexibility in facilitating different cell densities and volumes of processes in multi-product plants.

Due to these disadvantages, use of centrifugation as part of the sustainable trends of manufacturing is being phased out where feasible especially in clarifying mammalian cell cultures and low-density suspensions. The assessment of non-centrifugal solutions is also gaining pace because of the added advantages of lower energy consumption, single-use ability and less demanding on cell and product manipulation.

26.3.2 Alternatives to Centrifugation: Acoustic Wave Separation and Depth Filtration

26.3.2.1 Acoustic Wave Separation (AWS)

AWS is a cutting edge and non-mechanical technique that utilizes ultrasonic standing waves in manipulating and separating cells or particles depending on their density and compressibility differences. Applying a standing acoustic field across a microchannel or flow through chamber will focus cells and debris material to the nodes or the antinodes of the wave, depending upon the acoustic property of the material. This allows continuous, low-shear and mild separation without centrifugation or filters membranes [38].

Advantages of AWS in sustainable DSP include:

- Energy efficiency: AWS uses very little energy as compared to conventional centrifuges.
- Low-shear environment: Reduces cell lysis and minimizes product degradation.
- Single-use compatibility: Is compatible with disposable flow paths so that cleaning is not necessary.
- Scalability: Modular design allows scalability for various production volumes.

Albeit that AWS remains in an early, commercialization stage of large-scale biologics, its adoption on a continuous bioprocessing platform, and its capability of operating in perfusion-based upstream systems, makes it a promising candidate when it comes to sustainable facilities in the future [39, 40].

26.3.2.2 Depth Filtration

The depth filtration is another scalable and practical substitute of centrifugation especially when dealing with mammalian cell cultures. Depth filters unlike surface filters trap cells and debris at the depth of the whole matrix thickness, which has the inherent advantage of having greater dirt holding capacity and extended service life [41].

Advantages of Depth filtration in Sustainable DSP include:

- Low energy needs: Uses low pressure and has no moving parts.
- Less downtime and fouling: The current depth filters have layered media or gradient media, decreasing clogging.
- Integration with single-use systems: Filters are usually built into closed single-use assemblies.
- Modular scalability: Easily adjusted to accommodate variable production volumes.

The development of the depth filter media, whose examples are synthetic polymers, cellulose composites or charged surfaces, has enhanced the selectivity, capacity as well as throughput. Depth filtration is considered as a component of the hybrid approaches to clarification. It is frequently applied along with prefiltration, or flocculation, to achieve optimum performance and conserved buffer consumption [18, 42].

26.3.3 Single-Use and Closed Systems: Reducing Waste and Contamination

Rising use of single-use (SU) and closed systems, especially in multiproduct factories or manufacturing of products in hospitals, represents one of the key characteristics of sustainable cell harvesting and clarification plans. Such systems have pre-sterilized, disposable items (bags, tubing, filters, and connecters) that are put together into compactable structures [43–45].

Advantages of SU and Closed Systems in Harvesting, Clarification:

- Reduction in cleaning and sterilization demand
- Single-use systems radically cut down on water, and chemical intake because they do away with cleaning-in-place (CIP), steam-in-place (SIP). This results in great savings of energy particularly in large facilities.
- Lower cross-contamination risk

- Better containment and control occur in closed systems, especially multi-product or high-potency manufacturing of biologics, where the contamination of one product by another may have severe consequences.
- Reduced facility footprint and infrastructure needs
- Single-use harvest skids are also small and modular, so smaller infrastructures of utility installations or stainless-steel hardware are not necessary.
- Decreased downtime and faster turnaround
- The ease of changing over batches using disposable parts is most useful to contract development and manufacturing organizations (CDMOs).

Environmental trade-offs and mitigation

There is no benefit in the use of single-use systems since they produce plastic waste; however, life cost assessments have demonstrated that they frequently result in fewer overall environmental impacts because of fewer water, electricity, and chemicals used. Incineration of waste-to-energy and developments in bioplastic recycles enhance that sustainability rating further.

Single-use components are available through harvest bag designs that incorporates filters, depth filter capsules and completely enclosed acoustic separation modules. This combined with their ease of use with flexible manufacturing strategies also means they are becoming popular on many levels of biomanufacturing.

26.3.4 Energy-Efficient and Low-Shear Methods

Besides substituting energy-intensive mechanical processes associated with mechanical clarification, the ultimate goal of sustainable clarification systems is maintaining product quality, which is achievable once avoiding induction of shear stress, temperature changes, and residence time. These parameters in the process have great effects on the product stability and impurity profiles. Key strategies include:

- ***Gravity settling combined with flocculants:*** Gravitational sedimentation with natural of biodegradable flocculants can be used to increase the removal of biomass in the case of microbial cultures without the need of the high-speed centrifugation. The technique is energy-efficient and scalable on specific process streams which are clarified passively [46, 47].

- ***Continuous low-pressure filtration:*** As compared to the batch-mode clarification, the continuous clarification processes through low-pressure, cross-flow arrangements result in less overall energy intensity and is a perfectly suitable clarification mode to perfusion bioreactors [48].
- ***Precipitation-assisted clarification:*** By addition of precipitating agents (e.g., calcium salts, polyethylene glycol) some impurities or cells can be specifically flocculated to be easily removed by simple filtration. These agents can be well optimized to ensure high recovery of the products with few processing steps [49].
- ***Shear-controlled pumping systems:*** When pumps are chosen, type of pumps such as peristaltic or diaphragm pumps can be selected instead of centrifugal pumps or gear pump to maintain product integrity during transfer. Such pump types have particular significance in the harvesting of expensive biologics like monoclonal antibodies or therapeutic vectors, especially gene-therapy vectors [50].

Adopting these low-shear, energy-frugal approaches to the clarification schemes in an integrated way may help manufacturers minimize protein aggregation, ease the downstream chromatography, and eventually decrease the downstream processing cost and environmental impact.

26.3.5 Industrial Examples and Integration with Continuous Bioprocessing

Some biopharmaceutical producers are already moving to use sustainable cell harvesting and clarification techniques on the commercial scale:

- The full closed, single-use harvest platforms with the integration of disposable bioreactors have been said to have been successful at Genentech and Amgen as it has saved substantial water and downtimes [51].
- Single-use, water-saving harvest trains with depth filtration in KUBio modular bioprocessing facilities allow significant saving in water usage in comparison with classic stainless-steel plants [45].

- Pfizer and Sartorius have been working on acoustic wave systems to perfuse harvest and show their applicability to high-density cultures in continuous systems [52, 53].

Integration with continuous bioprocessing further amplifies sustainability gains. In another example, steady-state operation and constant feed to downstream chromatography can be achieved by the use of continuous clarification through either AWS or depth filtration; this is a more efficient purification process using less hold tanks and less amounts of buffer.

26.4 Green Alternatives in Chromatography and Separation

Chromatography continues to be the heart of biologics downstream processing (DSP) for purification of complex molecules including monoclonal antibodies, therapeutic enzymes and nucleic acid-based products. Nevertheless, the traditional chromatographic processes also happen to be the resource-intensive operations in DSPs: they consume large quantities of buffer, high-grade water, disposable plastics, and occasionally even hazardous solvents. Also, the conventional use of packed-bed column, especially in the capture and polishing processes needs high operating pressure, extensive cycle rates, and frequent cleaning-in-place (CIP) actions, which add significantly to the biomanufacturing environmental load [32].

Due to high demand of the sustainable bioprocessing, green analogs of conventional chromatography are gaining traction. Such advances have been the use of new separation media using membranes, monoliths and reusable resins, elimination of hazardous solvents in favor of aqueous-based or green solvents, and the transfer of magnetic-based separation and affinity-based separation technology, which reduces processing steps, energy and waste products. The following section discusses these options in detail as the viable, scalable options of minimizing environmental impact of chromatography and at the same time keeping products with high recovery and purity.

26.4.1 Membrane Chromatography: High Throughput with Low Impact

The membrane chromatography has presented a strong argument against packed resin columns especially during flow-through polishing. Rather than depending on diffusion of the particles inside the porous beads, membrane systems take advantage of porous structure but with modified surfaces that allow convective flow. This is a large-scale mass transfer enhancing high rate, low pressure drop and time of processing [54]. The key sustainable benefits of membrane chromatography include:

- Reduced amount of buffer: Membranes use low amount of buffer per cycle as compared to columns due to the high throughput and rapid flow.
- Decreased cycle time: The decreased equilibration time and elution time reduce processing time and energy requirement.
- Disposable format: The cartridges made of membranes that are placed in single use systems, dispense with cleaning and sterilization, thereby saving water and chemicals.
- Low pressure services: Energy savings made by decreasing the pump loads.

The membrane adsorbers excel in the process of adsorbing host cell proteins (HCPs), DNA, viruses, and endotoxins in flow-through [55, 56]. Sartobind(r) Q and Mustang(r) Q are most commonly used in ion-exchange polishing steps. Due to the ability to run such membranes at a higher flow rate and shorter residence time, they are excellent to continuous processing and facility intensification strategies [57]. In addition, the compatibility of membrane chromatography with modular, closed systems contributes to process flexibility and minimizes cross-contamination possibilities, which are highly important considerations to greener, multiproduct facilities [58].

26.4.2 Monolithic Chromatography: Continuous Pores, Continuous Sustainability

Another novelty in the chromatographic media is the monoliths, which are a single, continuous block of porous material instead of packed resin beads. This architecture aids the convective mass movement of molecules and is especially useful with separating large biomolecules like the plasmid DNA, viruses, and virus-like particles (VLPs) that do not diffuse as fast on traditional resins [59]. Various types and their working as sustainable components are discussed in Table 26.2 and the advantages of monoliths from a green processing perspective are given here:

Table 26.2 Monolith types, working and applications.

Type	Material example	Working principle	Typical applications	References
Polymeric Monoliths	Poly (glycidyl methacrylate-co-ethylene dimethacrylate) (GMA-EDMA)	Convective flow through continuous polymer network; surface functionalization enables specific binding	Protein purification, plasmid DNA	[60]
Silica-Based Monoliths	Continuous porous silica rods	Convective transport in rigid silica matrix; high surface area for adsorption	Peptide/protein separation, small molecule HPLC	[61, 62]
Hybrid Organic–Inorganic Monoliths	Silica backbone with polymer coating	Combines silica rigidity with polymer selectivity; enhances pH tolerance	Wide biomolecule range	[63]
Bio-Based Monoliths *(Emerging)*	Cellulose, chitosan	Convective flow in biopolymer framework; natural functional groups for binding	Early-stage protein capture, environmental bioprocessing	[64]

- High permeability: the availability of open pore structure leads to low pressures of operation, mechanical strength and energy input.
- Fast throughput: Enhanced convective flow shorten cycle times, increasing productivity without lowering the resolution.
- Better cleaning: Due to its open pore structure, fouling agents are removed easily and less amount of buffers are utilized.
- Regenerability: Several monoliths may be cleaned and withstanding many usages without losing their performance.

Commercial platforms like CIMmultus by BIA Separations have proven effective in applications ranging from vaccine to gene therapy and mRNA purification. They can be candidates for sustainable biomanufacturing facilities due to their low footprint, low usage of water and solvents, and possibility of being integrated into continuous DSP platforms [65]. Further, research into biodegradable or hybrid polymer monoliths is ongoing, aiming to further reduce the environmental burden associated with end-of-life disposal.

26.4.3 Recyclable and Bio-Based Resins: Optimizing Its Use and Minimizing Waste

Conventional chromatographic resins such as Protein A and ion exchange resins are both expensive and have short lifetime characterized by fouling, bacterial growth or chemical decomposition. The manufacture and end of life of these resins are related to material footprint of DSP. Green chemistry initiatives are currently centered trying to come up with resins that can be recycled, renewable sources or with durability in their use.

a) Recyclable Resins: To be used in repeated CIP, some manufacturers optimize their resins formulations to assure hundreds of CIP cycles without loss of binding capacity and selectivity. Indeed, current Protein A resins like MabSelect PrismA 124 are expected to be used more than 200 times in a product lifecycle, cutting costs and minimizing resin waste by 50 per cent or more [66]. There is also a development of regeneration protocols with less severe cleaning reagents (e.g. sodium citrate, acetic acid) to substitute severe caustic chemicals, and the benefits to the operator and less emissions of the chemicals in the supply are applied [67].
b) Bio-Based Resins: Resin matrices made up of agarose, cellulose or chitosan are biodegradable and have renewable origins. The natural polymers can offer a more environmentally friendly alternative to synthetic, acrylamide- or polystyrene-based supports. Such resins, functionalized with ion-exchange or affinity ligand, can be used similarly to conventional resin with the difference that they are comparable especially during the polishing stages [32]. To be even more advanced one, enzymatic cross-linking of support matrices and use of plant-based affinity ligands has been proposed so that the entire separation system is fully sustainable in source through sink [68].

26.4.4 Aqueous-Based Mobile Phases and Green Solvents

Employment of organic solvents, especially in hydrophobic interaction (HIC) and reverse-phase chromatography as well as in the cleaning procedures, is one of the major contributors to the environmental burden of chromatography. Solvents such as ethanol, methanol, and isopropanol, despite being effective, are dangerous in the sense of flammability, toxicity, and cost of disposing of them [69, 70]. As their solution to this, green DSP strategies are turning towards using mobile phases that are based on water, reusable solvents, and non-toxic replacements of solvents.

a) Aqueous Mobile Phases: Usually, aqueous buffers are used in ion exchange, affinity and size exclusion chromatography and new formulations are being made to use less salt and to work under less harsh pH conditions. Histidine, citrate and phosphate buffers are examples, that provide buffering capabilities, with less corrosivity and fewer downstream neutralization needs [71]. Water and chemical use are also reduced by minimizing volumes of buffers using streamlined gradient programs and inline conditioning systems.
b) Green Solvent Alternatives: Volatile organic compounds will be phased out as innovative solvent systems are currently being developed. Examples include:

- *Ionic liquids:* Concentrated, non-volatile and adjustable solvents which can be created to fit specific protein characteristics; hence, selectively solvable [72].
- *Supercritical CO_2:* Used in solvent-assisted precipitation and extraction steps [73].
- *Bio-based solvents*: e.g. ethyl lactate, is based on renewable raw materials, has comparable elution power to classical alcohols [74].

Closed-loop systems of solvent recovery and reuse are becoming even more popular in larger facilities and minimize emissions as well as cutting the costs.

26.4.5 Low Impact Magnetic and Affinity Based Separation

Besides resin-based chromatography, magnetic particle technology and affinity separation provide high levels of selectivity, using minimal resource inputs. Such technologies are particularly applicable to those applications using low-titer feed streams or more niche products, such as enzymes, gene therapies or exosomes.

a) Magnetic Separation: Magnetic separation removes targeted molecules directly out of mixtures by using magnetized beads or particles, which are ligand (e.g., antibodies, peptides, protein A analog) functionalized. The particles are then separated through a magnetic field, washed and eluted once attached [75, 76]. Sustainability advantages of this magnetic separation include:

- Elimination of columns and pressure systems
- Minimal buffer usage
- Low shear, preserving product integrity
- Reduced energy consumption
- Compatibility with small-volume, high-value products

Antibody purification by magnetic separations is already employed at laboratory and clinical level and improved large-scale magnetic separation units (e.g., Mag Sepharose) are allowing other more extensive industrial applications [77].

b) Alternative Affinity Ligands: Protein A though highly effective is prone to leaching, highly expensive and needs several polishing steps. The Green alternatives to this are the use of synthetic peptides, engineered protein scaffolds, aptamers, metal-chelate ligands. These substitutes are usually more stable, less immunogenic and can be eluted under a mild condition, which implies the implications of less use of denaturing or of extreme pH conditions [78–80]. Besides, cell-free expression system or microbial-based fermentation-based ligands increase independence of animal products and enhance reproducibility of the process.

There are ethical and technological reasons as to why downstream bioprocessing should take on sustainable chromatography. Separation processes using biologics are being transformed to innovative membrane adsorbers, monolithic columns, recyclable and biodegradable polymers and greener solvents. These alternatives are already environmentally beneficial in reducing use of the buffers and water, minimizing emissions, streamlining operations and even becoming commercially viable. These modular and continuous biomanufacturing methods use the flexible and process-easing methods as well as the magnetic and novel affinity-based techniques of separation. Alternatives to green chromatography offer a better performance in production costs, compliance, and environmental impact of biologics manufacturing.

26.5 Integration and Process Intensification Approaches

Downstream processing (DSP) continues to shift away as a batch series of discrete serial steps that consume resources to a more continuous, integrated and intensified model of manufacturing. It is made possible due to the requirement to create a more efficient process, decimate the environmental burden, increase flexibility, and enable speedier and more cost-effective scale production capabilities of high-quality biologics. A key role towards the achievement of the green biomanufacturing goals is played by integration and process intensification (PI). These practices will support green engineering practices as they have the advantage of reducing equipment redundancy, streamlining business operations, limiting inter-stage buffering, and energy utilization. They also improve consistency and compliance to products achieved by improved control and automation. Three essential strategies at the forefront of green DSP, which are continued and integrated platforms, inline buffer preparation with real-time observation, and upstream-downstream coupling, are described in this section. Integration with intensification work together and create the bedrock of the less-wasteful and sustainable future-proofed biomanufacturing systems [81].

26.5.1 Continuous and Integrated DSP Platforms

The traditional DSP is mainly batch-wise in nature, whereby consecutive processes are separated by holding steps. Such an arrangement needs high volumes of buffer storage, huge plant areas, long cycles, and manual interventions. Continuous DSP, in contrast, combines several steps into a smooth workflow, eliminating wastes, and variability, as well as, simplifying operations. The benefits of continuous DSP for sustainability are shown in Figure 26.5. Continuous DSP is all about modular unit operations that can be interconnected. Examples include:

- Multicolumn chromatographic assays like periodic counter-current chromatography (PCC) to carry out continuous capture [82].
- Single pass tangential flow filtration (SPTFF) concentrates or diafiltration [83].
- In-line virus filtration and membrane chromatography in removal of impurities [84].

These modules enable product and facility-specific flexibility between operations either fully continuous or a hybrid semi-continuous workflow. One of the more striking examples is the one that involves the use of continuous Protein A capture and the inline viral inactivation and buffer exchange; this reduces the capturing and conditioning steps into just a single and continuous combination [32]. This cuts down on hold time,

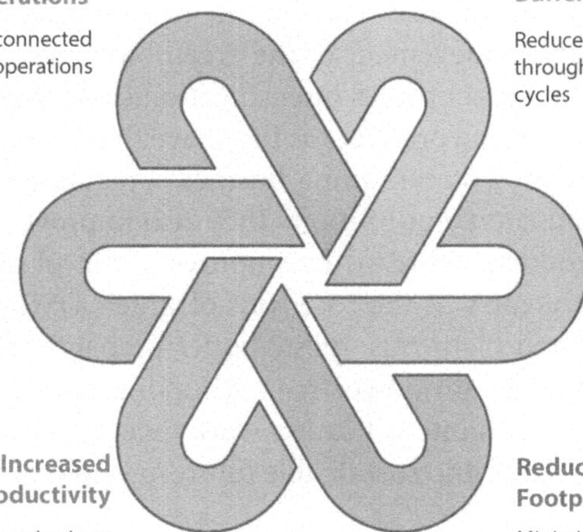

Figure 26.5 Benefits of continuous DSP.

increases yield and it avoids the use of a number of stainless-steel vessels and transfer processes.

Also, a few facilities are adopting end-to-end continuous manufacturing, in which upstream perfusion bioreactors connect directly to continuous DSP modules. The model is not only more sustainable but can also favor smaller and more nimble manufacturing footprints which is quite a benefit to personalized and regional therapies.

26.5.2 Inline Buffer Preparation and Real-Time Monitoring

The preparation of buffer is one of the most expensive elements in DSP both in terms of resources and manpower. Quantities of buffer need to be added in large amounts throughout chromatographic and filtration procedures, as well as large amounts of water and chemicals to ascertain that the facilities are correctly prepared, stored, and cleaned. The solution is available in prep of inline buffers (IBP) or inline conditioning which is the real time production of the process buffers on demand [85]. The following are the advantages of inline buffer preparation:

- Effective decrease of the volume of necessary buffers due to the reduction of the overproduction and decrease of waste.

- Reduced storage space and equipment, freeing facility footprint.
- Reduced cleaning and validation which reduces use of water and chemicals.
- Better work place safety given reduced manual handling procedures.
- Enhanced flexibility to switch between products and recipes.

The mechanism of inline conditioning systems is to take the concentrated stock solutions and dilute it manually to the desired pH and conductivity with water for injection (WFI) using automated control systems. These are connected to chromatography skids and they can change the recipes quickly and provide high quality of buffers. Combined with IBP, real-time process monitoring tools constitute the digital horsepower of enhanced DSP. These instruments are collectively termed as process analytical technologies (PAT) and they can be used to control key quality attributes and performance properties including-pH, conductivity, and concentration (UV); turbidity and particle counts (clarification); pressure and flow monitoring (filtration and chromatography) purity or titer of protein through spectroscopic or biosensor based techniques [86–88]. Incorporating PAT into inline operations ensures:

- Better consistency across runs, reducing out-of-spec batches.
- Early fault detection, enabling rapid corrective action.
- Minimized rework and resource wastage.
- Facilitating adaptive control, with the system manipulating parameters on a real time basis to keep the process conditions.

Relatively, both IBP and PAT are a keystone of green, smart DSP, decreasing not only the environmental tolls, but also the burden on operations to manage the number of buffers, and do process control.

26.5.3 Coupling Upstream and Downstream Operations for Efficiency

Historically, the processes of upstream and downstream operation were regarded as distinct and quite independent spheres. Hold tanks, clarification and downstream purification follows fermentation or cell culture. The consequence of this segregation is idle time, underutilization of equipment, waste and use of more energy and water in cleaning, sterilization, and cold storage. Bioreactor upstream-downstream coupling universally is encouraged in modern sustainable bioprocessing as exemplified by continuous or perfusion bioreactors fed into integrated DSP modules. Benefits of Upstream–Downstream Integration:

- It avoids intermediate holding, and lessens the exposure to contamination and degradation risk involved.
- Reduces the need of cold chain, which reduces energy consumption on chilled storage.
- Enables steady-state operation, enhancing yield and reducing variability.
- Synchronizes equipment utilization, maximizing productivity and process economy.

As an example, in a perfusion-driven upstream process, cells are retained when the product-rich media is harvested continuously and channeled into clarification module such as acoustic wave separation, depth filtration or centrifuge-alternatives. This clarified stream can directly be taken into continuous protein A chromatography and then inline virus inactivation with inline UF/DF and final polishing. Single-use technologies and cleanroom designs that are modular further provide support in the process integration and allow high deployability and flexibility of equipment installed. This enables multiproduct facilities to interchange campaigns with limited use of resource and downtime. Digital twins and integrated control systems have the potential of tracking and coordinating upstream and downstream parameters to allow a process to run continuously, foresee problematic areas and ensure the balance of flow. The digital twin of an entire bioprocess is simulated in these systems, which allows to make proactive changes and make sound decisions [89–91].

26.5.4 Industrial Adoption and Case Studies

Some of the industry giants have adopted process intensification and integration with substantive sustainability consequences:

- Amgen's Singapore site (ASM) combines warehouse capacity optimization with transport risk mitigation to enhance downstream processing. By employing a Warehouse Capacity Model, they enhanced material flow and lessened storage requirements, while the qualification of 2000 L disposable bioreactor bags decreased safety stock and contamination hazards. This efficient, risk-conscious strategy guarantees dependable supplies, reduced expenses, and sustained biologics production [92].

- Framingham's Sanofi digital manufacturing facility uses fully integrated, continuous upstream-downstream processing and real-time digital monitoring to speed batch release and reduce environmental impact. Digital, data-centric operations save energy, water, chemical, and carbon emissions while improving efficiency and agility. This innovative design shows how digital biomanufacturing makes high-performance biologics sustainable [93].
- KUBio is the modular-based facility integrated into GE Healthcare, which performs the intensification of upstream and downstream processes to reduce the carbon footprint of facilities and allow their production on a regional basis [45].

These case studies illustrate that sustainable intensification is not an idea that is impractical, but it is a possibility that can be scaled up and most likely to be essential in the biopharmaceutical activities of the world. DSP is becoming an efficient, intelligent and sustainable production system because of integration and process intensification. Continuous DSP platforms and inline buffer preparation and upstream coupled downstream are helping biomanufacturers improve yield, flexibility and facility efficiency by cutting energy levels, use of buffers and process reproducibility. Such techniques encourage a more sustainable ways of bio-making and flexible, economical production that meets the requisites of biologics market share, particularly in the age of high-mix, low-volume pharmaceuticals. Next-generation biomanufacturing, with its focus on process integration as regulatory frameworks and digital infrastructure allow bringing these breakthroughs to fruition, will become environmentally responsible.

26.6 Life Cycle Assessment (LCA) and Environmental Impact Analysis

With the current wave of sustainable biomanufacturing, there is an increasing requirement to shift past those who focus on process-level performance and begin to system-level environmental analysis. The positive impact to operation illustrated by the adoption of a membrane chromatography, inline buffer preparation, or a single- use technology could have a clear benefit; whereas the overall environmental benefit has to be proven according to the comprehensive assessment tools. And that is where Life Cycle Assessment (LCA) becomes a vital thing.

LCA is a preset approach (according to ISO 14040/14044) of evaluating environmental consequences of all of the stages of a product, process or system, including the epoch of extraction of uncooked material and end-of-life disposal. Within the topic of downstream processing (DSP), LCA incurs a systematic guideline in terms of the analysis of conventional purification versus green processing and a tradeoff quantifying sustainability process design decision making.

This section shows an in-depth discussion on LCA to the DSP systems highlighting some comparative analyses, the carbon and water footprints, waste generation measures, and incorporating the sustainability indicators in the formulation of purification strategies.

26.6.1 Downstream Processing LCA Principles and Scopes

Sustainable biomanufacturing requires environmental performance assessments beyond process improvements. Membrane chromatography, inline buffer preparation, and single-use systems improve operations, but Life Cycle Assessment should determine their environmental impact. According to ISO 14040/14044, LCA examines impacts from raw material extraction to end-of-life, comparing traditional and green DSP methods, evaluating trade-offs, and guiding sustainable purification strategy design [94]. A Life Cycle Analysis requires a set of four standard steps as depicted in Figure 26.6:

In DSP, the scope may cover:

- Upstream effect on DSP (e.g., harvest yield on downstream load)
- Primary purification steps (chromatography, filtration, virus inactivation)
- Buffer and solvent usage
- Cleaning cycles and CIP/SIP utilities
- Waste handling and emissions
- End-of-life treatment of single-use components

A good LCA will yield comparative information, e.g., whether it makes sense to adopt a single-use system over stainless-steel, or whether to choose a batch or a continuous chromatography platform.

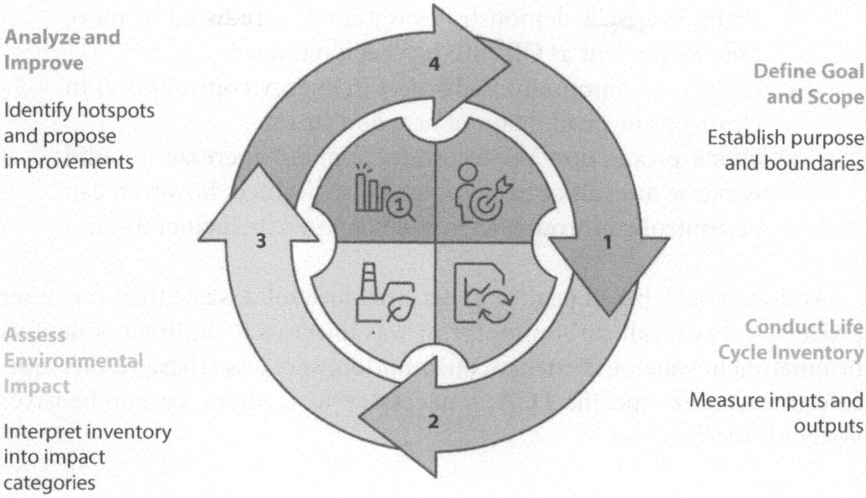

Figure 26.6 Steps of LCA.

26.6.2 Comparative LCA: Conventional vs. Sustainable DSP Strategies

Case 1: Batch Chromatography vs. Continuous Chromatography- Comparatively in terms of LCA, a study of the production of monoclonal antibodies employed conventional batch Protein A chromatography, and was compared to periodic counter-current chromatography (PCC):

- Resin consumption: PCC enhances a utilization of resin by up to 90 percent, as the total aggregate consumption of resin is taken into consideration.
- Buffer requirements: Continuous systems reduce buffer volume by 40–60%.
- Energy consumption: The use of pumps with reduced loads and reduced cleaning cycles translates to less energy consumption.

The LCA demonstrated that the PCC-based capture chromatography significantly reduced the global warming potential (GWP) per gram of mAb in addition to savings in process time and water used [95].

Case 2: Stainless Steel vs. Single-Use DSP Systems- A different study involving comparing stainless-steel systems against single-use technologies (SUT) within a hybrid DSP train found out:

- Water use: SUT demonstrates water use is reduced by more than 80 percent as CIP/SIP has been removed.
- Energy consumption: Reduction in energy consumption in clean-ups by production of heat and steam.
- Waste production Waste production: An increase in solid waste as a result of disposable plastics, which, however, can be controlled through energy-efficient waste incineration.

Although SUT-based process generates more solid waste than the other processes, its overall carbon footprint was lower as its indirect emissions originating in water and energy consumption were less. These results indicate that context-specific LCA is necessary to facilitate comprehensive-system trade-offs [96].

26.6.3 Carbon Footprint, Water Usage, and Waste Generation

Life Cycle Assessment (LCA) studies demonstrate that downstream processing (DSP) can have significant environmental implications, making carbon footprint, water use, and waste creation important. Sustainable technologies and process innovations can reduce these consequences while retaining efficiency [97].

a) **Carbon Footprint (GWP):** Energy-intensive operations, solvent and chemical manufacture, and utilities emit DSP. For stainless-steel chromatography, CIP/SIP is important, whereas continuous and single-use capture methods reduce GWP and boost efficiency.
b) **Water Use:** WFI, buffer prep, and cleaning use lots of water. Inline buffer preparation and single-use technologies eliminate buffer overproduction and cleaning cycles, saving water.
c) **Solid and liquid waste production:** Used buffers, consumables, and disposables are waste. Single-use systems increase plastic trash, although waste-to-energy incineration and recyclable bioplastics lessen environmental effect.

26.6.4 Sustainability Indicators for DSP Design

To measure the process of implementing sustainability into process development, quantifiable indicators are employed during the process development alongside LCA in the course of intensifying the process and selecting the material [98]. Key sustainability indicators for DSP include:

- ***E-Factor (Environmental Factor):*** The ratio of the waste mass produced amount to mass of product and is defined as such. Lower E-factors indicate cleaner processes. E-factors of traditional DSP are large because of overuse of buffer and cleaning waste.
- ***Process Mass Intensity (PMI):*** PMI DSP benchmarks are being formulated whereby improvement over the baseline (e.g., a switch to continuous operations lowers PMI by 4060 percent). PMI is given as

$$PMI = \frac{\text{(Mass of all raw materials, water, solvents)}}{\text{(Mass of the product)}}$$

- ***Water Intensity:*** It is the total amount of water consumed on a unit product (L/g). It is possible to reduce this indicator by half with the help of targeting water reduction *via* closed systems and inline preparation of buffers.
- ***Energy Intensity:*** Measured in kWh per gram of product. Energy intensity is reduced throughout DSP due to inline and continuous processes operations as well as low-pressure chromatography.
- ***GHG Emissions per Unit Output:*** The indicator facilitates the reporting of scope 1 and scope 2 emissions. Usable data that is monitored in a real-time facility is one that can be calculated easily and this can be used in the procurement and in the operations.

26.6.5 Tools and Software for LCA in Bioprocessing

Life Cycle Assessment (LCA) tools and databases help bioprocessors assess environmental performance and build sustainable processes. They help make efficient, environmentally friendly judgments [99–101]. Field-specific bioprocessing LCA modeling tools/databases:

- SimaPro with GaBi—Expert LCA software for cradle-to-grave examination.
- BioGrace—EU bio-based LCA tool for green biopharma feedstock analysis.
- Aspen Plus with Sustainability Toolbox — Compatible with over 1,000 environmental tools and process simulation for sustainability evaluations.

- Ecoinvent and GREET databases provide chemical, energy, and material life cycle inventory data.

These methods help prioritize environmental performance alongside economics and yield by selecting lower-impact chemicals, resins, and solvents early in process development.

Comparative LCAs show that intensive, continuous, and single-use DSP systems save carbon emissions, water and chemicals as well as increase facility flexibility and product uniformity. Such advantages should be considered during the life cycle, along with the procurement of materials, disposal of the wastes, and diversion of the energy, to benefit the environment. The metrics and digital process control of sustainability can assist the developers in enhancing the DSP system and making it environmentally friendly. To obtain green certification, compliance and sustainability of the corporation and the world, LCA should be included in the process design.

26.7 Challenges, Industrial Applications, and Future Outlook

Sustainable downstream processing (DSP) transition has emerged as a strategic focus in the biopharmaceutical sector, and has associated environmental, economic, and functioning advantages. Nevertheless, it is quite complicated to transfer the green technologies succeeding at the laboratory level into the proper scale of industry. This section presents the current challenges, the industrial success stories and the perspectives of the future, such as the potential innovative forms in digitalization, synthetic biology, and circular bioprocessing.

26.7.1 Obstacles and Scalability Barriers and Regulations Adoptions

Sustainable DSP technologies have distinct advantages; yet, their extensive implementation is frequently obstructed by practical, technological, and regulatory obstacles. These obstacles encompass expenses associated with process reform, insufficient data, and ambiguous compliance frameworks. Principal challenges encompass:

- ***Technical and Operational Complexity***: Sustainable techniques such as membrane chromatography or continuous processing necessitate process reconfiguration, novel control strategies, and operator training, which might incur significant costs in current facilities.

- ***Scalability and Compatibility***: Methods like magnetic separation or novel monoliths frequently lack extensive infrastructure and encounter challenges related to flow rate, pressure stability, and GMP compliance.
- ***Insufficient biologics***: specific life cycle assessment data hinders the measurement and comparison of environmental benefits.
- ***Regulatory Ambiguity***: Emerging processes such as continuous chromatography encounter ambiguous validation and control criteria, while global regulatory harmonization remains underway.
- ***Capital Expenditures and Return on Investment***: Substantial initial investments in automated or single-use systems may hinder uptake, particularly for smaller or nascent enterprises with brief product life cycles.

26.7.2 Industrial Case Studies and Success Stories

As previously said, firms such as Amgen and Sanofi have illustrated that the implementation of green DSP technologies can markedly diminish environmental impact while enhancing operational efficiency. In addition to these, several industry leaders have adopted new strategies to attain sustainable improvements. Nonetheless, other businesses have managed to adopt the green DSP technologies, which provides practical value when proving sustainability benefits:

- ***Case study of Lonza's Continuous DSP Implementation:*** PCC and single-pass TFF– The periodic counter-current chromatography (PCC) and single-pass TFF (SPTFF) are used in the commercial stage that Lonza has adopted. These enabled increased throughput of fewer resins, less footprint, and shorter cycles lowering the cost per gram of monoclonal antibody but requiring less utility [102].
- ***Case Study of WuXi Biologics' Greenfield Facility (Ireland):*** This unit is built literally on a modular single-use basis both at the upstream and downstream lines level. Better use of membrane-based polishing steps, inline buffer dilution

and disposable flow paths have allowed quicker batch turnaround and less use of cleaning resources [103].

- ***Case study of Roche:*** Roche has introduced energy recovery technologies and waste reduction measures in its biologics manufacturing facilities [104].
- ***Case study of Novartis:*** Novartis has invested in water recycling technologies and the integration of renewable energy into its biopharmaceutical plants [105].

The examples indicate that much environmental and fiscal advantage can be attained on sustainability when the design and selection of processes in facilities are sustainable through the design of the facility.

26.7.3 Future Perspectives; Digitalization, Synthetic Biology and Circular Bioprocessing

The following decade of sustainable DSP will be characterized by integration of engineering, digital intelligence, and biology. Several transformative trends are on the horizon:

a) Digitalization and AI-Driven Process Control: Using digital twins, machine learning, and predictive analytics helps one control the operations of the DSP dynamically. These tools can: Optimize chromatography gradients and buffer use in real time, detect process drift early and reduce batch failures, inform predictive maintenance and scheduling to reduce downtime and the Manufacturers can involve environmental KPIs (e.g., water use, CO_2 emissions) into the process models to make sustainability a control variable and not a reporting one [91].
b) Synthetic Biology for Simplified DSP: It is in the field of synthetic biology that the troublesome complexity of DSP can be ablated. For example: Genetically manipulating host cell to produce proteins of low levels of impurities (e.g., lower host cell protein burden); producing fusion proteins that bind reversible ligands, simplifying capture; developing self-assembling protein scaffolds which require no multi-step polishing. Such upstream strategies directly reduce downstream unit operations, energy, and buffer requirement [106].
c) Circular Bioprocessing and End-of-Life Solutions: Sustainability is incomplete without addressing end-of-life management [107]. Future DSP strategies will embrace circular bioprocessing, including:

- Chemical or mechanical reprocess of single-use plastics by means of recycling or reuse
- Recovery of solvents and buffer salts
- Processing of waste biomass by valorization (e.g., utilization of spent fermentation media as a substrate biogas production)
- Closed-loop systems that regenerate water and reduce external inputs

The likeliness of biomanufacturing campuses becoming more resource-efficient biorefineries, with a minimized environmental impact, and a reduced supply chain risk, is highly possible.

d) Green Metrics in Regulatory Submissions: With the increasing expectations of sustainability, regulators are likely to encourage or demand the information on green metrics--including carbon footprint, PMI, and the water intensity--be included in CMC submissions. This is able to move the industry towards transparency and environmental performance-driven competition which would further catalyze green innovation [108].

26.8 Conclusion

Sustainable downstream processing (DSP) is essential for harmonizing biomanufacturing with global environmental objectives, minimizing resource intensity while preserving product quality. Green technologies such as membrane chromatography, continuous processing, and bio-based resins substantially reduce water, energy, and waste footprints. Process intensification and digitalization improve efficiency, adaptability, and real-time sustainability oversight. Notwithstanding obstacles in scalability and regulatory acceptance, industry case studies illustrate viable, economical solutions. The future of DSP is rooted in circular bioprocessing, synthetic biology, and data-driven technologies, guaranteeing environmentally sustainable and economically feasible biologics manufacturing.

AI disclosure statement: The author acknowledges the use of AI-based tools for minor language editing or grammatical correction. However, all intellectual contributions, analyses, and conclusions presented in this chapter are entirely the author's own.

References

1. Gupta, S.K. and Shukla, P., Sophisticated cloning, fermentation, and purification technologies for an enhanced therapeutic protein production: a review. *Front. Pharmacol.*, 8, 419, 2017.
2. Forde, G.M., Rainey, T.J., Speight, R., Batchelor, W., Pattenden, L.K., Matching the biomass to the bioproduct: Summary of up-and downstream bioprocesses. *Phys. Sci. Rev.*, 1, 20160046, 2016.
3. Bhat, S.G., Thivaharan, V., Divyashree, M.S., Sustainable opportunities in the downstream processing of the intracellular biopolymer polyhydroxyalkanoate. *ChemBioEng Rev.*, 11, 79–94, 2024.
4. Figoli, A., Marino, T., Simone, S., Di Nicolò, E., Li, X.M., He, T., Tornaghi, S., Drioli, E.J., Towards non-toxic solvents for membrane preparation: a review. *Green Chem.*, 16, 4034–59, 2014.
5. Jungbauer, A., Ferreira, G., Butler, M., D'Costa, S., Brower, K., Rayat, A., Willson, R., Status and future developments for downstream processing of biological products: Perspectives from the Recovery XIX yield roundtable discussions. *Biotechnol. Bioeng.*, 121, 2524–41, 2024.
6. Jungbauer, A. and Walch, N., Buffer recycling in downstream processing of biologics. *Curr. Opin. Chem. Eng.*, 10, 1–7, 2015.
7. Yang, J., Novel Strategies for Cleaning-in-Place Operations, pp. 1–100, 2018.
8. Yabré, M., Ferey, L., Somé, I.T., Gaudin, K., Greening reversed-phase liquid chromatography methods using alternative solvents for pharmaceutical analysis. *Molecules*, 23, 1065, 2018.
9. Gottschalk, U., Disposables in downstream processing, in: *Disposable Bioreactors*, R. Eibl and D. Eibl (Eds.), pp. 171–183, Springer, Berlin, Heidelberg, 2009.
10. Elyasi, S.N., Marami, H., He, L., Kaab, A., Pan, J., Liu, H., Khoshnevisan, B., Up and downstream technologies of anaerobic digestion from life cycle assessment perspective, in: *Renewable Energy Technologies for Energy Efficient Sustainable Development*, A. Sharma and S. Kumar (Eds.), pp. 361–389, Springer, Cham, 2022.
11. Wilson, M.P. and Schwarzman, M.R., Toward a new US chemicals policy: rebuilding the foundation to advance new science, green chemistry, and environmental health. *Environ. Health Perspect.*, 117, 1202–9, 2009.
12. Glavič, P., Pintarič, Z.N., Levičnik, H., Dragojlović, V., Bogataj, M., Transitioning towards net-zero emissions in chemical and process industries: a holistic perspective. *Processes*, 11, 2647, 2023.
13. Arora, P., Benefits and Challenges of Green Marketing in the Pharmaceutical Industry, in: *Green Marketing Perspectives*, S. Kumar (Ed.), pp. 273–284, Emerald Publishing Limited, Bingley, 2025.
14. Koenig, S.G., Bee, C., Borovika, A., Briddell, C., Colberg, J., Humphrey, G.R., Kopach, M.E., Martinez, I., Nambiar, S., Plummer, S.V., Ribe, S.D., A green

chemistry continuum for a robust and sustainable active pharmaceutical ingredient supply chain. *ACS Sustain. Chem. Eng.*, 7, 16937–51, 2019.
15. Schaub, J., Ankenbauer, A., Habicher, T., Löffler, M., Maguire, N., Monteil, D., Püngel, S., Stepper, L., Stiefel, F., Thoma, J., Unsöld, A., Process intensification in biopharmaceutical process development and production–an industrial perspective. *Phys. Sci. Rev.*, 9, 2989–3041, 2024.
16. Ündey, C., Ertunç, S., Mistretta, T., Looze, B., Applied advanced process analytics in biopharmaceutical manufacturing: Challenges and prospects in real-time monitoring and control. *J. Process Control*, 20, 1009–18, 2010.
17. Jorge, A.M. and Pereira, J.F., Aqueous two-phase systems–versatile and advanced (bio) process engineering tools. *Chem. Commun.*, 60, 12144–68, 2024.
18. Schmidt, A., Hengelbrock, A., Strube, J., Continuous biomanufacturing in upstream and downstream processing. *Phys. Sci. Rev.*, 9, 3167–222, 2024.
19. Wurzer, C., Tziambazis, E., Richenhagen, E., From Talk to Action: How Biopharma and Medtech Can Drive Sustainability, in: *Curious Future Insight: Science for a Better Tomorrow*, M. Branca (Ed.), pp. 171–190, Springer, Cham, 2023.
20. Papalexi, M., An investigation into the aspects of innovation within the downstream domain of the pharmaceutical supply chain, pp. 1–200, University of Huddersfield, Huddersfield, 2017.
21. Aryanfar, Y., Castellanos, H.G., Hammoodi, K.A., Farifteh, M., Keçebaş, A., Ghasemlou, S.M., Mammadova, A., Yengejeh, E.A., Neghabi, Z.H., Cleaner and sustainable circular economy approaches for bio-based product recovery from industrial effluents in a biorefinery. *Environ. Prog. Sustain. Energy*, 43, e14483, 2024.
22. Abdussalam-Mohammed, W., Ali, A.Q., Errayes, A.O., Green chemistry: principles, applications, and disadvantages. *Chem. Methodol.*, 4, 408–23, 2020.
23. Boodhoo, K. and Harvey, A. (Eds.), *Process Intensification Technologies for Green Chemistry: Engineering Solutions for Sustainable Chemical Processing*, pp. 1–400, John Wiley & Sons, Hoboken, 2013.
24. Xu, J., Li, C., Dai, L., Xu, C., Zhong, Y., Yu, F., Si, C., Biomass fractionation and lignin fractionation towards lignin valorization. *ChemSusChem*, 13, 4284–95, 2020.
25. Farinha, S., Sá, J.V., Lino, P.R., Galésio, M., Pires, J., Rodrigues, MA., Henriques, J., Spray freeze drying of biologics: a review and applications for inhalation delivery. *Pharm. Res.*, 40, 1115–40, 2023.
26. Misra, N.N., Sullivan, C., Cullen, P.J., Process analytical technology (PAT) and multivariate methods for downstream processes. *Curr. Biochem. Eng.*, 2, 4–16, 2015.
27. Gopi, S.P., Development and characterization of thermoplastic biopolyester compounds and their processing by 3D bioplotting, pp. 1–200, Queen's University, Kingston, 2023.

28. Pareek, A., Buddhiraju, V.S., Masampally, V.S., Premraj, K., Runkana, V., Comparison of multi-column chromatography configurations through model-based optimization. *Biotechnol. Progr.*, 39, e3376, 2023.
29. Madsen, E., Kaiser, J., Krühne, U., Pinelo, M., Single pass tangential flow filtration: Critical operational variables, fouling, and main current applications. *Sep. Purif. Technol.*, 291, 120949, 2022.
30. Zarbis-Papastoitsis, G., Kuczewski, M., Frau, N., Hirai, M., Hydrophobic Membrane Adsorbers for Large-Scale Downstream Processing. *BioPharm Int.*, 22, 2009.
31. Kaul, R. and Mattiasson, B., Extractive bioconversions in aqueous two-phase systems, in: *Extractive Bioconversions*, B. Mattiasson and O. Holst (Eds.), pp. 173–188, CRC Press, Boca Raton, 2021.
32. Singh, N. and Herzer, S., Downstream processing technologies/capturing and final purification: Opportunities for innovation, change, and improvement. A review of downstream processing developments in protein purification, in: *New Bioprocessing Strategies: Development and Manufacturing of Recombinant Antibodies and Proteins*, B. Kiss, U. Gottschalk, M. Pohlscheidt (Eds.), pp. 115–178, Springer, Cham, 2017.
33. Székely, G., Gil, M., Sellergren, B., Heggie, W., Ferreira, F.C., Environmental and economic analysis for selection and engineering sustainable API degenotoxification processes. *Green Chem.*, 15, 210–25, 2013.
34. Perera, C.G., Life Cycle Assessment (LCA) and Techno-Economic Assessment (TEA) of Emerging Technologies: A Case Study of Laboratory Based Carbon Fiber Production from Alberta Oilsands Asphaltene, pp. 1–200, 2023.
35. Li, J., Albrecht, J., Borovika, A., Eastgate, M.D., Evolving green chemistry metrics into predictive tools for decision making and benchmarking analytics. *ACS Sustain. Chem. Eng.*, 6, 1121–32, 2018.
36. Dryden, W.A., Larsen, L.M., Britt, D.W., Smith, M.T., Technical and economic considerations of cell culture harvest and clarification technologies. *Biochem. Eng. J.*, 167, 107892, 2021.
37. Kwang, T.W., Zeng, X., Wang, S., Manufacturing of AcMNPV baculovirus vectors to enable gene therapy trials. *Mol. Ther. Methods Clin. Dev.*, 3, 16002, 2016.
38. Xie, Y., Mao, Z., Bachman, H., Li, P., Zhang, P., Ren, L., Wu, M., Huang, T.J., Acoustic cell separation based on density and mechanical properties. *J. Biomech. Eng.*, 142, 031004, 2020.
39. Gao, Y., Wu, M., Lin, Y., Xu, J., Acoustic microfluidic separation techniques and bioapplications: a review. *Micromachines*, 11, 921, 2020.
40. Banerjee, S., Jesubalan, N.G., Nikita, S., Rathore, A.S., Multi-objective optimization of continuous CHO cell clarification using acoustic wave separation. *Sep. Purif. Technol.*, 338, 126579, 2024.
41. Nejatishahidein, N. and Zydney, A.L., Depth filtration in bioprocessing—new opportunities for an old technology. *Curr. Opin. Chem. Eng.*, 34, 100746, 2021.

42. Gronemeyer, P., Ditz, R., Strube, J., Trends in upstream and downstream process development for antibody manufacturing. *Bioengineering*, 1, 188–212, 2014.
43. Karnieli, O., Bioreactors and downstream processing for stem cell manufacturing, in: *Stem Cell Manufacturing*, J.M. Croop and O. Karnieli (Eds.), pp. 141–160, Elsevier, Amsterdam, 2016.
44. Whitford, W.G., Petrich, M.A., Flanagan, W.P., Environmental Impacts of Single-Use Systems, in: *Single-Use Technology in Biopharmaceutical Manufacture*, R. Eibl and D. Eibl (Eds.), pp. 169–179, John Wiley & Sons, Hoboken, 2019.
45. Eibl, R. and Eibl, D. (Eds.), *Single-Use Technology in Biopharmaceutical Manufacture*, pp. 1–400, John Wiley & Sons, Hoboken, 2019.
46. Shokravi, Z. and Shokravi, H., Implementation of gravity sedimentation in microalgae downstream processes: Principles, advances, and challenges, in: *Algal Bioreactors*, J. Pruvost (Ed.), pp. 59–70, CRC Press, Boca Raton, 2025.
47. Fawell, P.D., Nguyen, T.V., Solnordal, C.B., Stephens, D.W., Enhancing gravity thickener feedwell design and operation for optimal flocculation through the application of computational fluid dynamics. *Miner. Process. Extr. Metall. Rev.*, 42, 496–510, 2021.
48. Patil, R. and Walther, J., Continuous manufacturing of recombinant therapeutic proteins: upstream and downstream technologies, in: *New Bioprocessing Strategies: Development and Manufacturing of Recombinant Antibodies and Proteins*, B. Kiss, U. Gottschalk, M. Pohlscheidt (Eds.), pp. 277–322, Springer, Cham, 2017.
49. Duan, Y., Zhong, Z., Tong, Z., Cheng, Z., Xiong, R., Pilot-scale study on seeded precipitation assisted nanofiltration for flue-gas desulfurization wastewater softening. *Water Sci. Technol.*, 85, 862–71, 2022.
50. Kristensen, M.F., Leonhardt, D., Neland, M.L., Schlafer, S., A 3D printed microfluidic flow-cell for microscopy analysis of *in situ*-grown biofilms. *J. Microbiol. Methods*, 171, 105876, 2020.
51. Joshi, A.K. and Gupta, S.K., Single-use continuous manufacturing and process intensification for production of affordable biological drugs, in: *Process Control, Intensification, and Digitalisation in Continuous Biomanufacturing*, A.K. Joshi (Ed.), pp. 179–208, Wiley-VCH, Weinheim, 2022.
52. Rezk, A.R., Ahmed, H., Ramesan, S., Yeo, L.Y., High frequency sonoprocessing: A new field of cavitation-free acoustic materials synthesis, processing, and manipulation. *Adv. Sci.*, 8, 2001983, 2021.
53. Pierce, T.T., Ozturk, A., Sherlock, S.P., Moura Cunha, G., Wang, X., Li, Q., Hunt, D., Middleton, M.S., Martin, M., Corey, K.E., Edenbaum, H., Reproducibility and repeatability of US shear-wave and transient elastography in nonalcoholic fatty liver disease. *Radiology*, 312, e233094, 2024.
54. Lavoie, J., Fan, J., Pourdeyhimi, B., Boi, C., Carbonell, R.G., Advances in high-throughput, high-capacity nonwoven membranes for chromatography in downstream processing: A review. *Biotechnol. Bioeng.*, 121, 2300–17, 2024.

55. Nadar, S., Shooter, G., Somasundaram, B., Shave, E., Baker, K., Lua, L.H., Intensified downstream processing of monoclonal antibodies using membrane technology. *Biotechnol. J.*, 16, 2000309, 2021.
56. Schneider, A., Herlevi, L.M., Guo, Y., Fernandez Lahore, H.M., Perspectives on adsorption technology as an effective strategy for continuous downstream bioprocessing. *J. Chem. Technol. Biotechnol.*, 97, 2305–16, 2022.
57. Trnovec, H., Impurity removal by ion-exchange membrane absorbers, pp. 1–200, Univerza v Ljubljani, Fakulteta za kemijo in kemijsko tehnologijo, Ljubljana, 2020.
58. Sharma, V., Mottafegh, A., Joo, J.U., Kang, J.H., Wang, L., Kim, D.P., Toward microfluidic continuous-flow and intelligent downstream processing of biopharmaceuticals. *Lab Chip*, 24, 2861–82, 2024.
59. González-González, M., González-Valdez, J., Mayolo-Deloisa, K., Rito-Palomares, M., Monolithic chromatography: insights and practical perspectives. *J. Chem. Technol. Biotechnol.*, 92, 9–13, 2017.
60. Kodermac, MS., Rotar, S., Dolenc, D., Božič, K., Štrancar, A., Černigoj, U., Grafting chromatographic monoliths with charged linear polymers for highly productive and selective protein or plasmid DNA purification. *Sep. Purif. Technol.*, 353, 128253, 2025.
61. Ma, S., Li, Y., Ma, C., Wang, Y., Ou, J., Ye, M., Challenges and advances in the fabrication of monolithic bioseparation materials and their applications in proteomics research. *Adv. Mater.*, 31, 1902023, 2019.
62. Vuignier, K., Fekete, S., Carrupt, P.A., Veuthey, J.L., Guillarme, D., Comparison of various silica-based monoliths for the analysis of large biomolecules. *J. Sep. Sci.*, 36, 2231–43, 2013.
63. Baydemir Peşint, G., Eren Yüngeviş, B., Perçin Demirçelik, I., Enhanced invertase binding from baker's yeast *via* cryogels included boronic acids. *World J. Microbiol. Biotechnol.*, 39, 267, 2023.
64. Chen, J., Yu, B., Zhu, J., Gao, Y., Deng, W., Chen, R., Wang, H.L., Electrospinning of biodegradable, monolithic membrane with distinct bimodal micron-sized fibers and nanofibers for high efficiency PMs removal. *ACS Appl. Mater. Interfaces*, 15, 35507–15, 2023.
65. Grinsted, J., Liddell, J., Bouleghlimat, E., Kwok, K.Y., Taylor, G., Marques, M.P., Bracewell, D.G., Purification of therapeutic & prophylactic mRNA by affinity chromatography. *Cell Gene Ther. Insights*, 8, 335–49, 2022.
66. Ravi, N., Huerta, J., Ferreira, G., Evaluating multiproduct chromatography Protein A resin reuse for monoclonal antibodies in biopharmaceutical manufacturing. *Biotechnol. Progr.*, 39, e3333, 2023.
67. Agarwal, C. and Pandey, A.K., Remediation and recycling of inorganic acids and their green alternatives for sustainable industrial chemical processes. *Environ. Sci. Adv.*, 2, 1306–39, 2023.
68. Sheldon, R.A., Cross-linked enzyme aggregates as industrial biocatalysts. *Org. Process Res. Dev.*, 15, 213–23, 2011.

69. Shukla, A.A., Sunasara, K.M., Rupp, R.G., Cramer, S.M., Hydrophobic displacement chromatography of proteins. *Biotechnol. Bioeng.*, 68, 672–80, 2000.
70. Tseng, W.Y. and van Dam, R.M., Compact microfluidic device for rapid concentration of PET tracers. *Lab Chip*, 14, 2293–302, 2014.
71. Mant, C.T. and Hodges, R.S. (Eds.), *High-Performance Liquid Chromatography of Peptides and Proteins: Separation, Analysis, and Conformation*, pp. 1–960, CRC Press, Boca Raton, 2017.
72. Almeida, C., Pedro, A.Q., Tavares, A.P., Neves, M.C., Freire, M.G., Ionic-liquid-based approaches to improve biopharmaceuticals downstream processing and formulation. *Front. Bioeng. Biotechnol.*, 11, 1037436, 2023.
73. Santana, A.L., Santos, D.T., Meireles, M.A., Perspectives on small-scale integrated biorefineries using supercritical CO2 as a green solvent. *Curr. Opin. Green Sustain. Chem.*, 18, 1–2, 2019.
74. Abdullah, M.A., Hussein, H.A., Alshajrawi, O.M., Ethyl lactate as a green solvent in the pharmaceutical industry, in: *Green Sustainable Process for Chemical and Environmental Engineering and Science*, I. Ahmad, S. Ahmad, M.A. Quraishi (Eds.), pp. 185–194, Elsevier, Amsterdam, 2021.
75. Fields, C., Li, P., O'Mahony, J.J., Lee, G.U., Advances in affinity ligand-functionalized nanomaterials for biomagnetic separation. *Biotechnol. Bioeng.*, 113, 11–25, 2016.
76. Iranmanesh, M. and Hulliger, J., Magnetic separation: its application in mining, waste purification, medicine, biochemistry and chemistry. *Chem. Soc. Rev.*, 46, 5925–34, 2017.
77. Ebeler, M., Lind, O., Norrman, N., Palmgren, R., Franzreb, M., One-step integrated clarification and purification of a monoclonal antibody using Protein A Mag Sepharose beads and a cGMP-compliant high-gradient magnetic separator. *New Biotechnol.*, 42, 48–55, 2018.
78. Kruljec, N. and Bratkovič, T., Alternative affinity ligands for immunoglobulins. *Bioconjugate Chem.*, 28, 2009–30, 2017.
79. Schmitz, F.R., Valério, A., de Oliveira, D., Hotza, D., An overview and future prospects on aptamers for food safety. *Appl. Microbiol. Biotechnol.*, 104, 6929–39, 2020.
80. Santos, M.A., Irto, A., Buglyó, P., Chaves, S., Hydroxypyridinone-based metal chelators towards ecotoxicity: Remediation and biological mechanisms. *Molecules*, 27, 1966, 2022.
81. Patrașcu, I., Bîldea, C.S., Kiss, A.A., Eco-efficient downstream processing of biobutanol by enhanced process intensification and integration. *ACS Sustain. Chem. Eng.*, 6, 5452–61, 2018.
82. Gomis-Fons, J., Andersson, N., Nilsson, B., Optimization study on periodic counter-current chromatography integrated in a monoclonal antibody downstream process. *J. Chromatogr. A*, 1621, 461055, 2020.

83. Brinkmann, A., Elouafiq, S., Pieracci, J., Westoby, M., Leveraging single-pass tangential flow filtration to enable decoupling of upstream and downstream monoclonal antibody processing. *Biotechnol. Progr.*, 34, 405–11, 2018.
84. Göbel, S., Pelz, L., Reichl, U., Genzel, Y., Upstream processing for viral vaccines-Process intensification, in: *Bioprocessing of Viral Vaccines*, A. Noth and U. Reichl (Eds.), pp. 137–173, CRC Press, Boca Raton, 2022.
85. Gibson, K., Oliveira, J.C., Ring, D., Optimisation of biopharmaceutical buffer management to maximise the benefits of inline preparation systems. *Results Eng.*, 25, 104080, 2025.
86. Andersson, N., Fons, J.G., Isaksson, M., Tallvod, S., Espinoza, D., Sjökvist, L., Andersson, G.Z., Nilsson, B., Methodology for fast development of digital solutions in integrated continuous downstream processing. *Biotechnol. Bioeng.*, 121, 2378–87, 2024.
87. Vojinović, V., Cabral, J.M., Fonseca, L.P., Real-time bioprocess monitoring: Part I: *In situ* sensors. *Sens. Actuators B Chem.*, 114, 1083–91, 2006.
88. Wasalathanthri, D.P. and Patel, B.A., The role of process analytical technology (PAT) in biologics development, in: *Continuous Pharmaceutical Processing and Process Analytical Technology*, A.S. Rathore and Z.K. Nagy (Eds.), pp. 339–354, CRC Press, Boca Raton, 2023.
89. Drobnjakovic, M., Hart, R., Kulvatunyou, B., Ivezic, N., Srinivasan, V., Current challenges and recent advances on the path towards continuous biomanufacturing. *Biotechnol. Progr.*, 39, e3378, 2023.
90. Samaras, J.J., Micheletti, M., Ding, W., Transformation of biopharmaceutical manufacturing through single-use technologies: current state, remaining challenges, and future development. *Annu. Rev. Chem. Biomol. Eng.*, 13, 73–97, 2022.
91. Canzoneri, M., Horner, M., Banerjee, P., Digital twins in biomanufacturing, in: *Bioconvergence Revolution: Challenges and Opportunities in Research Productivity*, M. Branca (Ed.), pp. 59–81, Springer, Cham, 2024.
92. Yang, M., Optimization of warehouse operations and transport risk mitigation for disposable bioreactor bags to support launch of Amgen Singapore Manufacturing, pp. 1–200, Massachusetts Institute of Technology, Cambridge, 2015.
93. Demesmaeker, M., Kopec, D., Arsénio, A.M., Bioprocessing 4.0–Where Are We with Smart Manufacturing in 2020? *Pharm. Outsourcing*, 2020.
94. Connelly, E.B., Colosi, L.M., Clarens, A.F., Lambert, J.H., Life cycle assessment of biofuels from algae hydrothermal liquefaction: the upstream and downstream factors affecting regulatory compliance. *Energy Fuels*, 29, 1653–61, 2015.
95. Bunnak, P., Allmendinger, R., Ramasamy, S.V., Lettieri, P., Titchener-Hooker, N.J., Life-cycle and cost of goods assessment of fed-batch and perfusion-based manufacturing processes for mAbs. *Biotechnol. Progr.*, 32, 1324–35, 2016.

96. Pietrzykowski, M., Flanagan, W., Pizzi, V., Brown, A., Sinclair, A., Monge, M., An environmental life cycle assessment comparing single-use and conventional process technology. *BioPharm Int.*, 24, 30–8, 2011.
97. Jiménez-González, C. and Overcash, M.R., The evolution of life cycle assessment in pharmaceutical and chemical applications–a perspective. *Green Chem.*, 16, 3392–400, 2014.
98. Veleva, V.R., Cue Jr., B.W., Todorova, S., Thakor, H., Mehta, N.H., Padia, K.B., Benchmarking green chemistry adoption by the Indian pharmaceutical supply chain. *Green Chem. Lett. Rev.*, 11, 439–56, 2018.
99. Dunn, J.B., Adom, F., Sather, N., Han, J., Snyder, S., He, C., Gong, J., Yue, D., You, F., Life-cycle analysis of bioproducts and their conventional counterparts in GREET, Argonne National Lab., Argonne, IL, 2015.
100. Prado-Rubio, O.A., Gasca-González, R., Fontalvo, J., Gómez-Castro, F.I., Pérez-Cisneros, E.S., Morales-Rodriguez, R., Design and evaluation of intensified downstream technologies towards feasible lactic acid bioproduction. *Chem. Eng. Process. Process Intensif.*, 158, 108174, 2020.
101. Samson-Bręk, I., Smerkowska, B., Filip, A., Environmental aspects in the life cycle of liquid biofuels with biocomponents, taking into account the storage process, in: *Storage Stability of Fuels*, K. Biernat (Ed.), pp. 1–20, IntechOpen, London, 2015.
102. Macdonald, G.J., CDMOs Embrace Industry 4.0 to Satisfy Customer Demand: The digitalization of bioprocessing technology isn't happening for its own sake, but to ensure that outsourced manufacturing tasks, however diverse, will be part of highly integrated processes. *Genet. Eng. Biotechnol. News*, 41, 43–5, 2021.
103. Zhou, H., Fang, M., Zheng, X., Zhou, W., Improving an intensified and integrated continuous bioprocess platform for biologics manufacturing. *Biotechnol. Bioeng.*, 118, 3618–23, 2021.
104. Kelley, B., Renshaw, T., Kamarck, M., Process and operations strategies to enable global access to antibody therapies. *Biotechnol. Progr.*, 37, e3139, 2021.
105. Behera, B., *Biopharmaceuticals: Challenges and Opportunities*, pp. 1–300, CRC Press, Boca Raton, 2020.
106. Shi, T., Han, P., You, C., Zhang, Y.H., An *in vitro* synthetic biology platform for emerging industrial biomanufacturing: Bottom-up pathway design. *Synth. Syst. Biotechnol.*, 3, 186–95, 2018.
107. Aranda-Barradas, J.S., Guerrero-Barajas, C., Ordaz, A., Addressing Challenges in Large-Scale Bioprocess Simulations: A Circular Economy Approach Using SuperPro Designer. *Processes*, 13, 2025.
108. Fisher, A.C., Kamga, M.H., Agarabi, C., Brorson, K., Lee, S.L., Yoon, S., The current scientific and regulatory landscape in advancing integrated continuous biopharmaceutical manufacturing. *Trends Biotechnol.*, 37, 253–67, 2019.

Part 7

FUTURE DIRECTIONS AND ADVANCES

27

Next-Gen Nanocarriers for Transdermal Transport of Therapeutic Biologics

Gurugubelli Sowjanya*, Varri Ambica Taruni and Kommula L. N. S. S. Dharani Devi

School of Pharmacy, Aditya University, Surampalem, Andhra Pradesh, India

Abstract

The transdermal drug delivery system has no permeability for biologics. As they are large molecules, our skin, the epidermal layer, acts as an impediment. Next generation of nanocarriers is being developed in the field of nanotechnology. It enhances the transdermal transport of biologics like amino acid chains, oligopeptides, and polynucleotides and also overcomes the skin's natural barrier function with an efficient on-site specific drug delivery system. Polymeric nanoparticles, liposomes, and metallic nanoparticles develop various nanocarriers. These nanocarriers can be designed by incorporating them with specific properties like size, charge, and surface modification for easy facilitation. The steps in the methodology involve selection of biologics, designing, surface modification, and loading the biologic drug, *in vitro* skin penetration tests, evaluation of stability and release, and data analysis and optimization. The nanocarriers have various features, like they specify the target, improving the bioavailability, reducing the toxicity, and enhancing patient compatibility. The major disadvantages include the biological instability, low drug loading, and manufacturing complexity. Thus, the NEXT-GEN nanocarriers can be revolutionized, offer enhanced efficacy, safety, and patient compliance. This innovative approach provides treatment for various diseases like cancer, autoimmune disorders, and other infectious diseases, enhancing patient compliance and development in the field of biomedical research.

***Keywords*:** Stratum corneum, liposomes, penetration, bioavailability, autoimmune disorder, nanoparticles

**Corresponding author*: Vekshijaasri@gmail.com

Sandip Kunar, Jagadeesha T. and D. Sathis Kumar (eds.) Biomanufacturing: Transforming Production of Biologics and Biopharmaceuticals, (887–938)

27.1 Introduction

27.1.1 Overview of Transdermal Drug Delivery Systems (TDDS)

Transdermal drug delivery systems (TDDS) illustrate the critical enhancement in controlled release and non-intrusive therapeutic administration, offering an appealing manifest to per-oral and injectable routes. TDDS allows drugs to be absorbed straight away through the skin into blood i.e.; systemic circulation, thereby bypassing hepatic pre-systemic metabolism and gastrointestinal disintegration [1]. These systems release drugs in small amounts in a continuous manner by using adhesive patches applied to the skin. From the first transdermal patch invention which is having scopolamine in 1979, TDDS has seen significant improvement in technologies in dosage forms [2].

The outer layer of body, which is known as skin, produces a largest surface area to apply the medication. Anyways, exclusively a small range of drugs typically small particles having low molecular weight (less than 500 Daltons), moderate lipophilicity, and more potency can passively diffuse *via* the stratum corneum which is the outermost layer of the skin [3]. Conventional TDDS platforms include reservoir-type and matrix-type patches, gels, sprays, and films. These systems have found clinical utility in the management of chronic effects like hypertension, endocrine replacement therapy, analgesia, and smoking cessation [4]. Recent innovations have seen the appearance of advanced TDDS that use chemical enhancers, iontophoresis, sonophoresis, and microneedles to avoid the shielding nature of the skin and improve drug penetration [5]. Nevertheless, delivering big, water loving, and unstable biological therapeutics remains a formidable challenge.

27.1.2 Importance of Biological Therapeutics

It involves proteins, peptides, and monoclonal immunoglobulin's, nucleic acids which comprises of siRNA and mRNA, and gene-editing tools have improved modern medicine due to their high specificity, potency, and ability to target diseases at a molecular level [6]. These biologics form the backbone of therapeutic strategies for complex disorders along with hypersensitivity diseases, diabetes, unproliferated cell division, and infections. The global market of bio pharmaceuticals was appreciated at over USD 400 billion in 2022, and it is expected in the future that there is an enhancement of a yearly growth rate (CAGR) of over 8% in the next 5 years [7].

In spite of their efficacy, biologics face significant delivery problems because of their increased molecular weight, hydrophilicity, along with structural sensitivity with environmental conditions. The majority of these agents are taken through injections, which are induced with poor patient compliance, pain, infection risks, and the need for professional healthcare assistance [8]. Oral delivery of biologics is largely unsuccessful because of enzymatic deterioration in the pathway of stomach and intestine and poor intestinal absorption [9]. In this context, transdermal release of biologics offers a hopeful, patient-friendly choice. However, successful implementation is limited to the behavior of the outermost layer of body which is known as skin. Therefore, the development of advanced nanocarriers capable of facilitating the transdermal delivery of biologics is gaining increasing attention in pharmaceutical research and development [10].

27.1.3 Limitations of Classic Delivery Methods

Classic drug delivery approaches, while efficient for small-molecule treatment, are ill-suited for the delivery of drugs due to various major drawbacks. Intravenous or subcutaneous injections, even though commonly employed, needs trained healthcare professionals and carry the risk of infection and needle-stick injuries [11]. Furthermore, the common need for repeated dosing in chronic conditions leads to decreased adherence and poor therapeutic outcomes [12]. Oral administration of biologics is largely ineffective due to the presence of proteolytic enzymes, acidic pH in the stomach, and tight epithelial junctions in the intestines, all of which impede drug stability and absorption [13]. Consequently, the systemic bioavailability of orally administered drugs is usually below 2% [14].

Moreover, the use of chemical enhancers or physical methods (such as iontophoresis or sonophoresis) in conventional TDDS to improve the permeability of biological molecules often results in skin irritation, inflammation, or cytotoxicity [15]. These methods may temporarily disturb the stratum corneum but are restricted in their capability to distribute high molecular weight, hydrophilic drugs efficiently.

In this context, nanotechnology produces a transformative solution. Nanocarriers such as liposomes, ethosomes, solid lipid nanoparticles, and dendrimers are prepared to encapsulate biologics, protect them from degradation, and facilitate their passage across the skin barrier [16]. They can be prepared with tuneable surface chemistry, size, and responsiveness to physiological stimuli, enabling targeted and controlled drug release [17]. Inducing nanocarriers into transdermal systems not only improves the permeability of biologics but also produces opportunities for easy delivery mechanisms, such as stimuli-responsive release or ligand-mediated targeting. These next-generation systems embrace the potential to overcome the drawbacks of traditional methods and provide effective, safe, and patient-compliant delivery of biologics.

27.2 The Skin Barrier and Challenges in Transdermal Delivery

27.2.1 Anatomy and Physiology of the Skin

The skin is an advanced and complicated organ, acting as an obstruction and interface within the outer environment and the inner environment. Structurally, the skin consists of three major layers: the epidermis, dermis, and hypodermis. Among these layers, the stratum corneum (SC) which is the extreme layer of the epidermis is the important obstacle to transdermal delivery of biomedicals [17]. This layer is comprised of dead, flattened keratinocytes (corneocytes) incorporated in a lipid-rich matrix, often likened to a "brick-and-mortar" structure: the corneocytes being the bricks and the intercellular lipids forming the mortar [18].

The below Figure 27.1 illustrates the diagram of skin labeled with epidermis - stratum corneum as a primary barrier, viable epidermis, dermis with vasculature, tight junctions, lipid bilayers that block a macromolecule entry.

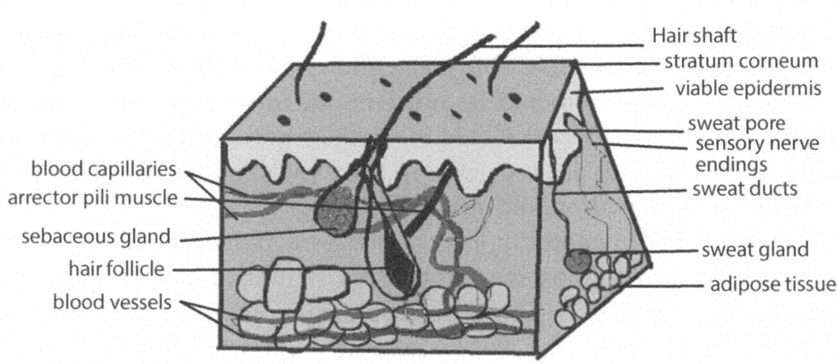

Figure 27.1 Structural layers of the skin and associated barriers.

The outer most layer of epidermis is nearly 10–20 μm width in most body areas and is highly lipophilic in nature. It is prepared to stop trans epidermal water loss and prevent the entry of foreign particles, which involves pathogens and therapeutic molecules. Beneath the SC lies the viable epidermis (~100 μm), which is relatively more permeable and metabolically active, followed by the dermis, which houses blood vessels, lymphatics, and nerve endings [19].

27.2.2 Physicochemical Barriers to Transdermal Delivery

Transdermal transport is maintained primarily by Fick's first law of diffusion, which describes states that a molecule must exhibit appropriate lipophilicity, less molecular weight (<500 Da), and sufficient potency to diffuse across the SC [20]. This creates significant constraints for the delivery of biological drugs, which usually exceed 1,000 Da and contains poor lipid solubility. Biologics such as insulin (~5.8 kDa), erythropoietin (~34 kDa), and monoclonal antibodies (~150 kDa) face extreme resistance from the SC due to their size and hydrophilicity [21]. In addition, these molecules are susceptible to disintegration by skin enzymes like esterases and proteases, which can compromise bioavailability before the therapeutic reaches systemic circulation [22].

Furthermore, drug absorption is affected by different physiological factors including kin hydration and pH (typically ~5.5), Anatomical site (e.g., face and scrotum are more permeable than palms and soles), age (neonatal and geriatric skin exhibit altered barrier functions), skin diseases (e.g., psoriasis or eczema can either enhance or hinder drug absorption) [23].

27.2.3 Immunological and Enzymatic Barriers

Apart from physical barriers, the skin also works as an immunological protector. Langerhans cells, located within the epidermis, engaged in a major activity in MHC presentation and immune triggering. The involvement of these immune cells can sometimes lead to undesired immunogenicity when foreign therapeutic proteins are delivered transdermally [24].

Moreover, enzymatic activity in the epidermis can lead to the disintegration of sensitive biologics before they enter into deeper layers. The skin expresses a wide arrangement of enzymes which involves aminopeptidases, esterases, and cytochrome P450 isoforms [25]. These enzymes act as a major barrier especially for gene- and peptide-based therapies.

27.2.4 Mechanical and Structural Resistance

The lipid matrix of the stratum corneum contains ceramides, cholesterol, and free fatty acids which are arranged in highly ordered lamellar forms. These lipids stops the permeation of polar and large molecules [26]. In addition, tight junctions located in the stratum granulosum maintains paracellular transport and further prevent drug passage through intercellular spaces.

A vital structural barrier also arises from the lack of aqueous pores in the SC, which means hydrophilic macromolecules must rely on either physical disruption of the skin or engineered delivery systems like nanocarriers to breach this layer [27].

27.2.5 Methods to Overcome the Epidermal Barrier

To improve transdermal transport, many chemical and physical improvers have been examined: Chemical enhancers like ethanol, oleic acid temporarily disturb the SC lipid matrix, enhancing permeability [28]. Physical methods such as iontophoresis i.e., electric current, sonophoresis i.e., ultrasound and microneedles develop microchannels or use energy to support transport [29]. However, both methods have drawbacks. Chemical enhancers may cause irritation or long-term barrier disturbance, while physical methods usually need devices that may not be user-friendly or cost-effective [30].

Thus, the development of nanocarrier systems offers a promising alternative. These carriers—engineered at the nanoscale—can interact with the skin on a molecular level to facilitate drug penetration, either by fusing with lipid domains or by forming temporary disruptions in the SC [31].

27.2.6 The Need for Next-Generation Nanocarriers

Given the complex barriers outlined above, it is evident that traditional TDDS are insufficient for distributing the drugs across the skin. Nanoparticles such as lipid vesicles, niosomes, solid lipid nanoparticles, and dendrimers have been prepared to avoid these limitations. These systems can: improve drug solubility, block the cargo enzymatic deterioration, enhance less and sustained release, interact directly with skin components to enhance permeation [32]. Some systems are also engineered to be stimuli-responsive, releasing their payload upon exposure to pH, temperature, enzymes, or light. This responsiveness in future enhances bioavailability while preventing systemic side effects [33].

Moreover, surface changed of nanocarriers with cell- penetrating peptides (CPPs), specific ligands, or stealth coatings (e.g., PEGylation) has shown to significantly enhance skin penetration and reduce immune clearance [34]. The convergence of these technologies offers a novel avenue for the transdermal route of administration of macromolecular therapeutics, mainly in chronic diseases where patient compliance is essential.

27.3 Overview of Nanocarrier Systems

27.3.1 Introduction to Nanocarriers in Transdermal Delivery

Nanocarriers are an advanced class of drug administration systems prepare to avoid the restrictions of conventional transdermal techniques. These nanoscale delivery platforms—ranging from 1 to 1000 manometers in size can summarize therapeutic agents, enhance their physio-chemical stability, and improve skin permeability by reacting directly with the stratum corneum [34]. They also provide reduced and sustained delivery, which is specifically beneficial for delivering sensitive biological medications such as peptides, proteins, and nucleic acids [35]. The capability of nanocarriers to avoid enzymatic degradation, enhance skin retention time, and reduce systemic toxicity has rendered them indispensable in transdermal therapeutics [36].

27.3.2 Classification of Nanocarrier Systems

Nanocarriers for transdermal delivery are divided into different types based upon their composition, structure, and mechanism of delivery.

27.3.2.1 Lipid-Based Nanocarriers

These involves the following:

Liposomes: Spherical vesicles with amphipathic bilayers skilled of embedded in both hydrophilic and hydrophobic drugs. They enhance skin hydration and fuse with the lipid grid of the stratum corneum to improve medicament permeation [37].

Niosomes: Non-ionic surfactant-based vesicles gives an improved stability and skin penetration over conventional liposomes [38].

Solid Lipid Nanoparticles (SLNs): Consist of biocompatible solid lipids, SLNs to enhance stability of drug and provide slow release [39].

Nanostructured Lipid Carriers (NLCs): Modified SLNs involving a mix of solid and liquid lipids, offering enhanced drug incorporation and producing kinetics [40].

27.3.2.2 Polymeric Nanoparticles

These involve decomposable and biopolymer substances like

PLGA (poly (lactic-co-glycolic acid)) nanomaterials, chitosan nanoparticles, polycaprolactone nanoparticles. These systems protect medicines from breaking and can be coated using some particles which are used to enhance transdermal specification [41].

27.3.2.3 Dendrimers

Dendrimers are more divided, tree-like large molecules having well-defined structure. They can carry drugs either through covalent conjugation or imprisoned within internal spaces. Polyamidoamine (PAMAM) dendrimers are the ones which are used to study for transdermal applications [42].

27.3.2.4 Nano Emulsions and Microemulsions

These are energetically stable suspension systems composed of oil, water, and surfactant/co-surfactant mixtures. Their small droplet size allows improved penetration and bioavailability of drugs [43].

27.3.2.5 Metallic and Inorganic Nanoparticles

Examples include Au nanoparticles, silica nanoparticles, and quantum dots. Even though still in early stages for transdermal drug delivery, these carriers provide high surface area and the possibility for changes with targeting moieties [44].

27.3.3 Mechanisms of Nanocarrier Penetration through the Skin

Nanocarriers improve skin penetration through several routes:

Intercellular pathway: Between corneocytes, exploiting the lipid lamellae.
Transcellular pathway: Directly with the corneocytes.
Trans appendageal route: Through sweat glands or hair follicles.

Different studies propose that nanosized materials (<300 nm) primarily use the trans follicular route, while lipid-based nanocarriers manage to mix with the SC to produce temporarily disruption in the barrier [45]. Dendrimers, with their nanoscale size particles and surface charge, have shown efficient skin absorption *via* both follicular and intercellular pathways [46].

27.3.4 Advantages of Nanocarriers in Biological Drug Delivery

Nanocarriers produces many advantages mainly for therapeutic biologicals:
Prevention from enzymatic degradation: By embedding or complexing the payload, nanocarriers inhibit degradation by proteases and nucleases [47]. Enhanced transdermal permeation: By interacting along with skin lipids or promoting temporary diffusion pathways [48]. Targeted delivery: By modification with ligands, antibodies, or peptides which binds to skin-specific receptors [49]. Stimuli-responsive release: pH-sensitive or temperature-sensitive systems permit precise spatiotemporal drug delivery [50]. These advantages are mainly essential for biologics such as insulin, calcitonin, siRNA, monoclonal antibodies, and growth hormones which are poorly absorbed *via* skin when delivered in their native form.

27.3.5 Examples of Nanocarrier-Based Transdermal Biological Systems

Different nanocarrier-based preparations have been interrogated for biologicals:

Insulin-loaded nanostructured lipid carriers (NLCs) depict improved skin permeability and sustained glucose-lowering effects in diabetic animal models [51]. Chitosan-coated liposomes were used for transdermal transportation of interferon-alpha, attaining systemic levels with very less skin irritation [52].

Dendrimer-siRNA complexes have produced successful demolish of target genes in skin cancer models upon topical administration [53]. Nano emulsion-based vaccines transported through skin patches have been determined as non-invasive immunization strategies [54].

27.3.6 Safety and Regulatory Considerations

Despite their promising features, nano carriers must be evaluated for:

Cytotoxicity: Especially for cationic dendrimers and metallic nanoparticles.
Immunogenicity: Unusual immune stimulation or hypersensitivity reactions.

Long-range accumulation: Nanoparticles may gather in the skin or systemic organs if not properly cleansed [55]. Regulatory agencies like the FDA (Food and Drug administration) and EMA (European Medicines Agency) are developing frameworks to produce nano medicine safety. Key measurements such as particle size distribution, zeta potential, surface chemistry, and biodegradability are regarded critical quality features [56].

27.4 Keen: Stimuli-Responsive Nano Carriers

Stimuli-responsive nano carriers allow precise control over when drugs are released by responding to internal changes like pH, redox reactions, chemicals, or external factors like light, magnetic fields, and ultrasound. These nanoparticles help to make sure whether the drugs and detecting agents are using exactly when and where they're needed. Improved detecting methods can also detect how drugs travel through the body, how they're delivered, and how well they are absorbed, helping to calculate their delivery efficiency [57]. Nanocarriers have been very efficient at producing effects to specific tissues and cells, which show improved bioavailability, higher drug loading capability, better cell uptake, and more efficient drug outcome. They show great effect for specific site drug delivery in conditions where reduced release is necessary based on particular body conditions or stimulators. In recent years, multifunctional nanocarrier systems have been developed by attaching various targeting ligands, such as folic acid, peptides, and antibodies, to the surface of the nanomaterials [58].

The compounds like quercetin, curcumin, and resveratrol are structured through stimuli-responsive carriers, and basically, they show fundamental triggered-release component, and restorative potential properties. Shrewd for long-term: Stimuli-responsive smart/intelligent materials play a surprising part in hardware, sensors and restorative applications owing to changes of physico-chemical properties and mechanical changes [59].

27.4.1 pH-Responsive Nanocarriers

pH-responsive nano systems show a path way for transporting and delivering anticancer agents. These systems also have applications in producing metal particles and biomolecules like proteins and nucleic acids, as well as in co-delivering different types of cargo. They include double pH-sensitive nanocarriers, dual or multi-stimuli-responsive nano systems, and are being explored for new treatment approaches for diseases such as Alzheimer's. To produce a system with optimized characteristics, it is important to understand different pH-responsive micro and nanoparticles. Tools like long-circulating PEGylated liposomes and PEG-phosphatidylethanolamine (PEG-PE)-based micelles have been developed with various functionalities [60, 61]. pH-responsive nanocarriers can be made from stimuli-responsive polymers that detect little differences in the surrounding pH, which in turn causes changes in the polymer's physical properties such as size, shape, or hydrophobicity.

27.4.2 Thermo-Responsive Nanocarriers

Temperature-sensitive nanocarriers are among the very necessary types of intelligent nano systems which have been studied over the old few decades. Temperature can act as an outside trigger when heat is triggered from outside, or as an inside trigger when there is an abnormal increase in temperature due to injury or disease. A lower critical solution temperature (LCST) is a unique property of some polymeric materials, and most temperature-sensitive nanocarriers are modified with the help of this property [62].

Mitochondria have become a key target in cancer treatment. By using the slightly higher temperature of mitochondria, a carefully designed thermos responsive nanocarrier was improved to particularly deliver the anticancer agents, paclitaxel (PTX), to mitochondria in proliferating cells [63]. A process was described to create thermos responsive polymer coatings on the surface of magnetic nanocarriers prepared using many iron oxide superparamagnetic nanoparticles impregnated in poly(maleic anhydride-alt-1-octadecene) polymer nanobeads [64].

27.4.3 Enzyme-Sensitive Systems

Most of the chemicals, such as proteases, phospholipases, glycosidases, and others, are known to be present in tumor and inflamed tissues. The substrates of these chemicals can be used as parts of a controlled delivery system to achieve drug release that is triggered by proteins at the target tissue. Enzymes help with normal body functions and can become imbalanced in many disorder -related environments and unusual cell processes [65].

Enzymes with inherent properties, like high related for various disorders, particular substrate selection, and high catalytic efficiency, have been efficiently used in the design of responsive particles.

Matrix metalloproteinase (MMPs) are the most well-established enzymes used as triggers for enzyme-responsive systems. MMPs are a family of zinc-dependent endopeptidases that play a role in the breakdown of extracellular matrices (ECM) and the regulation of bioactive molecules on the cell surface [66].

27.4.4 Magnetic and Light Triggered Responsive Nanocarriers

Surface-functionalized magnetic iron oxide nanoparticles offer new strategies for MNP-based drug delivery in cancer imaging and therapy, addressing key limitations [67].

The FDA-approved superparamagnetic iron oxide nanoparticles [IONPs], known for their good compatibility and high magnetization, have become a focus in biopharmaceutical fields, especially for drug delivery systems that combine magnetic tumor targeting with MR imaging and chemo-phototherapy. HMCuS nanoparticles were used to carry the chemotherapy drug doxorubicin [DOX] and then coated with multifunctional IONPs to enhance features such as improved photothermal therapy effect, controlled drug release, magnetic targeting ability, and MR imaging. This created a smart and highly versatile all-in-one drug delivery system. Additionally, the hydrophilic PEG coating improved the nanoparticles' biocompatibility and extended circulation time in the body [68].

27.5 Sorts of Nanocarriers for Transdermal Delivery

27.5.1 Liposomes and Transferosomes

27.5.1.1 Liposomes

Liposomes are round, lipid-based structures, usually between 50 to 500 nano meters in size, made up of one or more layers of lipids. They are formed when normal or specially designed lipids are mixed with water, creating a stable structure. Structure of liposome is shown in Figure 27.2.

It highlights the liposome's aqueous core and lipid bilayer, which shows its versatile nature in transdermal drug delivery system. It is the structure of liposomes which were first discovered in the 1960s by Bingham. Because they are biocompatible, stable, easy to make, and effective at delivering drugs, they are widely used. Their structure makes them useful in transdermal drug delivery systems. Liposomes are categorized into four types based on their size and the number of layers they have: small unilamellar vesicles (SUV), large unilamellar vesicles (LUV), multilamellar vesicles (MLV), and multivesicular vesicles (MVV) [71].

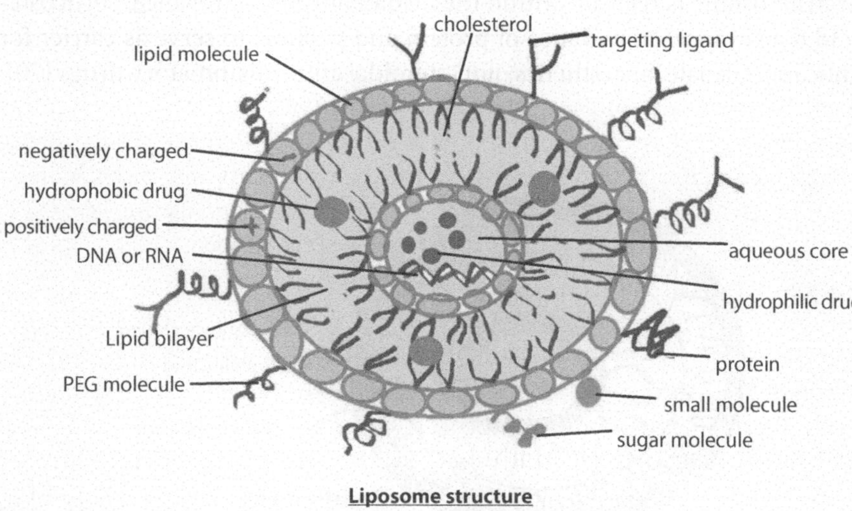

Figure 27.2 Structure of liposome.

27.5.1.2 Transferosomes

Transferosomes, developed by Thought AG, are flexible vesicles that were first introduced in the 1990s. Initially, these vesicles were prepared by using soy phosphatidylcholine, sodium cholate, and a small amount of ethyl alcohol. They have been used to deliver various materials, including small particles, peptides, proteins, and vaccines, both *in vitro* and *in vivo.* Thought AG claims that these transferosomes can pass through the outer layer of the skin and the deeper layers into the bloodstream [69].

Figure 27.3 contains the components like edge activator, phosphatidylcholine, hydrophilic, and hydrophobic drugs. It is used for drug delivery that is encapsulated with both water soluble and fat-soluble drugs.

Transferosomes has a few potential preferences such as the capacity to misshape and pass through limit pores, serve as a carrier for both moo as well as tall atomic drugs.

The component behind the transferosome entrance into skin is the advancement of osmotic angle since of vanishing of surface water due to body warm. There are different strategies existing for the planning of transferosome such as film scattering strategy, switch stage vanishing strategy, high-pressure homogenization strategy, and ultrasonic scattering strategy. Transferosome is relevant within the field of affront conveyance, corticosteroid conveyance, conveyance of protein and peptide, to serve as carrier for anticancer sedate, anaesthetics, non-steroidal anti-inflammatory drugs [70].

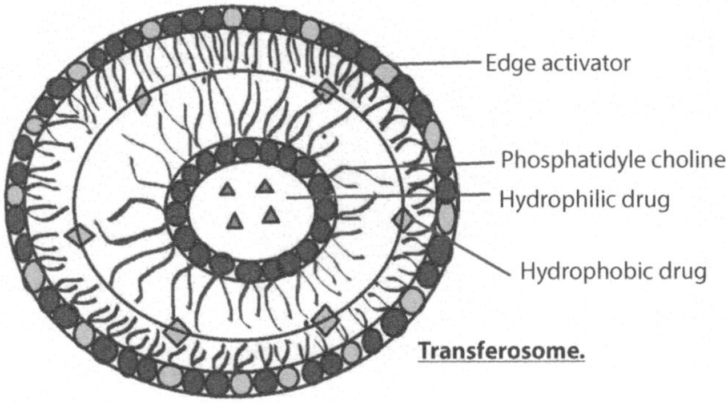

Figure 27.3 It illustrates a transfer some, a deformable liposome.

27.5.2 Ethosomes and Niosomes

27.5.2.1 Ethosomes

Ethosomes are the bilayer lipid vesicles which permit the exchange of medicament to the profound outer body layers and into the systemic circulation. It contains expensive sum of ethanol; entrance enhancers are joined into it to upgrade the rate of entrance. It contains the phospholipid bilayer with polar head and nonpolar tail.

At the center of ethanolic depth is a display which is composed of the hydroethanolic solution [72]. Touitou designed an unused vesicular system, which was named ethosomes, because of the nearness of ethyl alcohol within the structure of vesicle.

The below Figure 27.4 represents a detailed illustration of an ethosome, which shows its layered composition and highlighting its potential applications in drug delivery systems. The ethosome structure is characterized by its lipid molecules, phospholipid bilayers, and separate hydrophilic and hydrophobic regions.

<u>Types of ethosomal frameworks</u>:-There are three sorts of ethosomal frameworks based on their composition.

Classical ethosomes: Classical ethosomes are a type of adjusted ethosome made up of phospholipids, water, and a high amount of ethanol, up to 45% weight per volume. Drugs with an atomic weight between 130.077 Da and 24,000 Da that work best when encapsulated in classical ethosomes.

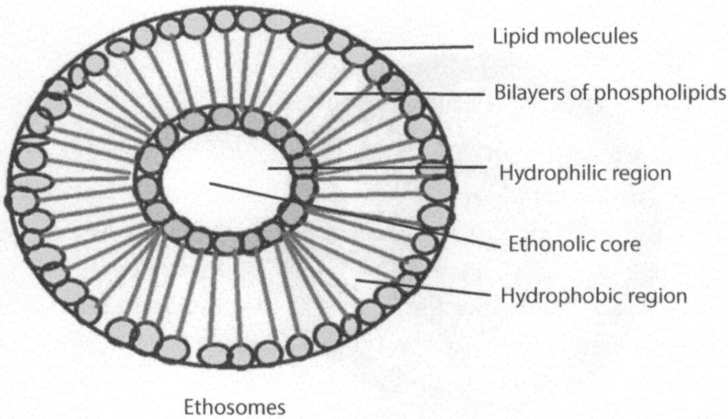

Figure 27.4 Structure of ethosomes.

Double ethosomes: They were introduced by Zhou *et al.* They are made by adding another type of alcohol to the classical ethosome formula. Commonly used alcohols include propylene glycol (PG) and isopropyl alcohol (IPA). Transethosomes are a newer form of ethosomal systems designed to combine the benefits of classical ethosomes and transferosomes into one formulation. Their composition includes the basic components of classical ethosomes along with a penetration enhancer or a surfactant, which helps in increasing drug delivery efficiency [73].

27.5.2.2 Niosomes

Niosomes are new types of delivery systems made mainly from non-ionic surfactants. These surfactants naturally form small, bubble-like structures in water. Their ability to carry more substances and better spread on the skin makes them useful for delivering medicines through the skin. Niosomes are round-shaped vesicles made of surfactants. They come in different types based on their size and how many layers they have.

The below Figure 27.5 illustrates the structure of a niosome, a type of vesicle that is used for transdermal drug delivery system with its various components. It shows the encapsulation of both hydrophilic and lipophilic drugs within the niosome.

Niosomes are made into different types of vesicles. There are minute unilamellar vesicles with sizes ranging from 10 to 100 nanometers, larger unilamellar vesicles from 100 to 300 nanometers, and multilamellar vesicles that are larger than 1000 nanometers. The number of layers in the

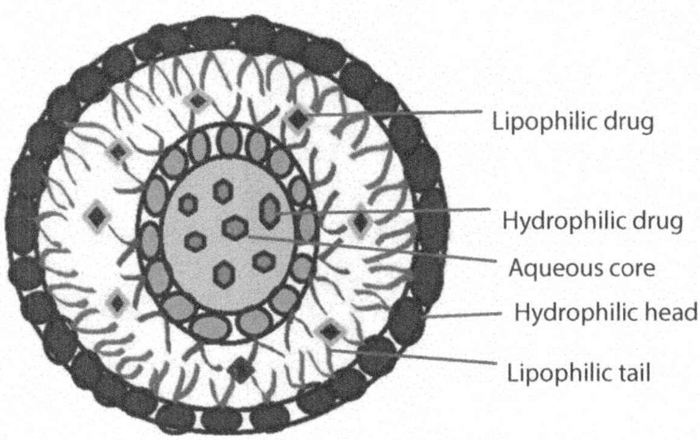

Figure 27.5 Structure of niosomes.

niosomes, known as lamellarity, is a necessary factor that divides them into unilamellar, which has one surfactant layer, and multilamellar, which has more than one. Common methods used in labs to make niosomes include thin-film hydration, ether infusion, phase transition, transmembrane pH gradient, and the bubble method. Microfluidics is also used to create niosomes with specific physical and chemical properties. The other method contains using proniosomes to form niosomes, which is a way to make the final dosage form. In most of these approaches, organic solvents are used, so the bubble method is commonly used a greener approach for lab-scale production. Recent uses of niosomal nanocarriers include improving the delivery of drugs through the skin, achieving a controlled release of medication, and enhancing the stability of the formulation during storage. The drugs used in these systems are used for treating various conditions like hormonal disorders, diabetes, pain management, wound healing with bioactive ingredients, psoriasis, fungal infections, and anaesthesia.

27.5.3 Solid Lipid Nanoparticles and Nanostructured Lipid Carriers (NLCs)

27.5.3.1 Solid Lipid Nanoparticles

Solid lipid nanoparticles (SLNs) are presented as a carrier framework in successfully water dissolvable pharmaceutical [74].

Figure 27.6 provides a visual representation of the SLN's structure. It consists of an active pharmaceutical ingredient core surrounded by a solid lipid matrix and a surfactant layer.

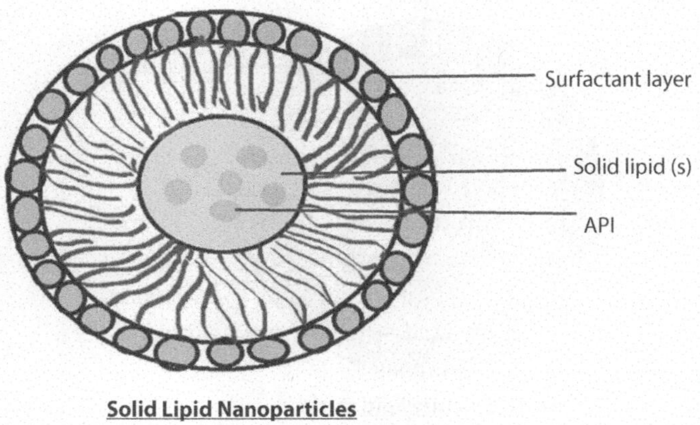

Figure 27.6 Structure of solid lipid nanoparticle.

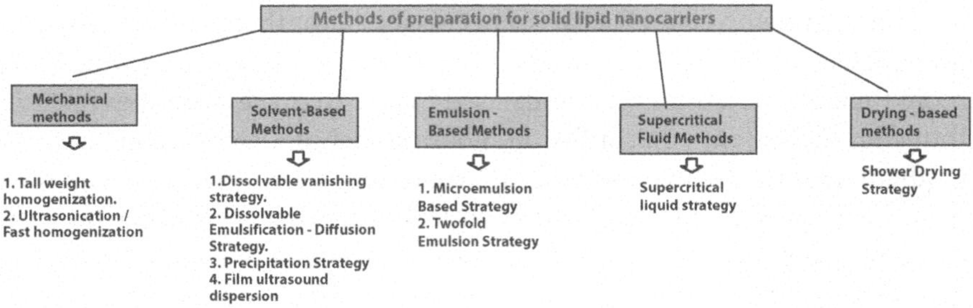

Flow chart 27.1 It depicts the methods of preparation for solid lipid nanocarriers.

Flow chart 27.1 shows various methods like mechanical methods, solvent-based methods, emulsion-based methods, supercritical fluid methods and drying-based methods for the preparation of solid lipid nanocarriers.

The SLN is a type of nanoparticle used for transdermal drug delivery, with the surfactant layer stabilizing the particle and preventing aggregation.

27.5.3.2 Nanostructured Lipid Carriers (NLCs)

Lipid nanosubstance's, such as solid lipid nanoparticles and nanostructured lipid vesicles, are tiny transport systems which are made up of a lipid structure. These substances are standardized using surface active agents in a liquid form. They are important for use in the lungs because they can be easily introduced into nasal formulations. Their small size allows them to reach too deep into the lungs, and they produce slow release and less toxicity, making them an assurance option for pulmonary uses.

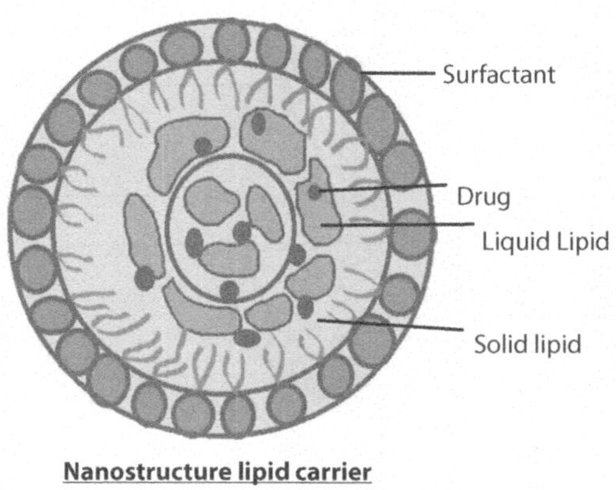

Figure 27.7 Nanostructure lipid carrier.

Figure 27.7 illustrates a brief structure of NLC with its labeled components that include surfactant, drug, liquid lipid, and solid lipid. The NLC is a circular structure with a liquid lipid outer layer and a liquid lipid inner core, where the drug is dispersed. The diagram is accompanied by explanatory text describing the composition and characteristics of NLCs.

In the 1990s, three research groups—Müller *et al.*, Gasco *et al.*, and Westesen—developed the first generation of lipid nanoparticles, known as Strong Lipid Nanoparticles (SLN). SLN consist of 0.1 to 30% (weight/weight) lipid dispersed in a water-based solution containing 0.5 to 5% (weight/weight) surfactant as a stabilizing agent. The size of SLN typically ranges from about 40 to 1000 nanometers. SLN shows different advantages, along with physical stability, controlled release, low cytotoxicity, and the use of well-tolerated adjuvants. NLC, which is similar to SLN, has a solid lipid matrix at room and body temperature, made up of a combination of a solid lipid and oil. Common methods for its production include hot or cold high-pressure homogenization, microemulsion technique, emulsification-solvent evaporation, or dispersion methods.

Great tolerance of SLN and NLC within the aviation routes can be guaranteed, in the event that lipids are chosen which are breaking down and coming about in non-toxic, frequently indeed endogenous corruption items. Besides, because of their measure, nanoparticles can be effortlessly captured into particles or aerosolized into beads with efficiently appropriate characteristics [75]. Cosmetics and cosmeceuticals.- Since lipids can help with skin hydration and moisturization, using a lipid nanocarrier system might be a good way to improve makeup products [76].

27.5.4 Polymeric Nanoparticles

Polymeric nanoparticles (PNPs) are created using biocompatible and biodegradable polymers, and their size typically ranges between 10 to 1000 nanometres. Based on how the nanoparticles are structured, they can form either nanospheres or nano capsules [77].

Figure 27.8 shows a polymeric nanoparticle with labeled components, including lipid shell, polymer core, and targeting ligands. This gives a detailed structure highlighting its potential applications in transdermal drug delivery system.

Nano capsules are structures where the medicine is contained within a small pocket surrounded by a layer of the same kind of polymer film. Nanospheres are frameworks where the drug is evenly spread out throughout the structure. The field of polymer nanoparticles (PNPs) is growing

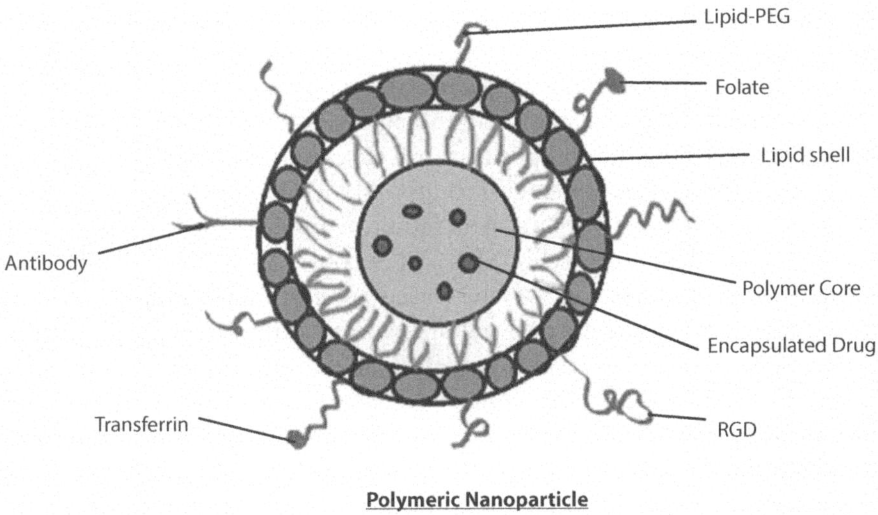

Figure 27.8 Polymeric nanoparticle.

quickly and is becoming essential in many areas such as electronics, photonics, conductive materials, sensors, pharmaceuticals, biotechnology, pollution control, and environmental technology. Polymer-based nanoparticles can efficiently deliver drugs, proteins, and DNA to specific cells and organs. Methods such as soluble vanishing, salting-out, dialysis, and supercritical fluid technology are used. On the other hand, PNPs can be made directly by joining monomers through different polymerization methods like micro-emulsion, mini-emulsion, surfactant-free emulsion, and interfacial polymerization [78].

27.5.5 Dendrimers

Dendrimers are tiny, branched substances that have a tree-like shape. They are uniform in size and have the capability to form bonds with other molecules. They are known to help skin absorb particles better and enhance how well certain compounds become soluble. Dendrimers are broadly used in medicine and biotechnology to deliver vital biological molecules. They have three main parts: (i) a central starting point, (ii) layers of repeating units that branch out from the center, and (iii) the outermost part that can connect to other molecules.

The below Figure 27.9 shows a structure of dendrimer, a highly branched molecule, with labeled components such as surface group, void space, core, and interior branching.

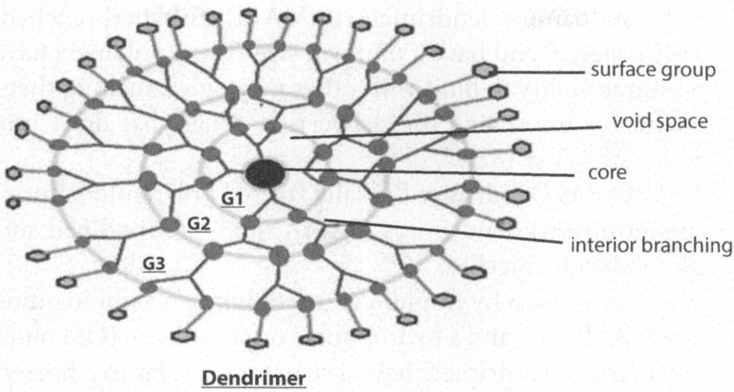

Figure 27.9 Structure of dendrimer.

There are broadly two components for medicate conveyance:

1. One way to start is by using *in vivo* debasement of medicated dendrimers that are covalently attached, which relies on the presence of suitable proteins or an environment capable of breaking down the bonds.
2. Another method involves releasing the medication because of changes in the physical conditions, like pH or temperature. This method doesn't require external factors and occurs either inside the receptor's core (endo-receptor) or on its outer layer (exo-receptor).

Recently, there has been significant research on dendrimers, focusing on developing new transdermal drug delivery systems using polyhydroxyalkanoate and starburst polyamidoamine (PAMAM) dendrimers.

Dendrimer-mediated transdermal delivery improves the drug entering capacity into blood circulation of indomethacin and enhances the skin permeation of 5-fluorouracil [79–81].

Types of Dendrimers

- PAMAM Dendrimer Poly (amidoamine) dendrimers (PAMAM) are made using a special method starting with smelling salts or ethylenediamine as the starting point. The name comes from the star-like shape you see when staring at the outline of these high-generation dendrimers in two dimensions.

- Polyamidoamine dendrimers (PAMAM) are hyperbranched, well-ordered, and have a uniform size. These polymers have a unique ability to bind with other molecules, making them useful for improving the delivery of drugs that don't mix well with water [82].
- PAMAMOS Dendrimer Radially layered poly(amidoamine-organosilicon) dendrimers (PAMAMOS) are modified single-molecule micelles.
- They comprise a hydrophilic, nucleophilic polyamidoamine (PAMAM) core and a hydrophobic organosilicon (OS) outer layer. These dendrimers have a structure similar to a honeycomb, with tiny PAMAM and OS sections.
- PPI Dendrimer PPI-dendrimers stand for "Poly (Propylene Imine)" and describe the propylamine groups in the oldest known type of dendrimer first developed by Vogtle.

  These dendrimers are mostly poly-alkyl amines with primary amines at the ends, and the inside contains many tertiary tris-propylene amines.
- Tecto Dendrimer These dendrimers are made up of a central dendrimer surrounded by other dendrimers in multiple steps to perform specific functions in smart nanodevices.
- Multilingual Dendrimers These dendrimers have multiple copies of a specific functional group on their surface.
- Chiral Dendrimers – The chirality in these dendrimers comes from the creation of a naturally unique but chemically similar branch that acts like a chiral center.
- Hybrid Dendrimers Direct Polymers – These are combinations of dendritic and straight polymers, having properties of both.
- Amphiphilic Dendrimers – These are made with two separate ends of the chain—one end gives electrons and the other takes them.
- Micellar Dendrimers – These are single-molecule micelles made from water-soluble hyper-branched polyphenylenes.
- Multiple Antigen Peptide Dendrimers – These are dendron-like structures based on a polylysine framework.
- Lysine, with its alkyl amino side chain, is a good building block for creating multiple branching points.
- Frechette-Type Dendrimers – These are a newer type of dendrimer developed by Seller and Frechet, based on a poly-benzyl ether hyper-branched structure [83].

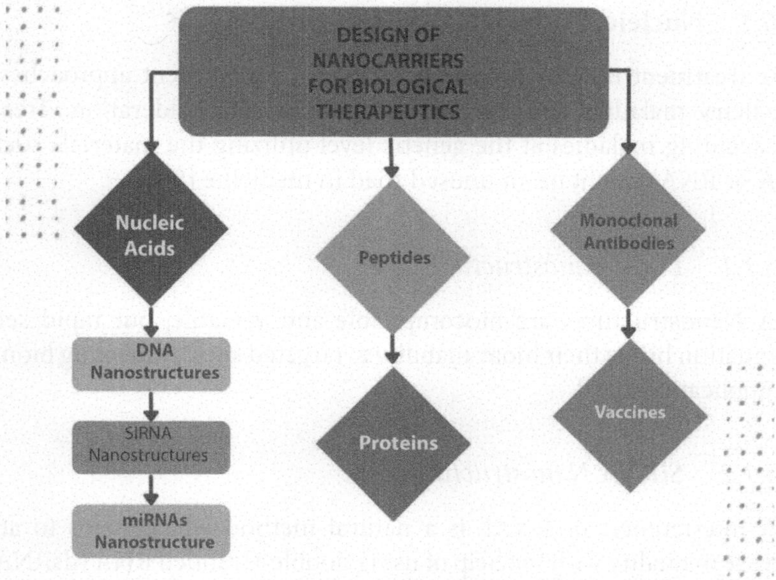

Figure 27.10 Design of nanocarriers for biological therapeutics.

Figure 27.10 illustrates the different types of biological therapeutics that can be delivered using nanocarriers in transdermal drug delivery system.

27.6 Design of Nanocarriers for Biological Therapeutics

Plan of Nanocarriers for Natural Therapeutics

An assortment of nano sized materials has been created for biomedical applications taking after the revelation of the EPR wonder, of which the foremost striking cases are different inorganic nanocarriers, such as mesoporous silica nanocarriers, graphene oxide carbon nanotubes, gold nanocarriers and attractive nanocarriers, etc.; and an assortment of natural nanocarriers, such as polymer nanocarriers, nano gel, liposomes, dendritic polymers and micelle, etc. In addition, several composite nanocarriers have been created too, and proposed for pharmaceutical applications through the hybridization of inorganic and natural materials.

27.6.1 Nucleic Acids (DNA, SiRNA, mRNA)

Gene treatment may be a basic constituent of treatment approaches for hereditary maladies and has picked up colossal consideration. Treating and avoiding maladies at the genetic level utilizing the materials such as DNA or RNAs might be an unused road in medicine [84].

27.6.1.1 DNA Nanostructures

DNA Nanostructures are biocompatible and versatile, but rapid serum degradation limits their bioavailability at targeted sites, restricting biomedical applications [85].

27.6.1.2 SiRNA Nanostructures

RNA interference, or RNAi, is a natural method which helps to attain increase in quality with the help of using double-stranded RNA (dsRNA) as a signal to disintegrate specific mRNA molecules. This approach is mostly used for gene silencing and drug enhancement due to its being highly specific, effective, has fewer side effects, and is easy to use. RNAi promotes the precise targeted reduction of problematic proteins. The method happens naturally through microRNAs (miRNAs) and can also be stimulated artificially using small interfering RNAs (siRNAs).

27.6.1.3 MiRNAs Nanostructures

miRNAs adversely direct quality expression, basically through coordinate interaction with the 3^1 untranslated locale (3^1-UTR) of comparing target delivery person RNAs (mRNAs) [86].

27.6.2 Peptides and Proteins

Proteins and peptides are efficient and strong in their actions, making them the preferred choice for treating many disorders. Essential areas of studies involve issues related to delivering proteins, such as keeping them stable in delivery systems and designing carriers that target particular areas in the body. When a drug is unstable in the digestive system or when patients need long-term or frequent doses, the transdermal route of delivery is seen as a good alternative. To get past the barriers that stop cells from letting substances in, a new method has been developed recently. This method is called cell-penetrating peptides (CPPs), also known as protein transduction domains (PTDs). These CPPs are usually short, up to 30 amino acids long, water-soluble, and often have a positive charge or are amphipathic. They can transport large molecules across cell layers. One big advantage of CPPs is that they can enter cells at very low concentrations, both in the lab and in the body, without needing receptors or causing damage to cell membranes. These peptides can take in active drugs that are either electrically bound or chemically linked to them, with high efficiency and low toxicity. There are two main methods to attach CPPs to their cargo. The first involves producing the CPP and cargo together in E. Coli and then purifying the resulting protein. The second method is to chemically link the CPP to the cargo to form a complex. They suggested that macro pinocytosis and changes in actin structures are important in improving the delivery of topical medicines [87].

27.6.3 Monoclonal Antibodies and Vaccines

Monoclonal counter acting agent invention, the mix of antibodies focused on to a particular foreign body is basically hybridoma invention, which is present in the hybridization of mouse B cells with myeloma cells. The future cell hybridoma technique leads to cancer cell life span. The release of monoclonal antibodies is used in the treatment.

Immunization innovation is one of the union of antibodies to a huge degree by recombinant DNA innovation and microbial generation. An immunization, is a process that produces a resistant reaction without disease of the body, may be a bacterium, infection, an antigen or poison, a recombinant vector or engineered peptide, or DNA [88].

27.7 Surface Building and Functionalization

New eras of built nanoparticles, moreover, also indicated as outlined and functionalized nanomaterials, have opened up modern conceivable outcomes within the uses of biomedical methods such as biomolecular identification, sedate conveyance, natural detecting agent, and treatment. Many nanomaterials are best possible to be used in biopharmaceuticals, counting nanotubes, nanoparticles, nanoplates and nanowires. The approaches for making and controlling functionalized nanomaterials [FNMs] open energizing unused openings for creating novel multifunctional natural equipment's.

Additionally, the little evaluation of FNMs upgrades uncommon physicochemical highlights like solubility, diffusivity, immunogenicity and the capacity to specific the assigned local with least dissemination to its encompassing target [89]. Besides, functionalization anticipates nanoparticles from agglomeration and makes them congruous in ensuing stages. Finally, FNMs can transport more effectively after systemic infusion and have better pharmacokinetic properties *in vivo*. FNMs can be profoundly driven into tissues *via* contract capillaries and epithelial filming.

27.7.1 PEGylation and Stealth Strategies

Pegylated proteins use polyethylene glycol (PEG) that ranges in size from about 40 kDa, which is similar in size to a protein, and it's attached directly to the protein to help it. PEG particles are safe for the body to break down through a process called phagocytosis and don't really trigger the body's complement system. This makes them useful for storing drugs and in devices that are placed inside the body. The major uses of pegylated proteins are that they dissolve better in the body, are less likely to be disintegrated by enzymes, produce fewer immune reactions, and stay in the body longer. Polylactic acid and polyethylene glycol nanoparticles as researched in many ways carry plasmid DNA, and they have shown promising results. Hormone proteins are usually PEGylated prodrugs, which means the ester or amide bond must be eliminated to get the active drug. These micellar systems are used to safely and effectively deliver compounds that don't dissolve well in water. PEGylation is a method where the surface of nanocarrier systems used in drug delivery is coated with a polymer like polyethylene glycol (PEG), which is a water-loving, non-toxic, and electrically neutral compound with the formula (C2nH4n+2On+1). PEGylation of lipid carriers can be done in two ways: either by involving PEG-lipids while the production of liposomes (pre-insertion method) or by mixing a PEG lipid solution with previously formed liposomes (post-insertion method). The most usually used PEG lipid is methoxy PEG-distearoyl phosphoethanolamine with a molecular weight of 2000 (mPEG-DSPE2000). It helps make liposomes less likely to interact with the mononuclear phagocyte system (MPS), which in turn enhances the time the drug delivery system stays active in the body. Many drugs based on PEGylated therapeutics are already on the market, and new ones are being developed to make drugs last longer in the body [90, 91]. The advantages of PEGylation involve increasing the time a variation of drugs, such as proteins, chemicals, small molecule drugs, nanoparticles, and liposomes, stay in the bloodstream. It has been referred that nanocarrier-based drugs, like PEGylated and non-PEGylated liposomes, are importantly eliminated from the blood by the reticuloendothelial system (RES) through a process called phagocytosis.

27.7.2 Cell-Penetrating Peptides (CPPs)

Protein transduction spaces (PTDs) or cell-penetrating peptides (CPPs) or film transduction peptides (MTPs) consist of 30 or less amino acids, are divided as either cationic or amphipathic and have capability to cross the cell layer and enter into cells.

Antioxidant proteins like superoxide dismutase and catalase are exceptionally viable against the illnesses intervened by responsive oxygen species such as skin aggravation, skin cancer, cutaneous immune system infections, phototoxicity and skin enhancing. Polylysine-9 (K9 coupled antioxidant proteins can translocate into epidermis and dermis.

It was recommended that CPPs can translocate over the cellular film by a receptor and vitality autonomous instrument, i.e., endosomal free pathway; a single CPP may utilize numerous pathways such as vitality subordinate endocytosis, vitality free coordinate translocation over films or utilize these substitute pathways concurrently. The CPP-based conveyance framework into skin has amplified from proteins to nanocarriers. CPP can translocate nanoparticles with their payloads into the reasonable layers of skin [92].

27.8 Crossover and Multi-Layered Nanocarriers

By looking at the challenges of both liposomal carriers and polymer-based nanoparticles, a new type of drug delivery system was developed. This system, called lipid polymer nano assemblies, combines the best features of both. These structures have a polymer core surrounded by self-assembled phospholipid layers that are connected to a PEG-containing layer, usually a PEG-lipid, which helps the system stay in the body longer. Despite the advantages of these nanostructures, to the best of our knowledge, there hasn't been much research on using them for targeting drugs to the brain [93].

In the lipid-based nanocarriers, multi-layered cochleae's have been improved as a new and innovative delivery system.

They are particularly useful for securing both water loving and water hating drugs from oxidation, enhancing their solubility, and reducing their size. These structures also improve bioavailability and increase drug safety by delivering the medication to a particular location; therefore, it reduces the side effects. Nanostructured cochleates are used as carriers for delivering water-hating drugs with anticancer, antiviral, and anti-inflammatory effects [94].

27.8.1 Core-Shell Nanostructures

Core-shell structured nanoparticles are ensuring nano-transporter frameworks for reduced and particular drug delivery uses [95]. They serve as extended-release carriers for both topical and systemic drug delivery. In the different skin that nanocarriers produced, liposomes, transferosomes, ethosomes, niosomes, dendrimers, lipid and polymer nanoparticles, and nano-emulsions are the most usually used [95]. Figure 27.11 which is given below shows the types of crossover and multi-layered nanostructures which are used for enhancing properties like optical, electrical, or magnetic behaviors in material science and nanotech.

Usually, drug-loaded microneedles (MN's), which are part of core-shell nanostructures, can be divided into two types: coated microneedles and solubilizing microneedles (DMNs).

Coated microneedles are made by filming the surface of a strong MN base, which may be made from substances like metal, silicon, or biocompatible polymers. The coating process for coated microneedles can induce methods such as immersion or layered coating techniques. Solubilizing microneedles, on the other side, can be made to dissolve into the skin upon application, after delivering the drug. These microneedles are made from bio composable materials such as saccharides like sucrose and maltose, as well as biopolymers such as carboxymethyl cellulose, polyvinyl alcohol, and hyaluronic acid. The fabrication process often includes a hybrid method that combines micro-molding with solvent removing process.

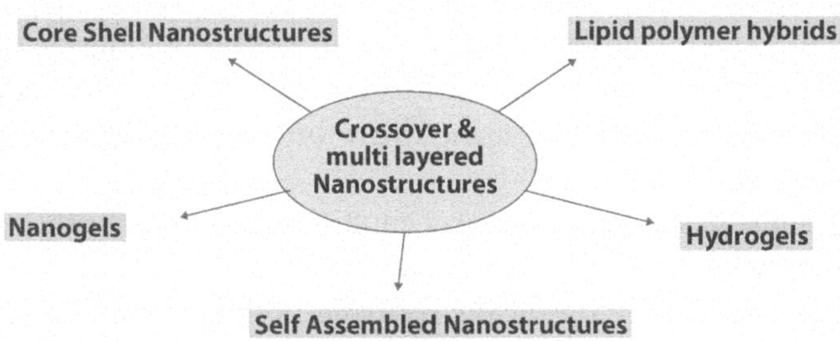

Figure 27.11 A schematic diagram representing the various types of crossover and multi-layered nanostructures.

To manage heat sensitivity, the production process can be adapted to use a very little amount of heat. Since these microneedles are made from less cost using materials and present in a relatively simple manufacturing process, they are more helpful for mass production compared to coated micro needles [96].

27.8.2 Lipid-Polymer Hybrids

Polymeric lipid hybrid nanoparticles, or PLNs, are structures that have a core made of a polymer and a shell made of lipid or lipid combined with PEG. These nanoparticles combine the physical stability and biocompatibility of both polymer-based nanoparticles and liposomes. Many different types of polymers, both natural and partially synthetic, have been used to make them. A new type of delivery system called polymeric lipid hybrid nanoparticles was created to use the best features of both polymeric nanoparticles and liposomes, which helped make them effective in the beginning.

This system also solves problems like structural weakness, short time the drug stays in the body, and leakage that happens with traditional methods. This hybrid system can be a helpful way to deliver drugs, offering great release of the medication, high efficiency in targeting specific areas, and better absorption by the body's tissues and cells. To overcome the limits of both lipid nanoparticles and polymeric nanoparticles, polymeric lipid hybrid nanoparticles (PLNs) have been developed. The word "hybrid" describes these nanoparticles that have features of both polymeric and lipid materials. The polymer helps control how the drug is delivered, while the lipid part improves how well the drug moves through cell membranes and increases how much of it is wrapped inside. Polymeric lipid hybrid nanoparticles [PLNs] can improve the compatibility of drugs with the body and make them more physically stable, making them a good option for delivering drugs to specific areas. PLNs are made up of three parts that coincide with the qualities of both polymeric nanoparticles and lipoid substances. These parts are involved in a hydrophobic polymeric core which holds the active ingredients, a phospholipid monolayer, and an outer PEG layer that acts as a protective covering and helps the PLNs stay in the body longer. This layer also provides stabilization by preventing the nanoparticles from sticking together. These nanoparticles are grouped based on how the lipids and polymers are arranged, such as polymer core-lipid shell hybrid NPs, lipid-bilayer-coated NPs, polymer-caged nanobins, mixed polymer-lipid hybrid NPs [Monolithic PLN], and lipid-polymer-lipid NPs [Hollow-core NP] [97].

27.8.3 Nanogels and Hydrogel

Drug nanocarriers, including crosslinked hydrogel nanoparticles, enhance therapeutic efficacy by controlling release, prolonging circulation, and protecting against premature degradation. Nanogels have emerged as versatile biomaterials with broad biomedical applications.

The characteristics of nanogels are mainly shaped by their large surface area, stable particle structure, ability to adjust size, and the use of water-based preparations. These features have made nanogels popular as drug delivery systems, as shown by many *in vivo* studies. The gel structure that holds the nanoparticles shows a significant increase in drug delivery when used for transdermal and topical applications [98]. It is possible that nanogels can be changed to have either lipid loving or water loving properties, making them suitable for drugs with different affinities. This helps them connect more effectively with specific groups of cells and the outline of cells. Smart nanogels can be prepared to respond to external triggers such as near-infrared (NIR) light, pH changes, temperature, and mechanical stress. Recently, nanogels have been changed by adding nanofillers, creating composite materials with multiple, versatile properties. For example, carbon nanotubes, fullerene, magnetic nanoparticles, and plasmonic materials have been added to nanogels to improve their use in image-guided therapies and on-demand drug release [99].

27.8.4 Self-Assembled Nanostructures

Self-assembly, also known as self-organization, could be an unconstrained procedure utilized for the manufacture frameworks right now, utilized in medicate conveyance and tissue building. This procedure, based on readily available components, is getting to be more attractive because of the availability, reproducibility, and low manufacturing cost. Self-assembly may be a wonder watched in amphipathic atoms, which have a 'built-in" fondness for two sorts of situations: solvent-loving and solvent-hating. The two affinities come from the covalent connecting of two or more joining parts, each with diverse chemical properties. The widely used amphiphilic particles incorporate surfactants, phospholipids, and square copolymers. When these molecules are blended with certain solvents and/or their physicochemical properties are tuned (e.g., composition, shape, surface properties), the distinctive components self-organize to make steady structures, such as, micelles, emulsions, liposomes/vesicles, hydrogels, strong lipid nanoparticles, and polymersomes. The most recent innovations that are used in transdermal biomolecule transporters include electrical frameworks (physical infiltration enhancers) a few chemical infiltration improvers and nanoparticles. All of them permit the support of the physical and chemical characters of the most proteins and peptides with these clinical medicines, permitting their proficient capacity, transport, and discharge and guaranteeing the accomplishment of their target and superior comes about within the treatment of numerous infections [100].

27.9 Components of Nanocarriers Infiltration through Skin

The transdermal approach of drug delivery shows various advantages with respect to traditional methods like oral or injectable forms for preventing various diseases and using beauty care products. The outer layer of the body acts as a reservoir which allows the drug to be absorbed over a longer period in a controlled manner. This reduces the risk of dangerous side effects and local irritation due to the medication absorbed through multiple points. Both lipid-soluble and water-soluble drugs can be transported across the outer layer of the skin, called the stratum corneum, with the help of nanocarriers, enabling them to reach both local and systemic areas to treat different conditions. The stratum corneum plays a nearly complete and surprisingly effective role in protecting the body, functioning as a unique bio-membrane in terms of its structure and composition. Within the epidermis, this layer forms a covering made up of tightly packed, keratin-filled dead skin cells held together in a lipophilic (oil-like) matrix. The lipids in this outer layer have distinct features: (i) they serve as the only pathway for substances to move from the skin's surface to the deeper layers; (ii) the skin's composition is unique among bio-membranes, as it does not contain phospholipids; (iii) despite lacking phospholipids, the stratum corneum lipids can form multiple layers; and (iv) the long, saturated hydrocarbon chains in these lipids create a highly ordered, interdigitated arrangement and form a gel-like structure that is more restricted than the typical fluid crystalline structure found in other membranes [101].

27.9.1 Passive Dissemination vs. Dynamic Transport

27.9.1.1 Passive Methods

The normal ways of applying drugs to the skin which involves using products like treatments, creams, and gels. In recent days, these forms have been enhanced or modified to develop the ability of the medicine to move into the skin (thermodynamic movement) and/or increase the skin's permeability. Some of these methods include using penetration enhancers, supersaturated systems, prodrugs, metabolic approaches, liposomes, and other types of vesicles. However, the amount of medicine that can be delivered with these methods is still limited because the skin's barrier properties are not significantly changed.

27.9.1.2 Active Methods

The advent of biotechnology within the last-mentioned half of the 20th century has driven to the era of therapeutically active, expansive atomic weight (>500 Da) polar and hydrophilic particles, generally peptides and proteins. With the help of a driving force, they can be penetrated into the skin layers. Such approaches guarantee to lead to propels within the productivity of transdermal delivery. Recent advance in these innovations has happened because of propels in exactness building (bioengineering), computing, chemical designing, and fabric sciences, which have all made a difference the creation of scaled down effective gadgets that can produce the specified clinical reaction [102]. The two modes of particle transportation include - detached and dynamic transport. Inactive transport, it includes dissemination, in which particles move from a high-to-low concentration angle. Cases will incorporate the dissemination of gasses over alveolar films and the dissemination of neurotransmitters such as acetylcholine over the synapse or neuromuscular intersection. Osmosis may be a shape of detached transport when water particles move from low solute concentration (high water concentration) to tall solute or moo water concentration over a film that's not penetrable to the solute. There's a type of inactive transport called encouraged dissemination. It takes place when atoms such as glucose or amino acids move from high concentration to low concentration, encouraged by carrier proteins or pores within the layer. Dynamic transport requires vitality for the method by transporting atoms against a concentration or electrochemical angle. Dynamic transport is an energy-driven system in which layer proteins transport atoms over cells, basically classified as essential or auxiliary, based on how energy is coupled to fuel these components. The previous constitutes implies by which a chemical response, e.g., ATP hydrolysis, powers the coordinate transport of atoms to set up particular concentration slopes, as seen with sodium/potassium-ATPase and hydrogen-ATPase pumps.

27.9.2 Endocytosis and Transcytosis

27.9.2.1 Endocytosis

Endocytosis is a process where materials from outside the cell are wrapped into small pouches called vesicles and brought inside the cell. These materials then follow specific paths to reach the right parts of the cell. The materials can be reused, like transferring, broken down, such as LDL and other developmental components, moved back through the same pathway that sends materials out (like bacterial toxins), or passed to the other side of the cell in specialized cells, such as immunoglobulins. Different ways of taking up nanoparticles can be grouped into phagocytosis, pinocytosis, and passive entry. Phagocytosis happens when specialized cells, like macrophages, engulf larger particles. Smaller particles are taken in through pinocytosis, which has subtypes including clathrin-mediated endocytosis (CME), caveolae-mediated endocytosis (CavME), and macropinocytosis (MP). CME is used by all mammalian cells and plays a key role in how cells take in substances. Vesicles taken in through CME are usually about 100 nm in size. CavME is also a common way for cells to take in materials. This pathway avoids the lysosomes, which are the cell's waste disposal units, and is often used by pathogens like viruses and bacteria to avoid being destroyed by the lysosomes.

27.9.2.2 Transcytosis

Transcytosis is the method by which cargo internalized at one plasma layer space of a polarized cell is transported by means of vesicular intermediates to the contralateral plasma layer 4. With regard to the conveyance and transport of pharmaceuticals, characterization of this pathway ought to lead to propel within the improvement of transcellular transepithelial and trans endothelial conveyance systems [103].

Types of Transcytoses
Solute Carrier-Mediated Transcytosis (CMT)
The point-by-point instrument of inspired solute carrier-mediated transcytosis (CMT) is accepted to continue by means of carrier-vector

acknowledgment at the luminal side taken after by a conformational alter of the carrier from outward to inward-facing states, coming about within the transport of the substrate down to its concentration angle over luminal and abluminal films of the ECs.

Receptor-Mediated Transcytosis (RMT)

In receptor-mediated transcytosis (RMT), a receptor-vector acknowledgment occasion happens at the luminal side. To begin with, point of passage is taken after by endosome arrangement, which squeezes the luminal film and transports the cargo over the endothelial cell to the inverse conclusion.

Adsorptive-Mediated Transcytosis (AMT)

Similarly to RMT, adsorptive-mediated transcytosis (AMT) continues with an invagination of the AMT-substrate taken after by means of endosomal arrangement and transport from the luminal to the abluminal side, finishing with exocytosis of the endocytosed cargo. AMT is activated by an electrostatic interaction between the AMT-vector and anionic proteoglycans, such as syndecan-4 [104, 105].

27.9.3 Disruption of Tight Intersections and Lipid Bilayers

Tight intersections (TJs) are intercellular contacts that seal the space between the person cells of an epithelial sheet or stratifying epithelia, such as the epidermis, so that they can collectively isolated tissue compartments. Intercellular intersections, such as adherens and TJs, play a significant part within the arrangement and support of epithelial and endothelial boundaries [106].

Tight intersection is a powerfully controlled structure. Cytoskeletal reorganization, especially myosin light chain phosphorylation induced actomyosin withdrawal, has progressively been recognized as an arbiter of physiological and pathophysiological tight intersection control. At resting conditions, the tight intersection is kept up by energetic protein to protein intuitive, which may give a stage for quick tight intersection control [107].

In transdermal sedate conveyance, tight intersections (TJs) and lipid bilayer disturbance play vital parts in controlling the entry of drugs over the skin. TJs, found between cells, act as obstructions, whereas lipid bilayers within the stratum corneum (the furthest skin layer) offer another noteworthy obstruction to medicate saturation. Disturbing these obstructions can upgrade medicate conveyance, but it too presents dangers that require cautious administration.

27.10 Preclinical and Clinical Evaluation

Transdermal therapy has more popularity in the present market as it is widely preferred for transporting active pharmaceutical ingredients into the brain. This way of approach provides various benefits like minimal invasiveness, administration of drug individually, preventing first-pass metabolism, optimum plasma concentrations, target specific delivery, safety and improved patient compatibility to the therapy [108]. The World Health Organization [WHO] and other global health agencies has established a clear bridge between UV light exposure in the occupational area and the increased risk of non–melanoma skin cancer. This highlighted text interprets the requirement of new innovative ideas with solutions of various ways for drug delivery, enhancement in tumor penetration and to decrease systemic toxicity [109].

27.10.1 *In Vitro* Models (Franz Cells, Manufactured Skin)

The Franz diffusion cell plays as a crucial tool for examining the permeability and stability of different pharmaceutical formulations, which also includes topical formulations such as gels and creams and transdermal formulations including lipid nanoparticles and other applications. By calculating the release of active pharmaceutical ingredients from various topical formulations such as liposomes, gels, creams, treatments, they have more value regarding pharmacokinetics of these preparations. It also includes model and assessment of strategies which give detail information regarding quality and safety of potent active toxic ingredients. By using these models and devices, it is easy to enable the advancements of novel topical therapeutic conditions [110].

Furthermore, *ex vivo* human subjects skin models or creature skin models are commonly used to perform analysis regarding skin penetration ability; with thorough regulatory standard procedures, these models act as surrogates for *in vivo* human skin to provide assessment for the bioequivalence studies for transdermal products.

27.10.2 Extracorporeal *Homo Sapiens* Skin Models

The human skin that is used in the research is generally obtained from plastic surgeries, or cadavers and in some cases from both with proper ethical approval. The dermis layer can be taken from the parts like stomach, breast or back as it provides required surface area as skin characters vary across different body parts regarding the stratum corneum, hydration level, thickness, and composition of lipids. According to the study conducted by Barbero and Frasch, carefully preserved human dermis is most suitable for evaluating the penetration of chemical agents. But this should not include factors like skin integrity and metabolic activity.

27.10.3 *Ex Vivo* Human Skin: Utilize in Bioequivalence Studies

When examining the bioequivalence of topical medications, researchers consider comparative clinical endpoint studies to compare their activity with a standard or reference product. Even though, the outcome of these studies is insufficient due to individual variations, and also factors regarding the drug, which have limit regarding their reliability and usefulness, there is an alternative for this. It includes the utilization of drug kinetic studies for examining the bioequivalence, in which we can determine the absorption and distribution of drug in the body. The other few methods used widely for bioequivalence testing are: *in vitro* permeation testing, *in vivo* tape stripping and dermatopharmacokinetics which also include *in vivo* micro dialysis and micro perfusion.

27.10.4 *Ex Vivo* Creature Skin Models

Rodent skin includes mice, rats, guinea pigs which are common models for *in vitro* skin penetration examining studies due to their ready availability and affordability. Specific rodent strains are chosen by researchers for their similarity to human skin in terms of permeability. Notably, rat skin is mostly used in research due to its structural similarities to human skin. Comparison of studies interpret that rodent skin shows more permeability than human skin, as they exhibit various degree of porosity among different physicochemical properties.

27.10.5 Artificial and Remade Skin Models

In specific research scenarios, artificial and reconstructed skin models play a vital role as an alternative to well established routine testing methods involving human and creature skin. This alternative substitute is used widely by researchers in traditional *in vivo* and *in vitro* tests as they produce accurate and more précised results in a consistent environment. The complexity and compatibility of these models vary from straight forward polymer-based systems to more intricate models which replicate the properties of human skin regarding its structure and functions. By decreasing the complexity of human and animal skin, researchers can examine the major principles that are unknown regarding the passive transport [111].

27.10.6 Clinical Studies for Transdermal Nanoparticle Delivery System

Innovation in sedate conveyance frameworks is a key technique utilized to make strides. The bioavailability of dynamic pharmaceutical fixings (APIs) energy-driven or electrically helped strategies of transcutaneous delivery uses electronic equipment for enhancing dermal uptake of the drugs.

Based on Prausnitz and Langer's classification, the energy-driven approaches are classified into two eras. The moment era of electrically helped approaches are iontophoresis and noncavitational ultrasound, the third era are electroporation, sonophoresis and warm deletion. Prausnitz and Langer classified the iontophoresis as the moment era, since this innovation does not grant an expansive effect on the stratum corneum, such as micro channel creation the third era of upgrade innovation requires stronger disturbance of the SC, since the most target of these strategies is to disturb or evacuate the SC, without causing harm to the more profound skin tissue. In the survey, the second-generation energy-driven strategy, iontophoresis, and the third-generation strategies of sonophoresis and electroporation will be portrayed as cases of these approaches.

Iontophoresis is described as a method for increasing penetration of drugs into the skin using little power-related streams. The electrical current connected in iontophoresis changes from 0.5 to 20 mA. The rule of this framework is based on the various charges of the anodes used, should be particularly considered the anode and cathode.

The increasing transdermal method is sonophoresis, moreover also called as phonophoresis. This framework employs ultrasound at frequencies of 20 kHz to 16 MHz for adjusting lipid bilayer course of action of the SC. There are two conceivable improvement components of activity by means of sonophoresis, specifically, warm impacts and cavitation (counting steady and transient cavitation). By applying ultrasound, the skin temperature increments, coming about in higher drug diffusivity into the skin.

The final depiction of an energy-facilitated delivery strategy is electroporation. Electroporation is a technique which creates micron sized pores deep in the skin with the help of applying a high voltage for a very short duration, may be less than a few hundred milliseconds. The procedure here involves placing an arrangement on the skin such a way that it generates thermally heat waves by creating aqueous pores in the lipid bilayer of stratum corneum (SC). As a result, it allows for the infiltration of substances deep into the skin layers through these pores. Electroporation has the potential that easily facilitates the transportation of macromolecules across the skin [112].

27.11 Administrative, Fabricating, and Adaptability Aspects

Advancement of nano therapeutics and related treatments based on Safe-by-Design and Quality-by-Design approaches empowered improved viability with a way better resilience due to assessment of frameworks to more secure forms. Issues emerging amid scale-up and the issues to be covered for administrative acknowledgment can be recorded as the foremost common challenges for nano medicine fabrication. Fabricating strategies effortlessly versatile to existing offices, and inevitably gear, are required to empower bigger generation volumes with less vitality and less fabric utilization but with higher control of item quality and security [113].

27.11.1 Regulatory Systems (FDA, EMA)

The FDA, a government office entrusted with defending open wellbeing, directs a wide cluster of items, such as nourishment, drugs, restorative devices, vaccines, biological, and beauty care products. It is the essential for managing the improvement, assessment, and endorsement of nano formulation items guaranteeing they meet meticulous safety, efficacy, and quality as a worldwide pioneer in nanotechnology direction, the FDA coordinates universally to set up rules.

The FDA directs the arrangement and explains the industry partners with clear suggestions and administrative desires for creating and assessing nano medicine items. These covers topics like characterization, security appraisal, fabricating forms, and administrative entries particularly to nanoscale technology-based products. By following regulatory guidelines, it advances the straightforwardness, streamlines administrative forms, and encourages the convenient endorsement and marketization of nano pharmaceuticals items, subsequently improving healthcare and therapeutic effects. FDA survey shows an FDA-regulated item that utilizes nanotechnology based on the factors like: The fabric or conclusion item is outlined to have measurements within the nano scale extend either remotely, inside, or on its finishing.

The fabric or conclusion item is designed to show attributes or impacts, such as physicochemical, or organic, that are affected by its measurements; in case these measurements expand past the nano scale extend, up to 1mm (1000 nm).

Guidance for Industry: Security of Nano materials in Corrective Products Published on June 15, 2020, this direction record presents proposals for evaluating the security of nanomaterial's utilized in cosmetics Guidance for Industry: Liposome Sedate Products (400). Issued in April 2018, this rule examines the important information and details necessary for potential candidates submitting an unused sedate application (NDA) or shortened modern sedate application (ANDA) for a liposome-based medicine under audit by the Centre for Sedate Assessment and Investigate at the FDA.
Labeling Requirements for New Drug Approval and Advanced New Drug Approval: This provides a detailed information regarding the content of the compound labeling which are particularly custom fitted for liposome sedate items by ensuring the compliance with the regulatory standards.
Guidance for Industry: Medicate items, counting natural items, that include nano materials.

Issued as of late April 2022, this rule gives indications on the enhancement of human sedate items, including organic items, which incorporated with nanomaterials. Risk-based system (Regarding Potential hazard components): This part includes in depth review of the contemplations basic for assessing the dangers related with compounds that are formulated by nanoparticles.

Nonclinical ponders for sedate items: This area gives bits of knowledge into the preclinical evaluation of sedate items joining nanomaterials. It emphasizes adjusting by the regulatory guidelines such as USFDA and ICH recognizing the requirement for custom-made approaches.

Quality Suggestions: This area gives a point-by-point system for guaranteeing the quality of sedate items containing nanomaterial's, tending to basic angles of characterization, fabricating, soundness, and post market changes.

Clinical advancement: This area offers nitty gritty bits of knowledge into the administrative system and methodological contemplations for assessing sedate items in clinical trials in which nano carriers are being used by equal compatibility with the rules for clinical trial application, Marketing Authorization Application, and Generic Drug Applications.

Environmental Affect Contemplations –This segment underscores an administrative commitments and subtleties in assessing the natural consequences of medicate items containing nanomaterials [114].

European Medications Authority

EMA, headquarters is located in Amsterdam, in which the key component known as European Union [EU] endowed with assessing, overseeing, and guaranteeing the quality and efficacy of medications planning for utilization inside the EU. Its obligations include reviewing submissions for showcasing accreditation of medications, advertising logical direction to healthcare companies, and directing medication safety tracking endeavors to screen the security of drugs after releasing into the market [108]. The EMA plays a main role in recognizing the challenges and opportunities that arise in the nano pharmaceutical companies, advancement in developing an innovative characteristics and upgrowing importance of nanotechnology in the healthcare field.

- The EMA conveys the taking of mandates for the assessment of surface modified nanoparticles.
- Comprehensive analysis of the layer fabric enveloping its formulations and regulatory compliances.
- Ancillary assessment for guaranteeing homogeneity and reliability, particularly in case of the sheathing operator or the expansion focusing on ligand involves an intricate particle (e.g., polypeptides or immunoglobulins).
- Meticulous approval of the surface finishing's preparation, depicting the chemical structure supporting anchorage of adsorption-based plating and coupling of permanently attached sheathing compounds.
- Examination of the possible impact on the interface, scope, variability on the security and viability of the result.
- Reliable delineation on the introduction and ligand bound conformation, especially for items highlighting a dynamic focusing on moiety on the surface.
- Inclusive evaluation of the surface treatments, soundness amid capacity and execution of the drug delivery systems.

- Lab based experimentation of the material properties, soundness of the protective layer regarding its expected utility beneath the circumstances under the intelligent course of organization, drug kinetics, targeting and localization, and disease mechanism.
- Assessment of possible implications on viability and security within the occasion of untimely separation or corruption of the coverings, driving to vulnerability of modern useful bunches on the nanoscale therapeutics.
- Thoughtful examination of the biological repercussions of distinctive barriers and exterior scope on drug kinetics and drug circulation
- To focus on understanding the distribution of coating materials within the body and their metabolic fate after releasing into the systemic circulation [115, 120].

27.11.2 Scale-Up Challenges and Solutions

Reproducibility – numerous nano therapeutics don't effectively reach the showcase due to challenges in accomplishing reproducibility and assembly the prerequisites for versatile synthesis. Complexity, the fabricating of nanoparticles (NPs) for large-scale medicate arrangements is challenging due to their complexity. Furthermore, in nanocarrier development, increasing formulation complexity requires in-process quality control at each unit operation, making the overall process lengthy and labor-intensive.

Consistent soundness– Nano medicine is challenging due to the need of keeping up reliable physicochemical properties over diverse clumps. Minor varieties in generation strategies or composition can incredibly impact the structure and work of the nano medicine. Cost, in large-scale generation, is fundamental to figure within the costs of crude materials and the necessity for endless multistep forms, driving to noteworthy time and budgetary costs within the mechanical fabricating of nano medicines. Regulatory angles: The need of administrative systems and measures administering fabricating hones, quality control, security, and adequacy evaluation presents a noteworthy obstruction to the creation of nano therapeutics.

Embracing green eco-friendly blend procedures and economical fabricating hones is significant for diminishing the dependence on unsafe substances and bringing down squander generation. A safe-by-design procedure– This methodology incorporates the utilization of naturally secure solvents, renewable vitality sources, and empowers reusing and squandering minimization. Green blend alludes to the utilization of clean, secure, and eco-friendly strategies, counting normal extricates from plants, organisms, biopolymers, and bio wastes. Its objective is to diminish the hurtful impacts regularly connected to conventional chemical and physical strategies for creating nanoparticles [85, 86]. Focal points of green union: 1) sustainability, 2) biosafety 3) biocompatibility [115].

27.12 Challenges and Future Directions

In this new era achievements in surface functionalization of nanoparticles have provided great opportunities for nano carriers in specific site drug delivery system. The function of nanotechnology in medical field is proving to be transformative, significantly impacting how we diagnose, prevent, and treat various illnesses. A number of nano molecules have given approval into therapeutic trials phase. Thus, they have been actively modified to detect and cure different types of cancers. Moreover, various compositions are currently undergoing trials for therapeutic safety. When it comes to stability and storage, it's crucial for nano carriers to be consistent, tolerant assemblies, and hold on to the compound they reach their receptor site. Any shelf-life concerns could be undermined by their effectiveness in the pharmacological activity [116].

Stability and storage: As stability plays a vital role in pharmacological activity, storing of nano carriers at suitable temperatures also play a very crucial role; moreover, we need to focus on designing stronger designs, improving the storage procedures and finding new ways to preserve their integrity [117].

Long-Term Toxicity and Biocompatibility: By studying the detail and standard information regarding the toxicology and bioavailability, we can undergo novel inventions for reducing their long tern toxicity effects and more improved bioavailability by developing biodegradable nano carriers which are essential to reduce potential risks [118].

Personalized Nano medicine for Transdermal Delivery: Customizing nano carriers in such a way that it is compatible for every individual skin type, conditions, or diseases can help simultaneously to boost both the significant effectiveness and safety of treatments.

- AI and Machine Learning in Nanocarrier Design:– capitalizing automatic learning and machine intelligence in the concept of molecular engineering can help us design the structure, function, and performance of nanocarriers. Future efforts should focus on using these technologies to predict behaviors, stimulate interactions, and pinpoint the best design parameters [119]; approaching and resolving these challenges is key to bring nanocarrier technology into the most advanced world and unlocking its full prospects in the sector of medicine and beyond. Continuous research and development are crucial to overcome these obstacles and accruing the benefits of nanocarriers.

27.13 Conclusion

The final brushstroke reveals that this discussion has paved the way for the change in the upcoming generation of nanocarriers in natural remedies. The increase in the advancement of the nanocarriers in different fields, like transdermal nanomedicine, enables more efficient and targeted drug delivery. The enhancement of these nanocarriers will overcome the traditional walls to medicine delivery and improve treatment issues. As the study continues further, the unborn outlook for transdermal nano medicine is promising, with implicit operations in a various range of diseases and different conditions. The inclusion of the upcoming nano carriers into clinical practice is awaited, which enhances patient results and standard of life. Similarly, continuous advancement of the nano carrier technology is likely to drive innovation in the field, which enables the creation of further effective treatments. In the upcoming era it is similar to the transdermal nano medicine which act as a crucial step in changing the healthcare system and furnishing new and innovative results for complex medical challenges. Finally, along with the ongoing changes and enhancement of transdermal nano medicine, is large and instigative.

AI disclosure statement: The author acknowledges the use of AI-based tools for minor language editing or grammatical correction. However, all intellectual contributions, analyses, and conclusions presented in this chapter are entirely the author's own.

References

1. Prausnitz, M.R. and Langer, R., Transdermal drug delivery. *Nat. Biotechnol.*, 26, 11, 1261–8, 2008 Nov.
2. Barry, B.W., Novel mechanisms and devices to enable successful transdermal drug delivery. *Eur. J. Pharm. Sci.*, 14, 2, 101–114, 2001.
3. Elias, P.M., Skin barrier function. *Curr. Allergy Asthma Rep.*, 8, 4, 299–305, 2008.
4. Benson, H.A.E., Transdermal drug delivery: penetration enhancement techniques. *Curr. Drug Delivery*, 2, 1, 23–33, 2005.
5. Ita, K., Transdermal delivery of drugs with microneedles—potential and challenges. *Pharmaceutics*, 7, 3, 90–105, 2015.
6. Leader, B., Baca, Q.J., Golan, D.E., Protein therapeutics: a summary and pharmacological classification. *Nat. Rev. Drug Discovery*, 7, 1, 21–39, 2008.
7. Market Research Future, Biologics Market Research Report - Forecast to 2030, 2022.
8. Stroh, M., *et al.*, Parenteral drug administration: challenges and solutions. *Clin. Pharmacokinet.*, 49, 9, 573–582, 2010.
9. Kumar, R., *et al.*, Oral delivery of therapeutic proteins and peptides: a review. *Pharmaceutics*, 12, 2, 117, 2020.
10. Larrañeta, E., *et al.*, Transdermal delivery of macromolecules using microneedle arrays. *Int. J. Pharm.*, 512, 1, 62–73, 2016.
11. Mitragotri, S., Devices for overcoming biological barriers: the use of physical forces to disrupt the barriers. *Adv. Drug Delivery Rev.*, 65, 1, 100–103, 2013.
12. Osterberg, L. and Blaschke, T., Adherence to medication. *N. Engl. J. Med.*, 353, 5, 487–497, 2005.
13. Shaji, J. and Patole, V., Protein and peptide drug delivery: oral approaches. *Indian J. Pharm. Sci.*, 70, 3, 269–277, 2008.
14. Mahato, R.I. and Narang, A.S., *Pharmaceutical dosage forms and drug delivery: revised and expanded*, CRC Press, Boca Raton, FL, USA, 2017 Nov 22.
15. Kalia, Y.N., *et al.*, Iontophoretic drug delivery. *Adv. Drug Delivery Rev.*, 56, 5, 619–658, 2004.
16. Jain, S., *et al.*, Nanocarrier-based transdermal drug delivery: current status and future prospects. *Ther. Deliv.*, 6, 5, 601–612, 2015.
17. Chen, Y., *et al.*, Stimuli-responsive nanocarriers for drug delivery. *Int. J. Nanomed.*, 8, 2951–2967, 2013.
18. Madison, K.C., Barrier function of the skin: "la raison d'être" of the epidermis. *J. Invest. Dermatol.*, 121, 2, 231–241, 2003.
19. Elias, P.M., Epidermal lipids, barrier function, and desquamation. *J. Invest. Dermatol.*, 80, Suppl, 44s–49s, 1983.
20. Proksch, E., Brandner, J.M., Jensen, J.M., The skin: an indispensable barrier. *Exp. Dermatol.*, 17, 12, 1063–1072, 2008.
21. Mitragotri, S., Blankschtein, D., Langer, R., Transdermal drug delivery: past, present, and future. *Nat. Rev. Drug Discovery*, 13, 11, 815–833, 2014.

22. Prausnitz, M.R., Mitragotri, S., Langer, R., Current status and future potential of transdermal drug delivery. *Nat. Rev. Drug Discovery*, 3, 2, 115–124, 2004.
23. Wiechers, J.W., Formulating for efficacy: a historical perspective. *Int. J. Cosmet. Sci.*, 30, 1, 1–8, 2008.
24. Bouwstra, J.A. and Honeywell-Nguyen, P.L., Skin structure and mode of action of vesicles. *Adv. Drug Delivery Rev.*, 54, Suppl 1, S41–S55, 2002.
25. Kaplan, D.H., Ontogeny and function of murine epidermal Langerhans cells. *Nat. Immunol.*, 8, 8, 784–792, 2007.
26. Ryman, J.T. and Meibohm, B., Pharmacokinetics of monoclonal antibodies. *CPT Pharmacometrics Syst. Pharmacol.*, 6, 9, 576–588, 2017.
27. Potts, R.O. and Francoeur, M.L., Lipid biophysics of water loss through the skin. *Proc. Natl. Acad. Sci. U.S.A.*, 87, 10, 3871–3873, 1990.
28. Karande, P. and Mitragotri, S., Enhancement of transdermal drug delivery *via* synergistic action of chemicals. *Biochim. Biophys. Acta*, 1788, 11, 2362–2373, 2009.
29. Walters, K.A., Transdermal drug delivery: pharmaceutical and pharmacokinetic considerations. *Drug Delivery*, 9, 1, 19–26, 2002.
30. Prausnitz, M.R., Microneedles for transdermal drug delivery. *Adv. Drug Delivery Rev.*, 56, 5, 581–587, 2004.
31. Williams, A.C. and Barry, B.W., Penetration enhancers. *Adv. Drug Delivery Rev.*, 64, Suppl, 128–137, 2012.
32. Lopes, L.B., Overcoming the cutaneous barrier with microemulsions. *Pharmaceutics*, 6, 1, 52–77, 2014.
33. Jain, S. and Tiwary, A.K., Recent trends on the use of lipid-based nanoformulations for dermal/transdermal delivery. *Curr. Nanosci.*, 4, 4, 370–388, 2008.
34. Torchilin, V.P., Multifunctional and stimuli-sensitive pharmaceutical nanocarriers. *Eur. J. Pharm. Biopharm.*, 71, 3, 431–444, 2009.
35. Wang, J., *et al.*, Cell-penetrating peptides for enhanced transdermal drug delivery. *Adv. Drug Delivery Rev.*, 65, 4, 570–587, 2013.
36. Barras, A., Mezzetti, A., Richard, A., Lévêque, P., Wolf, C., Boukherroub, R., Lipid-based nanocarriers for skin delivery of drugs and cosmetics: a review. *Adv. Colloid Interface Sci.*, 279, 102139, 2020.
37. Patel, D.B., *et al.*, Nanoparticle-based drug delivery for cancer therapy, in: *Nanostructures for Cancer Therapy*, pp. 127–146, Elsevier, Amsterdam, Netherlands, 2017.
38. Cevc, G., Lipid vesicles and other colloids as drug carriers on the skin. *Adv. Drug Delivery Rev.*, 56, 5, 675–711, 2004.
39. Honeywell-Nguyen, P.L. and Bouwstra, J.A., Vesicles as a tool for transdermal and dermal delivery. *Drug Discovery Today Technol.*, 2, 1, 67–74, 2005.
40. Jain, C.P., *et al.*, Niosomes: a novel approach for drug delivery. *Int. J. Pharm. Sci. Rev. Res.*, 5, 3, 41–44, 2010.
41. Pardeike, J., *et al.*, Lipid nanoparticles (SLN, NLC) in dermal and transdermal drug delivery. *Eur. J. Pharm. Biopharm.*, 71, 2, 247–256, 2009.

42. Muller, R.H., Radtke, M., Wissing, S.A., Nanostructured lipid matrices for improved microencapsulation of drugs. *Int. J. Pharm.*, 242, 1–2, 121–128, 2002.
43. Prego, C., *et al.*, Chitosan-based nanoparticles for improving immunization against hepatitis B infection. *Vaccine*, 28, 14, 2607–2614, 2010.
44. Svenson, S. and Tomalia, D.A., Dendrimers in biomedical applications–reflections on the field. *Adv. Drug Delivery Rev.*, 57, 15, 2106–2129, 2005.
45. Shakeel, F., *et al.*, Nanoemulsions as vehicles for transdermal delivery of aceclofenac. *AAPS PharmSciTech*, 8, 4, E104–E109, 2007.
46. Kohli, A.K. and Alpar, H.O., Potential use of nanoparticles for transcutaneous vaccine delivery: effect of particle size and charge. *Int. J. Pharm.*, 275, 1–2, 13–17, 2004.
47. Vogt, A., *et al.*, Follicular targeting—a promising tool in selective dermatotherapy. *J. Invest. Dermatol. Symp. Proc.*, 10, 3, 252–255, 2005.
48. Chauhan, A.S., Dendrimers for drug delivery. *Molecules*, 23, 4, 938, 2018.
49. Ma, P. and Mumper, R.J., Paclitaxel nano-delivery systems: a comprehensive review. *J. Nanomed. Nanotechnol.*, 4, 2, 1000164, 2013 Feb 18.
50. Verma, D.D., *et al.*, Particle size of liposomes influences dermal delivery of substances. *Int. J. Pharm.*, 258, 1–2, 141–151, 2003.
51. Zhang, L., Li, Y., Jin, C., Xia, Y., Chen, Q., Liu, Q., *et al.*, Ligand-modified nanocarriers for targeted delivery of therapeutic agents: recent advances and perspectives. *Small*, 17, 42, 2101424, 2021.
52. Raza, A., *et al.*, Stimuli responsive nanocarriers for improved drug delivery. *Curr. Pharm. Des.*, 25, 3, 326–346, 2019.
53. Zhang, Z., *et al.*, Transdermal delivery of insulin using a novel nanostructured lipid carrier. *Int. J. Pharm.*, 454, 1, 104–111, 2013.
54. Yu, Z., *et al.*, Enhanced transdermal delivery of interferon alpha 1b by ethosomes. *Int. J. Nanomed.*, 7, 4789–4797, 2012.
55. Fakhraei Lahiji, S., *et al.*, Topical gene silencing using dendritic siRNA carriers in skin cancer. *Mol. Ther.*, 21, 4, 763–772, 2013.
56. Chen, X., *et al.*, Transdermal immunization *via* nanoemulsion-based vaccine. *Biomaterials*, 33, 32, 8166–8176, 2012.
57. Chen, X., Lei, S., Lin, J., Huang, P., Stimuli-responsive image-guided nanocarriers as smart drug delivery platforms. *Expert Opin. Drug Deliv.*, 19, 11, 1487–504, 2022 Nov 2.
58. Majumder, J. and Minko, T., Multifunctional and stimuli-responsive nanocarriers for targeted therapeutic delivery. *Expert Opin. Drug Deliv.*, 18, 2, 205–27, 2021 Feb 1.
59. Mrinalini, M. and Prasanthkumar, S., Recent advances on stimuli-responsive smart materials and their applications. *ChemPlusChem*, 84, 8, 1103–21, 2019 Aug.
60. Sawant, R.M., Hurley, J.P., Salmaso, S., Kale, A., Tolcheva, E., Levchenko, T.S., Torchilin, V.P., "SMART" drug delivery systems: double-targeted

pH-responsive pharmaceutical nanocarriers. *Bioconjugate Chem.*, 17, 4, 943–9, 2006 Jul 19.
61. Manchun, S., Dass, C.R., Sriamornsak, P., Targeted therapy for cancer using pH-responsive nanocarrier systems. *Life Sci.*, 90, 11–12, 381–7, 2012 Mar 10.
62. Karimi, M., Sahandi Zangabad, P., Ghasemi, A., Amiri, M., Bahrami, M., Malekzad, H., Ghahramanzadeh Asl, H., Mahdieh, Z., Bozorgomid, M., Ghasemi, A., Rahmani Taji Boyuk, M.R., Temperature-responsive smart nanocarriers for delivery of therapeutic agents: applications and recent advances. *ACS Appl. Mater. Interfaces*, 8, 33, 21107–33, 2016 Aug 24.
63. Wang, D., Huang, H., Zhou, M., Lu, H., Chen, J., Chang, Y.T., Gao, J., Chai, Z., Hu, Y., A thermoresponsive nanocarrier for mitochondria-targeted drug delivery. *Chem. Commun.*, 55, 28, 4051–4, 2019.
64. Pernia Leal, M., Torti, A., Riedinger, A., La Fleur, R., Petti, D., Cingolani, R., Bertacco, R., Pellegrino, T., Controlled release of doxorubicin loaded within magnetic thermo-responsive nanocarriers under magnetic and thermal actuation in a microfluidic channel. *ACS Nano*, 6, 12, 10535–45, 2012 Dec 21.
65. Hu, Q., Katti, P.S., Gu, Z., Enzyme-responsive nanomaterials for controlled drug delivery. *Nanoscale*, 6, 21, 12273–12286, 2014.
66. Mu, J., Lin, J., Huang, P., Chen, X., Development of endogenous enzyme-responsive nanomaterials for theranostics. *Chem. Soc. Rev.*, 47, 15, 5554–73, 2018.
67. Yang, H.Y., Li, Y., Lee, D.S., Multifunctional and stimuli-responsive magnetic nanoparticle-based delivery systems for biomedical applications. *Adv. Ther.*, 1, 2, 1800011, 2018 Jun.
68. Feng, Q., Zhang, Y., Zhang, W., Hao, Y., Wang, Y., Zhang, H., Hou, L., Zhang, Z., Programmed near-infrared light-responsive drug delivery system for combined magnetic tumor-targeting magnetic resonance imaging and chemo-phototherapy. *Acta Biomater.*, 49, 402–13, 2017 Feb 1.
69. Benson, H.A., Transfersomes for transdermal drug delivery. *Expert Opin. Drug Deliv.*, 3, 6, 727–37, 2006 Nov 1.
70. Pawar, A.Y., Transferosome: A novel technique which improves transdermal permeability. *Asian J. Pharm. (AJP)*, 10, 04, 2016 Dec 21.
71. Nsairat, H., Khater, D., Sayed, U., Odeh, F., Al Bawab, A., Alshaer, W., Liposomes: structure, composition, types, and clinical applications. *Heliyon*, 8, 5, 2022 May 1.
72. Monisha, C., Ganesh, G.N., Mythili, L., Radhakrishnan, K., A review on ethosomes for transdermal application. *Res. J. Pharm. Technol.*, 12, 7, 3133–43, 2019.
73. Kumar, N., Dubey, A., Mishra, A., Tiwari, P., Ethosomes: A novel approach in transdermal drug delivery system. *Int. J. Pharm. Life Sci.*, 11, 5, 2020 May 1.
74. Mehnert, W., Mäder, K., Solid lipid nanoparticles: production, characterization and applications. *Adv. Drug Delivery Rev.*, 64, 83–101, 2012 Dec 1.

75. Weber, S., Zimmer, A., Pardeike, J., Solid lipid nanoparticles (SLN) and nanostructured lipid carriers (NLC) for pulmonary application: a review of the state of the art. *Eur. J. Pharm. Biopharm.*, 86, 1, 7–22, 2014 Jan 1.
76. Salvi, V.R. and Pawar, P., Nanostructured lipid carriers (NLC) system: A novel drug targeting carrier. *J. Drug Delivery Sci. Technol.*, 51, 255–67, 2019 Jun 1.
77. Rao, J.P. and Geckeler, K.E., Polymer nanoparticles: Preparation techniques and sizecontrol parameters. *Prog. Polym. Sci.*, 36, 7, 887–913, July 2011.
78. Geckeler, K.E. and Stirn, J., Polyreaktionen – Mechanismen, Systematik, Relevanz. *Naturwissenschaften*, 80, 487–500, 1993.
79. Huang, B., Dong, W.J., Yang, G.Y., Wang, W., Ji, C.H., Zhou, F.N., Dendrimer-coupled sonophoresis-mediated transdermal drug-delivery system for diclofenac. *Drug Des. Devel. Ther.*, 3867–76, 2015 Jul 23.
80. Kalomiraki, M., Thermos, K., Chaniotakis, N.A., Dendrimers as tunable vectors of drug delivery systems and biomedical and ocular applications. *Int. J. Nanomed.*, 1–2, 2016 Dec 22.
81. Venuganti, V.V. and Perumal, O.P., Effect of poly(amidoamine) (PAMAM) dendrimer on skin permeation of 5-fluorouracil. *Int. J. Pharm.*, 361, 1–2, 230–8, 2008.
82. Chauhan, A.S., Sridevi, S., Chalasani, K.B., Jain, A.K., Jain, S.K., Jain, N.K., Diwan, P.V., Dendrimer-mediated transdermal delivery: enhanced bioavailability of indomethacin. *J. Controlled Release*, 90, 3, 335–43, 2003 Jul 31.
83. Tripathy, S. and Das, M.K., Dendrimers and their Applications as Novel Drug Delivery Carriers. *J. Appl. Pharm. Sci.*, 3, 09, 142–149, 2013.
84. Singh, P., Muhammad, I.J., Nelson, N.E., Tran, K.T., Vinikoor, T., Chorsi, M.T., D'Orio, E., Nguyen, T.D., Transdermal delivery for gene therapy. *Drug Delivery Transl. Res.*, 12, 11, 2613–33, 2022 Nov.
85. Ita, K., Dermal/transdermal delivery of small interfering RNA and antisense oligonucleotides-advances and hurdles. *Biomed. Pharmacother.*, 87, 311–20, 2017 Mar 1.
86. Jain, A., Jain, A., Gulbake, A., Shilpi, S., Hurkat, P., Jain, S.K., Peptide and protein delivery using new drug delivery systems. *Crit. Rev.™ Ther. Drug Carrier Syst.*, 30, 4, 2013.
87. Stryjewska, A., Kiepura, K., Librowski, T., Lochyński, S., Biotechnology and genetic engineering in the new drug development. Part II. Monoclonal antibodies, modern vaccines and gene therapy. *Pharmacol. Rep.*, 65, 5, 1086–101, 2013 Sep.
88. Díez-Pascual, A.M., Surface engineering of nanomaterials with polymers, biomolecules, and small ligands for nanomedicine. *Materials*, 15, 9, 3251, 2022 Apr 30.
89. C., Sanchez, A., Putnam, D., Ting, D., Langer Perez, R., Alonso, M.J., Poly (lactic acid)-poly (ethylene glycol) nanoparticles as new carriers for the delivery of plasmid DNA. *J. Controlled Release*, 75, 1–2, 211–24, 2001.

90. Mohamed, M., Alaaeldin, E., Hussein, A., Sarhan, H.A., Liposomes and PEGylated liposomes as drug delivery system. *J. Adv. Biomed. Pharm. Sci.*, 3, 2, 80–8, 2020 Apr 1.
91. Desai, P., Patlolla, R.R., Singh, M., Interaction of nanoparticles and cell-penetrating peptides with skin for transdermal drug delivery. *Mol. Membr. Biol.*, 27, 7, 247–59, 2010 Oct 1.
92. Ishak, R.A., Mostafa, N.M., Kamel, A.O., Stealth lipid polymer hybrid nanoparticles loaded with rutin for effective brain delivery–comparative study with the gold standard (Tween 80): optimization, characterization and biodistribution. *Drug Delivery*, 24, 1, 1874–90, 2017 Jan 1.
93. Shende, P., Khair, R., Gaud, R.S., Nanostructured cochleates: A multi-layered platform for cellular transportation of therapeutics. *Drug Dev. Ind. Pharm.*, 45, 6, 869–81, 2019 Jun 3.
94. Kumar, R., Mondal, K., Panda, P.K., Kaushik, A., Abolhassani, R., Ahuja, R., Rubahn, H.G., Mishra, Y.K., Core–shell nanostructures: perspectives towards drug delivery applications. *J. Mater. Chem. B*, 8, 39, 8992–9027, 2020.
95. Wang, Q.L., Zhang, X.P., Chen, B.Z., Guo, X.D., Dissolvable layered microneedles with core-shell structures for transdermal drug delivery. *Mater. Sci. Eng. C*, 83, 143–7, 2018 Feb 1.
96. Sivadasan, D., Sultan, M.H., Madkhali, O., Almoshari, Y., Thangavel, N., Polymeric lipid hybrid nanoparticles (plns) as emerging drug delivery platform—A comprehensive review of their properties, preparation methods, and therapeutic applications. *Pharmaceutics*, 13, 8, 1291, 2021 Aug 18.
97. Tariq, L., Arafah, A., Ali, S., Beigh, S., Dar, M.A., Dar, T.U., Dar, A.I., Alsaffar, R.M., Masoodi, M.H., Rehman, M.U., Nanogel-based transdermal drug delivery system: A therapeutic strategy with under discussed potential. *Curr. Top. Med. Chem.*, 23, 1, 44–61, 2023 Jan 1.
98. Vashist, A., Alvarez, G.P., Camargo, V.A., Raymond, A.D., Arias, A.Y., Kolishetti, N., Vashist, A., Manickam, P., Aggarwal, S., Nair, M., Recent advances in nanogels for drug delivery and biomedical applications. *Biomater. Sci.*, 2024.
99. Peña-Juárez, M.C., Guadarrama-Escobar, O.R., Escobar-Chávez, J.J., Transdermal delivery systems for biomolecules. *J. Pharm. Innovation*, 17, 2, 319–32, 2022 Jun.
100. Yu, Y.Q., Yang, X., Wu, X.F., Fan, Y.B., Enhancing permeation of drug molecules across the skin *via* delivery in nanocarriers: novel strategies for effective transdermal applications. *Front. Bioeng. Biotechnol.*, 9, 646554, 2021 Mar 29.
101. Brown, M.B., Martin, G.P., Jones, S.A., Akomeah, F.K., Dermal and transdermal drug delivery systems: current and future prospects. *Drug Delivery*, 13, 3, 175–87, 2006 Jan 1.
102. Okamoto, C.T., Endocytosis and transcytosis. *Adv. Drug Delivery Rev.*, 29, 3, 215–28, 1998 Feb 2.

103. Nagai, N., Ogata, F., Ishii, M., Fukuoka, Y., Otake, H., Nakazawa, Y., Kawasaki, N., Involvement of endocytosis in the transdermal penetration mechanism of ketoprofen nanoparticles. *Int. J. Mol. Sci.*, 19, 7, 2138, 2018 Jul 23.
104. Abdul Razzak, R., Florence, G.J., Gunn-Moore, F.J., Approaches to CNS drug delivery with a focus on transporter-mediated transcytosis. *Int. J. Mol. Sci.*, 20, 12, 3108, 2019 Jun 25.
105. Sapra, B., Jindal, M., Tiwary, A.K., Tight junctions in skin: new perspectives. *Ther. Deliv.*, 3, 11, 1297–327, 2012 Nov 1.
106. Henzler Wildman, K.A., Lee, D.K., Ramamoorthy, A., Mechanism of lipid bilayer disruption by the human antimicrobial peptide, LL-37. *Biochemistry*, 42, 21, 6545–58, 2003 Jun 3.
107. Nguyen-Thi, P.T., Vo, T.K., Le, H.T., Nguyen, N.T., Nguyen, T.T., Van Vo, G., Translation from Preclinical Research to Clinical Trials: Transdermal Drug Delivery for Neurodegenerative and Mental Disorders. *Pharm. Res.*, 41, 6, 1045–92, 2024 Jun.
108. Rane, B., Mhatre, N., Jain, A., Next-Generation Nanocarriers for Skin Cancer Therapy *via* Transdermal Delivery. *Nano Biomed. Eng.*, 17, 2, 2025 Jun 1.
109. Kumar, M., Sharma, A., Mahmood, S., Thakur, A., Mirza, M.A., Bhatia, A., Franz diffusion cell and its implication in skin permeation studies. *J. Dispersion Sci. Technol.*, 45, 5, 943–56, 2024 Apr 1.
110. Abd, E., Yousef, S.A., Pastore, M.N., Telaprolu, K., Mohammed, Y.H., Namjoshi, S., Grice, J.E., Roberts, M.S., Skin models for the testing of transdermal drugs. *Clin. Pharmacol.: Adv. Appl.*, 163–76, 2016.
111. Ramadon, D., McCrudden, M.T., Courtenay, A.J., Donnelly, R.F., Enhancement strategies for transdermal drug delivery systems: Current trends and applications. *Drug Delivery Transl. Res.*, 12, 4, 758–91, 2022 Apr.
112. Türeli, N.G. and Türeli, A.E., Upscaling and GMP production of pharmaceutical drug delivery systems, in: *Drug delivery trends*, pp. 215–229, Elsevier, Amsterdam, Netherlands, 2020 Jan 1.
113. Souto, E.B., Silva, G.F., Dias-Ferreira, J., Zielinska, A., Ventura, F., Durazzo, A., Lucarini, M., Novellino, E., Santini, A., Nanopharmaceutics: Part I—Clinical trials legislation and good manufacturing practices (GMP) of nanotherapeutics in the EU. *Pharmaceutics*, 12, 2, 146, 2020 Feb 11.
114. Desai, N., Rana, D., Patel, M., Bajwa, N., Prasad, R., Vora, L.K., Nanoparticle Therapeutics in Clinical Perspective: Classification, Marketed Products, and Regulatory Landscape. *Small*, 2502315, 2025 Jun 2.
115. Abdelmonem, R., Ayman, M., Lofty, S.E., Ezzat, Y., Nabeah, J.M., Mohamed, H., Safwat, M., Abdelzaher, A., AboAlfa, A., Abdelkhalek, M., Ali, R.M., Advancing Nanoparticle Production: Scaling Up Techniques, Challenges, and Future Perspectives in Pharmaceutical Applications. *J. Pharm. Sci. Drug Manuf.-Misr Univ. Sci. Technol.*, 2, 2, 26–39, 2025 Jun 1.
116. Alshawwa, S.Z., Kassem, A.A., Farid, R.M., Mostafa, S.K., Labib, G.S., Nanocarrier drug delivery systems: characterization, limitations, future

perspectives and implementation of artificial intelligence. *Pharmaceutics*, 14, 4, 883, 2022 Apr 18.
117. Zhang, X., Servos, M.R., Liu, J., Ultrahigh nanoparticle stability against salt, pH, and solvent with retained surface accessibility *via* depletion stabilization. *J. Am. Chem. Soc.*, 134, 24, 9910–3, 2012 Jun 20.
118. Nel, A., Xia, T., Madler, L., Li, N., Toxic potential of materials at the nanolevel. *Science*, 311, 5761, 622–7, 2006 Feb 3.
119. Gupta, N.V., Murugan, M.D., Shanbhag, S.J., AI/ML-driven nanocarriers for the management of neurodegeneration, in: *The neurodegeneration revolution*, pp. 361–372, Amsterdam, Netherlands: Elsevier, 2025 Jan 1.
120. U.S. Food and Drug Administration, What we do [Internet], U.S. Food and Drug Administration, Silver Spring (MD), 2023 [cited 2024 Apr], Available from: https://www.fda.gov/about-fda/what-we-do.

28

Recent Advances in Biocatalysis and Metabolic Engineering for Biomanufacturing

Saivenkatesh Korlam[1], Sankara Rao Miditana[2]* and Satheesh Ampolu[3]

[1]*Department of Botany, SVA Government Degree College (M), Srikalahasti, A.P., India*
[2]*Department of Chemistry, Government Degree College, Puttur, A.P., India*
[3]*Department of Chemistry, Centurion University of Technology and Management, Vizianagaram, A.P., India*

Abstract

Biocatalysis and metabolic engineering have emerged as transformative technologies in biomanufacturing, enabling sustainable and efficient production of chemicals, pharmaceuticals, biofuels, and bioplastics. This chapter provides a comprehensive overview of recent advancements in enzyme engineering, synthetic biology, and systems biology, which have significantly enhanced the efficiency, scalability, and specificity of biocatalytic processes and metabolic pathways. Enzyme discovery innovations that are driven by high-throughput screening, computational tools like molecular dynamics and machine learning, have opened new doors in the design of more biocatalysts with special functionalities. Simultaneous utilization of metabolic engineering by adopting tools like CRISPR/Cas systems, omics approaches, and systems biology, has promoted the development of microbial cell factories destined for producing high-value compounds. The synergy between biocatalysis and metabolic engineering has involved the creation of complex cascade reactions and multi-enzyme frameworks, improving both *in vitro* and *in vivo* bioprocesses. However, obstacles such as enzyme activity in industrial conditions, substrate specificity, pathway resilience, and regulatory hurdles take part. This review also highlights emerging trends, such as the utility of artificial intelligence and automation facilitated bioprocess refinement, the trending movements toward circular bioeconomy models and the biomanufacturing

**Corresponding author*: sraom90@gmail.com

Sandip Kunar, Jagadeesha T. and D. Sathis Kumar (eds.) Biomanufacturing: Transforming Production of Biologics and Biopharmaceuticals, (939–956) © 2026 Scrivener Publishing LLC

reliability in personalized healthcare and sustainable chemistry. By understanding the existing challenges and expecting the future possibilities, this review describes the revolutionary impact of biocatalysis and metabolic engineering in delivering eco-friendly and cost-effective solutions for industrial biotechnology.

Keywords: Biocatalysis, metabolic engineering, enzyme engineering, CRISPR/Cas, bioprocesses, circular bioeconomy, industrial biotechnology

28.1 Introduction

Biomanufacturing utilizes biology to make useful products, and it is an important tool for sustainable production in future industries. It addresses problems such as resource exhaustion, adverse environmental impacts, and energy efficient technologies. Biocatalysis and metabolic engineering are core concepts in this branch of science. Biocatalysis utilizes enzymes to catalyze chemical reactions with high specificity and efficiency, whereas metabolic engineering reengineers cellular pathways to improve synthesis of molecules we desire [1]. This is revolutionizing how pharmaceuticals, biofuels and green chemicals are made - replaced by less energy-intensive and environmentally destructive chemical processes. This chapter describes recent developments in these areas, explaining their combined impact on biomanufacturing and identifying future challenges and opportunities.

The transition to sustainable manufacturing are more urgent than ever. As the world is expected to see an increase in carbon emissions in the absence of substantial search for decarbonization, biomanufacturing provides an alternative route— to lessen the dependence on fossil-derived feedstocks, reduce waste [2]. Biocatalysis is efficient in synthesizing complex molecules with high stereospecificity under optimum conditions. It reduces energy consumption and waste compared to the traditional chemical methods. Enantioselective production of chiral intermediates for pharmaceuticals, such as statins, by biocatalysis has greatly alleviated the environmental burden incurred by drug synthesis [3]. Likewise, metabolic engineering has facilitated the scale-up economically efficient microbial production of valuable compounds, for example, artemisinin (an antimalarial drug) when compared to plant-based extraction [4]. These results indicate the feasibility of combining biocatalysis and metabolic engineering for industrial biotechnology. Advancements in new technologies have prompted development in both areas.

Synthetic biology tools such as CRISPR-Cas9 allows precision control of genome editing to optimize metabolic pathways and enzyme performance [5].

Computational tools, such as AlphaFold and Rosetta, have revolutionized enzymology by enabling protein structure prediction and rational design [6]. Systems biology has been advanced through integrating omics technologies (Genomics, Transcriptomics, Proteomics and Metabolomics) which has led to comprehensive models of cellular metabolism to address the challenges and achieving engineering goals [7]. These developments increase the range surface of biomanufacturing by machine production of commodity including new compounds and the use of abundant, renewable feedstock s such lignocellulosic as biomass and CO_2 [8].

Biocatalysis has evolved based on recent breakthroughs in enzyme-based engineering, immobilization, and cascade reactions. It has been a Nobel prize awarded technology utilizing random mutagenesis and high-throughput screening to enhance the activity and stability of enzyme [9]. Dialkylketone reductases (DKRs) have been engineered to allow these enzymes to work at industrial conditions (e.g., in the presence of organic solvents and at higher temperatures), to produce chiral alcohols [3]. Immobilization strategies, such as on MOFs, provide an enhanced enzyme recyclability and scalability, thereby, reducing the cost of biocatalysis for continuous processing [10]. Multi-enzyme cascade systems involve the sequential functioning of enzymes, improve the simplified complex syntheses, reducing purification steps and boosting yields [11]. These recent developments have established biocatalysis as an important center of green chemistry, with immense applications in pharmaceuticals and pure chemicals.

Metabolic engineering transforms microbes into efficient platforms of production. By altering the direction of metabolic pathways, engineers optimized the synthesis of biofuels, bioplastics and therapeutic compounds. Engineered or transformed strains of *Escherichia coli* and *Saccharomyces cerevisiae* have achieved high yields of isoprenoids and fatty acids, respectively [5, 12]. Non-traditional hosts like *Yarrowia lipolytica* and *Clostridium thermocellum*, expanded the range of products, enabling the synthesis of omega-3 fatty acids and second-generation biofuels from lignocellulosic biomass [11, 13]. Synthetic microbial consortia of multiple collaborative species, improved the processes efficiency, as seen in co-cultures that degrade biomass to produce biofuels [8]. These developments demonstrate the ability of metabolic engineering to leverage microbial diversity for sustainable production.

The integration of biocatalysis and metabolic engineering has formed efficient combinations. Combining biocatalytic reactions with engineered microbial pathways has facilitated the production of complex molecules, including polyketides [7]. Such a combined effect is particularly relevant in the context of the bioeconomy, where renewable feedstocks (e.g. glycerol or agricultural residues) are converted into value-added products in line with the principles of circular economy [3]. Moreover, it enabled enzyme [8] and pathway design [9] by artificial intelligence (AI) and machine learning (ML), to predict best strategies and reduce experimental trials. These interdisciplinary methodologies have the potential to move biomanufacturing closer to sustainability and scale.

Now, each of these has its disadvantages. Enzyme stability in industrial environments, pathway complexity in engineered microorganisms and bioprocess scalability are the major challenges [5]. Economic and regulatory limitations, such as the high costs of enzymes production and the edition of transgenic organisms have also effects in the successful implementation of the technology [12]. Addressing these challenges requires continual innovation, collaboration between academia and industry, and policies that promote sustainable technologies.

In this chapter, the recent progresses of biocatalysis and metabolic engineering and their applications in biomanufacturing are discussed. It discusses the new features in enzyme optimization, iterative design for pathway optimization and design, multi-pathway design, and strain design and improvement, and includes new case studies for example, in pharmaceutical, biofuel and chemical productions.

28.2 Advances in Biocatalysis

Biocatalysis emerged as a natural process of catalyst that is used in various chemical reactions with an advantage of being environmentally friendly compared with classical chemical processes. Recent advances of enzyme engineering, immobilization, and cascade reaction systems have promoted biocatalysis to a highly relevant technology for the synthesis of pharmaceuticals, biofuels, and sustainable materials. The various advances necessary for improving the enzyme-based processes are discussed in this section.

28.2.1 Enzyme Based Engineering

Enzyme engineering applies selected enzymes that are engineered to perform more efficiently, being faster, more robust, or easily adaptable to industrial needs. Directed evolution, a key strategy, speeds up natural selection by creating random mutations that turn out as a variety of enzyme variants and select those with better activities. This is exemplified by the case of a thermostable keto reductase reported by Sheldon and Pereira that can be

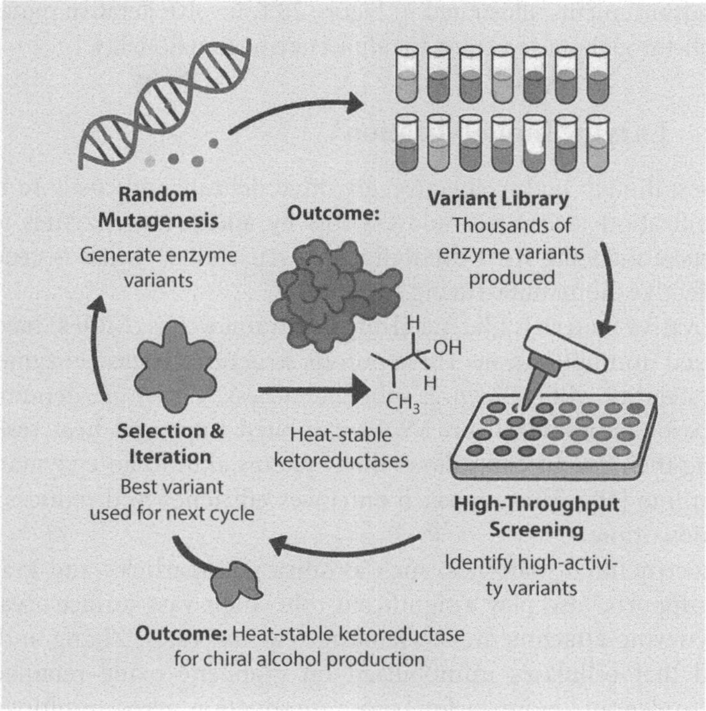

Figure 28.1 Directed evolution process for engineering a heat-stable ketoreductase, enabling efficient chiral alcohol synthesis for pharmaceuticals.

applied to the synthesis of pharmaceuticals chiral alcohols at high temperatures, high pressures conditions [3].

In addition to this, computational assets can be used to apply rational design to create accurate mutations in enzymes. High-throughput software such as Rosetta and AlphaFold can predict protein folding and substrate interactions for targeted improvements. Li *et al.*, extended this concept to optimize cytochrome P_{450} enzymes. This transformed them into highly effective catalysts for hydroxylation reactions critical to drug molecule synthesis [14]. These reactions ensure the precise 3D configurations required for therapeutic efficacy.

De novo enzyme design takes innovation further by creating entirely new enzymes. Kan *et al.*, achieved a milestone by designing an enzyme capable of forming carbon-silicon bonds, a reaction not found in nature [15]. This breakthrough paves the way for novel chemical classes. Additionally, machine learning has emerged as a powerful tool in *de novo* design, predicting enzyme sequences that catalyze reactions with exceptional efficiency [16].

These advancements, illustrated in Figure 28.1, involve iterative mutagenesis and high-throughput screening to refine enzyme functionality.

28.2.2 Enzyme Immobilization

Enzymes, though highly effective, are often delicate and costly to replace. Immobilization techniques address this by anchoring enzymes to solid supports, enhancing their durability and reusability, which is crucial for cost-effective biomanufacturing.

Innovative materials like metal-organic frameworks (MOFs) have revolutionized immobilization. These porous structures encase enzymes, providing stability while allowing substrate access. Wu *et al.*, demonstrated that lipases immobilized on MOFs exhibited improved heat resistance, enabling their use in continuous-flow systems akin to an enzymatic production line [10]. This approach enhances efficiency and reduces operational downtime.

Nanostructured materials, such as silica nanoparticles and graphene-based supports, also play a significant role. Their vast surface areas maximize enzyme attachment, accelerating reaction rates. Zhang and Wang showed that cellulases immobilized on graphene oxide retained their ability to degrade biomass for biofuel production across multiple cycles [17]. Another technique, cross-linked enzyme aggregates (CLEAs), chemically binds enzymes into robust clusters. Sheldon and Brady explored this with laccases, enabling them to withstand harsh solvents and supporting greener chemical processes [18]. These developments make biocatalysis both practical and economically viable for industrial applications.

28.2.3 Cascade Reactions

Enzyme cascades resemble a seamless relay, where multiple enzymes collaborate in a single system, passing intermediates without requiring purification. This approach minimizes waste and enhances sustainability that make it ideal for complex syntheses.

A notable example is Zhang *et al.*'s cascade involving glucose oxidase and catalase to produce gluconic acid from glucose [11]. The first enzyme generates an intermediate, while the second eliminates byproducts, ensuring a smooth reaction. This efficiency reduces energy consumption and simplifies processes for applications like food additives or bio-based chemicals.

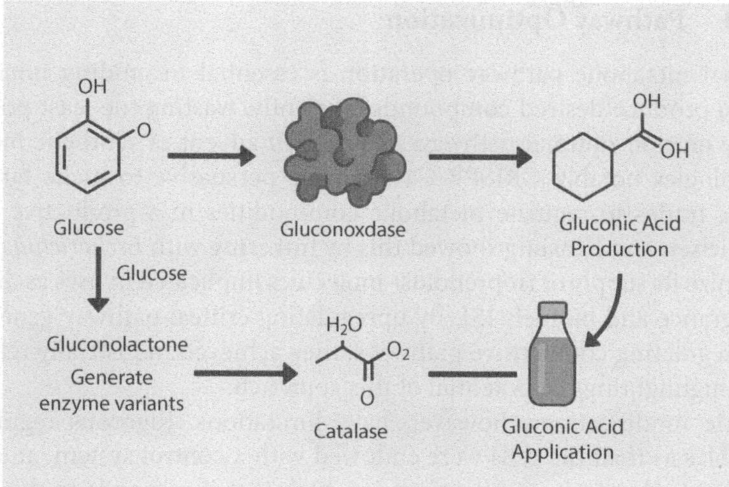

Figure 28.2 Enzyme cascade involving glucose oxidase and catalase for efficient gluconic acid production, a key food additive, with byproduct removal enhancing sustainability.

In pharmaceutical manufacturing, cascades streamline the synthesis of complex molecules, such as erythromycin precursors, reducing steps and costs [19]. Combining cascades with immobilized enzymes in flow reactors creates continuous production systems. Turner and Humphreys demonstrated this with chiral amine synthesis for pharmaceuticals, achieving yields comparable to traditional methods but with a significantly lower environmental impact [20]. By optimizing reactions and reducing waste, enzyme cascades are a cornerstone of sustainable chemistry, as depicted in Figure 28.2.

28.3 Advances in Metabolic Engineering

Metabolic engineering is like giving cells a new set of instructions to churn out the stuff we want like biofuels, drugs, or sustainable chemicals faster and better. It's all about tweaking the inner workings of microbes to boost product titers, rates, and yields (often called TRY in the biz). Over the past few years, this field has exploded—thanks to some seriously cool tools and ideas. Let's dive into the big strides in pathway optimization, microbial host design, and systems biology integration, which are making biomanufacturing more efficient and sustainable than ever.

28.3.1 Pathway Optimization

Optimal metabolic pathway operation is essential in guiding microbial cells to produce desired compounds efficiently, wasting the least possible energy on non-optimal pathways. The recent advent of synthetic biology technologies notably CRISPR-Cas9 can be persuasive to make targeted genetic trades to regulate metabolic commodities in a productive manner. Nielsen and Keasling showed this by tinkering with *Escherichia coli* to maximize its supply of isoprenoids–molecules implicated in uses as diverse as fragrance and biofuels [5]. By upregulating critical pathway genes and downregulating competitive pathways, they achieved industrially relevant titers, highlighting the potential of this approach.

Static modifications, however, have limitations. Rigorous regulation resembles as if all the cells were endowed with a control system, an adaptive system that levies gene expression such that it responds to the inner information. Gupta *et al.*, demonstrated this in brewer's yeast, saccharomyces cerevisiae, with the use of biosensor-driven feedback loops controlling fatty acid synthesis [12]. These biosensors track changes in metabolism and adapt gene activity in real time- a method that produces 30% more fatty acids than traditional static cultures. This kind of adaptive control is crucial for optimal pathway operation.

A second newly developed strategy is modular pathway design that allows researchers to break down intricate metabolic pathways into manageable pieces. Wang *et al.*, extended this approach to improving lycopene production, an important antioxidant, in *E. coli* [21]. By fine-tuning individual modules similar to adjusting components on a control panel they achieved a 40% improvement in lycopene yield. This modular framework simplifies scaling production and adapting pathways for diverse products, from vitamins to bioplastics, offering a versatile solution for biomanufacturing.

28.3.2 Microbial Host Design

Choosing the right microbe for the desirable task is a big question in biomanufacturing. It's like picking the perfect actor for a role. Different microbes bring different strengths to the table. *E. coli* and *S. cerevisiae* are the superstars—thanks to how well we understand them and how easy they are to tweak. But lately, non-conventional hosts are stealing the spotlight because of their unique metabolic tricks. Take *yarrowia lipolytica*, a yeast that's a wizard at handling lipids. Xie *et al.*, engineered it to produce omega-3 fatty acids those heart-healthy compounds in fish oil by tapping into its natural knack for fat metabolism [13]. This kind of specialized host can make all the difference for certain products.

Then there's *clostridium thermocellum*, a microbe that thrives in hot environments and loves breaking down tough plant material. Zhang *et al.*, harnessed it to produce isobutanol, a biofuel, directly from cellulose, skipping expensive preprocessing steps [11]. These non-traditional hosts are opening up new possibilities, especially for sustainable feedstocks like agricultural waste.

Synthetic microbial consortia, think of them as microbial dream teams are shaking things up. By combining different species, each with its own strengths, you can tackle complex tasks. Li *et al.*, paired *E. coli* with *Pseudomonas putida* in a co-culture that breaks down lignocellulosic biomass and turns it into biofuels [8]. It's like one microbe does the heavy lifting, and the other polishes the final product. This teamwork approach is super promising for handling tricky substrates or multi-step processes.

Another cool trend is engineering hosts for extreme conditions, like high acidity or salt levels, which are common in industrial settings. A study by Chen *et al.*, engineered a salt-tolerant bacillus subtilis to produce itaconic acid, a building block for biodegradable plastics, in harsh saline conditions [22]. This kind of resilience makes microbes more practical for real-world bioreactors, where conditions can get rough.

28.3.3 Systems Biology Integration

If metabolic engineering is about rewiring cells, systems biology is about understanding the whole wiring diagram. It pulls together data from genomics, transcriptomics, proteomics, and metabolomics to create a big-picture view of what's going on inside a cell. This holistic approach helps spot bottlenecks those pesky roadblocks that slow down production and figure out how to fix them. Machine learning has become a trusty sidekick here, crunching massive datasets to predict where things are going wrong and suggest smart tweaks.

A standout example is Kim *et al.*, who used a random forest model a type of machine learning algorithm to dig into polyketide biosynthesis in streptomyces. Polyketides are complex molecules used in antibiotics and other drugs, but their production is tricky [7]. The model pinpointed rate-limiting steps, leading to a twofold increase in output. It's like having a super-smart assistant who can scan a cell's entire metabolic network and say, "Fix this, and you're golden."

Beyond machine learning, multi-omics integration is helping engineers design more robust pathways. For instance, Liu *et al.*, combined transcriptomics and metabolomics to optimize shikimate production in corynebacterium glutamicum [23]. By mapping out how gene expression and metabolite levels interact, they boosted yields by 25%. This kind of data-driven approach is making metabolic engineering less of a trial-and-error game and more of a precision science.

What's really exciting is how systems biology is bridging the gap between lab experiments and industrial-scale production. A recent paper by Park *et al.* (2025), used systems biology to model how *E. coli* behaves in large bioreactors, accounting for things like nutrient gradients and oxygen levels [24]. Their model predicted optimal gene edits that kept production stable even at massive scales, which is a remarkable event for taking lab results to the factory floor.

28.4 Applications in Biomanufacturing

Biomanufacturing is revolutionizing industries by producing pharmaceuticals, biofuels, and sustainable chemicals with enhanced efficiency and environmental benefits. By understanding biocatalysis—using enzymes to accelerate chemical reactions—and metabolic engineering—modifying microbial metabolic pathways—innovative, eco-friendly production methods are emerging. This section examines the transformative impact of these technologies across pharmaceuticals, biofuels, and sustainable chemicals, supported by practical examples that highlight their potential.

28.4.1 Pharmaceuticals

Pharmaceutical production was revolutionized with biocatalysis and metabolic engineering allowing the synthesis of the complex molecules, with high stereoselectivity, and ensuring they had the preferred chiral 3D structure in order to be effective. A well-known example would be artemisinin, an important antimalarial medicine. Once obtained from sweet wormwood using expensive, time-consuming filtration schemes, artemisinin can now be manufactured directly *via* genetically modified *Escherichia coli*, thereby reducing cost and availability in malaria-endemic areas [4]. This development has significant implications for global health.

Biocatalysis is also optimized statin production, critical for cholesterol management. Enzymatic catalysis of key synthesis steps reduces waste and environmental impact compared to traditional chemical methods [1]. Similarly, engineered yeast strains have advanced insulin analog production. It offers tailored properties like rapid starting or extended duration, enhancing patient outcomes in diabetes management [25]. These innovations improve efficiency while making medicines more sustainable and accessible.

Recent developments have further expanded biocatalysis into complex natural products like cannabinoids, which show impact for pain relief and neurological disorders. By engineering *saccharomyces cerevisiae* to synthesize cannabinoids such as CBD and THC from simple sugars, this approach eliminates the need for resource-intensive *cannabis* cultivation, ensuring consistent, high-purity products at lower costs [26]. These advancements illustrate how biocatalysis and metabolic engineering are redefining pharmaceutical manufacturing.

28.4.2 Biofuels

Biofuels are an important component in addressing climate change, and metabolic engineering is deciding their evolution as viable alternatives to fossil fuels. Unlike first-generation biofuels like corn-based ethanol, which competed with food production, second-generation biofuels utilize non-food sources such as lignocellulosic biomass (e.g., crop residues or wood chips). A breakthrough example is the modification of *clostridium thermocellum* to produce isobutanol directly from cellulose is a bypassing expensive pretreatment steps [11]. Isobutanol's higher energy density and lower corrosiveness make it a superior gasoline substitute.

Further innovations include engineering microbes to produce advanced biofuels, such as drop-in hydrocarbons compatible with diesel or jet fuel. For instance, *Yarrowia lipolytica* has been modified to convert waste oils into long-chain hydrocarbons, offering a sustainable aviation fuel alternative [27]. This aligns with circular economy principles by repurposing waste into valuable products, particularly for industries like aviation where electrification remains challenging.

Synthetic microbial consortia represent another leap forward. A study demonstrated that a co-culture of *Trichoderma reesei* and *E. coli* could simultaneously degrade lignocellulose and produce butanol, enhancing efficiency and yield [28]. These collaborative systems are making biofuels more environmentally and economically competitive with fossil fuels.

28.4.3 Sustainable Chemicals

Biomanufacturing is redirecting the production of commodity chemicals, providing eco-friendly alternatives to petrochemicals. Polylactic acid (PLA), a biodegradable polymer used in packaging and medical implants, is an example. By optimizing lactate dehydrogenase pathways in *Lactobacillus* strains, researchers have enhanced PLA production, creating a cost-effective, sustainable alternative to conventional plastics [5]. PLA's biodegradability helps mitigate plastic waste in landfills and oceans.

Another achievement is the biocatalytic production of 1, 3-propanediol, which is a key component in polymers like polytrimethylene terephthalate (PTT) used for textiles and carpets. Engineered microbes convert glycerol, which is a biodiesel byproduct into 1, 3-propanediol with high efficiency and minimal energy use [3]. This process transforms waste into a valuable resource while reducing reliance on fossil fuels.

Bio-based succinic acid which is used in plastics, solvents and food additives, is another success. Engineered *corynebacterium glutamicum* strains produce succinic acid from renewable feedstocks like corn stover is achieving yields comparable to petroleum-based methods [29]. This supports sustainable supply chains even in growing pressure to reduce carbon footprints.

Additionally, bio-based vanillin is a widely used flavor compound and it is now produced by engineered *E. coli* from glucose. It is offering a scalable, sustainable alternative to vanilla bean extraction or synthetic processes [30]. These examples demonstrate how biomanufacturing delivers smarter, greener solutions for everyday chemical needs.

Table 28.1 summarizes key applications in pharmaceuticals, biofuels, and sustainable chemicals, highlighting specific examples, enabling technologies, and their contributions to sustainable production.

28.5 Challenges and Limitations

Despite significant progress in biocatalysis and metabolic engineering, several challenges persist. Enzyme stability remains a critical bottleneck, particularly in continuous-flow reactors where high temperatures or organic solvents can denature biocatalysts, limiting scalability for applications like biofuel production [31]. For instance, cellulases used in biomass breakdown often lose activity after repeated cycles, increasing costs. Metabolic engineering faces issues of pathway robustness, with engineered microbial strains, such as *Escherichia coli*, often exhibiting reduced fitness or genetic instability due to metabolic burden [32]. Scalability hurdles also arise, as laboratory-optimized processes may falter in industrial bioreactors due to mass transfer limitations or inconsistent substrate availability, as seen in large-scale isobutanol production [33].

Table 28.1 Applications of biocatalysis and metabolic engineering in biomanufacturing.

Application area	Description	Examples	Enabling technologies	Impact	References
Pharmaceuticals	Precise, efficient, greener drug production	Artemisinin (*E. coli*), Statins, Insulin analogs, Cannabinoids	Biocatalysis, Metabolic Engineering	Reduced costs by up to 70%, lower environmental footprint	[4, 25, 26]
Biofuels	Sustainable alternatives from non-food sources	Isobutanol (C. thermocellum), Drop-in hydrocarbons (Y. lipolytica), Butanol (consortia)	Metabolic Engineering, Synthetic Microbial Consortia	Higher energy content, reduced pretreatment costs	[11, 27, 28]
Sustainable Chemicals	Eco-friendly petrochemical replacements	PLA, 1,3-Propanediol, Succinic acid, Vanillin	Biocatalysis, Metabolic Engineering	Biodegradable products, waste valorization	[3, 5, 29, 30]

The high costs associated with enzyme production and purification is often comprising 50–80% of biocatalytic process expenses that pose significant obstacles to competing with conventional chemical synthesis methods [34]. Furthermore, the regulatory landscape for genetically modified organisms (GMOs) which varies across countries is creating complexities for commercialization, especially in pharmaceutical applications [12]. Public attitudes toward GMOs also play an important role, as societal acceptance shapes regulatory policies. Overcoming these hurdles demands is possible by adopting innovative approaches, such as enhanced enzyme immobilization techniques and simplified regulatory frameworks, to facilitate broader involvement of biomanufacturing.

28.6 Future Perspectives

The convergence of artificial intelligence (AI) and automation offers transformative opportunities for biocatalysis and metabolic engineering. AI-powered protein engineering can demand the creation of customized enzymes with specific functionalities, while automated high-throughput screening enables rapid optimization of metabolic pathways on a large scale [35]. Additionally, progress in synthetic biology such as cell-free systems could answer the constraints of living organisms by providing greater control and efficiency in *in vitro* biocatalytic processes [36].

Sustainability will continue to drive innovation, with research emphasizing waste valorization and circular bioeconomy principles. For instance, engineering microbes to convert carbon dioxide or plastic waste into valuable products could tackle environmental issues while generating useful compounds [37]. Collaborative efforts among academia, industry, and policymakers will be essential to address regulatory and economic challenges, paving the way for the widespread implementation of biomanufacturing.

28.7 Conclusion

Biocatalysis and metabolic engineering are transforming biomanufacturing, delivering breakthroughs like a 70 % cost reduction in artemisinin production and sustainable biofuel synthesis from lignocellulosic waste. These technologies leverage advanced enzyme engineering, pathway optimization, and systems biology to produce pharmaceuticals, biofuels, and eco-friendly chemicals with unprecedented efficiency. Despite challenges such as enzyme stability and scalability, innovations in AI-driven design and automation are paving the way for robust, scalable bioprocesses. By fostering collaboration across academia, industry, and policymakers, these fields will drive a greener, more resilient bioeconomy, revolutionizing sustainable industrial processes for a brighter future.

AI disclosure statement: The author acknowledges the use of AI-based tools for minor language editing or grammatical correction. However, all intellectual contributions, analyses, and conclusions presented in this chapter are entirely the author's own.

References

1. Bornscheuer, U.T., Huisman, G.W., Kazlauskas, R.J., Lutz, S., Moore, J.C., Robins, K., Engineering the third wave of biocatalysis. *Nature*, 485, 185, 2012.
2. IPCC, *Climate Change 2022: Mitigation of Climate Change*, Cambridge University Press, Cambridge, UK and New York, NY, USA, 2022.
3. Sheldon, R.A. and Pereira, P.C., Biocatalysis engineering: The big picture. *Chem. Soc. Rev.*, 46, 2678, 2017.
4. Paddon, C.J. and Keasling, J.D., Semi-synthetic artemisinin: A model for the use of synthetic biology in pharmaceutical development. *Nat. Rev. Microbiol.*, 12, 355, 2014.
5. Nielsen, J. and Keasling, J.D., Engineering cellular metabolism. *Cell*, 164, 1185, 2016.
6. Jumper, J., Evans, R., Pritzel, A., Green, T., Figurnov, M., Ronneberger, O., Hassabis, D., Highly accurate protein structure prediction with AlphaFold. *Nature*, 596, 583, 2021.
7. Kim, J., Lee, S., Park, J.H., Machine learning-guided optimization of polyketide biosynthesis in Streptomyces. *Biotechnol. J.*, 19, e2300456, 2024.
8. Li, Z., Wang, Y., Zhao, H., Microbial consortia for lignocellulosic biofuel production: Advances and challenges. *Biotechnol. Adv.*, 62, 108076, 2023.
9. Arnold, F.H., Directed evolution: Bringing new chemistry to life. *Angew. Chem. Int. Ed.*, 57, 4143, 2018.
10. Wu, X., Yang, Y., Zhang, Y., Enzyme immobilization on metal-organic frameworks: Efficient biocatalysts for continuous-flow processes. *ACS Catal.*, 11, 8793, 2021.
11. Zhang, L., Liu, Y., Yang, J., Multi-enzyme cascades for efficient biomanufacturing: Design and applications. *Biotechnol. Biofuels*, 16, 45, 2023.
12. Gupta, A., Reizman, I.M., Reisch, C.R., Prather, K.L.J., Dynamic regulation of metabolic flux in engineered bacteria using a pathway-independent quorum-sensing circuit. *Nat. Biotechnol.*, 40, 3, 355–363, 2022.

13. Xie, D., Jackson, E.N., Zhu, Q., Sustainable production of omega-3 fatty acids by metabolically engineered Yarrowia lipolytica. *Microb. Cell Fact.*, 18, 1, 121, 2019.
14. Li, A., Casali, S., Zhang, F., Engineering cytochrome P450 enzymes for pharmaceutical biosynthesis. *Curr. Opin. Biotechnol.*, 65, 111, 2020.
15. Kan, S.B.J., Lewis, R.D., Chen, K., Arnold, F.H., Directed evolution of cytochrome c for carbon–silicon bond formation: Bringing silicon to life. *Science*, 354, 1048, 2016.
16. Yang, K.K., Wu, Z., Arnold, F.H., Machine learning-guided enzyme design for novel biocatalytic functions. *Nat. Chem. Biol.*, 20, 312, 2024.
17. Zhang, W. and Wang, J., Nanostructured supports for enzyme immobilization in biofuel production. *Renewable Energy*, 215, 119012, 2023.
18. Sheldon, R.A. and Brady, D., Cross-linked enzyme aggregates (CLEAs): Advances in carrier-free immobilization. *ChemSusChem*, 15, e202200123, 2022.
19. Liu, Y., Chen, X., Zhao, H., Multi-enzyme cascades for antibiotic precursor synthesis: Design and optimization. *Biotechnol. Adv.*, 68, 108234, 2024.
20. Turner, N.J. and Humphreys, L., Biocatalytic cascades in flow: Continuous production of chiral amines. *Green Chem.*, 25, 1789, 2023.
21. Wang, X., Zhang, Y., Li, T., Modular pathway engineering for enhanced lycopene production in Escherichia coli. *Synth. Syst. Biotechnol.*, 9, 456, 2024.
22. Chen, X., Zhang, H., Liu, Y., Engineering salt-tolerant Bacillus subtilis for itaconic acid production in high-salinity conditions. *Appl. Microbiol. Biotechnol.*, 109, 12, 2025.
23. Liu, Q., Wang, L., Zhao, J., Multi-omics integration for enhanced shikimate production in Corynebacterium glutamicum. *Metab. Eng.*, 81, 101, 2024.
24. Park, S., Kim, H., Lee, J., Systems biology-guided scale-up of Escherichia coli for industrial biomanufacturing. *Biochem. Eng. J.*, 205, 109245, 2025.
25. Baeshen, N.A., Baeshen, M.N., Sheikh, A., Bora, R.S., Ahmed, M.M., Ramadan, H.A., Al-Hejin, A.M., Cell factories for insulin production. *Microb. Cell Fact.*, 19, 182, 2020.
26. Luo, X., Reiter, M.A., d'Espaux, L., Wong, J., Denby, C.M., Lechner, A., Keasling, J.D., Complete biosynthesis of cannabinoids and their unnatural analogues in yeast. *Nature*, 567, 123, 2019.
27. Liu, H., Xu, W., Li, X., Engineering Yarrowia lipolytica for the production of drop-in biofuels from waste oils. *Biotechnol. Biofuels*, 15, 89, 2022.
28. Minty, J.J., Singer, M.E., Scholz, S.A., Bae, C.H., Ahn, J.H., Foster, C.E., Lin, X.N., Design and optimization of a lignocellulosic biomass-to-butanol microbial consortium. *Nat. Commun.*, 12, 4567, 2021.
29. Chung, S.C., Park, J.S., Lee, J.W., Metabolic engineering of Corynebacterium glutamicum for succinic acid production from lignocellulosic biomass. *Bioresour. Technol.*, 315, 123863, 2020.

30. Li, X., Chen, Y., Zhang, H., Biosynthesis of vanillin using engineered Escherichia coli from renewable feedstocks. *J. Agric. Food. Chem.*, 72, 1456, 2024.
31. Wu, H., Chen, Q., Zhang, W., Mu, W., Overview of strategies for developing high thermostability industrial enzymes: Discovery, mechanism, modification and challenges. *Crit. Rev. Food Sci. Nutr.*, 63, 14, 2057–2073, 2023.
32. Daboussi, F. and Lindley, N.D., Challenges to Ensure a Better Translation of Metabolic Engineering for Industrial Applications. *Methods Mol. Biol. (Clifton, N.J.)*, 2553, 1–20, 2023.
33. Crater, J.S. and Lievense, J.C., Scale-up of industrial microbial processes. *FEMS Microbiol. Lett.*, 365, 13, fny138, 2018.
34. Arnau, J., Yaver, D., Hjort, C.M., Strategies and Challenges for the Development of Industrial Enzymes Using Fungal Cell Factories, in: *Grand Challenges in Fungal Biotechnology*, Nevalainen, H. (eds.). Grand Chall. Biol. Biotechnol., Springer, Cham, 2020, https://doi.org/10.1007/978-3-030-29541-7_7.
35. Merlicek, L.P., Neumann, J., Lear, A., Degiorgi, V., de Waal, M.M., Cotet, T.S., Mulholland, A.J., Bunzel, H.A., AI.zymes: A Modular Platform for Evolutionary Enzyme Design. *Angew. Chem. Int. Ed. Engl.*, 64, 27, e202507031, 2025.
36. Meyer, C., Nakamura, Y., Rasor, B.J., Karim, A.S., Jewett, M.C., Tan, C., Analysis of the Innovation Trend in Cell-Free Synthetic Biology. *Life (Basel, Switzerland)*, 11, 6, 551, 2021.
37. Aggarwal, N., Pham, H.L., Ranjan, B., *et al.*, Microbial engineering strategies to utilize waste feedstock for sustainable bioproduction. *Nat. Rev. Bioeng.*, 2, 155–174, 2024.

Index

Wiley The manufacturer's authorized representative according to the EU General Product Safety Regulation is Wiley-VCH GmbH, Boschstr. 12, 69469 Weinheim, Germany, e-mail: Product_Safety@wiley.com.

Printed and bound by CPI Group (UK) Ltd, Croydon, CR0 4YY

03/08/2026

02185160-0001